REVIEWS in MINERALOGY
Volume 19

HYDROUS PHYLLOSILICATES
(exclusive of micas)

S.W. Bailey, editor

Series Editor: Paul H. Ribbe

MINERALOGICAL SOCIETY of AMERICA

T.m

HYDROUS PHYLLOSILICATES
(exclusive of micas)
S.W. Bailey, editor

The authors and series editor:

S.W. Bailey
Dept of Geology & Geophysics
University of Wisconsin
Madison, Wisconsin 53706

R.G. Berman
Dept of Geological Sciences
University of British Columbia
Vancouver, BC, Canada V6T 1W5

L.T. Bryndzia
The James Franck Institute
University of Chicago
Chicago, Illinois 60637

D.M. Burt
Dept of Geology
Arizona State University
Tempe, Arizona 85287

Cristina de la Calle
Instituto de Ciencia de Materiales
C.S.I.C.
Serrano 113, 28006 Madrid, Spain

J.V. Chernosky, Jr.
Dept of Geological Sciences
University of Maine
Orono, Maine 04469

R.A. Eggleton
Geology Dept
Australian National University
Canberra, ACT, Australia

B.W. Evans
Dept of Geological Sciences
AJ20 University of Washington
Seattle, Washington 98195

E. Galan
Dept de Geología
Universidad de Sevilla
Sevilla, Spain

R.F. Geise, Jr.
Dept of Geological Sciences
State University of New York
Buffalo, New York 14226

S. Guggenheim
Dept of Geology
University of Illinois at Chicago
Chicago, Illinois 60680

N. Güven
Dept of Geology
Texas Tech University
Lubbock, Texas 79409

B.F. Jones
U.S. Geological Survey
432 National Center
Reston, Virginia 22092

Jo Laird
Dept of Earth Sciences
University of New Hampshire
Durham, New Hampshire 03824

M. Lee
Mobil Research & Development Corp
P.O. Box 819047
Dallas, Texas 75381

H.H. Murray
Dept of Geology
University of Indiana
Bloomington, Indiana 47405

D.S. O'Hanley
Royal Ontario Museum
100 Queens Park
Toronto, ON, Canada M5S 2C6

R.C. Reynolds, Jr.
Dept of Earth Sciences
Dartmouth College
Hanover, New Hampshire 03755

S.M. Savin
Dept of Earth Sciences
Case Western Reserve University
Cleveland, Ohio 44106

Hélène Suquet
Laboratoire de Chimie des Solides
Université Pierre et Marie Curie
4 Place Jussieu, 75230 Paris, France

F.J. Wicks
Royal Ontario Museum
100 Queens Park
Toronto, ON, Canada M5S 2C6

Series Editor: **P.H. Ribbe**
Dept Geological Sciences
Virginia Polytechnic Inst & State Univ
Blacksburg, Virginia 24061

REVIEWS in MINERALOGY

(Formerly: SHORT COURSE NOTES)

ISSN 0275-0279

Volume 19: *HYDROUS PHYLLOSILICATES*

(exclusive of micas)

ISBN 0-939950-23-5

ADDITIONAL COPIES of this volume as well as those listed below
may be obtained from the MINERALOGICAL SOCIETY of AMERICA
1625 I Street, N.W., Suite 414, Washington, D.C. 20006 U.S.A.

Reviews in Mineralogy

*Volume 1: Sulfide Mineralogy, 1974; P. H. Ribbe, Ed. 284 pp.
Six chapters on the structures of sulfides and sulfosalts; the crystal chemistry and chemical bonding of sulfides, synthesis, phase equilibria, and petrology. ISBN# 0-939950-01-4.

*Volume 2: Feldspar Mineralogy, 2nd Edition, 1983; P. H. Ribbe, Ed. 362 pp. Thirteen chapters on feldspar chemistry, structure and nomenclature; Al,Si order/disorder in relation to domain textures, diffraction patterns, lattice parameters and optical properties; determinative methods; subsolidus phase relations, microstructures, kinetics and mechanisms of exsolution, and diffusion; color and interference colors; chemical properties; deformation. ISBN# 0-939950-14-6.

*Volume 3: Oxide Minerals, 1976; D. Rumble III, Ed. 502 pp.

OUT of PRINT

*Volume 4: Mineralogy and Geology of Natural Zeolites, 1977; F. A. Mumpton, Ed. 232 pp. Ten chapters on the crystal chemistry and structure of natural zeolites, their occurrence in sedimentary and low-grade metamorphic rocks and closed hydrologic systems, their commercial properties and utilization. ISBN# 0-939950-04-9.

*Volume 5: Orthosilicates, 2nd Edition, 1982; P. H. Ribbe, Ed. 450 pp. Liebau's "Classification of Silicates" plus 12 chapters on silicate garnets, olivines, spinels and humites; zircon and the actinide orthosilicates; titanite (sphene), chloritoid, staurolite, the aluminum silicates, topaz, and scores of miscellaneous orthosilicates. Indexed. ISBN# 0-939950-13-8.

*Volume 6: Marine Minerals, 1979; R. G. Burns, Ed. 380 pp. Ten chapters on manganese and iron oxides, the silica polymorphs, zeolites, clay minerals, marine phosphorites, barites and placer minerals; evaporite mineralogy and chemistry. ISBN# 0-939950-06-5.

*Volume 7: Pyroxenes, 1980; C. T. Prewitt, Ed. 525 pp. Nine chapters on pyroxene crystal chemistry, spectroscopy, phase equilibria, subsolidus phenomena and thermodynamics; composition and mineralogy of terrestrial, lunar, and meteoritic pyroxenes. ISBN# 0-939950-07-3.

*Volume 8: Kinetics of Geochemical Processes, 1981; A. C. Lasaga and R. J. Kirkpatrick, Eds. 398 pp. Eight chapters on transition state theory and the rate laws of chemical reactions; kinetics of weathering, diagenesis, igneous crystallization and geochemical cycles; diffusion in electrolytes; irreversible thermodynamics. ISBN# 0-939950-08-1.

*Volume 9A: Amphiboles and Other Hydrous Pyriboles—Mineralogy, 1981; D. R. Veblen, Ed. 372 pp. Seven chapters on biopyribole mineralogy and polysomatism; the crystal chemistry, structures and spectroscopy of amphiboles; subsolidus relations; amphibole and serpentine asbestos—mineralogy, occurrences, and health hazards. ISBN# 0-939950-10-3.

*Volume 9B: Amphiboles: Petrology and Experimental Phase Relations, 1982; D. R. Veblen and P. H. Ribbe, Eds. 390 pp. Three chapters on phase relations of metamorphic amphiboles (occurrences and theory); igneous amphiboles; experimental studies. ISBN# 0-939950-11-1.

*Volume 10: Characterization of Metamorphism through Mineral Equilibria, 1982; J. M. Ferry, Ed. 397 pp. Nine chapters on an algebraic approach to composition and reaction spaces and their manipulation; the Gibbs' formulation of phase equilibria; geologic thermobarometry; buffering, infiltration, isotope fractionation, compositional zoning and inclusions; characterization of metamorphic fluids. ISBN# 0-939950-12-X.

*Volume 11: Carbonates: Mineralogy and Chemistry, 1983; R. J. Reeder, Ed. 394 pp. Nine chapters on crystal chemistry, polymorphism, microstructures and phase relations of the rhombohedral and orthorhombic carbonates; the kinetics of $CaCO_3$ dissolution and precipitation; trace elements and isotopes in sedimentary carbonates; the occurrence, solubility and solid solution behavior of Mg-calcites; geologic thermobarometry using metamorphic carbonates. ISBN# 0-939950-15-4.

*Volume 12: Fluid Inclusions, 1984; by E. Roedder. 644 pp. Nineteen chapters providing an introduction to studies of all types of fluid inclusions, gas, liquid or melt, trapped in materials from the earth and space, and their application to the understanding of geological processes. ISBN# 0-939950-16-2.

*Volume 13: Micas, 1984; S. W. Bailey, Ed. 584 pp. Thirteen chapters on structures, crystal chemistry, spectroscopic and optical properties, occurrences, paragenesis, geochemistry and petrology of micas. ISBN# 0-939950-17-0.

*Volume 14: Microscopic to Macroscopic: Atomic Environments to Mineral Thermodynamics, 1985; S. W. Kieffer and A. Navrotsky, Eds. 428 pp. Eleven chapters attempt to answer the question, "What minerals exist under given constraints of pressure, temperature, and composition, and why?" Includes worked examples at the end of some chapters. ISBN# 0-939950-18-9.

*Volume 15: Mathematical Crystallography, 1985; by M. B. Boisen, Jr. and G. V. Gibbs. 406 pp. A matrix and group theoretic treatment of the point groups, Bravais lattices, and space groups presented with numerous examples and problem sets, including solutions to common crystallographic problems involving the geometry and symmetry of crystal structures. ISBN# 0-939950-19-7.

*Volume 16: Stable Isotopes in High Temperature Geological Processes, 1986; J. W. Valley, H. P. Taylor, Jr., and J. R. O'Neil, Eds. 570 pp. Starting with the theoretical, kinetic and experimental aspects of isotopic fractionation, 14 chapters deal with stable isotopes in the early solar system, in the mantle, and in the igneous and metamorphic rocks and ore deposits, as well as in magmatic volatiles, natural water, seawater, and in meteoric-hydrothermal systems. ISBN #0-939950-20-0.

*Volume 17: Thermodynamic Modelling of Geological Materials: Minerals, Fluids, Melts, 1987; H. P. Eugster and I. S. E. Carmichael, Eds. 500 pp. Thermodynamic analysis of phase equilibria in simple and multi-component mineral systems, and thermodynamic models of crystalline solutions, igneous gases and fluid, ore fluid, metamorphic fluids, and silicate melts, are the subjects of this 14-chapter volume. ISBN # 0-939950-21-9.

HYDROUS PHYLLOSILICATES (exclusive of micas)

FOREWORD

The authors of this volume presented a short course by the same title to about 120 participants in Denver, Colorado, October 29-30, 1988, just prior to the 100th anniversary meeting of the Geological Society of America. This was the seventeenth in the series of courses that the Mineralogical Society of America began sponsoring in 1974, and this book joins the long list of those published by the Society for the geoscience community (see opposite page for details).

S.W. ("Bull") Bailey convened the course and edited this volume, his second for *Reviews in Mineralogy*. Because he is retiring at the end of this academic year after 38 years' teaching at the University of Wisconsin (Madison), his colleagues, friends and I (a diligent student of "Bull" thirty years ago) agreed that it would be appropriate to dedicate this volume to him, odd though it seems to have him editing a book honoring himself. He had no advance knowledge of this dedication.

DEDICATION

Sturges Williams Bailey was born February 11, 1919 in Waupaca, Wisconsin. He stuck close to home, eventually obtaining from the University of Wisconsin (Madison) a B.A. in Geology (1941) and -- after service in World War II -- an M.A. in Geology (1948). He was a Fulbright scholar at the Cavendish Laboratory, University of Cambridge, where he received his Ph.D. in Physics (Crystallography) in 1954. From 1951 on he was employed by the Department of Geology and Geophysics at his Alma Mater, becoming full professor in 1961, serving as Chairman of the department from 1968 to 1971, and being designated the Roland D. Irving Professor of Geology in 1976.

Bailey was Councillor of the Clay Minerals Society from 1964 to 1984 and its President in 1971-72. He edited *Clays and Clay Minerals* from 1964 to 1970 and was presented the Distinguished Member Award of the Society in 1975.

S.W. Bailey

Bailey also served the Mineralogical Society of America as Councillor (1970-72) and President (1973-74), the Association Internationale pour L'Etude des Argiles as Councillor (1972-89) and President (1975-78), and the International Mineralogical Association and the International Union of Crystallography as Chairman of the Joint Committee on Nomenclature (1971-77). His other accomplishments and services to the scientific community and his university are too numerous to elaborate, but they include more than 90 papers and chapters in books and the editorship of two volumes in the *Reviews* series, "Micas" (Vol. 13) and this volume.

Thus it is with greatest pleasure that we dedicate "Hydrous Phyllosilicates" to S.W. Bailey, who has dedicated most of his very highly productive career to their study.

Paul H. Ribbe
Blacksburg, Virginia
September 16, 1988

TABLE OF CONTENTS

Chapter 5   F.J. Wicks & D.S. O'Hanley
SERPENTINE MINERALS: STRUCTURES AND PETROLOGY

Chapter 6   S.W. Bailey

STRUCTURES AND COMPOSITIONS
OF OTHER TRIOCTAHEDRAL 1:1 PHYLLOSILICATES

Chapter 7   S.M. Savin & M. Lee

ISOTOPIC STUDIES OF PHYLLOSILICATES

Chapter 8   B.W. Evans & S. Guggenheim

TALC, PYROPHYLLITE, AND RELATED MINERALS

Chapter 9   J.V. Chernosky, Jr., R.G. Berman & L.T. Bryndzia
STABILITY, PHASE RELATIONS, AND THERMODYNAMIC
PROPERTIES OF CHLORITE AND SERPENTINE GROUP MINERALS

Chapter 10   S.W. Bailey
CHLORITES: STRUCTURES AND CRYSTAL CHEMISTRY

Chapter 11   J. Laird
CHLORITES: METAMORPHIC PETROLOGY

Chapter 12   C. de la Calle & H. Suquet

VERMICULITE

Chapter 13   N. Güven

SMECTITES

x

Chapter 14   D.M. Burt
VECTOR REPRESENTATION OF PHYLLOSILICATE COMPOSITIONS

Chapter 15   R.C. Reynolds, Jr.
MIXED LAYER CHLORITE MINERALS

Chapter 16
B.F. Jones & E. Galan
SEPIOLITE AND PALYGORSKITE

Chapter 17
S. Guggenheim & R.A. Eggleton
CRYSTAL CHEMISTRY, CLASSIFICATION, AND IDENTIFICATION OF MODULATED LAYER SILICATES

PREFACE AND ACKNOWLEDGMENTS

Volume 13 of *Reviews in Mineralogy* presented much of our present-day knowledge of micas. At the time of that volume (1984), I mentioned that there was too much material available to attempt to cover all of the hydrous phyllosilicates in one volume. The micas were treated first because of their abundance in nature and the fact that more detailed studies had been carried out on them than on the rest of the phyllosilicates. The serpentines, kaolins, smectites, chlorites, etc. would have to wait their turn. Now, four years later, that turn has come. Hence the peculiar nature of the title of this volume.

We know less about the rest of the phyllosilicates than we do about the micas, primarily because many of them are of finer grain sizes and lower crystallinities than most of the micas. As a result, we have been unable to determine as much detail regarding their structures, crystal chemistries, and origins. Nevertheless, there is a considerable body of literature about them, and this volume will attempt to collate and evaluate that literature. One compensating factor that has helped greatly in the accumulation of knowledge about these minerals is that some of them occur in large deposits that are of great economic value and thus stimulate interest. For this reason considerable emphasis in this volume will be related to the occurrence, origin, and petrology of the minerals.

I thank the authors of the various chapters for their enthusiasm in undertaking the writing of this volume and for their diligence in [almost] adhering to the timetable needed to produce the volume prior to the Short Course. The series editor, Paul Ribbe, with the help of Marianne Stern and Margie Sentelle, has expedited this process by his usual fast and efficient processing of the manuscripts.

Scholarships and a reduced registration fee to encourage graduate student attendance at the Short Course were made possible by grants from

Shell Development Company	Southern Clay Products
Chevron Oil Field Research Company	Oil-Dri Corporation of America
J.M. Huber Corporation	Thiele Kaolin Corporation
Conoco, Incorporated.	

These grants are gratefully acknowledged.

S.W. Bailey
Madison, Wisconsin
September 1, 1988

Chapter 1

INTRODUCTION

STRUCTURAL DEFINITIONS

The phyllosilicates to be considered in this chapter ideally contain continuous two-dimensional *tetrahedral sheets* of composition T_2O_5 (T = tetrahedral cation, normally Si, Al, Fe^{3+}, and rarely Be^{2+} or B^{3+}), in which individual tetrahedra are linked with neighboring tetrahedra by sharing three corners each (the basal oxygen atoms) to form an hexagonal mesh pattern (Fig. 1a). In platy, non-modulated structures, all of the unshared corners (the apical oxygen atoms) point in the same direction to form part of an immediately adjacent *octahedral sheet* in which individual octahedra are linked laterally by sharing octahedral edges (Fig. 1b). The common plane of junction between the tetrahedral and octahedral sheets consists of the apical oxygen atoms plus OH groups that lie at the center of each tetrahedral 6-fold ring at the same z-level as the apical oxygens. F, Cl, and S may substitute for OH in some species. The octahedral cations normally are Mg, Al, Fe^{2+}, and Fe^{3+}, but other medium-sized cations such as Li, Ti, V, Cr, Mn, Co, Ni, Cu, and Zn also occur in some species. In some layer silicate minerals the apical oxygen atoms of the tetrahedral sheet do not all point in the same direction. Instead, as shown later, certain tetrahedra may be inverted so that their apical oxygen atoms are shared with other tetrahedra or octahedra of a different sheet. This inversion and relinkage usually produces a regular *modulated* structure in which either or both of the resultant lateral repeat distances (*a,b*) become *commensurate* or *incommensurate* superlattice dimensions relative to their non-modulated values.

The assemblage formed by linking one tetrahedral sheet with one octahedral sheet is known as a *1:1 layer*. In such layers the uppermost, unshared plane of anions in the octahedral sheet consists entirely of OH groups. Although isomorphous substitutions can take place in both component sheets, 1:1 layers must be electrostatically neutral overall. A *2:1 layer* links two tetrahedral sheets with one octahedral sheet. In order to accomplish this linkage, Figure 1c shows that the upper tetrahedral sheet is inverted so that its apical oxygens point down and can be shared with the octahedral sheet below. Both octahedral anion planes then are of the same O,OH composition. A 2:1 layer may be electrostatically neutral or may have an excess negative charge. Any excess layer charge is neutralized by various *interlayer* materials, including individual cations, hydrated cations, and hydroxide octahedral groups and sheets. The total assemblage of a layer plus interlayer is referred to as a *unit structure*. It contains one or more chemical formula units (Z). It should be noted that the terms plane, sheet, layer, and unit structure as used here have precise meanings and are not interchangeable. They refer to increasingly thicker parts of the layered arrangement.

2

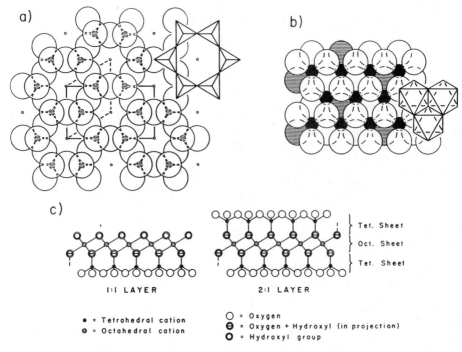

Figure 1. Idealized tetrahedral sheet with hexagonal (dashed line) and orthogonal (full line) cells. (b) Idealized octahedral sheet with OH groups of lower anion plane (dashed circles) shown shaded. For a 1:1 layer upper anion plane (full circles) will be entirely OH groups. For a 2:1 layer upper anion plane will be of same composition as lower anion plane. (c) Combination of sheets to form 1:1 and 2:1 layers. Modified from Bailey (1980a).

Layer charge (x) is the net negative charge per formula unit that may exist on a 2:1 layer when the total anionic negative charge per formula unit is greater than the total cationic positive charge, assuming complete ionization. A net negative charge can arise from

  (1) substitution of R^{3+} or R^{2+} for Si^{4+} in tetrahedral coordination,
  (2) substitution of R^{1+} or R^{2+} for R^{2+} or R^{3+}, respectively, in octahedral coordination,
  (3) presence of octahedral vacancies, or
  (4) dehydroxylation of OH to O.

The net charge may originate entirely within the tetradral sheet or entirely within the octahedral sheet in some species, or may come partly from both sheets. Sometimes a positive charge on the octahedral sheet (usually due to substitution of a higher charged cation for a lower charged cation or for a vacancy) may offset a negative charge on the tetrahedral sheet, either exactly as in 1:1 type layers or partly as in some 2:1 layers. Hydroxylation of O to OH also can lead to reduction of the magnitude of a negative layer charge. Whatever the source of net negative charge, its magnitude must be balanced by a correponding positive charge on the interlayer so that the crystal as a whole is electrostatically neutral.

In non-modulated layers the smallest structural unit contains three octahedral sites. If all three octahedral sites are occupied, the sheet is classified as *trioctahedral*. If only two octahedra are occupied and the third octahedron is vacant, the sheet is classified as *dioctahedral*. Most natural specimens do not contain exactly 3.0 or 2.0 octahedral cations per formula unit, but 2.5 will be used here as a boundary for classification because few homogeneous layer silicate minerals have octahedral cation totals near this value. In modulated structures portions of the structure can be identified as part of a 1:1 or 2:1 layer. The geometry of relinkage of the inverted tetrahedra is likely to change the tetrahedral-octahedral cation ratio, however, and thereby make a structural allocation of the chemical analysis difficult without knowledge of the structural details.

LATERAL FIT OF SHEETS WITHIN A LAYER

In order to fit together in a common layer the tetrahedral and octahedral sheets must have similar lateral dimensions. Sheets that fit together only with difficulty will have strained interfaces that influence greatly the resultant crystal size, morphology, and structure. The relative degree of lateral fit between the tetrahedral and octahedral sheets can be estimated by determining the lateral dimensions such sheets would have in the free state. Free octahedral sheets occur in some hydroxide minerals and can be used as analogues to estimate the ideal lateral dimensions for the octahedral sheets in layer silicates of similar composition. Thus, gibbsite and bayerite have $b = 8.64$ Å and 8.67 Å, respectively, for the dioctahedral $Al(OH)_3$ composition and brucite has $b = 9.43$ Å for the trioctahedral $Mg(OH)_2$ composition. These observed values are 0.5 Å smaller than those calculated on the basis of regular octahedra. Although tetrahedral sheets do not exist in an unconstrained state in nature, it is possible to estimate their ideal lateral dimensions in the free state by assuming an average T--O bond length and ideal tetrahedral angles:

$$a = \frac{4\sqrt{6}}{3} (T\text{--}O); \quad b = a\sqrt{3} = (4\sqrt{2})(T\text{--}O) \qquad (1)$$

Thus, for undistorted hexagonal geometry and Si--O $= 1.62\pm.01$ Å the resulting lateral dimensions are $a = 5.28$ Å and $b = 9.15$ Å. Substitution of the larger Al for Si in the tetrahedral sites increases these values. Assuming Al^{IV}--O $= 1.75\pm.01$ Å, then $b = 9.89$ Å and $b(Si_{1-x}Al_x) \cong 9.15$ Å $+ 0.74x$. This relationship should be considered as only approximate, not only because of the uncertainties in the assumed bond lengths but also because the tetrahedral angles are known to vary somewhat from structure to structure.

Despite the uncertainties noted above it is evident that a tetrahedral sheet containing only Si has an ideal lateral b dimension that is intermediate between those of the ideal octahedral sheets containing only Mg or only Al. The magnitude of the misfit is approximately 3-5% for these compositions. Structural adjustments of some type are necessary in order to eliminate this misfit and to articulate the tetrahedral and octahedral sheets into a layer with common planes of junction if the structure is to be of the non-modulated type.

Tetrahedral rotation

In the great majority of layer silicates the ideal lateral dimensions of the tetrahedral sheet are larger than those of the octahedral sheet either because the latter are dioctahedral or as a result of tetrahedral substitution of Al or Fe^{3+} for Si. Mathieson and Walker (1954), Newnham and Brindley (1956), Zvyagin (1957), Bradley (1959), and Radoslovich (1961) have shown that it is relatively easy to reduce the lateral dimensions of a tetrahedral sheet by rotating adjacent tetrahedra in opposite senses (clockwise and anticlockwise) in the (001) plane. The amount of rotation necessary to relieve the misfit is given by

$$\cos\alpha = b(\text{obs})/b(\text{ideal}) \quad , \tag{2}$$

where α is the rotation angle. This expression is only an approximation in practical application because of the difficulty in estimating b(ideal). It predicts a non-linear lateral reduction that ranges from 1.5% for $\alpha = 10°$ to 6.0% for $\alpha = 20°$ and 13.4% for $30°$, which is the maximum possible rotation because it brings three basal oxygens into contact at the center of each hexagonal ring. Toraya (1981) proposed a linear relation between the rotation angle (α) and the dimensional misfit (Δ) between the tetrahedral and octahedral sheets

$$\alpha = 35.44\Delta - 11.09 \quad , \tag{3}$$

where $\Delta = [2\sqrt{3}e_b - 3\sqrt{2}d_0]$, e_b = the mean of all basal tetrahedral edge lengths, and d_0 = the mean of all M-0,OH,F octahedral bond lengths. Although equation (3) was formulated for micas, it should be approximately applicable to other phyllosilicates as well.

The direction of tetrahedral rotation is governed by attraction of the basal oxygen atoms by one or more of three sets of atomic neighbors, namely

(1) hydrogen bonding to adjacent OH or H_2O surfaces,
(2) octahedral cations within the same layers, and
(3) octahedral cations in the next layer or interlayer.

In micas, talc, and pyrophyllite only factor (2) is important. In layer silicate minerals with 1:1 layers or with 2:1 layers having appreciable interlayer material present, however, all three factors may be important. The relative strengths of the attractions then are as listed above, with shortening of 0--OH contact distances strongest. During the rotation every other basal oxygen atom within a 6-fold ring is constrained to rotate toward the center of that ring, and the alternate oxygen atoms move away from the center of that ring in directions that are toward the centers of adjacent rings. This creates a ditrigonal rather than hexagonal symmetry for the tetra- hedral sheet. For compositions most abundant in nature, ditrigonal rings are to be expected because the tetrahedral sheets will have lateral dimensions that are larger than those of the octahedral sheets to which they are joined. Peterson et al. (1979)

calculated that an isolated 6-fold ring would have minimum energy for
$\alpha = 16°$ rather than $0°$, suggesting that even the "ideal" tetrahedral
sheet for a layer silicate should be shown as having ditrigonal
symmetry.

Other structural adjustments

Because the sheets are semi-elastic some adjustment for lateral
misfit where the tetrahedral sheet is larger than the octahedral
sheet can be accomplished by thickening the tetrahedral sheet and
thinning the octahedral sheet. This has the effect of decreasing
and increasing, respectively, their lateral dimensions. It should
be emphasized, however, that this mechanism plays a minor role in
compensating for lateral misfit because of the limitations possible
in the distortion of individual tetrahedra and octahedra. Also,
other factors can be important in affecting these polyhedral shapes.
The adjustments that do take place can be expressed either in terms
of the measured sheet thicknesses (t_o) or of the τ_{tet} and ψ_{oct}
angles. An ideal tetrahedron has an angle $\tau = (O_{apical}\text{--}T\text{--}O_{basal})$
$= 109°28'$, which is increased by thickening and decreased by
thinning. An ideal octahedron has an angle ψ, which is defined as
the angle between an apex-to-apex body diagonal and the vertical.
For a sheet of linked octahedral, ψ will be calculated here as

$$\cos \psi = \text{oct. thickness}/2(M\text{--}O,OH) \quad , \qquad (4)$$

where M--O,OH is the mean of all octahedral cation-anion distances,
including vacancies. The ideal ψ angle is $54°44'$, which increases
with octahedral thinning.

It is much more difficult to stretch the lateral dimensions
of a tetrahedral sheet in order to fit with a larger octahedral
sheet, and the structural adjustments may be more drastic as a
result. Mechanisms observed in layer silicate minerals include
adjustments of sheet thicknesses in the reverse sense to
that described above, tetrahedral tilting (to increase their
$O_{apical}\text{--}O_{apical}$ distances) that may lead to curling of 1:1 layers
(as in asbestiform chrysotile), and inversion of tetrahedral apical
directions and relinkage in some periodic pattern to form a modulated
structure. Examples of these kinds of lateral misfits are restricted
to species with little or no tetrahedral substitution for Si and with
trioctahedral sheets populated with relatively large cations of the
size of Mg, Ni, Co, Zn, Fe^{2+}, and Mn^{2+}. Some of the resulting
structures are discussed in Chapters 5, 16, and 17 of this volume.

UNIT CELLS

The individual sheets making up a layer have hexagonal symmetry
when undistorted (Figs. la,b). In an actual crystal, however, the
resultant 3-dimensional symmetry may be lowered either because of the
geometry of stacking of successive layers along the Z direction or
because of layer distortions due to lateral misfit of tetrahedral
and octahedral sheets, ordering of cations, or ionic interactions.
If the resultant symmetry remains hexagonal or trigonal, the smallest
unit cell is Primitive and has an hexagonal shape with three
equivalent X axes 120° apart lying in the horizontal plane of the

layer and with the Z axis normal to the layer (dashed line cell in Fig. 1a). The axial repeat distances for this cell are $a = b \cong$ 5.15-5.4 Å and $c \cong n$ x unit structure thickness, where n = number of 1:1 or 2:1 layers in the unit cell. A one-layer Primitive cell of this shape contains one chemical formula unit (Z), provided there is no superperiodicity in the plane of the layer.

A trigonal or hexagonal structure, alternatively, can be described in terms of a larger C-centered cell with an orthogonal-shaped base in which the Y axis is chosen to be normal to the Z axis and to one of the X axes (solid line cell in Fig. 1a). The axial repeat along this Y axis is $b \cong a\sqrt{3}$, and the cell volume is twice that of the Primitive cell for the same periodicity along Z. Because of the greater ease of indexing X-ray powder patterns on an orthogonal (sometimes termed orthohexagonal) rather than hexagonal cell, the larger C-cell often is used for indexing until the true symmetry of the mineral is known. C-cells normally are used to describe structures with orthorhombic, monoclinic, or triclinic resultant symmetries, although for the latter two cases the Z-axis is not necessarily normal to the layer.

The number of chemical formula units (Z) present in a unit cell of a hydrous phyllosilicate will be unity for a one-layer Primitive cell of hexagonal shape and two for the corresponding orthogonal C-cell. This number must be multiplied by the number of layers in the cell for regular-stacking multi-layer structures. Because of this variability in the unit cell size and shape the writer recommends that all chemical formulas be expressed in terms of one formula unit. This is a safe procedure, whatever the unit cell, and enables consistent correlation of chemistry with structure within the entire family of hydrous phyllosilicates. Statements often found in the literature that the structural allocation of a chemical analysis is based on a full cell or a half cell presuppose a knowledge of the shape and volume of the true cell that may not be justified.

CLASSIFICATION

Hydrous phyllosilicate minerals can be classified conveniently on the basis of layer type (1:1, 2:1, or modulated), layer charge (x) per formula unit, and type of interlayer into several major groups. Further subdivision into subgroups is made on the basis of the octahedral sheet type (dioctahedral or trioctahedral) and into species by chemical composition or, occasionally, by the geometry of superposition of individual layers and interlayers. Such a classification scheme for non-modulated species is given in Table 1. The structures of the major groups of non-modulated hydrous phyllosilicates are shown diagrammatically in Figure 2. The names used for groups, subgroups, and species are those approved by the AIPEA Nomenclature Committee (Bailey, 1980b) or by the IMA Commission on New Minerals and Mineral Names. A classification scheme for modulated species is given in Chapter 17 of this volume.

The following chapters in this volume will cover all of these minerals groups and sub-groups (excluding the micas and brittle micas) approximately in the order listed in Table 1.

Table 1. Classification of non-modulated hydrous phyllosilicates

Layer type	Interlayer	Group*	Subgroup	Species
1:1	None or H_2O only	Serpentine-kaolin $x \sim 0$	Serpentines	Chrysotile, lizardite, amesite, berthierine, cronstedtite, etc.
			Kaolins	Kaolinite, dickite, nacrite, halloysite
2:1	None	Talc-pyrophyllite $x \sim 0$	Talcs	Talc, willemseite
			Pyrophyllites	Pyrophyllite, ferripyrophyllite
	Hydrated exchangeable cations	Smectite $x \sim 0.2$-0.6	Saponites	Saponite, hectorite, sauconite, stevensite, etc.
			Montmorillonites	Montmorillonite, beidellite, nontronite, volkonskoite, etc.
	Hydrated exchangeable cations	Vermiculite $x \sim 0.6$-0.9	Trioctahedral vermiculites	Trioctahedral vermiculite
			Dioctahedral vermiculites	Dioctahedral vermiculite
	Non-hydrated cations	True mica $'x \sim 0.5$-1.0	Trioctahedral true micas	Phlogopite, biotite, lepidolite, zinnwaldite, annite, etc.
			Dioctahedral true micas	Muscovite, illite, glauconite, tobelite, paragonite, etc.
	Non-hydrated cations	Brittle mica $x \sim 2.0$	Trioctahedral brittle micas	Clintonite, bityite, anandite, kinoshitalite
			Dioctahedral brittle micas	Margarite
	Hydroxide sheet	Chlorite x variable	Trioctahedral chlorites	Clinochlore, chamosite, nimite, pennantite, baileychlore
			Dioctahedral chlorites	Donbassite
			Di,trioctahedral chlorites	Cookeite, sudoite

* x is charge per formula unit.

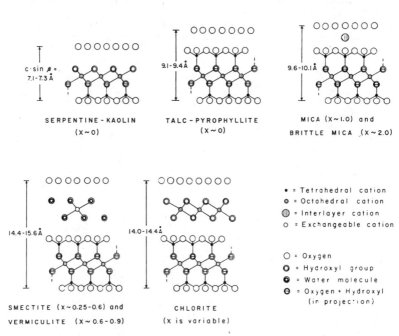

Figure 2. [010] view of structures of major non-modulated hydrous phyllosilicate groups. Modified from Bailey (1980a).

8

Bailey, S.W. (1980a) Structures of layer silicates. Ch. 1 in: Crystal Structures of Clay Minerals and their X-ray Identification, G.W. Brindley and G. Brown (Eds.), Mono. 5 Mineralogical Soc., London, 1-124.

_____ (1980b) Summary of recommendations of AIPEA Nomenclature Committee. Clay Minerals 15, 85-93.

_____ (1984) Micas. (Ed.), Reviews in Mineralogy 13, Mineral. Soc. Am., 584 p.

Bradley, W.F. (1959) Current progress in silicate structures. Clays & Clay Minerals 6, 18-25.

Mathieson, A.McL. and Walker, G.F. (1954) Crystal structure of magnesium-vermiculite. Am. Mineral. 39, 231-255.

Newnham, R.E. and Brindley, G.W. (1956) The crystal structure of dickite. Acta Crystallogr. 9, 759-764.

Peterson, R.C., Hill, R.J. and Gibbs, G.V. (1979) A molecular-orbital study of distortions in the layer structures brucite, gibbsite, and serpentine. Canadian Mineral. 17, 703-711.

Radoslovich, E.W. (1961) Surface symmetry and cell dimensions of layer lattice silicates. Nature 191, 67-68.

Toraya, H. (1981) Distortions of octahedra and octahedral sheets in 1M micas and the relation to their stability. Z. Kristallogr. 157, 173-190.

Zvyagin, B.B. (1957) Determination of the structure of celadonite by electron diffraction. Soviet Phys. Crystallogr. 2, 388-394 (English transl.).

Chapter 2   S.W. Bailey

POLYTYPISM OF 1:1 LAYER SILICATES

INTRODUCTION

Hydrous phyllosilicates of the 1:1 type have layers, about 7 Å thick, that consist of one tetrahedral sheet joined to one octahedral sheet. The common plane of junction between the sheets consists of the apical oxygen atoms of the tetrahedral sheet, which serve also as octahedral corners, plus unshared OH groups that are positioned at centers of the 6-fold rings at the same z coordinates as the apical oxygen atoms. The octahedral cations are positioned on top of this plane of junction, where they may occupy either the set of three positions labeled I in Figure 1a or the set of three positions labeled II. But the cations may not be intermixed between these sets in the same plane. The upper anion plane of the octahedral sheet consists entirely of OH groups that are positioned above the cations to complete the 6-fold, octahedral coordination. Small amounts of F and Cl are known to substitute for OH in some specimens, and much more can be introduced synthetically.

Depending on the set of positions (I or II) the octahedral cations occupy, the octahedra themselves and the octahedral sheet have slants (Fig. 1b) that differ by ±60° or 180° relative to fixed axes. Because the set of positions labeled I can be transformed to

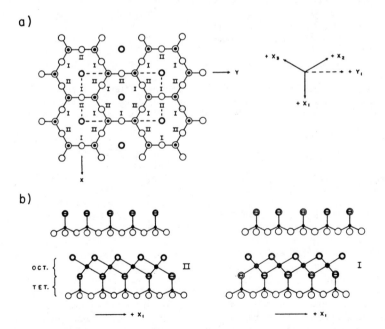

Figure 1. (a) Definition of I and II sets of octahedral cation positions above a tetrahedral net for a 1:1 layer, relative to a fixed set of hexagonal axes. (b) Octahedral sheets slant in opposite directions relative to a fixed X_1 direction for occupancy of I or II set of octahedral cation positions (from Bailey, 1969).

10

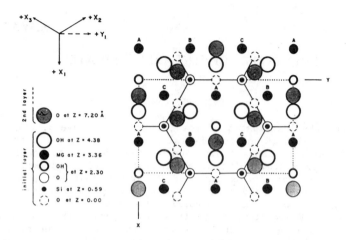

Figure 2. Pairing of tetrahedral oxygen atoms (stippled) at the base of an upper 1:1 layer with the octahedral OH groups (large double circles) on the upper surface of the layer below (modified from Bailey, 1980). Other geometrical patterns of pairing also are possible.

the set of positions labeled II by rotation of ±60° or 180°, these sets of positions are not really different in the case of regularly stacked one-layer structures. We are concerned not with which set of positions is occupied by octahedral cations, therefore, but whether the same set of three positions is occupied in each successive layer or whether there is a regular alternation between the I and II sets in successive layers. This variable is defined as the first degree of freedom for hydrous phyllosilicates having either 1:1 or 2:1 type layers. It will be recognized from Figure 1 that alternation of octahedral cations between I and II sets of positions in successive layers can be described also as relative rotations of layers.

In all planar 1:1 layer structures studied in detail, the positions of adjacent layers are determined by the pairing of each oxygen of the basal tetrahedral surface of one layer with an OH group of the upper octahedral surface of the layer below. One possible pairing is shown in Figure 2. This O--OH pairing results in the formation of long and relatively weak hydrogen bonds, approximately 2.9-3.0 Å between the anion centers, that bond the layers together. Curling of the layers into cylinder-like coils, as in chrysotile, prevents ideal hydrogen bonds of this sort (Wicks and Whittaker, 1975). Because O--OH pairing can be obtained by several different ways of superimposing one layer on top of another, this variable represents the second degree of freedom in 1:1 layer structures. There is some controversy as to whether all O--OH pairs actually participate in the interlayer bonding in the kaolin materials and as to the correlation of their IR absorption spectra with the orientation of the O...H vectors of the OH groups. This controversy is considered in more detail in Chapter 3. Additional electrostatic attractions between layer surfaces also may exist as a result of tetrahedral and octahedral substitutions. Although a 1:1 layer ideally must be electrostatically neutral, coupled substitutions can create balanced negative and positive charges on the tetrahedral and octahedral sheets, respectively. These charges in turn can interact between sheets across the interlayer space.

The name *septechlorite* was proposed by Nelson and Roy (1954, 1958) as a name for all trioctahedral layer silicates based on a 1:1 layer. The name has not found favor among most mineralogists because the structure of chlorite is quite different from that of a 1:1 layer. The name implies that the chlorite group should be subdivided into 7 Å and 14 Å structural subgroups, which is at odds with the structural rationale behind the approved classification scheme in Table 1 of Chapter 1. The name has not been approved by any nomenclature committee and should not be used.

DERIVATION OF STANDARD POLYTYPES

Assumptions

In deriving the different structural sequences (polytypes) of 1:1 layers that are theoretically possible, some limiting assumptions must be made in order to obtain a group of structures that is small but still representative of natural specimens. It will be assumed here that the individual sheets are undistorted, have trioctahedral compositions with no ordering of cations, and that the layer stacking is regular. Correlation of these theoretical trioctahedral structures with the natural dioctahedral kaolin minerals will be considered separately. For the case of interlayer displacements along X (see below), it is assumed that the same interlayer stacking angle, e.g., $\pm 60°$ or $\pm 120°$, is found between all layers.

Bailey (1963, 1969) has shown for the case of planar structures that interlayer hydrogen bonds can be formed for the following three relative positions of adjacent 1:1 layers.

(1) Shift of the second layer by $a/3$ along the fixed hexagonal axes X_1, X_2, or X_3 of the initial layer, with or without rotation of the second layer. Because of the different slants of the I and II octahedra, the interlayer shift must be in the negative direction to give hydrogen bonding if the II octahedral set is occupied in the initial layer, as shown in Figure 2, or in the positive direction if the I octahedral set is occupied. This is the reverse of the *interlayer* shifts of $a/3$ within 2:1 layers.

(2) No shift of the second layer. The hexagonal rings in the tetrahedral sheets of successive layers are exactly superimposed along Z (aside from possible rotations of $\pm 60°$ or $180°$, equivalent to a change of octahedral cation sets occupied in the layers).

(3) Shift of the second layer by $\pm b/3$ along any of the three hexagonal Y axes of the initial layer (defined as normal to the three X axes), with or without rotation of the second layer. If the layers are assumed to be undistorted, shifts along each of the three Y axes give identical results so that the particular Y axis used is immaterial. Interlayer shifts along the positive or negative directions of Y are both possible, regardless of the octahedral set occupied in the initial layer.

A critical assumption may now be made that will limit the total number of theoretical polytypes substantially. It is assumed that the three types of relative layer superpositions listed above are not intermixed in the same crystal. With this assumption Bailey

(1969) derived 12 standard trioctahedral 1:1 polytypes, plus four enantiomorphic structures, with c-periodicities between one and six layers.

Other derivations for planar 1:1 type polytypes have been given by Newnham (1961), Zvyagin (1962, 1967), Steadman (1964), Zvyagin et al. (1966), Dornberger-Schiff and Durovic (1975a,b), and Durovic et al. (1981). Bailey (1969) discusses the differences between his derivation and those of the earlier authors. The original references should be consulted for details. Wicks and Whittaker (1975) give a notation for cylinder-like or coiled structures that is utilized in Chapter 5.

Derivation of 12 standard polytypes

Variability both in the positioning of adjacent layers and in the occupancy of octahedral sets within each layer (equivalent to layer rotations) leads to different permissible stacking sequences of layers. Structures resulting from different stacking sequences of layers are termed *polytypes*, provided the layer compositions are similar. The 12 polytypes that are derived based on the assumptions listed above are termed here the "standard" polytypes. The great majority of natural platy 1:1 specimens will be shown to adopt one or the other of these structures.

The stacking sequences of layers for the 12 standard trioctahedral polytypes are shown diagrammatically in plan view in Figure 3. In all cases the initial 1:1 layer is oriented so the II set of octahedral cation positions is occupied; for the case of interlayer X shifts the first $a/3$ shift is directed along $-X_1$; for the case of interlayer Y shifts the first $b/3$ shift is directed along $-Y_1$. [The sense (+ or -) of shift is governed by the octahedral cation set occupied in the lower layer for $a/3$ shifts, whereas the X axis used is optional. For $b/3$ shifts, both the sense of shift and the Y axis used are optional for occupancy of either octahedral cation set, but shifts along different Y axes do not lead to different results]. In Figure 3 the interlayer shifts that give rise to valid interlayer hydrogen bonding are shown as vectors directed from the center of an hexagonal ring (circles) in the tetrahedral sheet of a lower layer to the center of an hexagonal ring in the layer above. These vectors have projected lengths on (001) of $a/3$ in Figure 3a, zero in Figure 3b, and $b/3$ in Figure 3c. The *interlayer stacking angle* will be defined as the angle between two adjacent vectors in projection on (001), where a clockwise measurement (from a lower to a higher vector) denotes a positive angle. The six possible interlayer stacking angles are 0°, 60°, 120°, 180°, 240°, and 300°. Because of the plane of symmetry that is parallel to the interlayer shift vector, angles 60° and 300° are considered equivalent (= ±60°), as are angles 120° and 240° (= ±120°).

Each standard polytype in Figure 3 is designated by a Ramsdell structural symbol consisting of a number to indicate the number of layers in the repeating unit along Z and letters to indicate the resulting lattice symmetry (M = monoclinic, T = trigonal, H = hexagonal, R = rhombohedral, Or = orthorhombic, and Tc = triclinic).

13

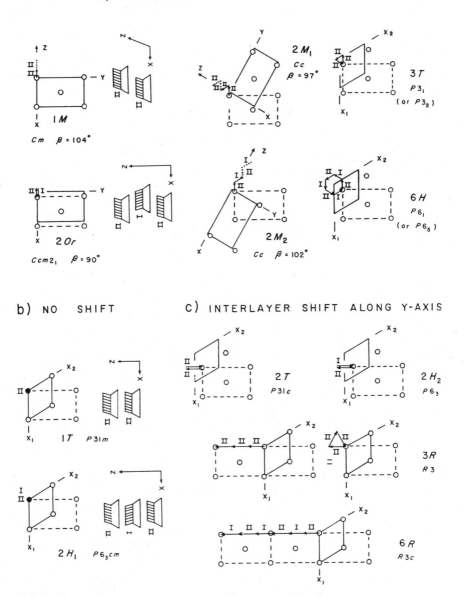

Figure 3. Derivation of 12 standard 1:1 polytypes (from Bailey, 1980). The orientation of Figure 1a is assumed for the first layer of each polytype with the II set of octahedral cation positions occupied. The base of the resultant unit cell (solid lines) may or may not coincide with the fixed cell (dashed lines) used for the first layer. The origin of each resultant unit cell has been placed as close to the center (circles) of an hexagonal ring in the first layer as the symmetry permits. A solid circle at the origin indicates zero interlayer shift. Two equivalent interlayer shift sequences using different Y axes are illustrated for the 3R polytype. Enantiomorphic structures are not shown.

Subscripts 1 and 2 are used to differentiate structures that have the same periodicity and symmetry.

For interlayer shifts of $a/3$ (Fig. 3a) six standard vector patterns are derived that are identical to the vector patterns of the six standard mica polytypes, so these will be given the same structural symbols as for the micas. There is a fundamental difference in the two cases, of course, in that the vectors for the micas (as shown in Fig. 3 of Bailey, 1984) represent shifts of $a/3$ of the upper tetrahedral sheet relative to the lower tetrahedral sheet *within* the same 2:1 layer whereas for the 1:1 layers in Figure 3a the vectors represent actual shifts *between* layers of $a/3$. It should be noted also that a 1:1 layer is not centrosymmetric and is thinner than a centrosymmetric 2:1 layer, so the resulting space groups and crystallographic β angles will be different than those for the micas. Following the style of derivation given for the micas by Bailey (1984), the two vectors patterns at the left of Figure 3a represent structures in which interlayer shifts are restricted to one X axis (selected as X_1). For the two vector patterns in the middle part of the diagram shifts are allowed along two X axes (selected as X_1 and X_2). On the right of the diagram shifts occur along all three X axes. The upper member of each of the three pairs of structures has the octahedral cations in the same set of positions in each layer, whereas in the lower member there is a II,I regular alternation of cations in successive layers along Z. These two groups (of three structures each) will be termed groups A and B, respectively, in the following discussion. They will be shown to be the least stable of the standard polytypes. The choices of the X axes used here for the interlayer shift directions are based solely on convenience of illustration and do not affect the polytypes derived; only the resultant orientations of the polytypes relative to the fixed axes assumed for the first layer are affected.

In Figure 3b, for the case of zero interlayer shift, the two possibilities are either to have all of the layers the same ($2T$ polytype) or to alternate octahedral cation sets regularly ($2H_1$ polytype). In Figure 3c, interlayer shifts of $b/3$ along Y_1 are shown. At the top, alternating - and + directions of shift lead to two structures. Below, shifts of $b/3$ in the same direction lead to another two structures. Group C is defined to include the $1T$, $2T$, and $3R$ polytypes, all having octahedral cations the same in each layer. In group D the $2H_1$, $2H_2$, and $6R$ polytypes have a regular alternation of II,I cation sets in successive layers. All six structures in groups C and D are shown as based on an hexagonal-shaped P cell.

Figure 4 illustrates the 12 standard polytypes diagrammatically in XZ and YZ projections. It also shows the four enantiomorphic structures that were ignored in the derivation above. The differences in projection give clues as to the classes of X-ray reflections that may be best used for differentiation.

15

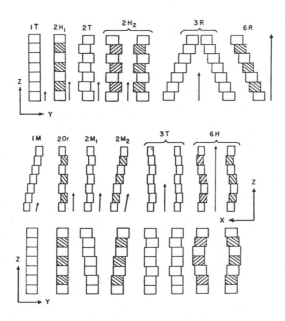

Figure 4. Diagrammatic *XZ* and *YZ* projections of 12 standard polytypes plus 4 enantiomorphic structures (linked by brackets). Blank and ruled rectangles represent layers with octahedral cations in sets II and I, respectively. Thin arrows indicate layer periodicity along Z.

Table 1. Structure amplitudes (F_c) for strong reflections of four groups of standard polytypes

| Interlayer Shifts of $a/3$ | | | | Shifts of 0 or $b/3$ | |
| Oct. same all layers Group A $(1M$-$2H_1$-$3T)$ | | Oct. alternate I,II Group B $(2Or$-$2M_2$-$6H)$ | | Oct. same all layers Group C $(1T$-$2T$-$3R)$ | Oct. alternate I,II Group D $(2H_1$-$2H_2$-$6R)$ |
d(Å)	F_c	d(Å)	F_c	F_c	F_c
2.649	25	2.670	11	12	8
2.590	17	2.624	17	0	11
2.387	54	2.499	24	18+51	34
2.262	13	2.326	32	0	22
2.007	33	2.134	17	19+39	17
1.886	18	1.945	15	0	27
1.665	47	1.771	25	9+45	18
1.568	31	1.615	33	0	25
1.396	16	1.478	8	44+30	29
1.322	44	1.358	28	0	20
1.191	14	1.254	21	32+10	19
1.134	6	1.162	3	0	12

Calculations made with B = 1.2, 0.8, and 2.0 for Mg, Si, and O,OH respectively, and normalized to one formula unit. Indices are of type 20ℓ for monoclinic and orthorhombic C cells (except $2M_2$) and 11ℓ for hexagonal P cells. In group C, $F_c(11\bar{\ell}) \neq F_c(11\ell)$, and each is listed (e.g., 18+51).

IDENTIFICATION OF NATURAL SPECIMENS

Trioctahedral species

The sequences of layers derived above lead directly to atomic coordinates from which theoretical X-ray diffraction patterns can be calculated. For identification purposes it is best to divide the 12 standard planar polytypes into the four groups (A-D) given above. The most intense X-ray reflections ($k = 3n$ for monoclinic- and orthorhombic-shaped C-cells or $hh\ell$ and equivalents for hexagonal-shaped P-cells) have identical calculated intensities for the three structures within each group, assuming identical compositions, but quite different intensities between groups (Table 1). Thus, classification of a good X-ray diffraction pattern of an unknown planar 1:1 layer silicate into one of four groups can be accomplished easily by means of the most intense reflections. Knowledge of the group defines the directions of interlayer shift and the occupancy pattern of octahedral cation sets in the structure. Further differentiation between the three structures within each group is more difficult because it requires the use of weaker reflections, some of which are sensitive to the precise layer stacking sequence. If these reflections are very weak or are diffuse and streaked out due to stacking disorder, the three structures within each group may be indistinguishable. But the group affinity will still be evident from the stronger reflections.

X-ray powder patterns. Bailey (1969) gives calculated powder patterns for all 12 standard polytypes for a Mg-rich composition. These will not be repeated here. Figure 5 illustrates observed powder patterns for representatives of Groups A, C, and D. The X-ray powder patterns of the group A polytypes ($1M, 2M_1, 3T$) differ from those of the other three groups in the d-values of the strongest reflections (Table 1). As in the micas, the 1:1 trioctahedral $1M$ and $3T$ powder patterns are identical so that differentiation of these two polytypes must be made by single crystal study. The $2M_1$ powder pattern can be recognized both by additional weak reflections not present in the $1M$ and $3T$ patterns and by a different ratio of $hk\ell$ ($k \neq 3n$) to 00ℓ intensities due to multiplicity differences. In group B ($2M_2, 2Or, 6H$) the powder patterns of the trioctahedral $2M_2$ and $6H$ polytypes are also identical, as in the micas, and single crystals are needed for differentiation. The $2Or$ pattern can be recognized because it has $k \neq 3n$ reflections that are fewer in number and occur at different d-values than for the $2M_2$ and $6H$ patterns.

In group C ($1T, 2T, 3R$) about one-half of the weak $h \neq k$ reflections (hexagonal indexing) are similar for the $1T$ and $2T$ patterns, but the $2T$ pattern has additional weak reflections that verify its 2-layer nature. The $3R$ weak reflections (of index $h \neq k$ on hexagonal axes) have different d-values than those for $1T$ and $2T$. A similar situation exists in group D ($2H_1, 2H_2, 6R$), in which there are additional weak $h \neq k$ reflections for $2H_2$ that differentiate it from the otherwise similar pattern of $2H_1$. The $6R$ pattern has d-values for its weak $h \neq k$ reflections that are different from those of the $2H_1$ and $2H_2$ patterns.

1.1   1.5   2.0   3.0   7.0 Å

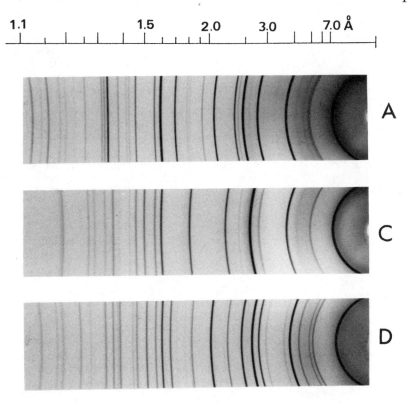

Figure 5. X-ray powder patterns of representative examples of group A, C, and D polytypes of cronstedtite. All patterns taken in Gandolfi cameras of diameter 114.6 mm with graphite-monochromatized Fe$K\alpha_1$ radiation.

The powder patterns of specimens from groups B and D tend to have a greater number of intense lines than do patterns of specimens from groups A and C. This is due to the alternation of octahedral cations in sets I and II in the former groups and a consequent two-layer periodicity in the strong reflections (see below). The weaker reflections needed to distinguish the three structures within each group tend to be very weak for trioctahedral specimens and may not be visible on X-ray powder patterns, especially if the layer stacking sequences have been distorted by grinding. Examples are given in Table 1 of Chapter 6 of patterns in which these weak reflections are visible.

Single crystal patterns. When single crystals are available, X-ray precession photographs can provide more certain identification of individual polytypes. A convenient procedure is to mount the crystal with its (001) cleavage normal to the dial axis. The reciprocal $Z^* = [001]^*$ axis then is horizontal and parallel to the film, and any desired lateral reciprocal axis can be made parallel to the film and approximately vertical by an appropriate rotation around Z^* on the dial axis. In order to avoid confusion between indexing on hexagonal and orthohexagonal axes, only the latter will be used in this subsection. Films of the $h0\ell$ zone (X^* approximately vertical) are strikingly different for the four groups A-D and serve to identify the groups (Fig. 6). Dial rotations of ±60° to bring the [1$\bar{3}$0]* or [130]* zone axes vertical produce precession films identical to that with the X^* = [100]* axis vertical for all polytypes and are not illustrated here. Because all octahedral cations and anions repeat at $b/3$ intervals in the structure, they contribute strongly to X-ray reflections of index k = $3n$ (for orthohexagonal indexing). The periodicity along Z^* for these reflections thus indicates whether the octahedral cations are always in the same set of positions in every layer (repeat of 7 Å, as in groups A and C) or whether they alternate regularly between the II and I sets of positions in successive layers (repeat of 14 Å, as in groups B and D). These periodicities in the structures are illustrated by the left hand arrows for each diagram of Figure 7. For groups B and D, therefore, the spots on the film (in reciprocal space) will be twice as close together and the number of strong k = $3n$ reflections approximately doubled relative to groups A and C. Weaker k = $3n$ reflections indicating a multi-layer periodicity may occur as a result of layer distortion for any polytype.

Precession films of the $0k\ell$ zone (Y^* = [010]* axis vertical) serve to identify individual polytypes within each group (Fig. 8). The basal oxygen atoms do not repeat at intervals of $b/3$ and are principal contributors to the intensities of reflections of index $k \neq 3n$. The periodicity along Z^* for these reflections is that of the basal oxygen planes, therefore, and this is a function of the interlayer shifts and rotations that exist in each polytype (right hand arrows in Fig. 7). This periodicity can be observed along the 02ℓ and 04ℓ row lines in most cases. But if a c-glide plane parallel to (100) is present in the structure, as in $2Or$, $2H_1$, and $6R_1$, the diagnostic $0k\ell$ spots with ℓ odd will be systematically absent and other $k \neq 3n$ reflections (such as 11ℓ) must be examined for the periodicity. The same is true of the $2M_2$ structure but in this case the c-glide plane has been designated as parallel to (010) in order to conform with the rules of monoclinic symmetry. This requires reversal of the customary X and Y directions. Figure 8 shows observed and calculated $0k\ell$ precession photographs for the polytypes in groups A and B, and Figure 9 shows those in groups C and D.

The resultant unit cell shape and ideal symmetry can be obtained from inspection of the three films taken with the reciprocal [010]*, [110]*, and [1$\bar{1}$0]* zone axes (as indexed on orthohexagonal axes) approximately vertical (taken 60° apart around the dial axis). For a monoclinic-shaped cell only the Y^* = [010]* axis is normal to Z^* = [001]* and is in the plane of the film. For an hexagonal-shaped cell (trigonal or hexagonal ideal symmetry) all three of these axes are normal to Z^* and are in the plane of the film when rotated to a vertical position. For a rhombohedral-shaped cell none of the three axes are in the plane of the film when vertical, and white radiation streaks through reflections such as 060, 330, and 3$\bar{3}$0 (030, 300, and 3$\bar{3}$0 for hexagonal indexing) do not extend to the center of the film and are positioned asymmetrically relative to reflections on adjacent row lines (see $6R$ pattern in Fig. 9).

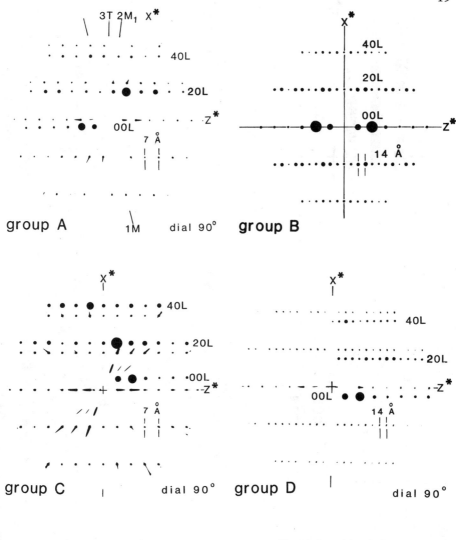

group A   1M   dial 90°   group B

group C | dial 90°   group D | dial 90°

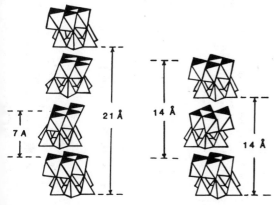

Figure 6. (above) X-ray precession photographs showing observed $h0\ell$ zones of Group A, C, and D polytypes, taken with filtered $MoK\alpha$ radiation. Solid circles of different radii on superimposed strips represent calculated intensities. Group B shows calculated intensities only. Periodicities of the octahedral cations along Z^* are indicated by tick marks. See text for details. From Bailey (1988).

Figure 7. (left) Diagrammatic views of 1:1 structures to illustrate different repeat values of octahedral cation sets (left hand arrows) and of basal oxygen planes (right hand arrows). From Bailey (1988).

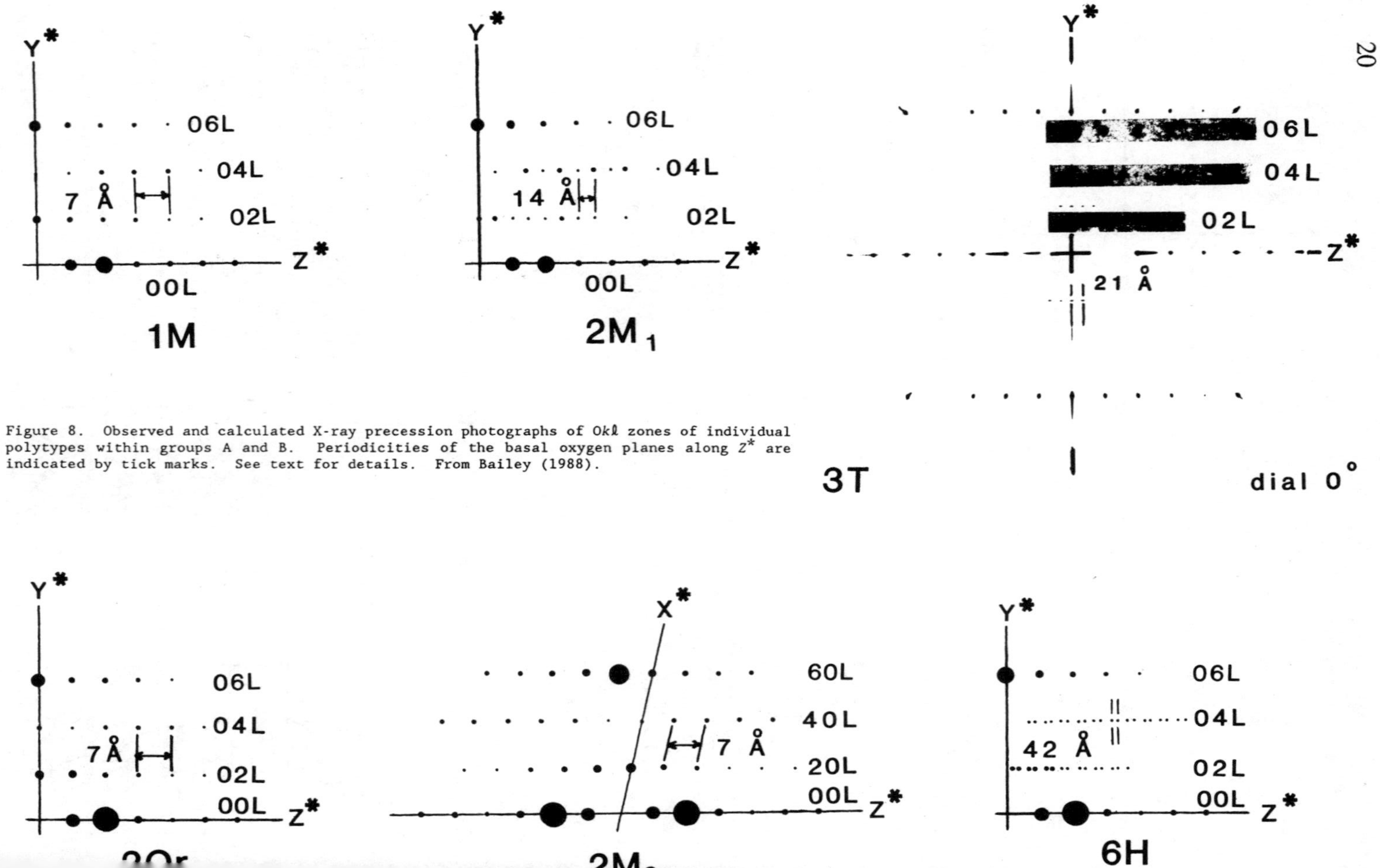

20

Figure 8. Observed and calculated X-ray precession photographs of 0kℓ zones of individual polytypes within groups A and B. Periodicities of the basal oxygen planes along Z* are indicated by tick marks. See text for details. From Bailey (1988).

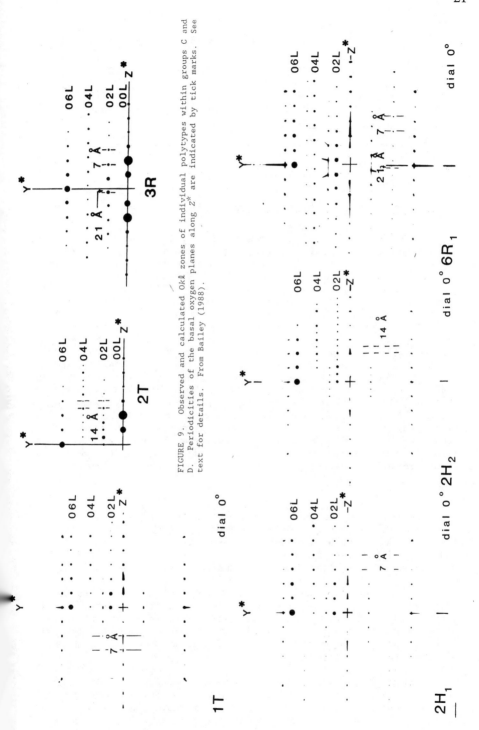

FIGURE 9. Observed and calculated Okl zones of individual polytypes within groups C and D. Periodicities of the basal oxygen planes along Z^* are indicated by tick marks. See text for details. From Bailey (1988).

Table 2. Natural examples of standard 1:1 structures

Species	Trioctahedral	Species	Dioctahedral
lizardite*	$1T$ (+$2H_1$ rare)	kaolinite	$1Tc$ - [$1M_{ord}$]
Al-lizardite	$1T,1M$-$3T$	dickite*	$2M$ - [$1M_{ord}$]
amesite*	$2H_2$(+$2H_1$,$6R$ rare)	nacrite*	$2M$ - [$6R_{ord}$]
kellyite*	$2H_2$		
berthierine	$1T,1M$-$3T$		
brindleyite	$1T,1M$-$3T$		
fraipontite	$1T,1M$-$3T$		
nepouite	$2Or,1T$		
cronstedtite*	$1T,3T,2H_1,2H_2$, $2T,1M,2H_1,3R$		
garnierite	group C + group D		

*Identification by single crystal study

Trioctahedral 1:1 polytypes found in nature. Table 2 lists the correlation of natural 1:1 species with standard polytypes, based on comparison of observed and calculated X-ray patterns. Where single crystals were available, identification was based on single crystal X-ray precession photographs. Examples of all four major groups (A-D) exist in nature, including at least 10 of the 12 individual standard structures. Only the $2M_2$ and $6H$ polytypes have not been identified. There is a strong tendency for certain compositions to prefer specific stacking sequences so that most natural specimens are neither truly polytypic nor truly polymorphic with one another. They may be described instead as *polytypoids*. Only a few natural multi-layer specimens of aluminian serpentine, amesite, and cronstedtite have structures more complex than those of the 12 standard polytypes in that they involve intermixing of different interlayer shift mechanisms (e.g., zero and b/3) within the same crystal. These are discussed in Chapters 5 and 6. Only planar structures have been considered here. Chrysotile and antigorite with curled and modulated layers, respectively, are discussed in Chapter 5, and the modulated structures of greenalite and caryopilite are covered in Chapter 17.

Dioctahedral species

Previous derivations of standard polytypes for trioctahedral micas (Smith and Yoder, 1956) were found to apply without change to the dioctahedral species as well. There were no changes in the space groups or layer periodicities for the latter because the octahedral vacancy was always located on the symmetry plane of the 2:1 layer.

For the three non-hydrated kaolin minerals, however, the vacant octahedral site does not lie on the symmetry plane. Its location requires symmetry changes for all three minerals relative to those of the corresponding trioctahedral 1:1 structures. In addition, the location of the vacancy changes the layer periodicity along Z for dickite and nacrite.

Bailey (1963) noted that kaolinite and dickite are both based on the $1M$ stacking sequence of layers and would be identical if they were trioctahedral (space group Cm). In kaolinite the octahedral vacancy is at the same position in every layer, at position B of Figure 2 (or at its mirror image C). This imposes triclinic symmetry on the structure due to loss of the symmetry plane that relates B to C. The structure is also distorted slightly to triclinic geometry. In dickite the vacant site alternates regularly between sites B and C in adjacent layers to create a two-layer superlattice of monoclinic symmetry (space group Cc). In nacrite the sequence of layers is that of the standard $6R$ polytype (space group $R3c$). The pattern of vacant octahedral sites causes the loss of the three-fold axes, which reduces the symmetry from rhombohedral to monoclinic and allows selection of an inclined Z axis that has true two-layer periodicity. The conventional X and Y axes must be reversed in this structure (as well as in the $2M_2$ structure) because the resulting symmetry plane [which is indexed as (010) in the monoclinic system] is parallel to the 9 Å lateral axis rather than to the 5 Å axis. The symbols in brackets in Table 2 are used to indicate that the observed periodicities and symmetries for these three kaolin minerals are a result of ordering of the octahedral cations and vacancy within the layer sequences of the standard polytypes that are listed in the brackets.

RELATIVE STRUCTURAL STABILITIES

Insufficient synthesis data exist to rank the relative stabilities of the different layer sequences, especially in view of the diverse chemical compositions involved. Following the method used by Newnham (1961) for the dioctahedral kaolin minerals and by Bailey and Brown (1962) for the trioctahedral chlorites, relative structural stabilities of the 12 standard polytypes may be predicted approximately by considering the relative amounts of attraction and repulsion between the ions in the structures. Different stacking sequences of layers lead to different patterns of superposition of ions in adjacent layers and, therefore, to different amounts of electrostatic attraction and repulsion.

Bailey and Brown (1962) and Shirozu and Bailey (1965) found the presence or absence of repulsion between superposed tetrahedral and interlayer octahedral cations to be the major factor in the stability of chlorites (see Chapter 10 for details). In 1:1 layers the analogous situation refers to tetrahedral cation positions in one layer relative to octahedral positions in the layer below. Minimum repulsion exists between these cations for all structures in groups C and D and maximum repulsion for all structures in groups A and B, where these cations exactly superimpose in projection along the normal to (001). Structures in groups C and D, therefore, are expected to be considerably more stable than those in groups A and B.

The direction of tetrahedral rotation in response to lateral misfit between the tetrahedral and octahedral sheets is a second important structural factor. Bailey (1966) analyzed the forces that determine the direction of rotation in 1:1 layers and attempted to rank these in terms of the strengths of their attractive forces on the basal oxygens. Subsequent structural refinements showed that some of the earlier structures on which that ranking was based were incorrect, and a revised ranking of the forces is given below in decreasing order of attraction.

(1) Attraction of basal oxygens by surface hydroxyl groups in the layer below to form interlayer hydrogen bonds as short as possible.
(2) Attraction of basal oxygens by octahedral cations in the same layer.
(3) Attraction of basal oxygens by octahedral cations in the layer immediately below.

For different layer sequences these three forces may act either in unison to attract the basal oxygens to rotate in one particular direction or may act partly or wholly in opposition to one another. In well refined structures of both 1:1 layer structures and chlorites, however, the rotation direction is always such as to shorten the interlayer O--OH contacts, even when this direction moves the basal oxygen atoms away from the octahedral cations both above and below. As a consequence, factor (1) must be stronger than factors (2) and (3) combined. In group C structures the latter two forces combine to act in opposition to that of (1). Factor (3) is not operative in group A and B structures because tetrahedral cations always project onto octahedral cations in the layer below. In all group D structures refined to date the direction of rotation of basal oxygen atoms is always toward the octahedral cation in the same layer (factor 2) and away from the cation in the layer below (factor 3). Because the rotation direction is determined primarily by hydrogen bond contacts (factor 1), however, this is not a true test of the relative strengths of factors (2) and (3), and the distances over which the forces must operate may be the best criterion of relative strengths.

In Table 3 the several forces that may influence the relative structural stabilities have been assigned weights in accord with the structural observations discussed above. Numerical positive values divided by negative values give a resulting structural stability index. Group D is indicated as most stable by this procedure, followed by groups C, A, and B. The same relative order is obtained by use of other linear and exponential weighting schemes, provided the relative weights of the cation-cation repulsion (at 4.5 Å) and the hydrogen bond contacts (at 3.0 Å) are maintained. Unfortunately, the criteria used in this ranking scheme predict identical or similar stabilities for the three structures within each group.

The predicted relative structural stabilities of the four groups are only in moderate agreement with the relative abundances of the structures in nature, as listed in Table 2. Group C and D structures are certainly more abundant than those of groups A and B. This confirms the importance of repulsion between the tetrahedral cations and the octahedral cations in lowering the stability of group A and group B structures. Group B has the lowest ranking and is represented in nature only by rare occurrences of the $2Or$ structure and none of the $2M_2$ and $6H$ structures. It is more difficult to judge the relative stabilities of the structures of groups C and D. Those of group C are adopted by the largest number of species and are volumetrically most abundant because of their importance in lizardite serpentine. This contradicts the structural stability predicted by Table 3 if one uses abundance as the measure of stability. Another point to be noted is that there is no distinctive pattern of

Table 3. Summation of structural factors that influence stability. Features that cause repulsion or weakening of a bond marked by (-), those that cause strengthening of a bond or enhancement of stability marked by (+)

Feature	Operative distance	Relative Weight	Group A $1M$-$2M_1$-$3T$	Group B $2M_2$-$2Or$-$6H$	Group C $1T$	Group C $2T$-$3R$	Group D $2H_1$	Group D $2H_2$-$6R$
Repulsion tet. cations and oct. cations below	4.5 Å	5	-	-	+	+	+	+
Rot. to shorten OH--O bond	3.0	3.5	+	+	+	+	+	+
to oct. cation same layer	3.5	1.8	+	-	-	-	+	+
to oct. cation layer below	4.0	1.5	0	0	-	-	-	-
Repulsion tet cations of adjacent layers	7.2	0.2	0	0	-	1/2-	-	1/2-
Repulsion oct. cations of adjacent layers	7.2	0.2	0	0	-	-	0	0
Repulsion tet. cations and oct. cations in layers above	10.2	0.05	0	-	0	0	0	0
Stability index			$\frac{5.3+}{5.0-}$ -1.06	$\frac{3.5+}{6.85-}$ -0.51	$\frac{8.5+}{3.7-}$ -2.30	$\frac{8.5+}{3.6-}$ -2.36	$\frac{10.3+}{1.7-}$ -6.06	$\frac{10.3+}{1.6-}$ -6.44

25

metastable 1:1 structures changing to more stable 1:1 structures with increasing depth of burial in diagenesis or with increasing grade of metamorphism, as has been found for the polytypes of muscovite and chlorite. This may well be due to most of these 1:1 structures being metastable relative to some different structural arrangement, such as chlorite, biotite, stilpnomelane, etc., as temperature is increased. The high temperature species amesite and kellyite do have the predicted most stable group D structures. Their compositions are also unique, leading to substantial interlayer ionic attractions, and the importance of these factors in determining the structure adopted has not been evaluated. The preference of certain compositions for specific layer stacking sequences, as listed in Table 2, is certainly impressive and warrants the attention of future researchers for explanation.

ACKNOWLEDGMENTS

Previously unpublished research in this chapter was sponsored in part by grant EAR-8614868 from the National Science Foundation and in part by grant 17966-AC2-C from the Petroleum Research Fund, administered by the American Chemical Society. Dr. S. Guggenheim provided constructive criticism of the text.

REFERENCES

Bailey, S.W. (1963) Polymorphism of the kaolin minerals. Am. Mineral. 49, 1196-1209.
_____ (1966) The status of clay mineral structures. Clays & Clay Minerals 14, 1-23.
_____ (1969) Polytypism of trioctahedral 1:1 layer silicates. Clays & Clay Minerals 17, 355-371.
_____ (1980) Structures of layer silicates. Ch. 1 in: Crystal Structures of Clay Minerals and their X-ray Identification, G.W. Brindley and G. Brown (Eds.), Mono. 5 Mineralogical Soc., London, 1-124.
_____ (1984) Classification and structures of the micas. Ch. 1 in: Micas, S.W. Bailey (Ed.), Vol. 13 Reviews in Mineralogy, Mineral. Soc. America, 1-12.
_____ (1988) X-ray identification of the polytypes of mica, serpentine, and chlorite. Clays & Clay Minerals 36, (in press).
_____ and Brown, B.E. (1962) Chlorite polytypism: I. Regular and semi-random one-layer structures. Am. Mineral. 47, 819-850.
Dornberger-Schiff, K. and Durovic, S. (1975a) OD-interpretation of kaolinite-type structures--I: Symmetry of kaolinite packets and their stacking possibilities. Clays & Clay Minerals 23, 219-229.
_____ and _____ (1975b) OD-interpretation of kaolinite-type structures--II: The regular polytypes (MDO-polytypes) and their derivation. Clays & Clay Minerals 23, 231-246.
Durovic, S., Miklos, D. and Dornberger-Schiff, K. (1981) Polytypism of kaolinite-type minerals: an aid to visualize the stacking of layers. Crystal Research Tech. 16, 557-565.
Newnham, R.E. (1961) A refinement of the dickite structure and some remarks on polymorphism in kaolin minerals. Mineral. Mag. 32, 683-704.

27

Shirozu, H. and Bailey, S.W. (1965) Chlorite polytypism. III. Crystal structure of an orthohexagonal iron chlorite. Am. Mineral. 50, 868-885.

Smith, J.V. and Yoder, H.S. (1956) Experimental and theoretical studies of the mica polymorphs. Mineral. Mag. 31, 209-235.

Steadman, R. (1964) The structure of trioctahedral kaolin-type silicates. Acta Crystallogr. 17, 924-927.

Wicks, F.J. and Whittaker, E.J.W. (1975) A reappraisal of the structures of the serpentine minerals. Canadian Mineral. 13, 227-243.

Zvyagin, B.B. (1962) Polymorphism of double-layer minerals of the kaolinite type. Soviet Phys. Crystallogr. 7, 38-51 (English transl.).

_____ (1967) Electron-Diffraction Analysis of Clay Mineral Structures. Plenum Press, New York, 364 p.

_____, Mishchenko, K.S. and Shitov, V.A. (1966) Ordered and disordered polymorphic varieties of serpentine-type minerals, and their diagnosis. Soviet Phys. Crystallogr. 10, 539-546 (English transl.).

Chapter 3   R.F. Giese, Jr.

KAOLIN MINERALS: STRUCTURES AND STABILITIES

INTRODUCTION

The exploitation of white clay for the fabrication of ceramic ware seems to have originated in China. A major source of that material was a mine at Kauling, Kiangsi Province (Grim, 1968), which was in operation from the 11th century until it was closed in 1964 (Keller et al., 1980). The material that was mined took its name from the locality, viz. kaolin. Commercially, kaolins, with kaolinite as the dominant mineral, are presently mined from major deposits in the Cornwall region of England and the south eastern United States, principally Georgia (Jepson, 1984). Two types of deposit have been identified: primary (or residual), and transported (or secondary) (Keller, 1978a). Major deposits are found in South Carolina and Georgia (secondary) and Cornwall, England (residual) with smaller deposits distributed throughout the world (Jepson, 1984). See Chapter 4 of this volume for details of occurrences and origins of kaolins. The major uses of kaolins are in the paper industry where they serve as fillers and coatings. A secondary use is in ceramic ware of various kinds (Jepson, 1984).

The minerals that occur in kaolin are typically very fine-grained and therefore difficult to purify and study. Chemical analyses, optical properties, and reactions due to heating were, for a long time, the principal methods of study. The modern era of research on the kaolin-group minerals began with Ross and Kerr (1930) and their demonstration that X-ray diffraction could identify and characterize individual minerals in a straight-forward manner. They distinguished three minerals: (1) kaolinite, the most abundant of the kaolin-group, (2) dickite, named after Dick (1888), and (3) nacrite, named by Brongniart (1807) and established as different from the other kaolin-group minerals by Des Cloizeaux (1862) and Dick (1908). The hydrated analogue of the kaolin-group minerals, halloysite, presents other problems that will be discussed in a separate section.

In terms of the number of published papers, kaolinite has received most attention. Doubtless this reflects the fact that dickite and nacrite are easier to study (single crystals are available), they are not as common as kaolinite (this is particularly true of nacrite), and both seem to be less complex structurally. Kaolinite, on the other hand, exhibits a wide variability due to the common occurrence of defects in the structure.

KAOLIN-GROUP MINERALS

Unlike the other phyllosilicate minerals, the members of the kaolin group show a nearly uniform chemistry, 46.54 wt % SiO_2, 39.50 wt % Al_2O_3, 13.96 wt % H_2O, corresponding to $Al_2Si_2O_5(OH)_4$. Typical chemical analyses do show some variation in the major elements, but these are generally small (see Table 32 in Deer et al., 1971, for representative analyses). Small quantities of iron, titanium, potassium, and magnesium, normally are found also. The presence of ancillary minerals in small quantities is to be expected, and sometimes these can be identified in X-ray diffraction patterns. The fact that deviations from the ideal stoichiometry are small suggests that ionic substitutions in the structure are few. Bulk chemical or other analytical methods are of little use because it is not possible to eliminate the contamination problem. Improvements in instrumentation, particularly in electron microscopy, both scanning and transmission, along with energy dispersive analysis has allowed the examination of individual particles of these minerals. Very careful work has shown: (1) statistically significant departures from the ideal Si/Al atomic ratio of 2 (Jepson and Rowse, 1975) and (2) the presence of iron (Jepson and Rowse, 1975; Malden and Meads, 1967), small amounts of titanium (Rengasamy, 1976; Jepson and Rowse, 1975), and potassium (Angel and Hall, 1972; Malden and Meads, 1967; 1974) substituting in the structure, as surface coatings, or originating as random interleaving of other phyllosilicates (Lee et al., 1975). Of these, the presence of iron and its possible relation to defects in the crystal structure has attracted the most interest.

The topology of the crystal structure of kaolinite was outlined by Pauling (1930) by analogy with the other phyllosilicates. The picture he presented was of an elementary layer composed of a sheet of edge-sharing octahedra with every third site vacant (dioctahedral) sharing a plane of oxygens and hydroxyls with a sheet of corner-sharing tetrahedra (Fig. 1). Unlike the micas, smectites, and vermiculites, the layer of the kaolin-group minerals has either no charge or, at most, a very small one, no interlayer cation, and necessarily a different kind of bonding holding the individual layers together to form a crystal. There are two surfaces for such a layer: the basal oxygens of the silica tetrahedra and a surface of hydroxyls which are part of the octahedra. These latter hydroxyls are termed here *surface* hydroxyls (S in Fig. 1) to distinguish them from the hydroxyl which lies in the shared oxygen plane, termed the *inner* hydroxyl (I in Fig. 1). The nature of the interlayer bonding became clear with the determination of the crystal structures of the kaolin-group minerals. Hendricks (1938) first pointed out that all three structures have layer stackings such that the surface hydroxyls of one layer are close to the tetrahedral basal oxygens of the adjacent layer. This arrangement allows long hydrogen bonds to form between layers, and these bonds were assumed to account for the stability of these minerals (Fig. 1).

Crystallography

Historically, there have been three phases in the study of the structures of the kaolin-group minerals; the first phase established the space group, unit cell parameters and idealized crystal structure (typically in this literature no positional parameters are given for the atoms in the unit cell), the second phase involved refinements of the crystal structures to establish the details of the atomic positional parameters and to locate the hydroxyl hydrogen atoms, and the third phase is the study of the defect structure, with principal emphasis on kaolinite. The transition from the first to the second phase was marked by the single crystal refinement of dickite (Newnham and Brindley, 1956) while the beginning of the study of defects dates to the paper by Brindley and Robinson (1946a) in which they reported the existence of stacking faults. The following is a brief account of the early research that established the unit cell, space group, and idealized structures of the kaolin-group minerals. Brindley (1961) has given a more detailed account of this.

It was not long after Pauling described the kaolinite layer that attempts were made to verify the structures of the kaolin-group minerals by determining the unit cell parameters and space group of each and comparing observed with calculated X-ray diffraction patterns. All of this early work was done with structure models based on idealized polyhedra. Chronologically, kaolinite was the first to be examined. As early as 1929, Hendricks had observed a 7.1 Å spacing in oscillation diagrams from kaolinite particles. It was realized early on that the possibility of obtaining single crystals of kaolinite suitable for structure studies was unlikely, so the structure verification and subsequent refinement had to be done using powder patterns. Gruner (1932a) published the first detailed structural interpretation of the kaolinite powder pattern and concluded (incorrectly) that the space group was Cc with $d_{001} \approx 14.3$ Å, corresponding to a two-layer structure. Hendricks (1936) came to the same conclusions using X-ray and electron diffraction. Brindley and Robinson (1945, 1946b) examined a number of

31

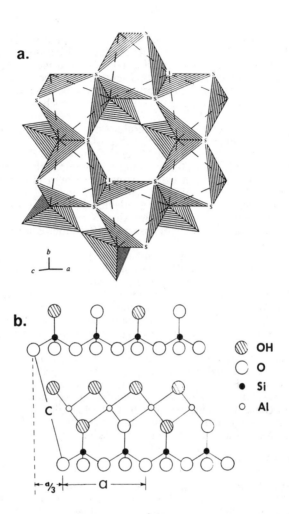

a.

b.

⬚ OH
○ O
● Si
○ Al

Figure 1 (above). A view (a) of the 1:1 layer in the kaolin-group minerals projected onto (001). The layer structure is that of kaolinite (Bish and Von Dreele, 1988). The pseudo-mirror plane characteristic of the layer is seen to approximately follow the [1,1] direction. The surface hydroxyls are marked S, the inner hydroxyl shown in the figure is marked I. A diagrammatic projection down the b-axis is shown in (b). The offset of the 1:1 layers for the kaolinite stacking is indicated; the juxtaposition of surface hydroxyls of one layer and oxygens of the adjacent layer allowing hydrogen bond formation is evident.

Figure 2 (below). Three views projected onto (001) of the octahedral sites in kaolinite and dickite showing the possible placement of the vacant octahedral site. In each of these views the X direction is down and the Y direction is across from left to right. (after Bailey, 1963)

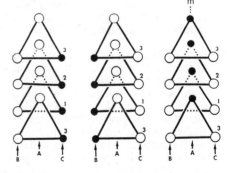

kaolinites and observed more reflections than reported by Gruner (1932a). More importantly, some reflections could not be indexed with a monoclinic unit cell but were consistent with a triclinic unit cell having a one-layer structure ($d_{001} \simeq 7.1$ Å). Brindley and Robinson obtained sufficiently good agreement between observed and calculated diffraction intensities to confirm the essential correctness of the triclinic structure. The structure is built up of a stacking of identical layers, each with a translation of $-a/3$.

The position of the vacant octahedral site is important, and this has been discussed by Bailey (1963). Figure 2 shows the disposition of the octahedral sites in three successive 1:1 layers with either the C site (left view), the B site (middle view) or the A site (right view) vacant. In the kaolin-group minerals, the vacant site is not A as in the micas, rather it has been assumed that the site is either B or C, these being enantiomorphic (Bailey, 1963; 1988). However, Bookin et al. (1988) argue that if the real structure is considered (and not an idealized one) the vacant site must be B and not C because the two do not give equivalent X-ray diffraction patterns and only the vacant B site structure agrees with observed diffraction patterns.

In contrast to kaolinite, the structures of dickite and nacrite have proved to be easier to determine. Gruner (1932b) examined the powder diffraction pattern of dickite and determined the structure as monoclinic with two possible ways of stacking the layers. The stacking directions were shown by Newnham and Brindley (1956) to be incorrect; the layers are stacked so that the interlayer shifts are in the same direction, that found in kaolinite ($-a/3$), but the vacant octahedral site alternates between B and C (Fig. 2) in successive layers giving a monoclinic two-layer structure.

The nacrite structure was examined first by Gruner (1933), but the correct structure was determined by Hendricks (1938) as having a stacking sequence based on a six-layer repeat. Bailey (1963) pointed out that because of the distribution of octahedral vacancies in successive layers, a smaller unit cell could be chosen which had a two-layer stacking. It is of interest that the conventional X and Y axes of phyllosilicates are interchanged for nacrite. The structure has shifts of $-a/3$ but this is now the Y axis as opposed to the X axis for kaolinite and dickite.

Structure refinements

Re-examination of the kaolinite structure following the discovery by Newnham and Brindley (1956) that the 1:1 layer in dickite is distorted from the ideal, showed that the layer in kaolinite also is distorted (Brindley and Nakahira, 1958). The nature of these distortions has been described several times (e.g. see Bailey, 1980). They arise from two fundamental characteristics of the 1:1 kaolin layer: the filling of two-thirds of the octahedral sites and the lateral misfit between the tetrahedral and octahedral sheets. These result in rotation of the tetrahedra, shortening of the shared octahedral edges, counter-rotation of the upper and lower triads of the octahedra, tilt of the tetrahedra, and changes to the thickness of both sheets.

Much of the research since 1956 has been directed at determining accurately the distortions that are present and to locate the hydrogen atoms. Since 1960, nineteen structural refinements of kaolin-group minerals have appeared (Table 1). Two additional studies listed in the table give only positional parameters for the hydrogen atoms (Adams, 1983; Adams and Hewat, 1981). This effort has not been uniformly distributed; only two refinements of nacrite are in print (although a third is underway; Bailey, personal communication), six are of dickite, six are of kaolinite, two are of organic intercalates of dickite, and three are of organic intercalates of kaolinite. Thus, we have nine independent views of the layer in kaolinite or its intercalates, eight of the layer in dickite or its intercalates, and two of nacrite.

Table 1. A tabulation of the crystal structure refinements for the kaolin-group minerals exclusive of halloysites. The refinements by Adams (1979) and Adams and Hewat (1983) report positional parameters for hydrogens only. The code is used to identify each refinement in other tables and figures. The structure identifiers are: K = kaolinite, D = dickite, N = nacrite, I = intercalate. Technique identifiers are: X = X-rays, N = neutrons, E = electrons, M = magic angle spinning nuclear magnetic resonance.

Reference	code	structure	space group	refinement type	technique	H positions
Adams and Jefferson, 1976	AFMA	DI	Cc	single crystal	X	
Adams, 1979	ANMF	DI	Cc	single crystal	X	
Adams, 1983	-	K	C1	Rietveld	N	Y
Adams and Hewat, 1981	-	D	Cc	Rietveld	N	Y
Bish and Von Dreele, 1988	BISH	K	C1	Rietveld	X	
Blount et al., 1969	BLNT	N	Cc	single crystal	X	
Drits and Kashaev, 1960	DRIT	K	C1	single crystal	X	
Joswig and Drits, 1986	JSWG	D	Cc	single crystal	X	Y
Newnham, 1961	NEWN	D	Cc	single crystal	X	
Raupach et al., 1987	RSEO	KI	C1	Rietveld	X	
Raupach et al., 1987	RMSO	KI	C1	Rietveld	X	
Rozdestvenskaya et al., 1982	ROZ	D	Cc	single crystal	X	Y
Sen Gupta et al., 1984	GUPT	D	Cc	single crystal	X	Y
Suitch and Young, 1983	SYDK	D	Cc	Rietveld	XN	
Suitch and Young, 1983	SYB	K	P1	Rietveld	XN	Y
Suitch and Young, 1983	SYZ	K	P1	Rietveld	XN	Y
Thompson and Cuff, 1985	THB	KI	P1	Rietveld	XN	Y
Thompson and Cuff, 1985	THZ	KI	P1	Rietveld	XN	Y
Young and Hewat, 1988	YNG	K	P1	Rietveld	XN	Y
Zvyagin, 1960	ZVY	K	C1	oblique texture	E	
Zvyagin, et al., 1972	ZNA	N	Cc	oblique texture	E	

The techniques that have been used to carry out these refinements have ranged from trial-and-error to least squares, based on powder diffraction or single crystal intensities and using diffraction of electrons, X-rays, and neutrons. Two recent developments have rekindled interest in refining these structures: one is the proliferation of computer controlled X-ray diffractometers, allowing very precise intensity measurements of the entire powder diffraction pattern (not just the peaks), and the second is the application of the Rietveld structure refinement technique to these materials, and particularly to kaolinite. The combination of improved intensity measurements and the ability to do a least-squares refinement of a low symmetry structure would seem to be all that is needed to finally determine the structure of kaolinite. In fact, this seems not be the case for reasons that will appear later.

A major problem that has plagued crystallographers and mineralogists has been the difficulty in determining the hydrogen positions in these structures. A number of techniques in addition to diffraction have been applied to this problem, and the results and current status of this effort will be treated in a separate section.

Refined structures. Normally in crystal structure studies, one begins with a model and refines it using the diffraction data until some sort of criterion is met that terminates the refinement. In several of the Rietveld refinements of the kaolin-group minerals (Suitch and Young, 1983; Thompson and Cuff, 1985), the same set of diffraction data has been used to refine different starting models. These refinements do not converge to the same final structure, and hence they are treated as independent determinations in the following discussion. In a sense, these multiple refinements act a bit like replicates and this may allow one to estimate the reliability of the Rietveld refinement procedure as it has been applied to kaolinite and dickite.

The early studies established that kaolinite is triclinic and dickite and nacrite are monoclinic. The unit cell dimensions have been agreed upon for some time, with, of course, some variations due to experimental errors, the use of different techniques and the occasional choice of a different setting for the unit cell. The following are typical values:

kaolinite: $a = 5.15560 \pm 0.00010$ Å (Bish and Von Dreele, 1988)
$b = 8.94460 \pm 0.00017$ Å
$c = 7.40485 \pm 0.00017$ Å
$\alpha = 91.697 \pm 0.002°$
$\beta = 104.862 \pm 0.002°$
$\gamma = 89.823 \pm 0.002°$

dickite: $a = 5.1375 \pm 0.0014$ Å (Joswig and Drits, 1986)
$b = 8.9178 \pm 0.0019$ Å
$c = 14.389 \pm 0.002$ Å
$\beta = 96.74 \pm 0.02°$

nacrite: $a = 8.908 \pm 0.002$ Å (Blount et al., 1969)
$b = 5.146 \pm 0.001$ Å
$c = 15.697 \pm 0.002$ Å
$\beta = 113.70 \pm 0.08°$

There are numerous ways to compare topologically related crystal structures; in phyllosilicate structures these traditionally have been measurements of the various parameters involving tetrahedra and octahedra, such as cation-oxygen distances, tetrahedral rotation, sheet thickness, quadratic elongation, and angle variance (the latter two are described by Robinson et al., 1971). These parameters are shown in Table 2 for the nineteen structure refinements. One problem in comparing structures with different symmetries is that the number of crystallographically unique polyhedra depends on the space group symmetry. For example, in the kaolinite:DMSO intercalate structure (space group $P1$) there are four unique tetrahedra and octahedra while the $C1$ structure of kaolinite has only two of each. To simplify comparisons, only average values are shown in Table 2.

Given so many parameters (7 for the tetrahedra and 8 for the octahedra) it is not a simple task to compare the different structures. This is a useful thing to do, however, because, in spite of the chemical similarity and topological identity of the layers in each of these minerals and their intercalates, one would like to know if there is any difference between the structure of the 1:1 layers in kaolinite, dickite, and nacrite. A major difference between kaolinite and dickite has been reported for the inner hydroxyl orientations (Young and Hewat, 1988), but it is not clear that differences exist in the heavy atom framework . Further, when organic molecules are intercalated into these minerals, the original interlayer bonding changes to a completely new one. It is at least possible that this change in interlayer bonding would distort the structure of the layer.

In order to compare all fifteen parameters of Table 2 for the nineteen structure refinements, a multivariate statistical treatment is appropriate. Principal components analysis (PCA) is a useful procedure for such a problem because one makes essentially no assumptions about the structure inherent in the data, as must be done in factor analysis. Many different types of measurement are represented in Table 2: distances, angles, and dimensionless numbers. In order to put these on an equal basis, the values were first standardized, then a correlation matrix was calculated, and the eigenvectors and eigenvalues were extracted. The correlation matrix is listed in Table 3, and the first eight of the eigenvalues are shown in Table 4. These eigenvalues indicate that the variance present in the fifteen parameters of the data set can be represented adequately by the first four (> 94% of the total variance) eigenvectors. The components of these eigenvectors are shown in Table 5, and the loadings are shown in Table 6. The first three eigenvectors (>86% of the variance) are plotted in Figures 3 and 4. The plot of eigenvector 1 versus eigenvector 2 shows that the structure refinements fall into essentially three groups: a cluster near the origin, a few structures that are dispersed along eigenvector 1 from the origin toward positive values (ending with the kaolinite:DMSeO structure of Raupach et al., 1987), and a few structures (kaolinites of Drits and Kashaev (1960) and Zvyagin (1960) with the dickite by Suitch and Young (1983)) dispersed along eigenvector 2. Examination of eigenvector 1 (Table 5) shows that there are five important contributors to this eigenvector: T-O distance, quadratic elongation, and angle variance (for the tetrahedra), and M-O distance and sheet thickness (for the octahedra). All five factors are positively correlated. For eigenvector 2, the major contributors relate to descriptions of the octahedra: counter-rotation of the upper and lower triangular faces, quadratic elongation, and angle variance; there is a smaller contribution from the tetrahedral rotation angle. Eigenvector 3 largely reflects variation in the tetrahedra: T-O distance, sheet thickness, tetrahedral tilt (Δz), and tetrahedral volume. Finally, eigenvector 4 is dominated by tetrahedral rotation and ψ with a smaller contribution from quadratic elongation (the latter two factors are octahedral distortions).

If one were making repeated measurements of the same object, in this case refinements of the kaolin-group minerals, the random errors in the measurement of diffraction intensities along with the normal uncertainties encountered in least-squares refinements should yield a plot such as shown in Figure 3 with a random distribution of structure refinements scattered about the origin. Figure 3 shows the refinements of kaolinite by Zvyagin (1960) and Drits and Kashaev (1960) to lie well above the cluster at the origin while the dickite refinement of Suitch and Young (1983) lies below. The question arises as to whether there is reason to believe that these are inaccurate refinements or whether they simply represent the extent of errors normally encountered. Some observations can be made: both kaolinite refinements are old, the oldest in Table 1. The kaolinite refinement of Drits and Kashaev (1960) has been criticized

Table 2. A summary of structural parameters for refined structures of the kaolin-group minerals. The data are segregated into three groups: the top group contains kaolinites, the middle one dickites, and the bottom is nacrite. All the values are averages over the crystallographically unique polyhedra.

Structure code	α (°)	T-O (Å)	sheet thickness (Å)	Δz (Å)	volume (Å³)	quad. elong.	angle variance (°²)
BISH	7.3	1.622	2.20	0.11	2.183	1.002	8.49
DRIT	11.2	1.626	2.20	0.19	2.190	1.009	19.37
RMSO	6.0	1.591	2.05	0.32	2.043	1.011	35.62
RSEO	6.4	1.677	2.33	0.29	2.296	1.041	136.18
SYB	6.9	1.622	2.21	0.15	2.168	1.010	28.09
SYZ	7.4	1.619	2.20	0.17	2.158	1.011	27.10
THB	5.8	1.643	2.26	0.33	2.236	1.015	51.85
THZ	6.8	1.623	2.18	0.32	2.160	1.015	49.39
YNG	8.4	1.629	2.22	0.24	2.174	1.017	52.90
ZVY	13.4	1.619	2.14	0.14	2.151	1.009	32.10
AFMA	6.3	1.606	2.15	0.17	2.107	1.007	27.14
ANMF	5.3	1.611	2.19	0.12	2.132	1.006	20.99
GUPT	6.8	1.613	2.19	0.19	2.148	1.002	10.26
JSWG	6.8	1.612	2.20	0.19	2.143	1.002	9.60
NEWN	7.5	1.621	2.21	0.17	2.175	1.003	11.52
ROZ	7.6	1.617	2.20	0.17	2.161	1.002	10.21
SYDK	4.6	1.616	2.22	0.16	2.156	1.004	9.47
BLNT	7.3	1.618	2.21	0.16	2.159	1.011	17.35
ZNA	3.2	1.617	2.24	0.18	2.157	1.005	16.52

The bottom row identifies each parameter as it appears in the PCA statistical analysis. Parameters 1-7 are tetrahedral parameters, 8-15 are octahedral parameters.

Structure code	ψ (°)	M-O (Å)	sheet thickness (Å)	rotation (°)	volume (Å³)	quad. elong.	angle variance (°²)	shared edge (Å)
BISH	57.1	1.915	2.079	15.67	9.143	1.017	61.54	2.408
DRIT	57.3	1.941	2.094	10.75	9.602	1.012	41.60	2.498
RMSO	57.0	1.927	2.097	15.90	9.294	1.024	76.74	2.424
RSEO	55.9	1.958	2.197	16.60	9.609	1.029	90.41	2.503
SYB	57.8	1.911	2.034	14.43	9.038	1.026	81.56	2.389
SYZ	57.8	1.913	2.036	13.99	9.086	1.026	79.08	2.399
THB	56.8	1.940	2.123	14.57	9.495	1.032	90.87	2.466
THZ	57.2	1.931	2.093	14.70	9.086	1.028	84.96	2.440
YNG	57.7	1.908	2.037	14.76	9.028	1.026	81.57	2.388
ZVY	58.7	1.932	2.007	7.66	9.440	1.015	49.86	2.480
AFMA	57.6	1.918	2.057	15.04	9.159	1.021	70.97	2.407
ANMF	57.4	1.909	2.058	15.29	9.050	1.020	68.05	2.394
GUPT	57.5	1.902	2.044	16.02	8.925	1.021	72.03	2.371
JSWG	57.6	1.904	2.040	16.33	8.873	1.021	72.73	2.365
NEWN	57.6	1.900	2.039	16.34	8.893	1.022	76.00	2.362
ROZ	57.2	1.911	2.073	15.95	9.087	1.020	68.29	2.398
SYDK	57.6	1.870	2.005	21.60	8.308	1.039	119.58	2.259
BLNT	58.0	1.914	2.027	12.41	9.138	1.019	65.09	2.414
ZNA	57.2	1.924	2.085	13.2	9.302	1.015	51.96	2.448
				11	12	13	14	15

Table 3. The correlation matrix for the fifteen structural parameters listed in Table 2.

	1	2	3	4	5	6	7	8	9	10	11	12	13	14	15
1	1.000														
2	0.064	1.000													
3	-0.229	0.851	1.000												
4	-0.172	0.327	0.047	1.000											
5	0.077	0.961	0.911	0.173	1.000										
6	-0.025	0.793	0.478	0.591	0.604	1.000									
7	-0.005	0.796	0.466	0.619	0.600	0.980	1.000								
8	0.478	-0.552	-0.446	-0.572	-0.468	-0.547	-0.603	1.000							
9	0.276	0.534	0.169	0.520	0.414	0.665	0.673	-0.471	1.000						
10	-0.224	0.632	0.398	0.633	0.520	0.683	0.730	-0.923	0.773	1.000					
11	-0.693	0.002	0.193	0.114	-0.001	-0.048	0.002	-0.428	-0.555	0.068	1.000				
12	0.363	0.424	0.098	0.355	0.338	0.517	0.515	-0.342	0.957	0.659	-0.680	1.000			
13	-0.443	0.284	0.276	0.450	0.217	0.350	0.360	-0.278	-0.261	0.084	0.689	-0.442	1.000		
14	-0.489	0.229	0.262	0.363	0.170	0.281	0.298	-0.268	-0.371	0.031	0.785	-0.551	0.984	1.000	
15	0.390	0.452	0.122	0.372	0.366	0.541	0.538	-0.314	0.973	0.648	-0.713	0.980	-0.438	-0.550	1.000

Table 4. The first eight eigenvalues from the principal components analysis of the correlation matrix in Table 3.

	eigenvalue	percent of trace	cumulative percent
1	6.73	44.9	44.8
2	4.51	30.0	74.9
3	1.72	11.5	86.4
4	1.19	7.9	94.3
5	0.34	2.2	96.6
6	0.32	2.1	98.7
7	0.12	0.9	99.6
8	0.026	0.2	99.8

Table 5. The first four eigenvectors for the fifteen parameters estimating structural variation (Table 2).

Parameter	1	2	3	4
1	0.0107	0.3221	-0.2417	0.4651
2	0.3309	-0.0978	-0.3531	0.0299
3	0.2198	-0.1785	-0.4917	-0.2456
4	0.2369	-0.1133	0.4380	0.2616
5	0.2847	-0.0894	-0.4675	-0.0773
6	0.3401	-0.0758	-0.0027	0.2844
7	0.3452	-0.0882	0.0303	0.2500
8	-0.2774	0.1816	-0.2308	0.4058
9	0.3266	0.2232	0.1548	0.0167
10	0.3417	-0.0339	0.2293	-0.2866
11	-0.0587	-0.4341	0.1013	-0.1965
12	0.2825	0.2995	0.1124	-0.0474
13	0.0543	-0.4156	0.0484	0.3574
14	0.0206	-0.4405	0.0341	0.2982
15	0.2883	0.3027	0.0874	-0.0089

Table 6. The loadings for the first four eigenvectors for each of the structure refinements.

Structure code	1	2	3	4
AFMA	-1.214	0.540	1.092	-0.089
ANMF	-1.332	0.046	0.117	-1.051
BISH	+0.576	0.661	-0.667	-1.675
BLNT	-0.873	1.152	-0.853	0.249
DRIT	1.627	3.880	-0.461	-0.519
GUPT	-1.741	-0.396	-0.009	-0.455
JSWG	-1.888	-0.504	-0.094	-0.401
NEWN	-1.553	-0.718	-0.900	-0.251
RSEO	8.540	-2.084	-0.746	-0.103
RMSO	-0.530	0.352	4.594	0.650
ROZ	-0.811	0.139	-0.094	-0.961
SYDK	-3.525	-5.598	-0.846	0.702
SYB	-0.779	-0.482	-0.854	0.533
SYZ	-0.685	-0.122	-0.495	0.763
THB	3.867	-1.289	0.369	0.189
THZ	1.510	-0.717	1.437	0.997
YNG	0.308	-0.768	-0.671	1.468
ZVY	-0.644	4.796	-1.204	2.300
ZNA	0.300	1.112	0.285	-2.345

38

before (Bailey, 1980), and the refinement by Zvyagin (1960) had a large R-factor, indicating an incompletely refined structure. On the other hand, the Suitch and Young refinement is recent, and therefore less obviously suspect. Examination of the interatomic distances calculated from the published structural parameters for the dickite of Suitch and Young (1983) shows that the two independent octahedra are unusually small and have Al-O distances as small as 1.64 Å for Al3 and 1.74 Å for Al2. Normal values (Table 2) are near 1.9 Å. A plot of the octahedral sheet (not shown) indicates an extreme distortion of the polyhedra. These distortions show up in the data in the form of a small and highly distorted average octahedron, with a very short shared edge, and a thin octahedral sheet; all of these fall well outside the range of variation observed for other dickite refinements.

If one eliminates these three anomalous refinements (SYDK, ZVY, and DRIT) from consideration, the dispersion along eigenvector 2 becomes much smaller, and all the structures give the appearance of belonging to the same population. Eigenvector 3 (Fig. 4) has a single outlying point, the structure of the kaolinite:DMSO intercalate (Raupach et al., 1987). The reliability of this refinement is more difficult to estimate, as is the reliability of the structures responsible for the dispersion along eigenvector 1 toward positive values. The dispersion along eigenvector 1 (see Fig. 3 or 4, ignoring the three points DRIT, SYDK and DRIT) is largely due to three structures, all intercalates of kaolinite (RSEO, THB, and THZ). Thus, the four remaining outlying points along eigenvectors 1 and 3 are intercalate structures. One implication of this is that the intercalation of kaolinite by DMSO or DMSeO (Thompson and Cuff, 1985; Raupach et al., 1987) changes the dimensions and/or distortions of the layer from those in the unintercalated kaolinite. According to the PCA, the change lies largely in an increase in the sizes of the octahedra and tetrahedra that result in increases in the octahedral and tetrahedral sheet thicknesses, an increase in the length of the shared octahedral edge, and greater tetrahedral tilt. If these structure refinements are accurate, the observed changes in the polyhedral dimensions suggest that intercalation can distort the 1:1 layer structure, either by allowing the layer to relax or as the result of a strong interaction between the intercalating molecule and the atoms of the layer. Such strong interactions have been inferred to exist in some kaolinite intercalates (Lipsicas et al., 1986).

Hydroxyl hydrogen positions. Because hydrogen atoms are not easy to locate in crystal structures by diffraction (they scatter X-rays weakly and have a strong incoherent scattering for neutrons), other techniques were first applied to the problem of locating either the hydroxyl hydrogens themselves or of determining the hydroxyl orientation. These were infrared spectroscopy (IR) and potential energy calculations.

IR stretching frequencies of the hydroxyl groups of the kaolin minerals lie generally in the region between 3600 and 3700 cm^{-1} where they appear as sharp, well-defined bands. The ease with which these bands can be observed coupled with the ready availability of IR spectrometers has led many researchers to apply the technique to a wide variety of samples, and the literature on this subject is large. The hydroxyl group is unusual in that it consists of one atom, oxygen, that is immobile (at least for the OH stretching frequencies) and an atom, hydrogen, that is to a considerable extent free to move, with the constraint that the O-H distance remain reasonable constant. Thus, the OH is able to reorient and will do so in response to changes in the local structural environment.

Given such a situation, it is not surprising that the frequencies, intensities and number of bands are different for each of the three polytypes (Farmer and Russell, 1964), suggesting that their different stacking is reflected in the IR spectrum. Disorder in the stacking of the layers can often be seen as a broadening and loss of intensity of the bands. Intercalation has a profound effect on the hydroxyl vibrational frequencies (Ledoux and White, 1964a,b). Because of the dipolar nature of the hydroxyl, the OH

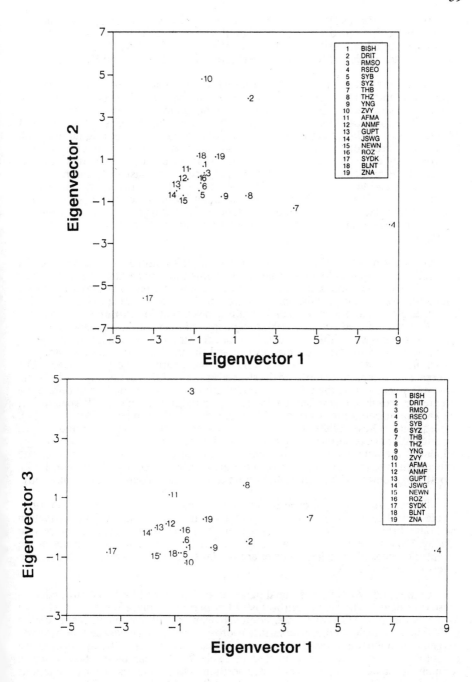

Figure 3 (top). A plot of eigenvector 1 versus eigenvector 2 for the PCA analysis of the structural data shown in Table 2. The codes (e.g. SYDK) for each structure are given in Table 1.

Figure 4. A plot of eigenvector 1 versus eigenvector 3 as in Figure 3.

40

couples to and adsorbs IR radiation most when the O-H vector lies perpendicular to the spectrometer beam and not at all when the beam and the hydroxyl are colinear. By tilting oriented samples in the IR beam it is possible to observe the pleochroism of the absorption bands and to infer hydroxyl orientations (Farmer and Russell, 1964).

While the recording of the IR spectrum of hydroxyl groups is simple, the derivation of structural information from the spectrum is not. Kaolinite has received the most attention. There are four bands and four crystallographically unique hydroxyls so that initially it was supposed that each OH gave rise to a specific band (see Farmer, 1964 for a discussion of this). However, Farmer (1964) proposed a set of orientations with the three surface hydroxyls oriented normal, or nearly so, to the 1:1 layer and the inner hydroxyl oriented more or less in the plane of the 1:1 layer. The stretching vibrations of each of the surface hydroxyls couple and produce an in-phase vibration normal to the 1:1 layer and two out-of-phase vibrations in the plane of the layer. This would assign the lowest frequency (3620 cm⁻¹) to the inner OH and the three high frequency bands to the surface OHs. The model was supported by changes in the bands resulting from intercalation and from selective deuterium substitution (Ledoux and White, 1964a,b).

Given the ability of the OH group to reorient, it is reasonable to assume that its orientation corresponds to a potential energy minimum. If one had an accurate way to calculate the potential energy, it would be a simple matter to determine the OH orientation by searching for the energy minimum. Giese and Datta (1973) did this for kaolinite, dickite, and nacrite. Their results showed that for all three minerals, the inner OH was directed out of the (001) plane toward the vacant octahedral site. The surface hydroxyls were differently oriented for each structure: in dickite all three were nearly vertical, in nacrite two were vertical and the third had an intermediate inclination, in kaolinite two were vertical and one was nearly in the (001) plane. The differences in orientations between the structures were related to the way in which one layer was stacked upon the next. The observation that each mineral had differently oriented hydroxyls was consistent with the known differences in their IR spectra, but the kaolinite orientations were not in agreement with the model of Farmer (1964). Subsequently (Giese, 1982), it became clear that the kaolinite structure upon which the calculations were based (Zvyagin, 1960) was inaccurate, and this inaccuracy strongly influenced the potential energy calculations. A recalculation using a more accurate structure showed that the three surface OHs are nearly normal to (001) (Giese, 1982), in better agreement with neutron diffraction refinements (Young and Hewat, 1988).

There have been eight experimental determinations of the hydrogen positions (Table 1) in kaolinite (3), dickite (4), and a kaolinite intercalate (1) by diffraction methods. In line with the similarity between the structures of kaolinite and dickite (Bailey, 1963) the environments about the surface hydroxyls and the positions of the acceptor oxygens of the adjacent layer are similar (Fig. 5a, 5b), while that for nacrite is very different (Fig. 5c).

Comparing hydrogen positional parameters determined by X-ray diffraction to those from neutron diffraction can be misleading because X-ray refinements are known to underestimate the O-H distances. For that reason, the only meaningful comparisons of OHs involve angles and, for the surface hydroxyls, the distance between the hydroxyl oxygen (the donor) and the oxygen of the adjacent layer that is the hydrogen bond acceptor. These values are tabulated in Table 7. Neither of the two structure refinements of nacrite reported hydrogen positions. As a comparison, the orientations determined by Giese and Datta (1973) for nacrite and by Giese (1982) for kaolinite are listed in the table. Not all reports of hydrogen positions are listed; those of Rozdestvenskaya et al. (1982), Suitch and Young (1983), and Thompson and Cuff (1985) are not included in the table. The hydrogen positions of Young and Hewat (1988) supersede the earlier refinement by Suitch and Young, and the work of Thompson and

41

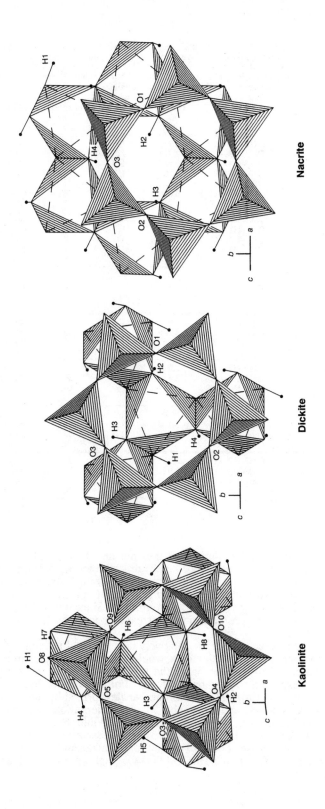

Figure 5. Projections of the structure of kaolinite (a), dickite (b), and nacrite (c) onto (001) showing the octahedral sheet of one layer and the tetrahedral sheet of the layer lying above. The hydroxyl hydrogen positions for kaolinite are from Young and Hewat (1988), for dickite are from Joswig and Drits (1986), while for nacrite the positions are those calculated by Giese and Datta (1973) from electrostatic energy calculations.

Table 7. A comparison of the environments about each of the hydroxyls in structures where hydrogen positions have been located by neutron diffraction, X-ray diffraction, or potential energy minimization techniques. The atom designations are those of the original papers. Od is the donor (hydroxyl) oxygen and Oa is the acceptor oxygen, where one exists. The "kaolinite" studied by Giese (1982) was a hypothetical structure with a 1:1 layer similar to that found in dickite by Newnham (1961).

Reference	Atoms	Od-H -- Oa (Å)	Od - Oa (Å)	Od-H/ab plane (deg)	Type
Young and Hewat, 1988	OH2-H2 -- O4	2.915	153	62	surface
kaolinite	OH3-H3 -- O3	2.927	136	48	surface
	OH4-H4 -- O5	2.765	143	66	surface
	OH6-H6 -- O9	3.205	158	74	surface
	OH7-H7 -- O8	3.022	173	75	surface
	OH8-H8 -- O10	3.010	145	48	surface
	OH1-H1	-	-	12	inner
	OH5-H5	-	-	-22	inner
Adams, 1983	OH1-H1 -- O8	2.954	168	69	surface
kaolinite	OH2-H2 -- O9	2.943	147	67	surface
	OH3-H3 -- O7	3.077	146	61	surface
	OH6-H6	-	-	-36	inner
Giese, 1982	H1	-	-	70	surface
"kaolinite"	H2	-	-	72	surface
	H3	-	-	61	surface
	H4	-	-	15	inner
Joswig and Drits, 1986	O7-H2 -- O1	2.937	174	76	surface
dickite	O8-H3 -- O3	3.126	149	53	surface
	O9-H4 -- O2	2.957	172	69	surface
	O6-H1	-	-	0	inner
Sen Gupta et al. 1984	OH2-H2 -- O1	2.936	177	75	surface
dickite	OH3-H3 -- O3	3.125	148	56	surface
	OH4-H4 -- O2	2.953	169	66	surface
	OH1-H1	-	-	-22	inner
Adams and Hewat, 1981	O7-H2 -- O8	2.98	141	65	surface
dickite	O8-H3 -- O9	3.106	155	67	surface
	O9-H4 -- O5	2.965	148	57	surface
	O6-H1	-	-	-18	inner
Blount et al., 1969	OH2-H2 -- O1	3.058	1.9	37	surface
Giese and Datta, 1973	OH3-H3 -- O2	2.971	166	80	surface
nacrite	OH4-H4 -- O3	2.982	173	72	surface
	OH1-H1	-	-	20	inner

Cuff (1985) is for an intercalated kaolinite and thus not directly comparable. The dickite refinement of Rozdestvenskaya et al. (1982) reported an orientation for the inner OH that lies very far from the pseudosymmetry plane passing through the shared edge of two octahedra. In this orientation the hydrogen is 1.916 Å from one aluminum and 2.607 Å from the other, hardly a reasonable orientation. Since this hydrogen is so badly placed, all of the hydrogen positions must be treated as suspect.

Perhaps it would be expected that the relatively new Rietveld refinement, particularly using the noisy neutron diffraction data, would yield hydrogen positions with a fair amount of scatter. On the other hand, the results from X-ray single crystal refinements would be expected to be more uniform. For the surface hydroxyls of kaolinite and dickite, the experimentally determined angles formed by the OHs and the (001) plane vary between 48° and 75° (Table 7). The orientations derived from the potential energy calculations lie well within this range. While there is some variation, the consensus is that all of the surface hydroxyls are essentially normal to the (001) plane and form long hydrogen bonds to oxygens of the adjacent layer. Similarly, the data agree that the hydrogen bonds (the Od-H -- Oa angle) are bent, as is to be expected. Neither the OH orientations nor the distances in the table suggest why the IR spectra of kaolinite, dickite, and nacrite are different. It is likely that the errors in the determinations are such as to mask any true differences between the three polytypes.

While the orientations of the surface hydroxyls are reasonably well known now, the orientation of the inner hydroxyl is much less clear. The electrostatic calculations of Giese and Datta (1973) and Giese (1982) agree that the inner OH of each of the polymorphs is directed by a small angle toward the vacant octahedral site (a positive Od-H/ab plane angle in Table 7). On the other hand, all of the diffraction studies prior to 1983 indicated an orientation away from the vacant octahedral site, and, in an extreme case (an angle with the ab plane of -36°) reported by Adams (1983), it was proposed that the inner OH forms a hydrogen bond with a basal tetrahedral oxygen of the same layer. It is not clear why the electrostatic calculations give seemingly good orientations for the surface hydroxyls but not for the inner hydroxyl. The contradiction between the calculated and experimentally determined orientations led Giese (1982) to conclude that the ionic model is not accurate enough to be useful for the inner hydroxyl. However, the refinements of the kaolinite structure by Suitch and Young (1983) and Young and Hewat (1988) have confused what previously seemed to be a clear situation. They determined that the two inner hydroxyls are independent and have different orientations; one is directed into the vacant octahedral site (as predicted by the electrostatic energy calculations) while the other is directed away from the vacant site (as indicated by previous diffraction studies). The existence of two unique hydroxyls in the unit cell results from a choice of a non-centered lattice ($P1$ as opposed to the usual $C1$). The positional parameters of the non-hydrogen atoms in these two refinements are nearly all consistent with the C-face centering. Refinements based on accurate X-ray intensity data, indicate no good reason to choose the primitive lattice (Bish and Von Dreele, 1988). The choice of a primitive lattice relies principally on the inner hydroxyl hydrogen positions as determined by the neutron diffraction refinements by Suitch and Young (1983) and Young and Hewat (1988). Thompson and Withers (1987) have tested two structural models of kaolinite, one with the atoms related by C-face centering, the other with no such relation. Their calculated intensities for transmission electron diffraction seem to favor the C-centering of the non-hydrogen atoms and they argue that if that is so, there is little reason to expect the hydrogen atoms not to obey the same centering translations. The correct choice of the Bravais lattice of kaolinite is presently a problem.

Defect structure

It was early noticed that the quality of the diffraction patterns of kaolinites from different localities varied considerably, some having sharp, narrow peaks and others with less well-defined, broad and unsymmetrical peaks. In general it is observed that the hkl reflections with $k = 3n$ (where n is an integer) are less affected than those with k not equal to $3n$ (Brindley and Robinson, 1946a; Murray, 1954). In extreme cases, peaks lose their identity and merge to form a two-dimensional modulated band of diffracted intensity. Figure 6 shows diffraction patterns of Keokuk kaolinite (a $de facto$ standard for a low defect kaolinite; Keller et al., 1966) and a kaolinite from Warren County, Georgia (sample IV-L from Brindley et al., 1986). The IV-L kaolinite differs, for example, in having lost the major peaks in the region between 19° and 24° 2θ (the 02,11 band).

Even though the structure of kaolinite is triclinic, the isolated layer has a great deal of pseudosymmetry. Hendricks (1940) and Hendricks and Jefferson (1938) noted that this allows the layers to be translated by ±nb/3 while still maintaining the juxtaposition of hydroxyls of one surface with oxygens of the adjacent surface. For Bragg reflections with $k = 3n$, the shift of layers by ±nb/3 has no effect on the intensity (for an idealized layer), while there would be changes for other reflections.

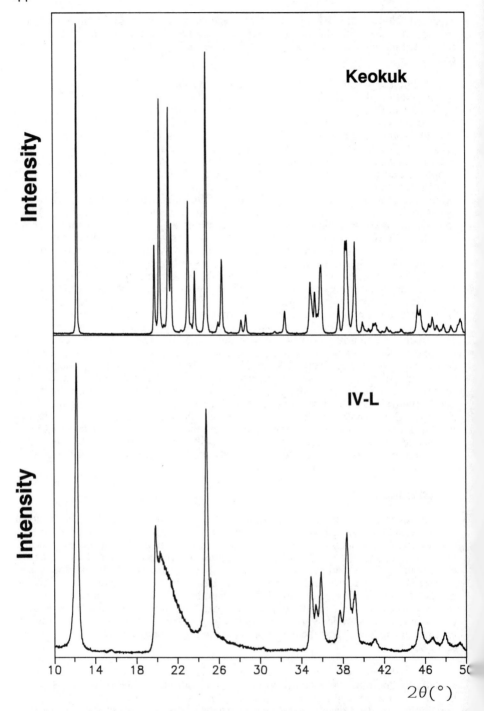

Figure 6. Observed XRD profiles for Keokuk kaolinite (upper) and a kaolinite from Warren County, Georgia. The latter has considerably more defects as shown by the loss of definition in the 02,11 band (19 to 24° 2θ). The differences between the two profiles are less marked for the 20,13 band (34 to 40° 2θ).

Murray (1954) pointed out that rotations of ±2π/3 about an axis perpendicular to the *a-b* plane and passing through the ditrigonal arrangement of tetrahedra would also not change the hydrogen bonding pattern. This type of defect would cause the vacant octahedral site in the rotated layer to be in a different position with respect to the layer next to it. As with the ±*nb*/3 translations, reflections with *k* = 3*n* would also not be affected (for an idealized layer structure). These rotations are in agreement with ideas about the polytypism of the kaolin-group minerals in which dickite can be treated as having rotations of +2π/3, -2π/3, etc., and those in nacrite +2π/6, -2π/6, etc.

Given the common occurrence of defect-containing kaolinites and the observation that the properties of kaolinites may be related to the defect structure (Murray and Lyons, 1956), it is not surprising that many attempts have been made to determine the specific types of defects and their abundances in kaolinites. The development of a defect model and its application to kaolinite samples has been attempted by several groups, each with a different approach. In the first, Mitra (1963) and Mitra and Bhattacherjee (1969a,b, 1970) proposed a model based on ±*nb*/3 translations between layers, using peaks of the type *hk*0 (with *k* ≠ 3*n*) with a Stokes deconvolution method. The procedure was limited to cases where the asymmetry of the diffraction peak was small (few defects) and where the peak did not overlap with other peaks (Plançon, 1976).

Noble (1971) took a very different approach. His idea was that each crystal consisting of M layers was composed of domains of m layers, each domain being perfectly ordered. The domains were separated by ±*nb*/3 translations or ±2π/3 rotations. Noble supposed that the coherent diffraction region for *hkl* reflections with *k* ≠3*n* would be bounded by these defects but that the 00*l* reflections would see a much larger domain. The model assumed that the diffraction from the ordered blocks would be added to the diffraction from the isolated single layers representing the places where faults occurred. By adjusting M and m, Noble calculated diffraction patterns which showed a modulation rather than discrete peaks. However, the assumptions underlying the model are incorrect (Plançon, 1976).

The first comprehensive defect model was developed by Tchoubar and Plançon (Plançon, 1976; Plançon and Tchoubar, 1977a,b) and is based on a matrix representation of the diffraction intensity (Kakinoki and Komura, 1952). The defects included in their model were: translations of ±*nb*/3 and arbitrary translations between layers, rotations of ±2π/3 and other values, and displacements of the vacant octahedral site. Each type of defect has associated with it a probability for its occurrence. In addition to these defects, the model also required the specification of the average number of layers in a crystallite (determined by broadening of the 001 peak), the radius of the diffracting domain (in the *a-b* plane; this changes somewhat with diffraction angle), and the so-called monoclinic character of the sample. The latter factor takes into account the fact that kaolinite is triclinic but, with an increasing proportion of defects, the γ angle appears to shift towards 90°

This defect model has been used to study several different kaolinites (Plançon and Tchoubar, 1977b; Tchoubar et al., 1982). In this work reasonably good agreement was achieved between calculated and observed diffraction patterns for the 02,11 and 20,13 bands. The principal conclusions of the two studies were that the primary defects in kaolinites are random displacements of the vacant octahedral site and not the ±*nb*/3 translations nor the ±2π/3 rotations as had been supposed previously. This defect model has also been applied to a kaolinite which was artificially disordered by intercalating it with hydrazine and then driving off the hydrazine by heating (Barrios et al., 1977). The process of separating and recombining the layers resulted in an increase in the proportion of ±*nb*/3 defects compared to the original clay.

46

The model proposed by Tchoubar and Plançon has generally been regarded as accurately describing the defect structure of kaolinites, largely because the experimental and calculated diffraction profiles matched reasonably well. More recently, however, a collaboration between the group in Orleans (Tchoubar and Plançon) and in Moscow (Drits, Bookin) has produced a new model (Bookin et al., 1988). They begin with the observation that the structure of an isolated 1:1 layer has a pseudo-symmetry mirror plane (Fig. 1). The interlayer translation (termed t_1), when projected onto the a-b plane, has coordinates of (-0.369, -0.024). Reflection across the mirror plane creates a second translation (termed t_2) with projected coordinates of (-0.349, 0.305). The difference between the two vectors is (0.020, 0.329) a value remarkably close to the $\pm nb/3$ translation originally suggested by Brindley and Robinson (1946a), but modified to account for the real structure of the kaolinite layer. The pseudo-trigonal symmetry of the 1:1 layer suggests that a third translation (termed t_0) can also be chosen with projected coordinates of (-0.315, -0.315). The translation t_0 is apparently not common in kaolinites (Plançon et al., 1988a). Giese (1982) has shown that there are three stable interlayer translations (for a one-layer structure), stability being defined as a net attraction between adjacent layers. This can be seen in a potential energy map (Fig. 7) in which a topographic depression indicates a stable stacking. The three minima seen in the figure correspond to the three translations of the defect model. Support for this proposal is found in the MASNMR study of Komarneni et al. (1985) who observed essentially the same ^{27}Al in highly disordered and highly ordered kaolinites. Disorder due to the placement of the vacant octahedral site would be expected to have a strong effect on the MASNMR since the local environment would change with the frequency of the defect. Bookin et al. (1988), based on arguments related to the dimensions of the unit cells and the stacking of the layers in kaolinite and dickite, concluded that completely ordered kaolinite has the vacant octahedral site in the B site.

Plançon et al. (1988a,b) applied this model to a number of kaolinite samples with differing degrees of disorder. The samples were chosen from a collection studied by Brindley et al. (1986) in which an attempt was made to correlate the disorder, as expressed by the Hinckley index (HI) (Hinckley, 1963), with a number of spectroscopic, chemical, and physical parameters. In all the kaolinites examined by Plançon et al. (1988a,b), the interlayer translations were the predominant defect with the t_1 being the most prevalent and t_0 the least common. No C layers were present in five of the nine kaolinites, and in the others the proportion of C layers was very small. This result is contrary to the previous studies of Tchoubar and Plançon.

For the kaolinites with a moderate- to low-frequency of structural defects (HI > 0.43), the model predicted a much lower background between peaks of the 02,11 band than was observed. It was not clear whether this discrepancy was due to shortcomings of the defect model or whether it was due to some other factor. The problem which is encountered is illustrated in Figure 8 for the 02,11 band. The two curves in this figure are the observed XRD scan from sample II-D (Twigg County, Georgia, irregular trace) compared with a calculated curve (smooth trace, termed BC) for a low defect kaolinite. Sample II-D has a HI of 0.59 while the calculated pattern yields a HI of 1.76. It can be seen from Figure 8 that the position and breadth of each of the peaks is reasonably well matched by the calculated curve, but the inter-peak intensities are predicted to be too weak, and the calculated intensity for the 020 peak is also too weak. All attempts to increase the inter-peak intensities by increasing the frequency of the defects resulted in a rapid loss of agreement between the peak positions, breadths, and amplitudes. It was not possible to improve the agreement between calculated and observed intensities for the entire 02,11 trace using the present model.

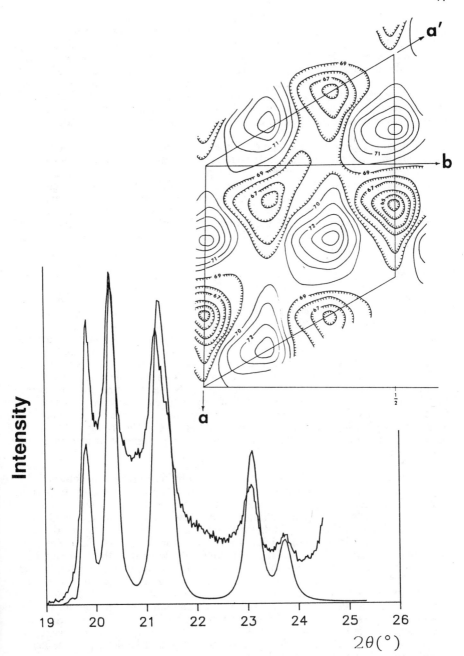

Figure 7. A map of the potential energy of stackings of 1:1 layers, at a constant interlayer distance, and translated parallel to (001). The three minima correspond to stable structures. Random mixing of these three translations constitute the major defects in kaolinites.

Figure 8. Observed (irregular) and calculated (smooth) profiles of the 02,11 band of the Twigg County, Georgia kaolinite (II-D) from the collection of Brindley et al. (1986). The calculated profile is for the low defect kaolinite referred to in the text as the BC kaolinite. The agreement in peak position, intensity (except for 020), and width for the two profiles is reasonably good; the background between the peaks is calculated to be much lower than observed.

A possible solution to this problem (Plançon et al., 1988a,b) is to suppose that this kaolinite is a mixture of a low-defect kaolinite (termed the BC kaolinite) and a moderate-defect kaolinite (termed the MC kaolinite). The former contributes primarily to the intensities of the peaks (see Fig. 9) while the latter contributes to the inter-peak intensity. The major differences between the MC and the BC are the existence of a small number of C layers (as envisaged by Brindley and Porter, 1978) and a larger proportion of the t_2 translation. The HI for this kaolinite is 0.30. The BC and MC were mixed in different proportions until a good fit with the II-D diffraction pattern was obtained. The best fit was for 30% BC and 70% MC (Fig. 10).

The good agreement between calculated and observed intensities shown in this figure supports the proposal that this material is a mixture of two kaolinites with different defect structures. The argument in support of a mixture of two different kaolinites does not rest on this single sample: a total of five kaolinites (Plançon et al., 1988a) with very different HIs showed good agreement between observed and calculated diffraction patterns, based on different proportions of the BC and MC kaolinites. It is not clear from the work done so far whether these kaolinites are simple physical mixtures of two different kaolinites or whether the two occur as domains in individual kaolinite particles.

Spectroscopic studies

All spectroscopic techniques are based on the observation of one or more transitions between states with different energy levels. Usually the energy differences between states are related to the local environment, and thus are in contrast to diffraction effects which average the structure over large distances. As mentioned earlier, spectroscopic observations often allow one to examine a crystalline material in a manner not available to traditional diffraction methods, but it is difficult to extract structural information unambiguously. For the kaolin-group minerals, and largely for kaolinite, three techniques have been used to study the structures and the types of defects present in them: infrared spectroscopy (and especially Fourier transform infrared spectroscopy, FTIR), electron paramagnetic resonance (EPR), and magic angle spinning nuclear magnetic resonance (MASNMR). This aspect of the study of the kaolin-group minerals has been aided greatly by improvements in instrumentation and techniques.

Infrared spectroscopy. The investigations of kaolinite, and more recently of dickite, have shown that in order to understand these minerals it is essential to study groups of related samples rather than isolated specimens as was done for so long. The most comprehensive examination of kaolinite and dickite was done by Brindley et al. (1986). They assembled a collection of minerals that spanned a wide range of defects as estimated by the Hinckley index, with values ranging from 0.18 to 1.72. They used a combination of IR, EPR, and X-ray diffraction. Underlying both the design of the research and the interpretation of the data was the assumption, following the early work of Tchoubar and Plançon, that defects in kaolinite are largely due to errors in the position of the vacant octahedral site and that this constitutes the introduction of dickite layers. Similarly, defects in dickite would amount to the random introduction of kaolinite layers, and the logical conclusion would be that a continuous series should exist between defect-free dickite and defect-free kaolinite by way of increasingly disordered material. At the boundary between dickite and kaolinite, the defect-rich samples of each should be indistinguishable.

By deconvolution of FTIR spectra of these samples, they determined the frequency, the intensity, and the width at half height of the four bands of kaolinite and the three bands of dickite. The high frequency band, ν_1, at approximately 3695 cm^{-1} is generally agreed to be due to an in-phase combination of the surface hydroxyls while the low frequency band, ν_4, is the pure vibration of the inner hydroxyl (see earlier

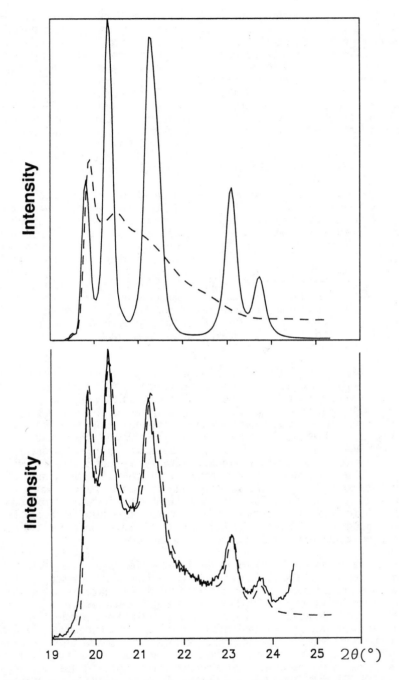

Figure 9 (top). A comparison of the calculated profiles for the BC and MC kaolinites (not to the same intensity scale).

Figure 10 (bottom). The final plot of observed and calculated diffraction profiles for the 02,11 band of the Twigg County, Georgia kaolinite. The calculated profile is a mixture of 30% BC and 70% MC.

50

discussion). In plotting the frequency of ν_1 versus the Hinckley index, it was observed that all eighteen samples fell on a smooth curve, as expected on the assumption of a continuous series. Figure 11 is a plot of the kaolinite and dickite samples from their study showing the nearly linear relation between frequency and HI for the kaolinites and an exponential relation for the dickites. The curves are drawn differently than in the paper of Brindley et al. who emphasized the continuum idea. The low defect dickites lie at higher frequency than the low defect kaolinites, suggesting that the interlayer hydrogen bonding is stronger in dickites than in kaolinites. For the kaolinites, as the number of defects increases, the high frequency band shifts slowly toward higher frequencies while the reverse is true of the dickites. This again, is in agreement with the idea of dickite layers in kaolinite as the major source of defects. As the number of dickite layers increased (a decrease in HI), the frequency of the ν_1 band, now a composite of kaolinite and dickite layers, would slowly increase in frequency.

This simple picture, however, is not capable of explaining the relation between the ν_4 width and the HI (Fig. 12). First of all, the kaolinites fall into two distinct groups; three low defect kaolinites lie at higher frequencies than the main group of six kaolinites. Both groups follow a linear trend of increasing width as the proportion of defects increases. The dickites show a single trend of decreasing width with an increasing proportion of defects. It seems reasonable to assume that an error in the position of the vacant octahedral site would change the environment about the inner hydroxyls, thereby perturbing the stretching frequency. The increasing proportion of vacant site defects would then result in a steadily increasing line width. The problem is that the line width for the dickites decreases as the Hinckley index decreases. Finally, the assumption by Brindley et al. (1986) that the major defect in kaolinites is due to errors in the placement of the vacant octahedral site is now no longer a tenable idea as shown in a previous section.

Electron paramagnetic resonance spectroscopy. Iron is present in small quantities in the kaolin-group minerals and the presence of Fe^{3+} substituting for Al in octahedral sites has been examined by a number of workers using EPR (Angel et al., 1974; Jones et al., 1974; Meads and Maulden, 1975; Mestdagh et al., 1982). In the g ≈ 4 region there are four overlapping lines; the singlet I line and a triplet of E lines. The I line is generated by iron in a site with octahedral symmetry while the E lines are due to a site with higher symmetry. Presumably, the distortions in the octahedral sites result from the local structure, including interlayer hydrogen bonding. As such, EPR should be a useful technique for examining changes in the structure as the number of defects vary. The strengths of the EPR signals change as a function of disorder in kaolinites and Mestdagh et al. (1980, 1982) concluded that the I line intensity correlated with the proportion of vacant site defects while the E line intensities did not correlate. However, both Jones et al. (1974) and Komusinski et al. (1981) saw evidence that both the I and E lines were related to the defect structure of kaolinites.

There are two difficulties with the work of Mestdagh et al. (1980, 1982): the relative areas of the EPR (derivative) signals were determined by assuming a triangular line shape, and the disorder estimate for the kaolinites was taken from the calculations of Tchoubar and Plançon. Brindley et al. (1986) deconvoluted the four overlapping peaks of the EPR signal using a Gaussian line shape. For example, Figure 13 (left) shows the integrated g ≈ 4 signal fitted with four Gaussian peaks and the derivative signal (Fig. 13, right). As a check for the accuracy of the deconvolution, both the signal and the derivative were calculated and are shown in the figure as dashed lines. The close match of the observed and calculated curves attests to the accuracy of this approach. The disorder of the kaolinites was estimated by the Hinckley index. With these data, Brindley et al. observed what they referred to as a functional relationship between disorder and the areas of both the I line and the E lines. The samples they examined fall into two groups: one with a Hinckley index centered around 0.4 and another group lying between 1.0 and 1.6. There are no samples between, so it is not

51

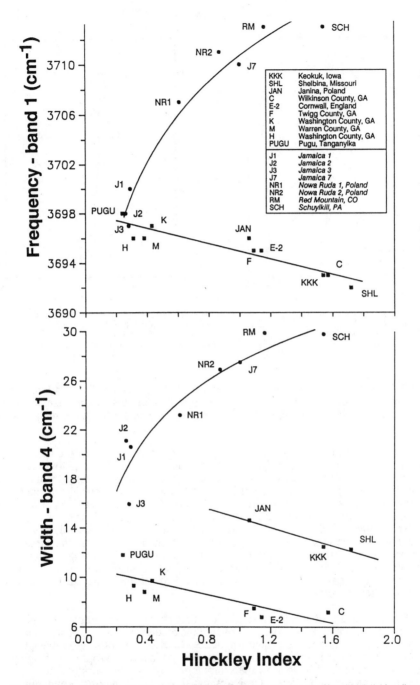

Figure 11 (top). A plot of the high frequency IR band of kaolinites (upper part of inset) and dickites (lower part of inset in italics) versus Hinckley index using the data of Brindley et al. (1986). The lines through the data are fitted independently to the dickite points (upper line) and the kaolinite points (lower line).

Figure 12 (bottom). The width of the ν_4 band due to the inner hydroxyl of kaolinites (lower two lines) and dickites (upper curve) versus the Hinckley index using the data of Brindley et al. (1986). The lines through the three groups are fitted independently.

52

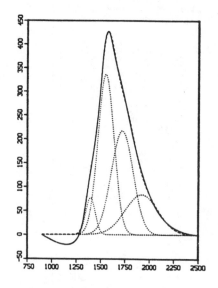 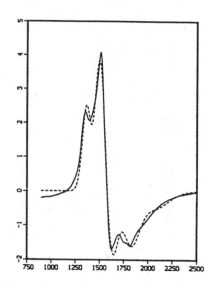

Figure 13. The EPR signal (left) and its derivative (right) for a kaolinite from Warren County, Georgia (sample L from Brindley et al., 1986). The singlet I line and the triplet E lines have been extracted from the signal by a Gaussian deconvolution. The individual components (dotted lines) are shown in the figure to the left along with their sum (dashed line). The calculated derivative signal is shown at the right as a dashed line. (after Brindley et al., 1986)

clear if there is a real functional relation between the iron seen by EPR and the defect structure or whether there are two distinct groups of kaolinites. In any case, the areas of both the I and E lines were observed to be affected by disorder (as reflected in the HI).

Nuclear magnetic resonance. In common with other spectroscopic techniques, NMR is sensitive to the local environment and, thus, is potentially useful for examining short-range order. This sensitivity to the immediate environment arises from the fact that each nuclear spin can interact with local entities such as neighboring nuclear spins, paramagnetic ions (e.g. Fe^{3+}), circulating electronic currents, etc. In solids, the dominant local interaction is usually that due to neighboring nuclear spins and gives rise to a significant (*spin-spin*) line broadening. However, the most useful information about the local environment is usually derived from the effect of the circulating electronic currents (the *chemical shift*), or from the effect of neighboring paramagnetic ions. For example, Stone and Torres-Sanchez (1988) studied the relation between structural disorder and the spatial distribution of Fe^{3+} through measurements of T_1, the spin-lattice relaxation time.

For chemical shift studies in solids, it is first necessary to remove the spin-spin line broadening which masks the shift of the line frequency. It is also advantageous to work at high magnetic field (high frequency NMR) since both signal strength and line shift are increased. These requirements have led to the development of high field magic angle spinning (MASNMR) spectrometers, based on the work of Andrew et al. (1958) and Lowe et al. (1959). These authors showed that the line broadening in a solid that is due to the spin-spin interaction can be removed by spinning the sample in the magnetic field at the magic angle ($\approx 54°$). The resulting spectrum consists of one or more chemically shifted lines and a series of equally spaced side bands which are artifacts of the spinning. Each narrow line corresponds to a different electronic (chemical) environment.

The ability to observe the NMR signal of ^{29}Si and ^{27}Al in solids has attracted much attention, particularly for studies of zeolites, but also for other silicates. Magi et al. (1981) observed a single ^{29}Si line at 39.75 MHz in kaolinite, implying that the silicon occurred in a single environment. This is consistent with the small differences observed between the two (or four, depending on whether the cell is centered) unique tetrahedral sites. However, Barron et al. (1983), working at higher field and with enhanced resolution detected a clear splitting of ^{29}Si signal into two nearly equal bands separated by about 27 Hz. Such doublets were also observed for dickite and nacrite. Unfortunately, no information was provided as to the origin of these minerals.

Barron et al. (1983) suggested that the splitting, which is not observed in comparable 2:1 structures, indicated the existence of two populations of silicon in the structure. The two populations were thought to represent distortions of the layer structure resulting from interlayer hydrogen bonding and from the polar nature of the 1:1 layer. Thompson, in a series of papers, has taken up this topic and examined a range of kaolinites with different degrees of order (as indicated by the HI), in the form of intercalates and as materials which have been artificially disordered by hydrazine intercalation followed by deintercalation (Thompson, 1984; Thompson, 1985; Thompson and Barron, 1987). His conclusions are that any distortions which may exist in the layer cannot be responsible for the splitting; rather it is the position of the surface hydroxyl hydrogens that creates two differently shielded silicon environments. Because of the uncertainty in the actual positions of the hydrogens in the kaolin-group minerals, this proposal cannot be definitively tested.

HALLOYSITES

Berthier (1826) identified a mineral similar in chemical composition to the kaolin-group minerals but different in that it incorporated water between its layers and called it *halloysite*. Hofmann et al. (1934) and Ross and Kerr (1934) demonstrated that this mineral diffracted X-rays and Hofmann et al. showed that the stoichiometry is $Al_2Si_2O_5(OH)_4.2H_2O$ for the fully hydrated phase and that the water between the layers, as shown by an increase in d_{001} from approximately 7 Å to 10 Å. It was observed that the interlayer water was unstable at room temperature and the material could be kept fully hydrated only by keeping the sample in a sealed container in contact with water or a water saturated atmosphere. Without these precautions halloysites rapidly lose much or substantially all of their interlayer water. To distinguish these two situations, the term *metahalloysite* was introduced (Mehmel, 1935) for the dehydrated form while using halloysite for the hydrated form. This was the beginning of many attempts to choose a nomenclature that would be both convenient and informative. There is still discussion about what constitutes an acceptable nomenclature (Brindley, 1980; Keller and Johns, 1976). The difficulty arises, as is often the case, because so much remains unknown about halloysites. A useful approach is to use the name halloysite as a generic term (MacEwan, 1947) and to specify the state of hydration of a specific sample by appending the d_{001} value, as in "halloysite(10 Å)" (Churchman and Carr, 1975). In this usage, specification of d_{001} as 7 Å or 10 Å refers to materials that are substantially dehydrated (7 Å) or hydrated (10 Å), and intermediate states are identified by d_{001} to a tenth of an Angstrom (e.g. halloysite(7.9 Å)). The flexibility of this approach is shown by its utility in naming artificially hydrated kaolinites, as described in a later section.

Given the similar chemical composition (excluding the interlayer water) of halloysite and the kaolin-group minerals the question of why one would form and not the other has attracted much attention. The generally accepted view is that kaolinites are formed either by low temperature weathering below the water table and over long

54

periods of time (Keller, 1978b) or by hydrothermal alteration of igneous rocks, especially granites (Bristow, 1977). Conditions of alternating wet and dry periods along with rapid growth, and conditions that promote disorder favor the formation of halloysite(10 Å) (Hurst and Kunkle, 1985).

At one time, it was thought that a continuous series existed between halloysite(7 Å) and kaolinite (Brindley and Robinson, 1946a). In this view, halloysite(7 Å) is simply a kaolinite with a very high proportion of translational defects. The origin of these defects was thought to lie in the misfit between adjacent layers that results from the rolling of the layers into tubes. With such a morphology, it is impossible to maintain a coherence between layers and this will lead to a highly disordered X-ray diffraction pattern (Spruce Pine halloysite, Fig. 14). Later work on the electron diffraction patterns of tubular halloysites suggested that halloysites have a unique two-layer structure that is different from that of kaolinite and that there cannot be a continuous series (Chukhrov and Zvyagin, 1966).

Similarly, there have been disagreements about the relation between the hydrated and dehydrated forms of halloysite. Mehmel (1935), Hendricks (1938), Alexander et al. (1943), Brindley and Goodyear (1948), and others proposed that only two forms existed; hydrated and non-hydrated. Others (MacEwan, 1947; Harrison and Greenberg, 1962; Churchman et al., 1972) saw evidence of a continuous sequence between the two extremes. The studies of Costanzo and Giese (1985) examining the dehydration of the synthetically hydrated kaolinite equivalent to halloysite(10 Å) strongly support the continuous series concept.

In contrast to the kaolin-group minerals, halloysites exhibit a variety of morphologies, including tubes, spheres, plates, oblate spheroids, stubby cylinders, and irregular shapes (Bates et al., 1950; Dixon and McKee, 1974; Kirkman, 1981). Most attention has been paid to the tubes, whose long axis is most frequently coincident with the b-axis, and less frequently with the a-axis or other directions in the a-b plane (Honjo et al., 1954). The tubes appear to be formed by kaolin layers that have been rolled up, with little or no space been adjacent layers when fully hydrated. Dehydration leaves the rolls unchanged except that shrinkage along the perpendicular to the a-b plane (normal to the surface of the tubes) leaves large voids between the different parts of the rolls (Kohyama et al., 1978). The origin of the rolling has been puzzling. Bates et al. (1950) advanced the idea that the dimensional mismatch between the octahedral and tetrahedral sheets placed a strain on the 1:1 layer causing the layer to curve and roll up around the a-axis. This presumably occurred only for very thin crystallites. A number of observations argue against this simple explanation: (1) most tubes roll up around the b-axis, (2) the dimensional mismatch has been shown to be reduced by a number of distortions of the structure, tetrahedral rotation being the most common, (3) intercalating organic compounds should lead to the formation of tubes and concomitant stacking disorder that is not observed, and (4) finally one would expect to find rolled up tubes more commonly in kaolinite deposits where thin crystallites are common. There is a general consensus now that the tubular morphology, along with the other shapes, is a primary feature created as the halloysite(10 Å) grew and is not secondary resulting from hydration of a 7 Å material (Kohyama et al., 1978).

The common observation is that halloysite(10 Å) easily and rapidly loses its interlayer water under ambient conditions and even samples stored under saturated water vapor or in liquid water may still dehydrate. The synthetic kaolinite(10 Å) is a special case in that it holds on to its interlayer water more tenaciously (Costanzo et al., 1982). The phase relations of kaolinite and halloysite have been examined in hydrothermal experiments by Hurst and Kunkle (1985).

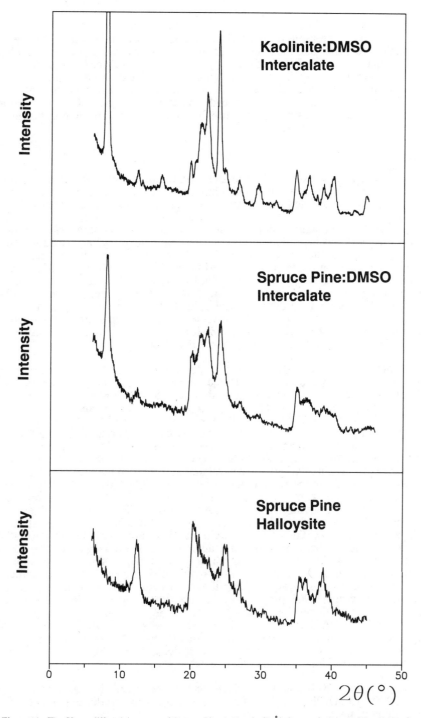

Figure 14. The X-ray diffraction traces of Spruce Pine halloysite(7 Å) (bottom), Spruce Pine halloysite intercalated with dimethylsulfoxide (middle), and, for comparison, a Cornwall kaolinite intercalated with dimethylsulfoxide (top).

There are many reasons for our lack of understanding of halloysites: (1) the ease with which water is lost and the impossibility of directly rehydrating a sample that has lost water, making it difficult to study fully hydrated samples, (2) the fact that different hydrated halloysites lose their water at different rates and to different degrees, (3) the high proportion of defects in the halloysite structure, partly structural and partly due to the interlayering of hydrated and dehydrated layers, giving an X-ray diffraction pattern that contains little structural information, and (4) the variety of morphological forms that are commonly observed.

Structure

X-ray diffraction patterns of halloysites do not contain much information, and therefore no quantitative structure studies have been based on this technique. On the other hand, electron diffraction has proved to be more useful. A number of workers have observed that selected area electron diffraction of individual tubes or particles of halloysite(7 Å) and halloysite(10 Å) give evidence of a three-dimensional order. Most of this work has been done on the dehydrated halloysite(7 Å) because of the rapid loss of water in the vacuum of the electron microscope and the heating of the sample in the electron beam (Honjo and Mihama, 1954; Honjo et al., 1954; Souza Santos et al. 1965; Chukhrov and Zvyagin, 1966), but recently hydrated specimens have also been examined using an environmental sample cell (Kohyama et al., 1978). The reasons for the greater apparent structural order seen in electron diffraction experiments is probably related to the smaller volume of the coherent scattering domain of electrons than for X-rays (see Brindley, 1980 for a discussion of this). The structure in a small volume is more likely to be ordered than in a larger volume.

All of the electron diffraction studies of the structure of halloysite agree that the structure has a two-layer periodicity. Honjo and Mihama (1954) first proposed that halloysite(7 Å) was monoclinic with dimensions; a = 5.14 Å, b = 8.93 Å, c = 14.7 Å, β = 104°. Subsequently, Honjo et al. (1954) proposed a triclinic arrangement of layers with a = 5.14 Å, b = 8.93 Å, c = 14.40 Å, α = 91.8°, β = 83°, γ = 90° based on examination of a material that they termed *tubular kaolinite*, probably because of the idea at that time that halloysite did not have a three-dimensional structure. This structure has been discussed in some detail by Chukhrov and Zvyagin (1966). In the terminology of Zvyagin (1962) the structural unit according to the setting of Honjo et al. (1954) is $\sigma_4\tau_-\sigma_4\tau_0\sigma_4$ or the enantiomorphic $\sigma_2\tau_+\sigma_2\tau_0\sigma_2$. Chukhrov and Zvyagin point out that this structure, if rotated by 2π about the normal to (001), yields the stacking $\sigma_1\tau_+\sigma_1\tau_0\sigma_1$ or the enantiomorph $\sigma_5\tau_-\sigma_5\tau_0\sigma_5$ with a β of 97°. There are difficulties with this structure: Honjo et al. (1954) did not examine all possible ways to construct a two-layer structure, they did not demonstrate that their model gave better calculated diffraction intensities that other possible structures, and the stacking created by the τ_0 translation is not a stable one (Chukhrov and Zvyagin, 1966). It should be mentioned that the description of this structure in the review of Brindley (1980) is incorrectly ascribed to Chukhrov and Zvyagin; he describes the structure of Honjo et al. In contrast, Chukhrov and Zvyagin (1966) proposed a structure with the layer stacking of $\sigma_3\tau_+\sigma_3\tau_-\sigma_3$ that has the unit cell dimensions of a = 5.14 Å, b = 8.90 Å, c = 14.7 Å, and β = 96° and is monoclinic. According to Chukhrov and Zvyagin, this stacking avoids the energetically impossible τ_0 interlayer translation and gives calculated intensities that are in reasonable agreement with observed intensities (but see Figure 7 and the discussion above of the Bookin et al. (1988) defect model).

More recently, Kohyama et al. (1978) have re-examined the structure of both halloysite(10 Å) and halloysite(7 Å). For both they deduced a two-layer structure with the cell dimensions

halloysite(10 Å) a = 5.14±0.04 Å
b = 8.90±0.04 Å
c = 20.7±0.1 Å
β = 99.7°

halloysite(7 Å) a = 5.14±0.04 Å
b = 5.90±0.04 Å
c = 14.9±0.1 Å
β = 101.9°

and a space group of Cc for both forms. They point out that their β-angle does not agree with that observed by Chukhrov and Zvyagin (1966) and, as a consequence, they suggest that halloysite might occur with more than one structure, although, while possibly true, there is little other evidence to support this. Kohyama et al. (1978) indicate that they are refining the structure to decide which of the proposed models is correct, but, to date, this publication has not appeared.

A completely different approach to the halloysite structure is that of Edelman and Favejee (1940). The structure is based on a layer that is different from the kaolinite layer in that it has half the silica tetrahedra inverted, pointing into the interlayer volume and terminated by hydroxyl groups. This structure, when fully hydrated, has a stoichiometry of $Al_2Si_2O_4(OH)_6.H_2O$. There is little evidence to support this structure, and the observation by Costanzo and Giese (1986) that intercalating tubular halloysite with dimethylsulfoxide (DMSO) produces a highly ordered structure similar to the ordered kaolinite:DMSO (Fig. 14) strongly argues against the validity of the structure.

Interlayer water

Hendricks and Jefferson (1938) proposed a structure for halloysite(10 Å) in which the water molecules adopted an ice-like structure with different types of hydrogen bonds: (1) between water molecules, (2) from water molecules to the oxygens of a bounding 1:1 layer, and (3) from the hydroxyls of the opposite bounding layer to water molecules. A key feature of the Hendricks and Jefferson model is the structural equivalence of the water molecules, in that, even though they are not all hydrogen bonded identically, they all occupy the interlayer volume at the same z-coordinate. The details of the arrangement of the interlayer water molecules is presented in a single oblique view of the water and parts of the adjacent layers (Fig. 8 in Hendricks and Jefferson, 1938). It is not easy to understand the structural arrangement from this fragment; a more conventional view is shown in Figure 15a, a projection onto the (001) plane of the octahedral sheet, water molecules lying above the octahedra, and a partial sheet of tetrahedra from the next higher layer in the stacking. The water molecules (the water oxygens are shown as open circles in the figure) are divided into two equal populations that play two roles; one set, visible in the figure through the ditrigonal holes, forms hydrogen bonds with adjacent water molecules (the hydrogens are shown diagrammatically as short dashed lines with a small filled circle at the end, pointing in the direction of the hydrogen bond). The second type of water molecules have one hydrogen bond directed toward the oxygens of the tetrahedral sheet (shown as a small filled circle within the oxygen atom) and the other hydrogen is directed laterally toward a water molecule. The first type of water molecule sits above a hydroxyl of the octahedral surface and receives a hydrogen bond from that hydroxyl. Curiously, two-thirds of the surface hydroxyls are not involved in hydrogen bonding to the interlayer water molecules; they are assumed to hydrogen bond to neighboring hydroxyls of the same octahedral sheet (Hendricks and Jefferson, 1938). This follows the arrangement found in gibbsite but not in any of the kaolin-group mineral structures.

58

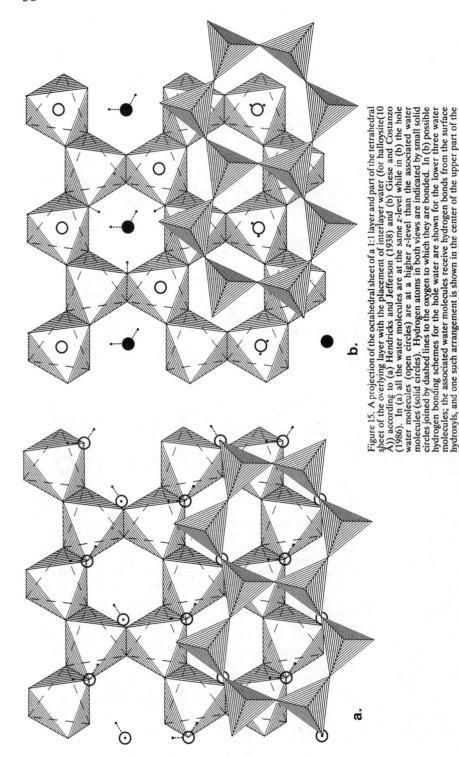

Figure 15. A projection of the octahedral sheet of a 1:1 layer and part of the tetrahedral sheet of the overlying layer with the placement of interlayer water (for halloysite(10 Å)) according to (a) Hendricks and Jefferson (1938) and (b) Giese and Costanzo (1986). In (a) all the water molecules are at the same z-level while in (b) the hole water molecules (open circles) are at a higher z-level than the associated water molecules (solid circles). Hydrogen atoms in both views are indicated by small solid circles joined by dashed lines to the oxygen to which they are bonded. In (b) possible hydrogen bonding schemes for the hole water are shown for the lower three water molecules; the associated water molecules receive hydrogen bonds from the surface hydroxyls, and one such arrangement is shown in the center of the upper part of the figure. See the text for a discussion of these models.

The geometry of the Hendricks and Jefferson model was in agreement with the general principles of hydrogen bond formation as they were known in 1938, and the simplicity and elegance of the structure were such as to convince many mineralogists of its correctness. But, there are difficulties with the model: (1) it would lead necessarily to a high degree of structural order, something not observed in X-ray diffraction patterns, and (2) the structure is based on orientations of the surface hydroxyls that are not in agreement with those observed in the kaolin-group minerals, and (3) the water molecules are essentially all of one type which is not consistent with the complex and variable dehydration phenomena often observed in halloysites. '

Part of the difficulty in working with halloysites lies in their lack of uniformity so that it is difficult to compare one sample with another. As a way out of this dilemma, Costanzo et al. (1980) succeeded in introducing water into the interlayer region of several different kinds of kaolinite to form a hydrated material, kaolinite(10 Å), with properties similar to halloysite(10 Å). These hydrated kaolinites are much better materials to study because they are made from planar 1:1 layers with a high degree of regularity. In a series of papers, Costanzo and coworkers examined the properties of the interlayer water with infrared spectroscopy, nuclear magnetic resonance, thermal gravimetry, and X-ray diffraction (Costanzo, 1984; Costanzo et al. 1980; Costanzo et al. 1984a, 1984b; Lipsicas et al. 1985; Giese and Costanzo, 1986). In addition to the 10 Å product, two other lower hydrates were prepared; one with a spacing of about 8.6 Å and the other a spacing of 8.4 Å. Evidence for these spacings in natural halloysites was found by Kohyama et al. (1978) in electron diffraction patterns.

Examination of the various kaolinite hydrates indicated clearly that the water in the 10 Å hydrate is of two types in equal quantity. One type (termed *hole water*) is keyed into the ditrigonal holes of the silica tetrahedra and forms hydrogen bonds to the tetrahedral basal oxygens. The second type (termed *associated water*) lies at a different level in the structure that is farther from the tetrahedral sheet than the hole water. The arrangement of the water molecules is shown in Figure 15b, where the hole water molecules are shown as open circles (not to size) and the associated water molecules are shown as filled circles (Giese and Costanzo, 1986). The hydrogen bonding from the hole water to the basal oxygens is seen from the figure to be necessarily disordered since there are two water hydrogens and three possible acceptors around the ditrigonal hole. The associated water molecules donate hydrogen bonds to adjacent hole water molecules and receive hydrogen bonds from three hydroxyls of the octahedral sheet that lies directly below in the figure.

The nonequivalence of the hole and associated water molecules is demonstrated by the fact that the kaolinite(10 Å) hydrate easily loses half the interlayer water and produces the 8.6 Å hydrate. From this it is concluded that the hole water is more stable than the associated water and is lost more slowly. The two rates are the key to understanding the dehydration of natural halloysites. Those halloysites that are more ordered will exhibit a more rapid loss of the associated water than the hole water so the d_{001} will drop from 10 Å to something below 8.6 Å and will ultimately stabilize (if not heated) at spacings around 7.9 Å or perhaps a bit smaller. More disordered halloysites where there is less distinction between the associated and hole water will lose the interlayer water in an apparently continuous manner and stabilize at d_{001} values nearer to 7.2 Å. These two scenarios are extreme situations; halloysites would be expected to exhibit behaviors lying between.

The advantages of this model of the interlayer water are that it is based on a large body of experimental evidence whereas that of Hendricks and Jefferson relied solely on geometric arguments. Further, it explains the dynamic properties of the interlayer water in halloysite(10 Å) and kaolinite(10 Å) and also offers a rationalization for the stacking disorder. This comes from the observation that the associated water molecules have a high degree of mobility at normal temperatures (Costanzo et al., 1982; Lipsicas et al., 1985) and thus the coherence from one layer to the next is easily lost. This mobility involves the movement of associated water molecules from one set of sites to another (Giese and Costanzo, 1986).

Identification

At one time it was assumed that the tubular morphology of halloysites was a property that distinguished them from kaolinites and thus identification was easy regardless of the hydration state of the halloysite (Churchman et al., 1984). This distinction has blurred with the discovery of platy halloysites (Noro, 1986), fibrous kaolinites (Souza Santos et al., 1965) and spherical kaolinites (Tomura et al., 1985). In reality, the fundamental distinction between the kaolin-group minerals and halloysites lies in the interlayer water present in the latter and this can be used to distinguish the two groups (Churchman and Carr, 1975). The presence of even a few interlayer water molecules disrupts both the hydrogen bonding across the interlayer volume and the regularity of the layer stacking, both of which allow easier access to intercalating molecules. Churchman et al. (1984) have shown that halloysites intercalate both formamide and N-methylformamide at a faster rate than does kaolinite and, thus, allow the two to be distinguished.

SUMMARY

It should be pointed out that what follows represents the opinion of the writer, and his opinions may not be greeted with universal enthusiasm and approbation. In fact, some points may turn out to be incorrect, some salient items may have been omitted, or some may be of no significance. For any or all of these, the author apologizes in advance.

The kaolin-group minerals and halloysites have been examined often enough and by a sufficiently large number and variety of theoretical and experimental techniques that many aspects of these minerals can be concluded to be known with some confidence. Based on the discussion presented above one can conclude that:

1.   the structure of the layers (excluding hydrogen atoms) in kaolinite, dickite, and nacrite are substantially the same,

2.   the refinements (excluding hydrogen atoms) of kaolinite by Zvyagin (1960) and Drits and Kashaev (1960); of dickite by Sen Gupta et al. (1984) and Suitch and Young (1983); and the hydrogen positions of Rozdestvenskaya et al. (1982) seem unreliable in one or more aspects,

3.   the surface hydroxyls in the kaolin-group minerals are normal or nearly so to the layer,

4.   on the basis of X-ray diffraction intensity data, there seems no reason to conclude that kaolinite is not C-face centered,

5.   the structure of halloysite (both hydrated, dehydrated, and intermediate states) is different from the other kaolin-group minerals,

6.   the defect structure of kaolinite is largely due to interlayer translations, of which there are three possibilities, and errors in the placement of the vacant octahedral site are of minor importance,

7. halloysite(10 Å) loses water in a continuous manner with halloysite(7 Å) as the final product, even though many samples probably will not reach that state,

8. there are two types of interlayer water in halloysite(10 Å) that have different positions in the interlayer volume and they participate in hydrogen bonds differently,

9. the 1:1 layer in halloysite is essentially (overlooking questions of distortions) the same as in the kaolin-group minerals,

10. at least some low-defect (large HI) kaolinites are mixtures of two different kaolinites each having a different defect structure.

There remain a number of uncertainties and open questions. These are, in no particular order:

1. the orientation of the inner hydroxyl and the attendant question of the Bravais lattice of kaolinite are unclear,

2. the structures of halloysite(10 Å) and halloysite(7 Å) are not known,

3. it is unclear what the relation (if any) is between iron (as seen by EPR) and the defect structure of kaolinite,

4. the relation between the hydroxyl stretching bands of the IR spectrum and the defect structure of kaolinite and dickite is not clear,

5. we know next to nothing about the defect structure of dickite,

6. does intercalation of kaolinite really distort the 1:1 layer?

7. can the defect structure of a kaolinite tell us something about the conditions of formation and subsequent history of the material?

8. does the defect structure of kaolinites influence the properties of the clay (e.g. rheology, optical properties, thermal dehydroxylation, intercalation)?

It is satisfying to look back and marvel at all the progress that has been made in the study of the kaolin minerals, especially in the last few years, but it is also sobering to consider how much remains to be done. The writer has the distinct impression that the study of these minerals is not a mature field of research but, rather, one that is just developing and that will blossom in the future.

ACKNOWLEDGMENTS

The writer is indebted to a number of colleagues, particularly A. Plançon, for their collaboration and tutelage in many aspects of kaolin mineralogy. The manuscript benefited from comments by S. W. Bailey, M. Lipsicas, and especially P. M. Costanzo. David Bish kindly furnished diffraction data for the Keokuk kaolinite. This effort, and much of the contributing research in which the author participated, was funded by the National Science Foundation.

62

REFERENCES

Adams, J. M. (1979) The crystal structure of a dickite:*N*-methylformamide intercalate $Al_2Si_2O_5(OH)_4.HCONHCH_3$: Acta Crystallogr., B35, 1084-1088.

———— (1983) Hydrogen atom positions in kaolinite by neutron profile refinement: Clays & Clay Minerals 31, 352-356.

———— and Hewat, A. W. (1981) Hydrogen atom positions in dickite: Clays & Clay Minerals 29, 316-319.

———— and Jefferson, D. A. (1976) The crystals structure of a dickite:formamide intercalate $Al_2Si_2O_5(OH)_4.HCONH_2$: Acta Crystallogr., B32, 1180-1183.

Alexander, L. T., Faust, G. T., Hendricks, S. B., Insley, H. and McMurdie, H. F. (1943) Relationship of the clay minerals halloysite and endellite: Am. Mineral., 28, 1-18.

Andrew, E. R., Bradbury, A., and Eades, R. G. (1958) Nuclear magnetic resonance spectra from a crystal rotated at high speed: Nature 182, 1659.

Angel, B. R. and Hall, P. L. (1972) Electron spin resonance studies of kaolins: Int'l Clay Conf., Madrid, 1, 71-86.

————, Jones, J. P. E., and Hall, P. L. (1974) Electron spin resonance studies of doped synthetic kaolinite. I: Clay Minerals 10, 247-255.

Bailey, S. W. (1963) The polymorphism of the kaolin minerals: Am. Mineral., 48, 1196-1209.

———— (1980) Structure of layer silicates: in Crystal Structures of Clay Minerals and Their X-ray Identification, ed. G. W. Brindley and G. Brown, Monograph No. 5, Mineralogical Society, London, 1-124.

———— (1988) X-ray diffraction identification of the polytypes of mica, serpentine, and chlorite: Clays & Clay Minerals 36, 193-213.

Barrios, J., Plançon, A., Cruz, M. I., and Tchoubar, C. (1977) Qualitative and quantitative study of stacking faults in a hydrazine-treated kaolinite - relationship with the infrared spectra: Clays & Clay Minerals 25, 422-429.

Barron, P. F., Frost, R. L., Skjemstad, J. O., and Koppi, A. J. (1983) Detection of two silicon environments in kaolins by solid-state ^{29}Si NMR: Nature 302, 49-50.

Bates, T. F., Hildebrand, F. A., and Swineford, A. (1950) Morphology and structure of endellite and halloysite: Am. Mineral., 35, 463-484.

Berthier, P. (1826) Analyse de l'halloysite: Ann. Chim. Phys., 32, 332-335.

Bish, D. L. and Von Dreele, R. (1988) Rietveld refinement of the crystal structure of kaolinite: Annual Meeting of the Clay Minerals Society, East Lansing, Michigan (Abstr.).

Blount, A. M., Threadgold, I. M., and Bailey, S. W. (1969) Refinement of the crystal structure of nacrite: Clays & Clay Minerals 17, 185-194.

Bookin, A. S., Drits, V. A., Plançon, A., and Tchoubar, C. (1988) Stacking faults in kaolin-group minerals in the light of real structural features: Clays & Clay Minerals accepted.

Brindley, G. W. (1961) Kaolin, serpentine, and kindred minerals, Ch. 2 in The X-ray Identification and Crystal Structures of Clay Minerals (G. Brown, Ed.), 2nd. ed., Mineralogical Soc., London.

———— (1980) Order and disorder in clay mineral structures, Ch. 2 in Crystal Structures of Clay Minerals and their X-ray Identification (G. W. Brindley and G. Brown, Ed.) Monograph No. 5, Mineralogical Soc., London.

————, and Goodyear, J. (1948) X-ray studies of halloysite and metahalloysite. Part II. The transition of halloysite to metahalloysite in relation to relative humidity: Mineral. Mag., 28, 407-422.

————, Kao, C. C., Harrison, J. L., Lipsicas, M., and Raythatha, R. (1986) Relation between structural disorder and other characteristics of kaolinites and dickites: Clays & Clay Minerals 34, 239-249.

———— and Nakahira, M. (1958) Further considerations of the crystal structure of kaolinite: Mineral. Mag., 31, 781-786.

———— and Robinson, K. (1945) Structure of kaolinite: Nature 156, 661-663.

———— and ———— (1946a) Randomness in the structures of kaolinitic clay minerals: Trans. Faraday Soc., 42B, 198-205.

———— and ———— (1946b) The structure of kaolinite: Mineral. Mag., 27, 242-253.

_____, Kao, C. C., Harrison, J., Lipsicas, M., and Raythatha, R. (1986) Relation between structural disorder and other characteristics of kaolinites and dickites: Clays & Clay Minerals 34, 239-249.

_____ and Porter, A. R. D. (1978) Occurrence of dickite in Jamaica: ordered and disordered varieties: Am. Mineral., 63, 554-562.

Bristow, C. M. (1977) A review of the evidence for the origin of the kaolin deposits in S. W. England: In Proc. 8th Int'l Kaolin Symposium and Meeting on Alunite (ed. E. Galen), Madrid-Rome: Servicio de Ministerio de Industrie y Energia, 1-19.

Brongniart, A. (1807) Traité élémentaire de minéralogie, 1, 506.

Chukhrov, F. V. and Zvyagin, B. B. (1966) Halloysite, a crystallochemically and mineralogically distinct species: Proc. Int'l Clay Conf. 1966 Jerusalem, 1, 11-25.

Churchman, G. J., Aldridge, L. P., and Carr, R. M. (1972) The relationship between the hydrated and dehydrated states of an halloysite: Clays & Clay Minerals 20, 241-246.

_____ and Carr, R. M. (1975) The definition and nomenclature of halloysites: Clays & Clay Minerals 23, 382-388.

_____, Whitton, J. S., Claridge, G. C. C., and Theng, B. K. G. (1984) Intercalation method using formamide for differentiating halloysite from kaolinite: Clays & Clay Minerals 32, 241-248.

Costanzo, P. M. (1984) Synthesis and characterization of hydrated kaolinites and the chemical and physical properties of the interlayer water: Ph.D. dissertation, State University of New York at Buffalo, Buffalo, New York, 113 pp.

_____, Clemency, C. V., and Giese, R. F. (1980) Low-temperature synthesis of a 10-Å hydrate of kaolinite using dimethylsulfoxide and ammonium fluoride: Clays & Clay Minerals 28, 155-156.

_____, Giese, R. F., and Clemency, C. V. (1984a) Synthesis of a 10-A hydrated kaolinite: Clays & Clay Minerals 32, 29-35.

_____, Giese, R. F., and Lipsicas, M. (1984b) Static and dynamic structure of water in hydrated kaolinites: I. The static structure: Clays & Clay Minerals 32, 419-428.

_____ and Giese, R. F. (1985) Dehydration of synthetic hydrated kaolinites: A model for the dehydration of halloysite(10 Å): Clays & Clay Minerals 33, 415-423.

_____, _____, Lipsicas, M., and Straley, C. (1982) Synthesis of a quasi-stable hydrated kaolinite and heat capacity of interlayer water: Nature 296, 549-551.

_____, _____ (1986) Ordered halloysite:dimethylsulfoxide intercalate: Clays & Clay Minerals 34, 105-107.

Deer, W. A., Howie, R. A., and Zussman, J. (1971) Rock-Forming Minerals, Vol 3, Sheet Silicates, Longman, London, 270 pp.

Des Cloizeaux, A. (1862) Supplement to Manuel de Minéralogie, 1, 548-549.

Dick, A. (1888) On kaolinite: Mineralog. Mag., 8, 15-17.

_____ (1908) Supplementary note on the mineral kaolinite: Mineralog. Mag., 15, 127.

Dixon, J. B. and McKee, T. R. (1974) Internal and external morphology of tubular and spheroidal halloysite particles: Clays & Clay Minerals 22, 127-137.

Drits, V. A. and Kashaev, A. A. (1960) An X-ray study of a single crystal of kaolinite: Soviet Phys. Crystallogr. 5, 207-210.

Edelman, C. H. and Favejee, J. C. L. (1940) On the crystal structure of montmorillonite and halloysite: Z. Kristallogr., 102, 417-431.

Farmer, V. C. (1964) Infrared absorption of hydroxyl groups in kaolinite: Science 145, 1189-1190.

_____ and Russell, J. D. (1964) The infrared spectra of layer silicates: Spectrochim. Acta 20, 1149.

Giese, R. F. (1982) Theoretical studies of the kaolin minerals: Electrostatic calculations: Bull. Mineral. 105, 417-424.

_____ and Costanzo, P. M. (1986) Behavior of water on the surface of kaolin minerals: In Geochemical Processes at Mineral Surfaces, ACS Symposium Series 323 (J. A. Davis and K. F. Hayes, eds.), Chap. 3, 37-53.

_____ and Datta, P. (1973) Hydroxyl orientations in kaolinite, dickite, and nacrite: Am. Mineral. 58, 471-479.

Grim, R. E. (1968) Clay Mineralogy: McGraw-Hill, New York, 596 pp.

Gruner, J. W. (1932a) The crystal structure of kaolinite: Z. Kristallogr., 83, 75-88.

64

_____ (1932b) The crystal structure of dickite: Z. Kristallogr., 83, 394-404.

_____ (1933) The crystal structure of nacrite and a comparison of certain optical properties of the kaolin group with its structures: Z. Kristallogr., 85, 345-354.

Harrison, J. L. and Greenberg, S. S. (1962) Dehydration of fully hydrated halloysite from Lawrence County, Indiana: Clay Minerals 9, 374-377.

Hinckley, D. N. (1963) Variability in "crystallinity" values among the kaolin deposits of the coastal plain of Georgia and South Carolina: Clays & Clay Minerals 11, 229-235.

Hendricks, S. B. (1929) Diffraction of X-radiation from some crystalline aggregates: Z. Kristallogr., 71, 273-275.

_____ (1936) Concerning the crystal structure of kaolinite, $Al_2O_3.2SiO_2.2H_2O$, and the composition of anauxite: Z. Kristallogr., 95, 247-252.

_____ (1938) The crystal structure of nacrite $Al_2O_3.2SiO_2.2H_2O$ and the polymorphism of the kaolin minerals: Z. Kristallogr., 100, 509-518.

_____ (1940) Variable structures and continuous scattering of X-rays from layer silicate lattices: Physical Rev. 57, 448-454.

_____ and Jefferson, M. E. (1938) Structures of kaolin and talc-pyrophyllite hydrates and their bearing on water sorption of the clays: Am. Mineral. 23, 863-875.

Hofmann, U., Endell, K., and Wilm, D. (1934) X-ray and colloid-chemical studies of clay: Angew. Chem., 47, 539-547.

Honjo, F., Kitamura, N., and Mihama, K. (1954) A study of clay minerals by means of single-crystal electron diffraction diagrams - The structure of tubular kaolin: Clay Minerals Bull., 2, 133-141.

_____ and Mihama, K. (1954) A study of clay minerals by electron-diffraction diagrams due to individual crystallites: Acta Crystallogr., 7, 511-513.

Hurst, V. J. and Kunkle, A. C. (1985) Dehydroxylation, rehydroxylation, and stability of kaolinite: Clays & Clay Minerals 33, 1-14.

Jepson, W. B. (1984) Kaolins: their properties and uses: Phil. Trans. R. Soc. Lond. A 311, 411-432.

_____, Rowse, J. B. (1975) The composition of kaolinite - an electron microscope microprobe study: Clays & Clay Minerals 23, 310-317.

Jones, J. P. E., Angel, B. R., and Hall, P. L. (1974) Electron spin resonance studies of doped synthetic kaolinite. II: Clay Minerals, 10, 257-270.

Joswig, W. and Drits, V. A. (1986) The orientation of the hydroxyl groups in dickite by X-ray diffraction: N. Jb. Mineral. Mh., H1, 19-22.

Kakinoki, J. and Komura, J. (1952) Intensity of X-ray diffraction by one dimensionally disordered crystals: J. Phys. Soc. Japan 7, 30-35.

Keller, W. A. (1978a) Classification of kaolins exemplified by their textures in scan electron micrographs: Clays & Clay Minerals 26, 1-20.

_____ (1978b) Kaolinization of feldspars as displayed in scanning electron micrographs: Geology 6, 184-188.

_____, Cheng, H., Johns, W. D., and Meng, C. S. (1980) Kaolin from the original Kauling (Gaoling) mine locality, Kiangsi Province, China: Clays & Clay Minerals 28, 97-104.

_____ and Johns, W. D. (1976) "Endellite" will reduce ambiguity and confusion in nomenclature of "halloysite": Clays & Clay Minerals 24, 149.

_____, Pickett, E. E., and Reesman, A. L. (1966) Elevated dehydroxylation temperature of the Keokuk geode kaolinite - a possible reference material: Proc. Int'l Clay Conf., 1966 Jerusalem 1, 75-85.

Kirkman, J. H. (1981) Morphology and structure of halloysite in New Zealand tephras: Clays & Clay Minerals 29, 1-9

Kohyama, N., Fukushima, K. and Fukami, A. (1978) Observation of the hydrated form of tubular halloysite by an electron microscopy equipped with an environmental cell: Clays & Clay Minerals 26, 25-40.

Komarneni, S., Fyfe, C. A., and Kennedy, G. J. (1985) Order-disorder in 1:1 type clay minerals by solid state [27]Al and [29]Si magin-angle-spinning nmr spectroscopy: Clay Minerals 20, 327-334.

65

Komusinski, J., Stoch, L., and Dubiel, S. M. (1981) Application of electron paramagnetic resonance and Mössbauer spectroscopy in the investigation of kaolinite-group minerals: Clays & Clay Minerals 29, 23-30.

Ledoux, R. L. and White, J. L. (1964a) Infrared study of the OH groups in expanded kaolinite: Science 143, 244-246.

_____ and _____ (1964b) Infrared study of selective deuteration of kaolinite and halloysite at room temperature: Science 145, 47-49.

Lee, S. Y., Jackson, M. L., and Brown, J. L. (1975) Micaceous occlusions in kaolinite observed by ultra-microtomy and high resolution electron microscopy: Clays & Clay Minerals 23, 125-129.

Lipsicas, M., Straley, C., Costanzo, P. M., and Giese, R. F. (1985) Static and dynamic structure of water in hydrated kaolinites: Part II. The dynamic structure: J. Coll. Interface Sci. 107, 221-230.

Lipsicas, M., Raythatha, R., Giese, R. F., and Costanzo, P. M. (1986) Molecular motions, surface interactions, and stacking disorder in kaolinite intercalates: Clays & Clay Minerals 34, 635-644.

Lowe, I. J. (1959) Free induction decays of rotating solids: Phys. Rev. Letters 2, 285-287.

MacEwan, D. M. C. (1947) The nomenclature of the halloysite minerals: Mineral. Mag., 28, 36-44.

Magi, M., Samoson, A., Tarmak, M., Engelhardt, G. and Lipmaa, E. (1981) Investigations into the structure of silicate minerals using high-resolution solid-state ^{29}Si NMR spectroscopy: Dokl. Akad. Nauk. SSSR 261, 1169-1174.

Malden, P. J. and Meads, R. E. (1967) Substitution by iron in kaolinite: Nature 215, 844-846.

Meads, R. E. and Malden, P. J. (1975) Electron-spin resonance in natural kaolinites containing Fe^{+3} and other transition metal ions: Clay Minerals 10, 313-345.

Mehmel, M. (1935) Uber die Struktur von Halloysit und Metahalloysit: Z. Kristallogr., 90, 35-43.

Mestdagh, M. M., Vielvoye, L., and Herbillon, A. J. (1980) Iron in kaolinite: II. The relationship between kaolinite crystallinity and iron content: Clay Minerals 15, 1-13.

_____, Herbillon, A. J., Rodrique, L., and Rouxhet, P. G. (1982) Evaluation du rôle du fer structural sur la cristallinité des kaolinites: Bull. Mineral., 105, 457-466.

Mitra, G. B. (1963) Structure defects in kaolinite: Z. Kristallogr. 119, 161-175.

_____ and Bhattacherjee, S. (1969a) X-ray diffraction studies of the transformation of kaolinite into metakaolin: I. variability of interlayer spacings: Am. Mineral. 54, 1409-1418.

_____ and _____ (1969b) Layer disorders in kaolinite during dehydration: Acta Crystallogr. B25, 1668-1669.

_____ and _____ (1970) X-ray diffraction studies on the transformation of kaolinite into metakaolin. II. study of layer shift: Acta Crystallogr. B26, 2124-2128.

Murray, H. H. (1954) Structural variations of some kaolinites in relation to dehydrated halloysite: Am. Mineral. 39, 97-108.

_____ and Lyons, S. C. (1956) Correlation of paper-coating quality with degree of crystal perfection of kaolinite: Clays & Clay Minerals 4, 31040.

Newnham, R. E. and Brindley, G. W. (1956) The crystal structure of dickite: Acta Crystall., 9, 759-764.

Noble, F. R. (1971) A study of disorder in kaolinite: Clay Minerals 9, 71-81.

Noro, H. (1986) Hexagonal platy halloysite in an altered tuff bed, Komaki City, Aichi Prefecture, Central Japan: Clay Minerals 21, 401-415.

Pauling, L. (1930) The structure of chlorites: Proc. Nat'l. Acad. Sci. U.S.A. 16, 578-582.

Plançon, A. (1976) Phénomène de diffraction produits par les systemes stratifiés comportant simultanément des feuillets de nature différente et des fautes d'empilements. Thèse, Docteur ès Sciences Physiques, Université d'Orléans, 89 pp.

_____ and Tchoubar, C. (1977a) Determination of structural defects in phyllosilicates by X-ray diffraction - I. Principle of calculation of the diffraction phenomenon: Clays & Clay Minerals 25, 430-435.

_____ and _____ (1977b) Determination of structural defects in phyllosilicates by X-ray diffraction - II. Nature and proportion of defects in natural kaolinites: Clays & Clay Minerals 25, 436-450.

_____, Giese, R. F., Snyder, R., Drits, V. A., and Bookin, A. S. (1988a) Stacking faults in the kaolin-group minerals: The defect structures of kaolinite: Clays & Clay Minerals accepted.

_____, _____, and _____ (1988b) The Hinckley index for kaolinites: Clay Minerals, in press.

Raupach, M., Barron, P. F., and Thompson, J. G. (1987) Nuclear magnetic resonance, infrared, and X-ray powder diffraction study of dimethylsulfoxide and dimethylselenoxide intercalates with kaolinite: Clays & Clay Minerals 35, 208-219.

Rengasamy, P. (1976) Substitution of iron and titanium in kaolinites: Clays & Clay Minerals 24, 265-266.

Robinson, K., Gibbs, G. V., and Ribbe, P. H. (1971) Quadratic elongation: a quantitative measure of distortion in coordination polyhedra: Science 172, 567-570.

Ross, C. A. and Kerr, P. F. (1930) The kaolin minerals: U.S. Geol. Surv. Prof. Paper 165E, 151-175.

Rozdestvenskaya, I. V., Drits, V. A., Bookin, A. S., and Finko, V. I. (1982) Location of protons and structural peculiarities of dickite: Mineral. Zh., 4, 52-58.

Sen Gupta, P. K., Schlemper, E. O., Johns, W. D., and Ross, F. (1984) Hydrogen positions in dickite: Clays & Clay Minerals 32, 483-485.

Stone, W. E. E. and Torres-Sanchez, R-M. (1988) Nuclear magnetic resonance spectroscopy applied to minerals Part 6 - Structural iron in kaolinites as viewed by proton magnetic resonance: J. Chem. Soc., Faraday Trans. I 84, 117-132.

Souza Santos, P. de, Brindley, G. W., and Souza Santos, H. de (1965) Mineralogical studies of kaolinite-halloysite clays: Part III. A fibrous kaolin mineral from Piedade, Sao Paulo, Brazil: Am. Mineral., 50, 619-628.

Suitch, P. R. and Young, R. A. (1983) Atom positions in well-ordered kaolinite: Clays & Clay Minerals 31, 357-366.

Tchoubar, C., Plançon, A., Ben Brahim, J., Clinard, C., and Sow, C. (1982) Caractéristiques structurales des kaolinites désordonnées: Bull. Minéral., 105, 477-491.

Thompson, J. G. (1984) Two possible interpretations of ^{29}Si nuclear magnetic resonance spectra of kaolin-group minerals: Clays & Clay Minerals 32, 233-234.

_____ (1985) Interpretation of solid state ^{13}C and ^{29}Si nuclear magnetic resonance spectra of kaolinite intercalates: Clays & Clay Minerals 33, 173-180.

_____ and Barron, P. F. (1987) Further consideration of the ^{29}Si nuclear magnetic resonance spectrum of kaolinite: Clays & Clay Minerals 35, 38-42.

_____ and Cuff, C. (1985) Crystal structure of kaolinite:dimethylsulfoxide intercalate: Clays & Clay Minerals 33, 490-500.

_____ and Withers, R. L. (1987) A transmission electron microscopy contribution to the structure of kaolinite: Clays & Clay Minerals 35, 237-239.

Tomura, S., Shibasaki, Y., Mizuta, H. and Kitamura, M. (1985) Growth conditions and genesis of spherical and platy kaolinite: Clays & Clay Minerals 33, 200-206.

Young, R. A. and Hewat, A. W. (1988) Verification of the triclinic crystal structure of kaolinite: Clays & Clay Minerals 36, 225-232.

Zvyagin, B. B. (1960) Electron-diffraction determination of the structure of kaolinite: Soviet Phys. Crystallogr., 5, 32-42.

_____ (1962) Polymorphism of double-layer minerals of the kaolinite type: Soviet Physics - Crystallography 7, 38-51.

_____, Soboleva, S. V., and Fedotov, A. F. (1972) Refinement of the structure of nacrite by high-voltage electron diffraction: Soviet Phys. Crystall., 17, 448-452.

Chapter 4   H.H. Murray

KAOLIN MINERALS: THEIR GENESIS AND OCCURRENCES

INTRODUCTION

The word kaolin is derived from the Chinese word "Kauling" which means high ridge. Kaolin was mined for many centuries from a high ridge near Jauchau Fu, China (Zheng et al., 1981), hence the derivation of the anglisized word kaolin. As now used kaolin has more than one meaning, including a clay mineral group, a rock term, and an industrial mineral commodity. In many parts of the world and particularly in Europe the term "China Clay" is used synonomously with kaolin in the sense of an industrial mineral commodity (Murray, 1988). The origin of the expression "China Clay" is unclear. Some say that it originated with Marco Polo who brought large ceramic urns on his return from China, and he and his contingent referred to these articles as "Chinaware". Later when kaolin was discovered at Cornwall in southwestern England it was described as similar to the clay that was used in China to make Chinaware and hence was called "China Clay". When the term kaolin is used it should be defined because of its multiple meanings.

The kaolin minerals kaolinite, halloysite, dickite, and nacrite have essentially the same chemical composition but as has been described in the preceding chapter there are important structural differences. The most common of the kaolin minerals, kaolinite $Al_2Si_2O_5(OH)_4$, is formed as a weathering product, by hydrothermal alteration, and as a authigenic sedimentary mineral. Electron micrographs (Fig. 1) show platey, fine particle size kaolin and large vermicular stacks or books of plates. Halloysite is an elongate mineral (Fig. 2) that occurs in hydrated and dehydrated forms. The hydrated form has the formula $Al_2Si_2O_5(OH)_4$, $2H_2O$ and the dehydrated form has the same formula as kaolinite. Halloysite occurs in saprolites as a weathering product and as a hydrothermal mineral. It is uncommon in sedimentary deposits. Dickite and nacrite are two polymorphic forms of kaolinite that have been described by Bailey (1963). Dickite occurs as a hydrothermal mineral and occasionally as an authigenic sedimentary mineral (Rex, 1966). Nacrite is a rare kaolin mineral and occurs only in hydrothermal environments.

Kaolinite exhibits varying degrees of order in its crystal structure. Brindley and Robinson (1947) first described structural variation in an X-ray study of some kaolinitic fireclays. Murray and Lyons (1956) showed that the crystallinity of the Georgia sedimentary kaolins ranged from poor to very well ordered forms. Hinckley (1963) proposed a procedure whereby the crystallinity of kaolinite could be quantified, and his method has now become rather widely known as the Hinckley index. Thus the X-ray diffraction patterns of kaolinite vary considerably because of translations of the unit cells parallel to the b-axis.

68

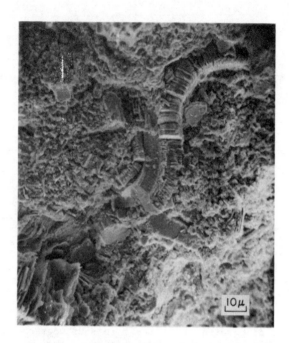

Figure 1. SEM showing a large vermicular kaolinite stack embedded in a fine platy
kaolinite matrix from Wilkinson County, Georgia.

Figure 2. SEM of halloysite from Bauxite, Arkansas.

Kaolin is an important economic industrial mineral. In the United States in 1987 more than 8,000,000 tons were produced and in the world over 20,000,000 million tons were produced (Murray, 1988). Approximately 60% of this total was used by the paper industry, where kaolin is used as a filler in the interstices of the cellulose fibers that form the paper sheet and as a coating on the surface of the sheet to improve the fidelity of printing using colored inks (Murray, 1984). Other major uses of kaolin are in ceramics, paint, rubber, plastics, catalysts, and many other minor uses which are restricted essentially to kaolinite because halloysite, dickite, and nacrite are relatively rare. Halloysite is produced in New Zealand primarily for use as a ceramic raw material (Murray et al., 1977). Dickite is used as a refractory raw material in Mexico (Hansen et al., 1981) and in Japan (Sudo and Shimoda, 1978). Nacrite is used as a refractory material in Mexico and occurs in the same hydrothermal deposit as the dickite (Hansen et al., 1981).

Kaolinite is an important mineral in world trade. In 1987 the United States exported more than 1,000,000 tons, primarily to Europe and Japan. The value per ton at the producing plant ranges from about $35 to as high as $400 per ton, depending upon the physical properties and the application. The important physical properties are brightness, particle size distribution, abrasion, and viscosity (Murray, 1984). The average value per ton is about $90.

This discussion is not concerned with ball clays, underclays, flint clays, or fireclays that are kaolinitic sedimentary clays with special characteristics and are used primarily in the ceramic and refractory industries. For those interested in these clays see the discussion by Patterson and Murray (1984). This chapter focuses on the genesis of the kaolin minerals and describes many of the important occurrences, and in the case of halloysite, dickite, and nacrite, the rare occurrences.

GENESIS

Kaolin occurs in hydrothermal, residual, and sedimentary deposits. The hydrothermal and residual occurrences are classed as primary and the sedimentary occurrences as secondary. Primary kaolins are those that have formed in situ by the alteration of crystalline rocks such as granites and rhyolites. The alteration results from surface weathering, groundwater movement below the surface, or hydrothermal action. Secondary kaolins are sedimentary and were eroded, transported, and deposited as beds or lenses associated with other sedimentary rocks. Most kaolin deposits of secondary origin have been formed by the deposition of kaolinite formed elsewhere. Some deposits have been formed from arkosic sediments that were altered after deposition primarily by groundwater. There are far more deposits of primary kaolins in the world than secondary kaolins because special geologic conditions are necessary for the deposition and preservation of the sedimentary kaolins.

The physical and chemical conditions under which the kaolin minerals form are relatively low temperatures and pressures. The most common parent minerals from which the kaolin minerals form are feldspars and muscovite, both of which contain the necessary alumina and silica. The theoretical composition of a pure kaolinite is 46.3 wt.% SiO_2, 39.8

wt.% Al_2O_3, and 13.9 wt.% H_2O, and the chemical formula is $(OH)_4Al_2Si_2O_5$. The transformation of potassium feldspar into kaolinite occurs by intense weathering of the feldspar and leaching of K and SiO_2 according to the equation:

Potash Feldspar	Water	Kaolinite	Silica	Potash

$$2KAlSi_3O_8 \;+\; 3H_2O \;\rightarrow\; Al_2Si_2O_5(OH)_4 \;+\; 4SiO_2 \;+\; 2K(OH)$$

All of the potassium must be lost in solution because if some is retained illite will form rather than kaolinite. Solution is essential to chemical weathering and to the formation of kaolinite. Solubilities are pH dependent, and the solubilities of Al_2O_3 and SiO_2 usually are plotted (Fig. 3) against pH (Mason, 1952). Alumina is insoluble between pH 4 and pH 10 whereas silica solubility increases with increasing pH. According to Mason (1952) the pH values of natural waters normally lie between 4 and 9. The alumina is not soluble in this range, but the alkalies and alkaline earth elements are soluble and mobile. Silica is only slightly soluble, and thus kaolinite is one of the most widespread minerals in soils (Dixon, 1977). The stability relations of some phases (Fig. 4) in the system K_2O-Al_2O_3-SiO_2-H_2O as functions of $[K^+]/[H^+]$ and $[H_4SiO_4]$ are given by Garrels and Christ (1965). Kaolinite is bounded by the stability fields of gibbsite, K-mica, K-feldspar, and amorphous silica.

Granites and rhyolites (Fig. 5) weather readily to kaolinite and quartz under favorable conditions of high rainfall, rapid drainage, temperate to tropical climate, a low water table, and adequate ground water movement to leach the soluble components. The more immobile components are alumina and silica whereas alkalies and alkaline earths are mobile. Arrows indicate the path taken by alteration products of (1) mica and (2) feldspar in successive stages of weathering. Arrow (3) shows the common reaction in tropical climates where the most intense weathering removes silica and forms gibbsite. Plagioclase feldspars are relatively unstable and alter before either potash feldspar or muscovite (Murray et al., 1978).

The exact conditions of formation of halloysite versus kaolinite are not defined. Halloysite is an elongate mineral (Fig. 2) and is hydrated. A recent study (Merino et al., 1988) suggested that the concentration of aqueous tetrahedrally coordinated aluminum is pH-dependent and that tetrahedral aluminum for silica substitution causes bending of the layers into a tubular morphology. For most halloysites the Al/Si ratio is greater than unity implying the presence of tetrahedral aluminum (Deer et al., 1966; Weaver and Pollard, 1973; Bates, 1959; and Newman and Brown, 1987). Dickite and nacrite are rarely found in residual deposits.

Primary kaolins of hydrothermal origin generally are comprised of kaolinite with lesser amounts of halloysite and occasionally dickite and nacrite. The kaolin mineral that forms under hydrothermal conditions is largely dependent upon temperature. Nacrite and dickite may form as a result of high temperature hydrothermal solutions (Hanson et al., 1981; Sudo and Shimoda, 1978) although they are relatively rare and the temperatures are probably of the order of $100°$ to $400°C$ (Kerr, 1955).

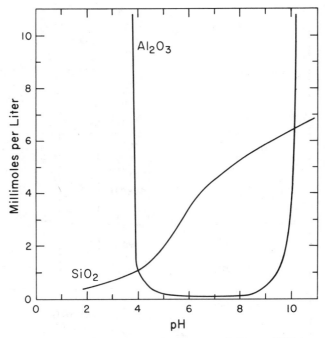

Figure 3. Solubility of Al_2O_3 and SiO_2 with respect to pH (Mason, 1952).

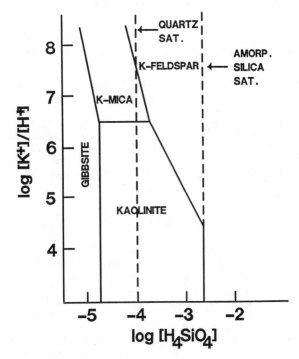

Figure 4. Stability relations in the system $K_2O-Al_2O_3-SiO_2-H_2O$ as functions of $[K^+]/[H^+]$ and $[H_4SiO_4]$ (from Garrels and Christ, 1965).

72

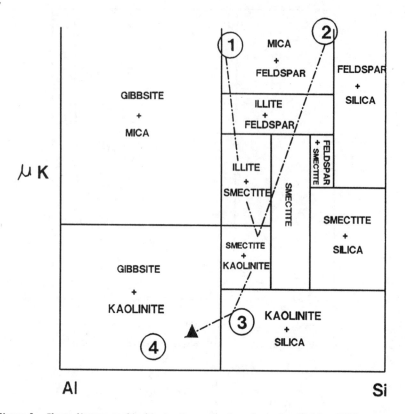

Figure 5. Phase diagram applicable to the weathering of granite (Velde, 1985).

Sedimentary or secondary kaolin deposits are not as common as primary deposits, however large sedimentary deposits are found in the United States, Brazil, and Australia. The largest sedimentary deposits that are preserved in the rock record are Cretaceous or younger in age and were deposited in deltaic, lagoonal, estuarian, or lacustrine environments. Kaolinite is the dominant kaolin mineral in the secondary deposits with halloysite rare, dickite very rare, and nacrite not present. Authigenic halloysite (Dixon, 1977) and dickite (Rex, 1966) have been reported in sedimentary rocks.

OCCURRENCES OF SELECTED KAOLIN DEPOSITS

Some of the more important and unusual deposits in the world are described in some detail in this section. They will be classed as hydrothermal, residual, mixed hydrothermal and residual, and sedimentary. Kaolin has many industrial uses and new uses are still being discovered (Murray, 1963; 1986). Historically, kaolin was used only in ceramics, and this application is perhaps still the best known even though the total tonnage sold for this use is small compared with that marketed to the paper industry. The physical and chemical properties of a kaolin determine its potential uses. Some deposits are good for ceramic uses, others for fillers, a few for paper coating clays, and a very few for all the aforementional uses. Diagnostic tests are used to evaluate kaolin properties in order to determine its eventual uses (Murray, 1980).

The important properties for kaolin are particle size distribution, brightness, screen residue (percent greater than 44 microns), viscosity, pH, and mineralogy. For ceramic useage there are additonal special properties including plasticity, shrinkage, modulus of rupture, absorption, fired color, and fusion point. The most difficult specification for kaolins that are used for coating paper is viscosity. The kaolin must flow readily at a solids level of 70 percent. At the present time only four deposits in the world meet the rigorous specifications for producing coating-grade kaolins. These four are located in Georgia in the United States, Cornwall in southeastern England, Rio Jari in Brazil, and Weipa in northeast Australia.

HYDROTHERMAL KAOLIN DEPOSITS

China

One of the largest and highest quality kaolin deposits in China is located near Suzhou in Jingsu province (Zheng et al., 1982). The geology of the Suzhou kaolin mining area is very complex. Carboniferous, Permian, and Triassic carbonate and clastic rocks occur in the area along with igneous rocks including granites, quartz latites, adamellite intrusives, and multistage acid and alkaline dikes which fill large high angle faults. As a result of multiple intrusions of granitic rocks and Jurassic volcanism, intermediate and low temperature hydrothermal alteration is very widespread. Kaolinization of many rock types result in several types of kaolin that are classed as compact, banded, sandy, and kaolin with relict phenocrysts. The thickness of the kaolin zone ranges from 5 to 30 meters with local thicknesses reaching as much as 80 meters. The dominant minerals are kaolinite and halloysite. Accessory minerals are alunite, pyrite, sericite, quartz, and chalcedony. The processed kaolin is used in ceramics, as fillers for paper, rubber, and paint, and as an ingredient in cement. The Suzhou kaolin deposit is very large, and the indicated reserves are over 20 million tons.

Italy

The largest kaolin mine in Italy is at Locchera on the island of Sardinia (Lombardi and Mattias, 1987). The kaolin is derived from hydrothermal alteration of Tertiary rhyolite ignimbrites and andesitic-basaltic lavas. The major kaolin mineral is well-crystallized kaolinite along with minor amounts of dickite, halloysite, and allophane. Cristobalite, quartz, opal, alunite, and sulfates are the most common impurities, and a small amount of residual feldspar is present. Smectite is common at the periphery of some of the kaolin deposits. This hydrothermal kaolin is used for cement, ceramics, and refractories.

Japan

Hydrothermal kaolin deposits are very common in Japan. The largest kaolin clay mine in Japan is the Itaya located in the Okayama Prefecture where "Rosecki" is mined. The name "Rosecki" is applied to compact clays comprised largely of minerals such as pyrophyllite, kaolinite, and

74

diaspore. The deposits were formed by the alteration of Pliestocene
volcanic and pyroclastic rocks (Sudo and Shimoda, 1978). Dickite and
alunite are also found at the Itaya mine (Togashi, 1976), and nacrite
associated with dickite has also been reported (Honda et al., 1964).
The kaolin clays from the Itaya mine are used as fillers for paper and
for ceramics and refractories.

Mexico

Kaolin is mined in at least 9 districts (Fig. 6) in Mexico and is
used as fillers in paper, rubber, and plastics, as extenders in paint,
and in ceramic and refractory products. The production of kaolin in
Mexico is greater than 100,000 tons per year (Patterson and Murray,
1984). All the deposits in Mexico that are mined are hydrothermal, and
most of them are related to upper Cenozoic volcanic rocks. Refractory
kaolin of hydrothermal origin occurs in a large district located 7 to 10
km west of Guanajuato. The refractory clay consists mainly of
disordered kaolinite and very fine quartz (Hanson and Keller, 1966).
The hydrothermal alteration developed along nearly vertical fracture
zones in a medium grained greenschist.

Another hydrothermal kaolinite clay deposit that is used for
refractories is located near San Luis Potosi (Keller and Hanson, 1968).
The mine is commonly known as the General Zaragosa. The kaolin was
formed by hydrothermal argillation of rhyolite flow-breccia and probably
welded tuff. The end product of the hydrothermal alteration is a white,
relatively pure, poorly ordered kaolinite with a small amount of
halloysite. Some cristobalite and alunite are mixed sporadically in the
kaolin.

Nacrite, dickite, and kaolinite were reported in one deposit in
Nayarit, Mexico (Hanson et al., 1981). Cenozoic volcanic rocks of
andesitic to rhyolitic composition are hydrothermally altered to
kaolinite and dickite along with minor quantities of nacrite. Quartz
and cristobalite are also present in minor amounts along with
halloysite.

United States

Hydrothermal kaolin deposits occur at several localities in the
western United States. In the Little Antelope Valley District in
California rhyolites have been altered to kaolinite and quartz along
with small amounts of cristobalite, opal, and alunite. This kaolin is
used as fillers in paint, plastics, rubber, and paper and in portland
cements and stucco. Cleveland (1962) believes that most of the
alteration was by sulfuric acid. The typical alteration zone consists
of a core of clay that grades outward into country rock. Deposition of
silica by the thermal solutions has caused localized contamination of
clay.

Almost all of the catalytic-grade halloysite that has been mined in
the United States came from the Dragon Mine, Tintic District, Utah. The
halloysite bodies are located along conspicious fissures at the
northeast end of a monzonite porphyry stock (Kildale and Thomas, 1957).
They are hydrothermal replacements of lower Paleozoic limestone near the
contact with the porphyry. The typical fresh clay is nearly pure
halloysite containing very fine grained disseminated pyrite that is

75

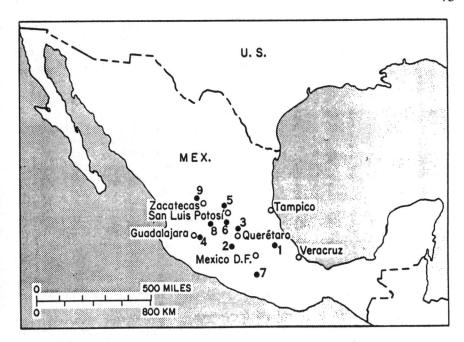

Figure 6. Location of kaolin districts in Mexico.

rarely large enough to be observed in hand specimens. The Dragon Mine
is no longer operating because all of the accessible reserves of
halloysite have been mined.

RESIDUAL KAOLIN DEPOSITS

Argentina

Extensive residual kaolins altered from rhyolitic volcanic and
related pyroclastic rocks of Jurassic age are present in the Southern
Patagonia region of Argentina (Fig. 7) in the provinces of Chubut and
Santa Cruz (Romero et al., 1974). The deposits located along the Chubut
River valley are relatively thick with some deposits as thick as 20
meters. The dominant mineral present in these kaolins is kaolinite
along with the accessory minerals halloysite and quartz. Minor amounts
of smectite and illite are present in some of the deposits. These
kaolins are processed for use as raw materials for the ceramic industry
including wall tiles, whitewares, sanitaryware, and refractories. Some
of the better quality kaolin is used as fillers in paper and paint.

Australia

Residual kaolinitic clays altered from granite are mined and
processed in the region of Ballarat and Pittong northwest of Melbourne.
The parent granitic rocks are probably late Devonian in age (Gaskin,
1944), and the residual kaolin deposits are found in gently undulating
topography. The depth of alteration of the acidic rocks ranges from 15

to more than 30 m. English China Clays, Ltd. mine this residual kaolin at Pittong for use as a filler in paper, paint, plastics, and rubber and for use in various types of ceramics. Kaolinite is the major clay mineral present along with quartz and minor amounts of halloysite.

China

There are many residual kaolin deposits in China (Zheng et al., 1981), and most of these deposits were formed from weathering of granites. An example of this type of deposit is located in Gaoling Village near Jingdezhen in Jiangxi Province. The depth of weathering is 30 to 50 m. The highest quality kaolin is generally found in the upper portion of a medium-grained granite. The major mineral component is kaolinite along with halloysite, quartz, and muscovite. Some feldspar is present in the lower portion of the alteration zone. This kaolin is used primarily for the production of fine china in a major ceramic manufacturing center at Jingdezhen.

Czechoslovakia

The largest kaolin deposits were formed in the Karlovy Vary areas by alteration of granites (Fig. 8). The kaolin is used as a filler in paper and rubber and in the manufacture of fine and heavy ceramics and refractories. The major mineral constituent is kaolinite along with quartz and minor amounts of halloysite and illite. Proved depth of kaolinization is more than 50 meters (Kuzvart, 1969) and the areal extent is approximately 85 square kilometers. The kaolinization of this area took place during Cretaceous and Paleocene time.

German Democratic Republic

Residual kaolins are found in several locations in the southern part of the GDR. The most important mines are in the Kemmlitz area of northwest Saxony between Dresden and Leipzig. The parent rocks of the kaolin in the Kemmlitz area are volcanic ignimbrites and porphyritic andesites (Storr, 1975). The thicknesses of the kaolins range between 10 and 40 m. The major minerals are kaolinite and quartz along with a minor amount of illite-smectite mixed-layer material. Other rock types that are altered in the southern part of GDR are granite, granodiorite, and graywackes. All of these rocks are part of the Bohemian Massif that has been deeply weathered since Cretaceous time. The kaolins are used as fillers in paper, paint, plastics, and rubber and as ceramic raw materials. In 1764 the kaolin deposits at Seilitz, near Meissen, were discovered and have been exploited up to the present time, thus being the oldest kaolin mine in Europe still working. The kaolin is altered from pitchstone, felsite, and quartz porphyry.

Indonesia

Kaolin is mined on the islands of Belitung and Banka in the Java Sea (Fig. 9). A porphyritic biotite granite on Belitung is altered to kaolinite (Murray et al., 1978). The mineralogical and chemical changes indicate that the feldspars alter to kaolinite and halloysite, the biotite alters to hydrobiotite to vermiculite to smectite, and quartz is unaltered except for solution pitting. This setting is a classic chemical weathering area where the high rainfall, the low relief, the water table level, the presence of granite, and the movement of ground

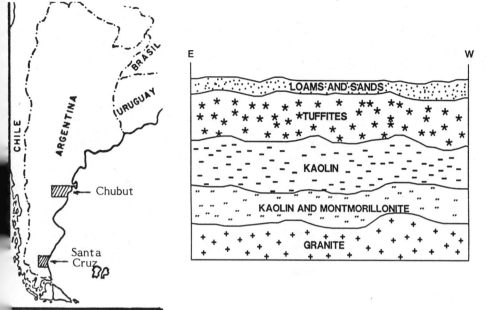

Figure 7 (left). Location of two kaolin districts in Argentina.

Figure 8 (right). Cross section of a portion of the Karlovy-Vary kaolin deposit (from Kuzvart, 1969).

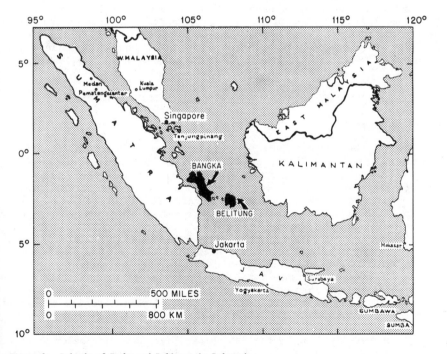

Figure 9. Islands of Banka and Belitung in Indonesia.

waters through the granite have altered it sequentially to kaolin (Fig. 10). The minerals present in the kaolin that is mined are dominantly kaolinite and quartz along with a minor amount of halloysite (Fig. 11). The kaolin is used as a filler for paper and for whiteware and sanitaryware ceramics.

USSR

The largest deposits of kaolin in the USSR were formed on an ancient weathering crust largely comprised of granites or other feldspathic or micaceous acid rocks (Petrov and Chukhrov, 1977). The most extensive area of kaolinization is in the Ukraine where the weathering took place in the early Mesozoic. Four deposits in the Ukraine (Prosyanovskoe, Glukhovetskoe, Veliko-Gradominetskoe, and Turbovskoe) are exploited. The geological cross-section of the Prosyanovskoe deposits (Fig. 12) is typical of the Ukrainian kaolins. Kaolinite is the major mineral along with variable quantities of mica and quartz. Lower in the weathering profile feldspar and smectite are present. The weathering depth generally ranges between 40 and 80 m thick. The processed kaolins are used as fillers in paper and rubber, and for all types of ceramics.

South Africa

An interesting residual kaolin deposit in South Africa is located near Grahamstown in southeastern Cape Province. Here kaolin is derived from Permian tillites and Carboniferous shales (Murray and Smith, 1973). The kaolin is fine particle size and may contain appreciable quartz, feldspar and musovite impurities. The area of alteration on the Grahamstown peneplain is large and the reserves are extensive. The processed kaolin is used primarily for ceramics, but some of the whiter materials that are derived from the Carboniferous shales are used as fillers in rubber, paper, paint, and plastics. A sizable quantity is also used as carriers for insecticides in agriculturial applications.

MIXED HYDROTHERMAL AND RESIDUAL KAOLIN DEPOSITS

Cornwall, England

One of the largest and highest quality primary kaolin deposits in the world is located in the Cornwall region of southeastern England. The parent rock from which the kaolin is derived is the St. Austell granite (Bristow, 1977). The kaolinite content in the altered granite ranges between 10 and 20 percent. Bristow (1977) states that a very special combination of conditions were required to produce the kaolin bodies in Devon and Cornwall, which are:
1) a low iron parent granite virtually free of biotite,
2) a hydrothermal phase of alteration that yields a metastable matrix which is extremely susceptible to later kaolinization,
3) a mesogene or supergene phase that is sufficiently intense to alter completely the metastable matrix to the kaolin we see today, and
4) a fairly gentle relief with lack of rapid erosion either by water or ice to preserve the kaolin matrix in place, once formed.

Extensive drilling by English China Clays, Ltd., the major producer in the area, has shown that the typical kaolin body is funnel or trough-

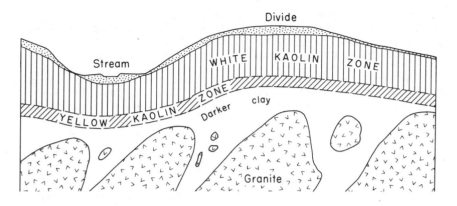

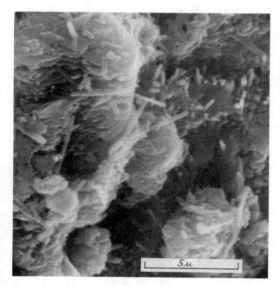

Figure 10 (above). Profile of kaolin altered from granite on the island of Belitung.

Figure 11 (left). SEM showing kaolinite stacks and plates and halloysite elongates from Belitung Island, Indonesia.

Figure 12 (below). Cross section of the Prosyanovskoe kaolin deposit in the Ukraine (from Petrov and Chukhrov, 1977).

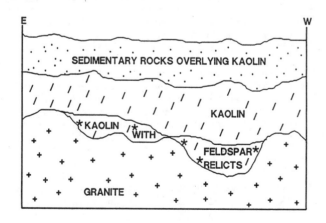

like in form, narrowing downwards (Fig. 13). This feature along with the fact that the granites in Cornwall are extensively mineralized with tin, copper, and tungsten indicates hydrothermal activity. Most mineralized veins are composed dominantly of quartz and tourmaline.

The processed kaolin is a well crystallized kaolinite with minor quantities of muscovite, quartz, and feldspar. The fine particle portion of the kaolinite is very white and has practically no impurities. The kaolin is used for paper coating and filling, for ceramics, and for a filler in paints, plastics, and rubber.

New Zealand

The Maungaparerua deposit is located on the North Island of New Zealand some 210 km north of Auckland. The halloysite deposit was formed by hydrothermal alteration of rhyolitic flow rocks on which residual weathering has been superimposed (Murray et al., 1977). The area at Maungaparerua is a complex system of fractured coalescing rhyolite domes with associated flow volcanics covering large areas (Fig. 14). The alteration zone is ovate and covers about 350 acres. The altered rocks contain about 50% quartz, fine amorphous silica, and cristobalite and 50% of a mix of halloysite, kaolinite, and allophane with a small amount of plagioclase feldspar in the coarse fraction (Fig. 15). The upper 8 to 30 m of the halloysite weathering zone consists of relatively soft clay. Below this soft clay the material becomes hard and dense. The processed halloysitic clay is a very high quality ceramic raw material. Whiteware ceramics are produced in New Zealand, and the clay is exported to Japan, England, and the United States because of its fired whiteness and translucency. This clay is also used as a filler for paper in New Zealand and in paints, plastics, and rubber. This deposit is unique in that it contains a very high proportion of halloysite in the clay fraction.

SEDIMENTARY KAOLIN DEPOSITS

Australia

The Weipa kaolin deposit is located on the west coast of the Cape York Peninsula in northeast Queensland. The sedimentary kaolinitic sands at Weipa are overlain by extensive lateritic bauxites of Tertiary age. Below the bauxite is an interval called the "pallid zone" that contains iron stained and mottled kaolin, bauxitic kaolin, and quartz sand (Murray and Harvey, 1982). The kaolin occurs in a sand matrix containing 30 to 60 percent coarse quartz. The exact geologic age of this sedimentary sand kaolin is uncertain with some geologists indicating a late Cretaceous age and others believing it is early Tertiary. The kaolinite in the pallid zone is 8 to 12 m thick, fine in particle size, and platy in shape. The fine clay fraction consists of kaolinite with a very small amount of smectite in some areas. It is used primarily for paper coating because it has good viscosity characteristics at high solids concentration (70%).

Brazil

A large sedimentary kaolin deposit is located in northern Brazil on the Jari River, a tributary of the Amazon, in the territory of Amapa

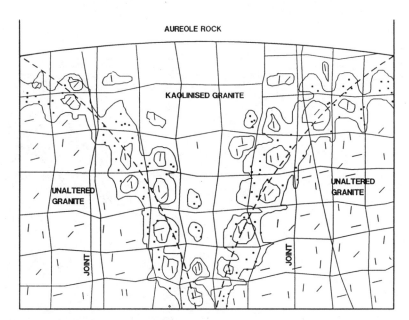

Figure 13. Diagrammatic cross section of a typical kaolinized granite "funnel" in the St. Austell granite (from Bristow, 1977).

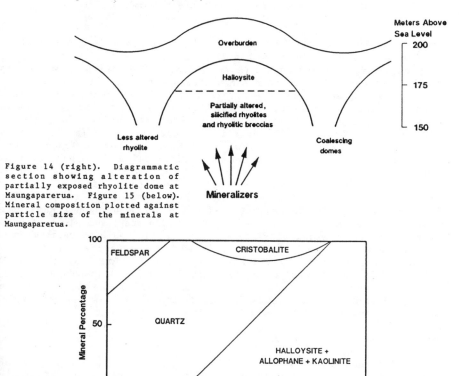

Figure 14 (right). Diagrammatic section showing alteration of partially exposed rhyolite dome at Maungaparerua. Figure 15 (below). Mineral composition plotted against particle size of the minerals at Maungaparerua.

82

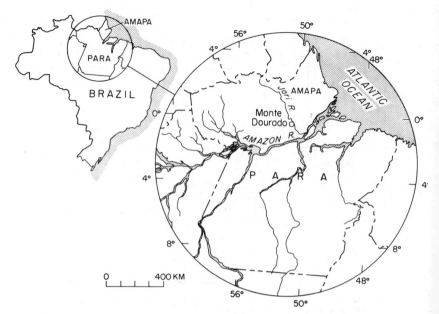

Figure 16. Location of Jari kaolin deposit at Monte Dourado in Brazil.

(Fig. 16). It is on the east side of the Jari River in the Barreiros
Series, a sequence of unconsolidated sands, sandy clays, kaolin, and
conglomerates, of Pliocene age (Murray, 1981). The kaolin is at the top
of this sequence and is called the Belterra Clay. In the area where it
is being mined, kaolin reaches a thickness of 85 m. Weathering in this
area of Brazil is intense; rainfall is heavy and the temperature is
hot. Jungle growth is abundant and provides organic acids that
intensify the laterization near the ground surface. The kaolin is
laterized to a depth of 7 to 8 m, and extensive bauxite deposits are
being mined in the Amazonia region of Brazil. Below the highly stained
lateritic zones is a light colored kaolin layer which averages about 30
m. The dominant clay mineral is kaolinite with only traces of smectite
and illite in some areas and a small percentage of quartz. The
kaolinite is fine in particle size with the pseudo-hexagonal kaolinite
crystals being small (2 microns or less in diameter) and platy with no
thick booklets or vermicular stacks. The kaolin has been derived from
weathered granites on the Guyana Shield and deposited in a large
lacustrine basin (Murray and Partridge, 1982). The kaolin is used
primarily for paper coating and is shipped to markets in Europe, Japan,
and South America. The viscosity characteristics of this kaolin are
very good at 70 percent solids, and it is one of the few kaolins in the
world that can be used for coating paper.

Federal Republic of Germany

      A sedimentary sand kaolin is located in Bavaria near the village of
Hirschau and Schnaittenbach (Koster, 1977). An arkosic sand of Triassic
age has been kaolinized and is being mined in an area approximately 6 km
long. The arkoses contain 75 to 85% quartz, about 10% kaolinite, and
the remainder a partially kaolinized feldspar. There is a small amount

83

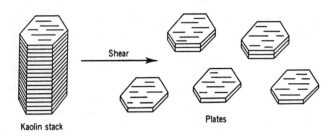

Kaolin stack

Shear

Plates

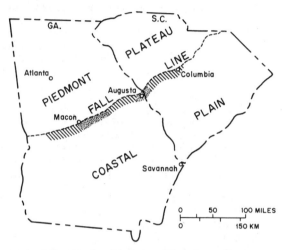

Figure 17 (above). Illustration of
delamination process. Figure 18
(right). Kaolin mining belt in
Georgia and South Carolina (hachured).

of smectite and illite associated with the kaolinite. The deposits are
mined, and glass sand, feldspar, and kaolin are the products. The
kaolin is used as a filler in paper and for high quality ceramic
whitewares. Some of the coarse kaolin booklets are separated and
sheared to produce large thin plates (Fig. 17) which are used to mix
with paper coating clays (Murray, 1984).

Spain

Kaolinitic sands of Cretaceous age extend from the Guadalajara
region to Valencia and are an important source of kaolin in central
Spain (Martin-Vivaldi, 1969). These kaolinitic sands are an important
source of silica sands for the glass industry, and the kaolin is used
for ceramics and as fillers in paper, paint, plastics, and rubber. The
thicknesses in the Wealdian and Utrillas formations range from 10 to 30
m. The kaolinite content generally ranges from 5 to 20 percent. These
kaolinitic sands probably were originally deposited as arkosic sands
derived from deeply weathered crystalline rocks of "Meseta Ubérica"
during early Cretaceous time. The feldspar content sands decreases from
bottom to top, indicating alteration by percolating ground water. These
Cretaceous kaolinitic sands cover a large area of northeast and east
central Spain, and the reserves are very large even though the kaolinite
content is small.

Suriname

The Onverdacht Series in the Moengo area of eastern Suriname is

84

divided into the Upper Onverdacht which is composed of bauxite underlain
by kaolins and the Lower Onverdacht which is composed of kaolinite and
quartz sands (Montagne, 1964). The age of the Onverdacht Series is
Eocene, and the kaolin and sands have been derived from the Guyana
Shield to the south. Subsequent to deposition in a coastal deltaic
environment intense lateritic weathering produced the bauxites which
overlie the kaolin. The kaolin averages about 6 to 8 m thick. The
upper two meters of the kaolin immediately under the bauxite contains
gibbsite dispersed throughout the kaolin and along bedding planes.
Below this gibbsitic kaolin the kaolinite contains up to 10 percent
quartz and minor amounts of mica. The kaolin in the Moengo area is high
quality and has excellent viscosity characteristics. At the present
time, however, no kaolin has been mined in Suriname. There are large
reserves, and a plentiful water supply so that at some future time these
kaolins could be mined and processed to produce coating grade products.
The kaolin is the parent material on which the high grade bauxites were
developed (Aleva, 1965).

United States

The sedimentary kaolins in Georgia and South Carolina (Fig. 18) are
the most extensive secondary kaolin deposits in the world. These
kaolins, which occur in Late Cretaceous and early Tertiary rocks (Fig.
19) were derived from deeply weathered crystalline rocks, granites, and
gneisses on the Piedmont plateau. Beginning in late Cretaceous time the
residual weathering products including kaolinite, quartz, partially
altered feldspar, muscovite, smectite, and minor amounts of magnetite,
ilmenite, rutile, and other heavy minerals, were stripped off and
transported to the coastline located along a line extending from western
Georgia through Macon and Augusta, Georgia, to Columbia, South Carolina.
The kaolin was deposited in estuaries, lagoons, oxbow lakes, and ponds
in a coastal area about 30 to 50 km wide.

The Cretaceous kaolins are generally lenticular-to saucer-shaped
bodies that are found in relatively unconsolidated, variably kaolinitic,
micaceous, cross-bedded sands. These kaolin beds range to as much as 12
m thick and more than 2 km long. The Tertiary kaolins also occur in
relatively unconsolidated, micaceous, cross-bedded sands and are
generally more persistant and thicker than the Cretaceous kaolins.

Kaolinite is the dominant mineral in both the Cretaceous and
Tertiary beds and usually comprises from 90 to 95 percent of the
minerals present. Other minerals commonly present as accessories in the
kaolins are quartz, muscovite, biotite, smectite and small amounts of
magnetite, ilmenite, tourmaline, zircon, goethite, and occasionally
other heavy minerals. In general the Cretaceous kaolins are coarser in
particle size than the Tertiary kaolins. The Cretaceous kaolins
generally range from 50 to 70 percent less than two microns and the
Tertiary kaolins 80 to 90 percent less than two microns. The Cretaceous
kaolins almost always contain a large number of coarse vermicular
crystals interspersed in a fine matrix of kaolinite books and plates
(Fig. 1) whereas the Tertiary kaolins are comprised of thin platy
particles with no vermicular kaolin particles (Fig. 20). Generally the
iron and titanun values of the Tertiary kaolins are higher than these
values in the paper grade Cretaceous kaolins. Dombrowski and Murray
(1984) found that the average thorium content of the Cretaceous kaolins
are higher than the average thorium content of the Tertiary kaolins.

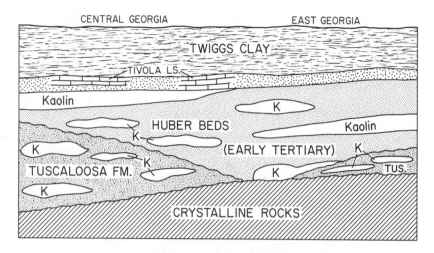

Figure 19. Diagrammatic section showing Tuscaloosa (Cretaceous) and Huber (Tertiary) kaolins in Georgia (Murray, 1976).

Figure 20. Tertiary kaolinite with thin, platy, very fine particles.

Virtually all who have investigated the kaolin deposits of Georgia and South Carolina agree that they were derived from granites and gneisses on the Piedmont plateau. Some believe that the kaolins were transported and deposited as kaolin and others suggest that the feldspar, quartz, and muscovite were originally deposited as such and that post-depositional alteration and recrystallization produced the kaolin deposits. A summary of these ideas is given by Patterson and Murray (1984). A study using oxygen isotopes as an indicator of kaolin genesis (Murray and Janssen, 1984) revealed that the large vermicular kaolinite crystals in the Cretaceous kaolins have a very low value compared to the platy matrix of particles surrounding these vermicular crystals. They suggested that the large vermicular kaolin books are altered from feldspar grains, and that the transformation from these detrital feldspars to kaolinite involved very little contribution from the $\delta^{18}O$ in meteoric water so that the low $\delta^{18}O$ value of the vermicular books reflect the low feldspar values.

The author believes that many of the Tertiary kaolins are reworked Cretaceous kaolins (Murray, 1976) and thus are second cycle, which accounts for their increased fineness, lack of vermicualr books, lower thorium content, and other physical and chemical characteristics.

The resources of kaolin in Georgia are estimated to be over 5 billion tons (Murray, 1983). These kaolins comprise the largest reserve of high quality kaolins in the world. They are used for paper coating at paper mills all over the world because of their excellent viscosity characteristics. They are also used for ceramics and as quality fillers in paper, plastics, paint, ink, and many other uses.

SUMMARY

The kaolin minerals kaolinite, halloysite, dickite, and nacrite have essentially the same chemical composition. Dickite and nacrite are relatively rare and are normally restricted to hydrothermal occurrences. Kaolinite is by far the most common mineral of the group and is found as a residual weathering product, in hydrothermal alteration zones, and in sedimentary rocks. Halloysite is found in hydrothermal and residual deposits but is very rare in kaolins of sedimentary origin.

Kaolins are generally derived from the alteration of feldspars and muscovite but can be derived from other aluminous minerals under intense alteration conditions. The most common origin of kaolins is the result of surface weathering of granites and other acidic crystalline rocks, and are classed as primary residual deposits. Another occurrence of kaolins is primary hydrothermal deposits, which are generally small in extent and contain limited quantities of exploitable reserves. Two well known primary kaolins, Cornwall in England and Maungaparerua in New Zealand, are mixed hydrothermal and residual kaolins.

Sedimentary or secondary kaolin deposits are found in lacustrine, paludal, deltaic, and lagoonal environments but are not found in marine environments. The largest sedimentary kaolin occurrence in the world is in the southeastern United States in Georgia and South Carolina. Other large sedimentary occurrences are in Brazil and Australia.

Kaolin is a very versatile industrial mineral that has many industrial uses. The largest user by far is the paper industry where it is ulitized as a filler in the paper sheet and as a coating on the surface. The specifications for its use as a paper coating material are very rigid, and only four areas in the world now produce large quantities for this purpose. These areas are the United States, England, Brazil, and Australia. Only one halloysite deposit is being mined at present, and it is in New Zealand.

87

REFERENCES

Aleva, G.J.J. (1965) The buried bauxite deposit of Onverdacht, Suriname, South America. Geologie en Mijnbouw 44, 45–58.
Bailey. S.W. (1963) Polmorphism of the kaolin minerals. Am. Mineral. 48, 1196–1209.
Bates, T.F. (1959) Morphology and crystal chemistry of the layer lattice silicates. Am. Mineral. 44, 78–114.
Brindley, G.W. and Robinson, K. (1947) An x-ray study of some kaolinitic fireclays. British Ceram. Soc. Trans. 46, 49–62.
Bristow, C.M. (1977) A review of the evidence for the origin of the kaolin deposits in S.W. England. Proc. 8th Int'l Kaolin Symp., Madrid–Rome, 19 p.
Cleveland, G.B. (1962) Geology of the Little Antelope Valley clay deposit, Mono County, California. Calif. Div. of Mines Geol. Spec. Rpt. 72, 28 p.
Deer, W.A., Howie, R.A. and Zussman, J. (1966) Introduction to the Rock Forming Minerals. Wiley, New York, 528 p.
Dixon, J.B. (1977) Kaolinite and serpentine group minerals, in Minerals in Soil Environments, Dixon, J.B. and Weed, S.B. eds. Soil Science Soc. of Amer., Madison, Wisconsin. 367–403.
Dombrowski, T. and Murray H.H. (1984) Thorium – A key element in differentiating Cretaceous and Tertiary kaolins in Georgia and South Carolina. Proc. 27th Int'l Geol. Cong. 15, 305–317.
Garrels, R.M. and Christ, C.L. (1965) Solutions, Minerals and Equilibria. Harper and Row, Inc., N.Y. 450 p.
Gaskin, A.J. (1944) Kaolinized granodiorite in the Bulla-Broadmeadows area. Proc. Royal Soc. Victoria 56, 1–18.
Hanson, R.F. and Keller, W.D. (1966) Genesis of refractory clay near Guanajuato, Mexico. In Bailey, S.W. ed., Clays & Clay Minerals, Proc. 14th Nat. Conf., Pergamon Press, 259–267.
Hanson, R.F., Zamora, R. and Keller, W.D. (1981) Nacrite, dickite, and kaolinite in one deposit in Nayarit, Mexico. Clays & Clay Minerals, 29 451–453.
Hinckley, D.N. (1963) Variability in "crystallinity" values among the kaolin deposits of the coastal plain of Georgia and South Carolina. Clays & Clay Minerals 11, 229–235.
Honda, S., Miura, T., Ohira, Y. and Tamanoi, M. (1964) Industrial mineral resources in the Tohoku district 4, 87.
Keller, W.D. and Hanson, R.F. (1968) Hydrothermal alteration of a rhyolite flow breccia near San Luis Potosi, Mexico, to refractory kaolin. Clays & Clay Minerals 16, 223–229.
Kildale, M.B. and Thomas, R.C. (1957) Geology of the halloysite deposits at the Dragon mine. Utah Geol. Soc. Guidebook 12, 94–96.
Koster, H.M. (1977) A contribution to the geochemistry and the genesis of the kaolin-feldspar deposits of eastern Bavaria. Proc. 8th Int'l Kaolin Symp. Madrid–Rome, 6 p.
Kuzvart, M. (1969) Kaolin deposits of Czechoslovakia. Proc. Symp. I, Kaolin Deposits of the World. A–Europe 23rd Int'l Geol. Congress, 47–73.
Lombardi, G. and Mattias, P. (1987) The kaolin deposits of Italy. Estratto da L'industria Mineraria 6, 1–34.
Martin-Vivaldi, L. (1969) Kaolin deposits of Spain. Proc. Symp. I. Kaolin Deposits of the World, A–Europe, 23rd Int'l Geol. Congress, 225–262.

88

Mason, B. (1952) Principles of Geochemistry. John Wiley and Sons, N.Y., 274 p.

Merino, E., Harvey, C.C. and Murray, H.H. (1988) Aqueous-chemical control on the tetrahedral-aluminum content of quartz, halloysite, and other low-temperature silicates. Clays & Clay Minerals (submitted).

Montagne, D.G. (1964) New facts on the geology of the young unconsolidated sediments of northern Suriname. Geologie Mijnbouw 43, 499-514.

Murray, H.H. (1963) Industrial applications of kaolin. In Swineford, A., ed. Clays & Clay Minerals, Proc. 10th Nat. Conf. MacMillan, N.Y., 291-298.

_____ (1976) The Georgia sedimentary kaolins. Int'l Geol. Correlation Program Project 23, 7th Symp. on Genesis of Kaolin, Tokyo, Japan, 114-125.

_____ (1980) Diagnostic tests for evaluation of kaolin physical properties. Acta Mineralogica-Petrographica XXIC, Proc. 10th Int'l Kaolin Symp., Budapest, Hungary, 67-76.

_____ (1981) Kaolin project in the Amazonia area, Brazil. Soc. of Mining Engineers Preprint 81-57, 7 p.

_____ (1983) Nonbauxite alumina resources, in Shanks, W.C. ed., Cameron Volume on Unconventional Mineral Deposits. Soc. Mining Engrs., 111-119.

_____ (1984) Clay. In Hagemeyer, R.W., ed., Pigments for Paper. Tappi Press, Atlanta, GA., 95-143.

_____ (1986) Clays. Ullmann's Encyclopedia of Industrial Chemistry A7, 109-136.

_____ (1988) World kaolins - diverse quality needs permit different resource types. Proc. 8th Ind. US. Minerals Int'l Congress. Metal Bulletin, London, 127-130.

_____ and Harvey, C.C. (1982) Australian and New Zealand clays. Soc. Mining Engineers Preprint 82-383, 5 p.

_____, Harvey, C.C. and Smith, J.M. (1977) Mineralogy and geology of the Maungaparerua halloysite deposit in New Zealand. Clays & Clay Minerals, 25, 1-5.

_____ and Jansen, J. (1984) Oxygen isotopes - indicators of kaolin genesis? Proc. 27th Int. Geol. Congress 15, 287-303.

_____ and Lyons, S.C. (1956) Correlation of papercoating quality with degree of crystal perfection of kaolinite. In Swineford, A., ed., Clays & Clay Minerals. Proc. 4th Nat'l Conf., Nat'l Res. Council Pub., 456, 31-40.

_____ and Partridge, P. (1982) Genesis of Rio Jari kaolin. Int. Clay Conf. 1981, Van Olphen, H. and Veniale, F., eds., Dev. in Sedimentology 35, Elsevier, 279-291.

_____, Partridge, P. and Post, J.L. (1978) Alteration of a granite to kaolin - Mineralogy and Geochemistry. Schriftenr. Geol. Wiss. 11, Berlin, 197-208.

_____ and Smith, J.M. (1973) The geology and mineralogy of the Grahamstown, South Africa kaolin deposit. Abstr., 22nd Ann. Clay Minerals Conf., Clay Minerals Soc., p. 48.

Newman, A.C.D. and Brown, G. (1987) The chemical constitution of clays In Chemistry of Clays and Clay Minerals, A.C.D. Newman, ed., Mineral. Soc. Monograph 6, Wiley, 1-128.

Patterson, S.H. and Murray, H.H. (1984) Kaolin, refractory clay, ball clay, and halloysite in North America, Hawaii, and the Caribbean Region. U.S. Geol. Survey Prof. Paper 1306, 56 p.

Petrov, V.P. and Chukhrov, F.V. (1977) Kaolin deposits in the USSR. Proc. 8th Int'l Kaolin Symp., Madrid, Rome, 15 p.

Rex, R.W. (1966) Authigenic kaolinite and mica as evidence for phase equilibria at low temperatures. Clays & Clay Minerals 13, 95-104.

Romero, A, Dominquez, E. and Whewell, R. (1974) El area caolinera de Gaiman, Prov. del Chubut. Fundacion Bariloche. Centro Nacional Patagonico, 423-444.

Storr, M. (1975) Kaolin deposits of the GDR in the northern region of the Bohemian Massif. Ernst-Moritz-Arndt-Univ. Greifswald. Sektion Geolog. Wiss., 243 p.

Sudo, T. and Shimoda, S. (1978) Clays and Clay Minerals of Japan. Dev. in Sedimentology 26, Elsevier, 326 p.

Togashi, Y. (1976) Geology of the Seto, Shokozan, and Itaya kaolin deposits. Int'l Geol. Correlation Program Project 23. 7th Symp. on Genesis of Kaolin, Tokyo, Japan, 16 p.

Velde, B. (1985) Clay Minerals. A Physics-Chemical Explanation of their Occurrence. Dev. in Sedimentology 40, Elsevier, Amsterdam, 427 p.

Weaver, C.E. and Pollard, L.D. (1973) The Chemistry of Clay Minerals. Dev. in Sedimentology 15, Elsevier, Amsterdam, 213 p.

Zheng, Z., Lu, D., Feng, M., Feng, B. and Jin, T. (1982) Kaolin deposits of China. Int'l Clay Conf. 1981, Van Olphen, H. and Veniale, F., eds., Dev. in Sedimentology 35, Elsevier, Amsterdam, 719-731.

Chapter 5   F.J. Wicks & D.S. O'Hanley

SERPENTINE MINERALS: STRUCTURES AND PETROLOGY

INTRODUCTION

The serpentine minerals, lizardite, chrysotile, and antigorite are
trioctahedral hydrous phyllosilicates based on 1:1 layer structures.
Lizardite has a planar structure with similarities to other
trioctahedral 1:1 layer structures discussed in Chapter 6. Chrysotile
has a cylindrical structure with some similarities to halloysite
discussed in Chapter 3, and antigorite has a curved modulated structure
related to other modulated structures discussed in Chapter 17. The
three serpentine minerals have similar chemical compositions, with the
occupancies at the octahedral sites dominated by magnesium. The lateral
dimensions of an ideal magnesium-occupied octahedral sheet ($b \approx 9.43$ Å,
Chap. 1) are larger than the lateral dimensions of an ideal silicon-
occupied tetrahedral sheet ($b \approx 9.1$ Å). The misfit between sheets is
significant, and leads to the three serpentine structures and the
recently postulated carlosturanite structure (Chap. 17), each a
different solution to the misfit problem. In lizardite, the misfit is
accommodated within the normal, planar 1:1 layer structure. In
chrysotile, the misfit is partly overcome by the cylindrical curvature
of the layers. In antigorite, the misfit is overcome by the curvature
of the alternating wave modulation. In carlosturanite, the misfit is
overcome through ordered vacancies and modifications of the tetrahedral
sheet within a planar structure. This is one of the few examples in
layer silicates in which four different structures occur in material of
the same or similar chemical compositions. Thus the serpentine minerals
and carlosturanite form an important group for study, as each structure
is a different solution to the same problem.

The serpentine minerals discussed here are common rock-forming
minerals that occur in similar geological environments. The other
species of the serpentine sub-group, amesite, berthierine, cronstedtite,
etc. occur much less frequently in a wide variety of geological
environments, and are more conveniently discussed separately (Chap. 6).

CRYSTAL STRUCTURES

Our understanding of the serpentine minerals has been hampered by
the lack of three-dimensional crystal structure refinements. The rarity
of suitable single crystals of lizardite and antigorite, and the
cylindrical structure of chrysotile, have limited structure refinements
to two dimensions. These refinements have provided the basic features
of the structures but none of the fine details. High-Resolution
Transmission Electron Microscopy (HRTEM) has verified these basic
structures, but again, none of the fine structural detail. The

information available from the two-dimensional refinements was compiled and discussed in detail by Wicks and Whittaker (1975).

Recently, three-dimensional refinements of three lizardite structures (Mellini, 1982; Mellini and Zanazzi, 1987) have provided a better understanding of the lizardite structure and have implications for the other serpentine structures. The HRTEM studies of antigorite, lizardite, chrysotile, and carlosturanite have broadened our understanding of these structures and their relationships to one another, but have revealed new problems to be solved. The recognition of a large class of modulated layer structures (Chap. 17) has provided new insight into the misfit problem and how it can be accommodated.

Lizardite

Lizardite is the most abundant of the serpentine minerals. It occurs commonly in retrograde serpentinites as the principal serpentine mineral, and is formed by pseudomorphic replacement of pre-existing silicates. It is usually apple green, but its color is often masked by fine-grained disseminated magnetite which colors the rock gray, black or brown, or less frequently by hematite which colors the rock red. Lizardite in veins is usually green (often apple green), but can be pale yellow to white. The habit of vein lizardite varies from massive to platey to pseudofibrous. It may be associated with chrysotile as coexisting but separate phases, or occur as a polygonal overgrowth of planar lizardite over a cylindrical chrysotile core. Euhedral crystals of lizardite are rare, but have been described by Midgley (1951), Krstanovic (1968), Wicks and Whittaker (1977); recently, two occurrences yielded crystals suitable for structure determination (Mellini, 1982; Mellini and Zanazzi, 1987).

Lizardite was first recognized as a distinct mineral species by Whittaker and Zussman (1956), on the basis of X-ray diffraction studies of a variety of specimens. These included single crystals of lizardite from a vein in the Lizard serpentinite at Kennack Cove, Cornwall, originally described by Midgley (1951). The stacking sequence of the lizardite crystals from Kennack Cove was studied by single-crystal methods by Rucklidge and Zussman (1965). The crystals are considerably disordered, and are composed of domains of 1-layer and 2-layer structures. Krstanovic (1968) did the first two-dimensional refinement of atomic positions in a (010) projection of 1-layered lizardite crystals with $\pm$ $b/3$ disorder from the Radusa chromite mine, Yugoslavia.

A number of multilayer serpentine structures have been discovered in naturally occurring serpentines (Brindley and von Knorring, 1954; Zussman and Brindley, 1957; Bailey and Tyler, 1960; Olsen, 1961; Müller, 1963; Krstanovic and Pavlovic, 1967; Coats, 1968; Jahanbagloo and Zoltai, 1968) and in synthetic hydrothermal serpentines (Roy and Roy, 1954; Gillery, 1959; Nelson and Roy, 1958; Shirozu and Momoi, 1972; Jasmund and Sylla, 1971, 1972). Bailey (1969) developed a polytype nomenclature scheme for 1:1 layer structures that provides a systematic basis for describing multilayer serpentines (see Chap. 2). Wicks and Whittaker (1975) pointed out that these multilayer serpentines are stacking variations of the same structure, disregarding compositional variations, and recommended that they be considered as polytypes of lizardite. Determining the stacking details of a multilayer polytype is not trivial, as can be seen by the study of Jahanbagloo and Zoltai (1968) on lizardite $9T$ from the north shore of Lake Superior in Minnesota, and by the study of Hall et al. (1976) on lizardite $6T$, from Unst, Shetland Islands, Scotland.

Wicks and Whittaker (1975) reviewed the available structural information on the serpentine minerals and this paper is still a useful source paper. However, the details of the lizardite structure given by Mellini (1982) and Mellini and Zanazzi (1987) have superseded the lizardite discussion in the earlier paper. Also, some of the polytype designations given by Wicks and Whittaker (1975) have been corrected by Hall et al. (1976). These authors have recommended that full polytype designation be given only to known structures. Thus the polytype designation $6R_2$, used by Wicks and Whittaker (1975) for a theoretical polytype developed by Steadman (1964) should be deleted because it has not been observed in natural crystals. Furthermore, Hall et al. (1976) already have used the $6R_2$ designation for a different polytype, (the one designated $6R_3$ by Wicks and Whittaker, 1975) observed in amesite and cronstedtite (Steadman and Nuttall, 1962, 1964) and in kellyite (Peacor et al., 1974). This use of $6R_2$ for observed structures has precedence over the $6R_3$ designation used by Wicks and Whittaker (1975) for a theoretical polytype, and thus the $6R_3$ designation should be deleted. These changes are listed in Table 4, of Hall et al. (1976, p. 320).

Structural refinements. Mellini (1982) determined the structure of lizardite $1T$ from Val Sissone, Italy. The lizardite occurs as light yellow, euhedral crystals with a truncated trigonal pyramidal habit, in a vein that cuts through a dolomite-clinohumite rock that is believed to be from the contact metamorphic aureole of the Bergell granite. The formula is $(Mg_{2.79}Fe^{2+}_{0.04}Fe^{3+}_{0.10}Al_{0.07})(Si_{1.83}Al_{0.17})O_5(OH)_4$.

Later Mellini and Zanazzi (1987) collected data for the refinement of the structures of lizardite $1T$ and lizardite $2H_1$ from Coli, Italy. The two types of structures occur together as light-green, euhedral crystals associated with calcite in a vein in the ophiolite complex of Monte dei Tre Abati. The $1T$ polytype occurs as truncated trigonal pyramidal crystals with a formula $(Mg_{2.82}Fe_{0.07}Al_{0.09})(Si_{1.94}Al_{0.06})O_5(OH)_4$. The $2H_1$ polytype occurs as truncated hexagonal pyramidal crystals and as hexagonal plates with a formula unit $(Mg_{2.83}Fe_{0.05}Al_{0.10})(Si_{1.93}Al_{0.07})O_5(OH)_4$. Thus at Coli the polytypes can be identified in hand specimen by their trigonal or hexagonal habit.

The $1T$ structures were refined in space group $P31m$ and seem to be trigonal within the limits of the refinements (R = 0.031 for Val Sissone, R = 0.074 for Coli). Thus neither the monoclinic space group

Cm used by Krstanovic (1968) in his two-dimensional refinement nor the pseudotrigonal orthorhombic symmetry suggested by Wicks and Whittaker (1975) is correct. However, if required, Mellini's structures can be described in terms of a *C*-centered orthorhombic unit cell for comparison with other serpentines. The $2H_1$ structure was refined in space group $P6_3cm$ to R = 0.024.

The three structures are essentially similar in all major features. Deviations from the ideal geometry are small, in spite of the fact that misfit between sheets is significant for the compositions of these crystals. The octahedral and tetrahedral sheets have well-defined trigonal symmetry, with the basal O(2) oxygen atoms, the silicon atoms, the magnesium atoms and the O(3) oxygen atoms that form the outer hydroxyl groups each lying on specific planes within the structure (Fig. 1). Thus the possible buckling of the plane of magnesium atoms into two separate planes discussed by Wicks and Whittaker (1975) does not occur.

Deviations from the ideal positions do occur. In the plane common to the octahedral and tetrahedral sheets, the apical O(1) oxygens of the tetrahedral sheet and the O(4) hydroxyl oxygens of the octahedral sheet are separated by 0.064 Å along Z in the Val Sissone lizardite 1*T*, by 0.116 Å in the Coli lizardite 1*T* (this may not be completely real but in part an artifact of the anisotropic thermal ellipsoid of O(1) in this refinement), and by 0.058 Å in the Coli lizardite $2H_1$. The plane of the magnesium atoms is also shifted from the center of the octahedral sheet away from the tetrahedral sheet by 0.082 Å in the Val Sissone lizardite 1*T*, by 0.088 Å in the Coli lizardite 1*T*, and by 0.089 Å in the Coli lizardite $2H_1$. The tetrahedra are rotated about Si-O(1) from hexagonal to ditrigonal symmetry (Fig. 2) by α = -3.5° in the Val Sissone lizardite 1*T*, by α = -1.7° in the Coli lizardite 1*T*, and by α = +6.4° in the Coli lizardite $2H_1$ (α is defined in Chap. 1, + and - rotation are defined in Fig. 2).

The tetrahedral sheet, measured from the weighted mean position of O(1) and O(4) is 2.22 Å thick in the two 1*T* structures and 2.21 Å thick in the $2H_1$ structure. Silicon forms three long bonds, 1.646 Å in Val Sissone 1*T*, 1.651 Å in Coli 1*T*, 1.648 Å in Coli $2H_1$, with the bridging O(2) oxygens and a single short bond, 1.616 Å in Val Sissone 1*T*, 1.577 Å in Coli 1*T* (accuracy limited by the strongly anisotropic O(1) thermal ellipsoids), 1.602 Å in Coli $2H_1$, with the non-bridging O(1) oxygen. The mean Si-O bond lengths are 1.639 Å in Val Sissone 1*T*, 1.632 Å in Coli 1*T*, and 1.636 Å in Coli $2H_1$, and generally reflect the Al-content of the tetrahedra. The Si-O(1) bond length seems to be the more sensitive to Al-content (Mellini and Zanazzi, 1987).

Measured from the weighted mean position at O(1) and O(4), the octahedral sheet is 2.10 Å thick in the Val Sissone 1*T* structure, and 2.12 Å thick in the two Coli structures. This is similar to chrysotile (2.08 Å, Wicks and Whittaker, 1975) and brucite (2.11 Å) but is thinner than previously reported for lizardite (2.20 Å, Krstanovic, 1968). The magnesium atom occupies a trigonal antiprism (flattened octahedron).

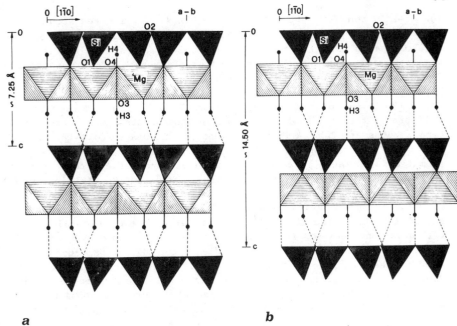

a   **b**

Figure 1. The crystal structure of lizardite, as seen along [110]: (a) $1T$ polytype; (b) $2H_1$ polytype. Octahedra are lined, tetrahedra are shaded, and H-bonds are schematically indicated (from Mellini and Zanazzi, 1987).

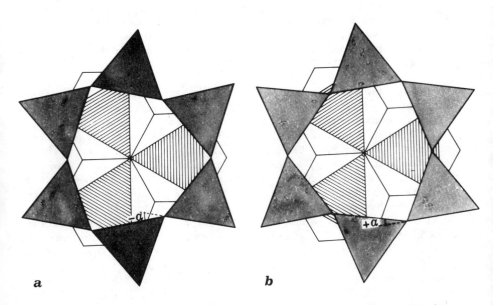

a   **b**

Figure 2. Schematic representation of (a) negative and (b) positive ditrigonal distortion in the T-O layer of lizardite. Opposite movement of the bridging oxygen atoms is shown, away from and toward the octahedral cations of the same layer (from Mellini and Zanazzi, 1987).

The Mg-O(3) bond lengths are 2.021 - 2.026 Å in all three structures, and when compared to the Mg-O(4) bond lengths of 2.082 Å in Val Sissone $1T$, 2.069 Å in Coli $1T$, and 2.087 Å in Coli $2H_1$, reflect the shift of the magnesium atom away from the center of the antiprism towards the O(3) plane. The bond between magnesium and O(1), the atom common to both the octahedral and tetrahedral sheets, is 2.121 Å in Val Sissone $1T$, 2.139 Å in Coli $1T$ and 2.125 Å in Coli $2H_1$, and most directly illustrates the effect of misfit between the sheets. The average Mg-O bond lengths are 2.067 Å in Val Sissone $1T$, 2.070 Å in Coli $1T$ and 2.067 Å in Coli $2H_1$. These are smaller than the value of 2.10 Å calculated from the ionic radii given by Shannon (1976) and reflect the effect of the substitution of trivalent ions, Al and Fe^{3+}, for Mg (Mellini, 1982).

Two hydrogen positions were located on the Fo map and refined by Mellini (1982). The H(4) is at the center of the ditrigonal ring of O(1) atoms and does not form hydrogen bonds. The second hydrogen, H(3), forms bonds between the donor O(3) and acceptor O(2) atoms, and helps to bond successive layers (Mellini, 1982). In all three structures the coupled rotation of tetrahedra to a ditrigonal geometry moves the O(2) atom towards the hydroxyl in the adjacent layer, promoting hydrogen bonding. This rotation is negative (as defined in Fig. 2) for the $1T$ structures but positive for the $2H_1$ structure. The $2H_1$ structure has a rotation of 180° between successive layers, so that a positive rotation of tetrahedra in this structure has exactly the same effect as a negative rotation in the $1T$ structure. The rotation of tetrahedra to ditrigonal symmetry is the principal mechanism of overcoming misfit when $b_{oct} < b_{tet}$ (Chap. 6). In the situation in which $b_{oct} > b_{tet}$, one can imagine the tetrahedral sheet being fully stretched to hexagonal symmetry. However, Mellini's three structure refinements indicate that tetrahedral rotation, albeit small ones, still occur in the $b_{oct} > b_{tet}$ situation.

These results can be used to re-assess the molecular orbital (MO) calculations of Peterson et al. (1979) on 1:1 layer structures. In this study, a cluster model consisting of a six-membered silicate ring with the Si-O bond lengths set at 1.62 Å was linked to three edge-sharing octahedra with an Mg-O bond length of 2.10 Å. The MO calculations indicated a total energy minimum at a tetrahedral rotation of $\alpha = -2.1°$ (Fig. 3a). Using a second cluster model of three tetrahedra and four octahedra that better modelled the effects of moving the bridging oxygens of the tetrahedral sheet over the filled octahedral sites, the total minimum energy occurred at $\alpha = -3.9°$. Further MO calculations with an Si-O bond length of 1.68 Å to approximate the amesite composition gave two total energy minima at $\alpha = -12.3°$ and $+12.7°$ for the first cluster model (Fig. 3b) and at $\alpha = -13.8°$ and $+10.6°$ for the second cluster model. Recent refinements of two amesite $2H_2$ structures indicate $\alpha = +14°$ or $+15°$ (Chap. 6), and support the utility of MO calculations in rationalizing real structures (Hall and Bailey 1976; Anderson and Bailey, 1981).

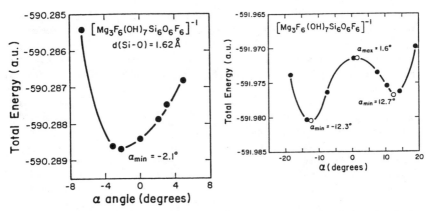

Figure 3. Total energy of a $[Mg_3F_6(OH)_7Si_6O_6F_6]^-$ cluster versus the tetrahedral rotation angle α. (left) Si-O = 1.62 Å representing the tetrahedra of the end-member Mg composition. (right) Si-O = 1.68 Å representing the tetrahedra of the amesite composition (from Peterson et al., 1979).

Lizardite   **Amesite**

$Mg_6Si_4O_{10}(OH)_8$ $(Mg_5Al)(Si_3Al)O_{10}(OH)_8$ $(Mg_4Al_2)(Si_2Al_2)O_{10}(OH)_8$

Figure 4. Approximate compositions of various lizardite, Al-serpentine and amesite specimens, illustrating a possible solid solution series. In order to plot the substitution and to maintain charge balance between sheets, minor amounts of Fe^{3+}, when it is present, has been added to Al. 1 = Wicks and Plant (1979), 2 = Mellini and Zanazzi (1987), 3 = Mellini (1982), 4 = calculated $b_{oct} = b_{tet}$ Chernosky (1975), 5 and 6 = Wicks and Plant (1979), 7 = Jahanbagloo and Zoltai (1968), 8 and 9 = Bailey and Tyler (1960), 10 = Taner and Laurent (1984), 11 = Hall and Bailey (1976), 12 = Anderson and Bailey (1981) and Gruner (1944).

The MO calculations of Peterson et al. (1979) and the structure refinements of Mellini (1982) and Mellini and Zanazzi (1987) suggest that a slight ditrigonal distortion is the preferred arrangement of the tetrahedral sheet. If this is the case, it should occur in chrysotile and other structures.

Peterson et al. (1975, p. 710) suggested that "the sense of tetrahedral rotation is not controlled by factors such as the interaction between the tetrahedral bridging oxygens and the octahedral cations, but by interlayer effects not modelled by these small clusters" used for MO calculations. Mellini's three structure refinements and other three-dimensional refinements with low R indicate that rotation of bridging oxygens is always towards the hydroxyls of the adjacent layer (Mellini, 1982). Thus the available evidence suggests that rotation always occurs, and in a direction that promotes hydrogen bonding between successive layers. This also has implications for chrysotile and other structures.

It would be interesting to have a structure refinement of end-
member Mg-lizardite. The Al-content at which the area of the octahedral
sheet equals the area of the tetrahedral sheet, $b_{oct} = b_{tet}$, has been
estimated by Chernosky (1975) to be at x = 0.6, where x is the Al-value
in the formula $(Mg_{6-x}Al_x)(Si_{4-x}Al_x)O_{10}(OH)_8$. The Val Sissone lizardite
1T with x = 0.34 (Fe^{3+} is combined with Al to determine x) lies
approximately halfway between the $b_{oct} = b_{tet}$ composition and the end-
member composition at which the $b_{oct} > b_{tet}$ misfit is at a maximum (Fig.
4). The Coli lizardite 1T and 2H_1 structures, with x ≈ 0.16, are closer
to the end-member composition, but still have all the essential features
of the Val Sissone lizardite 1T structure. A measure of how
representative the Mellini lizardite compositions are of lizardite in
general can be obtained by examining the microprobe analyses of the
large number of rock-forming lizardite 1T specimens analyzed by Wicks
and Plant (1979). By far the greatest number of lizardites have
aluminum contents of x <0.05 (0.9 wt % Al_2O_3); in most of the rest, Al
is in the range x = 0.05 - 0.2 (3.7 wt % Al_2O_3) (Fig. 2, p. 790, Wicks
and Plant, 1979). This means that the Val Sissone structure at x = 0.34
and the Coli structures at x = 0.16 are not representative of the
compositions of most lizardite specimens. The question then remains,
are the Mellini structures representative of the end-member Mg-
structure? It would be reasonable to assume that they are but it is not
a certainty.

In an attempt to answer this question, Bish (1981) and Wicks and
Hawthorne (1986) have used the Distance Least-Squares (DLS) method
(Meier and Villiger, 1969) to model the end-member Mg-structure. Both
the cell parameters of Krstanovic's Radusa mine lizardite 1T (Bish,
1981; Wicks and Hawthorne, 1986) and of Mellini's Val Sissone lizardite
1T adjusted to represent pure Mg-O and Si-O polyhedra (Wicks and
Hawthorne, 1986) were used as the basis for modelling. The results
indicate that the octahedral and tetrahedral sheets fit together without
any major distortions in the structure. Octahedral and tetrahedral
sheet thicknesses are similar to those observed in actual structures,
and buckling of the Mg position into two positions is not produced by
the DLS similations. Rotation of tetrahedra by α = 3 to 4° can also be
accommodated by the DLS modelling without significant distortion in the
structure. These modelling studies seem to indicate that end-member Mg-
lizardite will be structurally similar to the Val Sissone and Coli
structures refined by Mellini.

Electron microscopy. The planar nature of the lizardite structure
has been illustrated by Veblen and Buseck (1979, 1981), Wicks (1986) and
Livi and Veblen (1987) with TEM images showing regions of planar,
parallel (001) fringes of lizardite. Yada and Iishi (1974a) obtained
HRTEM images of the hexagonal array of 4.5 Å fringes of synthetic
lizardite lying on the (001) plane. Cressey (1979) and Livi and Veblen
(1987) have done electron diffraction studies of lizardite 1T in various
orientations. Wicks (1986) has used X-ray microbeam camera and single-
crystal orientation studies on a partly serpentinized enstatite to
estimate the distribution of lizardite 1T in various orientations, as a

control for the interpretation of subsequent TEM studies. Although most lizardite is planar, some of it is gently bent over hundreds of Å into broad "S" shaped curves (Wicks, 1986). Also, the structure at the edge of a planar region may be planar but is often curved. The amount of curvature varies from slightly over 20 to 30° of arc, (Fig. 5a) to nearly complete "chrysotile-like" rolls (Fig. 5b) (Veblen and Buseck, 1979; Wicks, 1986). The transition from planar region to curved region occurs without a break in the layers. However, no electron diffraction studies have been made on this curved material, so that its structural character is not yet characterized. Livi and Veblen (1987) found lizardite that contained occasional zones of paired antigorite offsets, crystallographically aligned along the [110], [210] and [120] axes of lizardite 1*T*. The possible combinations of lizardite and the other serpentine structures are discussed later in the chapter.

Composition. Caution must be used in attempting to extract crystal chemical information from chemical analyses of serpentine minerals. One must keep in mind that their chemical composition is controlled in part by the bulk composition of the ultrabasic rock in which the serpentine has formed. For example, nickel is always a minor or trace component of serpentine minerals in a serpentinite because it is a minor component of the initial rock and it remains dispersed throughout the rock during serpentinization. However, in a different geological environment, deep-weathering of a serpentinite can produce the Ni-serpentine minerals nepouite and pecoraite (Chap. 6) in response to changes in bulk chemistry brought about by the development of a nickel-laterite. The examination of the range in composition of serpentine minerals should be made with these factors in mind.

The principal substitutions in lizardite are Al (radius = 0.53 Å Shannon, 1976) for Mg(0.72 Å) in the octahedral sheet, usually coupled with substitution of Al (0.39 Å) for Si (0.26 Å) in the tetrahedral sheet. Ferric iron (0.645 Å [VI] and 0.49 Å [IV]) also can substitute for Mg and Si. Both the Al and Fe^{3+} substitutions help to relieve the misfit between sheets. Ferrous iron (0.78 Å) substitutes for Mg in the octahedral sheet, and increases the misfit. The coupled substitution of Al or Fe^{3+} for Mg and Si relieves the misfit in the planar serpentine structure even when Mg is replaced by Fe^{2+}, so that there can be solid solution between lizardite, $Mg_3Si_2O_5(OH)_4$, and amesite, $(Mg_2Al_1)(Si_1Al_1)O_5(OH)_4$ and between lizardite and cronstedtite $(Fe_2^{2+}Fe_1^{3+})(Si_1Fe_1^{3+})O_5(OH)_4$. The simple substitution of Fe^{2+} for Mg eventually will increase the size of the octahedral sheet past the point at which the planar serpentine structure can accommodate the misfit and it will give way to the greenalite structure (Guggenheim et al., 1982, and Chap. 17). The compositional limits for these two structures are not known.

The distribution of both natural (Fig. 4) and synthetic compositions (Gillery, 1959; Chernosky, 1975) suggests a solid solution series between lizardite and amesite. As already discussed, most lizardite has close to end-member Mg-composition. Amesite is a rare

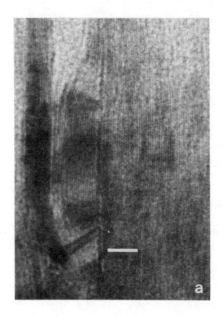 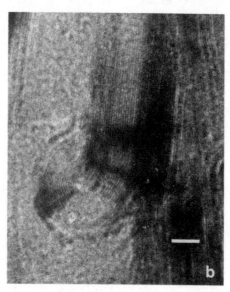

Figure 5. Curved features associated with lizardite after enstatite: (a) lizardite $1T$ grain ending in gentle curves. (b) lizardite $1T$ grain ending in a 270° "chrysotile-like" curl (from Wicks, 1986). Scale bar is 100 Å.

mineral, and the known specimens closely approximate the ideal end-member composition (Chap. 6), although the discovery of an iron-rich amesite (Taner and Laurant, 1984) suggests that substitutions in amesite may not be as limited as once thought. The composition of amesite seems to be controlled by the bulk composition of the rock in which it occurs (as are most serpentines), and not by crystal chemical restrictions of the crystal structure. Naturally-occurring crystals (usually called Al-serpentines) do span the gap between the two end-member compositions (Fig. 4). Some of these at the Al-poor end are rock-forming lizardite, x = 0.20-0.80, usually associated with the alteration of Al-rich pyroxenes (Wicks and Plant, 1979). Other more aluminous types (x = 1.2) occur in vugs in rhyolite (Jahanbagloo and Zoltai, 1968), and as accessory minerals (x = 1.5-1.75) in Lake Superior iron ores (Bailey and Tyler, 1960).

Further consideration other than chemical arguments must be given to this series as it contains at least three modifications of the basic serpentine structure. From x = 0-0.6, the lizardite structures refined by Mellini with $b_{oct} > b_{tet}$ are present. At x = 2, the amesite structure with the ordered distribution of Al at octahedral and tetrahedral sites is present (Hall and Bailey, 1976; Anderson and Bailey, 1981, Chap. 6). Between these compositions, from x = 0.6-2, the condition $b_{oct} < b_{tet}$ prevails and the main structural feature is the rotation of tetrahedra to produce $b_{oct} = b_{tet}$. However, little is known of the details of these structures. Approaching x = 2, there must be a tendency for ordering as in amesite, but in the direction of lower Al-contents, ordering is not possible. Wicks and Whittaker (1975) suggested that the Al-serpentine $9T$ structure (Jahanbagloo and Zoltai, 1968) be called amesite $9T$ because it is closer in composition to amesite than to lizardite (Fig. 4). However, with the lack of structural knowledge in this compositional range, it would be better to use the term Al-serpentine until new structural studies can resolve this problem.

Assessing the iron contents of serpentine minerals is hampered by the lack of knowledge of the oxidation state of iron for microprobe analyses, the largest source of information. The wet chemical analyses assembled from the literature and done by Faust and Fahey (1962) in their monumental work, show a greater range of compositions than most microprobe analyses. However, assessment (Whittaker and Wicks, 1970) of the reliability of mineral identification and analysis quality in this compilation suggested that much of these data, particularly the older analyses, can not be used with confidence.

In the large number of rock and vein forming lizardites analyzed by Wicks and Plant (1979) the total iron calculated as FeO ranged up to 16 wt % FeO although most lizardite contained less than 5 wt % FeO. Rozenson et al. (1979) showed that the occurrence of Fe^{3+} at the tetrahedral site is rare, so that Fe^{3+} is effectively restricted to the octahedral site in most lizardites, as is Fe^{2+}. Their Mössbauer studies indicate some significant differences in the proportions of Fe^{2+} and Fe^{3+} from that found in some of the wet chemical analysis, so that one should be careful when using wet chemical data, particularly when the total amount of iron is small. Substitution of other elements such as Ni, Co, Mn, Cr, and Zn is low in most lizardites found in serpentinites, except in unusual local environments (e.g. the titanian berthierine formed locally around altered spinel in an ultramafic dyke, Arima et al., 1985; and Cr_2O_3 contents up to 2 wt % in lizardite formed after pyroxenes, Dungan 1979; Golightly and Arancibia, 1979; Wicks and Plant, 1979). The effects of other elements on the serpentine structure is discussed by Wicks and Whittaker (1975).

Chrysotile

Chrysotile is probably the least abundant of the serpentine minerals but due to its distinctive fibrous habit it is the most easily recognized. It occurs most commonly in cross-fiber chrysotile veins in mildly prograde metamorphosed serpentinite hosts of asbestos deposits. It also occurs as slip fiber along shear zones, and less frequently as mass fiber formed by the replacement of the host serpentine in asbestos deposits. It is a minor component in retrograde lizardite serpentinites (Cressey, 1979) and in prograde antigorite serpentinites (Mellini et al., 1987). Chrysotile in serpentinized ultramafic rocks is usually some shade of green, but it is pale yellow to buff in serpentinized dolomites because the chrysotile is iron-free. In these occurrences, chrysotile is obviously fibrous and is usually called asbestos.

Chrysotile also occurs in a massive, macroscopically non-fibrous form that cannot be recognized as chrysotile without X-ray or electron diffraction studies. Chrysotile with this habit occurs as a non-fibrous component of cross-fiber and slip-fiber veins, as the sole component of veins or as coatings on shear surfaces in chrysotile asbestos deposits and other serpentinites. This chrysotile is usually green, commonly apple green, but can also be pale gray to white. It occurs in a variety of habits including massive, banded, splintery and pseudofibrous, and is commonly associated with lizardite (Wicks and Whittaker, 1977), or in the form of polygonal serpentine (Middleton and Whittaker, 1976; Cressey and Zussman, 1976). The occurrence of a non-asbestiform chrysotile has serious implications for environmental studies that estimate chrysotile asbestos contents based on X-ray powder diffraction methods. These methods will estimate a total chrysotile content but will not distinguish between asbestiform and non-asbestiform chrysotile.

Pauling (1930) suggested that the Mg-analogue of kaolinite would have a curved structure because of the misfit between the octahedral and tetrahedral sheets. The significance of this suggestion was not fully appreciated until Noll and Kircher (1951) and Bates et al. (1950) published electron micrographs showing cylindrical and apparently hollow chrysotile fibers. Once the cylindrical nature of chrysotile was recognized, two problems prevented solution of its crystal structure. The first problem was that the cylindrical structure produced a disorder between successive layers of the structure, and the second problem was that the theory of diffraction from a cylindrical lattice had not been developed. The first problem has not been solved, but the second was solved in a series of papers by Jagodzinski and Kunze (1954a,b,c) and Whittaker (1954, 1955a,b,c,d) and later by Toman and Frueh (1968a,b).

Structure refinements of well-formed chrysotile fibers were done by Whittaker, based on the theory of diffraction from cylindrical lattices and on the classification of cylindrical lattices given in Table 1. However, they were restricted by the circumferential disorder of the fibers to two dimensional refinements projected on to (010). Whittaker (1953, 1956a,b,c) defined three types of chrysotile: clinochrysotile, orthochrysotile, and parachrysotile. Clinochrysotile has a 2-layer monoclinic unit cell (disregarding the cylindrical nature of the structure), no rotation between layers, and has X parallel to the cylindrical axis. Orthochrysotile has a 2-layer orthorhombic unit cell with rotations of $180°$ between layers, and has X parallel to the cylindrical axis. Parachrysotile has a 2-layer orthorhombic unit cell with rotations of $180°$ between layers and Y parallel to the cylindrical axis. Later, Zvyagin (1967) recorded a clinochrysotile with a 1-layer monoclinic unit cell with X parallel to the cylindrical axis. Wicks and Whittaker (1975) pointed out that both 1- and 2-layer clinochrysotile and orthochrysotile are polytypes. Furthermore, parachrysotile with Y parallel to the cylindrical axis and the chrysotile structures with X

Table 1. Types of cylindrical latices (after Whittaker, 1955).

Symmetry	Kind	Interaxial angles*	Orientation of Y to right section**	Right section Circular	Spiral
anorthic	1st	$\gamma = \beta = 1/2\pi$	parallel	R	-
	2nd		inclined	H	H
monoclinic	1st	$\beta = \gamma = 1/2\pi$	parallel	R	-
	2nd	$\gamma = \beta = 1/2\pi$	parallel	R	R
	3rd	$\beta = \gamma = 1/2\pi$	inclined	H	H
	4th	$\gamma = \beta = 1/2\pi$	inclined	H	-
orthorhombic	1st	$\gamma = \beta = 1/2\pi$	parallel	R	R
			inclined	H	-

* X and Z axes interchanged from that given in the original paper.
** The right section displays either concentric, evenly spaced circles or an evenly spaced spiral. R = regular with Y lying in the plane at the right section. H = helical with Y lying at a low angle to the right section.

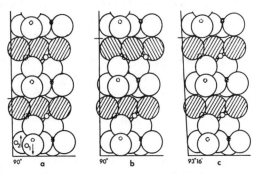

Figure 6. [010] projection of the structures of (a) lizardite $1T$; (b) chrysotile $2Or_{cl}$; (c) chrysotile $2M_{cl}$. The large open circles represent oxygen and the large shaded circles represent hydroxyl. The smallest circles represent silicon and the intermediate circles represent magnesium. The differences in the layer stacking and their effect on the relationship between the basal oxygens O_1 and O_2 and the underlying hydroxyls are shown (from Wicks, 1979). We now know (Mellini, 1982) that O_1 and O_2 in lizardite $1T$ are coplanar, so this adjustment should be made when using this figure.

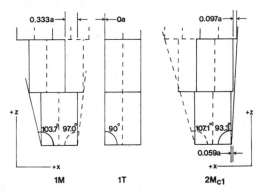

Figure 7. Stacking relationships between chrysotile $2M_{cl}$ and disordered $1T$ and $1M$ polytypes in the [010] projection. The standard orientation of $2M_{cl}$ reverses the polarity of the X-axis in comparison to the standard polytypes. All β values are calculated on the basis of $a = 5.34$ Å and $c = 7.32$ Å (from Wicks and Whittaker, 1975).

parallel to the cylindrical axis have a polymorphic relation to each other.

Cylindrical Polytypes. Before discussing the details of chrysotile structures, it is important to realize that the standard polytype scheme presented in Chapter 2, and those schemes developed by other researchers (Steadman, 1964; Zvyagin, Mischenko and Shitov, 1966; Zvyagin, 1967) cannot be applied directly to chrysotile. The basic assumption of these schemes is that successive layers are stacked in such a way that hydrogen bonding always develops between every basal oxygen of a given layer and every outer hydroxyl of the adjacent layers (Chap. 2). This stacking relationship does not occur in chrysotile, and considerable confusion has been generated by attempts to apply such schemes to chrysotile.

In order to distinguish the polytypes that arise in the cylindrical chrysotile structure from those applicable to flat-layer structures, Wicks and Whittaker (1975) proposed that the subscript c for cylindrical immediately follow the main symbol. Thus clinochrysotile becomes chrysotile $2M_{c1}$, where 2 = the numbers of layers in the unit cell, M = monoclinic (disregarding the curvature of the structure), c = cylindrical, and 1 = the polytype number used to differentiate structures that result in the same periodicity and symmetry. Similarly, the one-layer clinochrysotile becomes chrysotile $1M_{c1}$. Orthochrysotile becomes chrysotile $2Or_{c1}$ and the symbol D_c designates a disordered cylindrical chrysotile structure.

The structural refinements of both chrysotile $2M_{c1}$ (Whittaker, 1956a) and chrysotile $2Or_{c1}$ (Whittaker, 1956b) indicate that the basal oxygens O_1 and O_2 are separated in the radial direction by 0.2 Å with O_1 projecting from the layer structure and O_2 withdrawn into it (Fig. 6).

The cylindrical curvature of the chrysotile structure aligns the outer hydroxyl groups of the octahedral sheet so that they form rows, with grooves between the rows, running around the circumference of the cylindrical structure. Successive layers are stacked in such a way that the projecting O_1 atoms of the basal oxygen plane of the tetrahedral sheet fit into the underlying grooves between hydroxyl rows, and the withdrawn O_2 atoms lie approximately over the hydroxyl rows (Fig. 6). This stacking arrangement involves a shift of 0.4 Å (approximately $a/13$) away from the position of normal hydrogen bonding found in most 1:1 layer structures and assumed in the theoretical polytype schemes. If the initial 1:1 layer is oriented such that the II set of octahedral cation positions is occupied, the first $a/3$ shift in the standard polytype system (described in Chap. 2) is along $-X_1$. In the chrysotile structure, the first $a/13$ shift is along $+X_1$, the *opposite* direction from the standard polytype system. The effect of this is shown in Figure 7, which compares disordered standard $1M$ and $1T$ polytypes with the chrysotile $2M_{c1}$ structure. The cylindrical structure of chrysotile introduces disorder along the Y direction, the circumferential direction, and so for comparison, the $1M$ and $1T$ polytypes are also considered to be disordered along Y. This reduces the effective repeat unit along X to $a/2$, and makes it possible to choose alternate origins for a (represented by dashed lines in Fig. 7). The standard orientation of the chrysotile structure reverses the polarity of the X-axis in comparison to the standard polytypes, introducing another source of confusion. Examination of Figures 6 and 7 indicates that the chrysotile $2M_{cl}$ structure is significantly different from the standard polytypes. The same is true for chrysotile $1M_{cl}$ and chrysotile $2Or_{cl}$.

There is one other important structural detail that must be understood. In both the chrysotile $2M_{cl}$ and chrysotile $2Or_{cl}$ structures described by Whittaker (1953), the O_1 atoms not only project out of the layer in the radial direction but are displaced by 0.10 Å from their ideal position in the X-direction, i.e. parallel to the cylinder axis. This displacement of O_1 with respect to the upper part of its own layer occurs in either the $+$ or $-X$ direction, and has been designated as δ by Wicks and Whittaker (1975). As the O_1 atom is keyed into the groove of the layer below the shift between one layer and the next will be changed to $0.4 \pm \delta$ Å. Wicks and Whittaker (1975) defined the shift $0.4 + \delta$ Å as overshift and $0.4 - \delta$ Å as undershift. In the structure of chrysotile $2M_{cl}$ illustrated in Figure 7, δ is approximately 0.1 Å, and so overshift is 0.5 Å and the undershift is 0.3 Å, occurring in successive layers to produce a two-layer unit cell with $\beta = 93.3°$. If $\delta = 0$, the shift between all layers would be 0.4 Å, and a one-layer unit cell with $\beta = 93.3°$ would result; no such structure has been observed.

The two possible one-layer structures that can be developed with δ = 0.1 Å are shown in Figure 8. Overshift (0.5 Å) or undershift (0.3 Å) between successive layers would produce one-layer structures with β = 94.1° and β = 92.5°, respectively. The one-layer chrysotile discovered by Zvyagin (1967) has β = 94.2°, indicating it is an overshift structure which can be designated as chrysotile $1M_{cl}$. No one-layer structure with undershift has been found.

The chrysotile $2Or_{cl}$ structure described by Whittaker (1956b) has a rotation of 180° between successive layers, which produces a separation of $a/6$ between the rows of hydroxyls from one layer to the next (Fig. 6). In order to produce chrysotile $2Or_{cl}$ with O_1 keyed into the grooves between the rows of hydroxyls in the layer below, there must be a shift from the ideal stacking position of $0.4 \pm \delta$ Å in the opposite direction to the shift necessary in the standard polytype system (Chap. 2). This shift reduces the $a/6$ separation between hydroxyl rows in successive layers by $0.4 \pm \delta$ Å. Either undershift or overshift in both layers would be compatible with the orthorhombic character; however, the structures would be different and would give significantly different $20l$ intensities. The known structure of chrysotile $2Or_{cl}$ has undershift in both layers. The theoretically possible structure with overshift in both layers may be denoted $2Or_{c2}$ if it is discovered. Structures with alternate layers rotated by 180° but with alternating undershift and overshift can also be hypothesized, but they would be monoclinic with β = 90.4° for a value of δ = 0.1 Å, and would constitute an enantiomorphic

106

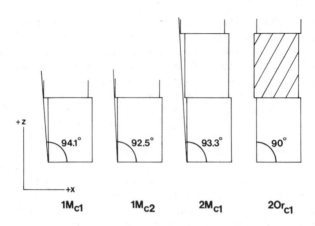

Figure 8. Stacking relationships among the chrysotile polytypes $1M_{c1}$, $1M_{c2}$, $2M_{c1}$ and $2Or_{c1}$ (from Wicks and Whittaker, 1975).

pair. Some chrysotile X-ray fiber diffraction patterns with X parallel to the fiber axis show considerable broadening of $20l$ reflections relative to the $00l$ reflections, indicating the presence of stacking faults. In extreme cases, the $20l$ reflections may be smeared out into an almost continuous streak, indicating a disordered structure designated chrysotile D_c (Wicks and Whittaker, 1975).

These polytype designations apply to chrysotile with X parallel to the fiber axis. Parachrysotile is a polymorph of these structures because the alignment of Y parallel to the fiber axis produces a different structure. A series of polytype symbols could be developed for parachrysotile should further work indicate that it is needed.

As it now stands, chrysotile is a generic name for all varieties. Those with X parallel to the fiber axis can be designated as chrysotile with a specific polytype symbol. Those with Y parallel to the fiber axis can be called parachrysotile.

Structural refinements. Whittaker (1956a) did a two-dimensional refinement of the chrysotile $2M_{c1}$ structure using a fiber bundle from the Bell mine at Thetford, Quebec, Canada. The composition of this specimen is not known, but chrysotile from the Quebec mines generally contains less than 0.5 wt % Al_2O_3 and 1-2 wt % total iron as FeO (Wicks and Plant, 1979).

Whittaker's (1956b) refinement of chrysotile $2Or_{c1}$ was done on a fiber bundle composed of 78% chrysotile $2Or_{c1}$ and 22% $2M_{c1}$ from Cuddapah, Madras, India. The intensities of the $h0l$ reflections of $2Or_{c1}$ had to be adjusted relative to the intensities of the $00l$ reflections to correct for the $2M_{c1}$ dilutant. The composition of this specimen of chrysotile asbestos is not known. The cylindrical nature of the fibers confined these structure determinations to two-dimensional Fourier syntheses projected on to (010). Recently, Yada (1979) has been able to resolve the Mg and Si positions of the Fourier projections to verify the stacking sequences in chrysotile $2M_{c1}$ and $2Or_{c1}$ by applying optical noise-filtering techniques to HRTEM images.

Chrysotile $2M_{c1}$ has a monoclinic cylindrical lattice of the 3rd kind (Table 1), and the structure was refined in the two-dimensional space group $p111$ to and R index of 0.14. Chrysotile $2Or_{c1}$ has an orthorhombic cylindrical lattice of the 1st kind (Table 1), and the structure was refined in space group $p11g$ to R = 0.21. The relationship of these structures with the other serpentine minerals can now be re-examined in the light of the three-dimensional refinements of the lizardite structure by Mellini (1982) and Mellini and Zanazzi (1987).

There are strong similarities between the two chrysotile structures and the newly determined lizardite structures. In the plane common to the octahedral and tetrahedral sheets, the apical oxygens and the hydroxyls are separated by 0.22 Å along Z in chrysotile, compared to 0.06 to 0.12 Å in the lizardites. The plane of the magnesium atoms is shifted from the center of the octahedral sheet away from the tetrahedral sheet by 0.07 Å in chrysotile compared to 0.08 and 0.09 Å in lizardite. Direct determinations of the rotation of tetrahedra similar to that found in the lizardite structures and indicated by the MO calculations of Peterson et al. (1979) cannot be obtained from the two-dimensional chrysotile refinements. However, as small tetrahedral rotation seems to be the preferred configuration in lizardite even where $b_{oct} > b_{tet}$, it is not unreasonable to assume it also occurs in chrysotile. The detected shift δ = 0.10 Å of the O_1 atoms with respect to the rest of the structure can be interpreted as evidence of rotation.

The tetrahedral sheet, measured from the weighted mean oxygen positions, is 2.26 Å thick in chrysotile $2M_{c1}$ and 2.13 Å thick in chrysotile $2Or_{c1}$. In comparison, the tetrahedral sheet of lizardite is 2.21 or 2.22 Å thick. The octahedral sheet is 2.08 Å thick in both chrysotile structures compared to 2.10 Å to 2.12 Å thick in the lizardite structures.

The basal oxygens O_1 and O_2 of the tetrahedral sheet in chrysotile are separated by 0.2 Å along Z (Fig. 6). These key into the underlying hydroxyl rows and grooves as discussed above, and give chrysotile its characteristic stacking arrangement. The equivalent oxygen atoms in lizardite are co-planar. Wicks and Whittaker (1975) suggested that buckling of the Mg-plane could account for the separation of O_1 and O_2 but it has been demonstrated by Mellini that buckling of the Mg-plane does not occur, so that the Wicks and Whittaker explanation is not valid and a new explanation must be sought. In lizardite, the rotation of tetrahedra promotes hydrogen bonding between layers. Presumably the importance of hydrogen bonding also applies to the movement of basal oxygens in the chrysotile structure. The detected shift δ = 0.10 Å along X is weak evidence for tetrahedral rotation. The interlayer shift of $0.4^{\pm}\delta$ Å along X_1 in the chrysotile structure is unique among 1:1

layer silicates. One can assume that this postion is adopted because it is the best position for hydrogen bonding in the cylindrical structure, even if the Y-axis disorder inhibits a strictly regular arrangement. Slight rotation about an axis parallel to X that passes through the apical oxygens will cause O_1 to project out of the structure and O_2 to be drawn into the structure. The combined movement of O_1 and O_2 will increase hydrogen bonding by allowing O_1 to extend into the groove of the hydroxyl row of the adjacent layer, and thus improve its position for hydrogen bonding in the cylindrical structure.

It is important to appreciate that the cylindrical structure does not completely compensate for the octahedral and tetrahedral sheet misfit. Only one layer will be at the ideal radius of curvature (calculated to be at 88 Å by Whittaker, 1957) for a perfect match of the octahedral and tetrahedral sheets along Y. As fibers are commonly 110 – 135 Å in outer radius (Yada, 1971), it can be seen that the large volume of the structure outside the ideal radius will have the misfit only partly relieved along Y, and that the small volume of the structure inside the ideal radius will be over compensated along Y. Furthermore, the misfit along X, the fiber axis, is not compensated by the curvature in any layers. Within the ideal radius, an octahedral sheet will be under tension along Y due to the over compensation, and under compression along X due to the misfit (the tetrahedral sheet will have the opposite stresses). This is the condition found throughout the antigorite structure and is discussed later; however, it is of minor importance in terms of the volume of layers in the chrysotile structure. Most of the octahedral sheets are outside the ideal radius, and will experience greater and greater compression along Y as the radius increases and misfit relief through curvature becomes less efficient. The anomalous growth patterns of large chrysotile fibers and polygonal-serpentine (discussed later in the chapter) provide some examples of the limits of misfit relief through the cylindrical structure. Compression will be constant along X in octahedral sheets throughout the structure, and there will be an equal tension in the tetrahedral sheets. As there is variable misfit relief along Y but not along X, it may be that the tetrahedral sheet does not have the ditrigonal symmetry found in lizardite, in which misfit is equal in all directions, but is slightly elongate along X to help relieve the misfit in this direction. However, without structural details, this remains speculation.

Details of the parachrysotile structure are not known to the same level as chrysotile $2M_{c1}$ and $2Or_{c1}$, as parachrysotile occurs only as a minor accessory with the other chrysotile structures. X-ray diffraction data were collected by Whittaker (1956c) from South African fiber bundles that contained approximately 9% parachrysotile, a high percentage for this mineral. Intensities were recorded from the first, second, third, fourth, and sixth layer lines, but the fifth and seventh layer lines were not detected due to their proximity to the third and fourth layer lines of chrysotile $2M_{c1}$ and $2Or_{c1}$. Parachrysotile has a helical orthorhombic cylindrical lattice of the second kind (Table 1). Whittaker's (1956c) study suggests general similarities of the parachrysotile structure with the chrysotile $2M_{c1}$ and $2Or_{c1}$ structure, but the details could not be confirmed. Due to the presence of a plane of symmetry perpendicular to Y, there is no possibility of a second parachrysotile structure related to parachrysotile in the same way as chrysotile $2M_{c1}$ is related to chrysotile $2Or_{c1}$ (Whittaker 1956c). Yada and Tanji (1980) applied optical noise-filtering to HRTEM images of parachrysotile and confirmed its stacking sequence as Yada (1979) had done for chrysotile $2M_{c1}$ and $2Or_{c1}$.

Electron microscopy. Chrysotile asbestos was one of the first minerals studied by HRTEM. The early work by Yada (1967, 1971) confirmed many of the earlier X-ray based studies and added many new details. These HRTEM studies provided images of fibers that can be related to a first approximation to the types of cylindrical lattice developed by Whittaker (1955c) and given in Table 1. Yada (1971) found that fibers sectioned perpendicular to the fiber axis show either one of the two structures, circular (Fig. 9a) or spiral (Fig. 9b), proposed by Whittaker (Table 1). Both types of structures occur in a given asbestos deposit, but the proportions of each vary from deposit to deposit.

When viewing fibers perpendicular to the fiber axis, Yada (1967, 1971) observed fibers with either regular or helical cylindrical structures (Table 1). In regular cylindrical fibers, the central part of a fiber has 4.5 Å fringes, representing the (020) planes, lying parallel to the fiber axis (Fig. 9c) and produces a normal fiber type SAD pattern (Fig. 9e). In fibers with helical cylindrical structures, the fringes are inclined a few degrees to the fiber axis (Fig. 9d) and the 0k0 and hk0 reflections of the SAD patterns are split into pairs that represent diffraction from either side of the twisted helical fiber (Fig. 9e). The tilt angle of the 4.5 Å fringes is commonly 2-3°, but ranges up to 10° (Yada, 1971), suggesting that the pitch of the helical path traced by the Y-axis is variable. Whittaker (1956c) noted that this is possible but that Y can not lie at any angle to the section perpendicular to the fiber axis (the right section), but must fulfil the condition that for every revolution of 360°, the origin must advance along the cylinder a distance equal to na (where n is an integer) or $na/2$ if the unit structure is centered. Individual fibers do not seem to contain combinations of concentric and spiral growth or of regular and helical growth. Similarly, combinations of chrysotile $2M_{c1}$, $2Or_{c1}$ and parachrysotile are rarely encountered in a single fiber, although Zvyagin (1967) has published an SAD pattern from a single fiber that shows the rare occurrence of both chrysotile $2Or_{c1}$ and parachrysotile.

The spiral growth fibers may be composed of a single layer, a double layer or multilayer spirals (Yada, 1971). Dislocations formed by the growth of a new layer occur in some multilayer fibers (Yada, 1967), and dislocations were also observed in longitudinal images (Yada, 1967). The double spiral and concentric circular fibers could possess a chrysotile $2M_{c1}$ structure or a chrysotile $2Or_{c1}$ structure. However, multispiral fibers with an odd number of layers, and those with numerous single-layer dislocations, cannot posses the chrysotile $2M_{c1}$ structure.

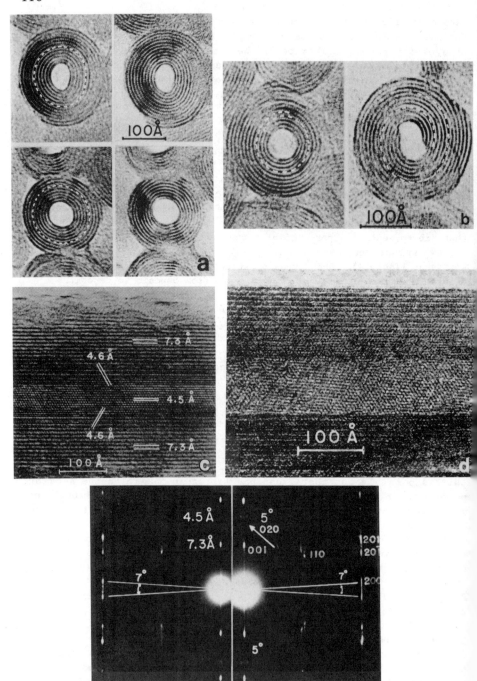

Figure 9. (a) Cross sections of chrysotile fibers. Pairs of through-focused images showing the concentric structure. (b) images showing the single-spiral structure (left) and multi-layer structure (right). (c) longitudinal image of a chrysotile fiber showing regular growth with (020) fringes (4.5 Å) parallel to the fiber axis. (d) helical growth with (020) fringes inclined to the fiber axis. (e) electron diffraction patterns from a regular cylindrical fiber (left) and a helical cylindrical fiber (right). Note splitting of 020 and 110 diffraction spots (from Yada, 1969, 1971).

The 2-layer structure cannot form because the alternate positive and negative displacement of layers with respect to one another would be put out of sequence by each extra layer added by dislocation, or by the odd number of layers in multispiral fibers. However, if the displacement is in the same direction in each layer, such that the chrysotile $1M_{c1}$ structure develops, these multilayer spiral and single-layer dislocation growths would be accommodated. Following the same reasoning, chrysotile $2Or_{c1}$ or parachrysotile cannot have multispiral growths with odd numbers of layers, or have a high number of single-layer dislocations. The dislocations offer an explanation for the streaking found along layer lines between $h0l$ reflections in some fiber diffraction patterns, and could lead to the disordered structure chrysotile D_c noted by Wicks and Whittaker (1975).

Yada (1967, 1971) found the inside diameter of the central hollow of fibers was most frequently 70-80 Å with quite uniform walls, but diameters of up to 100 Å were recorded. In some cases, the central core appears to be empty; in others, it appears to be filled with amorphous material. The most abundant outside diameter for normal fibers is 220-270 Å, but fibers greater than 650 Å in diameter were observed. Fibers over 350 Å in diameter usually have a second or even third stage of growth, not quite concentric with the first stage and often quite irregularly formed. This suggests a break in the growth process (Yada, 1971).

The outer surfaces of fibers appear smooth and clean, and Yada (1971) suggests that most inter-fibril sites are empty. Recently, Mellini (1986) observed chrysotile fibers in cross-section that show a regular hexagonal close-packed arrangement of parallel fibers. Occasionaly, a large chrysotile fiber 800 Å in diameter occupies the position of three small fibers in the array. Yada (1971) also recorded a variety of unusual growth patterns, including solid fibers without a central hollow, arc-shaped inter-fibril segments, and interlocking arcs. Anomalous growth patterns occur in some asbestos deposits but not in others (Yada, 1971).

Chrysotile fibers with many of the features observed by Yada in cross-fiber vein specimens also have been found in the host serpentinites. The principal feature used for identification of this chrysotile is the hollow cylindrical form of the fibers (Cressey and Zussman, 1976; Cressey, 1979; Veblen and Buseck, 1979; Veblen, 1980; Spinnler, 1985; Mellini et al., 1987). Yada and Iishi (1974b, 1977) discovered that hydrothermally synthesized chrysotile fibers often contain cone-in-cone structures not found in natural fibers, together with a much higher incidence of parachrysotile.

Polygonal serpentine. Most of the X-ray diffraction and TEM studies discussed above have been done on the parallel fibers of cross-fiber chrysotile asbestos. A lot of less well-aligned chrysotile occurs as macroscopically non-fibrous, often splintery, serpentine in both fracture-filling and slip veins. Early electron microscope studies of this type of material by Zussman et al. (1957), showed it to have a lath-like morphology, and anomalous chrysotile SAD patterns with a series of sharp hkl reflections on the odd layer-lines instead of $hk0$ reflections with diffuse tails. Krstanovic and Pavlovic (1964, 1967) noted similar features on X-ray rotation diffraction patterns of splintery chrysotile, and called this material Povlen-chrysotile. They suggested that it is less curved and has a greater degree of order than normal chrysotile. A detailed analysis of X-ray fiber diffraction patterns of other splintery serpentine led Middleton and Whittaker (1976) to propose that this serpentine is composed of a cylindrical core of chrysotile with a polygonal overgrowth of flat serpentine layers (Fig. 10). This model was almost immediately confirmed by the TEM studies of Cressey and Zussman (1976) and Cressey (1979), who recorded images of large fibers 1000-8000 Å in diameter, and composed of planar polygonal overgrowths on a cylindrical chrysotile core. Middleton and Whittaker (1976) found two types of polygonal serpentine, one has a cylindrical core of chrysotile $2M_{c1}$ and a polygonal overgrowth of what seems to be planar chrysotile $2M_{c1}$. The SAD patterns of Cressey and Zussman (1976) suggest the same configuration. How planar serpentine layers could stack in the chrysotile stacking position, but without curvature, is not understood and needs further study. The second type contains a core of chrysotile $2Or_{c1}$ and a polygonal overgrowth of lizardite $2H_1$. A third possible polygonal serpentine with a core of chrysotile $2M_{c1}$ and a polygonal overgrowth of multilayer lizardite is suggested by unpublished X-ray studies on splintery serpentine from the mineral collection at the Royal Ontario Museum (Wicks, unpub. data). This seems to be a common occurrence and also requires further study. Some cores contain a single chrysotile fiber (Mitchell and Putnis, 1988), groups of fibers (Cressey and Zussman, 1976; Mellini, 1986), no fibers, planar layers, or partly curved layers (Mellini, 1986). The polygonal overgrowth may be complete (Cressey and Zussman, 1976; Mitchell and Putnis, 1988) or incomplete and fan shaped (Cressey and Zussman, 1976; Cressey, 1979; Mellini, 1986). Wei and Shaoying (1984), Mellini (1986), Yada and Wei (1987) and Mitchell and Putnis (1988) have published images that clearly resolve the planar 7 Å layer of the polygonal sectors and a sharp angle between sectors (Fig. 10). Mellini (1986) also showed that some planar polygonal sectors are connected by curved layers, and Mitchell and Putnis (1988) have found "antigorite-like" offsets between planar sectors. Spinnler (1985) has observed possible antigorite polygonal overgrowths on chrysotile cores. Yada and Wei (1987) have noted that well-formed polygonal serpentines from Guangyuanpu, China, tend to be consistently composed of either 15 or 30 planar sectors (Fig. 10). Whittaker and Middleton (1979) have documented a polygonal parachrysotile.

Polygonal serpentine commonly occurs in serpentine veins. However, Cressey and Zussman (1976) and Cressey (1979) also found it in rock-forming serpentines from a variety of pseudomorphic and non-pseudomorphic textures. Recently, Mitchell and Putnis (1988) have discovered polygonal serpentine in kimberlites.

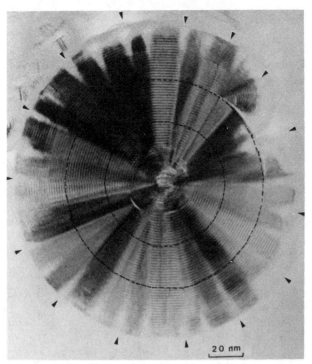

Figure 10. Cross-section of a polygonal-serpentine with a cylindrical chrysotile core and a polygonal overgrowth of 30 segments (from Yada, 1987).

Polygonal serpentine is a widely distributed configuration of serpentine. In spite of the intriguing details revealed by HRTEM images, we do not understand this complex material very well. In view of the uncertainties in our understanding, use of the term 'polygonal serpentine' proposed by Cressey and Zussman (1976) is strongly recommended unless detailed diffraction studies have been used to identify the structures present. It is incorrect to place polygonal serpentine on a par with the chrysotile, lizardite, and antigorite structures by calling it a fourth serpentine structure. Polygonal serpentine is a specific configuration of the known structures of chrysotile, lizardite, and antigorite. It is possible that the planar "chrysotile-like" layers identified by Middleton and Whittaker (1976) are composed of a new structural modification of the chrysotile structure, but we do not yet know the structure of this material.

Composition. Chemical substitution in the chrysotile structure is limited by the requirement that the misfit between octahedral and tetrahedral sheets must be maintained. Chernosky (1975) has calculated the composition of zero misfit at x = 0.6 (defined in the lizardite section) for aluminum substitution. At some compositon probably well before x = 0.6 (11 wt % Al_2O_3), the cylindrical structure can no longer form, but the composition at which this occurs is unknown. In the chrysotile analyses compiled by Whittaker and Wicks (1970) and obtained by electron microprobe by Wicks and Plant (1979) all samples, with one exception, contained less than x = 0.05 (0.9 wt % Al_2O_3). Lizardite coexisting with chrysotile often exceeds x = 0.05 (Wicks and Plant, 1979), so there is some evidence to suggest that aluminum is to a certain extent discriminated against by the chrysotile structure. The chrysotile with the exceptional composition, found by Wicks and Plant (1979), contains 3.2 wt % Al_2O_3 (x = 0.17) and 2.1 wt % Cr_2O_3 and is a polygonal chrysotile $20r_{c1}$ formed after enstatite. If normal cylindrical chrysotiles contain less than 1 wt % Al_2O_3 this higher value for the polygonal chrysotile $20r_{c1}$ suggests the possibility of a significant difference in composition between cores and polygonal overgrowths. Hydrothermal studies by Chernosky (1975) and earlier researchers (Nelson and Roy, 1958; Gillery 1959) showed that chrysotile forms at a bulk composition of up to x = 0.25 (4.6 wt % Al_2O_3). As the Al-content approaches this composition, the chrysotile fibers become shorter, thicker, and less abundant, but their actual composition is not known. Based on these natural compositions and synthetic studies, x = 0.2 (3.7 wt % Al_2O_3) is a possible upper limit for the aluminum content of chrysotile, although most chrysotiles contain much less.

Most chrysotile specimens contain less than 2 wt % total iron as FeO (Whittaker and Wicks, 1970; Wicks and Plant, 1979); however, the analyses of Wicks and Plant (1979) contain a significant number of chrysotile specimens with up to 6 wt % FeO, and two polygonal serpentines, composed of chrysotile $2M_{c1}$, chrysotile $20r_{c1}$ and minor parachrysotile, at 8 wt % FeO. Substitution of Fe^{2+} will slightly increase the misfit, but the limit of Fe^{2+} substitution is not known. Substitution of Fe^{3+} in chrysotile will decrease the misfit but its limit of substitution also is not known. Substitution of Fe^{3+} at tetrahedral sites is minor and infrequent, according to the Mössbauer studies of Rozenson et al. (1979) and the calculated formulae of available analyses by Whittaker and Wicks (1970) and Wicks and Plant (1979). Nonetheless, it is clear that the chrysotile structure can accommodate a few wt % FeO and/or Fe_2O_3. The existence of pecoraite, the Ni-analogue of chrysotile (Chap. 6), suggests the possibility that substitution of Ni for Mg is not limited in the chrysotile structure.

Antigorite

Antigorite is next in abundance to lizardite. It occurs most commonly as the principal mineral in prograde serpentinites, in which it usually forms non-pseudomorphic textures and only rarely pseudomorphs other minerals. In slip veins and on shear surfaces in these serpentinites, antigorite is foliated or splintery. It infrequently occurs in fracture-filling veins.

Antigorite is generally some shade of medium to pale green, but can be pale buff or gray. Antigorite serpentinites are usually tougher and less porous than lizardite or chrysotile serpentinites. The color of rocks composed of antigorite, as for the other serpentine minerals, is determined by the distribution of accessory minerals, principally magnetite. Antigorite occurs as fine to coarse, interpenetrating anhedral blades that give the rock a toughness not found in other serpentinites. Some of the coarse blades have been used for single-crystal studies.

Structural refinements. Aruja (1945) did the first single-crystal study of antigorite, on crystals from Mikonui, New Zealand, and suggested a basic similarity between the chrysotile and antigorite structures. He measured a large a parameter of 43.5 Å, and suggested a subcell with a of 5.4 Å. Weissenberg and optical transform studies by Zussman (1954) on antigorite from Mikonui, and by Kunze (1956, 1958) on a series of antigorite specimens including the Mikonui material, led to two-dimensional Fourier synthesis projected on (010) (Kunze, 1958) and considerable detail on the possible geometrical relationships within the basic structure (Kunze, 1959, 1961). This refinement of the Mikonui specimen gave a superstructure $A = 43.3$ Å and other cell dimensions, $b = 9.23$ Å, $c = 7.27$ Å, similar to the other serpentine minerals.

The structure given in Figure 11 is in the form of an alternating wave extending along X, the superstructure direction. The tetrahedral sheet is continuous through the structure, but reverses polarity at the midpoint where the wave changes its direction of curvature. The octahedral sheet is also continuous, but it too reverses at the midpoint and is bonded to different tetrahedral sheets in each half of the structure. An important feature of the reversals of curvature is that 3Mg and 6(OH) are omitted, relative to Si, in each unit cell. Thus the formula of antigorite is different from the ideal serpentine formula.

The antigorite superstructure A is not fixed at 43.3 Å, but varies amongst certain preferred values. The most commonly occurring values are $A = 33.7$-43.1 Å (Zussman et al., 1957), $A = 25.7$-51.5 Å (Kunze, 1961), and $A = 32.8$-51.4 Å (Uehara and Shirozu, 1985). A long superstructure group with $A = 80$-110 Å and a short superstructure at $A = 16$-19 Å were also recorded by Chapman and Zussman (1959). Mellini and Zussman (1986) have identified the specimen with the 16-19 Å fringes as carlosturanite, not antigorite, and suggest a lower limit of 33 Å for the superstructure of antigorite. Kunze (1961) defined the superstructure period in terms of m, the number of tetrahedra in the superperiod. When m is odd, the reversals in the tetrahedral sheet at PP' and PR' occur in normal 6-fold tetrahedral rings, but the reversals at QQ' occur in unusual 8- and 4-fold rings (Fig. 12). Under these conditions, the Mg-octahedra at all inversion points are only slightly distorted. When m is even, Kunze (1961) suggested that the 8- and 4-fold rings must occur at all inversion points; and complex Mg-bridges, composed of two magnesium and ten oxygens, are required at inversions PP' and PR'.

X-ray diffraction and electron microscope studies by Uehara and Shirozu (1985) on a series of Japanese antigorite specimens have produced a different solution. They defined the superstructure (A) in terms of the number (M) of subcells (a= 5.44 Å) making up the superstructure along X; thus $M = A/a$. This defines the superstructure in terms of the octahedra present (two per subcell) along X, rather than

116

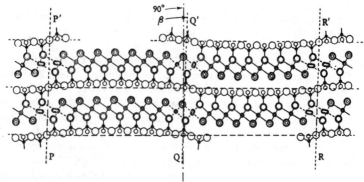

Figure 11. [010] projection of the structure of antigorite; the structure reverses polarity at PP', QQ' and RR' (from Kunze, 1956).

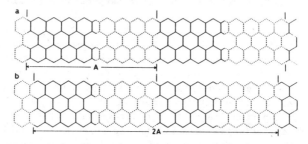

Figure 12. [001] projection of the tetrahedral sheet of (a) the $M = n$ structure which has the space group $P2/m$ with an A repeat. (b) the $M = (2n+1)2$ structure which is C—centered with a 2 A repeat (from Uehara and Shirozu, 1985).

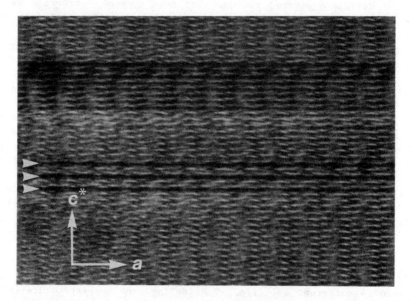

Figure 13. TEM images viewed down [010], showing beam damage at every other inversion point in the antigorite structure. Stacking faults indicated by arrows. Electron micrograph courtesy B. A. Cressey, Oxford Polytechnical Institute (unpub. data).

in terms of the tetrahedra present. The M of Uehara and Shirozu is related to the m of Kunze by the relationship $M = (m-1)/2$. (Compare Table 5 in Uehara and Shirozu 1985, p 307, to Table 5 in Kunze, 1961, p 239.)

Uehara and Shirozu (1985) interpreted their X-ray and electron diffraction patterns in terms of three structures. In the first structure, $M = n$ (where n is an integer), in the second structure, $M = (2n + 1)/2$, and in the third structure, $M \neq n/2$. The first structure contains an odd number of tetrahedra (m) and an even number of octahedra ($m - 1$) in the superstructure period A along X, and belongs to space group Pm (Fig. 12). This is the same as Kunze's structure when m = odd. The second structure is derived from, but is different from the structure proposed by Kunze (1961) with m = even and space group $P2/m$. The new structure is obtained by shifting alternate superstructure periods A of a Kunze structure along Y by $b/2$ (Fig. 12). This replaces the 8- and 4-fold tetrahedral rings at points PP' and RR' (Fig. 11) with normal 6-fold tetrahedral rings, and more importantly, replaces the energetically unfavorable Mg-bridges at these points with normal octahedra. The structure still contains an even number of tetrahedra (m) and an odd number of octahedra ($m - 1$), but the structure has a repeat period along X of $2A$, and has a C-centered space lattice. It seems to be preferred to the structure originally proposed by Kunze. Indirect evidence that supports the C-centered antigorite structure is found in the observations of Spinnler (1985). One of the inversion points in antigorite, thought to be the 6-fold inversion by Spinnler (1985), is much more susceptible to electron beam damage than the other 8-fold inversion (Fig. 13). In the many antigorite structures he examined, the damaged and undamaged inversions always occurred in pairs, suggesting the alternating sequence of 6-fold and 8-fold inversions found in the C-centered structure and in the primitive structure, but not the 8-fold and 8-fold sequence in the proposed Kunze structure with m = even. The third structure $M \neq n/2$ of Uehara and Shirozu (1987) is actually a mixture of the other two structures in poorly crystallized, disordered material.

The antigorite structures have been described by Spinnler (1985) in terms of a polysomatic series (Thompson, 1978; Ferraris et al., (1986). The series is composed of a half wave of n number of serpentine modules $[Mg_3Si_2O_5(OH)_4]$ with a = 2.7, b = 9.2, and c = 7.3 $\overset{\text{o}}{A}$, combined with a 6-fold module $[Mg_3Si_4O_8(OH)_2]^{4+}$ at one end of the series of serpentine modules, and an 8-fold (and 4-fold) module $[Mg_6Si_4O_{12}(OH)_8]^{4-}$ at the other end followed by a further series of n or ($n - 1$) serpentine modules in the other half of the wave. The b and c dimensions of all three modules are the same. The series begins at ideal serpentine composition (where n = ∞) and extends towards the ideal talc composition. This polysomatic series reproduces the variations in the idealized antigorite superstructure and the systematic change in Mg and OH with changes in the number of serpentine modules (n) in the series.

It is important to use idealized modules rather than real structures to build a polysomatic series. This has not always been done in the past. The use of lizardite modules instead of ideal serpentine modules is misleading and incorrect. The lizardite structure has variations from the ideal structure that may or may not be present in the antigorite structure. Furthermore, lizardite usually contains an excess of H_2O over the stoichiometric value, whereas antigorite usually contains the stoichiometric value (Whittaker and Wicks, 1970). Use of the idealized serpentine structure as a module in the antigorite polysomatic series facilitates comparisons among the real serpentine structures and the carlosturanite polysomatic series.

Kunze (1958) did a two-dimensional refinement of the antigorite structure (Fig. 11) on a small rectangular plate of antigorite from Mikonui, New Zealand, the same material used by Zussman (1954). The formula of this antigorite (calculated by Zussman, 1954) is $(Mg_{47.7}Fe^{2+}_{0.4}Fe^{3+}_{0.5})(Si_{33.2}Al_{0.8})O_{85}(OH)_{62.6}$, which can be reduced to $(Mg_{2.81}Fe^{2+}_{0.02}Fe^{3+}_{0.03})_{\Sigma=2.86}(Si_{1.95}Al_{0.05})_{\Sigma=2.00}O_5(OH)_{3.68}$ for comparison with the other serpentine minerals. The reduction in octahedral cations (from 3 to 2.86 atoms per forumula unit) and hydroxyls (from 4 to 3.68 apfu) relative to the tetrahedral cations (2 apfu) is easy to recognize in the reduced formula. This approximates a superstructure of $m = 17$ in the model of Kunze (1961), $M = 8$ in the model of Uehara and Shirozu (1985), and a polysome of 13 serpentine modules, seven in one half of the alternating wave and six in the other (Spinnler 1985).

The refined antigorite structure shows some features common to all serpentines as well as features found only in antigorite. The tetrahedral sheet is 2.22 Å thick, the same as found by Mellini in lizardite. At the accuracy level of this refinement, the basal oxygens are co-planar. The relationship of the basal oxygens to the hydroxyls of the adjacent layer changes along the superstructure direction so that hydrogen bonding is modified from the normal position. In the plane common to both sheets, the apical oxygens of the tetrahedral sheets and the hydroxyls of the octahedral sheet are co-planar.

At 2.44 Å, the octahedral sheet is anomalously thick in comparison with the lizardite structures (2.10-2.12 Å) and chrysotile (2.08 Å). The a and b parameter of the antigorite subcell are also anomalously large (5.42-5.46 Å, 9.24-9.26 Å; Uehara and Shirozu, 1985) compared to lizardite (5.31-5.33 Å; 9.21-9.24 Å; Mellini, 1982; Mellini and Zanazzi, 1987); and chrysotile (5.34 Å, 9.2 Å; Whittaker, 1956a,b). These values suggest that the octahedral sheet in antigorite has some unique features, but the two-dimensional structural refinement does not provide any of the details.

The antigorite structure is significantly different from the other serpentine structures in that each layer in the structure can attain the curvature along X that is most efficient for misfit relief. In contrast, the amount of misfit relief between sheets in the chrysotile structure depends on the radial position of a given layer in the cylindrical structure, with only one layer at the radius where misfit relief is ideal. Thus the antigorite structure is ideally the most efficient and most stable of the serpentine structures. Kunze (1961) has calculated a radius of curvature of 72 Å at one inversion in the structure and 50 Å at the next, with an average of 63 Å for the structure with $m = 17$ ($M = 8$). Wicks and Whittaker (1975) calculated that a radius of 190 Å would relieve the misfit, and thus the antigorite structure seems to have a smaller radius of curvature than necessary to match the misfit. This over-compensation position may be adopted to allow the structure greater freedom to compensate for the misfit along Y, the direction which is not directly relieved by the curvature. In a planar octahedral sheet, misfit will produce compression in all directions within the sheet. In an ideally curved octahedral sheet, the compression will be reduced to zero along the direction of curvature but still be present along the direction of the axis of curvature. Over-curvature will produce tension along the direction of curvature, and the balance between this tension and the compression along the axis of curvature seems to produce the most stable structure. The anomalously thick octahedral sheet and the slightly large a and b parameters of the antigorite subcell may be related to this situation.

Electron microscopy. Visual confirmation of the alternating wave structure of antigorite has been provided by Yada (1979), who first studied optically-enhanced images of antigorite viewed along Y. He experimentally determined the effects of both specimen thickness and focusing conditions. Spinnler (1985) did a systematic analysis of the effects of specimen thickness, defocusing (Fig. 14), and tilting (Fig. 15) on calculated images. This type of analysis is essential to provide a basis for confident interpretation of HRTEM images of antigorite (Fig. 16).

The most comprehensive HRTEM studies of antigorite have been done by Spinnler (1985) and Mellini et al. (1987). Spinnler (1985) studied antigorite in several samples from asbestos deposits, representing mildly prograde conditions. These samples, characterized by microbeam camera by Wicks and Whittaker (1977) and Wicks and Plant (1979), contained chrysotile and lizardite partly replaced by antigorite. Mellini et al. (1987) have studied a series of antigorite specimens from an area of known and increasing metamorphic grade. Well-formed antigorite with a uniform superstructure period does occur (Mellini et al., 1987, sample MG159), but a variety of anomalous features and growth defects were observed in both studies. These were particularly abundant in specimens of lower metamorphic grade, where the reaction to antigorite was incomplete. Twinning on (001) was common (Fig. 13) as were modulated dislocations (Mellini et al., 1987). Variations of the superstructure period occur within individual crystals. Two features, the non-coincidence of a^* in the subcell and A in the superlattice and the oscillating variations in A in successive layers along Z, are incommensurate and produce rotation of satellite diffraction spots, to give unusual *en echelon* SAD patterns. Long wave-length, non-periodic modulated material and disordered material were also observed; the same

120

THICKNESS

138.0 Å   220.8 Å   303.6 Å

DEFOCUS

-750 Å

-850 Å

-950 Å

-1050 Å

-1150 Å

-1250 Å

-1350 Å

-1450 Å

-1550 Å

-1650 Å

-1750 Å

-1850 Å

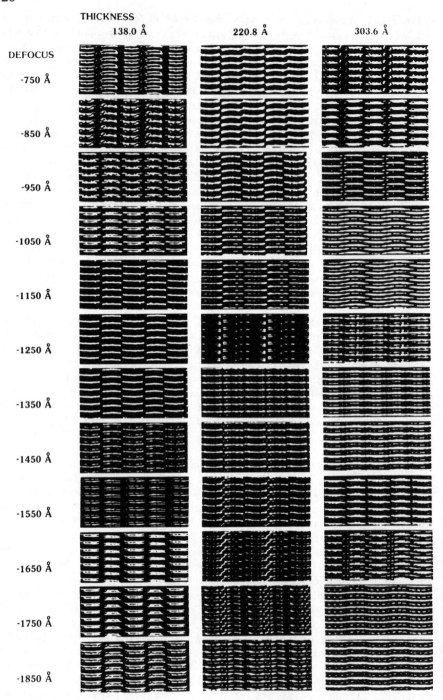

Figure 14. Calculated HRTEM images of antigorite as viewed down [010]. Calculations were made for the thicknesses and defoci indicated, a resolution of 3.0 Å, a C_s of 2.8 mm, and a divergence of 3.0 mrad. Figure courtesy G. E. Spinnler, Arizona State University, Tempe, Arizona (unpub. data).

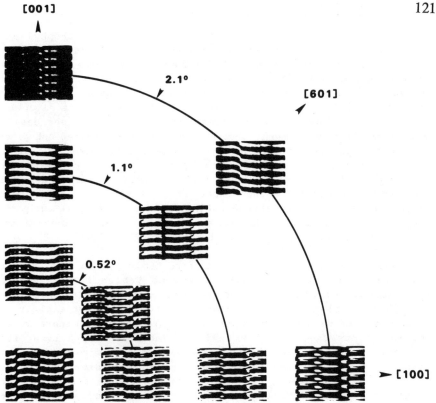

Figure 15. Calculated HRTEM images of antigorite as a function of crystal tilt. Calculations were performed for a crystal with a thickness of 220.8 Å, -900 Å defocus, 3.0 Å resolution, and 3.0 mrad divergence. The crystal orientation as indicated is the deviation of [010] from the direction of the electron beam. Figure courtesy G. E. Spinnler, Arizona State University, Tempe, Arizona (unpub. data).

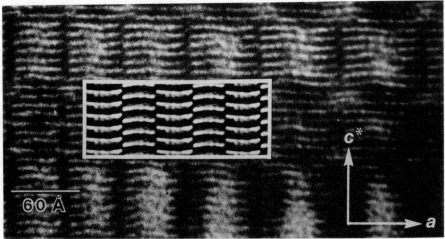

Figure 16. Match between an experimental HRTEM image of antigorite (sample W7035), taken parallel to [010], and a calculated image (inset). A crystal thickness of 220.8 Å and a defocus value of -950 Å were used in the calculation; all other parameters are the same as in Fig. 14. Figure courtesy G. E. Spinnler, Arizona State University, Tempe, Arizona (unpub. data).

features were seen by Veblen (1980) in altered anthophyllite, particularly with other serpentine minerals that were altering to antigorite. Mellini et al. (1987) found chrysotile altering to antigorite; Spinnler (1985) found chrysotile and lizardite altering to antigorite, as found in the microbeam camera studies by Wicks and Whittaker (1975) and Wicks and Plant (1979). Mellini et al. (1987) noted that the occurrence of these unusual features decrease as metamorphic grade increases. Thus the defects were annealed out of the antigorite crystals and the degree of crystallinity increases. It was further found that the superlattice period tends to decrease with increasing grade and becomes more uniform. This is expressed in thin section by a progression from interlocking textures of near-equant antigorite grains to interpenetrating textures of elongate antigorite blades (Mellini et al., 1987, Fig. 5a-c), and is similar to variations found by microbeam camera studies of Wicks and Plant (1979).

Features such as two-layer antigorite structures were observed by Yada (1979) and Veblen (1980). Livi and Veblen (1987) describe lizardite coupled with an oriented intergrowth of antigorite that formed through the replacement of phlogopite; this may in part explain why lizardite tends to form pseudomorphs after pre-existing minerals, and why antigorite does not. In Livi and Veblen's study, lizardite-phlogopite interfaces are parallel to (001) of both minerals and show little strain along the interface, so that lizardite can replace phlogopite without greatly disrupting the phlogopite structure. The formation of lizardite $1T$ pseudomorphs after phlogopite has been recorded by Wicks and Whittaker (1977), and is common in serpentinite that contains phlogopite or biotite. Livi and Veblen noted that the antigorite-phlogopite interfaces "exhibit marked strain contrast". The amount of antigorite in these intergrowths is small relative to lizardite, but if antigorite were abundant, it would severely disrupt the phlogopite structure, perhaps forming new antigorite grains unrelated to the original phlogopite structure. Antigorite is well-known for its lack of development of pseudomorphic textures (Wicks and Whittaker, 1977).

Composition. Substitution in antigorite is limited (as in chrysotile) to the condition that misfit between the octahedral and tetrahedral sheets must be maintained. The microprobe analyses of antigorites by Wicks and Plant (1979) show up to 2.5 wt % Al_2O_3. However, this is not an upper limit as Frost (1975) recorded 3 wt % Al_2O_3, Faust and Fahey (1962) up to 3.2 wt % Al_2O_3 (Whittaker and Wicks, 1970), and Uehara and Shirozu (1985) up to 4.1 wt % Al_2O_3. As antigorite is primarily a rock-forming mineral rather than a vein-forming mineral like chrysotile, this may be more an expression of the effect of the bulk chemistry of the rock rather than a fundamental difference between substitution in the antigorite and chrysotile structures. Whatever the reason, antigorite seems to accept greater substitution by aluminum than does chrysotile.

The substitution of iron in antigorite is subject to similar conditions as in chrysotile. However, the total iron content (expressed as FeO) of the antigorites analyzed by Wicks and Plant (1979) ranged up to 10 wt % FeO, 4 wt % greater than normal chrysotile and 2 wt % greater than polygonal chrysotile. Substitution for Si is usually low, and is by Al and rarely Fe^{3+} (Rozenson et al., 1979; Wicks and Plant, 1979).

The H_2O content of antigorite is significantly different from those of chrysotile and lizardite. The H_2O^+ contents in the antigorite specimens examined by Whittaker and Wicks (1970) are within the limits of error of the stoichiometric value based on the superstructure. Chrysotile and lizardite H_2O+ contents are invariably in excess of the stochiometric value (see Fig. 8, Whittaker and Wicks, 1970). Possible explanations for this feature have been discussed by Deer et al. (1962).

Uehara and Shirozu (1985) noted that the substitution of Al at both tetrahedral and octahedral sites decreased the c value of antigorite. Their series of specimens showed that the increasing substitution of Fe^{2+} at the octahedral sites increase the lateral dimensions of the octahedral sheet, particularly the b parameter. The effects on a were less obvious, but there is a trend for a smaller superstructure period with increasing Fe^{2+} content. However, samples from different localities had different superstructure values for similar Fe^{2+} contents, suggesting that other factors are also important in determining the superstructure parameter. Uehara and Shirozu (1985) observed that higher pressures and lower temperatures produced a long superstructure in specimens of the same Fe^{2+} content. In a complementary study, Mellini et al. (1987) found that increasing temperature shortened the superstructure in the antigorite serpentinites of the Swiss and Italian Alps.

Carlosturanite

Carlosturanite is a newly discovered asbestiform silicate (Compagnoni et al., 1985). It has a modulated structure and is discussed in detail in Chapter 17. However, as its structure represents a fourth mechanism for solving the serpentine misfit problem and as it may turn out to be a common mineral in serpentinites, it is included in this discussion of serpentine minerals.

Carlosturanite was described by Compagnoni et al. (1985) from a network of veins cutting an antigorite serpentinite near Sampeyre, Val Varaita, Italy. It is a low-grade metamorphic phase occurring in association with chrysotile, brucite, diopside, magnetite, and NiFe alloy, and locally with clinohumite, perovskite, and uvarovite. It occurs in [010] flexible, light-brown fibers up to several centimeters long, sometimes in parallel growth with [001] diopside and [100] chrysotile fibers. Individual carlosturanite fibers are usually less than 0.2 microns in diameter. The empirical formula derived by (Compagnani et al, 1985) for this specimen is $(Mg_{37.77}Fe^{2+}_{2.17}Ti^{4+}_{1.09}Mn^{2+}_{0.39}Cr^{3+}_{0.12})_{\Sigma 41.54}(Si_{22.92}Al_{0.81})_{\Sigma 23.73}H_{72.51}O_{1.26}.$

Mellini et al. (1985) have proposed a model for the carlosturanite structure based on available X-ray and electron diffraction data and TEM imaging. The structure is composed of a normal, planar octahedral sheet linked to a tetrahedral sheet modified by the systematic replacement along rows parallel to Y of one-seventh of the $[Si_2O_7]^{6-}$ groups by $[(OH)_6H_2O]^{6-}$ groups (see Fig. 7, Chap. 17). Each silicon tetrahedron is replaced by three hydroxyls and one water molecule in a hydrogen bonded tetrahedral arrangement. This results in a strip of a serpentine structure six tetrahedra wide, linked by the modified region to another serpentine strip. This produces an ideal formula for carlosturanite of $M_{42}T_{24}O_{56}(OH)_{68}(H_2O)_2$. The structure can be considered as part of a polysomatic series (Thompson 1978) built up of serpentine units $S = M_3T_2O_5(OH)_4$ (M = octahedral sites, T = tetrahedral sites) and the modified units, $X = M_6T_2O_3(OH)_{14}H_2O$, with a general formula $[M_3T_2O_5(OH)_4]_n \cdot M_6T_2O_3(OH)_{14} \cdot H_2O$. Carlosturanite has n = 5 and has a C-centered unit cell, as do all other polysomes with n = odd. Polysomes with n = even have a primitive cell with a shorter a parameter (Mellini, et al. 1985).

Carlosturanite illustrates a fourth mechanism of structural adjustment to the misfit problem. In this case, the octahedral sheet is essentially unmodified but the tetrahedral sheet is limited to strips parallel to Y, and misfit is overcome along the X direction by the modified zone in the tetrahedral sheet. However, this mechanism does not relieve the misfit along the Y direction. The b parameters in carlosturanite are 9.41 Å (Compagnoni et al., 1985) and 9.31 Å (Mellini and Zussman, 1986), slightly longer than in other serpentine minerals: lizardite, b = 9.21 –9.24 Å, chrysotile b = 9.19 Å, antigorite b = 9.24–9.26 Å. This suggests that the break in the tetrahedral sheet allows for near complete misfit relief along the X direction, so that the structure can expand slightly along the Y direction where misfit is not relieved, presumably slightly stretching the tetrahedral strips along Y.

Since the initial description of carlosturanite was published, it has been found in three more localities (Mellini and Zussman, 1986). It is possible that carlosturanite is a common mineral. Now that carlosturanite is known, it will undoubtedly be found in many serpentinites. Other minerals such as balangeroite (Compagnoni et al., 1983), although not a sheet silicate, may also be found in similar environments.

IDENTIFICATION

It is impossible to identify serpentine minerals without diffraction data of some kind, as they all have similar colors, habits and occurrences. The exception is cross-fiber asbestos, which can be identified with confidence as chrysotile. If the identification is taken a step further to chrysotile $2M_{c1}$, the distribution of chrysotile $2M_{c1}$, chrysotile $2Or_{c1}$ and parachrysotile established by X-ray and electron diffraction studies indicate that the odds strongly favor this answer. However, this is not an identification but a lottery. The geological conditions will indicate which serpentine mineral is likely to be present, or the most abundant, but reliable identification still depends on diffraction methods that will identify the structure and its stacking sequence. The recent recognition of carlosturanite, a mineral that has been in museum collections for some years and seems to be fairly wide-spread, illustrates the importance and difficulty of identification. There are several methods available.

X-ray powder diffraction

The X-ray powder diffraction patterns of representative serpentine minerals are given in Table 2. Where possible, the diffraction pattern from samples used for structure studies are listed; these include lizardite $1T$ and lizardite $6T$. The diffraction patterns from the chrysotile fibers used for structural refinements are not available, and new patterns were obtained from pure chrysotile $2M_{c1}$ and chrysotile $2Or_{c1}$ specimens at the Royal Ontario Museum. As chrysotile asbestos is difficult to grind, these samples were chopped in a microjet 10J ultracentrifugal mill with 0.2 mm sieve rings at liquid nitrogen temperatures. This produced a fluffy assemblage of short fibers that were dropped on a Guinier-de Wolff camera sample holder and lightly pressed into place. This procedure produced a minimum of preferred orientation of the fibers, mainly because the fibers are always handled and mounted in the dry state. The antigorite diffraction patterns of Uehara and Shirozu (1985) in Table 2 represent different superstructure periods, expressed as M, and different compositions.

The criteria for identifying the serpentine minerals by X-ray powder diffraction given by Whittaker and Zussman (1956) are still valid. The criteria for identifying lizardite polytypes are not included in Whittaker and Zussman (1956) but are available in Bailey (1969). Polygonal serpentine can not be conclusively identified by powder diffraction.

Lizardite. The calculated powder diffraction patterns given by Bailey (1969) for the 12 standard polytypes provide the basis for identifying lizardite and Al-serpentine polytypes. The basic information is also given in Chapter 2.

Lizardite $1T$ is by far the most abundant lizardite polytype, and has the simplest of the serpentine diffraction patterns (A, Table 2). An excellent powder pattern from the type lizardite is given by Rucklidge and Zussman (1965). The one-layer nature of lizardite $1T$ (in contrast to the two-layer structure of chrysotile) is clearly indicated by the 11ℓ series of reflections (Table 2).

The diffraction patterns of multi-layer poltypes are more complex, usually with a distinctive series of 01ℓ reflections (B, Table 2), but are encountered less frequently. They can be categorized into the four structure groups A, B, C and D (Bailey, Chap. 2), and sometimes the

Table 2. X-ray powder patterns of selected serpentine structures.

A. Lizardite 1T (Mellini, 1982)

hkl	I_{vis}	I_{meas}	d_{obs}	d_{calc}
001	s	60	7.216	7.238 Å
100	m	40	4.628	4.623
101	m	30	3.890	3.896
002	ms	50	3.624	3.619
102	vw	10	2.850	2.850
110	vw	10	2.668	2.669
111	s	100	2.506	2.504
003	vw	10	2.418	2.413
112	ms	60	2.152	2.148
202	vw	<5	1.948	1.948
113	ms	40	1.790	1.790
210	vw	10	1.748	1.747
211	vw	10	1.701	1.699
203	vw	<5	1.676	1.669
212	vw	5	1.575	1.574
300	m	50	1.539	1.541
301	m	{50	1.506	1.507
114	vw	}	1.498	1.498
302	vw	10	1.418	1.418
220	vw	10	1.335	1.335
221	vw	20	1.314	1.312

B. Lizardite 6T (Hall et al., 1976)

hkl	I_{obs}	I_{calc}	d_{obs}	d_{calc}
00.6	100	100.0	7.270	7.268 Å
01.0		15.8		4.609
01.1	(30.8)	7.6		4.584
01.2	34	7.4	4.585	4.509
01.3	7.3	27.6	4.402	4.393
01.4	3.9	6.2	4.251	4.245
01.5	5.0	5.4	4.074	4.075
01.6	6.1	19.8	3.890	3.892
01.7	(73.1)	4.0		3.705
00.12	84	69.1	3.637	3.634
01.8	3.9	3.3	3.521	3.520
01.9	3.4	11.1	3.170	3.340
01.10	2.8	2.3	3.004	3.168
01.11	2.2	1.9	2.859	3.006
01.12	1.7	6.2	2.710	2.854
01.13	2.2	1.3	2.659	2.712
11.0	14	3.1	2.617	2.661
11.3	v	10.1	2.585	2.618
11.6	72	90.4	2.502	2.580
00.18	2.8	3.2	2.457	2.499
11.9	1.1	5.3	2.424	2.459
02.3	26	30.7	2.333	2.423
02.6	v	1.0		2.332
11.12	(1.8)	0.5	2.274	2.276
01.18	(1.2)	1.3		2.197
02.9	(19.4)	17.7	2.200	2.197
01.19	1.6	1.6	2.090	2.147
02.10	(1.2)	0.4		2.144
02.11	0.4	0.4	2.020	2.081
11.15	31.8	2.054		2.054
02.13	0.4	0.4	1.962	2.038
01.21	(1.3)	0.9		1.992
00.24	27	1.5	1.895	1.963
11.18	10.7		1.817	1.899
12.0	(2.2)	1.4		1.893
12.1	3.4	0.8	1.817	1.817
12.3	4.5	2.8	1.792	1.791
12.9	1.4		1.742	1.742
11.21	18.8			1.741
03.0	12	28.3	1.728	1.730
11.24	(20.2)	13.0		1.639
02.23	35	19.3	1.638	1.637
00.30	20	(13.0)	1.536	1.536
03.12	(32.3)	(19.3)	1.502	1.503
12.18	0.3			1.501
11.27	2.2	0.3	1.462	1.464
22.0	3.9	(6.3)	1.453	1.454
	7.0		1.415	1.415
		6.3		1.414
	5.6	7.6	1.381	1.381
	(2.0)	2.0		1.331

C. Chrysotile 2M_{c1} (Wicks unpub.)

hkl	I	d_{obs}	d_{calc}
002	65	7.27	7.30 Å
020	40	4.55	4.59
004	50	3.65	3.65
200	25	2.655	2.657
201	20	2.588	2.588
202	35	2.540	2.544
$20\bar{2}$	100	2.454	2.452
203	10	2.370	2.390
$20\bar{3}$	<5	2.277	2.278
204	20	2.212	2.210
$20\bar{4}$	5	2.095	2.092
206	<5	1.829	1.826
008	20	1.743	1.746
$20\bar{6}$	55	1.531	1.531
060	30	1.321	1.320
$40\bar{2}$			

D. Chrysotile 2Or_{c1} (Wicks unpub.)

hkl	I	d_{obs}	d_{calc}
002	50	7.29	7.33 Å
020	40	4.56	4.60
004	40	3.66	3.66
200	15	2.659	2.656
201	20	2.617	2.614
202	100	2.500	2.497
006	25	2.451	2.442
$20\bar{2}$	30	2.332	2.333
203	5	2.145	2.150
205	10	1.966	1.968
206	<5	1.795	1.797
310	10	1.739	1.739
207	60	1.644	1.644
060	15	1.533	1.533
208	25	1.503	1.508
400	25	1.362	1.328
402	30	1.309	1.307

A. Specimen from Val Sissone, Italy. Film pattern of a single crystal recorded in a 57.3 mm Gandolfi camera with CuKα radiation. I_{vis} intensities estimated visually. I_{meas} intensities measured with a densitometer on a film pattern of an aggregate of single crystals from Coli, Italy (Mellini and Zanazzi 1987) recorded in a 114.6 mm Gandolfi camera with CuKα radiation (Wicks unpub.). Indexed on hexagonal axes with a = 5.338(2), c = 7.238(3) Å.

B. Specimen from Unst, Scotland. Film pattern of a natural aggregate of fine-grains recorded in a 114.6 mm Debye-Scherrer camera with CuKα radiation. Intensities measured by microdensitometer. Indexed on hexagonal axes with a = 5.332(1), c = 43.6(1) Å.

C and D. Film patterns of chopped chrysotile fibers recorded in a 57.3 mm Guinier-de Wolff camera with monochromated CuKα radiation. The slightly broad reflections limit the accuracy of the measurements. Intensities measured with a densitometer. C. Specimen from Johnson and Bell mine, Thetford, Quebec. ROM number M8569. Indexed on monoclinic axes with a = 5.32(1), b = 9.19(2), c = 14.63(3) Å, β = 93.29(7)°. D. Specimen from Sir R. Bond mine, Newfoundland, ROM number M15100. Indexed on orthorhombic axes with a = 5.31(1), b = 9.19(2), c = 14.65(3) Å.

Table 2 (cont.)

			E. Antigorite (M=8.52) (Uehara & Shirozu, 1985)			F. Antigorite (M=8.12) (Uehara & Shirozu, 1985)			G. Antigorite (M=7.44) (Uehara & Shirozu, 1985)			H. Antigorite (M=6.48) (Uehara & Shirozu, 1985)		
h	k	l	I_{obs}	d_{obs} Å	d_{calc} Å	I_{obs}	d_{obs} Å	d_{calc} Å	I_{obs}	d_{obs} Å	d_{calc} Å	I_{obs}	d_{obs} Å	d_{calc} Å
0	0	1	353	7.29	7.271	491	7.25	7.266	780	7.28	7.262	2200	7.27	7.258
0-1/M	0	1	40	7.21	{7.210	73	7.18	{7.199	103	7.17	{7.180	120	7.14	7.147
0+1/M	0	1			7.158}			7.141}			7.116}	110	7.10	7.074
0-2/M	0	1	8	6.98	6.987	6	6.92	6.953	11	6.88	6.892	11	6.77	{6.777
0+2/M	0	1	8	6.60	{6.895	6	6.54	{6.851	11		6.781			6.654}
0-3/M	0	1			6.645			6.580	8		6.465	9	6.25	6.255
0+3/M	0	1			6.526}			6.451}	6		6.329	9	6.14	6.111
1	1	0	6	4.673	4.676	8	4.671	4.684	10		4.692	4	4.685	4.703
0	2	0	8	4.618	4.619	11	4.615	4.621	5		4.624	7	4.618	4.627
1+1/M	0	0	8	4.293	4.299	8	4.272	4.286			4.256	6	4.214	4.213
0-1/M	0	2	229	3.623	3.631	254	3.620	3.628	330	3.622	3.625	520	3.611	3.620
0+1/M	0	2	30	3.589	{3.618	27	3.579	{3.600	13	3.581	3.608	10	3.574	{3.601
0-2/M	0	2			3.605			3.572			3.590			3.573}
0+2/M	0	2			3.579}			3.548}	9		3.558	14	3.538	3.537
0-3/M	0	2	14	3.557	3.558	11	3.537	3.507	5		3.529	9	3.485	3.494
0-4/M	0	2	13	3.530	3.521				8	3.527	3.484	11	3.448	3.443
												5	3.391	3.389
1-1/M	3	1	5	2.587	2.583	5	2.589	2.589		2.592	2.596	3	2.610	2.609
2	1	1	10	2.561	2.560	8	2.560	2.565		2.569	2.571	5	2.582	2.578
1	3	1	100	2.523	{2.522	100	2.525	{2.525	100	2.527	{2.527	100	2.532	{2.529
1+1/M	1	1			2.521}			2.524}			2.528}			2.534}
0	0	3	10	2.460	2.459	13	2.455	2.457	10	2.454	2.453	8	2.447	2.445
2+1/M	1	1	13	2.4232	2.4237	6	2.4213	2.4221	17	2.4207	2.4207	46	2.4169	2.4193
2-1/M	3	1	6	2.4032	2.3988	4	2.3952	2.3949	7	2.3903	2.3878	7	2.3757	2.3741
1-1/M	3	1	7	2.2336	2.2315	7	2.2373	2.2369	5	2.2459	2.2469	5	2.2637	2.2645
1	3	1	6	2.2064	2.2056	25	2.2085	2.2090	5		2.2134	7	2.2183	2.2207
2	1	3	25	2.1689	2.1680	20	2.1694	2.1695	22	2.1704	2.1707	38	2.1729	2.1722
1-2/M	3	1	16	2.1501	2.1493	3	2.1506	2.1500	22	2.1536	2.1519	29	2.1560	2.1551
2-1/M	3	1	2	1.8487	1.8475	8	1.8494	1.8500	7	1.8372	1.8402	3	1.8504	1.8594
1-1/M	3	1	8	1.8313	{1.8330	9	1.8330	{1.8352	10	1.8156	1.8325	10	1.8495	1.8495
0	0	4			1.8285}			1.8302}	33		1.8155	33	1.8368	1.8362
1+1/M	3	1	8	1.8166	1.8177	3	1.8166	1.8166				6	1.8139	1.8145
1+2/M	3	1	8	1.7835	1.7843	8	1.7831	1.7838		1.7802	1.7864	15	1.7786	1.7878
2+1/M	1	3									1.7820	3		1.7787
0	0	6	4	1.7592	1.7596	3	1.7567	1.7567		1.7520	1.7534	8	1.7461	1.7456
0	6	0	6	1.7405	1.7404	6	1.7380	1.7375		1.7347	1.7339	8	1.7274	1.7279
3	3	1	16	1.5583	1.5588	14	1.5605	1.5612	16	1.5638	1.5640	17	1.5687	1.5678
			11	1.5392	1.5397	11	1.5404	1.5405	12	1.5413	1.5413	12	1.5420	1.5423
			8	1.5298	1.5305	8	1.5346	1.5332	7	1.5364	1.5360	7	1.5383	1.5396

E, F, G and H. Patterns of fine-grained powders recorded on a Rigaku RAD-IIA diffractometer with CuKa radiation. Intensities are not normalized to 100. Indexed on monoclinic axes. Superlattice expressed in terms of M where $M = A/a$.
Composition in cations per $(7+2/M)$ oxygens.
E and F. Specimens from Nishisonogi area, Japan.
E. Subcell $a = 5.424$, $b = 9.238$, $c = 7.274$, $\beta = 91.32°$. Composition Al = 0.024, Fe^{2+} = 0.024, Fe^{3+} = 0.024 apfu.
F. Subcell $a = 5.435$, $b = 9.243$, $c = 7.270$Å, $\beta = 91.40°$. Composition Al = 0.026, Fe^{2+} = 0.047, Fe^{3+} = 0.114 apfu.
G and H. Specimens from Sasaguri area, Japan.
G. Subcell $a = 5.447$, $b = 9.248$, $c = 7.264$Å, $\beta = 91.47°$. Composition Al = 0.162, Fe^{2+} = 0.184, Fe^{3+} = 0.110 apfu.
H. Subcell $a = 5.463$, $b = 9.254$, $c = 7.260$Å, $\beta = 91.50°$. Composition Al = 0.044, Fe^{2+} = 0.051, Fe^{3+} = 0.195 apfu.

individual polytype can be established by comparison with the calculated powder patterns. In other specimens, a detailed study is required (Hall et al., 1976)

Chrysotile. The two-layer nature of the chrysotile structure may be established by the $20l$ reflections which require a doubling of c = 7.3 Å in order to be indexed. The monoclinic nature of chrysotile $2M_{c1}$ is indicated by these reflections, particularly the 202 and 202 reflections in the 2.6-2.4 Å region of the pattern. The chrysotile $20r_{c1}$ diffraction pattern could possibly be confused with those of lizardite $2H_1$ and $2H_2$. However, $20l$ with l = 3, 5 and 7 are stronger than $20l$ with l = 4, 6, and 8 in the chrysotile $20r_{c1}$ pattern whereas the equivalent to $20l$ with l = 3, 5, and 7 are weaker than those with l = 4, 6, and 8 in the lizardite $2H_1$ and $2H_2$ patterns.

Note that the chrysotile patterns (C and D in Table 2) were obtained with Guinier-de Wolff, not Debye-Scherrer, geometry. The lack of preferred orientation is illustrated by the decrease in the intensity of $00l$ reflections relative to $h0l$ reflections.

Antigorite. Whittaker and Zussman (1956) gave a composite antigorite powder diffraction pattern that included the essential diffraction features of antigorite. We now know through the work of Uehara and Shirozu (1985), Mellini et al. (1987), and several earlier workers that a given antigorite specimen can contain more than one superstructure period. This makes it impossible to uniquely index an antigorite powder pattern. The best that can be expected is an indexing representative of the average superstructure period. Uehara and Shirozu (1985) have used 132, 202, 004, 330, and 060 reflections to determine approximate a, b, c and β subcell values. They then use the $2+1/M$, 0, 3 reflection to estimate M using the formula

$$M = A/a = ua*/[\sqrt{l/d^2 - (k^2 b*^2 + l^2 c*^2 \sin^2\beta*)} - (lc*\cos\beta* + ha*)]$$

In all but exceptional specimens, the calculated M value will be irrational, as it represents more than one superstructure period. The final values are then refined using all reflections with d > 1.53 Å by least-squares program; representative results are given in E to H of Table 2. These diffraction patterns also show the effects of Al substitution on c and Fe^{2+} substitution on b.

The criteria of Whittaker and Zussman (1956) for distinguishing antigorite from other serpentine minerals are valid. The 330 reflection at 1.56-1.57 Å in antigorite is unique, not being found in any other serpentine diffraction patterns.

Single-crystal diffraction

The use of this method depends on the availability of suitable single crystals, so that it is applicable only to lizardite and antigorite. The method can be used with X-rays or electrons, although orientation and manipulation of the sample for electron diffraction is a problem.

Lizardite. In early electron microscope studies when samples were prepared as dispersed fragments, lizardite tended to lie on the (001) cleavage plane, producing a SAD pattern made up of a hexagonal array of spots (Zussman et al., 1957; Whittaker and Zussman, 1971). This type of diffraction pattern does not contain information related to the stacking of lizardite layers.

The single-crystal procedure for identifying lizardite and Al-serpentine polytypes was developed by Bailey (1988) (see Chap. 2). Bailey's method also can be applied to SAD patterns. Cressey (1979) has shown the effects of $b/3$ disorder on $0kl$ zones of lizardite $1T$. The disordered lizardite $1T$ pattern (Fig. 2b, Cressey, 1979) has streaks on the $02l$ and $04l$ reflections due to the $b/3$ disorder, but diffraction spots with no streaks on the $06l$'s which are unaffected by the disorder. The ordered lizardite $1T$ pattern (Fig. 9b, Cressey, 1979) has sharp diffraction spots on all layer lines (comparison of Fig. 7a, Chap. 2, with Fig. 6 of Bailey, 1988, illustrates this condition). More recently Livi and Veblen (1987) used oriented SAD patterns to identify a crystallographically controlled intergrowth of lizardite and antigorite.

Antigorite. Various single crystal methods (particularly Weissenberg) were used by Kunze (1958, 1959, 1961) and Zussman (1954) in the early studies on antigorite. Further work by Uehara and Shirozu (1985) included Weissenberg, rotation, and precession methods; however, the variability of the superstructure period, even within a single crystal, and the rarity of single crystals suitable for X-ray diffraction make electron diffraction the preferred method. Early electron diffraction studies were reviewed by Whittaker and Zussman (1971). Recently, Yada (1979), Uehara and Shirozu (1985), and Mellini et al. (1987) have used electron diffraction effectively on a variety of antigorite specimens. However, the most comprehensive study of the effect of various growth defects on antigorite electron diffraction patterns is that of Spinnler (1985).

Fiber Diffraction

The fibrous and splintery serpentine minerals, chrysotile, polygonal serpentine, and antigorite, are best studied by stationary or rotating oscillation or Weissenberg X-ray methods. By using equi-inclination rotation methods in a Weissenberg camera, up to the 10th layer line of chrysotile can be recorded. The electron fiber diffraction patterns are closely related to the X-ray patterns.

Chrysotile. Fiber diffraction patterns show the cylindrical nature of chrysotile (Whittaker, 1956a, 1957, 1963). Wicks (1979) has given indexed X-ray fiber diffraction patterns, which are reproduced in Figure 17a; these illustrate the configuration of chrysotile $2M_{c1}$ and chrysotile $2Or_{c1}$ for both normal and polygonal fibers, and for

130

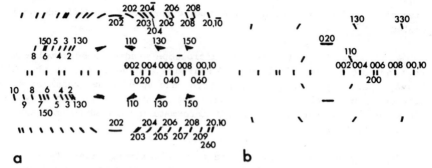

Figure 17. Diagramatic chrysotile fiber X-ray diffraction patterns. (a) the top half represent chrysotile $2M_{c1}$ and the bottom half chrysotile $2Or_{c1}$; the right half represents normal patterns and the left half polygonal chrysotile patterns; (b) parachrysotile (from Wicks, 1979).

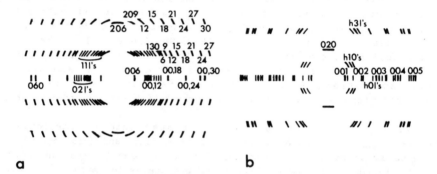

Figure 18. The X-ray fiber diffraction pattern of (a) lizardite $6T_1$, indexed for comparison with chrysotile, (b) antigorite (from Wicks, 1979).

parachrysotile (Fig. 17b). Middleton and Whittaker (1976) provide more detailed indexed patterns of the two types of polygonal serpentine that they studied. The diffraction pattern for a possible multilayer polygonal serpentine was given by Wicks (1979) and is reproduced in Figure 18a. The actual specimens also produce a weak superimposed chrysotile $2M_{c1}$ diffraction pattern.

Zussman et al. (1957) give indexed chrysotile $2M_{c1}$ and $2Or_{c1}$ SAD patterns out to the fourth layer line. Zvyagin (1967) published a variety of chrysotile SAD patterns including chrysotile $1M_{c1}$. Yada (1967, 1971) used SAD patterns of chrysotile to provide details of the helical or regular nature of the cylindrical structure. He has also presented type chrysotile $2M_{c1}$, $2Or_{c1}$ and parachrysotile SAD patterns (Yada, 1979).

Antigorite. Fiber diffraction was used to study splintery antigorite by Kunze (1958). An idealized fiber diffraction pattern for antigorite was given by Wicks (1979) and is reproduced in Figure 18b. The individual elongate antigorite grains that make up splintery antigorite are coarse enough at the electron optical level to produce single-crystal SAD patterns, not fiber patterns.

Microbeam camera

The microbeam camera with a 50 micron collimator is an ideal technique to examine to serpentine textures as they occur in thin section. The method and its resultant diffraction patterns are described in detail by Wicks and Zussman (1975). Each of the serpentine minerals produce distinctive diffraction patterns that have some of the features of rotation-type diffraction patterns, even though the specimen is stationary in the microbeam camera. The diffraction patterns show either or both of the effects of tilting a crystallographic axis out of the perpendicular to the X-ray beam, or lack of parallel orientation of adjacent crystals. The diffraction patterns have some of the features of the single-crystal diffraction patterns given in Chapter 2, but misorientation often has the affect of superimposing two diffraction patterns. The microbeam camera is applicable to all serpentine minerals in thin section, and is an ideal complementary technique to the electron microprobe (Wicks and Plant, 1979). Application of the microbeam camera to serpentine textures led to the establishment of criteria for identifying the serpentine mineral textures (Wicks and Whittaker, 1977), and forms the basis for the studies that are presented in the second part of this chapter.

The microbeam camera provides an overview of the mineralogy of a specimen that can be used to select samples for TEM studies and to assess the significance of the results obtained. TEM studies have been done by Cressey (1979), Spinnler (1985) and Wicks (1986) on specimens first studied by microbeam X-ray camera.

HRTEM studies

HRTEM studies have expanded our knowledge considerably. All three serpentine structures have been observed in combinations that had not been expected. The application of HRTEM to examination of the serpentine *in situ* has given us new insights to the serpentinization process. This is particularly true of studies of partly altered materials. Here, the reaction has stopped, equilibrium has not been reached, and many features are present that would not occur in an equilibrium assemblage. The demonstration of the effect of metamorphic grade on antigorite crystallinity by Mellini et al. (1987) also applies to the other serpentine minerals.

The ease with which serpentines suffer beam damage in HRTEM studies makes it difficult to obtain images and SAD patterns. The extensive imaging and electron diffraction studies of chrysotile by Yada (1967, 1971, 1979), of antigorite by Spinnler (1985) and Mellini et al. (1987), and of planar lizardite by Veblen and Buseck (1979, 1981), Spinnler (1985), Wicks (1986) and Livi and Veblen (1987) have provided a basis for visually identifying these minerals in HRTEM images. The many curved features often associated with the planar lizardite do present identification problems. These vary from gentle curves (Fig. 5a) to tight rolls (Fig. 5b) and hairpin curves. Certainly lizardite can bend, but how far can it curve and still remain lizardite? At what degree of curvature is the chrysotile type of stacking established? What is the planar chrysotile $2M_{c1}$ material in polygonal serpentine? There is no electron diffraction data from these materials, but this information is essential for the identification of these structures. Without SAD patterns we are back at the same level as we were when we held the piece of cross-fiber asbestos in our hands. The next major step in this field is the use of electron diffraction to identify the serpentine structures and the stacking of the layers in these structures. Yada (1969, 1971, 1979) and Livi and Veblen (1987) have demonstrated the power of electron diffraction to enhance our understanding of the visual images, but there remains much to be done.

TEXTURES IN SERPENTINITES

The serpentine minerals that make up serpentinites display a fascinating but bewildering array of textures and optical configuration in thin section. Even after the basic structures of the serpentine minerals were recognized by Whittaker and Zussman (1956) it was not possible to apply this knowledge directly to the identification of serpentine textures. Only the coarser textural units could be removed from thin section, ground up, and X-rayed but then the relationship of the serpentine structure to the optical orientation was lost. The application of the microbeam X-ray diffraction camera to this problem by Wicks and Zussman (1975) allows the serpentine minerals to be X-rayed *in situ* in thin section. Thus not only can the serpentine minerals in various textures and textural units down to 50 microns in diameter be identified, but the relationship of the crystal structure and optical orientation can be established as well.

Criteria for the identification of serpentine minerals in thin section were presented by Wicks and Whittaker (1977, Table 4), based on microbeam camera studies of a wide range of textures from numerous localities. They divide the textures into two groups: pseudomorphic textures that preserve the preserpentinization textures of the ultramafic rock, and non-pseudomorphic textures that do not preserve preserpentinization textures. The pseudomorphic process not only preserves the outline of the original grains and the original textures, but it also preserves the fracture patterns and cleavages of the parent minerals (Wicks, 1984c). This means that a considerable amount of information on the pre-serpentinization history is preserved despite complete serpentinization. The great majority of pseudomorphic textures, such as mesh and hourglass textures after olivine (Wicks et al., 1977) and bastites after pyroxenes, amphiboles and various layer silicates (Wicks and Whittaker, 1977), are composed of lizardite $1T$ in various orientations with, but usually without, brucite. Mesh and hourglass textures composed of antigorite or chrysotile are rare although they can be dominant in a given serpentinite. Bastites composed of chrysotile or antigorite were found to occur infrequently. The non-pseudomorphic textures are dominated by antigorite, usually as elongate grains that form an interpenetrating texture and obliterate previous textures. Chrysotile and lizardite non-pseudomorphic textures were also recorded, usually with more equant grains that form interlocking textures. The non-pseudomorphic textures may be massive or foliated. Some pseudomorphs may be present but they are usually indistinct and difficult to recognize, particularly when prograde serpentinization is intense.

Once the mineralogy of the textures was understood it became possible to subdivide the process of serpentinization based on the presence or absence of antigorite and the presence or absence of foliation (Table 3). From this it is possible to infer whether the serpentinite formed from simple retrograde hydration of peridotite or if it formed from prograde metamorphism of a previous existing serpentinite. The most common retrograde assemblage is composed of lizardite ± brucite ± magnetite in pseudomorphic textures (type 3) and forms one of the most abundant types of serpentinite. When sheared (type 4), the pseudomorphic textures are destroyed and chrysotile tends to dominate over lizardite (Wicks, 1984b). Retrograde assemblages composed of antigorite (types 1 and 2) occur at higher temperatures, appear to be rare, and are difficult to distinguish from prograde antigorite. Detailed work by Wicks (1984a) on the series of retrograde reactions in the Glen Urquhart serpentinite demonstrated the position of antigorite in the retrograde reaction sequence.

The most common prograde assemblage, antigorite ± magnetite, is usually found in massive serpentinites (type 7) but also occurs in foliated serpentinites (type 8, Wicks, 1984b). Antigorite is the high temperature serpentine mineral. It usually forms during prograde metamorphism, and it forms one of the more abundant types of serpentinite. Lower temperature prograde assemblages are composed of chrysotile and lizardite (types 5 and 6). These are commonly associated with chrysotile asbestos deposits and are much less wide spread than the higher temperature antigorite ± magnetite assemblages

Following the original microbeam camera studies, Wicks and Plant (1979) coupled microbeam X-ray diffraction with electron microprobe analysis to provide chemical data on the products of serpentinization. They discovered that lizardite ± brucite hourglass textures and chrysotile + lizardite ± brucite mesh textures, although closely related to lizardite ± brucite (type 3) retrograde textures, are promoted by prograde (type 5) serpentinization and thus form a bridge between type 3 and type 5 serpentinization. These textures are most commonly found in chrysotile asbestos deposits. Bastites in asbestos deposits are mineralogically more varied than those in retrograde (type 3) serpentinite. Whereas type 3 bastites are invariably lizardite $1T$ with Al and Cr-contents related to the parent pyroxene (Dungan, 1979), the bastites in asbestos deposits although still dominated by lizardite may also contain chlorite, chrysotile, or antigorite. Also the Al and Cr-contents of lizardite are lower, indicating that the elements are removed from the bastites (Wicks and Plant, 1979). All three serpentine

134

Table 3. Classifications of serpentinization.

Type	Wicks and Whittaker* (1977) T	Antigorite	Foliation	New Designation Regime	Location in P-T Space
1.	down	yes pseudomorphic**	no		
				A	above [BC] or [BL]
2.	down	yes	yes		
3.	down	no pseudomorphic	no		
				B	below [BC] or [BL]
4.	down	no	yes		
5.	up	no transitional	no		
				C	T below [BC] or [BL] P below [BF], [FT]
6.	up	no	yes		
7.	up	yes	no		
				D	above [BC] or [BL]
8.	up	yes	yes		

*T: Temperature increasing (up) or decreasing (down); Antigorite: nucleation (yes) or absence (no) Foliation: presence (yes) or absence (no) of shearing.
** Regimes that produce pesudomorphic textures are labelled. Regime 5 produces the transitional texture.

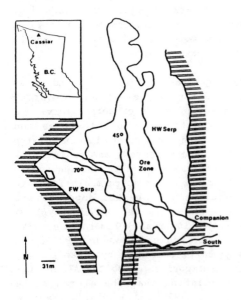

Figure 19. Bird's-eye view sketch map of the Cassiar asbestos deposit and the location of the Cassiar mine (inset). The ore zone is isolated from the metasediments and greenstones by the footwall (FW) serpentinite on the west, and the hangingwall (HW) serpentinite on the east. The faults of interest are the 45° and the 70° Shears, and the South and the Companion faults. The 45° Shear traverses the length of the pit but its boundaries are ill-defined in the northern portion of the pit.

minerals, chrysotile, lizardite, and antigorite, often are present within a particular chrysotile asbestos deposit. Chrysotile and antigorite were discovered within a single non-pseudomorphic texture (Wicks and Plant, 1979).

The complexity of the reactions and the potential for gaining a new understanding of the stability relationships of the serpentine minerals suggest that chrysotile asbestos deposits deserve intensive study. Fortunately, the opportunity for detailed mapping and structural study of the Cassiar chrysotile asbestos deposit at Cassiar, British Columbia, Canada, has provided the structural framework for interpreting the evolution of serpentine mineral textures, chrysotile asbestos veins, the distribution of stable isotopes, and the P/T relationships of the serpentine minerals. The following is a progress report of a study that is in the process of being published. It includes work by Dr. K. Kyser, University of Saskatchewan, Dr. J.V. Chernosky, University of Maine, as well as the authors of this chapter.

CASSIAR ASBESTOS MINE

The Cassiar asbestos deposit is located at Cassiar in north central British Columbia (Fig. 19, inset). The Cassiar serpentinite is located at the base of the Sylvester allochthon, which is emplaced onto Pre-Cambrian to Mississippian limestones, shales, and sandstones. The Cretaceous Cassiar batholith and the Eocene Cassiar stock are located to the west of the allochthon. The geology of the Sylvester Allochthon has been described by Harms (1986) and Gabrielse (1963) and the geology of the Cassiar asbestos deposit by Gabrielse (1960).

The Cassiar serpentinite is enclosed by argillites and cherts on its footwall and by argillites and greenstones on its hangingwall (Fig. 19). The serpentinite is dissected by several faults. The most important with respect to ore formation are the 70° shear, which strikes ESE and dips 70°NE, and the 45° shear, which strikes approximately NS and dips 70°E but contains zones of schistose serpentinite and of gouge that dip 55° to 45°E.

The Cassiar serpentinite was a harzburgite tectonite prior to serpentinization: both preserpentinization kink-bands and schistosity (Fig. 20a) can be seen in the pseudomorphically serpentinized samples (Wicks, 1984a). There is evidence to suggest that retrograde lizardite serpentinization took place at an early stage.

A variety of textures, many of them characteristic of asbestos deposits, occur in the Cassiar serpentinite. Most of the textures correspond to those described by Wicks and Whittaker (1977), but some variations occur that were not observed by Wicks and Whittaker. Knowledge of the structure within the serpentinite provides a framework for clarifying the relationships among pseudomorphic, non-pseudomorphic, and intermediate textures noted by Wicks and Plant (1979). It also

136

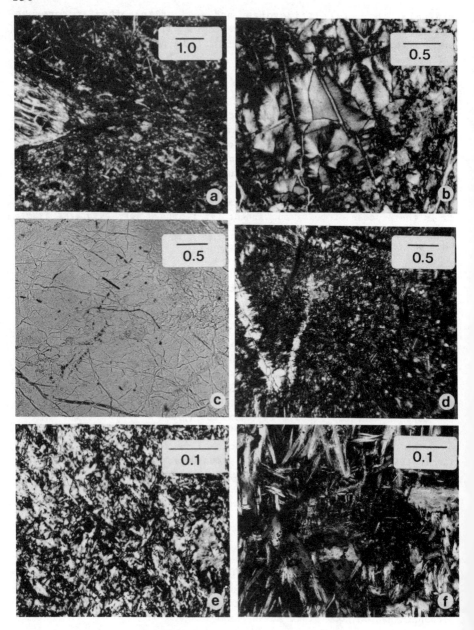

Figure 20. Photomicrographs of serpentine textures. Scale bars in mm are shown on the photomicrographs. (a) Preserpentinization schistosity wrapped around orthopyroxene. (b) lizardite 1T hourglass texture. (c) transition texture in plane-polarized light. (d) transition texture in crossed-nicols; chrysotile serrate veins on left side of photo, most of matrix is lizardite 1T. (e) Lizardite 1T and antigorite interlocking texture. (f) antigorite interpenetrating texture replacing lizardite 1T mesh texture.

provides the opportunity to classify the orthopyroxene bastites more fully than was done in the study of Wicks and Whittaker (1977)

Serpentine textures and their distribution

Pseudomorphic lizardite hourglass and mesh textures after olivine occur throughout the Cassiar serpentinite (Table 4). Lizardite $1T$ hourglass texture (Fig. 20b) dominates the footwall and extends to just east of the ore body. Lizardite $1T$ mesh textures sometimes associated with ribbon texture (Wicks, 1984c), dominate the hangingwall but occur in the ore zone and footwall as well. Lizardite mesh textures also occur between the Companion and the South Faults. The mesh centers are composed of isotropic serpentine or hourglass configuration. Relict olivine was not observed. Chrysotile $2M_{c1}$ mesh-rim textures with isotropic or hourglass centers are associated with the major faults.

Non-pseudomorphic textures, many formed by recrystallization of pre-existing serpentine textures, also are abundant in the serpentinite. A distinctive transition texture between a pseudomorphic and a non-pseudomorphic texture is present. When viewed in plane polarized light this serpentine is clearly pseudomorphic after forsterite (Fig. 20c) but under crossed-nicols it is isotropic, very fine-grained, and no pseudomorphs are visible (Fig. 20d). This transition texture may be composed of chrysotile $2M_{c1}$, lizardite $1T$, or less frequently antigorite, and it occurs near the ore zone. The transition texture passes into interlocking texture, and these two textures appear to represent successive stages in a recrystallization event. Interlocking textures of antigorite + chrysotile $2M_{c1}$ usually occur in association with chrysotile $2M_{c1}$ serrate veins.

Chrysotile $2M_{c1}$ serrate veins occur along fractures (Fig. 20d). They are characterized by length-slow fibers orientated perpendicular to the fracture and by irregular boundaries with the wall rock. Chrysotile serrate veins are associated with and form through the recrystallization of non-pseudomorphic textures. Occasionally the close arrangement and orientation of chrysotile-serrate veins can resemble a hourglass texture. Lizardite $1T$ serrate veins are similar in form but are composed of length-fast apparent fibers. They vary from true serrate veins formed through the recrystallization of non-pseudomorphic textures to lizardite mesh rims that form serrate veins through the recrystallization and replacement of isotropic or hourglass texture.

Lizardite $1T$ serrate veins are found at the margin of the ore body where they are associated with chrysotile $2M_{c1}$ serrate veins. Chrysotile serrate veins without lizardite serrate veins are found within the ore body. Interlocking antigorite + chrysotile texture (Fig. 20e) usually occurs in association with chrysotile serrate veins. Interpenetrating texture (Fig. 20f), consisting solely of antigorite at the microbeam camera level (50 μm), occurs mainly in association with the hangingwall alteration, and less commonly with the footwall alteration.

138

Table 4. Serpentine textures and mineralogy.

Texture	Optical Characteristics*	Mineralogy
Pseudomorphic		
Mesh rim	length-fast	lizardite
	length-slow	chrysotile
Mesh center, fine grained	isotropic	lizardite, antigorite or chrysotile
Mesh center, hourglass	length-fast	lizardite
	length-slow	chrysotile or antigorite
Hourglass	length-fast	lizardite
Ribbon	length-fast	lizardite
Non-pseudomorphic		
Transition	isotropic	lizardite, chrysotile or antigorite
Interlocking	length-fast	lizardite
	length-slow	antigorite, chrysotile or lizardite
Serrate veins	length-fast	lizardite
	length-slow	antigorite
Interpenetrating	length-slow	antigorite, rarely chrysotile or lizardite

* Based on elongation of fibers or apparent fibers.

The Cassiar serpentinite presents an excellent opportunity to study changes that occur in bastites associated with chrysotile asbestos deposits. Several textural types of orthopyroxene bastites have been observed together with the mineralogical changes. The textural types, in order of increasing degree of recrystallization, have been named uniform, domainal, patchy, shaggy, and indistinct (Table 5). The boundaries between textural types are not sharp, and a given bastite may exhibit more than one texture.

An uniform texture bastite is optically homogeneous and consists of only lizardite $1T$ (Fig. 21a), which can be either α or γ serpentine (defined in Table 5). Uniform lizardite bastites can be found in any serpentinite (Wicks and Whittaker, 1977). At Cassiar they occur in the SW corner of the serpentinite outside the ore zone. Their occurence is coincident with the pseudomorphic textures after olivine. Domainal texture is characterized by zones of optically distinct serpentine that are not controlled by fractures. The recrystallization appears to have begun at the edge of the bastite and proceeded inward (Fig. 21b). The recrystallized bastite consists of lizardite $1T$, but its strong parallel alignment is lost and a more random orientation has been developed. Bastites that have been cut by intergranular veins or fractures may contain a band of serpentine parallel to the edge of the fracture that is optically distinct from the rest of the bastite. In effect, these fractures subdivide the bastite and form circumferential bands of alteration around the smaller grains. Domainal textures occur around the ore zone.

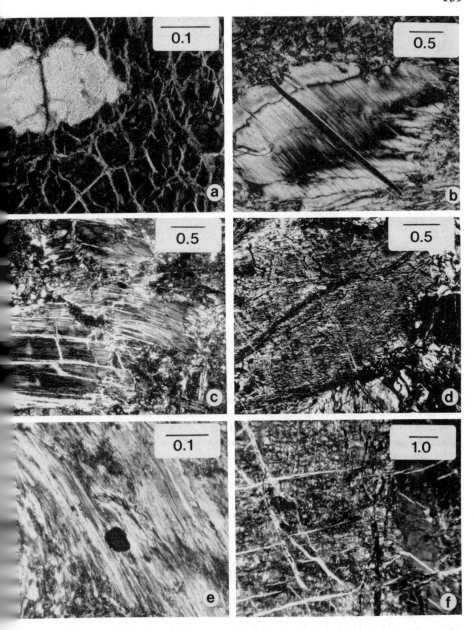

Figure 21. Photomicrographs of serpentine textures. Scale bars in mm are shown on the photomicrographs. (a) Lizardite $1T$ uniform bastite in lizardite $1T$ mesh rim with lizardite $1T$ hourglass centers. (b) Lizardite $1T$ domainal bastite in lizardite $1T$ hourglass texture. (c) Lizardite $1T$ and chrysotile $2M_{c1}$ patchy bastite in lizardite $1T$ and antigorite interlocking texture. (d) Lizardite $1T$, chrysotile $2M_{c1}$, and antigorite shaggy bastite in lizardite $1T$ and antigorite interlocking texture. (e) Antigorite and chlorite indistinct bastite. (f) Magnetite-free asbestos veins (horizontal) cutting magnetite-bearing asbestos veins (vertical) in lizardite $1T$ interlocking texture.

Table 5. Bastite textures and mineralogy.

Texture	Optical Characteristics*	Mineralogy
Uniform	Length-fast or length-slow Serpentine	lizardite
Domainal	Length-fast rims surrounding length-slow cores or Length-slow rims surrounding length-fast cores	lizardite
Patchy	Length-slow veins surrounding length-fast matrix	lizardite to chrysotile
Shaggy	Same as patchy	lizardite and chrysotile appearance of antigorite
Indistinct	Length-slow serpentine	antigorite, chrysotile, lizardite

*Sign of elongation when dominant cleavage (uniform, domainal, patchy) or lineation (shaggy, indistinct) is parallel to the slow ray of the accessory plate.

Patchy textures result from alteration of the bastites along intragranular fractures (Fig. 21c). This transition marks the formation of chrysotile $2M_{c1}$ from lizardite $1T$ without the formation of any other phase. Chrysotile bastites have been found only in asbestos deposits (Wicks and Plant, 1979). Patchy textures also occur around the ore zone.

Shaggy textures are dominated by lineations, altered boundaries (Fig. 21d), and sometimes the appearance of optically identifiable antigorite; the microbeam camera results indicate that both lizardite $1T$ and chrysotile $2M_{c1}$ are still present. The bastite is no longer a discrete grain but a patch of serpentine with a birefringence distinct from the matrix. Indistinct bastites contain antigorite as well and are no longer separate from the matrix (Fig. 21e). One may not recognize it as a bastite when first viewing it in the thin section. Shaggy and indistinct texture bastites occur in the ore zone and in the hangingwall alteration. Chlorite, magnetite, and brucite may be associated with these bastites.

Chlorite is associated with the ore zone and with the major faults. Its development is not related to the extent of bastite alteration nor to the type of serpentinization. Chlorite occurs in samples that have been recrystallized as well as in samples that have not. It is present in domainal bastites associated with lizardite hourglass textures as well as with indistinct bastites in non-pseudomorphic textures. Chlorite at Cassiar appears to be a prograde phase. It is crosscut by lizardite veins and by magnetite-free chrysotile asbestos veins.

TRANSITIONS AMONG SERPENTINE MINERALS AND TEXTURES

Lizardite is by far the most common serpentine mineral associated with retrograde serpentinization of peridotite (Wicks and Whittaker,

1977). The onset of serpentinization is due to the reaction olivine = lizardite $\pm$ brucite $\pm$ (magnetite and native metals). The presence or absence of brucite is related to the modal ratio of olivine to pyroxene (Hostetler et al., 1966) and may be related to the degree of recrystallization. This reaction invariably produces pseudomorphic mesh rims with olivine mesh centers in the early stages and randomly oriented lizardite $\pm$ brucite mesh centers in late stages.

Antigorite, formed after olivine, chlorite, and tremolite, occurs in retrograde alteration of ultramafic intrusions such as that found at Glen Urquhart (Wicks, 1984a) and the Hammett Grove meta-igneous complex in South Carolina (Mittwede et al., 1987). In this situation antigorite formation is one of a sequence of retrograde reactions (Thayer, 1967). The retrograde formation of antigorite pseudomorphs after olivine has been documented in two localities: the Trinity thrust sheet in California (Peacock, 1987) and a Precambrian sill in the Yukon Territories (Wicks and Whittaker, 1977; Wicks and Plant, 1979).

The transition from lizardite pseudomorphic texture to antigorite $\pm$ chlorite interpenetrating texture is common in prograde metamorphism (Springer, 1974; Pinsent and Hurst, 1977; Ikin and Harmon, 1983). Chrysotile has been found to be a common antigorite precursor in the Malenco serpentinite in the Italian Alps (Mellini et al., 1987). The transition from a lizardite pseudomorphic texture, to a very fine-grained transition texture, to a chrysotile interlocking texture and from chrysotile serrate veins to chrysotile-antigorite interlocking texture is characteristic of chrysotile asbestos deposits, as demonstrated in the Cassiar mine serpentinite. Antigorite is usually a minor phase in these deposits, although in several chrysotile asbestos deposits both slip- and cross-fiber chrysotile veins dissect an antigorite interpenetrating texture (Chidester et al., 1978; Compagnoni et al., 1980; Skarpelis and Dabitzias, 1987).

The transition from chrysotile-antigorite interlocking texture to antigorite $\pm$ brucite interpenetrating texture has been documented in prograde regimes in the Swiss Alps (Peretti et al., 1987). Mellini et al., (1987) showed that a correlation exists between the wavelength of the antigorite superstructure and temperature, as predicted by Kunze (1961) and by Trommsdorff (1983). As the superstructure period decreases with increasing grade, MgO and H_2O are expelled from the antigorite structure and form brucite. This process is accompanied by a coarsening of the antigorite, resulting in the change from interlocking to interpenetrating texture.

The sequence of serpentinization events in the Cassiar serpentinite helps define the mineral and textural transitions. Chrysotile asbestos veins postdate the development of some lizardite pseudomorphic textures. The reaction is lizardite = chrysotile + magnetite (Laurent, 1975). Recrystallization of early-formed asbestos veins accompanied by the development of other lizardite pseudomorphic textures suggests that lizardite was stablized again at the expense of chrysotile. In other areas, but within the same time span, we observe magnetite-bearing chrysotile veins cross-cutting lizardite pseudomorphic textures recrystallized to chrysotile. Thus, in one area of the serpentinite lizardite is replacing chrysotile while in another area chrysotile is forming from lizardite. In addition, chrysotile and antigorite occur within some interlocking textures. Textural relationships in various parts of the serpentinite indicate that the reaction lizardite = antigorite + chlorite is reversed.

The orthopyroxene bastites at Cassiar demonstrate similar mineralogical transitions. Thus serpentine prograde textures can be viewed as a continuum from retrograde lizardite pseudomorphic texture, to very fine-grained transitional texture of lizardite and chrysotile, to chrysotile-antigorite interlocking texture, and finally to antigorite interpenetrating texture. Not all serpentinites exhibit the same sequence of textural changes as the Cassiar body. Inferences regarding serpentine reactions based on the mineralogical and textural diversity of the Cassiar body suggest that Cassiar can be used as a model by which serpentinites with a lower mineralogical and textural diversity can be deciphered.

STABLE ISOTOPES

The fractionation of stable isotopes amongst the serpentine minerals has recently been reviewed by Kyser (1987) and need not be reviewed again here. Investigation of the geologic structure and the serpentine mineralogy of the Cassiar mine has presented the opportunity to analyse the stable isotopes from a well characterized serpentinite.

Changes in serpentine mineralogy and texture can result from three factors: a change in fluid composition, a change in fluid pressure, or a change in temperature. These changes may occur simultaneously in some instances, obscuring the relationship between cause and effect. One goal of investigations into the origin of serpentine is to decipher which factor is associated with a particular change in mineralogy and texture. Changes in mineralogy are accompanied by changes in texture. Analyses of hydrogen and oxygen isotopes in serpentine minerals indicate that the serpentine minerals lizardite and chrysotile can be distinguished from antigorite by their isotopic signature (Figs. 22 and 23) and that antigorite forms at a higher temperature than either lizardite or chrysotile (Wenner and Taylor, 1971). Given the difference between the isotopic signature of lizardite, chrysotile, and antigorite, there should also be a distinction between different textures as well. This is indeed the case. Partly serpentinized peridotite is invariably enriched in ^{18}O relative to unaltered peridotite (Margaritz and Taylor, 1974; Ikin and Harmon, 1983). Completely serpentinized peridotites are depleted in deuterium relative to partially altered peridotite because deuterium fractionates into the fluid phase with respect to serpentine (Kyser, 1987).

143

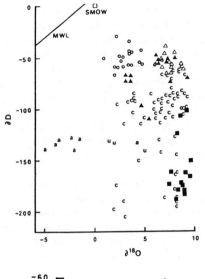

Figure 22. δD – δ^{18}O diagram for serpentinites, labeled by environment of formation. Oceanic (O), continental (C), S.E. Alaska zones intrusions (a), Urals (u), metamorphic antigorite (open triangles), chrysotile from greenstone belts (solid triangles), and samples from Cassiar serpentinite (squares). Data from: Wenner and Taylor (1974), Margaritz and Taylor (1974), Kyser (1987), and O'Hanley, Kyser, and Wicks (unpub. data).

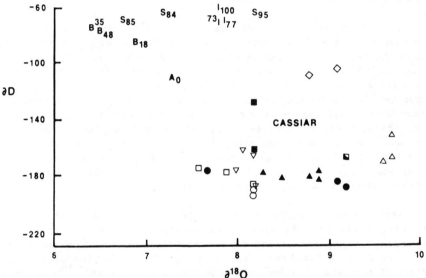

Figure 23. δD – δ^{18}O diagram for serpentinites from the Highland Border Suite (Ikin and Harmon, 1983), and the Cassiar serpentinite (Wenner and Taylor, 1974, O'Hanley, Kyser, and Wicks, unpub. data). For the Highland Border Suite samples the localities are: Aleyth (A), Balmaha (B), Innellan (I), and Scalpsie Bay (S). The subscript indicates the percent antigorite present in the sample. For Cassiar, the textures are: early-formed hourglass (open triangles), late-formed hourglass (open squares), antigorite-lizardite interlocking texture (solid squares), antigorite-lizardite-chrysotile interlocking texture (solid triangles), asbestos vein (solid circles), antigorite interpenetrating texture in hangingwall alteration (open diamonds). From Wenner and Taylor (1974): lizardite-chrysotile matrix (open inverted triangles), asbestos vein (open circles).

Oxygen and hydrogen stable isotope studies of the Cassiar serpentinite (O'Hanley, Kyser, and Wicks, unpub. data) point out further distinctions. The low deuterium values of the Cassiar serpentinite indicate that only meteoric water was involved in serpentinization (Fig. 23). The transition from one lizardite hourglass texture to another in the Cassiar serpentinite produced a rock depleted in both D and ^{18}O.

The magnitude of the change can only be accomplished by the introduction of another fluid. The replacement of the later lizardite hourglass by antigorite as a fault is approached has produced enrichment in both D and ^{18}O. These results suggest a slight rise in temperature. They also suggest that the fault was a fluid conduit. The transition from lizardite to chrysotile + antigorite $\pm$ chlorite interlocking textures occurs on the side of the fault opposite from the textures described above. This transition is different from the lizardite to antigorite transition described above. The discontinuity in serpentine mineralogy, texture, and isotope values across the fault indicates that movement occurred on the fault after the formation of the textures.

The hangingwall alteration is enriched in deuterium with respect to the rest of the serpentinite and suggests that it formed at a higher temperature than the mesh textures it replaces. This is consistent with the presence of antigorite, quartz, zoisite, and tremolite in the alteration (Gabrielse, 1963). The isotope values for the asbestos veins are similiar to the matrix they cut and this suggests that both were formed from the same fluid (Figs. 23 and 24).

The isotope ratios between chrysotile asbestos veins and the matrix they cut are complex and appear to reflect the differing mechanisms by which the relationship can be brought about (Fig. 24). Peacock (1986) found a difference in δD of 40% between a chrysotile vein and an antigorite matrix in the Trinity peridotite, California. This is accounted for by having chrysotile precipitate from a fluid (meteoric water) that is different from the metamorphic fluid that produced the antigorite. Smaller changes such as those found at Cassiar appear to indicate that the same fluid produced the chrysotile veins and the host rock. Thus, chrysotile may or may not be derived from the local host rock. Each case must be evaluated individually until a sufficient database is obtained for more general conclusions.

Ikin and Harmon (1983) documented the isotopic change accompanying the transition from lizardite pseudomorphic texture to antigorite interpenetrating texture in serpentinites from the Highland Border Fracture Zone in Scotland. There is a good correlation between the extent of recrystallization and heavier values for δD and $\delta^{18}O$ but this relationship breaks down when the different localities are treated separately (Fig. 23). For example, at Scalpsie Bay ^{18}O increases with the amount of antigorite present but at Balmaha ^{18}O decreases. However, in both instances the transition from lizardite to antigorite is due to prograde metamorphism. Deuterium appears to be the best indicator of temperature change.

The distribution of serpentine textures in alpine-type serpentinites is fault controlled. Changes in mineralogy, texture, and stable isotope values indicate that faults represent fluid conduits. The work at Cassiar indicates that the isotope values for samples from a particular alpine-type serpentinite can vary greatly. This implies that no one sample can be taken to represent the serpentinite as a whole.

145

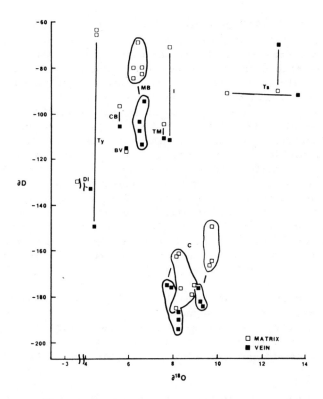

Figure 24. δD - δ^{18}O diagram for asbestos vein-matrix associations. localities are: Baie
Verte, Newfoundland (BV), Belvedere Mt., Vermont (MB), Carrity Burn, Scotland (CB),
Cassiar, British Columbia (C), Duke Island, Alaska (DI), Innellan, Scotland (I) Trinity
thrust sheet, California (Ty), and Troodos, Cyprus (Ts). Data taken from: Wenner and
Taylor (1974), Margaritz and Taylor (1974), Ikin and Harmon (1983), Peacock (1986), Kyser
(1987), and O'Hanley, Kyser, and Wicks (unpub. data).

The spatial variation in serpentine mineralogy, texture, and
isotopic values suggests that these samples reflect gradients in
pressure and temperature. If this premise is correct then an observed
sequence of serpentine mineral assemblages should correspond to a path
through the serpentine phase diagram. O'Hanley et al., (1988) have shown
that this is the case.

SERPENTINE PHASE DIAGRAM

The serpentine phase diagram has been of interest since the classic
debate between Bowen (1927) and Hess (1938) on the existence of a low
temperature primary peridotite magma and on the emplacement mechanism
for alpine peridotites. One result of this debate is that experimental
work on serpentine equilibria has focussed on the serpentine-out
reactions involving either lizardite, chrysotile, or antigorite with
various combinations of brucite, forsterite, olivine, talc, chlorite,
and water (Chernosky, 1975; Evans et al., 1976; Moody, 1976; Caruso and
Chernosky, 1979; Chernosky, 1982).

Evans et al. (1976) presented a phase diagram for a three-component ($MgO-SiO_2-H_2O$; MSH), six-phase system for chrysotile and antigorite with brucite, forsterite, talc, and water. Their model was based on field observations and on experimental reversals of the reactions antigorite + brucite = forsterite + water and antigorite = forsterite + talc + water. Caruso and Chernosky (1979) presented a different phase diagram for a four-component ($MgO-Al_2O_3-SiO_2-H_2O$: MASH), seven-phase system for an aluminian lizardite and an aluminum-free antigorite with brucite, forsterite, talc, clinochlore, and water. Their model was based on experimental reversals of the reaction lizardite (solid solution) = forsterite + talc + clinochlore + water for two lizardites of distinct aluminum content. The relationship between these two models was not clear because of the different bulk chemistry and of the different phases used in each model. The uncertainty remained until the use of dual networks by O'Hanley (1987a) permitted n-component, (n+4)-phase diagrams to be constructed readily. Now all three serpentine minerals can be included in one model along with other relevant phases (O'Hanley, 1987b; O'Hanley et al., 1988).

Serpentinization can be modelled in MSH because of the magnesium-rich composition of the serpentine protolith. O'Hanley (1987b) presented a model using MSH that included brucite, forsterite, lizardite, chrysotile, antigorite, enstatite, talc, and water. In this model lizardite and chrysotile are polymorphs. Figure 25 is taken from that model.

The presence of clinochlore in serpentinites and the observation of a chlorite-in reaction at Blue River in British Columbia (Pinsent and Hirst, 1976) suggests that Al_2O_3 may be important to serpentine stability. Caruso and Chernosky (1979) demonstrated that aluminum can have a dramatic effect on the thermal stability of lizardite. O'Hanley et al. (1988) constructed a model for MASH using brucite, forsterite, Al-lizardite, chrysotile, antigorite, talc, clinochlore, and water. Figure 26 is taken from that model.

None of the reactions involving two serpentine minerals have been calibrated experimentally. Linear programming has been used to optimize data obtained from calorimetry and from phase equilibrium studies (Berman et al., 1986). As a result the P-T coordinates of reaction chrysotile = antigorite + brucite are known. Otherwise, Figures 25 and 26 illustrate the topologic relationships among the phases only.

$MgO-SiO_2-H_2O$ system

Lizardite and chrysotile are polymorphs in MSH because they both can be synthesized in this system. In addition, microprobe investigations indicate that there is no consistent partitioning of elements between lizardite and chrysotile (Frost, 1975; Dungan, 1979; Wicks and Plant, 1979). There is indirect evidence of a role for aluminum in this capacity, but this point and the thermodynamic data available for minerals in this system are described by Chernosky et al.

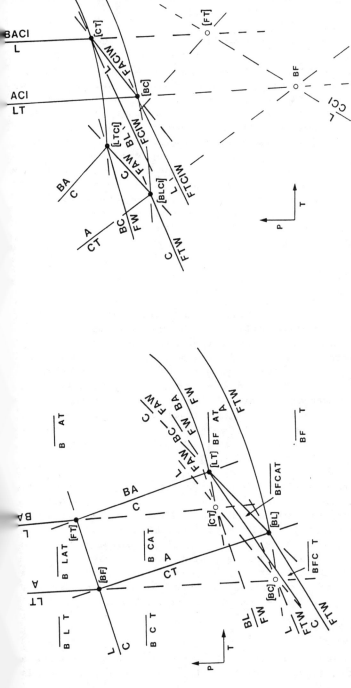

Figure 25. Orientated phase diagram for MSH system involving the phases brucite (B), forsterite (F), lizardite (L), chrysotile (C), antigorite (A), talc (T), and water (W). Stable and metastable invariant points are represented by solid and open dots respectively. Invariant points are labelled using the missing-phase convention. Stable, metastable, and doubly metastable univariant reactions are represented by solid, long-dash, and short-dash lines respectively. Divariant fields are represented by lines with divariant assemblages indicated by letters underneath the lines. Reaction L = C and invariant points [LT] and [BL] are stable. Reaction L = F + T + W and invariant points [CT] and [BC] are metastable. Stability field for chrysotile limited by reactions between invariant points [BF], [FT], [LT], and [BL]. See text for discussion.

Figure 26. Orientated phase diagram for MASH system involving the same phases and conventions as Figure 22, except that clinochlore (Cl); (L) is aluminian lizardite; and divariant fields are represented by triangles with divariant assemblages indicated by letters around the triangle. Reaction L = C + Cl and invariant points [LT] and [BL] are metastable. Reaction L = F + T + Cl + W and invariant points [CT] and [BC] are stable. See text for discussion.

in Chapter 9. The data indicate that lizardite is the high-P and low-T mineral, whereas chrysotile is the low-P, low-T mineral. Stable isotope data, petrographic and field observations, and phase equilibrium experiments all suggest that antigorite is the high-P, high-T mineral (Wenner and Taylor, 1974; Evans 1977; Berman et al., 1986).

Recent observations suggest that lizardite has a stability field (O'Hanley et al., 1988). In the Cassiar serpentinite lizardite has been observed replacing chrysotile asbestos veins, and it persists through the changes in lizardite hourglass textures as they recrystallized into a lizardite interlocking texture. In addition a lizardite hourglass texture and a lizardite interlocking texture are replaced by antigorite-chlorite interlocking texture. Lizardite also fills veins that formed after the antigorite-chlorite assemblage. Thus, lizardite forms after chrysotile and antigorite as well as being replaced by them. Observations of this type have now been documented for each serpentine mineral, so that it appears that each has a stability field in P-T space. A phase diagram that incorporates all of the information about the relationships between the serpentine minerals mentioned above is shown in Figure 25. Based on the calculations of Berman et al. (1986) invariant point [LT] is located below 500 bars and below 200°C.

Serpentine-out reactions

The antigorite-out reaction during prograde metamorphism is antigorite = forsterite + talc + water. Enstatite is never observed, which is consistent with its stability at very high temperatures (Berman et al., 1986). This implies that serpentine reactions involving enstatite are metastable. Frost (1975) has observed the persistence of lizardite microshears 30 meters beyond the antigorite-out reaction antigorite = forsterite + talc + water at Paddy-Go-Easy Pass, Washington, U.S.A. Microbeam studies of these microshears indicate that lizardite does persist beyond the antigorite-out reaction and that when it breaks down, lizardite alters to antigorite as well as to forsterite, talc, and chlorite, indicating that at least two reactions are present within these 30 meters (O'Hanley et al., 1988).

The stable forsterite-in reaction has been considered to be brucite + antigorite = forsterite + water (Evans et al., 1976). Recent work in the Malenco serpentinite suggests that the forsterite-in reaction may be the metastable reaction chrysotile = forsterite + antigorite + water (Peretti et al., 1987). This is not the first time a metastable reaction has been proposed for the serpentinites. A study of thermal metamorphism of the Blue River serpentinite in north-central British Columbia suggests that the serpentine-out reaction is spinel + serpentine + brucite = chlorite (Pinsent and Hirst, 1976).

$MgO-Al_2O_3-SiO_2-H_2O$ system

Microprobe data indicate that lizardite can accommodate a significant amount of Al_2O_3, although usually less than 1 wt % Al_2O_3 is present (Frost, 1975; Wicks and Plant, 1979; Dungan, 1979; Caruso and Chernosky, 1979). Petrographic observations and phase-equilibrium data indicate that chlorite is the stable Al-bearing phase formed from the breakdown of aluminian lizardite (Caruso and Chernosky, 1979). A model that incorporates chlorite into the serpentine phase diagram is shown in Figure 26. Lizardite is still the high-pressure mineral.

Caruso and Chernosky (1979) reversed the equilibrium lizardite = forsterite + talc + clinochlore + water for a lizardite containing 9 wt % Al_2O_3. This reaction intersects the reaction antigorite = forsterite + talc + water at 700°C, 35 kbar for $a(H_2O) = 1$ (Chernosky et al., in Chap. 9) and defines invariant point [BC]. Chernosky (1975) reversed the same lizardite reaction for a lizardite containing 3.5 wt % Al_2O_3. This reaction intersects the reaction antigorite = forsterite + talc + water at 550°C, 5 kbar for $a(H_2O) = 1$. These two reactions indicate that the stability of lizardite is a function of its Al-content and that invariant point [BC] moves to lower pressures and temperatures as the Al-content of lizardite decreases. In addition, if the reaction antigorite = forsterite + talc + water is stable, then the reaction aluminian lizardite = forsterite + talc + clinochlore + water must be stable as well.

$FeO-Fe_2O_3-MgO-Al_2O_3-SiO_2-H_2O$ system

The behavior of Fe in serpentine does not appear to be amenable to the type of generalization just used for aluminum. However, there are patterns in the behavior of Fe that are associated with specific sequences of serpentine minerals and textures. Cogulu and Laurent (1984) found that the transition from lizardite pseudomorphic texture to chrysotile interlocking texture to chrysotile asbestos vein was marked by a continuous decrease in the Fe content of chrysotile and the production of magnetite. The partitioning of Fe amongst minerals in the reaction antigorite + brucite = forsterite + water does appear to be a function of temperature in the Malenco serpentinite (Peretti et al., 1987). In contrast to these observations tabulations of microprobe data from pseudomorphic textures indicate that lizardite can be either richer or poorer in iron with respect to chrysotile (Wicks and Plant, 1979). The presence of certain sulfides, oxides, and native metals can be correlated to the changes in serpentine mineralogy (Eckstrand, 1975; Frost, 1985) and perhaps to texture. Frost (1985) explained this pattern by appealing to a systematic increase in $f(H_2O)$ and $f(S)$ that is associated with the change from olivine-bearing serpentinites to carbonate serpentinite to talc carbonate. This suggests that certain values for the fugacities of oxygen and sulfur may also be related to specific serpentine assemblages (Fig. 27). One important point made by Frost (1985) is that the fugacities of sulfur and oxygen may be locally controlled because both are minor fluid components. Thus, changes in the oxide and sulfide mineralogy may be out-of-step with the changes in serpentine mineralogy. This appears to be the case at the Balangero mine in which several opaque assemblages are present that together indicate inconsistent fugacities (Rossetti et al., 1987).

150

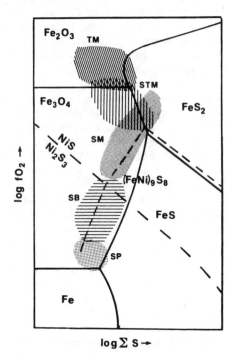

Figure 27. log fO_2 – log ΣS diagram showing the stability fields for major oxide and sulfide minerals and associated silicate or carbonate minerals. The shaded sections represent partially serpentinitized peridotite (SP), serpentinite ± brucite (SB), serpentine-magnesite (SM), serpentine-talc-magnesite (STM), and talc-magnesite (TM). Modified from Frost (1985) after Eckstrand (1975).

Other components

    Calcium is not soluble in serpentine and usually leaves the serpentinite to form rodingites (Coleman, 1977). If it remains in the serpentinite it does so by stabilizing either tremolite or diopside depending on temperature (Evans, 1977). Chromium is a minor component occurring in bastites after pyroxenes (Wicks and Plant, 1979) and in antigorite or chlorite associated with the breakdown of chromite (Bliss and MacLean, 1975). Titanium can be important on a limited scale, such as in titanian berthierine after spinel in the Picton ultramafic dike in Ontario (Arima et al., 1985) or in the occurrence of carlosturanite in the Balangero mine, Italy (Compagnoni et al., 1985). However, Ti is not an essential component of carlosturanite (Mellini and Zussman, 1987). The Ni content of olivine tends to be higher than that of the lizardite alteration product (Wicks and Plant, 1979). This is due to the low water fugacity during the serpentinization of olivine, which favors partitioning of Ni into oxides, sulfides, or alloys rather than silicates. However, during the formation of laterite, Ni rather than Mg preferentially partitions into serpentine (Golightly and Aranciba, 1979). The behavior of Fe, Ni, and Co also is related to the fugacities of sulfur and oxygen which determine the occurrence of sulfides and native alloys (Frost, 1985; Shiga, 1987).

Discussion of phase diagrams

The topology shown in Figure 25 does not agree with that in Figure 26. The topology in Figure 25 is valid for MSH whereas the topology in Figure 26 is valid for high-Al lizardites. That is, aluminian lizardite and chrysotile are not polymorphs if lizardite contains more than 3.5 wt % Al_2O_3. The existence of two possible topologies for lizardite with different amounts of Al indicates that lizardite can be classified according to its Al-content. Thus, low-Al lizardite contains less than 1 wt % Al_2O_3, intermediate-Al lizardite contains 1 to 3.5 wt % Al_2O_3, and high-Al lizardite contains greater than 3.5 wt % Al_2O_3.

Figure 26 is similiar to Figure 25. Note that in Figure 25 the positions of the lizardite-missing invariant points [LT] and [BL] relative to one another are the same as in Figure 26. The positions of the other invariant points are consistent relative to one another, but [CT] and [BC] occur stably at higher temperatures in Figure 26. The reason for this change is that the reactions involving lizardite shift towards higher temperatures with increasing amounts of Al. The lizardite-missing reactions will not shift at all, however, so that invariant point [FT], for example, will shift to lower pressure along reaction chrysotile = antigorite + brucite as lizardite = chrysotile + clinochlore and lizardite = antigorite + brucite + clinochlore shift to higher temperatures. With greater amounts of Al in lizardite [FT] moves past [LT], rendering [LT] metastable. Reaction lizardite = antigorite + brucite + clinochlore now intersects reaction antigorite + brucite = forsterite + water stabilizing [CT]. [CT] moved to higher temperatures along reaction antigorite + brucite = forsterite + water due to the effect of increasing amounts of Al in lizardite, which shifted the reactions lizardite = antigorite + brucite + clinochlore and lizardite = forsterite + antigorite + clinochlore + water towards higher temperatures.

There is still uncertainty with respect to intermediate-Al lizardites. Figure 28a illustrates a topology for a five-phase system brucite, lizardite, chrysotile, antigorite, and talc in MSH. If aluminian lizardite is substituted for lizardite and clinochlore is added as the other Al-bearing phase, one obtains Figure 28b, which is based on a six-phase system in MASH. Either Figure 28a or 28b may be appropriate for intermediate-Al lizardites, but we do not know at this time. As these lizardites are associated with bastites, observations on the relationship between lizardite, chrysotile, and clinochlore in patchy and shaggy bastites should solve this problem.

The geometry of these diagrams (Figs. 25, 26 and 28) will depend on the compositions of the various minerals and the activity of water. A comparison of the phase diagrams in Figures 25 and 26 indicates that for low-Al lizardite, lizardite, antigorite, and chrysotile each have their own region of stability in P-T space (Fig. 25). Antigorite is the high-temperature serpentine mineral, chrysotile is the low-temperature, low-pressure mineral, and lizardite is the low-temperature, high-pressure

152

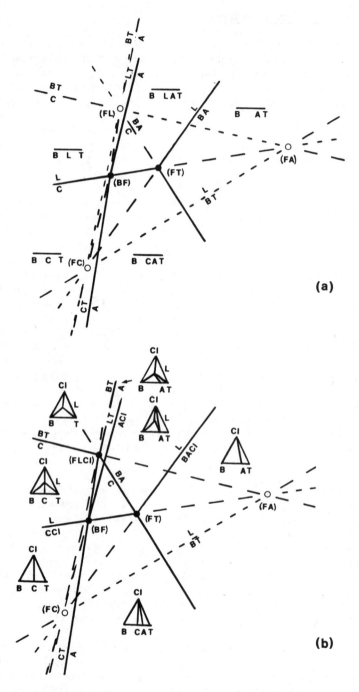

Figure 28. Orientated phase diagrams for low-Al lizardite (see Fig. 25 for symbolism).
(a) is for the reaction L = C in MSH. (b) represents a possible topology for the reaction
L = C + Cl and is obtained from (a) by adding aluminian lizardite for lizardite and adding
clinochlore as the other Al-bearing phase. See text for discussion.

mineral. For the high-Al lizardite phase diagram (Fig. 26), lizardite is stable to a higher temperature than antigorite below [BC] but antigorite is stable to a higher temperature than lizardite above [BC]. The stability field of chrysotile is the same as that in Figure 25. The relationships for intermediate-Al lizardite are not resolved.

The reaction lizardite = forsterite + talc + water $\pm$ clinochlore is located at low pressures in both diagrams. This reaction is metastable for low-Al lizardite but is stable for high-Al lizardite. This limit is based on the reaction studied by Chernosky (1975), and the actual limit is probably lower. Clearly the limits of the chrysotile stability field will depend on the location of the reaction lizardite = chrysotile $\pm$ clinochlore. The higher the Al-content of lizardite, the lower the pressure at which this reaction occurs. Thus the pressures and temperatures at which the reactions are located could be specific to a restricted range of composition. Knowledge of the P-T co-ordinates of the reaction lizardite = chrysotile $\pm$ clinochlore is also critical in deciding whether or not lizardite is stable in retrograde serpentinization because it controls the stability field of lizardite. If the reaction lizardite = chrysotile $\pm$ clinochlore is located at pressures above those of invariant points [CT] and [BC] (Fig. 25), it is a stable reaction and lizardite is metastable in retrograde serpentinization. If it is located at pressures below these invariant points, the reaction is metastable and lizardite is stable. Microprobe data from pseudomorphic textures indicate that they contain low-Al lizardite so that lizardite is probably metastable.

Despite this dependence on bulk composition we do know how the reactions change as these parameters change (see Chap. 9). As a result we can use the diagrams shown in Figures 25 and 26 in a qualitative manner to define paths characteristic of different sequences of serpentine mineral assemblages. The stability regions can be defined by the invariant points at which the reactions that limit the stability of each serpentine mineral intersect, and they provide a basis with which to describe the various paths of serpentinization that are recognised by petrographic observations and microbeam X-ray studies.

Importance of water pressure

Two conditions need to be considered in the discussion of the role of H_2O pressure in serpentinization. The first condition is H_2O pressure, $P(H_2O)$, less than geostatic pressure $P(g)$. This would pertain to H_2O under a hydrostatic head. The second condition is $P(H_2O)$ less than the fluid pressure $P(f)$. This condition would occur if more than one volatile is present, if solutes are present in the water or if the water is contained in the discontinuous pore structure of a sealed rock (Sanford, 1981).

Bruton and Helgeson (1983) have investigated the consequences of varying $P(H_2O)$ and $P(g)$. This condition assumes the presence of a distinct water fluid phase. If $P(H_2O) < P(g)$ then depth becomes an important parameter. The different densities of water and rock will affect the rate at which $P(H_2O)$ and $P(g)$ increase with depth, such that at any given depth $P(H_2O < P(g)$ until a depth is reached at which $P(H_2O) = P(g)$. The condition $P(H_2O) < P(g)$ does not affect the general geometry of the phase diagram because the $a(H_2O) = 1$. However, for $P(H_2O) < P(g)$ a reaction such as chrysotile = forsterite + talc + water would plot at a lower temperature for a given depth than that for $P(H_2O) = P(g)$ because the reaction has a positive slope. In contrast, the reaction chrysotile = brucite + antigorite will plot at a higher temperature at a given depth for $P(H_2O) < P(g)$ than for $P(H_2O) = P(g)$ because the reaction has a negative slope (Bruton and Helgeson, 1983).

Sanford (1981) has discussed the second condition $P(H_2O) < P(f)$ in which the $a(H_2O) < 1$. There are two variations to this condition. In the first the presence of other solvents such as CO_2 or S reduce the activity of water. In the second water is present as a discontinuous phase in the pore structure of the rock, which also reduces the activity of water. Both of these variations will affect the geometry of the phase diagram (see Chap. 9). The reactions that conserve vapor, such as chrysotile = antigorite + brucite, will not be affected by the change in the activity of water, but reactions that do not conserve vapor will shift to lower temperatures at a given pressure. Thus the invariant points will shift to higher pressures along the vapor-conserving reactions. Reactions such as chrysotile = forsterite + antigorite + water and chrysotile = forsterite + talc + water that were stable at less than 500 bars for $P(H_2O) = P(f)$ $(a(H_2O) = 1)$ are stabilized to higher pressure for $P(H_2O) < P(f)$ $(a(H_2O) < 1)$. The condition $P(H_2O) < P(f)$ is not important if CO_2 and S metasomatism are excluded from the discussion. The serpentine minerals are very sensitive to the presence of some solvents in the fluid. For example the presence of very little CO_2 is sufficient to stabilize a carbonate mineral.

The condition $P(H_2O) = P(g)$ and $a(H_2O) < 1$ could be very important in explaining partially serpentinized peridotites because a lack of water is a major factor in stopping serpentinization. The condition $P(H_2O) < P(g)$ does have implications for the development of serpentine minerals and textures. According to Wenner and Taylor (1974) lizardite and chrysotile form at temperatures less than 250°C. If $P(H_2O) < P(g)$, then for thermal gradients of 30°–100°C/km, this temperature would correspond to a fluid pressure of 800 bars at 7.5 km or 300 bars at 2.5 km for a fluid density of 1 g/cm^3. However, if $P(H_2O) = P(g)$, for pressure gradients of 200–300 bars/km with the same thermal gradients the pressures change to 6500–7500 bars at 7.5 km and 1800–2600 bars at 2.5 km for rock density of 2.5 g/cm^3 (Bruton and Helgeson, 1983).

Based on the calculations of Bruton and Helgeson (1983) and the discussion of Sanford (1981), the condition $P(H_2O) < P(g)$ can be used to explain the formation of pseudomorphic textures. The rate of serpentinization is controlled by the rate at which water gains access to the serpentinite. Both the temperature and the water pressure would be low because the rocks are relatively shallow and the water is under a hydrostatic head. The first serpentine reaction encountered is either chrysotile = forsterite + talc + water, chrysotile = forsterite + antigorite + water, lizardite = forsterite + talc + water, or lizardite = forsterite + antigorite + water, depending upon which invariant points are stable. The hydration reaction consumes water and buffers the water pressure at the reaction interface. Lizardite will coexist with olivine at the reaction front which migrates into the peridotite, yielding pseudomorphic textures.

Types of serpentinization

The eight types of serpentinization described by Wicks and Whittaker (1977) can be related to the serpentine phase diagrams by associating transitions between serpentine mineralogy and textures to reactions on the phase diagrams (Table 3). The eight types can be reduced to four in relation to the phase diagrams for the following reasons (O'Hanley et al., 1988). Firstly, penetrative deformation will certainly influence the texture developed, but in most cases the deviatoric stress does not appear to exert a great influence on mineralogy and thus does not need to be considered separately. Secondly, the presence of antigorite represents a specific region of P-T space rather than a problem of nucleation so that its appearance represents a change in the intensive variables rather than a kinetic effect. Thus, the paths through P-T space represented by transitions between serpentine textures depend on temperature and the relationship between water pressure and lithostatic pressure. The types of serpentinization of Wicks and Whittaker (1977) are re-defined as regimes in the right-hand column of Table 3 and are referred to by letters.

Regime A, serpentinization with decreasing temperature, occurs at pressures and temperatures above invariant point [LT] (Fig. 25) or [CT] (Fig. 26). This type of retrograde serpentinization, during which the hydration of olivine produces antigorite, is not common although it has been reported in the Glen Urquhart serpentinite (Wicks 1984a), the Hemmet Grove meta-igneous suite (Mittwede et al., 1987) and Trinity peridotite (Peacock, 1987) and has been inferred for the Fox River Sill in Manitoba (Wicks and Whittaker, 1977). This regime appears to be characterized by $P(H_2O) = P(g)$.

The retrograde regime B is located below invariant point [BL] (Fig. 25) or [BC] (Fig. 26) for $P(H_2O) < P(g)$, as discussed above. The alteration of forsterite and enstatite to lizardite occurs at low water pressure, and pseudomorphic textures are produced. The alteration of forsterite to lizardite is probably metastable because there is no consistent different between the chemistry of lizardite and chrysotile in pseudomorphic textures. The metastability could be due to the suppression of the forsterite to chrysotile reactions at low water pressures. The reaction of lizardite to chrysotile occurs once the influence of the olivine and pyroxene crystal structures is gone.

The prograde regime C, characterized by the reaction lizardite = chrysotile $\pm$ clinochlore, occurs below invariant points [BF] and [FT] when serpentinization is complete. Water pressure must not increase above invariant points [BF] and/or [FT] or chrysotile reacts to form lizardite. Regime C is characteristic of chrysotile asbestos deposits (Wicks and Whittaker, 1977).

At temperatures and pressures above invariant point [BL] or [BC] but below invariant point [LT] or [CT], lizardite reacts to produce antigorite and forsterite in regime D. This reaction has been inferred for the observations made by Frost (1975) at Paddy-Go-Easy Pass. X-ray microbeam films indicate the presence of antigorite after lizardite, which suggests that the reaction is lizardite = antigorite + forsterite + clinochlore + water rather than lizardite = forsterite + talc + clinochlore + water as inferred by Frost (1975). The stability of these reactions is not clear because the antigorite contains up to 3 wt % Al_2O_3. The common sequence, in regime D, occurs above invariant point [LT] (Fig. 25) or [CT] (Fig. 26) with increasing temperature (Wicks and Whittaker, 1977), in which lizardite reacts to produce antigorite which reacts to produce forsterite. This regime appears to be characterized by $P(H_2O) = P(g)$ as well.

THE ROLE OF FRACTURES AND DEFORMATION

The distinction between fracturing in a static regime and fracturing in a dynamic regime is important in serpentinization. A static regime is characterized by the lack of a penetrative fabric in the hand sample or thin section, and a dynamic regime is characterized by a penetrative fabric. In addition, a static regime is inferred for the isotropic, grid-like framework of fractures that surround olivine and pyroxene fragments and form the site for the initial lizardite alteration that eventually forms pseudomorphic textures. The grid-like framework is associated with the onset of serpentinization. MacDonald (1984), based on his water-diffusion experiments, concluded that this geometry is compatible with a water-diffusion controlled process. This process is closer to chemical corrosion rather than hydraulic fracturing. The large scale, continuous lizardite veins that have extensive alteration along their margin represent conduits by which the fluid gained access to the serpentinite and to the grid-like framework of fractures to form the pseudomorphic textures.

Chrysotile asbestos veins form a strong contrast to the grid-like framework of mesh rims and lizardite feeder veins because they have a preferred orientation. Different generations of chrysotile veins can be recognized in thin section and the veins of each generation have the same orientation across the thin section (Fig. 21f). In outcrop, asbestos veins can maintain their orientation for 5-10 meters. As the chrysotile asbestos veins show consistent orientations mesoscopically as well as in thin section, they must form in a dynamic regime.

Chrysotile veins do not form a conductivity network for fluids associated with serpentinization. Vein margins are very sharp, indicating that the fluid in the fracture did not alter the wall rock. The mineralogy of asbestos veins often reflects the rock that is being fractured. An asbestos vein is more likely to contain magnetite if the wall rock does, which suggests that the material comprising the vein minerals is locally derived (Cogulu and Laurent, 1984; Laubscher, 1986a, b). Nevertheless, the characteristics of the vein-wall interface of chrysotile veins is in marked distinction to large scale lizardite veins along which wall rock reaction, especially with bastites, is extensive.

Chrysotile asbestos veins may follow earlier fractures, but in most cases they occupy large scale fractures that cross-cut both pseudomorphic and non-pseudomorphic textures. The fractures and mesh rims of pseudomorphic textures are sealed and obliterated by the recrystallization that produces non-pseudomorphic textures. This interpretation is consistent with the observations of Raleigh and Paterson (1965) that serpentinites with non-pseudomorphic textures are stronger than those with pseudomorphic textures.

CHRYSOTILE ASBESTOS DEPOSITS

Chrysotile asbestos deposits can be grouped into six types based on the origin of the serpentine protolith and the mode of occurrence of the asbestos fiber (Table 6). The largest and most thoroughly studied deposits consist of cross-fiber chrysotile asbestos with the fiber oriented at a high angle to the vein wall. Slip-fiber deposits in which the fiber is oriented at a low angle to the vein wall occur less frequently (Mumpton and Thompson, 1975; Skarpelis and Dabitzias, 1987). Chrysotile serrate veins can develop to such an extent that they form a significant percentage of the rock and constitute a potential source of ore called mass-fiber asbestos. However, milling problems prevent its exploitation on a large scale.

Ross (1981) made a general review of asbestos deposits. Several papers describing the occurrence of chrysotile asbestos deposits in Southern Africa were published very recently (Anhaeusser, 1986; Barton, 1986; Laubscher, 1986a,b; Menell et al., 1986; Voight et al., 1986). These deposits are hosted by Archean ultramafic sills and thermally altered limestones. The mesoscopic characteristics of these cross-fiber desposits are similar to those of Phanerozoic cross-fiber deposits.

In all deposits the role of mechanical deformation is extremely important in that it controls the location and the orientation of the fractures that contain the fiber. Laubscher (1986a,b) invoked wrench faulting to rationalize the fault and talc-alteration patterns associated with Precambrian peridotite-hosted cross-fiber deposits, exclusive of those found in the Great Dyke. The asbestos veins at the Slade-Forbes deposit in the Abitibi Belt of Ontario appear to have

158

Table 6. Classification of chrysotile-asbestos deposits.

Lithology		Fiber Type	Example
Alpine-type	tectonized peridotites	cross-fiber	Cassiar, B.C.
peridotites		slip-fiber	Coalinga, California
Phanerozoic		mass-fiber	Asbestos, Quebec
Differentiated	sills	cross-fiber	Great Dike, Zimbabwe
Precambrian	komatiitic flows	cross-fiber	Abitibi Belt, Ontario
Serpentinized		cross-fiber	Globe, Arizona
limestones both			
Phanerozoic &			
Precambrian			

formed during folding (O'Hanley et al., 1988), an idea that has been presented before for other deposits (Anhaeusser, 1976).

Cross-fiber asbestos veins can be modelled successfully as tension fractures, and as such their kinematic interpretation should be consistent with that obtained from other kinematic indicators that formed at the same time (Butt and Glen, 1981; Laubscher, 1986a,b; O'Hanley, 1988). Studies done at the Woodsreef deposit in New South Wales (Butt and Glen, 1981), at the Cassiar deposit in British Columbia (O'Hanley and Wicks, 1987), and at several deposits in Black Lake, Quebec (O'Hanley, 1987c) indicate that this is the case.

The origin of the asbestos fiber is a problem distinct from that of the origin of the fractures (O'Hanley, 1987d). Ideas on the origin of the asbestos fiber fall into two groups: that the fiber fills space created by the opening of fractures (e.g., Cooke, 1936), or that the fiber replaces pre-existing serpentine minerals (e.g., Riordan, 1955). At present it is recognized that both fracture-filling and replacement fiber occurs, and that fracture-filling fiber is the dominant source of ore (Cogulu and Laurent, 1984). Laurent (1975) thinks that asbestos veins result from some variant of what is now called the crack-seal model (Ramsay, 1980), in which fracturing and mineralization proceeds in increments. Laurent (1975), based on work done in the Eastern Townships of Quebec, suggests that the orientation of the fiber contains information on the displacement path associated with the deformation, and this forms part of the basis on which cross-cutting relationships are determined. This is not the case with replacement fiber, serrate veins, nor fracture-filling fiber in which the orientation of the fiber is controlled by the substrate (O'Hanley, 1987c; Laubscher, 1986a,b).

Cross-cutting relationships among the chrysotile-asbestos veins indicate that they formed in a definite sequence. This information used in conjunction with the Laurent model for vein formation allows us to reconstruct partially the stress state at the time of vein formation. Detailed studies in progress on the cross-cutting relationships among asbestos veins in three mountain belts, in northern British Columbia,

Quebec, and New South Wales, show that the asbestos veins formed during two stages that occur before and after a change in the orientation of the regional stress field.

Studies completed to date indicate that the formation of a Phanerozoic, cross-fiber chrysotile-asbestos deposit within an alpine peridotite requires a specific set of conditions (O'Hanley 1988). The serpentinite must be massive; as long as a schistose fabric is not present, mineralogy is not important. The serpentinite must be in P-T-X conditions such that chrysotile is stable when the change in stress orientation occurs. Deformation must cease before the fiber is altered either by the development of a schistose fabric or by a change in temperature that destroys chrysotile and produces antigorite. Also chemical alteration, due to infiltration of silica, carbonate, or sulfur-bearing fluids, must not occur.

ACKNOWLEDGEMENTS

We thank Dr. Joseph V. Chernosky, Jr., University of Maine; Dr. Michael E. Fleet, University of Western Ontario; Dr. B. Ronald Frost, University of Wyoming and Dr. Frank C. Hawthorne, University of Manitoba for reviewing the manuscript thoroughly at short notice. Cassiar Mining Corporation kindly provided support for the study of their mine.

REFERENCES

Anderson, C.S. and Bailey, S.W. (1981) A new cation ordering pattern in amesite-$2H_2$. Am. Mineral. 66, 185-195.
Anhaeusser, C.R. (1976) The nature of chrysotile asbestos occurrences in Southern Africa: A review. Econ. Geol. 71, 96-116.
----- (1986) The geologic setting of chrysotile asbestos occurences in southern Africa. In C. R. Anhaeusser and S. Maske, eds., Mineral deposits of Southern Africa V. 2, 359-375, Geol. Soc. Johannesburg, Johannesburg, South Africa.
Arima, M., Fleet, E.E. and Barnett, R.L (1985) Titanian berthierine: A Ti-rich serpentine group mineral from the Picton ultramafic dyke, Ontario. Can. Mineral. 23, 213-220.
Aruja, E. (1945) An X-ray study of the crystal structure of antigorite. Mineral. Mag. 27, 65-74.
Bailey, S.W. (1969) Polytypism of trioctahedral 1:1 layer silicates. Clays and Clay Minerals 17, 355-371.
----- (1988) X-ray diffraction identification of the polytypes of mica, serpentine, and chlorite. Clays and Clay Minerals 36, 193-213.
----- and Tyler, S A (1960) Clay minerals associated with the Lake Superior iron ores. Econ. Geol. 55, 150-175.
Barton, C.M. (1986) The Havelock asbestos deposit in Swaziland, Barberton greenstone belt. In C.R. Anhaeusser and S. Maske, eds., Mineral deposits of Southern Africa V. 2, 395-407, Geol. Soc. Johannesburg, Johannesburg, South Africa.
Bates, T.F., Sand, L.B. and Mink, J.F. (1950) Tubular crystals of chrysotile asbestos. Sci. 3, 512.

160

Berman, R.G., Engi, M., Greenwood, H.J. and Brown, T.H. (1986) Derivation of internally-consistent thermodynamic data by the technique of mathematical programming: A review with application to the system $MgO-SiO_2-H_2O$. J. Petrol. 27, 1331-1364.

Bish, D.L. (1981) Distortions in the lizardite structure: A distance least-squares study, (Abstr.) EOS 62, 417.

Bliss, N.W. and MacLean, W.H. (1975) The paragenesis of zoned chromite from Central Manitoba. Geochim. Cosmochim. Acta 39, 973-990.

Bowen, N.L. (1927) The origin of ultrabasic and related rock. Am. J. Sci. 14, 89-108.

Brindley, G.W. and von Knorring, O. (1954) A new variety of antigorite (ortho-antigorite) from Unst, Shetland Islands. Am. Mineral 39, 794-804.

Bruton, C.J. and Helgeson, H.C. (1983) Calculation of the chemical and the thermodynamic consequences of differences between fluid and geostatic pressure in hydrothermal systems. Am. J. Sci. 283-A, 540-588.

Caruso, L. and Chernosky, J.V., Jr. (1979) The stability of lizardite. Can. Mineral. 17, 757-769.

Chapman, J.A. and Zussman, J. (1959) Further electron optical observations on crystals of antigorite. Acta Cryst. 12, 550-552.

Chernosky, J.V. Jr. (1975) Aggregate refractive indices and unit-cell parameters of synthetic serpentine in the system $MgO-Al_2O_3-SiO_2-H_2O$. Am. Mineral. 60, 200-208.

----- (1982) The stability of clinochrysotile. Can. Mineral. 20, 19-27.

Chidester, A.H., Albee, A.L. and Cady, W.M. (1978) Petrology, structure and genesis of the asbestos bearing ultramafic rocks of the Belvidere Mountain area in Vermont. U. S. Geol. Survey, Prof. Paper 1016, 83 pp.

Coats, C.J.A. (1968) Serpentine minerals from Manitoba. Can. Mineral. 9, 321-347.

Cogulu, C. and Laurent, R. (1984) Mineralogical and chemical variations in chrysotile veins and peridotite host rocks from the asbestos belt of southeastern Quebec. Can. Mineral. 22, 173-183.

Coleman, R.G. (1977) Ophiolites; ancient oceanic lithosphere. Springer-Verlag, N.Y., 229 pp.

Compagnoni, R., Ferraris, G. and Fiora, L. (1983) Balangeroite, a new fibrous silicate related to gageite from Balangero, Italy. Am. Mineral. 68, 214-219.

-----, ----- and Mellini M. (1985) Carlosturanite, a new asbestiform rock-forming silicate from Val Varaita, Italy. Am. Mineral. 70, 767-772.

-----, Sandrone, R. and Zucchetti, S. (1980) Some remarks on the asbestos occurrences in the western Alps with special reference to the chrysotile asbestos deposit of Belangero (Lanzo Valley, Piedmont, Italy. Proc. 4th Int'l. Conf. Asbestos, Torino, Italy, 49-71.

Cooke, H.C. (1936) Asbestos deposits of Thetford district, Canada. Econ. Geol. 31, 355-376.

Cressey, B.A. (1979) Electron microscope studies of serpentine textures. Can. Mineral. 17, 741-756.

----- and Zussman, J. (1976) Electron microscopic studies of serpentinites. Can. Mineral. 14, 307-313.

Deer, W.A., Howie, R.A. and Zussman J. (1962) Rock forming minerals. Vol. 3. Sheet Silicates, Longmans, London.

Dungan, M.A. (1979b) A microprobe study of antigorite and some serpentine pseudomorphs. Can. Mineral. 17, 711-784.

161

Eckstrand, O.R. (1975) The Dumont Serpentinite: A model for control of Nickeliferous opaque mineral assemblages by alteration reactions in ultramafic rocks. Econ. Geol. 70, 183–201.

Evans, B.W. (1977) Metamorphism of alpine peridotite and serpentinite. Ann. Rev. Earth Planet. Sci. 5, 397–447.

-----, Johannes, W., Oterdoom, H. and Trommsdorff, V. (1976) Stability of chrysotile and antigorite in the serpentine multisystem. Schweiz. Mineral. Petrog. Mitt. 56, 79–93.

Faust, G.T. and Fahey, J.J. (1962) The serpentine group minerals. U. S. Geol. Surv. Prof. Paper 384A, 1–92.

Ferraris, G., Mellini, M. and Merlino, S. (1986) Polysomatism and the classification of minerals. Rend. Soc. Ital. Mineral. Petrol. 41, 181–192.

Frost, B.R. (1975) Contact metamorphism of serpentinite, chloritic blackwall and rodingite at Paddy-Go-Easy Pass, Central Cascades, Washington. J. Petrol. 16, 272–313.

----- (1985) On the stability of sulfides, oxides, and native metals in serpentinite. J. Petrology, 26, 31–63.

Gabrielse, H. (1960) The genesis of chrysotile asbestos in the Cassiar asbestos deposit, northern British Columbia. Econ. Geol. 55, 327–337.

----- (1963) McDame map area, British Columbia. Geol. Survey Can. Memoir 319, 13, 96–138.

Gillery, F.H. (1959) The X-Ray study of synthetic Mg-Al serpentines and chlorites. Am. Mineral. 44, 143–152.

Glen, R.A. and Butt, B.C. (1981) Chrysotile asbestos at Woodsreef, New South Wales. Econ. Geol. 76, 1153–1169.

Golightly, J.P. and Arancibia, O.N. (1979) The chemical composition and infrared spectrum of nickel-and iron-substituted serpentine from a nickeliferous laterite profile, Soroako, Indonesia. Can. Mineral. 17, 719–728.

Gruner, J.W. (1944) The kaolinite structure of amesite $(OH)_8(Mg_1Fe)_4Al_2(Si_2Al_2)O_{10}$ and additional data on chlorites. Am. Mineral. 29, 422–430.

Guggenheim, S., Bailey, S.W. Eggleton, R.A. and Wilkes, P. (1982) Structural aspects of greenalite and related minerals. Can. Mineral. 20, 1–18.

Hall, S.H. and Bailey S.W. (1976) Amesite from Antarctica. Am. Mineral. 61, 497–499.

-----, Guggenheim, S., Moore, P. and Bailey, S.W. (1976) The structure of Unst-type 6-layer serpentines. Can. Mineral. 14, 314–321.

Harms, T. (1986) Structural and tectonic analysis of the Sylvester allochton, northern British Columbia: implications for paleogeography and accretion. Ph.D. thesis, University of Arizona, 80 pp.

Hess, M.M. (1938) A primary peridotite magma. Am. J. Sci. 35, 321–344.

Hostetler, P.B., Coleman, R.G., Mumpton, F.A. and Evans, B.W. (1966) Brucite in alpine serpentinites. Am. Mineral. 51, 75–98.

Ikin, N.P. and Harmon, R. S. (1983) A stable isotope study of serpentinization and metamorphism in the Highland Border Suite, Scotland, U.K. Geochim. Cosmochim. Acta 47, 153–167.

Jagodzinski, H. and Kunze, G. (1954a) Die Rollchen-struktur des Chrysotils. I Allgemeine Beugungstheorie und Kleinwinkelstreuung. N. Jahrb. Mineral. Mh, 95–108.

----- and ----- (1954b) Die Rollchenstruktur des Chrysotils. II Weitwinkelinterferenzen. N. Jahrb. Mineral. Mh., 113–130.

----- and ----- (1954c) Die Rollchenstruktur des Chrysotils. III Versetzungswachstum der Rollchen. N. Jahrb. Mineral. Mh., 137–150.

162

Jahanbagloo, I.C. and Zoltai, T. (1968) The crystal structure of a hexagonal Al-serpentine. Am. Mineral. 53, 14-24.

Jasmund K. and Sylla, H.M. (1971) Synthesis of Mg and Ni-antigorite. Contrib. Mineral. Petrol. 34, 84-86.

----- and ----- (1972) Synthesis of Mg and Ni-antigorite: A correction. Contrib. Mineral. Petrol. 34, 346.

Krstanovic, I. (1968) Crystal structure of single-layer lizardite. Zeit. Krist. 126, 163-169.

----- and Pavlovic, S. (1964) X-ray study of chrysotile. Am. Mineral. 49, 1769-1771.

----- and ----- (1967) X-ray study of 6-layer ortho-serpentine. Am. Mineral. 52, 871-876.

Kunze, G. (1956) Die gewellte Struktur des Antigorits, I. Zeit. Krist. 108, 82-107.

----- (1958) Die gewellte Struktur des Antigorits, II. Zeit. Krist. 110, 282-320.

----- (1959) Fehlordnungen des Antigorits. Zeit. Krist. 111, 190-212.

----- (1961) Antigorit. Strukturtheoretische Grundlagen und ihre praktische Bedeutung fur die weiters Serpentin-Forschung. Fortschr. Mineral. 39, 206-324.

Kyser, T.K. (1987) Equilibrium fractionation factors for stable isotopes. Chap. 1 in Stable Isotope Geochemistry of Low Temperature Fluids. T.K. Kyer, Ed., Mineral. Assoc. Can. Short Course 13, 1-84.

Laubscher, D.H. (1986a) Chrysotile asbestos in the Zvishavane (Shabani) and Mashava (Mashaba) areas, Zimbabwe. In C.R. Anhaeusser and S. Maske, eds., Mineral Deposits of Southern Africa Vol. 2, 377-393, Geol. Soc. Johannesburg, Johannesburg, South Africa.

----- (1986b) The new Amianthus chrysotile asbestos deposit, Kaapsehoop Barberton Greenstone belt. In C.R. Anhaeusser and S. Maske, eds., Mineral deposits of Southern Africa Vol. 2, 421-426, Geol. Soc. Johannesburg, Johannesburg, South Africa.

Laurent, R. (1975a) Occurrences and origin of the ophiolites of Southern Quebec, Northern Appalachians. Can. J. Earth Sci. 12, 443-455.

----- (1975b) Petrology of the alpine-type serpentinites of Asbestos and Thetford Mines, Quebec. Schweiz. Mineral. Petrog. Mitt. 55, 431-455.

Livi, K.J.T. and Veblen, D.R. (1987) "Eastonite" from eastern Pennsylvania: A mixture of phlogopite and a new form of serpentine. Am. Mineral. 72, 113-125.

MacDonald, A.H. (1984) Water-diffusion rates through serpentinized peridotites. Implications for reaction induced and chemical effects in ultramafic rocks. Ph.D. Thesis. Univ. Western Ontario, 226 pp.

Margaritz, M and Taylor, M.P., Jr. (1974) Oxygen and hydrogen isotope studies of serpentinization in the Troodos Ophiolite Complex, Cyprus. Earth Planet. Sci. Letters 23, 8-14.

Meier, W.M. and Villiger, G. (1969) A Fortran computer programme for the least-squares refinement of interatomic distances, Program Manual. Institut. Kristallogr. Eidg. Petrogr. Tech. Hochschule, Zurich, Switzerland.

Mellini, M. (1982) The crystal structure of lizardite $1T$: hydrogen bonds and polytypism. Am. Mineral. 67, 587-598

----- (1986) Chrysotile and polygonal serpentine from the Balangero serpentinite. Mineral. Mag. 50, 301-306.

-----, Ferraris, G. and Compagnoni, R. (1985) Carlosturanite: HRTEM evidence of a polysomatic series including serpentine. Am. Mineral. 70, 773-781.

-----, Trommsdorff, V. and Compagnoni, R. (1987) Antigorite polysomatism: behavior during progressive metamorphism. Contrib. Mineral. Petrol. 97, 147-155.

----- and Zanazzi, P.F. (1987) Crystal structures of lizardite-1T and lizardite-2H_1 from Coli, Italy. Am. Mineral. 72, 943-948.

----- and Zussman, J. (1986) Carlosturanite (not 'picrolite') from Taberg, Sweden. Mineral. Mag. 50, 675-679.

Menell, R.P., Brewer, T.H., Delve, J.R. and Anhaeusser, C.R. (1986) The Kalkloof chrysotile asbestos deposit and surrounding area, Barberton mountain land. In C. R. Anhaeusser and S. Maske, eds., Mineral Deposits of Southern Africa V. 2, 427-435, Geol. Soc. Johannesburg, Johannesburg, South Africa.

Midgley, H.G. (1951) A serpentine mineral from Kennack Cove, Lizard, Cornwall. Mineral. Mag. 29, 526-530.

Middleton, A.P. and Whittaker, E.J.W. (1976) The structure of Povlen-type chrysotile. Can. Mineral. 14, 301-306.

Mitchell, R.H. and Putnis, A. (1988) Polygonal serpentine in segregation-textured kimberlite. Can. Mineral. 26,

Mittwede, S., Idegard, M. and Sharp, W.E. (1987) Major chemical characteristics of the Hemmett Grove meta-igneous suite, northwestern South Carolina. Southeastern Geol. 28, 49-63.

Moody, J. (1976) An experimental study on the serpentinization of iron-bearing olivines. Can. Mineral. 14, 462-478.

Müller, P. (1963) 6-layer serpentin vom Piz Lunghim, Maloya, Schweiz. N. Jahrb. Mineral. Abh. 100, 101-111.

Mumpton, F.A. and Thompson, C.S. (1975) Mineralogy and origin of Coalinga asbestos deposit. Clays and Clay Minerals 23, 131-143.

Nelson, B.W. and Roy, R. (1958) Synthesis of the chlorites and their structure and chemical constitution. Am. Mineral. 43, 707-725.

Noll, W. and Kircher, H. (1951) Uber die Morphologie von Asbesten und ihren Zusammenhang mit der Kristallstruktur. N. Jahrb. Mineral. Mh. 10, 219-240.

O'Hanley, D.S. (1987a) The construction of phase diagrams by means of dual networks. Can. Mineral. 25, 105-119.

----- (1987b) A chemographic analysis of magnesian serpentinites using dual networks. Can. Mineral. 25, 121-133.

----- (1987c) Deformation-free growth of NaCl cross-fiber veins: Implications for syntectonic growth. North-Central region G.S.A., Geol. Soc. Am. Abstr. Program 19, #4, 237.

----- (1987d) The origin of the chrysotile asbestos veins in southeastern Quebec. Can. J. Earth Sci. 24, 1-9.

----- (1988) The origin of alpine peridotite hosted cross fiber chrysotile asbestos deposits. Econ. Geol. 83, 256-265.

-----, Chernosky, J.V. Jr. and Wicks, F.J. (1988) The stability of lizardite and chrysotile. Can. Mineral. 26, part 4.

----- and Wicks, F.J. (1987) Structural control of serpentine textures in the Cassiar Mining Corporation's open-pit mine at Cassiar, British Columbia. Geol. Assoc. Can.-Mineral. Assoc. Can. Joint Annual Meeting Program Abstr. 12, 77.

-----, Schandl, E.S. and Wicks, F.J. (1988) Time relationships between alteration and deformation at the Slade-Forbes asbestos deposit, Deloro Township, Ontario. Geol. Assoc. Can./Mineral. Assoc. Can./Can. Soc. Petroleum Geol. Program Abstr. 13, A92.

Olsen, E.J. (1961) Six-layer ortho-hexagonal serpentine from the Labrador Trough. Am. Mineral. 46, 434-438.

Pauling, L. (1930) The structure of the chlorites. Proc. Nat. Acad. Sci. USA 16, 578-582.

164

Peacock, S. (1987) Serpentinization and infiltration metasomatism of the Trinity peridotite, Klamath province, northern California. Implications for subduction zones. Contr. Mineral. Petrol. 95, 55–70.

Peacor, D.R., Essene, E.J., Simmons, W.B. and Bigelow, W.C. (1974) Kellyite, a new Mn-Al member of the serpentine group from Bald Knob, North Carolina, and new data on grovesite. Am. Mineral. 59, 1153–1156

Peretti, A., Trommsdorff, V. and Frost, B.R. (1987) Reactions governing the appearance of olivine in the Malenco serpentinite, Northern Italy. Geol. Soc. Am. Abstr. Programs 19, #7, p. 802.

Peterson, R.C., Hill, R.J., and Gibbs, G.V. (1979) A molecular-orbital study of distortions in the layer structures brucite, gibbsite and serpentine. Can. Mineral. 17, 703–711.

Pinsent, R.H. and Hirst, D.M. (1977) The metamorphism of the Blue River Ultramafic Body, Cassiar, British Columbia, Canada. J. Petrology 18(4), 567–594.

Raleigh, C.B. and Paterson, M.S. (1965) Experimental deformation of serpentinite and its tectonic implications. J. Geophys. Res. 70, 3965–3985.

Ramsay, John G. (1980) The crack-seal mechanism of rock deformation. Nature 284, 135–139.

Riordon, P.H. (1955) The genesis of asbestos in ultrabasic rocks. Econ. Geol. 50, 67–83.

Ross, M. (1981) The geologic occurrences and health hazards of amphibole and serpentine asbestos. Chap. 6 in Reviews in Mineralogy Vol. 9A. Amphiboles and other hydrous pyriboles-mineralogy (D.R. Veblen, ed.). Mineral. Soc. Am., 279–323.

Rossetti, P., Zucchetti, S. and Frost, B.R. (1987) Assemblages with native metals from serpentinites of the Balangero mine, Italian Western Alps. Geol. Soc. Am. Abstr. Programs 19, #7, p. 825.

Roy, D.M. and Roy, R. (1954) Synthesis and stability of minerals in the system $MgO-Al_2O_3-SiO_2-H_2O$. College of Mineral Industries No. 53-60, 147–178

Rozenson, I., Bauminger, E.R., and Heller-Kallai, L. (1979) Mössbauer spectra of iron in 1:1 phyllosilicates. Am. Mineral. 64, 893–901.

Rucklidge, J.C. and Zussman, J. (1965) The crystal structure of the serpentine mineral, Lizardite $Mg_3Si_2O_5(OH)_4$. Acta Cryst. 19, 381–389.

Sanford, R.F. (1981) Mineralogical and chemical effects of hydration reactions and applications of serpentinization. Am. Mineral. 66, 290–297.

Shannon, R.D. (1976) Revised effective ionic radii and systematic studies of interatomic distances in halides and chalcogenides. Acta Cryst. A32, 751–767.

Shiga, Y. (1987) Behavior of iron, nickel, cobalt and sulfur during serpentinization with reference to the Hayachine ultramaic rocks of the Kamaishi mining district, northeastern Japan. Can. Mineral. 25, 611–624.

Shirozu, H.S. and Momoi, H. (1972) Synthetic Mg-chlorite in relation to natural chlorite. Mineral. J. (Japan) 6, 464–476.

Skarpelis, N. and Dabitzias, S. (1987) The chrysotile asbestos deposit at Zidani, northern Greece. Ofioliti 12, 403–410.

Spinnler, G.E. (1985) HRTEM study of antigorite, pyroxene-serpentine reactions and chlorite. Ph.D. thesis, Arizona State Univ., Tempe, Arizona.

Springer, R.K. (1974) Contact metamorphosed ultramafic rocks in the western Sierra Nevada foothills, California. J. Petrol. 15, 160–195.

165

Steadman, R. (1964) The structure of the trioctahedral kaolinite-type silicates. Acta Cryst. 17, 924-927.

────── and Nuttall, P.M. (1962) The crystal structure of amesite. Acta Cryst. 15, 510-511.

────── and ────── (1964) Further polymorphism in cronstedtite. Acta Cryst. 17, 404-406.

Taner, M.F. and Laurent, R. (1984) Iron-rich amesite from the Lake Asbestos Mine, Black Lake, Quebec. Can. Mineral. 22, 437-442.

Thayer, T.P. (1967) Serpentinization considered as a constant volume metasomatic process: A reply. Am. Mineral. 52, 549-553.

Thompson, J.B., (1978) Biopyriboles and polysomatic series. Am. Mineral. 63, 239-249.

Toman, K. and Frueh, A.J. (1968a) Diffraction of X-rays by the faulted cylindrical lattice of chrysotile. I. Numerical computation of diffraction profiles. Acta Cryst. A24, 364-373.

────── and ────── (1968b) Diffraction of X-rays by the faulted cylindrical lattice of chrysotile. II. The form, position and width of some diffraction profiles. Acta Cryst. A24, 374-379.

Trommsdorff, V. (1983) Metamorphose magnesiumreicher gesteine: Kritischer Vergleich von Natur, Experiment und modynamischer Datenbasis. Fortschr. Mineral. 61, 283-308.

Uehara, S. and Shirozu, H. (1985) Variations in chemical composition and structural properties of antigorites. Mineral. J. (Japan) 12, 299-318.

Veblen, D.R. (1980) Anthophyllite asbestos: microstructures, intergrown sheet silicates, and mechanisms of fiber function. Am. Mineral. 65, 1075-1086.

────── and Buseck, P.R. (1979) Serpentine minerals: intergrowths and new combination structures. Science 206, 1398-1400.

────── and ────── (1981) Hydrous pyriboles and sheet silicates in pyroxene and uralite: intergrowth microstructures and reaction mechanisms. Am. Mineral. 66, 1107-1134.

Voight, J.C., Buttner, W. and Schaum, H.H. (1986) Chrysotile asbestos at the Msuali mine, Barberton Greenstone belt. In C.R. Anhaeusser and S. Maske, eds., Mineral Deposits of Southern Africa V. 2, 409-420, Geol. Soc. Johannesburg, Johannesburg, South Africa.

Wei, L. and Shaoying, J. (1984) Discovery and its significance of Povlen-type hydrochrysotile. Acta Mineral. Sinica 2, 136-142 (in Chinese).

Wenner, D.B. and Taylor, H.P., Jr. (1974) O/H and O^{18}/O^{16} studies of serpentinization of ultramafic rocks. Geochim. Cosmochim. Acta 38, 1255-1286.

────── and ────── (1971) Temperatures of serpentinization of ultramafic rocks based on O^{18}/O^{16} fractionation between co-existing serpentine and magnetite. Contrib. Mineral, Petrol. 32, 165-185.

Whittaker, E.J.W. (1953) The structure of chrysotile. Acta Cryst. 6, 747-748.

────── (1954) The diffraction of X-rays by a cylincrical lattice. I. Acta Cryst. 7, 827-832.

────── (1955a) The diffraction of X-rays by a cylindrical lattice. II. Acta Cryst. 8, 261-264.

────── (1955b) The Diffraction of X-rays by a cylindrical lattice. III. Acta Cryst. 8, 265-271.

────── (1955c) A classification of cylindrical lattices. Acta Cryst. 8, 571-574.

────── (1955d) The diffraction of X-rays by a cylindrical lattice. IV. Acta Cryst., 8: 726-729

────── (1956a) The structure of chrysotile. II. Clinochrysotile. Acta Cryst. 9, 855-862.

166

———— (1956b) The structure of chrysotile. III. Orthochrysotile. Acta Cryst. 9, 862–864.

———— (1956c) The structure of chrysotile. IV. Parachrysotile. Acta Cryst. 9, 865–867.

———— (1957) The structure of chrysotile. V. diffuse reflections and fibre texture. Acta Cryst. 10, 149.

———— (1963) Fine structure within the diffraction maxima from chrysotile. Acta Cryst. 16, 486–490.

———— and Middleton, A.P. (1979) The intergrowth of fibrous brucite and fibrous magnesite with chrysotile. Can. Mineral. 17, 699–702.

———— and Wicks, F.J. (1970) Chemical differences among the serpentine "polymorphs". Am. Mineral. 55, 1025–1047.

———— and Zussman, J. (1956) The characterization of serpentine minerals by X-ray diffraction. Mineral. Mag. 31, 107–126.

———— and ———— (1971) The serpentine minerals. In "The Electron-optical Investigation of Clays" (ed. by J.A. Gard) Chap. 5, Mineral. Society, London.

Wicks, F.J. (1979) Mineralogy, crystal chemistry and crystallography of chrysotile. In Ledoux, R.L., ed., Mineralogical Techniques of Asbestos Determination. Mineral. Assoc. Can. Short Course Handbook 4.

———— (1984a) Deformation histories as recorded by serpentinites. I. Deformation prior to serpentinization. Can. Mineral. 22, 185–195.

———— (1984b) Deformation histories as recorded by serpentinites. II. Deformation during and after serpentinization. Can. Mineral. 22, 197–204.

———— (1984c) Deformation histories as recorded by serpentinites. III. Fracture patterns developed prior to serpentinization. Can. Mineral. 22, 205–209.

———— (1986) Lizardite and its parent enstatite: a study by X-ray diffraction and transmission electron microscopy. Can. Mineral. 24, 775–788.

———— and Hawthorne, F.C. (1986) Distance least-squares modelling of the lizardite 1T structure. (Abstr.) Geol. Assoc. Can./ Mineral. Assoc. Can. Program Abstr. 11, 144.

———— and Plant, A.G. (1979) Electron microprobe and X-ray microbeam studies of serpentine minerals. Can. Mineral. 17, 785–830.

———— and Whittaker E.J.W. (1975) A reappraisal of the structures of the serpentine minerals. Can. Mineral. 13, 227–243.

———— and ———— (1977) Serpentine textures and serpentinization. Can. Mineral. 15, 459–488.

————, ———— and Zussman, J. (1977) An idealized model for serpentine textures after olivine. Can. Mineral. 15, 446–458.

———— and Zussman, J. (1975) Microbeam X-ray diffraction patterns of the serpentine minerals. Can. Mineral. 13, 244–258.

Yada, K. (1967) Study of chrysotile asbestos by a high resolution electron microscope. Acta Cryst. 23, 704–707.

———— (1971) Study of microstructure of chrysotile asbestos by high resolution electron microscopy. Acta Cryst. A27, 659–664.

———— (1979) Microstructures of chrysotile and antigorite by high resolution electron microscopy. Can. Mineral. 17, 679–691.

———— and Iishi, K. (1974a) Microstructures of synthetic serpentines observed by lattice imaging method. 8th Int'l. Congr. Electron Microscop. Canberra 1, 494–495.

———— and ———— (1974b) Serpentine minerals hydrothermally synthesized and their microstructures. J. Crystal Growth 24/25, 627–630.

———— and ———— (1977) Growth and microstructure of synthetic chrysotile. Am. Mineral. 62, 958–965.

167

----- and Tanji T. (1980) Direct observation of chrysotile at atomic resolution. Fourth Int'l. Conf. Asbestos, Torino, 335-346.

----- and Wei, L. (1987) Polygonal microstructures of Povlen chrysotile observed by high resolution electronmicroscopy. (Abstr.) Euroclay 87, Sevilla, Spain.

Zussman, J. (1954) Investigation of the crystal structure of antigorite. Mineral. Mag. 30, 498-512.

----- and Brindley, G.W. (1957) Serpentines with 6-layer ortho-hexagonal cells. Am. Mineral. 42, 666-670.

----- , ----- and Comer, J.J. (1957) Electron diffraction studies of serpentine minerals. Am. Mineral. 42, 133-153.

Zvyagin, B.B. (1967) Electron-Diffraction Analysis of Clay Mineral Structures. R.W. Fairbridge, ed., Plenum Press, New York.

----- , Mishchenko, K.S. and Shitov, V.A. (1966) Ordered and disordered polymorphic varieties of serpentine-type mineral and their diagnosis. Soviet Physics-Crystallogr 10: 539-546 (transl. from Kristallografiya) 10, 635-643.

Chapter 6

STRUCTURES AND COMPOSITIONS
OF OTHER TRIOCTAHEDRAL 1:1 PHYLLOSILICATES

This chapter includes those trioctahedral hydrous phyllosilicates based on planar 1:1 layers that have not been covered in Chapter 5, namely the Al-rich amesite and its compositional analogues, the Fe^{2+}-rich berthierine and its analogues, the Fe^{3+}-rich odinite, the Fe^{2+}- and Fe^{3+}-rich cronstedtite, and the Ni-rich nepouite and pecoraite. The modulated 1:1 structures of greenalite and caryopilite will be covered in Chapter 17.

AMESITE

Amesite is a rare trioctahedral layer silicate that typically occurs as white to green to pink hexagonal prisms that are elongate and tapered parallel to Z. Amesite has the maximum amount of tetrahedral and octahedral substitution of Al for Si and Mg found in 1:1 structures and is formed as a consequence of metamorphism in high Al environments. For many years the mineral was classified as an end member within the chlorite series, but Gruner (1944) showed that it is based on a 1:1 type layer. Gruner (1944) and Brindley et al. (1951) observed that amesite crystals often show streaking of $k \neq 3n$ reflections (indexed on an orthogonal C-cell) indicative of random layer displacements of $\pm b/3$. Most crystals of amesite that show more regular stacking are based on an ordered distortion of the standard $2H_2$ layer sequence, in which there are alternating interlayer shifts of $-b/3$ and $+b/3$ and alternating occupation of the I and II sets of octahedral positions in successive layers. Amesite is separated here from the aluminian lizardites of Chapter 5 because the structures are different, the presently known amesite specimens have compositions very close to the ideal end member $(Mg_2Al)(SiAl)O_5(OH)_4$ in terms of their tetrahedral Si:Al ratios, and they are separated by a compositional gap (at least in our present state of knowledge) from th aluminian lizardites.

Amesite has been well documented for four localities, namely in the emery deposit at Chester, Massachusetts, the Postmasburg Mn-ores at Gloucester, South Africa, the Saranovskoye chromite deposit in the North Urals, and the Dufek Massif in the Pensacola Mountains of Antarctica. An Fe-rich variety of amesite, in which Fe^{2+} substitutes for Mg, occurs in rodingite veins at the Lake Asbestos mine, Quebec. A Mn^{2+}-analogue, for which a different species name (kellyite) has been proposed, occurs in a manganese deposit at Bald Knob, North Carolina. Other reported occurrences of amesite have been shown to be either some other serpentine-type mineral or a 14 Å chlorite.

Steinfink and Brunton (1956) attempted the first single crystal structural refinement of amesite using a crystal from the North Urals locality. They refined the structure in the ideal space group $P6_3$ of the standard $2H_2$ polytype. Although ordering of octahedral cations cannot be determined in the ideal space group, they concluded that the cations in the tetrahedral sheet are completely disordered. They mentioned the observation of 6-fold biaxial sectors in the crystal, but attributed the biaxial nature to strain. Hall and Bailey (1976, 1979) identified this as sector twinning of truly biaxial material. Twinning of this sort plus more complex twinning is ubiquitous in all amesite crystals from the four known localities (Bailey, unpublished) and is attributed to cation ordering that lowers the symmetry from hexagonal or rhombohedral to triclinic.

Hall and Bailey (1979) excised an untwinned sector from a light green amesite-$2H_2$ crystal from the Antarctic locality and refined the structure on an orthogonal-based cell of symmetry $C1$. They found an optic angle 2V of 18°. In addition to the triclinic symmetry indicated by the X-ray diffraction intensities, they also found a slight distortion of the ideal orthogonal cell shape to monoclinic geometry with an observed crystallographic β angle of 90.27(3)°.

In subgroup symmetry $C1$ the two layers of the standard $2H_2$ polytype are no longer equivalent. The refinement by Hall and Bailey found that tetrahedral cations lying on the $pseudo$-6_3 axis are alternately Si-rich [site T(1)] and Al-rich [site T(11)] in adjacent layers. Of the three octahedra in each layer, one is smaller than the other two and is interpreted as Al-rich. These sites (Fig. 1) are M(3) in the first layer and M(11) in the second layer. The distribution of Al-rich and Mg-rich octahedra violates both the 3-fold axes within each layer and the 6_3 axes that relate one layer to the next in the ideal space group.

The mean T--O bond lengths of 1.639 and 1.729 Å for the two tetrahedra in layer 1, and 1.725 and 1.649 Å in layer 2 indicate that tetrahedral ordering of Si,Al is substantial but incomplete. Using a predictive linear equation based on the relative sizes of pure Si--O and Al--O bond lengths, they allocated Al^{IV} contents of 0.14, 0.76, 0.73, and 0.21 atoms, respectively, for the four mean bond lengths above. Octahedral cation ordering is nearly complete for the compositions involved, as indicated by four larger octahedra with mean O--O,OH bonds of 2.096, 2.087, 2.096, and 2.087 Å (indicating contents of essentially pure Mg plus a small amount of Fe) and two similar octahedra with mean M--O,OH bonds of 1.946 and 1.947 Å (slightly larger than pure Al values).

The pattern of distribution of Al-rich tetrahedra and octahedra in the structure is shown in Figure 1. In projection onto (001) the Al atoms form a slightly zigzag line parallel to true X and to one of the optical extinction directions (Fig. 1b). This is only one of several ordering patterns that are theoretically possible for the $2H_2$ structure in $C1$ symmetry. The 6-fold sector twinning is interpreted as operation of the 6_3 screw axis of the ideal structure as a twin axis in the lower symmetry to form the six ordered patterns of Figure 2, in which the optical extinction directions would be parallel and perpendicular to the projected line of ordered Al in each sector. Averaging the Al distributions in the six sectors gives an overall disordered cation distribution, which Steinfink and Brunton (1956) found by refinement when they used an entire twinned crystal to collect their intensity data.

171

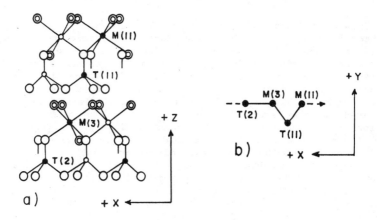

Figure 1.   (a) [010] projection of structure of amesite-$2H_2$ from Antarctica showing ordered positions of tetrahedral and octahedral Al atoms (solid circles) in two layers. (b) [001] projection shows zigzag line of Al atoms parallel to the resultant X axis.

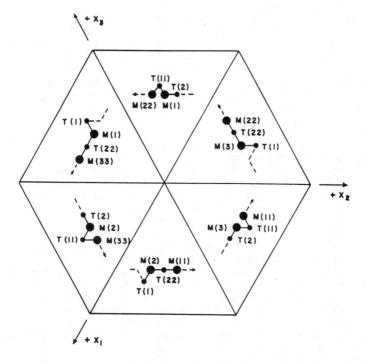

Figure 2.   Diagrammatic view of 6-fold sector twinning in amesite from Antarctica with postulated ordered positions of Al atoms, as generated by a 6_3 twin axis operating on one sector.   Arrow indicates direction of continuation of ordering pattern in the third layer along resultant -X axis in each sector.   From Hall and Bailey (1979).

The substantial substitution of Al into the tetrahedral and octahedral sheets creates an electrostatic attraction across the interlayer space in addition to the usual hydrogen bonding. Tetrahedral rotation of 14° of the basal oxygen atoms toward the adjacent OH groups combines with the attractive forces to give O--OH contact distances between layers of 2.7-2.8 Å that are appreciably shorter than for other 1:1 layer silicates. Adjacent layers also are keyed together sterically as a result of the cation ordering. Tilting of tetrahedra to enable shorter O_{apical}--O_{apical} lateral edges around the ordered small octahedral Al sites causes a downward buckling of one of the bridging basal oxygen atoms, which is paired with a similarly depressed OH group. This steric keying effect due to ordering is believed to be a positive factor in influencing the regularity of stacking of layers in amesite and in its resultant stability.

Anderson and Bailey (1981) refined the structure of an untwinned sector from a pink amesite crystal from the North Urals locality in order to resolve conflicting evidence as to the order or disorder of its cations. On one hand observed biaxial optics and complex twinning suggested ordering, but on the other hand IR spectra had been interpreted (Serna et al., 1977) as indicative of disorder. The resulting structural refinement showed nearly complete ordering of both tetrahedral and octahedral cations but arranged in a different pattern than for the Antarctic specimen. Instead of being aligned in a zigzag line parallel to X in projection on (001), as in Figure 1b, the sites of the ordered tetrahedral and octahedral Al in the amesite from North Urals form a spiral around the Z axis (Fig. 3). This pattern explains the more complex nature of the twinning, some of which is polysynthetic.

The Urals amesite also is based on the standard $2H_2$ polytype, reduced to triclinic symmetry and a unit cell quite similar in shape to that of the specimen from Antarctica but showing a small deviation of both crystallographic β and α from 90°. The reduction to triclinic symmetry for this ordering pattern is due to the distribution of the ordered octahedral Mg and Al, because the tetrahedral Si and Al ordering preserves the identity of the 6_3 screw axis so far as the tetrahedral compositions are concerned. There is a steric keying together of adjacent layers in the Urals specimen, similar to that in the Antarctic specimen but involving different oxygen and OH atoms because of the different ordering pattern. Anderson and Bailey (1981) showed that shifts of the octahedral and tetrahedral Al cations from their ideal positions and small offsets of adjacent layers affect the magnitudes of the triclinic β and α angles. The layer offsets are determined by a combination of (1) the ordered substitution of Al^{3+} for Mg^{2+} in the octahedral sites, which causes the H^+ protons of the surface OH groups to be repelled from the Al-rich sites, plus (2) the orientation of the corrugation of the plane of surface OH groups. Both of these factors influence the details of the geometry of the interlayer hydrogen bonding system, and thus the exact positions of adjacent layers.

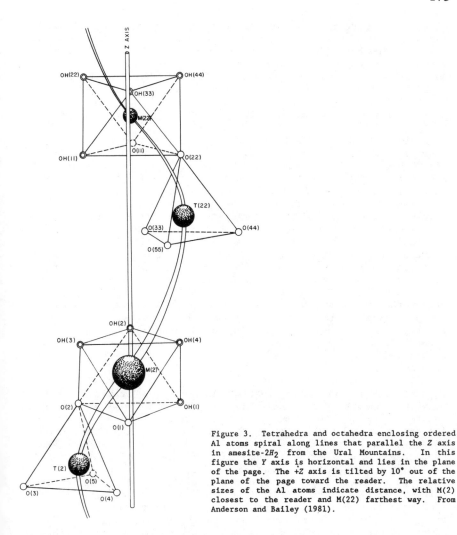

Figure 3. Tetrahedra and octahedra enclosing ordered Al atoms spiral along lines that parallel the Z axis in amesite-$2H_2$ from the Ural Mountains. In this figure the Y axis is horizontal and lies in the plane of the page. The +Z axis is tilted by 10° out of the plane of the page toward the reader. The relative sizes of the Al atoms indicate distance, with M(2) closest to the reader and M(22) farthest way. From Anderson and Bailey (1981).

Anderson and Bailey (1981) showed that there are 36 possible cation ordering patterns in space group $C1$ for the $2H_2$ structures, in which 2Si+2Al are distributed over four tetrahedral sites and 4Mg+2Al are distributed over six octahedral sites in the two layers. Only seven of these 36 patterns, however, would be expected to give the observed triclinic symmetry and geometry. The three most stable ordering patterns are judged to be those in which the predicted Al cation shifts (see above) are parallel in both layers so that the resulting distortions fit together well in adjacent layers. The Urals and Antarctic ordered patterns represent two of the three most stable arrangements.

Once a particular cation ordering pattern has been established for any mineral, it is often assumed that all specimens of that mineral will have the same ordering pattern, provided the environments of crystallization are similar. This is not the case for amesite-$2H_2$. It may be prudent to examine the geometry of other crystal structures to see if alternative ordering patterns are both possible and crystallochemically plausible. This applies particularly to structures involving ordering in subgroup symmetry.

Although the $2H_2$ standard polytype is most characteristic of amesite, other polytypes do occur in lesser amounts. Steinfink (in Oughton, 1957) reported a $2H_1$ crystal from the Urals locality. Steadman and Nuttall (1962) reported a $6R$ structure that is distorted to triclinic symmetry (presumably due to ordering). It is termed $6R_2$ here because it is different than the standard $6R_1$ polytype and involves intermixing of $b/3$ interlayer shifts with zero shifts. Hall and Bailey (1976) reported several $6R_1$ crystals in amesite from Antarctica. Anderson and Bailey (1981) mentioned that the $2H_2$ polytype is most abundant in the Urals sample but that the $2H_1$ standard polytype, both the $6R_1$ and $6R_2$ polytypes, and random stacking were observed also. The $6R_2$ structure is more abundant than $6R_1$ in this sample. Bailey (unpublished) observed the $2H_1$ polytype to be the dominant stacking sequence in crystals of amesite from South Africa. All natural crystals of amesite, regardless of polytypes, are observed to be optically biaxial, twinned, and of monoclinic or triclinic geometry. Thus, all crystals are likely to be ordered.

Identification of polytypes is made most easily by single crystal precession photographs. All of the regularly stacked polytypes reported for amesite belong to group D, as discussed in Chapter 2, and this seems to be a characteristic of amesite and its compositional analogues. The strong $k = 3n$ reflections, as a consequence, are the same for all polytypes. On precession photographs the 02ℓ, $11\bar{\ell}$, and 11ℓ row lines (as indexed on the resultant orthogonal axes) can be used to identify the polytypes by the type of lattice present and the intensities of the reflections. A rhombohedral lattice is identified on precession photographs by the fact that the three equivalent Y^* axes (normal to the three X axes) are not in the plane of any of the photographs taken with X_1, X_2, or X_3 as the precession axes so that the white radiation streaks through reflections such as 060, 330 and $\bar{3}30$ on these axes do not extend to the centers of the photographs (as they do for hexagonal lattices). The spots on adjacent row lines parallel to Z^* are offset from one another. An example is shown in Figure 7 of Chapter 2. For hexagonal lattices the three equivalent $Y*$ axes will be in the plane of the corresponding precession films, and spots on adjacent row lines are not offset along Z^*. The $k \neq 3n$ reflections that are diagnostic to distinguish $6R_1$ from $6R_2$ and $2H_1$ from $2H_2$ are weak and may not show up on X-ray powder patterns, especially if the sample has been ground. They do show up on Gandolfi powder patterns of single crystals, as tabulated in Table 1, and this permits identification also.

Table 1. X-ray powder patterns of selected 1:1 structures.

A. Cronstedtite-1T

hkℓ	I	d(obs.)	d(calc.)
001	90	7.1 Å	7.095 Å
100	20	4.75	4.751
101	7 B	3.96	3.948
002	60	3.54	3.548
102	2	2.85	2.843
110	15	2.75	2.743
111,11$\bar{1}$	100	2.560	2.558
003	10	2.362	2.365
112,11$\bar{2}$	60	2.170	2.170
103	1	2.122	2.124
113,11$\bar{3}$	50	1.792	1.791
004	2	1.773	1.774
121	2	1.740	1.741
300	40	1.582	1.584
301	30	1.544	1.546
114,11$\bar{4}$	45	1.489	1.489
302	40	1.446	1.446
005	10	1.419	1.419
220	2	1.370	1.372
221,22$\bar{1}$	40	1.348	1.347
303	20	1.318	1.318
222,22$\bar{2}$	30	1.280	1.279
115,11$\bar{5}$	40	1.261	1.260

B. Cronstedtite-2H₁

hkℓ	I	d(obs.)	d(calc.)
002	90	7.1 Å	7.100 Å
020,110	20	4.75	4.751
022,112	15	3.96	3.948
004	70	3.54	3.550
024.114	5	2.84	2.844
131,201	50	2.69	2.693
132,202	100	2.557	2.559
133,203 / 006 }	60	2.369	{2.373 / 2.368
134,204	30	2.168	2.170
026,116	2	2.12	2.118
135,205	70	1.972	1.973
136,206	25	1.792	1.792
008	5	1.773	1.775
152,242,312	2	1.742	1.741
137,207	40	1.633	1.631
330,060	40	1.583	1.584
332,062	35	1.544	1.546
138,208	45	1.492	1.490
334,064	25	1.447	1.446
00.10	15	1.420	1.420
139,209	50	1.367	1.368
262,402	30	1.348	1.347
263,403 / 336,066 }	35	1.319	{1.317 / 1.316

C. Cronstedtite-2H₂

hkℓ	I	d(obs.)	d(calc.)
002	90	7.1 Å	7.093 Å
110 / 020 }	25	4.75	{4.752 / 4.751
111 / 021 }	15	4.50	{4.506 / 4.505
112 / 022 }	20	3.95	{3.948 / 3.947
004	60	3.54	3.546
113 / 023 }	5	3.35	{3.352 / 3.351
114,024	10	2.84	2.842
201 / 131 }	60	2.693	{2.694 / 2.693
202,132	100	2.556	2.559
115,025	2	2.440	2.436
203,133 / 006 }	60	2.371	{2.373 / 2.364
204,134	30	2.170	2.170
116,026	3	2.119	2.117
205,135	50	1.970	1.972
117,027	1	1.865	1.864
225,045	1	1.820	1.821
206,136	20	1.791	1.791
151,241,311	3	1.778	1.782
152,242,312	7	1.740	1.741
153,243,313 / 226,046 }	5 / 2	1.680	{1.679 / 1.676
207,137	40	1.631	1.630
154,244,314	2	1.602	1.602
330,060	50	1.584	1.584
332,062	35	1.546	1.546
208,138	40	1.490	1.489
334,064	35	1.447	1.446
00.10	10	1.419	1.419
139,209 / 401 }	45	1.366	{1.367 / 1.366
261			1.365
262,402	35	1.347	1.347
403 / 263 }	35	1.319	{1.318 / 1.317
336,066			1.316

D. Cronstedtite-3T

hkℓ	I	d(obs.)	d(cal
001	80	7.1 Å	7.081
020	10 B	4.73	4.750
110	10 B	4.64	4.637
11$\bar{1}$	10 B	4.35	4.336
021	5 B	3.95	3.945
00$\bar{2}$,111	60	3.52	3.541
11$\bar{2}$	5	3.18	3.167
022	2	2.84	2.839
13$\bar{0}$,20$\bar{1}$	80	2.720	2.720
13$\bar{1}$,200	2	2.654	2.656
112	2	2.556	2.558
13$\bar{1}$,20$\bar{2}$	100	2.438	2.437
003	15	2.360	2.360
20$\bar{1}$ / 13$\bar{2}$ }	40	2.303	{2.305 / 2.304
04$\bar{1}$	2	2.550	2.252
20$\bar{3}$,132	80	2.035	2.035
11$\bar{3}$	2	1.941	1.940
20$\bar{2}$,13$\bar{3}$	30	1.910	1.908
31$\bar{1}$ / 150,24$\bar{1}$,11$\bar{4}$ }	7	1.80	{1.796 / 1.789
004	10	1.770	1.770
310,31$\bar{2}$	2	1.743	1.741
151,24$\bar{2}$	2	1.702	1.701
133,20$\bar{4}$	80	1.680	1.680
02$\bar{4}$	2	1.660	1.659
33$\bar{1}$ / 060 }	80	1.584	{1.584 / 1.583
134,20$\bar{3}$	40	1.580	1.579
330 / 33$\bar{2}$,06$\bar{1}$ }	60	1.545	{1.546 / 1.545
33$\bar{1}$ / 33$\bar{3}$,06$\bar{2}$ }	65	1.445	{1.446 / 1.445
005	15	1.416	1.416
134,20$\bar{5}$	15	1.405	1.404
26$\bar{1}$,40$\bar{1}$	20	1.370	1.369
02$\bar{4}$	2	1.360	1.359
135,20$\bar{4}$,26$\bar{2}$,400	80	1.328	1.328
134,06$\bar{3}$	20	1.315	1.315
26$\bar{1}$,40$\bar{3}$	20	1.305	1.305
26$\bar{3}$,40$\bar{1}$	20	1.250	1.250

Table 1, continued

All specimens from Pribram, Czechoslovakia. Data from Bailey (unpublished). Film patterns using single crystals in a Gandolfi camera, 114.6 mm diameter, graphite-monochromatized $FeK\alpha_1$ radiation. Indexing verified by comparison with single crystal intensities. Intensities estimated visually.

A. Cronstedtite-$1\underline{T}$ indexed on hexagonal axes with a = b = 5.486, c = 7.095 Å, γ = 120°

B. Cronstedtite-$2\underline{H}_1$ indexed on orthogonal axes with a = 5.486, b = 9.501, c = 14.200 Å

C. Cronstedtite-$2\underline{H}_2$ indexed on orthogonal axes with a = 5.488, b = 9.502, c = 14.186 Å

D Cronsedite-$3\underline{T}$ but indexed as $1\underline{M}$ with a = 5.486, b = 9.500, c = 7.313 Å,

	A. Amesite-$2\underline{H}_2$			B. Amesite-$2\underline{H}_1$		
hkℓ	I	d(obs.)	d(calc.)	I	d(obs.)	d(calc.)
002	100	7.0 Å	7.01 Å	100	7.0 Å	7.02 Å
020,110	25	4.60	4.588	35	4.59	4.583
021,111	25	4.38	4.361	--		
022,112	25	3.84	3.840	30	3.83	3.837
004	100	3.51	3.507	100	3.505	3.509
023,113	25	3.27	3.275	--		
024,114	10	2.790	2.786	25	2.790	2.786
130,200	1	2.65	2.648	1	2.65	2.646
131,201	40	2.600	2.602	40	2.600	2.602
132,202	80	2.476	2.478	80	2.475	2.477
025,115	12	2.392	2.394	--		
006	20	2.338	2.338	25	2.335	2.339
133,203	25	2.302	2.304	20	2.302	2.304
041,221	2	2.26	2.264	--		
042,222	1	2.18	2.180	1	2.18	2.178
134,204	25	2.112	2.113	20	2.110	2.113
026,116	1	2.08	2.083	5	2.085	2.084
043,223	2	2.06	2.060	--		
135,205	70	1.925	1.926	70	1.925	1.926
044,224	--			2	1.915	1.919
027,117	2	1.838	1.836	--		
045,225	2	1.778	1.776	--		
008,136,206	25	1.754	1.753	25	1.755	1.754
150,240,310	--			5	1.739	1.734
151,241,311	12	1.722	1.721	2	1.724	1.721
152,242,312	2	1.682	1.683	15	1.682	1.683
028,118, 046,226}	2	1.638	{1.638 1.637	2	1.639	1.639
153,243,313	5	1.625	1.626	--		
137,207	30	1.598	1.598	25	1.599	1.598
154,244,314	2	1.555	1.554	1	1.555	1.554
060,330	60	1.528	1.529	60	1.528	1.529
047,227	2	1.511	1.509	--		
062,332	12	1.495	1.494	20	1.494	1.494
029,119	2	1.477	1.476	--		
138,208	35	1.462	1.462	40	1.463	1.463
00.10 064,334}	25	1.402	{1.403 1.402	20 5	1.404 1.400	1.404 1.401
048,156+	--			1	1.39	1.393
139,209	25	1.343	1.343	30	1.343	1.344
261,401	2	1.319	1.318	5	1.319	1.318
262,402	15	1.301	1.301	20	1.302	1.301

Amesite specimens A and C from Saranovskoye, N. Urals, U.S.S.R., specimen B from Postmasburg, South Africa, and specimen D from Dufek Massif, Antarctica. All data from Bailey (unpublished) using a Gandolfi camera with single crystals, graphite-monochromatized $FeK\alpha_1$ radiation, 114.6 mm camera diameter. Intensities estimated visually. All indexing on basis of orthogonal axes with indexing confirmed by comparison with single crystal intensities. A. $2\underline{H}_2$ $\underline{a}$ = 5.229, $\underline{b}$ = 9.173, $\underline{c}$ = 14.028 Å. B. $2\underline{H}_1$ $\underline{a}$ = 5.290, $\underline{b}$ = 9.174, $\underline{c}$ = 14.036 Å. C. $6\underline{R}_2$ $\underline{a}$ = 5.298, $\underline{b}$ = 9.176, $\underline{c}$ = 42.194 Å. D. $6\underline{R}_1$ $\underline{a}$ = 5.297, $\underline{b}$ = 9.172, $\underline{c}$ = 42.047 Å.

Table 1, continued

hkl	C. Amesite-6R₂			D. Amesite-6R₁		
	I	d(obs.)	d(calc.)	I	(dobs.)	d(calc.)
006	100	7.0 Å	7.03 Å	100	7.0 Å	7.01 Å
02$\bar{1}$,11$\bar{1}$	40	4.58	4.561	--		
02$\bar{2}$,11$\bar{2}$	2	4.48	4.483	30	4.48	4.481
024,11$\bar{4}$	10	4.21	4.207	20	4.20	4.204
02$\bar{5}$,11$\bar{5}$	20	4.03	4.031	--		
027,117	20	3.65	3.651	--		
00.12	100	3.51	3.516	100	3.50	3.504
02$\underline{8}$,11$\bar{8}$	1	3.46	3.462	15	3.46	3.456
02.10,11.$\underline{10}$	2	3.11	3.106	10	3.10	3.099
02.$\underline{11}$.11.$\overline{11}$	2	2.94	2.943	--		
02.$\overline{13}$,11.13	5	2.65	2.650	--		
203,133	40	2.603	2.603	35	2.60	2.602
02.14,11.$\overline{14}$	--			5	2.51	2.513
206,136	80	2.478	2.479	90	2.474	2.474
00.18	20	2.345	2.344	15	2.335	2.336
209,$\underline{139}$	20	2.305		20	2.300	2.304
02.16,11.$\underline{16}$	--			1	2.28	2.280
02.17,11.$\overline{17}$	1	2.18	2.183	--		
20.12,13.12	20	2.115	2.116	20	2.110	2.113
20.15,13.15	45	1.929	1.929	40	1.922	1.925
20.18,13.18	20	1.756	1.755	20	1.752	1.752
241,151	5	1.732	1.733	2	1.73	1.732
02.23,11.$\overline{23}$	2	1.704	1.703	--		
247,157	1	1.665	1.667			
20.21,13.21	15	1.601	1.601	20	1.598	1.597
060,330	60	1.529	1.529	70	1.528	1.529
066,336	20	1.495	1.494	15	1.495	1.495
20.24,13.24	30	1.466	1.465	20	1.461	1.61
00.30 06.12,33.12 }	25	1.406	{ 1.406 1.402	20	1.402	{ 1.402 1.401
20.27,13.27	25	1.346	1.346	20	1.343	1.342
403,263	2	1.319	1.319	1	1.32	1.318
406,266	20	1.303	1.302	15	1.302	1.302
409,269	5	1.275	1.275	2	1.275	1.274
20.30,13.30	15	1.242	1.242	15	1.240	1.239
40.15,26.15	3	1.200	1.198	8	1.198	1.197
00.36	10	1.170	1.172	5	1.170	1.168

E. Kellyite-2H₂

hkl	I	d(obs.)	d(calc.)	hkl	I	d(obs.)	d(calc.)
002	85	7.0 Å	7.01 Å	20$\bar{5}$,135 }	25	1.960	{ 1.961 1.956
020,110 }	10	4.7	{ 4.705 4.703	205,13$\underline{5}$ }	25	1.940	{ 1.946 1.941
021,111 }	15	4.46	{ 4.461 4.447	20$\bar{6}$,136 }	10	1.778	{ 1.780 1.776
022,112 }	3	3.910	{ 3.907 3.890	136,206 }	10	1.760	{ 1.767 1.762
004	85	3.502	3.506	20$\bar{7}$,137 }	10	1.620	{ 1.620 1.616
023,113 }	2	3.32	{ 3.316 3.301	137,207 }	20	1.605	{ 1.608 1.604
024,114 }	1/2	2.81	{ 2.811 2.799	060,330	70	1.568	1.568
20$\bar{1}$			2.670	062,33$\underline{2}$ }	25	1.530	{ 1.530 1.527
13$\bar{1}$,131	50	2.665	{ 2.669 2.660	20$\bar{8}$,138 }	15	1.480	{ 1.480 1.476
201			2.660	138,208 }	20	1.466	{ 1.469 1.466
20$\bar{2}$,13$\bar{2}$,202 }	100 B	{ 2.540 2.530	{ 2.580 2.528 2.523	064,334 }	20	1.430	{ 1.432 1.426
20$\bar{3}$,13$\bar{3}$,203 }	40 B	{ 2.352 2.340	{ 2.358 2.354 2.343 2.337	00.10	5	1.402	1.402
20$\bar{4}$,13$\bar{4}$,134,204 }	20 B	{ 2.155 2.140	{ 2.157 2.152 2.142 2.136				

Kellyite-2H₂ from Bald Knob, North Carolina.
a = 5.430, b = 9.410, c = 14.025 Å, β = 90.58°.
Data from Bailey (unpublished).

178

Table 1, continued

Odinite (Bailey, 1988)

Int.	d(meas.)	d(calc.)	hkl:	Joint	1M	1T
100	7.15 Å	7.15 Å	001			
40	4.65	4.65-4.66	020			
20	4.53	4.55		110		
85	3.58	3.57	002			
1	2.84	2.84	022			
40	2.67	2.68			20$\overline{1}$,130	200,130
7	2.56	2.56	112			
3	2.51	2.51				201,131
30 B	2.42	2.40			20$\overline{2}$,131	
5	2.34	2.33	040			
1	2.14	2.15				202,132
10 B	2.02	2.02			20$\overline{3}$,13$\underline{2}$	
1	1.90	1.90			202,13$\overline{3}$	
7 B	1.74	1.74-1.75	150,240			
7 B	1.67	1.67-1.68			20$\overline{4}$,133	
2	1.62	1.63			15$\overline{2}$,241	
65	1.552	1.551-1.553	060			
10	1.492	1.491				204,134
5 B	1.43	1.42	062			
15 B	1.32	1.32-1.33			204,13$\overline{5}$, 400,40$\overline{2}$	

Sample from Islands of Los (off Guinea). Debye-Scheerer pattern taken with graphite-monochromatized FeKα1 radiation in camera of diameter 114.6 mm. Intensities estimated visually, 1T pattern indexed on orthogonal axes. Cell dimensions: 1M: a - 5.369, b - 9.307, c - 7.17 Å, β - 103.9°. 1T: a - 5.371, b - 9.316, c - 7.361 Å, β - 90°.

An Fe^{2+}-rich amesite was described from a rodingite dike in a metasomatically altered granite sheet emplaced in serpentinite of the Thetford Mines ophiolite complex at the Lake Asbestos mine in Quebec (Taner and Laurent, 1984). The amesite appears to have formed by replacement of biotite. The structural formula obtained by averaging electron microprobe analyses of two crystals is $(Mg_{1.140}Fe_{0.955}Al_{0.975})(Si_{1.045}Al_{0.945})O_5(OH)_{3.705}$. The crystals show sector twinning and are optically biaxial positive with 2V ≅ 20°. The X-ray powder pattern is that of a group D serpentine, but no lines diagnostic of a specific polytype are present. The crystallographic β angle is 90.4°, analogous to that of Mg-rich amesite. Single crystals have not been studied, due in part to the small grain size.

Kellyite is the yellow Mn^{2+}-analogue of amesite (Peacor et al., 1974). The mineral occurs as small tablets and laths interstitial to Ca-Mn carbonate grains at the manganese deposit at Bald Knob, near Sparta, North Carolina. The composition is $(Mn_{1.8}Mg_{0.2}Fe_{0.1}^{3+}Al_{0.9})$ $(Si_{1.0}Al_{1.0})O_5(OH)_4$. In addition to crystals showing stacking disorder, 2-layer and 6-layer forms with considerable regularity of stacking were recognized. The 2-layer form is the standard $2H_2$ polytype that is most abundant in amesite, and the 6-layer form is the $6R_2$ structure. Doublets on the powder pattern (Table 1) of kellyite-$2H_2$ require monoclinic or triclinic cell geometry to separate 20l and 20$\overline{l}$ reflections.

BERTHIERINE

Berthierine is the preferred term for the fine-grained iron-rich 1:1 layer silicate commonly present in unmetamorphosed sedimentary iron formations. In the past this mineral often has been called "chamosite," but this name has been used also for a 14 Å Fe-rich chlorite. The material from the type locality in Chamoson, France, is known now to be a true chlorite so that the term "chamosite" should be used for an iron-rich chlorite and not for the 1:1 layer material.

Berthierine is found in two structural forms, often intimately intermixed. One is of a apparent trigonal symmetry and the other of apparent monoclinic symmetry. The trigonal berthierine has the $1T$ structure (Brindley, 1951) with each layer identical and no inter-layer shift. Despite the trigonal symmetry it is convenient to index the powder pattern on a C-centered orthogonal-shaped cell. For this reason the compound sometimes is termed the orthohexagonal or orthorhombic variety in the literature. The monoclinic berthierine can be described as the $1M$ structure, in which each layer is displaced by $a/3$ along $-X_1$. The smallest unit cell for this structure is an inclined 1-layer cell with $\cos\beta = -a/3c$. The possibility of this mineral being the $3T$ structure cannot be dismissed, however, because the calculated powder patterns of the ideal $1M$ and $3T$ layer sequences are identical for trioctahedral compositions. Figure 4 illustrates the fact that any monoclinic n-layer unit cell with an interlayer shift of exactly $a/3$ can be described also by a larger orthogonal cell containing $3n$-layers. Because berthierine often shows randomness in the layer stacking, it is preferable to use the $1M$ description or $1M$-$3T$ in the absence of definitive evidence of a larger cell. Although most berthierine samples are a mixture of the $1M$ and $1T$ forms, the diagnostic powder X-ray lines necessary for the identification of each phase do not overlap because of the difference in the β angles (Table 1).

Brindley (1982) reviewed the best chemical analyses available for 14 different occurrences of berthierine. He used a formula of the type $(R_a^{2+}R_b^{3+}\square_c)(Si_{2-x}Al_x)O_5(OH)_4$, where $a + b + c = 3.0$, $b - x = 2c$, and a total valence of $+14$ is assumed for the cations. Tetrahedral Al is a characteristic and necessary feature of berthierine and ranges from $x = 0.45$ to 0.90 atoms per formula unit in the 14 samples. The R^{3+} octahedral cations exceed the Al^{3+} tetrahedral cations, and electrical neutrality of the layers is achieved by vacant octahedral sites ($\square$). A plot of octahedral R^{3+} vs. tetrahedral Al^{3+} shows a linear variation for 12 of the samples, with some scatter. Two points lie farther from the line (Fig. 5). The solid line through the data points corresponds to b (or R^{3+}) = $1.30x$, and thus also to c (or $\square$) = $0.15x$. Octahedral R^{3+} is primarily Al, ranging from 0.37 to 1.03 atoms per formula unit, and to a lesser extent Fe^{3+}, which varies from 0.01 to 0.27 atoms. The excess of octahedral R^{3+} over tetrahedral R^{3+} is not believed to be due to secondary oxidation of primary Fe^{2+} because the amount of Fe^{3+} is usually too small to account for the excess. Fe^{2+} is the dominant octahedral cation, ranging from 1.33 to 1.84 atoms per formula unit.

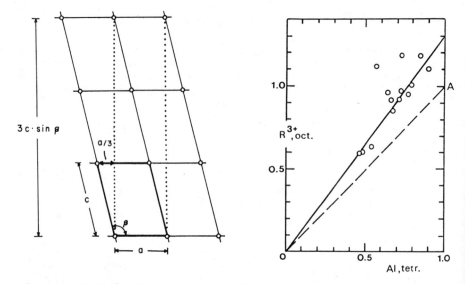

Figure 4. (left) Monoclinic 1-layer unit cell (heavy outline) with interlayer shift of a/3 also can be described by a larger orthohexagonal 3-layer cell (dotted lines). From Bailey (1980).

Figure 5. (right) Graph of octahedral R^{3+} vs. tetrahedral Al^{3+} for berthierine. Full line through data points has slope of 1.30. Dashed line corresponds to octahedral R^{3+} - tetrahedral Al^{3+} with no vacancies. Point A = amesite. From Brindley (1982a).

The amount of octahedral substitution of Mg is small, ranging from 0.08 to 0.66 atoms. Octahedral vacancies range from 0.04 to 0.27 per formula unit and thus provide a small but consistent deviation from ideal trioctahedral nature. The monoclinic form of berthierine appears to be more abundant in samples with high total Al_2O_3 contents and the trigonal form in those with low Al_2O_3 but high SiO_2 contents. In a few samples only the 1T form is present.

Berthierine can be described as the ferroan aluminian analogue of lizardite. Not enough data are available to determine if substitution of Al into the lizardite structure in the natural aluminian lizardites follows the same pattern as shown by the solid line for berthierine in Figure 5. An alternative trend would follow the dashed line leading to amesite at point A, which would mean that no octahedral vacancies exist.

A titanian berthierine replaces titanian magnesioferrite in a serpentinized ultramafic dike near Picton, Ontario (Arima et al., 1985). The substitution of Ti^{4+} for Fe^{2+} requires a greater charge compensation by octahedral vacancies than in Fe^{2+}-berthierine, and the resulting composition is intermediate between trioctahedral and dioctahedral. An average structural formula of 25 electron microprobe point analyses is $(Fe_{1.113}Ti_{0.773}Mg_{0.458}Al_{0.117}Cr_{0.010}Mn_{0.002}Ca_{0.035}Na_{0.011}K_{0.08}\square_{0.473})(Si_{1.29}Al_{0.71})O_5(OH)_4$.

A green Ni-analogue of berthierine occurring in the Marmara bauxite deposit in Greece was named *brindleyite* (Maksimović and Bish, 1978). Although originally believed to be an analogue of amesite, further study showed that it is more like berthierine in that the tetrahedral substitution is only about $Si_{1.5}Al_{0.5}$, vacancies exist in the octahedral sheet, and the natural samples are mixtures of both trigonal and monoclinic polytypes. Spectroscopic data indicate that both natural and synthetic samples have a disordered cation distribution and that the green color is due to octahedrally coordinated Ni^{2+} cations. The structural formula of brindleyite is approximately $(Ni_{1.75}Al_{1.0}\square_{0.25})(Si_{1.5}Al_{0.5})O_5(OH)_4$.

Cesaro (1927) gave the name *fraipontite* to a yellowish-white Zn-rich clay that he discovered in the Vieille Montagne mines at Moresnet in eastern Belgium. Fransolet and Bourguignon (1975) showed that fraipontite is the Zn-analogue of berthierine with a composition of $(Zn_{2.35}Al_{0.65})(Si_{1.35}Al_{0.65})O_5(OH)_4$. The X-ray powder pattern has broad lines, but can be interpreted as a mixture of $1T$ and $1M$-$3T$ polytypes (Table 1). Esquivin (1957) synthesized the mineral at low temperature.

Chukhrov (1956) used the term "zinalsite" to describe Zn-rich clays found in weathering zones around primary Zn-ores at several localities in the U.S.S.R. These clays were studied in more detail by Chukhrov et al. (1971) and shown to be similar to fraipontite in composition and in containing mixtures of trigonal and monoclinic phases. The name fraipontite has priority.

ODINITE

All of the berthierine samples studied by Brindley (1982) are from ancient Fe-rich sedimentary rocks, including the Lake Superior iron formations and several of the European oolitic ironstones. "Berthierine" also has been described by several authors as currently forming on continental shelf regions and reef lagoons close to the discharge areas of many river systems in tropical latitudes (23°S to 16°N). In these areas glauconite tends to form below about 150 m depth due to a greater oxidation potential, and "berthierine" forms at shallower depths of 15 to 60 m. Odin (1985, 1988) challenged the identification of these "berthierine" samples by showing that they are significantly more ferric (about 0.7 to 1.0 atoms per 3.0 octahedral sites) in composition than berthierine and have more compensating octahedral vacancies (0.5 to 0.7 atoms per 3.0 sites) so that the bulk composition is intermediate between trioctahedral and dioctahedral. There is also a much smaller amount of tetrahedral Al (0.00 to 0.20 atoms per two tetrahedral sites). A simplified formula for one of the purest samples (from the Islands of Los off Guinea) is $(R^{3+}_{1.35}R^{2+}_{1.05}\square_{0.60})(Si_{1.85}Al_{0.15})O_5(OH)_4$ with R^{3+} dominantly Fe and R^{2+} dominantly Mg. Odin (1985) originally considered the material to be a chlorite because of the presence of a 14 Å peak, but further work (Odin et al., 1988) showed that the primary green clay is a 1:1 layer silicate mineral that alters readily to Iba chlorite. Bailey (1988) gave the name *odinite* to the primary green clay to supersede the temporary designation of 7 Å phyllite V given by Odin (1988). Odinite is similar to berthierine in that it is usually a mixture of $1M$ and $1T$ polytypes (Table 1). Unlike berthierine, odinite has never been observed in oolitic habit nor crystallizing between detrital particles.

182

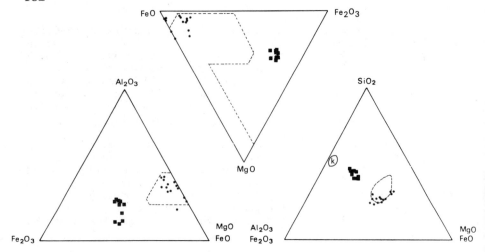

Figure 6. Comparison of compositional field of odinite (squares) with those of berthierine (asterisks from Brindley, 1982) and chlorite (dashed line areas enclosing 154 samples from Foster, 1962). Symbol k = kaolinite. From Odin et al. (1988).

The compositional relationship of odinite to berthierine and chlorite is shown in Figure 6. Odinite is an Fe^{3+}-rich 1:1 clay with 0 to 10% tetrahedral substitution of Al for Si. This charge is balanced by addition of divalent Mg and Fe in the third octahedral site of a normally dioctahedral structure. Total octahedral cations range from 2.30 to 2.54 atoms per formula unit for ten analyzed samples (without correction for small quartz and chlorite impurities). There is no tetrahedral Fe^{3+} according to Mössbauer study.

Odinite forms authigenically in marine waters near 25°C at pH values of 7.5 to 8.5 in sandy sediments that often have a significant bioclastic content. It occurs as infillings, impregnations, or replacements of biogenic or detrital porous grains or fecal pellets. It is believed to have formed in a micro-environment within pores and cavities that originally contained organic matter and were more reducing that the external marine environment. During maturation light green odinite becomes darker green, harder, and starts to recrystallize to I*ba* chlorite. It has not yet been recognized in rocks older than the Recent Quaternary.

CRONSTEDTITE

Cronstedtite typically occurs as jet black crystals of brilliant vitreous luster in low temperature hydrothermal sulfide veins. It has been reported also in carbonate chondrites. It is distinguished chemically in its ideal end member formula by the tetrahedral composition $Si:Fe^{3+}$ = 1:1 and by complete occupation of octahedral sites by Fe. Mg is known to substitute for Fe in octahedral coordination in natural specimens, and the tetrahedral $Si:Fe^{3+}$ ratio may be as high as 3:1. Brindley (1961) balanced the amount of tetrahedral Fe^{3+} with an equivalent amount of octahedral Fe^{3+}.

Both Gossner (1935) and Hendricks (1939) determined that cronstedtite is based on a 1:1 type layer. More detailed studies by Frondel (1962) and Steadman and Nuttall (1963, 1964) showed an extraordinary amount of variation in the layer stacking sequence. Frondel (1962) recognized crystals with periodicities of 1, 2, 3, and 9 layers as well as those with considerable stacking disorder in crystals from six different localities. Steadman and Nuttall (1963, 1964) found eight different stacking sequences in an X-ray examination of about 200 crystals from nine localities. These can be correlated with the standard polytypes $1T$, $1M$, $2T$, $2H_1$, $2H_2$, $2M_1$, and $3T$ plus the same hybrid $6R_2$ structure found in amesite and kellyite, in which interlayer shifts of $b/3$ along Y alternate regularly with zero interlayer shifts. Bailey (unpublished) identified the $2H_1$, $2H_2$, and $3T$ standard polytypes in cronstedtite crystals from Pribram, Bohemia, $1T$ from Kisbanya, Romania, and $1T$ and $3T$ polytypes in crystals from Llalaagua (Bolivia), Kuttenberg (Bohemia), and Wheal Jane (Truro, Cornwall). Most of the $3T$ crystals have a resultant apparent monoclinic symmetry in that one of the three lateral Y^* axes is unique, perhaps as a result of twinning or ordering. The identification of a true $3R$ cronstedtite polytype is doubtful. Although Hendricks (1939) gave intensity data for a crystal identified as $3R$ from Kisbanya, Romania, his published Weissenberg photograph has intensities corresponding to a group A structure ($3R$ is group C). This was recognized by Steadman and Nuttall (1963), who found a crystal with similar intensities from Wheal Jane, Cornwall, and determined the true space group as $P3_1$ with a structure corresponding to the standard $3T$ polytype. Frondel (1962) reported a $3R$ crystal from Kuttenberg, Bohemia, but Bailey (unpublished) examined the same crystal and identified both $1T$ and $3T$ polytypes to be present in different portions of the large crystal.

Identification of these cronstedtite structural types is best made by single crystal X-ray precession study. Different crystals from the same hand specimen often are different polytypes, and intermixing of these crystals by grinding to form a powdered sample may be misleading. To avoid this, a Gandolfi camera can be used to obtain a powder pattern from each crystal (see Fig. 5 of Chapter 2 and Table 1 of this chapter).

Steadman and Nuttall (1963) partially refined the atomic positions of the $1T$, $3T$, and 6-layer structures by electron density maps and trial-and-error adjustment of individual atomic parameters. They found that the large lateral dimensions of the tetrahedral sheet due to the presence of 50% Fe^{3+} are reduced by tetrahedral rotations between 2° and 8° to fit the slightly smaller dimensions of the Fe-rich octahedral sheet. For the partly refined structures, octahedral and tetrahedral cation ordering within the constraints of the symmetry of the ideal space groups is possible only in the 3-layer and 6-layer structures, and no evidence for such ordering was observed for these two crystals.

Geiger et al. (1983) refined the structure of a magnesian cronstedtite-$2H_2$ crystal from Pribram, Czechoslovakia. Examination of 1932 counter-collected X-ray reflections indicated a triclinic intensity distribution but a dimensionally hexagonal unit cell shape. The structure was refined in space group $C1$ to facilitate comparison with the amesite-$2H_2$ structures. The composition of the crystal was $(Fe_{1.64}^{2+}Fe_{0.49}^{3+}Mg_{0.71}Mn_{0.16})(Si_{1.51}Fe_{0.49}^{3+})O_5(OH)_4$, assuming tetrahedral Fe^{3+} is balanced by octahedral Fe^{3+}.

Electron density maps indicated complete ordering of tetrahedral Si and Fe^{3+} but disorder of all octahedral cations. Complete refinement was not possible due to the presence of domains. There are two kinds of $2H_2$ domains present that are shifted by $b/3$ relative to one another due to mistakes in the interlayer stacking sequence. The mistakes of zero shift (instead of the normal $\pm b/3$ shifts for a $2H_2$ polytype) create about 11% by volume of enantiomorphic domains of $2H_2$ packets in which the tetrahedral ordering pattern is reversed. Thus, at the domain interfaces small Si tetrahedra in one domain sit above large Fe^{3+}-rich tetrahedra in the other domain, and vice versa. Relief of strain by this arrangement may be the cause of the stacking mistakes. Stacking errors of zero shift affect only the tetrahedral sheets. A third kind of domain makes up about 26% by volume of the crystal and represents regions in which the sense of tetrahedral rotation (6.2°) is reversed, giving rise to split basal oxygen peaks on electron density maps for the crystal as a whole. This multiple domain model best explains the extra atomic positions observed on electron density maps, lack of streaking of $k \neq 3n$ reflections, and possible non-Bragg satellitic reflections observed on Weissenberg films.

The lowering of symmetry from $P6_3$ to $P1$ (or $C1$ as refined) is not due to cation ordering in this crystal as the distribution of tetrahedral Si and Fe^{3+} does not violate $P6_3$ symmetry. The lower apparent symmetry must be due to a combination of the contributions of the three kinds of domains to the overlapping reflections and of non-hexagonal distortions as a consequence of attempting to keep a regular tetrahedral sheet shape despite the alternation of cations of quite different sizes within the sheet. It is not known if all cronstedtite crystals have domains of this sort or whether a better refinement would be possible with a different crystal or a different polytype. The jet black color prohibits optical examination for macroscopic twins or domains.

Ni-RICH SPECIES

Nepouite

Faust (1966) pointed out that Ni-rich green clays commonly are mixtures of fine-grained, poorly crystallized 1:1 structures, talc, chlorite, smectite, or sepiolite. *Garnierite* is now used as a group name to include all hydrous Ni-Mg silicates. The name is particularly useful as a field term when a more detailed description cannot be given. "Genthite" from the type specimen was shown to be a mixture of 80% Ni-rich serpentine and 20% Ni-rich smectite (pimelite) and thus an unnecessary name. Maksimović (1973) showed that Ni^{2+} can substitute isomorphously for Mg^{2+} in any amount in the planar lizardite structure because of the similarity of their ionic radii. When Ni constitutes more than half of the octahedral cation total, adoption of the existing name *nepouite* was proposed. Brindley and Pham Thi Hang (1972) and Uyeda et al. (1973) made X-ray, chemical, and electron optical studies of 40 garnierite samples, concentrating on those showing predominantly 7 Å (lizardite-nepouite series) and 10 Å (kerolite-pimelite series of talc-like structures, as described in Chapter 8) periodicities. The 7 Å phases included tube- and rod-shaped particles as well as platy forms and fluffy aggregates. No correlation was noted between Ni content and morphology. Brindley and Wan (1975) showed for the planar 1:1 structures that the nepouite specimens studied showed less structural regularity than the Ni-bearing lizardites in general. Even the latter could not be correlated in detail with specific polytypes, except for two specimens that gave X-ray powder patterns close to that of the standard 2Or form. Some specimens also gave reflections suggestive of a monoclinic component. Bayliss (1981) indexed the X-ray powder patterns of several specimens as the 1T polytype.

Manceau and Calas (1986) used the extended X-ray absorption fine structure (EXAFS) technique to show that the Ni atoms in Ni-bearing lizardite, nepouite, kerolite, and pimelite specimens are not distributed randomly with Mg over the octahedral sites but tend to cluster together in domains. X-ray dispersive spectroscopy combined with transmission electron microscopy suggested that pure Ni octahedral sheets are associated with pure Mg sheets in the kerolite-pimelite series. In the lizardite-nepouite series, the extent of the Ni-enriched domains is variable and depends on the chemical composition of the specimen. But very large domains are suggested in some specimens.

Brindley (1980) noted that the compositions do not conform to a simple serpentine formula in the lizardite-nepouite series, even for carefully selected samples with no observable impurities in the X-ray patterns. Excess Si averages about 0.15 atoms per two tetrahedral sites, and octahedral R^{2+} is deficient by about 0.30 atoms if a total cation valency of +14 is assumed. Brindley (1980) suggested the deviations from ideal could be due to the presence of amorphous Si or to a leaching of the edges of the clay particles to remove octahedral R^{2+} and leave a residue of amorphous Si. The possibility should also be considered that the structures are not ideally planar and may contain modulations that change the structural formulae, as in antigorite (Chapter 5), greenalite, and caryopilite (Chapter 17).

Pecoraite

Electron microscopy shows the presence of rod- or tube-shaped particles, along with platy and irregular, fluffy particles, in many natural garnierite samples (Uyeda et al., 1973). The elongate particles have been shown to be Ni-bearing forms of chrysotile. Because the Ni^{2+} cation is only slightly smaller than Mg^{2+}, the same lateral misfit between the tetrahedral and octahedral sheets exists for Ni-rich compositions as was shown for Mg-rich chrysotile in Chapter 5. The misfit leads to curling of the 1:1 layers into cylinder-shaped coils and scrolls, although the Ni-rich forms are not as asbestiform as some chrysotiles.

Faust et al. (1969, 1973) gave the name *pecoraite* to the Ni-analogue of chrysotile found in a weathered meteorite from the Wolf Creek crater in Western Australia. The X-ray powder pattern of pecoraite (Table 1) is poor, but is best indexed on a monoclinic-shaped unit cell with $\beta \cong 92°$. Brindley (1980) suggested that the name pecoraite should be used for all minerals with fibrous 7 Å morphologies in which Ni is the dominant octahedral cation and that the name be prefixed by clino, ortho, or para, as in the corresponding forms of chrysotile, when the different structural forms can be recognized. X-ray powder patterns, if they are of poor quality, may not be sufficient to provide such identification or to differentiate pecoraite from nepouite.

Fibrous forms of pecoraite have been synthesized by several researchers, including Noll and Kircher (1952), Roy and Roy (1954), Noll et al. (1958), and Jasmund et al. (1976). The fibrous Fe- and Co-analogues of chrysotile have been synthesized also but not yet recognized in nature. The Ni-analogue of antigorite has not been synthesized or identified beyond doubt in nature. Jasmund and Sylla (1971) described the synthesis of fibrous pecoraite that converted to hexagonal platelets of a 6-layer Ni-analogue of "orthoantigorite" after holding at 320°-350°C and 200-250 atm for several weeks. The 6-layer form was later clarified as having a structure similar to that of the Unst serpentine (Chapter 5) and not that of antigorite (Jasmund and Sylla, 1972).

ACKNOWLEDGMENTS

Previously unpublished research in this chapter was sponsored in part by grant EAR-8614868 from the National Science Foundation and in part by grant 17966-AC2-C from the Petroleum Research Fund, administered by the American Chemical Society. Dr. S. Guggenheim provided constructive criticism of the text.

REFERENCES

Anderson, C.S. and Bailey, S.W. (1981) A new cation ordering pattern in amesite-2H_2. Am. Mineral. 66, 185-195.
Arima, M., Fleet, M.E. and Barnett, R.L. (1985) Titanian berthierine: a Ti-rich serpentine group mineral from the Picton ultramafic dyke, Ontario. Canadian Mineral. 23, 213-320.
Bailey, S.W. (1980) Structures of layer silicates. Ch. 1 in: Crystal Structures of Clay Minerals and their X-ray Identification, G.W. Brindley and G. Brown (Eds.). London, Mineralogical Soc. Mono. 5, 1-123.
_____ (1988) Odinite, a new dioctahedral-trioctahedral Fe^{3+}-rich 1:1 clay mineral. Clay Minerals 23, (in press).
Bayliss, P. (1981) Unit cell data of serpentine group minerals. Mineral. Mag. 44, 153-156.
Brindley, G.W. (1951) The crystal structure of some chamosite minerals. Mineral. Mag. 29, 502-530.
_____ (1961) Kaolin, serpentine, and kindred minerals. Ch. 2 in: The X-ray Identification and Crystal Structures of Clay Minerals, G. Brown (Ed.), 2nd ed. Mineralogical Soc., London, 51-131.
_____ (1980) The structure and chemistry of hydrous nickel-containing silicate and nickel-aluminium hydroxy minerals. Bull. Mineral. 103, 161-169.

_____ (1982) Chemical composition of berthierines--a review. Clays & Clay Minerals 30, 153-155.

_____, Oughton, B.M. and Youell, R.F. (1951) The crystal structure of amesite and its thermal decomposition. Acta Crystallogr. 4, 552-557.

_____ and Hang, Pham Thi (1973) The nature of garnierites-I. Structures, chemical compositions and color characteristics. Clays & Clay Minerals 21, 25-37.

_____ and Wan, H.M. (1975) Compositions, structures, and thermal behavior of nickel-containing minerals in the lizardite-nepouite series: Am. Mineral. 60, 863-871.

Cesaro, G. (1927) Sur la Fraipontite, silicate basique de zinc et d'aluminium. Ann. Soc. Geol. Belg. 50, 106-111.

Chukhrov, F.V. (1956) Zinc-clays from the Akdzhal deposits in Kazakhstan. Kora Vyvetrivaniya 2 (in Russian).

_____, Zvyagin, B.B., Gorshkov, A.I., Yermilova, L.P. and Rudnitskaya, E.S. (1971) To the nature of some zinc-clays and zinalsite. In: Problems of Mineral Homogeneity and Inhomogeneity (F.V. Chukhrov and N.V. Petrovskaya, Eds.), Nauka Publ. House, Moscow, 192-201 (in Russian).

Esquevin, J. (1957) Sur la composition minéralogique des moresnetites et l'existence probable d'une nouvelle phyllite zincifere. C.R. r. hebd. Séanc. Acad. Sci., Paris 244, 215-217.

Faust, G.T. (1966) The hydrous nickel-magnesium silicates-the garnierite group. Am. Mineral. 51, 279-298.

_____, Fahey, J.J., Mason, B. and Dwornik, E.J. (1969) Pecoraite, $Ni_6Si_4O_{10}(OH)_8$, nickel analog of clinochrysotile, formed in the Wolf Creek meteorite. Science 165, 59-60.

_____, _____, _____ and _____ (1973) The disintegration of the Wolf Creek meteorite and the formation of pecoraite, the nickel analog of clinochrysotile. U.S. Geol. Survey Prof. Paper 384-C, 107-135.

Fransolet, A-M. and Bourguignon, P. (1975) Donńees nouvelles sur la fraipontite de Moresnet (Belgique). Bull. Soc. fr. Mineral. Crystallogr. 98, 135-244.

Frondel, C. (1962) Polytypism in cronstedtite. Am. Mineral. 47, 781-783.

Geiger, C.A., Henry, D.L., Bailey, S.W. and Maj, J.J. (1983) Crystal structure of cronstedtite-$2H_2$. Clays & Clay Minerals 31, 97-108.

Gossner, B. (1944) Uber Cronstedtite von Kisbanya. Zentrallblatt Mineral., Abt. A, 195-201.

Gritsaenko, G.S., Bocharova, A.P. and Lyamina, A.N. (1943) On nepouite from the Tyulenevskoye deposit, middle Urals. Soc. Russe Minéral., Mém. 72, 7-28 (in Russian).

Gruner, J.W. (1944) The kaolinite structure of amesite, $(OH)_8$ $(M-g,Fe)-_4Al_2(Si_2Al_2)O_{10}$, and additional data on chlorites. Am. Mineral. 29, 422-430.

Hall, S.H. and Bailey, S.W. (1976) Amesite from Antarctica. Am. Mineral. 61, 497-499.

_____ and _____ (1979) Cation ordering pattern in amesite. Clays & Clay Minerals 27, 241-247.

Hendricks, S.B. (1939) Random structures of layer minerals as illustrated by cronstedtite $(2FeO.Fe_2O_3.SiO_2.2H_2O)$. Possible iron content of kaolin. Am. Mineral. 24, 529-539.

188

Jasmund, K. and Sylla, H.M. (1971) Synthesis of Mg- and Ni-antigorite. Contrib. Mineral. Petrol. 34, 84-86.

_____ and _____ (1972) Synthesis of Mg- and Ni-antigorite. Contrib. Mineral. Petro. 34, 346.

_____, _____ and Freund, F. (1976) Solid solution in synthetic serpentine phases. Proc. Internat. Clay Conf. 1975, S.W. Bailey (Ed.,) 267-274.

Maksimović, Z. (1973) The isomorphous series lizardite-nepouite. Zap. vses. mineral. Obshch. 102, 143-149.

_____ and Bish, D.L. (1978) Brindleyite, a nickel-rich aluminous serpentine mineral analogous to berthierine. Am. Mineral. 63, 484-489.

Manceau, A. and Calas, G. (1986) Nickel-bearing clay minerals: II. Intracrystalline distribution of nickel: an X-ray absorption study. Clay Minerals 21, 341-360.

Noll, W. and Kircher, H. (1952) Synthese des Garnierits. Naturwiss. 39, 233-234.

_____, _____ and Sybertz, W. (1958) Adsorptionsvermögen und specifische Oberfläche von Silikaten mit röhrenformig gebauten Primarkristallen. Koll. Zeit. 157, 1-11.

Odin, G.S. (1985) La "verdine", facies granulaire vert, marin et cotier distinct de la glauconie: distribution actuelle et composition. C.R. Acad. Sci. Paris 301, II(2), 105-108.

_____ (Ed.) (1988) Green Marine Clays. Developments in Sedimenology, Elsevier Publ., 439 pp. (in press).

_____, Bailey, S.W., Amouric, M., Fröhlich, F. and Waychunas, G.S. (1988) Mineralogy of the verdine facies. Chap. B5 in Green Marine Clays (G.S. Odin, Ed.), Developments in Sedimentology, Elsevier Publ., p. 159-206 (in press).

Oughton, B.M. (1957) Order-disorder structures in amesite. Acta Crystallogr. 10, 692-694.

Peacor, D.R., Essene, E.J., Simmons, W.B., Jr. and Bigelow, W.C. (1974) Kellyite, a new Mn-Al member of the serpentine group from Bald Knob, North Carolina, and new data on grovesite. Am. Mineral. 59, 1153-1156.

Roy, D.M. and Roy, R. (1954) An experimental study of the formation and properties of synthetic serpentines and related layer silicate minerals. Am. Mineral. 39, 957-975.

Serna, C.J., Velde, B.D. and White, J.L. (1977) Infrared evidence of order-disorder in amesites. Am Mineral. 62, 296-303.

Steadman, R. and Nuttall, P.M. (1962) The crystal structure of amesite. Acta Crystallogr. 15, 510-511.

_____ and _____ (1963) Polymorphism in cronstedtite. Acta Crystallogr. 16, 1-8.

_____ and _____ (1964) Further polymorphism in cronstedtite. Acta Crystallogr. 17, 404-406.

Steinfink, H. and Brunton, G. (1956) The crystal structure of amesite. Acta Crystallogr. 9, 487-492.

Taner, M.F. and Laurent, R. (1984) Iron-rich amesite from the Lake Asbestos mine, Black Lake, Quebec. Canadian Mineral. 22, 437-442.

Uyeda, N., Hang, Pham Thi and Brindley G.W. (1973) The nature of garnierites--II. Electron-optical study. Clays & Clay Minerals 21, 41-50.

ISOTOPIC STUDIES OF PHYLLOSILICATES

INTRODUCTION

Isotopic Fractionations

The stable isotopes of oxygen, ^{16}O, ^{17}O, and ^{18}O, have chemical and physical properties that differ from one another by small amounts. Larger differences exist between the properties of the stable isotopes of hydrogen, ^1H (H) and ^2H (D). As a result of these differences, the isotopes become separated to a small extent, or fractionated, during most geochemical reactions which involve oxygen and/or hydrogen. Consequently, the stable isotope ratios, $^{18}O/^{16}O$ and D/H, of oxygen- and hydrogen-containing compounds are, in general, different. Because the processes that cause isotopic fractionation are reasonably well understood, it is often possible to infer something about a geologic process from the isotopic composition of a rock, a mineral, or a suite of minerals. This paper is a review of the status of stable isotope geochemistry of phyllosilicates (other than the micas) and related phases. Most of the material presented deals with the stable isotopic behavior of mineral-water systems. A few geologic applications are mentioned. However, these have been reviewed amply elsewhere and are not discussed in detail here.

Most applications of stable isotope measurements to geological processes involve one or both of two approaches: (1) tracer studies, and (2) studies that are based upon the isotopic fractionations which occur in the environment under study. Tracer studies are used for identifying the origin of a phase which has acquired a distinctive isotopic composition in another geologic environment. Studies of the provenance of minerals in sediments are examples of this sort. More commonly, isotopic studies make use of isotopic fractionations that occur in the environment in which a process has occurred. Inferences drawn from the isotopic compositions of minerals about the origin of hydrothermal or diagenetic fluids usually involve both approaches.

Isotopic fractionations are either kinetic (i.e., arising out of different rates of reaction or transport of isotopically substituted compounds) or equilibrium fractionations (i.e., arising from the differences in the thermodynamic properties of isotopically substituted compounds). Most applications of isotopic data to geologic processes make use of the thermodynamic effects of isotopic substitution, although that does not imply that isotopic (i.e., thermodynamic) equilibrium is achieved in all cases.

Terminology and Notation

Oxygen and hydrogen isotope ratios are expressed in δ (delta) notation as deviations of the ratios in per mil or parts per thousand from the ratio of a standard.

$$\delta^{18}O = \left[\frac{(^{18}O/^{16}O)sample - (^{18}O/^{16}O)standard}{(^{18}O/^{16}O)standard} \right] \times 1000$$

and similarly for hydrogen. (In many earlier papers δD values were reported in percent rather than per mil deviations from the standard.) In all discussions in this paper the standard is S.M.O.W. (Standard Mean Ocean Water).

The magnitude of the isotopic fractionation between two phases, two compounds, or two isotopically substituted positions, A and B, within a single compound is frequently given in the form of the isotopic fractionation factor, α_{A-B}.

$$\alpha_{A-B} = \frac{(^{18}O/^{16}O)_A}{(^{18}O/^{16}O)_B}$$

The relationship between α^{ox}_{A-B} and $\delta^{18}O$ is

$$\alpha^{ox}_{A-B} = \frac{\left[1 + \dfrac{\delta^{18}O_A}{1000} \right]}{\left[1 + \dfrac{\delta^{18}O_B}{1000} \right]} \approx 1 + \frac{(\delta^{18}O_A - \delta^{18}O_B)}{1000}$$

and similarly for the hydrogen isotope fractionation factor, α^{hy}_{A-B}. Another useful approximation is

$$1000 \ln \alpha_{A-B} \approx \delta^{18}O_A - \delta^{18}O_B$$

The approximations above are especially useful when the fractionation is small, which is typically the case for oxygen isotope data in high temperature geological systems. However, when fractionations are greater than about 10 per mil, which is typical for oxygen isotope data in low temperature systems (i.e., those at surface and near-surface temperatures) and hydrogen isotope data in all geologic systems, the errors in using the approximations can be significant.

Factors Controlling Isotopic Compositions of Minerals

The theoretical basis of isotopic fractionation has been reviewed recently by O'Neil (1987a). Because the minerals discussed here form in hydrous systems, it is useful to consider isotopic fractionations between minerals and water. Fractionations between minerals can be readily calculated from the mineral water fractionations.

$$\alpha_{A-B} = \frac{\alpha_{A-water}}{\alpha_{B-water}}$$

The isotopic composition of a mineral that forms in isotopic equilibrium with the environmental water is a function both of the temperature of formation and of the isotopic composition of the water. Pressure has a negligible effect on the fractionation of oxygen isotopes between phases, at least within the pressure range encountered within the earth's crust

(Clayton et al., 1975). Hydrogen isotope pressure effects have not been identified, but tests for their existence have not been as thorough.

Isotopic compositions of natural waters. The isotopic composition of water in a geological process is strongly dependent on its origin and history. Water in the deep oceans has a $\delta^{18}O$ value within a few tenths of 0 per mil. Surface ocean waters are more variable, but those of normal salinity have $\delta^{18}O$ values between +1 and -1 per mil (Craig and Gordon, 1965). During Pleistocene glacial maxima, large volumes of ^{18}O- and D-depleted water were stored in continental icecaps. As a result, marine $\delta^{18}O$ values were approximately 1.3 to 1.5 per mil higher than at present (arguments summarized by Savin and Yeh, 1981), and δD values were approximately eight times higher. In the earlier parts of the Tertiary (pre-Oligocene), marine $\delta^{18}O$ values were approximately 1 per mil lower than today (Shackleton, 1967) reflecting the absence of large ^{18}O-depleted polar icecaps. $\delta^{18}O$ and δD values of meteoric waters (precipitation and fresh surface waters that have not been subject to intense evaporation) are correlated with one another (Craig, 1961), according to the relationship

$$\delta D = 8\ \delta^{18}O + 10,$$

which, when plotted on a graph of δD vs $\delta^{18}O$, is commonly called the *meteoric water line.* The isotopic composition of intensely evaporated sea water or meteoric water falls to the right of the meteoric water line, as a result of isotopic disequilibrium between vapor and liquid during evaporation (Craig, 1961). In evaporite facies brines, this effect can be marked (Fig. 1).

In the subsurface, interaction between water and rock is a key factor controlling the isotopic composition of water. Clayton et al. (1966) showed that the isotopic compositions of brines in four North American sedimentary basins plot to the right of the meteoric water line, and concluded that the brines were of meteoric origin and had undergone isotopic exchange with ^{18}O-rich sedimentary rocks. Because rocks contain much more oxygen than hydrogen, whereas water contains half as many oxygen as hydrogen atoms, interaction between water and rock ordinarily affects $\delta^{18}O$ value of the water to a greater extent than the δD value (Fig. 1). Knauth and Beeunas (1986) have argued that ^{18}O-enrichment of brines in sedimentary basins may also be the result of infiltration of evaporitic brines which have been strongly enriched in ^{18}O as the result of intense evaporation. In general, when water and rock interact, the lower the water:rock ratio the more the isotopic composition of the water is affected and the less the isotopic composition of the rock.

Temperature dependence of isotopic fractionations. Equilibrium isotopic fractionations vary with temperature. The theory of isotopic fractionation of simple diatomic gas molecules (Urey, 1947, Bigeleisen and Mayer, 1947) indicates that at high temperatures 1000 ln α should vary as T^{-2} (T is temperature in ^{0}K), and that at low temperatures it should vary as T^{-1}. What is "high" and "low" depends upon the vibrational frequencies of the isotopically substituted molecule as well as upon frequency shifts on isotopic substitution and the temperature itself, but in this regard, for all elements except hydrogen, temperatures of $0^{0}C$ and above can be considered to be high. Theoretical arguments indicate that the relation between 1000 ln $\alpha_{min-water}$ and T^{-2} for complex

192

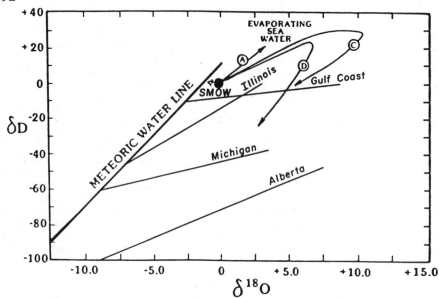

Figure 1. Summary of oxygen and hydrogen isotopic compositions of meteoric waters (after Craig, 1961), brines of the Gulf Coast, Illinois Basin, Michigan Basin and Alberta Basin (after Clayton et al., 1966) and highly evaporate sea water (along trajectories labeled A, C, and D, after Knauth and Beeunas, 1986).

molecules and/or condensed phases need not be linear (or even monotonic). However it has been found that for most mineral-water systems for which there are good data concerning oxygen isotopic fractionations over an extended temperature range, plots of $1000 \ln \alpha_{min-water}$ vs T^{-2} are linear or almost so. As a result, it is useful to display oxygen isotopic fractionation data on such plots, and in some instances, to extrapolate lines on those graphs as a way of estimating isotopic fractionations in temperature ranges for which experimental data are not available. As will be seen in subsequent sections, the variation of the hydrogen isotope fractionation between a mineral and water as a function of temperature is typically much more complex.

KINETICS OF MINERAL-WATER ISOTOPIC EXCHANGE

The isotopic composition of a mineral can provide information about a geologic process only if the mineral has retained the isotopic composition it acquired during that process. Rates of isotopic exchange between minerals and water are not well known, but data from isotopic exchange experiments in the laboratory and evidence from the isotopic compositions of naturally occurring samples provide some information. Massive exchange between the interlayer water of expandable clays and an aqueous vapor or liquid phase at room temperature occurs within a matter of hours, and hence the $\delta^{18}O$ and δD values of interlayer water provide no geologic information (Savin and Epstein, 1970a). Exchange of structural oxygen and hydrogen in clays is much slower, and in general, oxygen isotope exchange proceeds more slowly than hydrogen isotope exchange (O'Neil an Kharaka, 1976). In the absence of mineralogic reaction, exchange of both oxygen and hydrogen isotopes between phyllosilicates and water is slow in a geologic time frame at temperatures prevailing at the earth's surface. At higher temperatures rates of exchange may become significant. This topic recently been reviewed by O'Neil (1970b).

Exchange Rates Inferred from δD and δ¹⁸O Values in Natural Systems

Fine-grained minerals in nature are often out of isotopic equilibrium with the present-day aqueous environment, implying resistance to isotopic exchange between the mineral and water. When it is possible to estimate the length of time the mineral has been in the disequilibrium environment, limits can be set on the rate of isotopic exchange.

There is a great deal of data concerning the rate of exchange of clay minerals in the marine environment, in which the temperature for the past 15 million years has been within a few degrees of 0°C (Savin, 1977). Savin and Epstein (1970b) noted that most clay minerals in ocean sediments were detrital and of continental origin, and that in spite of thousands or millions of years' residence in the ocean they retained isotopic compositions they had obtained in their continental sources. Yeh and Savin (1976) compared δ¹⁸O values of different size fractions of detrital clay from a series of piston cores. They assumed that the initial δ¹⁸O values of all fractions from an individual core sample had been the same prior to deposition in the oceans, and that isotopic differences among fractions of a single sample were caused by isotopic exchange with sea water. The data indicate that if the assumptions were valid, less than 5 percent of the oxygen had exchanged in clay fractions coarser than 0.5 μm in times ranging from 14 thousand to 2.7 million years. Yeh and Savin concluded that as much as 42 percent of the oxygen of the <0.1μm fraction of a smectite rich sample may have exchanged. However, Yeh and Eslinger (1986), in a more refined study demonstrated that much smaller amounts of exchange had occurred in the <0.1 μm fraction of Gulf of Mexico sediments, suggesting that the higher values reported by Yeh and Savin (1976) may have resulted from contamination of the detrital smectite with ¹⁸O-rich authigenic smectite. In a similar study, Yeh and Epstein (1978) reported no evidence for hydrogen isotope exchange in detrital clays in ocean sediments in 2 to 3 m.y., except for the <0.1 μm fraction. They estimated that the maximum amount of exchange of hydrogen in that fraction was between 8 and 28 percent. Eslinger and Yeh (1981) reported no evidence for either oxygen or hydrogen isotopic exchange of even the fine-grained (<0.1 μm) clays in a 500 m column of Pleistocene sediments from Deep Sea Drilling Project (DSDP) Site 180 in the Aleutian trench. Eslinger and Savin (1976) found no evidence for isotopic exchange between <0.3 μm detrital mixed-layer illite/smectite and pore water in DSDP Site 323 in sediments as old as late Cretaceous and buried to depths as great as 700m, but that study was not sensitive to small amounts of exchange.

The refractory nature of detrital clay minerals in the marine environment is also apparent from other data. Detrital clays in the marine environment retain both the $^{87}Sr/^{86}Sr$ ratios (Dasch, 1969) and the K/Ar ages (Hurley et al., 1963) acquired in the continental source areas. As a result, both stable and radiogenic isotope measurements are useful as tracers of the origin of detrital clays in ocean sediments. This approach has been applied to the detrital clays of the South

Atlantic (Lawrence, 1979), the western Indian Ocean (Tsirambides, 1986), and the coast of northwestern Europe (Salomons et al., 1975).

Lack of isotopic equilibrium between a clay mineral and modern ambient water, or between two clay minerals in the same sample, or among different size fractions of the same mineral in a single sample can be taken as evidence for resistance to isotopic exchange in a variety of water-rich geologic settings: the weathering environment (Sheppard et al., 1969; Lawrence and Taylor, 1972; Hassanipak and Eslinger, 1985; Bird and Chivas, 1988a); buried shale sequences (Yeh and Savin, 1977); and diagenetically altered sandstones (Longstaffe, 1984, 1986; Burrows, 1985). In a study of the burial diagenesis of the shales of the U.S. Gulf Coast, Yeh and Savin (1977) found that the extent of internal isotopic disequilibrium, as indicated by the range of $\delta^{18}O$ values of illite/smectite of different particle sizes, decreased markedly in the depth range 2.5 to 4 km (temperature range 68 to 110°C) in which the conversion of smectite layers to illite layers was greatest (Fig. 2). By a depth of 5 km (155°C, upper Oligocene sediment), clay (predominantly illite/smectite) of several different particle sizes in the same sample had almost indistinguishable $\delta^{18}O$ values, and by inference, had undergone almost complete isotopic exchange with ambient fluids. This exchange was almost certainly facilitated by the accompanying mineralogic reaction.

In contrast to the resistance of clay minerals to isotopic exchange at low temperatures demonstrated in many studies, Bird and Chivas (1988b) argued that the isotopic compositions of a set of Permian kaolinites of weathering origin had undergone hydrogen, but not oxygen isotopic exchange at temperatures no higher than 60°C during the long period of time since their formation.

Exchange Rates Inferred from Laboratory Experiments

Data on the kinetics of oxygen and hydrogen isotopic exchange of phyllosilicates have been obtained in a number of experimental studies. O'Neil and Taylor (1969) reported that oxygen isotope exchange between muscovite and pure water was "prohibitively slow" in laboratory experiments conducted between 350° and 600°C at a pressure of 1 to 1.5 kbar. However, paragonite heated in KCl solutions underwent conversion to muscovite accompanied by oxygen isotopic exchange of about 60 to 85 percent of the oxygen in the mineral in time periods ranging from a few days to several weeks. While those data may reflect isotope exchange between a solid and water, it is more likely that one phase dissolved and another precipitated.

O'Neil and Kharaka (1976) measured rates of hydrogen and oxygen exchange between water and the clay minerals kaolinite, illite, and smectite at temperatures as high as 200°C (350°C for illite) and for times ranging from approximately two months to 1.5 years. Some exchange of both oxygen and hydrogen was observed in most of the experiments. At temperatures at or below 200°C the amount of oxygen isotope exchange of kaolinite and of illite was small, but 19 percent of the oxygen in the smectite was exchanged at 200°C in about 8 months. Between 33 and 86 percent of the oxygen in the kaolinite was exchanged at 350°C in about 1 year. Hydrogen isotopes exchanged much more rapidly than did oxygen isotopes in all experiments. Comparison of the extents of oxygen and hydrogen isotope exchange in different experiments strongly

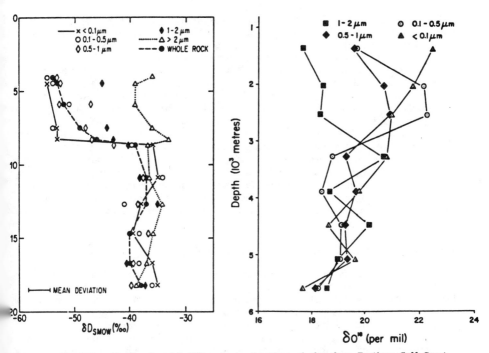

Figure 2. $\delta^{18}O$ and δD values of different size fractions of clay from Tertiary Gulf Coast shale sequence, CWRU Well 6 (after Yeh and Savin, 1967; Yeh 1980). Note that there are large intra-sample variations in both oxygen and hydrogen isotope ratios of shallow samples, but that samples become isotopically much more homogeneous in the depth range in which significant conversion of expandable mixed-layer illite/smectite to more illitic clay occurs.

suggests that the mechanism of hydrogen isotope exchange is replacement of -H rather than -OH. O'Neil and Kharaka concluded that the reason that smectite exchanged more rapidly than the other clay minerals might in part be the finer particle size of the smectite used in the experiments, but that it was almost certainly also a reflection of the greater access of water to the aluminosilicate structure in the expandable mineral. This conclusion is strengthened by the observation of James and Baker (1976) that illite which had been artificially expanded by intercalation of sodium tetraphenyl boron underwent significant oxygen isotopic exchange at room temperature in periods ranging from a few months to a few years. Similarly, Lawrence and Taylor (1972) reported that halloysite exchanges hydrogen isotopes rapidly with water. Thus in all of the cases noted in which rapid isotopic exchange at low temperature between a clay and water was noted, the mineral is either naturally expandable or has been rendered expandable by artificial means.

Additional information about rates of hydrogen isotope exchange between phyllosilicates and water at elevated temperatures comes from several studies (discussed in more detail in a later section) in which attempts were made to measure equilibrium isotopic fractionations in hydrothermal bomb experiments. Most hydrous minerals undergo rapid exchange of hydrogen isotopes at temperatures of 400°C and above. Suzuoki and Epstein (1976) reported values ranging between 17 and 100 percent exchange for a variety of phyllosilicates during experiments between 400 and 500°C and lasting between approximately two weeks and two months. Kaolinite and serpentine underwent more rapid exchange than did muscovite or biotite. Sakai and Tsutsumi (1978) measured extents of hydrogen isotope exchange between 18 and 25 percent between serpentine and water at 100°C in 2 to 4 months. Graham et al. (1987) exchanged the hydrogen of aluminous chlorite with water. At temperatures of 400°C and below they found the rate of exchange to be extremely slow, but at 500°C and above they obtained useful kinetic and isotopic equilibrium data. Treating their data with a diffusion model, and assuming a reasonable range of values for cooling of regionally metamorphosed terranes, they suggested that hydrogen isotope exchange between chlorite and water was sufficiently rapid to permit retrograde isotopic exchange at temperatures below those of greenschist facies. However, the rate of exchange is such that in terrains with greater cooling rates, such as hydrothermally altered oceanic crust, chlorite in greenschist facies rocks may retain its original hydrogen isotopic composition.

Discussion of mineral–water exchange kinetics

The data from natural systems as well as the data from laboratory exchange experiments indicate that at surface and near-surface temperatures the rate of oxygen and hydrogen isotope exchange between water and most phyllosilicates is slow, even on a geologic time scale. Both oxygen and hydrogen isotope ratios of clays appear to retain the isotope ratios they acquired when formed for periods of many millions, perhaps tens or hundreds of millions of years. When exchange does occur, hydrogen isotope ratios are likely to be altered to a greater extent than oxygen isotope ratios. At higher than surface temperatures, and especially above 100°C, the rate of isotopic exchange increases, but kinetics are, for the most part, insufficiently well known to predict rates of isotopic alteration of minerals in natural environments.

EQUILIBRIUM ISOTOPE FRACTIONATIONS

The approaches

Four approaches have been commonly used to estimate equilibrium oxygen and hydrogen isotopic fractionations between minerals and water. These are laboratory equilibration experiments, inferences from the isotopic compositions of naturally occurring samples, statistical mechanical calculations, and calculations based on an empirical bond-type model. Each of these has advantages and disadvantages.

Laboratory equilibration experiments. In laboratory equilibration experiments, a mineral and water of known isotopic composition are allowed to exchange isotopically at a constant temperature until they have equilibrated with one another. The isotopic fractionation factor, c mineral-water is calculated from the δ values of the two phases. This method, and several variants (including those involving three isotopes of oxygen, and those involving incomplete isotopic exchange) were reviewed recently by O'Neil (1987a). When it is possible to demonstrate that isotopic equilibrium between the mineral and water have been achieved, the laboratory equilibration method provides the most accurate estimates of isotopic fractionation factors.

There are, however, difficulties encountered with this approach. The rates of isotopic exchange reactions can be very slow, especially at low temperatures, preventing the achievement of isotopic equilibrium. The data of O'Neil and Kharaka (1976) suggest that a temperature of $200^{\circ}C$ may be a lower limit for oxygen isotopic exchange experiments between non-expandable clays and water, and that $100^{\circ}C$ may be a lower limit for hydrogen isotopic exchange experiments. To circumvent the problem of slow isotopic exchange between most minerals and water, experiments are sometimes designed in which the mineral of interest is synthesized from a glass or gel or from other minerals in the presence of an aqueous solution. While this approach can be useful, isotopic equilibrium can be rigorously demonstrated only in the case of a true isotopic exchange reaction between two phases. If the solid phase undergoes either a mineralogic transformation or a solution and reprecipitation process, kinetic isotopic fractionations may occur, and the resultant fractionation measured may not be an equilibrium one.

Naturally occurring samples. Many isotopic fractionations between low temperature minerals and water have been estimated from the isotopic compositions of naturally occurring minerals for which there is evidence of formation in the presence of water of well-defined temperature and isotopic composition. This is virtually the only experimental approach available in cases in which the rates of isotope exchange reactions are prohibitively slow, and/or in which minerals cannot be synthesized in the laboratory at appropriate temperatures. Difficulties with the approach are obvious. It is not possible to demonstrate that a mineral formed in isotopic equilibrium with the environment. Consistency of mineral-water isotopic fractionations obtained for the same mineral in a variety of deposits formed at the same temperature may be used as an argument in favor of isotopic equilibrium. However, consistency is not a conclusive argument for equilibrium. The argument is strengthened if both oxygen and hydrogen isotope fractionations are consistent for a suite of samples formed at similar temperatures. In the case of minerals formed from solutions of meteoric origin, a graph of δD vs. $\delta^{18}O$ can be useful. When different specimens of a mineral are formed at the same temperature in the presence of meteoric waters of different isotopic compositions, the relation between δD and $\delta^{18}O$ of the clay (Fig. 3) is, from Savin (1967)

$$\delta D = 8 \times \frac{\alpha^{hy}}{\alpha^{ox}} \times \delta^{18}O + 1000 \times 8 \times \left[\frac{\alpha^{hy}}{\alpha^{ox}} - 6.99 \, \alpha^{hy} - 1 \right]$$

Isotopic compositions of kaolinites of weathering origin are shown on such a graph in Fig. 4.

Additional uncertainties are encountered in this approach because minerals, especially in surficial and near-surface deposits may form over an extended period of time during which climate may have varied. Hence, it is often difficult to define either the temperature of formation or the isotopic composition of the ambient fluid within narrow limits.

Statistical mechanical calculations. Urey (1947) and Bigeleisen and Mayer (1947) described methods for calculating equilibrium fractionation factors using statistical mechanical techniques. O'Neil (1987a) has presented an excellent summary of the basic techniques. Application to

198

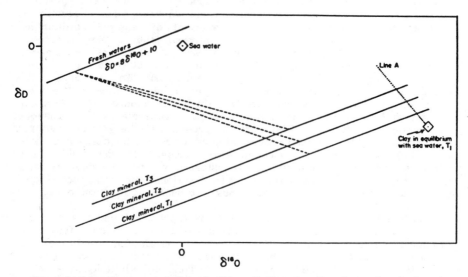

Figure 3. Schematic representation of $\delta^{18}O$ and δD systematics of a suite of clays in equilibrium with meteoric water. Each solid line represents the isotopic composition of a suite of clays in equilibrium with different meteoric waters at a single temperature $T_1 < T_2 < T_3$. Each dashed line connects the isotopic compositions of a single meteoric water with the clay in equilibrium with it at a single temperature: T_1, T_2, or T_3. The dotted line represents the isotopic composition of a suite of clays in equilibrium with sea water at a range of temperatures.

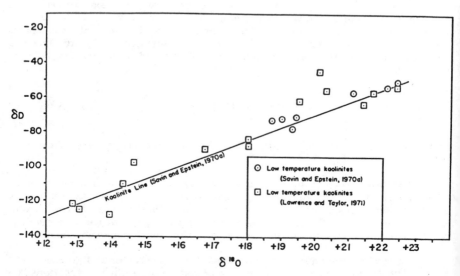

Figure 4. $\delta^{18}O$ and δD values of kaolinites of weathering origin from a variety of locations.

oxygen isotopic fractionations in mineral systems is somewhat complicated, but has been done by Bottinga and Javoy (1973), Kieffer (1982) and others. At present the technique has not been extensively used to calculate fractionations involving phyllosilicates, and there are few comparisons between calculated and experimentally measured mineral-water fractionations. However, the comparisons that have been made by Kieffer (1982) suggest that her technique may be useful for obtaining oxygen isotope fractionations at low temperatures between minerals and water. There have been no published successful attempts to calculate hydrogen isotope fractionations between minerals and water.

Empirical bond-type calculations. Given the lack of accurately determined isotopic fractionation factors for many silicates, the chemical variability of many phyllosilicates which form at low temperatures, and the sluggishness of low temperature isotope exchange reactions in the laboratory, it is useful to be able to approximate isotopic fractionation factors in some other fashion. Empirical bond models, in which it is assumed that oxygen in a chemical bond has similar isotopic behavior regardless of the mineral in which the bond is located, are useful in this regard. By assuming that the oxygen isotopic fractionation between two minerals is simply the weighted sum of the fractionations of the bonds they contained, Taylor and Epstein (1961) were able to make good approximations of the isotopic fractionations among a variety of igneous minerals. This approach to estimating oxygen isotopic fractionations among minerals has been refined and extended by Garlick (1966), Savin (1967), Wenner and Taylor (1971), and others. A further refinement, dealing primarily with phyllosilicates, is presented here. A semi-empirical approach, in which isotopic effects of vibrational frequencies, geometric arrangements, and bond strengths are considered, in addition to bond types, has been discussed by Schutze (1980).

While fractionation factors obtained using this approach must be regarded as only approximate, the approach is often the only one available. The basic assumption of the technique, i.e., that the isotopic behavior of oxygen in a mineral is dependent only upon the atoms to which the oxygen is bonded and not upon the structure of the mineral, is not strictly true. For example, there are small isotopic fractionations between calcite and aragonite. Still, the approach is useful for approximating oxygen isotope fractionations between minerals and water in the absence of better experimental or theoretical results.

The relative proportions of different oxygen-containing bonds in a mineral can be worked out with the aid of a crystal model. We have developed five conventions for work with phyllosilicates. (1) Interlayer cations that separate tetrahedral-octahedral-tetrahedral (T-O-T) layers from one another are ignored. (2) Bonding between T-O-T layers (or between T-O-T layers and brucite sheets in chlorite) is ignored. (3) Apical oxygen atoms (i.e., atoms shared by a tetrahedral and an octahedral sheet) are considered to participate only in bonds uniting an octahedral cation with a tetrahedral cation. While an apical oxygen can also be considered, in part, as a bridge between two octahedral cations, that is ignored. (4) In dioctahedral structures, bonds involving apical oxygens are considered as bonding a tetrahedral cation to an octahedral cation. In trioctahedral structures, a bond involving an apical oxygen is considered as bonding a tetrahedral cation to 1.5 octahedral cations. (Thus, 22.22 percent of the oxygen in kaolinite, $Al_2Si_2O_5(OH)_4$, is counted as being bonded in Al-O-Si bonds and 22.22 percent of the oxygen in serpentine, $Mg_3Si_2O_5(OH)_4$, is counted as being bonded in Mg-O-Si bonds.) (5) When a cationic site contains more than one element, the cations in the site are randomly distributed.

Two additional approximations are made in the absence of better data. (1) The isotopic behavior of Al- or Fe-containing bonds is independent of whether the atoms are octahedrally or tetrahedrally coordinated. (2) The isotopic behavior of all divalent and trivalent cations (denoted in what follows as M) other than Al is similar. (Thus, no differences are attributed to the oxidation state of Fe.) As additional data become available, these approximations should become unnecessary.

The relative proportions of various oxygen-containing bond types in a variety of minerals are given in Table 1. These may be used with the estimated fractionation equations for the different bond types given in Table 2 and shown in Figure 5 to compute a mineral-water oxygen isotope fractionation for any mineral. For example,

$$1000 \ln \alpha_{talc-water} = 0.5 \times 1000 \ln \alpha_{Si-O-Si \ water}$$

$$+ \ 0.333 \times 1000 \ln \alpha_{Si-O-M \ water} + 0.1667 \times 1000 \ln \alpha_{M-OH \ water}.$$

The accuracy of an oxygen isotope fractionation estimated in this way is no better than the assumption that bond-water fractionations in a mineral are additive. That is, each oxygen-containing bond fractionates oxygen with respect to water independently of all of the other bonds in the mineral. At this time there are not sufficient data to evaluate that assumption critically, and hence to place quantitative limits on the accuracy of mineral-water fractionations estimated using this technique.

The bond-type approach is also limited by the accuracy with which the bond-water fractionations given in Table 2 and shown in Figure 5 are known. The quality of these estimates varies greatly, both from bond-type to bond-type, and in some cases, as a function of temperature for a single bond type. For example, the Si-O-Si (i.e., quartz-water) fractionation is well-known at temperatures above about 250°C as the result of careful laboratory equilibration studies in the quartz-water system by Matsuhisa et al. (1979). The same fractionation is much less accurately known at 0°C, where we have assumed a linear extrapolation from the higher temperature data. While the value of the fractionation (1.0424) estimated in this way is reasonable in view of measured isotopic compositions of naturally occurring quartz and chert from the deep sea (Garlick, 1969; Savin, 1973; Knauth and Epstein, 1975), it certainly could be in error by as much as 2 or 3 per mil. The estimated value of the Al-OH fractionation factor at 17°C (1.0165) was derived in two ways from data for natural samples (see discussions of gibbsite and kaolinite fractionations below) and can be considered well defined (+/- about 1 per mil) at that temperature. However, the slope of the curve through the 17°C point is not well known, so the Al-OH fractionation is not as well known at higher and at lower temperatures. Probably the most poorly known of the bond-water fractionations is that for M-OH. The estimate of this fractionation in Table 2 is based on a variety of circuitous arguments. Ironically, it is a fractionation which is probably readily amenable to calibration in the laboratory by equilibration studies in the system brucite-water.

Table 1. Percentage abundances of various bond types in minerals

Mineral	Si-O-Si	Si-O-Al	Si-O-M	Al-O-Al	Al-O-M	M-O-M	Al-OH	M-OH
Quartz	100	....	....	....	....	....	....	....
K-feldspar	56.25	37.5	....	6.25	....	....	....	....
Kaolinite	33.33	22.22	....	....	....	....	44.44	....
Pyrophyllite	50	33.33	....	....	....	16.67	....	
Serpentine	33.33	....	22.22	....	....	....	....	44.44
Talc	50	....	33.33	....	....	....	....	16.67
Smectite[1]	50	27.83	5.58	....	....	....	13.92	2.75
Smectite[2]	37.84	22.18	18.13	2.47	2.71	....	6.25	10.42
Nontronite[3]	42.09	7.57	30.58	.35	2.75	....	....	16.67
Chlorite[4]	18.75	20.83	8.33	4.86	2.78	....	5.55	38.88
Chlorite[5]	22.69	27.96	....	4.91	....	....	18.89	25.56
Chlorite[6]	16.22	14.07	15.5	3.05	6.72	....	13.44	31.00
Illite[7]	36.25	37.54	3.54	5.5	0.63	....	14.58	2.08
Muscovite[8]	28.13	43.75		11.46	....	....	16.67	....
Biotite[9]	28.13	18.75	25	3.13	8.33	....	....	16.67

Notes:
M designates divalent and trivalent ions other than Al
[1] $R^+_{.33}(Al_{1.67}M_{.33})(Si_4O_{10})(OH)_2$
[2] $R^+_{.92}(Al_{.75}M_{1.25})(Si_{3.48}Al_{.52}O_{10})(OH)_2$
[3] $R^+_{.33}Fe_2(Si_{3.67}Al_{.33}O_{10})(OH)_2$
[4] $[Mg_{2.5}Fe_{.5}(OH)_6](Al_{1.5}Fe_{1.5})(AlSi_3O_{10})(OH)_2$
[5] $[Mg_{2.3}Al_{.7}(OH)_6](Al_2)(Al_{.7}Si_{3.3}O_{10})(OH)_2$
[6] $[Mg_{2.9}Fe_{1.5}Al_{1.21}(OH)_6](Mg_3)(Al_{1.21}Si_{2.79}O_{10})(OH)_2$
[7] $K_{.85}(Al_{1.75}M_{.25})(Si_{3.4}Al_{.6}O_{10})(OH)_2$
[8] $KAl_2(AlSi_3O_{10})(OH)_2$
[9] $KMg_3(AlSi_3O_{10})(OH)_2$

Table 2. Estimated bond-water oxygen isotope fractionations

Bond	Equation	Source
Si-O-Si	$1000 \ln \alpha = 3.34 \times 10^6 T^{-2} - 3.31$	1
Al-O-Si	$1000 \ln \alpha = -11.6 \times 10^3 T^{-1} + 5.40 \times 10^6 T^{-2} + 6.50$	2
Al-O-Al	$1000 \ln \alpha = -23.14 \times 10^3 T^{-1} + 7.455 \times 10^6 T^{-2} + 16.31$	2
M-O-M	$1000 \ln \alpha = -45.4 \times 10^3 T^{-1} + 24.3 \times 10^6 T^{-2} - 5.61 \times 10^9 T^{-3} + 0.504 \times 10^{12} T^{-4} + 22.38$	3
Si-O-M	$1000 \ln \alpha = -22.7 \times 10^3 T^{-1} + 13.82 \times 10^6 T^{-2} - 2.81 \times 10^9 T^{-3} + 0.252 \times 10^{12} T^{-4} + 9.54$	4
Al-O-M	$1000 \ln \alpha = -34.3 \times 10^3 T^{-1} + 15.88 \times 10^6 T^{-2} - 2.81 \times 10^9 T^{-3} + 0.252 \times 10^{12} T^{-4} + 19.33$	5
Al-OH	$1000 \ln \alpha = 29.6 \times 10^3 T^{-1} - 4.25 \times 10^6 T^{-2} - 35.28$	6
M-OH	$1000 \ln \alpha = 21.7 \times 10^3 T^{-1} - 2.492 \times 10^6 T^{-2} - 34.76$	7

Sources of fractionations
1. Quartz-water fractionation of Matsuhisa et al. (1979) between 250 and 500°C.
2. Calculated from quartz-water and albite-water fractionations (400 to 500°C) of Matsuhisa et al. (1979) and the assumption that $\alpha_{Al-O-Si} = (\alpha_{Si-O-Si} * \alpha_{Al-O-Al})^{1/2}$; and at low temperature from the estimated kaolinite-water fractionation at 17°C (1.0265), the quartz-water fractionation of Matsuhisa et al. (1979) and the Al-OH fractionation (this table)
3. Curve fitted to the magnetite-water results calculated by Becker (1971) and adjusted by Friedman and O'Neil (1977)
4. from $\alpha_{Si-O-M} = (\alpha_{Si-O-Si} * \alpha_{M-O-M})^{1/2}$
5. from $\alpha_{Al-O-M} = (\alpha_{Al-O-Al} * \alpha_{M-O-M})^{1/2}$
6. Curve fitted at the high temperature end to values inferred from the muscovite-water data of O'Neil and Taylor(1969) and the Si-O-Si, Al-O-Si, and Al-O-Al fractionations listed above, and at the low temperature end to a value of 1.0165 at 17°C (discussed in text).
7. Curve fitted to data from metamorphic biotite-muscovite assemblages at high temperatures. Estimated at low temperatures from a combination of the goethite-water data of Yapp (1987) and the M-O-M curve.

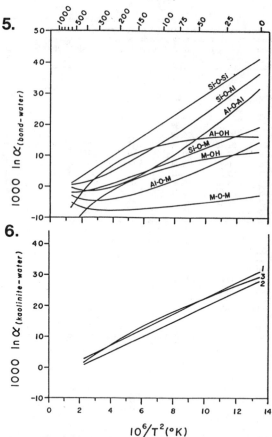

Figure 5. Estimated oxygen isotopic fractionations between oxygen-containing bonds and water as a function of temperature. M stands for any divalent or trivalent cation other than Al^{+3}. Equations of these curves are listed in Table 2.

Figure 6. Three estimates of the oxygen isotope fractionation between kaolinite and water as a function of temperature. Equations are listed in Table 3. Curve 1 is from Eslinger (1971) adjusted by Land and Dutton (1978). Curve 2 is from Kulla and Anderson (1978) modified by Anderson (personal communication). Curve 3 is calculated using the bond-type approach.

Table 3. Estimated oxygen isotope fractionations between kaolinite and water

Equation 1000 ln $\alpha^{ox}_{kaol-wat}$ =	Type of Data	Source
$2.50 \times 10^6 T^{-2} - 2.87$	a,b	1
$2.42 \times 10^6 T^{-2} - 4.45$	b	2
$10.6 \times 10^6 T^{-1} + 0.42 \times 10^6 T^{-2} - 15.337$	c	3

Notes:
a. based on naturally occurring materials
b. based on synthesis experiments
c. based on bond-model calculation
1. Eslinger (1971) adjusted by Land and Dutton (1978)
2. Kulla and Anderson (1978) modified by Anderson (pers. comm.)
3. this paper

We periodically revise the equations describing the bond-type fractionations in Table 2 to be consistent with new experimental data or data on naturally occurring samples. At the present time, oxygen isotope fractionations clearly cannot be estimated with sufficient accuracy using the bond-type approach to be useful for mineral-pair isotope geothermometry. Perhaps, even with greater refinement of bond-water fractionations, the failure of the basic assumption that bond-water fractionations are strictly additive will preclude sufficient accuracy for geothermometry. But, there are many applications of isotopic data in which knowledge of a mineral-water fractionation factor with an accuracy of 2 or 3 per mil is sufficient to permit meaningful geological conclusions to be drawn. An example is the study of the diagenetic alteration of sandstones. Formation of cogenetic mineral pairs in these rocks is very rare. However, from the isotopic composition of a single mineral it is possible to constrain the environment of its formation to lie along a trajectory on a graph of $\delta^{18}O_{water}$ vs. formation temperature, if the mineral-water fractionation has been calibrated. If it is possible to estimate the temperature of mineral formation by other than isotopic arguments (e.g., from fluid inclusion data), the $\delta^{18}O$ value of the formation fluid, and therefore its source, can then be constrained.

Another use of the bond-type approach is in evaluating the sensitivity of a mineral-water fractionation factor to variation in chemical composition.

Estimates of mineral-water fractionations

In this section the state of knowledge of oxygen and hydrogen isotope fractionations between several phyllosilicates and water and their variation with temperature is discussed.

Kaolinite. Oxygen and hydrogen isotope fractionations between kaolinite and water over a range of temperatures have been estimated on the basis of the isotopic compositions of naturally occurring samples as well as laboratory experiments. Kaolinites of weathering origin analyzed by Savin and Epstein (1970a) have $\delta^{18}O$ and δD values that plot along a line almost parallel to the line of meteoric waters (Fig. 4) and consistent with isotopic fractionation factors $\alpha \, ^{ox}_{kaol-water} = 1.0265$ and $\alpha \, ^{hy}_{kaol-water} = 0.970$ at surface temperatures (approximately 17°C). These values are similar to those obtained by Lawrence and Taylor (1971), and are probably fairly accurate. Eslinger (1971) estimated a kaolinite-water oxygen isotope fractionation of 12.3 per mil at 150°C on the basis of a single analysis of hydrothermal kaolinite from Broadlands, New Zealand. Experimental syntheses of kaolinite at elevated temperatures have been done by Kulla and Anderson (1978) between 170 and 320°C, and by Eberl and Savin (unpublished). Three estimates of the kaolinite-water oxygen isotope fractionation over a range of temperatures are given in Table 3 and shown in Figure 6. The difference between the kaolinite-water fractionations predicted by the two curves based on experimental studies and/or natural samples (Eslinger, 1971; Kulla and Anderson, 1978, as modified by Anderson, pers. comm.) is approximately 2.6 per mil at 0°C and 1.9 per mil at 200°C. The equation derived from the bond-model is closer to the curve of Eslinger than that of Kulla and Anderson at low temperatures, but that is not a completely independent result. That is, the estimate of the Al-O-Si fractionation at low temperatures is based in part on the

204

kaolinite fractionation factor (1.0265) at 17°C. We prefer the first or last equation in Table A for use at low temperatures. Choice of a fractionation equation for higher (above 200°C) temperatures must await the publication of experimental details of Kulla and Anderson.

There is good analytical evidence in support of major differences between the $\delta^{18}O$ value of Al-OH bonded oxygen and that of Al-O-Si and Si-O-Si bonded oxygen in the kaolinite structure. By reacting kaolinite in stepwise fashion with insufficient amounts of F_2 or at insufficient temperatures for complete reaction, Savin (1967) and Hamza and Epstein (1980) liberated O_2 with low $^{18}O/^{16}O$ ratios in amounts indicating that it was derived from the hydroxyl groups of the mineral. For a kaolinite of weathering origin (Langley, S. Carolina) for which Savin (1967) estimated 1000 ln $\alpha^{ox}_{kaol-wat}$ = 24.89, he estimated 1000 ln α whole min.- OH oxygen = 9.75, corresponding to a fractionation (1000 ln α) between the Al-OH bonded oxygen and water of approximately 15.14 (Fig. 7). Hamza and Epstein (1980) using a slightly different analytical technique, estimated 1000 ln α whole min.- OH oxygen = 12.6 for a kaolinite from Macon, Georgia (Fig. 8). These results suggest the possibility of development of a single-mineral oxygen isotope thermometer, in which the isotopic composition of the mineral would provide information about both the temperature and the $\delta^{18}O$ value of the water in which the mineral formed. Research into this is now underway.

Hydrogen isotope fractionations between kaolinite and water at temperatures higher than surface temperature have been derived both in experimental studies (Suzuoki and Epstein, 1976; Liu and Epstein, 1984) and in studies of naturally occurring samples (Lambert and Epstein, 1980; Marumo et al., 1980). The results of these studies are plotted in Figure 9. The results from three of the four studies, obtained using three different approaches, are consistent with one another. The results of Marumo et al. (which also include analyses of some mixtures of kaolinite and dickite) plot along a curve parallel to the other results, but are offset consistently by about 10 per mil. That could simply reflect a small error in the choice of the δD value assumed by Marumo et al for the water in which the kaolinites formed. Overall, the results in Figure 9 argue persuasively that the complex shape of the fractionation curve reflects equilibrium hydrogen isotope fractionation between kaolinite and water. The heavy curve through the data is the preferred curve at this time, but new data may require modification of the curve. The lack of any consistent isotopic difference between samples with and without dickite suggests that these polytypes, which are similar except for location of the octahedral vacancy, fractionate hydrogen isotopes similarly.

Pyrophyllite. No experimental studies of oxygen isotope fractionations between pyrophyllite and water have been published. The fractionation equation calculated using the bond-type approach,

$$1000 \ln \alpha^{ox}_{pyrophyllite-water} = 1.08 \times 10^3 T^{-1} + 2.76 \times 10^6 T^{-2} - 5.37$$

is plotted in Figure 10, where it is compared with the calculated talc-water fractionation. The calculations based on bond types indicate that at any temperature pyrophyllite should be the most ^{18}O-rich phyllosilicate discussed in this paper.

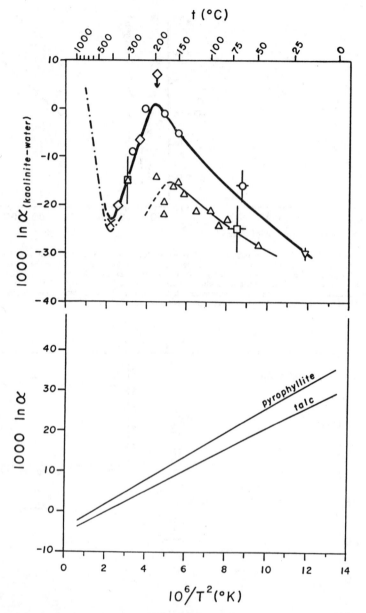

Figure 9 (top). Hydrogen isotope fractionations between kaolinite and water. Symbols are: circles from Lambert and Epstein (1980); diamonds from Liu and Epstein (1984); x from Suzuoki and Epstein (1976); squares from Sheppard et al., 1969; triangles from Marumo et al., 1980; inverse triangle from Savin and Epstein (1970a). The heavy curve is estimate by Liu and Epstein (1984) of the kaolinite-water fractionation. The thin line is drawn through the data of Marumo et al. (1980). The dash–dot line is the muscovite–water fractionation between muscovite and water of Suzuoki and Epstein et al. (1976), shown for comparison.

Figure 10 (bottom). Oxygen isotope fractionations between pyrophyllite and water and between talc and water, calculated using the bond-type approach and the data in Tables 1 and 2.

Gibbsite. While it is not a phyllosilicate, gibbsite is considered here because of its similarity to the octahedral sheet of dioctahedral phyllosilicates. Because of its common occurrence in high concentrations in lateritic deposits there are a number of analyses of gibbsite which formed under fairly well-defined conditions. Estimates of gibbsite-water oxygen isotope fractionation factors at weathering temperatures made from comparisons between gibbsite and meteoric water δ values in several studies are tabulated in Table 4.

Bernard (1978) argued that the results of Lawrence and Taylor (1971) are suspect because of possible gibbsite dehydration during heating as well as possible contamination with other phases. However, considering the uncertainties in the isotopic composition of the ambient water in which each of the gibbsites formed, and the variability of temperatures of formation, the agreement among the results of the different studies must be considered good. The oxygen isotope fractionation between gibbsite and water at weathering temperatures is similar to that estimated by Savin (1967) for the Al-OH oxygen in kaolinite and water (1.0153, see above). The hydrogen isotope fractionation between gibbsite and water is smaller than that between kaolinite and water.

Table 4. Estimated oxygen and hydrogen isotopic fractionations between gibbsite and water under weathering conditions

$\alpha^{ox}_{gibbsite-water}$	$\alpha^{hy}_{gibbsite-water}$	Source
1.018	0.985	Lawrence & Taylor (1971)
1.016+/-.001	0.992+/-.002	Chen et al. (1988)
1.0165+/-.008	0.9933+/-.0055	Bernard (1978)
1.0148	1.000	Bird (1988)

Smectite, mixed-layer illite/smectite, and illite. Early estimates by Savin and Epstein (1970a) of oxygen and hydrogen isotope fractionation factors in the smectite-water system, 1.027 and 0.94 respectively, at deep sea temperatures cannot be considered highly reliable. Those estimates were based on analyses of authigenic smectites from deep sea sediments, but the samples were not cleaned of contaminating oxide minerals (expected to have low $\delta^{18}O$ values) prior to isotopic analysis. In fact, the composition of tetrahedral and octahedral cations in smectites is highly variable, so different smectites ought to have different isotopic behavior. Estimates of smectite-water fractionation factors at low temperatures, based on analyses of naturally occurring samples, are listed in Table 5 with chemical compositional information where available.

McMurtry and Yeh (1981) reported a very ^{18}O-enriched (+30.9 per mil) authigenic smectite-rich sample from the Bauer Basin. If the phase were pure, it would suggest a minimum value of $\alpha^{ox}_{smectite-water}$ of 1.0311 at 0 to 2°C. They concluded, however, on the basis of structural formula calculations, that the sample was contaminated by (^{18}O-rich) biogenic amorphous silica, and that the $\delta^{18}O$ of the smectite component was 23.6+/-3.4 per mil. A $\delta^{18}O$ value of +30.62 per mil was reported by Yeh and Savin (1976) for an authigenic smectite, but there is no compositional data available for that sample. Hence it is not known

Table 5. Estimated smectite–water oxygen isotope fractionation factors at low temperatures

Composition	Occurrence	$\alpha^{ox}_{smectite-water}$	Ref.
3.5–4.5% Fe	weathering	1.0277	Lawrence & Taylor (1972)
6.5–8.5% Fe	weathering	1.0248	Lawrence & Taylor (1972)
not analyzed	deep sea	1.03083	Yeh & Savin (1976)
see below*	deep sea	1.0298	Hein et al. (1979)

*$(Ca_{.13}K_{.22}Na_{.41})(Ni_{.004}Cu_{.01}Zn_{.01}Mn_{.03}Ti_{.04}Mg_{.59}Fe^{+3}_{.53}Al_{.75})$
$(Al_{.52}Si_{3.48}O_{10})(OH)_2$

whether that sample was contaminated with biogenic silica. Hein et al. (1979) reported a value of 29.6 for one iron-rich dioctahedral smectite, and lower $\delta^{18}O$ values for a number of others. They interpreted those with lower values as being contaminated with terrigenous detritus, and estimated a maximum temperature of formation of 7°C. In discussions below, 30.6 per mil (corresponding to $\alpha^{ox}_{smectite-water}$ of 1.0308) is used as the equilibrium fractionation between smectite and water at 1°C, but it is obvious that there is a high degree of uncertainty in the value.

Bond-model calculations permit an estimate of the extent to which the smectite–water oxygen isotope fractionation factor varies with composition. Calculated curves for the three dioctahedral smectites given in Table 1 are shown in Figure 11. The most Si-rich smectite is calculated to have a $\delta^{18}O$ value approximately 5 per mil higher at 0°C than the nontronite which contains some tetrahedral Al. At higher temperatures differences among the $\delta^{18}O$ values of the smectites are smaller but remain significant (2.3 per mil at 150°C and 2.5 per mil at 300°C). While there is some uncertainty associated with the actual values of fractionation factors calculated using the bond-type approach, the calculations stress that Si/Al ratio and Fe content strongly affect the isotopic behavior of smectites.

Yeh (1974) and Yeh and Savin (1977) estimated the smectite–water oxygen isotopic fractionation at elevated temperatures using a series of assumptions: that $\alpha^{ox}_{smectite-water}$ is 1.0308 at 1°C; that both $1000 \ln \alpha^{ox}_{illite-water}$ and $1000 \ln \alpha^{ox}_{illite-smectite}$ vary linearly with T^{-2} over the entire temperature range represented on plots of $1000 \ln \alpha$ vs T^{-2} ; and that the illite–water fractionation equation of Eslinger and Savin (1973) and the quartz–water fractionation equation of Clayton et al. (1972) are valid. The resulting equation, as given by both Yeh (1974) and Yeh and Savin (1977) regrettably contained typographical errors. The correct version is

$$1000 \ln \alpha^{ox}_{smectite-water} = 2.60 \times 10^6 T^{-2} - 4.28,$$

A slightly modified form of the equation, which is calculated using the quartz–water fractionation curve of Matsuhisa et al. (1979), is preferable

$$1000 \ln \alpha^{ox}_{smectite-water} = 2.58 \times 10^6 T^{-2} - 4.19$$

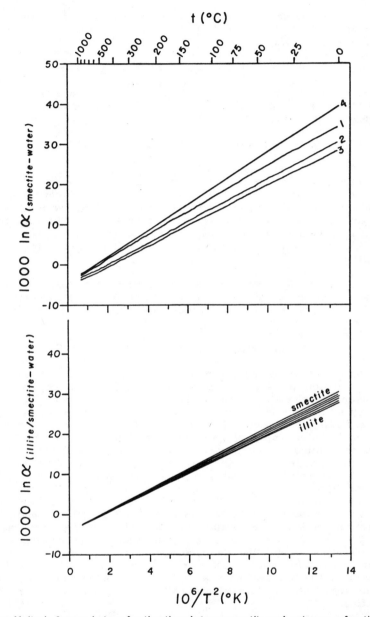

Figure 11 (top). Oxygen isotope fractionations between smectite and water as a function of temperature. Curves 1, 2 and 3 are calculated using the bond-type approach for three smectites of compositions given by footnotes 1, 2, and 3 in Table 1. Curve 4 was experimentally determined from syntheses of Mg smectites by Escande (1983). Samples for curve 4 were dehydroxylated prior to isotopic analysis. See text for further discussion.

Figure 12 (bottom). Estimates of the oxygen isotope fractionation between mixed-layer illite/smectite and water. Illite curve is adjusted from Eslinger and Savin (1973) and is derived from the isotopic compositions of hydrothermal quartz-illite pairs and the quartz-water fractionation of Matsuhisa et al. (1979). Smectite and mixed-layer (20, 40, 60, and 80% illite) curves are corrected from Yeh and Savin (1977). See text for details.

The illite-water fractionation equation calculated from the quartz-illite curve of Eslinger and Savin (1973) and the quartz-water curve of Matsuhisa et al., (1979) is

$$1000 \ln \alpha^{ox}_{illite-water} = 2.39 \times 10^6 T^{-2} - 4.19$$

By interpolating between the endmember illite and smectite fractionations, it is possible to estimate the fractionation factors for mixed-layer illite/smectites:

$$1000 \ln \alpha^{ox}_{illite/smectite-water} = (2.58 - 0.19 \times I) \times 10^6 T^{-2} - 4.19$$

where I is the proportion of illite layers in the clay. These fractionations are shown in Figure 12. However, because the curves were calculated on the basis of the isotopic composition of a single marine smectite, they cannot be considered to be representative of all illite/smectites. In fact, the range of fractionations of illite/smectites of varying illite content at most temperatures is smaller than the range of fractionations shown in Figure 11 for smectites of varying chemical composition.

Escande (1983) measured oxygen isotope fractionations between water and Mg-rich smectite (stevensite and saponite) synthesized over a range of temperatures. The fractionation equation he calculated for both smectite phases is:

$$1000 \ln \alpha^{ox}_{smectite-water} = 3.31 \times 10^6 T^{-2} - 4.82$$

This equation indicates significantly higher fractionations than does the revised equation of Yeh and Savin presented above, or any of the fractionations calculated using the bond-type model (Figure 11) especially at low temperatures. Escande's analytical technique involves dehydroxylating the smectite in vacuo at 1000°C after exchanging labile oxygen with water vapor of fixed isotopic composition at 200°C. This technique resulted in very good analytical reproducibility. Escande argued that dehydroxylation did not alter the isotopic composition of his samples but simply permitted him to achieve much greater analytical reproducibility. However, his data are not completely conclusive in this regard, and given the conclusion that hydroxyl oxygen is lower in [18]O than Al-O-Si bonded oxygen in kaolinite, the suspicion remains that dehydroxylation may have resulted in loss of low-^{18}O from the smectites. If so, the fractionations measured by Escande would be greater than those measured without the dehydroxylation step. On the other hand, Escande has pointed out that the fractionation curve of Yeh and Savin is defined in the low-temperature region by a single analysis of marine clay, the temperature, chemical composition and purity of which are not well-defined.

The oxygen isotope fractionation between illite and water is better known at high temperatures (above 150°C) than is the fractionation between water and smectite of any chemical composition. Confidence in the illite-water data comes from the similarity between the quartz-illite fractionation measured by Eslinger and Savin (1973) at Broadlands, New Zealand, and the quartz-muscovite fractionation calculated from the muscovite-water (O'Neil and Taylor, 1969) and quartz-water (Matsuhisa et al., 1979) curves experimentally determined at somewhat higher temperatures. The illite-water fractionation cannot be considered as well established at surface temperatures, however. A linear extrapolation of the high temperature curve indicates a fractionation at 0°C which is about 6 per mil lower than the fractionation indicated by the bond-type model. While that may be due in part to variation in the chemical composition of illite, it probably reflects both errors in the bond-type model and/or data and non-linearity of $1000 \ln \alpha\, {}^{ox}_{\text{illite-water}}$ as a function of T^{-2}. Lee (1984) proposed an alternative equation for $1000 \ln \alpha\, {}^{ox}_{\text{illite-water}}$ based on a combination of high temperature data and arguments that between 0 and 17°C illite and kaolinite fractionated oxygen isotopes similarly. His equation is

$$1000 \ln \alpha\, {}^{ox}_{\text{illite-water}} = -2.87 + 1.83xA + 0.0614xA^2 - 0.00115xA^3$$

where $A = 10^6T^{-2}$. The three illite-water curves discussed are plotted in Figure 13. Tentatively, we prefer the equation of Lee (1984), but it is clear that more work needs to be done in this system, especially at temperatures below 100°C.

Hydrogen isotope fractionations between dioctahedral T-O-T minerals and water are not well known. Most of the information about smectite, illite/smectite, illite, and muscovite hydrogen isotope fractionations is plotted in Figure 14. Lawrence and Taylor (1972) reported that the hydroxyl hydrogen of smectite partially exchanges with interlayer water during the dehydration procedure, permitting only approximate calculation of $\alpha\, {}^{hy}_{\text{smectite-water}}$. Other workers have not reported this problem with smectite. Lawrence and Taylor estimated $\alpha\, {}^{hy}_{\text{smectite-water}}$ to have a value of approximately 0.97 at weathering temperatures, and noted that iron-poor smectite from a weathering profile concentrated deuterium relative to the iron-rich smectite from the same profile. Yeh (1980) proposed the equation

$$\alpha\, {}^{hy}_{\text{illite/smectite-water}} = -19.6x10^3T^{-1} + 25$$

over the temperature range 29° to 120°C on the basis of the hydrogen isotope ratios of clays from a thick sequence of shales in the U.S. Gulf Coast. This equation is based upon a number of assumptions: that $1000 \ln \alpha$ varies linearly with $1/T$ over the temperature range, that the fractionation factor has a value of 0.96 at 25°C, and that within certain depth ranges in each of the wells studied, the δD value of the pore water is constant. While each of these assumptions is plausible, none is compelling. Yeh's equation is plotted in Fig. 14, as are the results of the high temperature (450° to 750°C) hydrogen isotope equilibration studies of Suzuoki and Epstein (1976) and estimates of sericite-water isotopic fractionations in the temperature range 130° to 250°C. While the curves and data in the figure are for three different minerals, their structures and compositions (and especially those of muscovite and sericite) are similar.

Serpentine. There is little experimental data on which to base an estimate of the serpentine-water oxygen isotope fractionation. Wenner and Taylor (1971) concluded that serpentine and chlorite fractionated oxygen isotopes similarly with respect to water, and proposed that their chlorite-water curve (see below) be used to calculate the serpentine-water fractionation.

$$1000 \ln \alpha\, {}^{ox}_{\text{serpentine-water}} = 1.56x10^6T^{-2} - 4.70.$$

212

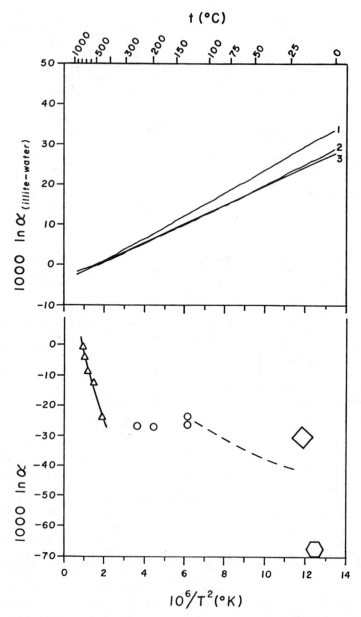

Figure 13 (top). Estimates of the oxygen isotope fractionation between illite and water. Curve 1 is calculated using the bond-type approach and the data in Tables 1 and 2. Curve 2 was proposed by Lee (1984) on the basis of a combination of bond-type arguments and data for naturally occurring samples.   Curve 3 is the illite curve adjusted from Eslinger and Savin (1973).

Figure 14 (bottom). Hydrogen isotope fractionations between dioctahedral T-O-T minerals and water as a function of temperature. Triangles are muscovite-water fractionations of Suzuoki and Epstein (1976).   Circles are fractionations estimated for naturally occurring sericites by Marumo et al. (1980).   The dashed line is inferred by Yeh (1980) for mixed-layer illite-smectite on the basis of the hydrogen isotope systematics of Gulf Coast shale sequences.   The diamond (Lawrence and Taylor, 1972) and hexagon (Savin and Epstein, 1970a) are estimates of fractionations between smectites of weathering origin and water.

The curve calculated from the bond water fractionation equations in Table 2 is

$$1000 \ln \alpha^{ox}_{serpentine-water} = 4.61 \times 10^3 T^{-1} + 3.08 \times 10^6 T^{-2}$$

$$- 0.623 \times 10^9 T^{-3} + 0.056 \times 10^{12} T^{-4} - 14.43$$

The two equations yield very different values, especially at temperatures below 100°C (Fig. 15). We tentatively prefer the latter equation, but at this time, neither one can be considered to have a very strong basis.

The hydrogen isotope fractionation between serpentine and water has been more thoroughly studied than the oxygen. It has been determined experimentally in the temperature range 100 to 500°C by Sakai and Tsutsumi (1978), who proposed the equation

$$1000 \ln \alpha^{hy}_{serpentine-water} = 2.75 \times 10^7 T^{-2} - 7.69 \times 10^4 T + 40.8.$$

There is considerable scatter of the experimental data about the curve given by this equation, for which Sakai and Tsutsumi quote a standard error of estimate of approximately 6 per mil. The equation (Figure 16) is consistent with a single data point at 400°C determined by Suzuoki and Epstein (1976). Wenner and Taylor (1973) proposed a tentative hydrogen isotope fractionation curve based upon the isotopic compositions of naturally occurring samples as well as the 400°C experimental point of Suzuoki and Epstein. The curve of Wenner and Taylor differs greatly from that proposed by Sakai and Tsutsumi at low and intermediate temperatures. Because of the unverified assumptions made by Wenner and Taylor, concerning isotopic compositions of serpentine-forming fluids and the resistance of serpentine to post-formational hydrogen isotope exchange, we consider the curve of Sakai and Tsutsumi (1978) to be preferable. Others, notably Ikin and Harmon (1981), have considered the partial exchange data of Sakai and Tsustsumi to be unreliable at temperatures below 400°C because of the partial exchange technique that they used, and prefer the curve of Wenner and Taylor (1973) at low temperatures.

Chlorite. Preliminary results of an experimental study of the chlorite-water oxygen isotope fractionation have been published by Cole (1985). In these experiments, which were not true isotope exchange experiments, biotite was altered to chlorite in NaCl solution at temperatures ranging from 170 to 300°C. Wenner and Taylor (1971) proposed an approximate chlorite-water oxygen isotope fractionation equation based upon the isotopic composition of naturally occurring chlorites and coexisting minerals in metasediments. The fractionation equations of Cole and of Wenner and Taylor are given in Table 6 and shown in Figure 17, as are fractionations for chlorites of three different chemical compositions estimated using bond-type calculations.

The oxygen isotope fractionations between chlorite and water calculated with the bond-type approach are very dependent upon chemical composition. The range of fractionations calculated for the three chlorites selected for their compositional differences increases from 1.5 per mil at 400°C to 4.6 per mil at 0°C. Those numbers are subject to revision because they are somewhat dependent on the rather poorly known fractionation between M-O-Si and water. The calculations do

214

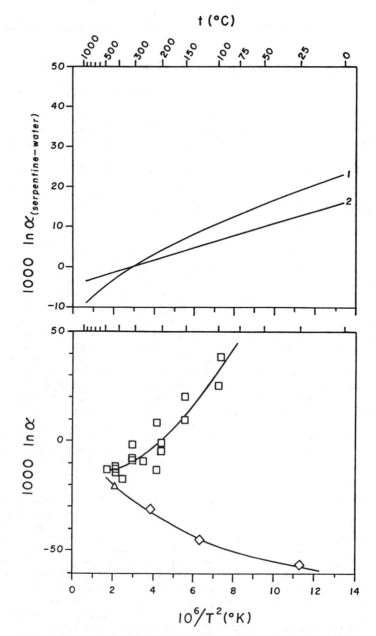

Figure 15 (top). Two estimates of the oxygen isotope fractionation between serpentine and water. Curve 1 is calculated using the bond-type approach and the data in Tables 1 and 2. Curve 2 is an approximate fractionation proposed by Wenner and Taylor (1971). Curve 2 was derived from the isotopic composition of coexisting minerals in chlorite-containing metamorphic assemblages and the assumption, based upon bond-type considerations, that chlorite and serpentine should fractionate oxygen isotopes in similar fashion.

Figure 16 (bottom). Two estimates of the hydrogen isotope fractionation between serpentine and water. The upper curve and the squares through which it passes are based upon experimental equilibration studies by Sakai and Tsutsumi (1978). The triangle is an experimental point of Suzuoki and Epstein (1976). The diamonds and the lower curve are from Wenner and Taylor (1973) and are based upon the isotopic compositions of naturally occurring serpentines.

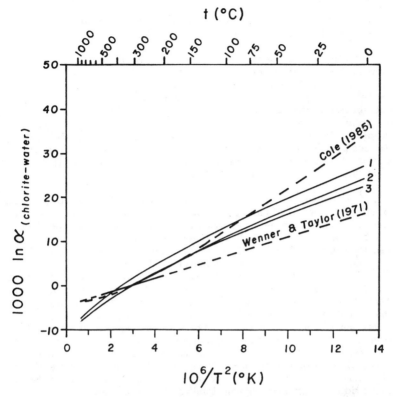

Figure 17. Estimates of the oxygen isotope fractionation between chlorite and water. The Wenner and Taylor (1971) curve was approximated on the basis of the isotopic compositions of coexisting minerals in chlorite-containing metamorphic assemblages. The Cole (1985) curve is based upon experimental formation of chlorite by hydrothermal alteration of biotite. The curves labeled 1, 2, and 3 are calculated using the bond-type approach and the chlorite compositions listed on footnotes d, c, and e, respectively, of Table 6.

===
Table 6. Estimated oxygen isotope fractionations between chlorite and water
===

Equation $1000 \ln \alpha^{ex}_{chlorite-water} =$	Type	Ref.
$-8.38 \times 10^3 T^{-1} + 4.81 \times 10^6 T^{-2}$	a	1
$1.56 \times 10^6 T^{-2} - 4.70$	b	2
$3.72 \times 10^3 T^{-1} + 2.50 \times 10^6 T^{-2} - 0.312 \times 10^9 T^{-3} + 0.028 \times 10^{12} T^{-4} - 12.62$	c	3
$6.78 \times 10^3 T^{-1} + 1.19 \times 10^6 T^{-2} - 13.68$	d	3
$2.56 \times 10^3 T^{-1} + 3.39 \times 10^6 T^{-2} - 0.623 \times 10^9 T^{-3} + 0.056 \times 10^{12} T^{-4} - 11.86$	e	3

Notes:
- a. based on hydrothermal synthesis experiments
- b. based on natural samples
- c. bond-type calculation $[Mg_{2.5}Fe_{.5}(OH)_6](Al_{1.5}Fe_{1.5})(AlSi_3O_{10})(OH)_2$
- d. bond-type calculation $[Mg_{2.3}Al_7(OH)_6](Al_3)(Si_{3.3}Al_7O_{10})(OH)_2$
- e. bond-type calculation $[Mg_{.39}Fe_{1.5}Al_{1.21}](Mg_3(Si_{2.79}Al_{1.21}O_{10})(OH)_2$
- 1. Cole (1985)
- 2. Wenner & Taylor (1971)
- 3. This study
===

clearly point out, however, the importance of knowing the chemical composition of chlorite before interpreting isotopic data.

In spite of a large number of assumptions on which the approximate fractionation curve of Wenner and Taylor (1971) is based, their equation yields fractionations which are similar to those of Cole (i.e., within 0.5 per mil) in the temperature range 250 to 375ºC. However, because chemical compositions are not given in either study, it is not clear whether or not this apparent internal consistency is coincidental. Pending publication by Cole of his experimental details and chlorite compositions, we are inclined to place most confidence on his chlorite-water curve at temperatures above 150ºC and on the calculated curves at lower temperatures.

An experimental equilibration study of hydrogen isotope fractionations between aluminous chlorite and water (Graham et al., 1987) indicated a constant value of $\alpha^{hy}_{chlorite-water}$ of about -28 per mil in the temperature range 500 to 700ºC. Those authors also synthesized (Fig. 18) fractionations inferred at lower temperatures by others from the isotopic compositions of natural chlorites. The curve drawn through the data is our estimate of the chlorite-water hydrogen isotope fractionation in the temperature range 130 to 700ºC. The fractionation cannot be considered well-defined, however, and probably has an uncertainty at least as great as 10 per mil, especially at temperatures below 250ºC. There is some disagreement in the literature about the effect of variation in the chemical composition of chlorite on the hydrogen isotope fractionation factor. Marumo et al. (1980) reported that $\alpha^{hy}_{chlorite-water}$ depends on the Fe/Mg of the mineral, while Graham et al. (1987) based in part upon a reconsideration of the same data, argued that the evidence for such a relationship was not clear.

Talc. No laboratory calibrations of isotopic fractionations between talc and water have been published. An oxygen isotope fractionation calculated on the basis of the bond types in the mineral is

$$1000 \ln \alpha^{ox}_{talc-water} = -3.94 \times 10^3 T^{-1} + 5.86 \times 10^6 T^{-2} - .934 \times 10^9 T^{-3}$$

$$+ .084 \times 10^{12} T^{-4} - 4.27$$

This fractionation, shown in Figure 10, must be considered to have a high degree of uncertainty because of the large role played by the uncertain fractionation between M-OH and water. The equation does, however yield reasonable results. Noack et al. (1986) reported $\delta^{18}O$ values of talc samples developed in a Quaternary weathering profile in Ghana to be between 18.6 and 23.2 per mil. Assuming a water $\delta^{18}O$ value of -5 per mil, the temperature of weathering can be estimated from the fractionation equation to be between 9º and 27ºC, a very reasonable range of values.

Brucite. There is no published calibration of the oxygen isotope fractionation between brucite and water. The oxygen in brucite can be considered to be bonded entirely as M-OH bonds, so the fractionation estimated on the basis of bond types can be read directly from Table 2.

$$1000 \ln \alpha^{ox}_{brucite-water} = 21.7 \times 10^3 T^{-1} - 2.49 \times 10^6 T^{-2} - 34.76$$

As noted earlier, however, this is the most poorly known of all of the bond-water fractionation equations.

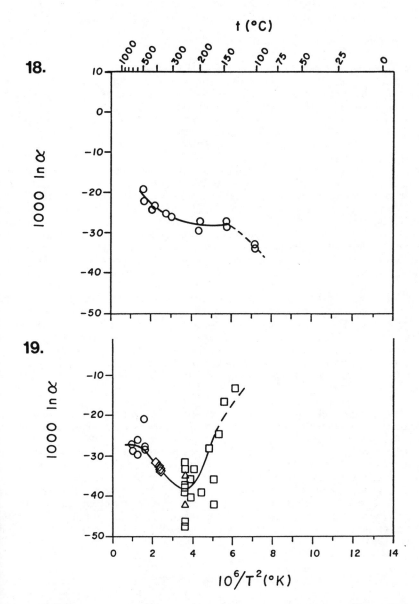

Figure 18. Estimates of the hydrogen isotope fractionation between chlorite and water as a function of temperature. Data sources are: circles, Graham et al. (1987); diamonds, Heaton and Sheppard (1977); triangles, Kuroda et al. (1976); squares, Marumo et al. (1980).

Figure 19. Hydrogen isotope fractionations between brucite and water measured experimentally by Satake and Matsuo (1984).

218

In an experimental equilibration study, Satake and Matsuo (1984) reported a brucite-water hydrogen isotope fractionation

$$1000 \ln \alpha \,^{hy}_{brucite-water} = 8.72x10^6T^{-2} - 3.86x10^4T^{-1} + 14.5$$

in the temperature range 144 to 510°C. At 100°C they obtained only a small amount (approx. 10 percent) of isotopic exchange, making the estimated value of $\alpha \,^{hy}_{brucite-water}$ unreliable, but suggestive (Fig. 19) of a marked change in the fractionation between 100 and 144°C.

CONCLUSIONS

Neither oxygen nor hydrogen isotope fractionations between most phyllosilicates and water are sufficiently well known over extended temperature ranges to permit accurate isotope geothermometry. For many phyllosilicates, however, there is sufficient information to make reasonable estimates of the fractionation factors. These estimates are sufficiently good for interpretation of stable isotope data obtained in the study of naturally occurring samples, especially studies in which attempts are made to use isotopic data to infer the origin of mineral-forming fluids.

An empirical bond model for estimating oxygen isotope fractionations has been discussed at some length. The model is useful for estimating fractionations in the absence of calibrations based upon laboratory equilibration studies or analyses of natural samples. The accuracy of the approach has, at this time, not been well tested, but where comparisons can be made between the bond model approach and accurately measured fractionations, agreement has usually been within 2 or 3 per mil or less.

The empirical bond-type approach has not proven useful in estimating hydrogen isotope fractionations between minerals and water. Other factors, in addition to the cation to which the -OH ion is bonded are important in determining the magnitude of hydrogen isotope fractionations. Note, for example, the significant difference between α hyserpentine-water and α hybrucite-water, in both of which minerals the -OH groups are in trioctahedral Mg sheets. Among the additional factors proposed as playing an important role in determining hydrogen isotope fractionations are hydrogen bonding (Suzuoki and Epstein, 1976; Graham et al., 1980) and distortion of octahedral sheets (Satake and Matsuo, 1984).

ACKNOWLEDGMENTS

We would like to thank Eric Eslinger and Thomas Anderson, each of whom pointed out the errors in the published smectite-water fractionation equation of Yeh and Savin (1977). This manuscript was greatly improved as the result of careful reviews by Eric Eslinger, Xiahong Feng, Jean-Pierre Girard, and an especially acerbic set of comments by James R. O'Neil. Linda Abel assisted in all phases of preparation of the paper. This work was supported by the National Science Foundation under Grant No. EAR 87-20387. Contribution No. 172 Department of Geological Sciences, Case Western Reserve University.

REFERENCES

Becker, R.H. (1971) Carbon and oxygen isotope ratios in iron-formation and associated rocks from the Hamersley Range of western Australia and their implications. Ph.D. thesis, Univ. of Chicago, Chicago, IL.

Bernard, C. (1978) Composition isotopique des mineraux secondaires des bauxites. Problemes de genese. Thesis, Univ. P. et M. Curie, Paris, France.

Bigeleisen, J. and Mayer, M.G. (1947) Calculation of equilibrium constants for isotopic exchange reactions. J. of Chem. Phys. 23, 2264-2269.

Bird, M.I. (1988) An isotopic study of the Australian regolith. Ph.D. thesis, Australian National Univ., Canberra.

---- and Chivas, A.R. (1988a) Oxygen isotope dating of the Australian regolith. Nature, 331, 513-516.

---- and ---- (1988b) Stable-isotope evidence for low temperature kaolinitic weathering and post-formational hydrogen-isotope exchange in Permian kaolinites. Chem. Geol. (Isotope Geosci. Sect.) 72, 249-265.

Bottinga, Y. and Javoy, M. (1973) Comments on oxygen isotope geothermometry. Earth Planet. Sci. Letters 20, 250-265.

---- and ---- (1975) Oxygen isotope partitioning among minerals in igneous and metamorphic rocks. Rev. Geophys. Space Phys. 13, 401-418.

Burrows, S.M. (1985) Oxygen isotope evidence for the conditions of diagenesis of the Muddy Formation, east flank of the Powder River Basin, Montana and Wyoming. M.Sc. thesis, Case Western Reserve Univ., Cleveland, OH.

Chen, C-H., Liu, K-K. and Shieh, Y-N. (1988) Geochemical and isotopic studies of bauxitization in the Tatun volcanic area, northern Taiwan. Chem. Geol. 68, 41-56.

Clayton, R. N., Friedman, I., Graf, D.L., Mayeda, T.K., Meents, W.F. and Shimp, N.F. (1966) The origin of saline formation waters. 1. Isotopic composition. J. Geophys. Res. 71, 3869-3882.

----, Goldsmith, J.R., Karel, K.J., Mayeda, T.K. and Newton, R.C. (1975) Limits on the effect of pressure on isotopic fractionation. Geochim. Cosmochim. Acta 39, 1197-1201.

----, O'Neil, J.R. and Mayeda, T.K. (1972) Oxygen isotope fractionation between quartz and water. J. Geophys. Res. 77, 3057-3067.

Cole, D.R. (1985) A preliminary evaluation of oxygen isotopic exchange between chlorite and water. Geol. Soc. Am. Abstr. Programs, 17, 550.

Craig, H. (1961) Isotopic variations in meteoric waters. Sci. 133, 1702-1703.

---- and Gordon, L. (1965) Deuterium and oxygen-18 variations in the ocean and the marine atmosphere, in Proc. Spoleto Conf. on Stable Isotopes in Oceanographic Studies and Paleotemperatures 2, 1-87.

Dasch, E.J. (1969) Strontium isotopes in weathering profiles, deep sea sediments and sedimentary rocks. Geochim. Cosmochim. Acta, 33, 1521-1552.

220

Escande, M-A. (1983). Echangeabilite et fractionnment isotopique de l'oxygene des smectites magnesiennes de syntheses. Etablissement d'un geothermometre. Thesis, Univ. de Paris-Sud, Centre d'Orsay, Paris, France.

Eslinger, E.V. (1971) Mineralogy and oxygen isotope ratios of hydrothermal and low-grade metamorphic argillaceous rocks. Ph.D. thesis, Case Western Reserve Univ., Cleveland, OH.

----- and Savin, S.M. (1973) Mineralogy and oxygen isotope geochemistry of the hydrothermally altered rock of the Ohaki-Broadlands, New Zealand geothermal area. Am. J. Sci. 273, 240-267.

---- and ---- (1976) Mineralogy and $^{18}O/^{16}O$ ratios of fine-grained quartz and clay from Site 323. In C.D. Hollister, C. Craddock et al., Initial Reports of the Deep Sea Drilling Project, 35, 489-496, Washington, D.C., U.S. Gov't. Printing Office.

---- and Yeh, H.-W., (1981) Mineralogy, $^{18}O/^{16}O$ and D/H ratios of clay rich sediments from Deep Sea Drilling Project Site 180, Aleutian Trench. Clays & Clay Minerals 29, 309-315.

Friedman, I. and O'Neil, J.R. (1977) Compilation of stable isotope fractionation factors of geochemical interest. Data of Geochemistry, U.S. Geol. Surv. Prof. Paper, 440K.

Garlick, G.D. (1966) Oxygen isotope ratios in coexisting minerals of regionally metamorphosed rocks. Ph.D. thesis, California Institute of Technology, Pasadena, CA.

---- (1969) The stable isotopes of oxygen. In K.H. Wedepohl (Ed.) Handbook of Geochemistry, 2. Springer Verlag, New York, NY., Part 1, Ch. 8B.

Graham, C.M. Viglino, J.A. and Harmon, R.S. (1987) An experimental study of hydrogen isotope exchange between aluminous chlorite and water. Am. Mineral. 72, 566-579.

Hamza, M.S. and Epstein, S. (1980) Oxygen isotopic fractionation between oxygen of different sites in hydroxyl-bearing silicate minerals. Geochim. Cosmochim. Acta 44, 173-182.

Hassanipak, A.A. and Eslinger, E.V. (1985) Mineralogy crystallinity $^{18}O/^{16}O$ and D/H of Georgia kaolins. Clays & Clay Minerals 33, 99-106.

Heaton, T.H.E. and Sheppard S.M.F. (1977) Hydrogen and oxygen isotope evidence for sea-water-hydrothermal alteration and ore deposition, Troodos complex, Cyprus. In Volcanic Processes in Ore Genesis, Spec. Pap. 7, 42-57, Geol. Soc. London.

Hein, J.R., Yeh, H-W. and Alexander, E. (1979) Origin of iron-rich montmorillonite from the manganese nodule belt of the equatorial Pacific. Clays & Clay Minerals 27, 185-194.

Hurley, P.M., Heezen, B.C, Pinson, W.H. and Fairbairn, H.W. (1963) K-Ar age values in pelagic sediments of the North Atlantic. Geochim. Cosmochim. Acta 27, 393-399.

Ikin, N.P. and Harmon, R.S. (1981) D/H and $^{18}O/^{16}O$ ratios and mineralogy of serpentinites from the highland border fracture zone, Scotland.Bull. Mineral. 104, 795-800.

James, A.T. and Baker, D.R. (1976) Oxygen isotope exchange between illite and water at 22ºC. Geochim. Cosmochim. Acta 40, 235-239.

Kieffer, S.W. (1982) Thermodynamics and lattice vibrations of minerals: 5. Applications to phase equilibria, isotopic fractionation, and high pressure thermodynamic properties. Rev. Geophys. Space Phys. 20, 827-849.

Knauth, L.P. and Epstein, S. (1975) Hydrogen and oxygen isotope ratios in silica from the JOIDES Deep Sea Drilling Project. Earth Planet. Sci.Letters 25, 1-10.

---- and Beeunas, M.A. (1986) Isotope geochemistry of fluid inclusions in Permian halite, with implications for the isotopic history of ocean water and the origin of saline formation waters. Geochim.Cosmochim.Acta 50, 419-434.

Kulla, J.B. and Anderson, T.F. (1978) experimental oxygen isotope fractionation between kaolinite and water. In Zartman, R.E. Ed., Short Papers of the 4th Int'l. Con., Geochronology, Cosmochronology, Isotope Geology: U.S. Geol. Surv., Open File Report No. 78-70, 234-235.

Kuroda, Y., Suzuoki, T., Matsuo, S. and Shirozu, H. (1976) A preliminary study of D/H ratios of chlorites. Contrib. Mineral. Petrol. 57, 223-225.

Lambert, S.J. and Epstein, S. (1980) Stable isotope investigations of an active geothermal system in Valles Caldera, Jemez Mountains, New Mexico. J. Volcanol. Geothermal Res. 8, 111-129.

Land, L.S. and Dutton, S.P. (1978) Cementation of a Pennsylvanian deltaic sandstone: Isotopic data. J. Sed. Petrol. 48, 1167-1176.

Lawrence, J.R. (1979) O^{18}/O^{16} of the silicate fraction of Recent sediments used as a provenance indicator in the South Atlantic. Marine Geol. 33, 1-7.

---- and Taylor, H.P., Jr. (1971) Deuterium and oxygen-18 correlation: Clay minerals and hydroxides in Quaternary soils compared to meteoric waters. Geochim. Cosmochim. Acta 35, 993-1003.

---- and ---- (1972) Hydrogen and oxygen isotope systematics in weathering profiles. Geochim. Cosmochim. Acta 36, 1377-1393.

Lee, M. (1984) Diagenesis of the permian Rotliegendes Sandstone, North Sea: K/Ar, O^{18}/O^{16} and petrologic evidence. Ph.D. thesis, Case Western Reserve Univ., Cleveland, OH.

Liu, K-K. and Epstein, S. (1984) the hydrogen isotope fractionation between kaolinite and water. Isotope Geology, 2, 335-350. Chem. Geol. 46, 335-350.

Longstaffe, F.K. (1984) The role of meteoric water in diagenesis of shallow sandstones. In D.A. McDonald and R.C. Surdam (Eds.)., "Clastic Diagenesis", Am. Assoc. Petrol. Geol. Memoir 37, 81-97.

---- (1986) Oxygen isotope studies of diagenesis in the basal Belly River sandstone, Pembina I-Pool, Alberta. J. Sed. Petrol. 56, 78-88.

Marumo, K., Nagasawa, K. and Kuroda, Y. (1980) Mineralogy and hydrogen, isotope geochemistry of clay minerals in the Ohnuma geothermal area, northeast Japan. Earth Plant. Sci. Letters 47, 255-262.

Matsuhisa,Y., Goldsmith, J.R. and Clayton, R.N. (1979) Oxygen isotopic fractionation in the system quartz-albite-anorthite-water. Geochim. Cosmochim. Acta 43, 1131-1140.

McMurtry, G.M. and Yeh, H-W. (1981) Hydrothermal clay mineral formation of East Pacific Rise and Bauer Basin sediments. Chem. Geo. 32, 189-205

Noack, Y., Decarreau, A. and Manceau, A. (1986) Spectroscopic and oxygen isotopic evidence for low and high temperature origin of talc, Bull. Mineral. 109, 253-263.

O'Neil, J.R. (1987a) Theoretical and experimental aspects of isotopic fractionation. In J.W. Valley, H.P. Taylor, Jr., and J.R. O'Neil, (Eds.), "Stable Isotopes in High Temperature Processes", Rev. Mineral. 16, 1-40.

222

---- (1987b) Preservation of H, C and O isotopic ratios in the low temperature environment. In T.K. Kyser (Ed.), "Stable Isotope Geochemistry of Low Temperature Fluids", Mineral. Assoc. Can. Short Course Handbook 13, 85-128.

---- and Kharaka, Y.K. (1976) Hydrogen and oxygen isotope exchange reactions between clay minerals and water. Geochim. Cosmochim. Acta, 40, 241-146.

---- and Taylor, H.P., Jr. (1969) Oxygen isotope equilibrium between muscovite and water. J. Geoph. Res. 74, 6012-6022.

Sakai, H. and Tsutsumi, M. (1978) D/H fractionation factors between serpentine and water at 100° to 500°C and 2000 bar water pressure, and the D/H ratios of natural serpentines. Earth Planet. Sci.Letters, 40, 231-242.

Salomons, W., Hofman, P., Boelens, R. and Mook, W.G. (1975) The oxygen isotopic composition of the fraction less than 2 microns (clay fraction) in Recent sediments from western Europe. Marine Geol. 18, M23-M28.

Satake, H. and Matsuo, S. (1984) Hydrogen isotopic fractionation factor between brucite and water in the temperature range from 100° to 510°C. Contrib. Mineral. Petrol. 86, 19-24.

Savin, S.M. (1967) Oxygen and isotope rations in sedimentary rocks and minerals. Ph.D. Thesis, California Institute of Technology, Pasadena, CA.

---- (1973) Oxygen and hydrogen isotope studies of minerals in ocean sediments. In Proceedings of Symposium on Hydrogeochemistry and Biogeochemistry, Sept. 1970, Tokyo 372-391. The Clarke Co., Washington, D.C.

---- (1977) The history of the earth's surface temperature during the past 100 million years. Ann. Rev. Earth Planet.Sci. 5, 319-355.

---- and Epstein, S. (1970a) The oxygen and hydrogen isotope geochemistry of clay minerals. Geochim. Cosmochim. Acta 34, 25-42.

---- and ---- (1970b) The oxygen and hydrogen isotope geochemistry of ocean sediments and shales. Geochim. Cosmochim. Acta, 34, 43-63.

---- and Yeh, H-W. (1981) Stable isotopes in ocean sediments. In C. Emiliani (Ed.) The Oceanic Lithosphere, 1521-1554. John Wiley & Sons, New York.

Schutze, V.H. (1980) Der isotopenindex--eine inkrementenmethode zur naherugsweisen berechnung von isotepenaustauschgleichgewichten zwischen kristallinen substanzen. Chem. Erde 39, 321-334.

Shackleton, N.J. (1967) The measurement of paleotemperatures in the Quaternary Era. Unpublished Ph.D. thesis, Univ. of Cambridge.

Sheppard, S.M.F., Nielsen, R.L., and Taylor, H.P., Jr. (1969) Oxygen and hydrogen isotope ratios of clay minerals from Porphyry Copper deposits. Econ. Geol. 64, 755-777.

Suzuoki, T. and Epstein, S. (1976) Hydrogen isotope fractionation between OH-bearing minerals and water. Geochim. Cosmochim. Acta 40, 1229-1240.

Taylor, H.P., Jr. and Epstein, S. (1961) Relationships between $^{18}O/^{16}O$ ratios in coexisting minerals of igneous and metamorphic rocks, Part 2. Application to petrologic problems. Geol. Soc. Am. Bull. 73, 675-694.

Tsirambides, A.E. (1986) Detrital and authigenic minerals in sediments from the western part of the Indian Ocean. Mineral. Mag. 50, 69-74.

Urey, H.C. (1947) The thermodynamic properties of isotopic substances. J. Chem. Soc. (London), 562-581.

Wenner, D. B. and Taylor, H.R., Jr. (1971) Temperatures of serpentinization of ultramafic rocks based on O^{18}/O^{16} fractionation between coexisting serpentine and magnetite. Contrib. Mineral. Petrol. 32, 165-168.

---- and ---- (1973) Oxygen and hydrogen isotope studies of the serpentinization of ultramafic rocks in oceanic environments and continental ophiolites. Am. J. Sci. 273, 207-239.

Yapp, C.J. (1987) Oxygen and hydrogen isotope variations among goethites (α-FeOOH) and the determination of paleotemperatures. Geochim. Cosmochim. Acta 51, 355-364.

Yeh, H-W. (1974) Oxygen isotope studies of ocean sediments during sedimentation and burial diagenesis. Ph.D. thesis, Case Western Reserve Univ., Cleveland, OH.

---- (1980) D/H ratios and late-stage dehydration of shales during burial. Geochim. Cosmochim. Acta 44, 341-352.

---- and Epstein, S. (1978) Hydrogen isotope exchange between clay minerals and sea water. Geochim. Cosmochim. Acta 42, 140-143.

---- and Eslinger, E.V. (1986) Oxygen isotopes and the extent of diagenesis of clay minerals during sedimentation and burial in the sea. Clays & Clay Minerals 34, 403-406.

---- and Savin, S.M. (1976) The extent of oxygen isotope exchange between clay minerals and sea water. Geochim. Cosmochim. Acta 40, 743-748.

---- and ----- (1977) The mechanism of burial metamorphism of argillaceous sediments: 3. oxygen isotopic evidence. Geol. Soc. Am. Bull. 88, 1321-1330.

Chapter 8
B.W. Evans & S. Guggenheim

TALC, PYROPHYLLITE, AND RELATED MINERALS

INTRODUCTION

General

The 2:1 layer silicate structures are composed of two types of sheets (see Chapter 1, Fig. 1). The *tetrahedral sheet* has a composition of T_2O_5 (where T is commonly Si, Al, and Fe^{3+}) in which individual TO_4 are linked by sharing three corners (*basal oxygens*) with TO_4 neighbors to form infinite two-dimensional sheets with a hexagonal-based pattern of linked six-fold rings. The fourth corner (*apical oxygen*) does not link to other TO_4 groups but does form a corner of the octahedral coordination unit around larger, M, cations (commonly Mg, Al, Fe^{2+}, and Fe^{3+}). These octahedrally-coordinated cations form what is commonly known as an *octahedral sheet*, in which individual octahedra are linked laterally by sharing octahedral edges. The *2:1 layer* is composed of two opposing tetrahedral sheets with the octahedral sheet between. Thus, the individual octahedral anion coordination unit is formed from four apical oxygens, two from each of the tetrahedral sheets, and two hydroxyl (OH) groups that are not shared with the tetrahedra. Therefore, the apical oxygens form a plane of anions common to adjacent tetrahedral and octahedral sheets. The OH group is at the same level as the apical oxygens in the center of each six-fold ring of tetrahedra.

The smallest structural unit contains three octahedral sites. Structures with all three sites occupied are known as *trioctahedral*, whereas if only two of these sites are occupied the structure is *dioctahedral*. The structural unit contains the 2:1 layer plus any material in the interlayer, the region between adjacent 2:1 layers. The 2:1 layer silicates are classified conveniently on the basis of the nature of the octahedral sheet (dioctahedral vs trioctahedral), layer charge, and the type of interlayer material. The interlayer may be vacant or may contain cations, hydrated cations, metal-hydroxyl octahedral sheets, or organic compounds.

Talc and pyrophyllite

Talc and pyrophyllite are the trioctahedral and dioctahedral forms, respectively, of the phyllosilicate group consisting of simple 2:1 layers. The terms talc and pyrophyllite may be used as either group terms or to designate the most common of species, $Mg_3Si_4O_{10}(OH)_2$ and $Al_2Si_4O_{10}(OH)_2$, respectively. In the ideal case, no interlayer material exists in these structures because the 2:1 layers are electrostatically neutral. The primary atomic forces holding adjacent 2:1 layers together are van der Waals bonds, which account for the very soft nature of talc and pyrophyllite. The weak cohesion between layers accounts also, at least in part, for the difficulty in obtaining quality crystalline material for mineralogic study, as these structures often exhibit varying degrees of stacking disorder of 2:1 layers. On the other hand, as the common members of the most simple 2:1 layer silicate group, talc and pyrophyllite represent important structures for comparison to the majority of other layer silicates.

NOMENCLATURE AND GENERAL CHEMICAL CLASSIFICATION

Pyrophyllite

Pyrophyllites do not vary greatly in composition (e.g., Kodama, 1958; Deer et al., 1962; Iwao and Udagawa, 1969). On the basis of 22 positive charges per formula unit, a small but ubiquitous substitution of aluminum for silicon occurs. This substitution ranges from 0.001 to 0.300 Al cations but is commonly closer to the low side of the range. Usually, octahedral sites are nearly filled with greater than 1.9 Al cations per two sites with remaining minor amounts of Fe^{2+}, Fe^{3+}, Mg, and Ti. Octahedral cation totals commonly vary between 1.95 to 2.05 (Newman and Brown, 1987). Sykes and Moody (1978) reported the occurrence of two samples with 0.35 wt % F substitution for (OH). Most pyrophyllite analyses show less then approximately 0.1 cations of (Ca+Na+K), although some wet chemical and microprobe analyses occasionally show more. Small amounts of alkali elements may be the result of contaminating interstratified phases, other impurities, or possible limited alkali interlayer occupancy (see further discussion under structure section). Kodama (1958) showed the existence of a 25 Å regularly interstratified pyrophyllite-montmoril-lonite mineral. Synthesis studies (Rosenberg, 1974; Rosenberg and Cliff, 1980) indicate a possible substitution of $Al^{3+} + H^+ \rightarrow Si^{4+}$ and a slightly expanded basal spacing. This substitution appears limited with the maximum observed variation at $Al_{2.0}(Si_{3.8}Al_{0.20})O_{9.8}(OH)_{2.2}$ assuming Si plus all Al = 6.0. Rosenberg (1974) suggested (OH) substitution for basal oxygens, which would produce interlayer hydrogen bonding and an increase in thermal stability. Generally, there is a trend in natural samples toward higher tetrahedral Al content with larger amounts of (OH) above 2.0, although a 1:1 correspondence is not observed, possibly because of the difficulty in analysis (Newman and Brown, 1987, p. 41). However, unlike synthetic samples which show a correlation between tetrahedral Al and (OH) content to basal spacing, natural samples have nearly identical basal spacings.

Ferripyrophyllite, the ferric iron analogue of pyrophyllite, has been identified from low temperature hydrothermal deposits at Strassenschacht, GDR, and central Kazakhstan (Chukhrov et al., 1979). Based on the analysis of Chukhrov et al. and additional Mössbauer data, Coey et al. (1984) determined the structural formula of the Strassenschacht sample as $Ca_{0.05}(Fe^{3+}_{1.96}Mg_{0.11})(Si_{3.80}Al_{0.13}Fe^{3+}_{0.07})O_{10}(OH)_2$.

Talc

Most talcs have typically near end-member compositions, although major Fe, minor Al and F, and trace Mn, Ti, Cr, Ni, Ca, Na, and K have been reported (e.g., Chidester, 1962; Moine et al., 1982; Hemley et al., 1977; Noack et al., 1986). McKie (1959) reported a talc sample with 3.95 wt % Al_2O_3 resulting in a formula of $(Mg_{2.49}Al_{0.27}Fe^{2+}_{0.03}Ca_{0.04})$ $(Si_{3.98}Al_{0.02})O_{10}(OH)_2$. Based on synthesis work, others (Yoder, 1952; Stemple and Brindley, 1960; Fawcett, 1962; Fawcett and Yoder, 1966) established that about 4.0 wt % is the apparent maximum Al substitution possible. For samples with silicon deficiency, Al (up to about 0.1 atoms per 4 tetrahedral sites) may occupy the tetrahedral sites. Ferric iron tetrahedral occupancy is rare and usually very minor (<0.05 atoms per 4 tetrahedral sites) in natural samples, and it may be related to analytical

227

inaccuracies or contamination. In a synthesis study, however, Forbes (1969) found a small substitution of the type $Fe^{3+} + H^+ \rightarrow Si^{4+}$ for his samples. Abercrombie et al. (1987) found up to 0.453 mole fraction of F in talc in coexisting talc-tremolite pairs in metamorphosed siliceous carbonates from Ontario, representing 4.47 wt % F (with a formula of $Mg_{2.98}Fe_{0.03}(Si_{3.95}Al_{0.05})(OH)_{1.10}F_{0.91}$; G.B. Skippen, Carleton Univ., pers. comm., 1988). In contrast, Ross et al. (1968) found a maximum of 0.48 wt % F substitution for (OH) in samples derived from the metamorphism of several types of sedimentary rocks and only trace amounts (<0.1) in samples derived from ultramafic rocks.

A moderate amount of iron substitution in talc (up to 0.45 cations Fe^{2+} per 3 octahedral sites) has been observed in high pressure pelitic blueschist deposits (Chopin,1981) and in (low temperature, high pressure) blueschist overprinted eclogites (Nisio and Lardeaux, 1987) of the Western Alps. Even higher amounts of iron substitution are possible in various types of hydrothermal deposits (e.g., Lonsdale et al., 1980, with apparently 0.75 cations; Kager and Oen, 1983, with 1.22 Fe^{2+} per 3 octahedral sites) and from iron formations (e.g., Floran and Papike, 1975; 1978; Klein and Gole, 1981). Iron-talc from the low temperature occurrence at the Sterling Mine, Antwerp, N.Y. has the apparent structural formula of $(Mg_{1.90}Fe^{2+}_{0.98}Fe^{3+}_{0.06}Al_{0.02})Si_{4.04}O_{10}(OH)_2$ based on chemical (Robinson and Chamberlain, 1984) and Mössbauer (R.G. Burns, M.I.T., pers. comm., 1985) data.

Mössbauer studies (Noack et al., 1986) of iron-containing talcs showed that hydrothermal talcs contain only ferrous iron, whereas talcs from weathering profiles contain both Fe^{2+} and Fe^{3+}. Noack et al. did not find any evidence of ferric iron in tetrahedral occupancy in these natural samples. When Fe^{3+} is present it is concentrated in the M(2) site only (see below for site nomenclature).

Determining the amount of iron that enters the talc structure as opposed to the minnesotaite structure (Gruner, 1944) is a considerable problem. Recently, Guggenheim and Bailey (1982) and Guggenheim and Eggleton (1986) determined that minnesotaite has complex structural modulations (see Chapter 17) and is not the ferrous analogue of talc. However, the structures and chemistries of both have strong similarities, thereby making their optical properties similar. As samples of Fe talc and minnesotaite tend to be fine-grained and impure, optical identification and/or probe analyses alone should not be used to differentiate between the two. Because earlier workers thought talc and minnesotaite represented a simple (Mg-Fe) solid solution series, the two phases were commonly separated arbitrarily by the Fe/Mg ratio (either by cation or by weight percent) of 1.0. Unless identification is based on diffraction data also, as was done by Chopin (1981), Lonsdale et al. (1980), Kager and Oen (1983), and for the Sterling Mine sample (Guggenheim, unpublished), the terms "Fe (ferrous) talc" and "minnesotaite" in the earlier literature do not necessarily designate true structural characteristics.

Ferrous talc and minnesotaite in close physical proximity have been reported by Floran and Papike (1978) on the basis of chemistry alone and by Kager and Oen (1983) using both chemical and diffraction data. Zoning of the spherulites for the sample studied by Kager and Oen indicates non-equilibrium conditions. Kagen and Oen gave compositions of ferroan talc, $(Mg_{1.80}Fe_{1.17})Si_{4.00}O_{10}(OH)_2$, in the interior, minnesotaite, approximately $(Fe_{2.62}Mg_{0.28}Mn_{0.10})Si_4O_{10}(OH)_2$, at the edges, and opal-chalcedony in the zone between, but considerable SiO_2 contamination and no water analysis makes interpretation of the minnesotaite data difficult.

Willemseite is defined as a talc structure with Ni as the major element in the octahedral sites. De Waal (1970) analyzed the type sample from Barberton Mountain Land, Transvaal: $(Ni_{2.115}Mg_{0.805}Fe^{2+}_{0.020}$ $Fe^{3+}_{0.102})_{\Sigma} = 3.042(Si_{3.945}Al_{0.034})_{\Sigma} = 3.979 \ O_{10}(OH)_{1.835}$. Stubican and Roy (1961) and Wilkins and Ito (1967) reported the synthesis and infrared spectra of Ni talcs. In addition, Wilkins and Ito synthesized talcs with reported octahedral ratios of up to $(Mg_{54}Co_{46})$, $(Mg_{86}Zn_{14})$, $(Mg_{52}Fe_{48})$, $(Mg_{92}Mn_8)$, and $(Mg_{55}Cu_{45})$.

Kerolite (sometimes *cerolite* in the older literature) is a fine-grained, often impure, material that is now believed to be chemically and structurally similar to talc (D'yakonov, 1963a, b; Maksimovic, 1966). Brindley and Thi Hang (1973) and Brindley et al. (1977) found the composition to be approximately $Mg_3Si_4O_{10}(OH)_2 \cdot nH_2O$ with n = 0.8-1.2. The H_2O molecules appear to be surface-held, although some may be in the interlayer region. Kerolite may be differentiated from common talc because it swells partially after long exposure to ethylene glycol. Kerolite has an enhanced basal spacing of 9.6 Å after the X-ray diffraction pattern (Brindley et al., 1977) is corrected for Lorentz-polarization effects (it has an apparent spacing of 10.0-10.1 Å before correction), and it has a general broadening of all reflections in comparison to talc. Because of this broadening, tabulation of the pattern is difficult and is not given here; diffraction patterns are presented in Brindley et al. (1977). This line broadening effect is consistent with a very small particle size (< 4-5 layers) and a turbostratic nature of the layer stacking, which expands the closeness of packing of oxygens along Z^* when compared to talc. Kerolite is of low temperature origin probably derived from a gel-like medium, and it commonly occurs in weathering profiles (Brindley et al., 1977). Stoessell and Hay (1978) found kerolite as an alteration product of sepiolite and as a direct precipitate of ground water with pH below 8. In dissolution experiments, some lasting as long as ten years, Stoessell (1988) found kerolite unstable with respect to sepiolite in solutions in equilibrium or supersaturated with SiO_2 at surface conditions, although reaction kinetics may be a major factor in the formation of kerolite relative to sepiolite. *Pimelites* are structurally similar to kerolites and are defined for the range in composition where Ni exceeds Mg.

A 10 Å phase similar to kerolite in composition was synthesized (Sclar et al., 1965; Bauer and Sclar, 1981) in the system $MgO-SiO_2-H_2O$ using a variety of starting materials (see also Yamamoto and Akimoto, 1977; Khodyrev and Agoshkov, 1986) between 32 and 95 kbar and 375-535°C. In contrast to kerolite, water is strongly bound in the interlayer either as H_2O or as H_3O^+; if it occurs as the latter, there must be considerable interaction with the hydroxyls. In the fully hydrated form, the basal spacing is 9.96 Å (a = 5.32 Å, b = 9.19 Å, c = 10.12 Å, ß = 100.25°), and the structure transforms to talc upon heating in the range of 475-675°C.

Schreyer et al. (1980) reported the occurrence of a sodian talc-like phase with tetrahedral Al substitution for Si. They suggested that Na resides between the 2:1 layers, thereby indicating that this phase may be classified also as an alkali and aluminum-deficient sodium phlogopite (aspidolite). Substitutions of this type were found to extend to 36 mol % aspidolite, and it could not be determined if it involved mixed layer effects or true solid solution. Regular mixed layer structures involving talc-like layers are well-known and range from the common chlorites (regularly interstratified talc-like and brucite-like layers) to more exotic arrangements such as kulkeite (e.g., chlorite/talc-like inter-stratifications, see Schreyer et al., 1982) and aliettite (e.g., saponite/talc-like interstratifications, see Alietti, 1956; Veniale and van der Marel, 1969; Alietti and Mejsner, 1980). Studies of structures containing 2:1 layers have shown that adjacent cations, sheets, or layers distort the component 2:1 layer sufficiently that it is advisable to call these layers "talc-*like*". Furthermore, these layers may have compositions distinct from talc. The mixed layer structures that contain talc-like layers are not discussed further here; chlorite structures are examined in detail elsewhere in this volume. Further discussion of the chemistry of pyrophyllite and talc is given in a petrologic context in the latter half of this chapter.

CRYSTAL STRUCTURE STUDIES

The stacking of adjacent 2:1 layers in talc or pyrophyllite is neither constrained by an interlayer cation as in the micas nor by optimum hydrogen bonding configurations as in the other 2:1 layer silicates. Instead, the major feature defining the stacking (Figs. 1A, 2A) appears to be related to the minimization of Si^{4+} to Si^{4+} repulsive forces across the interlayer region (Zvyagin et al., 1969). In this arrangement, Si^{4+} cations do not superpose over other Si^{4+} cations across the interlayer as they do in the micas but, rather, the adjacent 2:1 layers are shifted by approximately $0.3a$ along one of the three pseudohexagonal X directions (Bailey, 1984). Bailey (1984) described this arrangement in terms of the projection of four of the six basal oxygens within a hexagonal ring on one side of the interlayer onto the midpoint of tetrahedral edges of the hexagonal ring of the adjacent layer. The important consequence of this arrangement is that there are *no* six-fold or twelve-fold interlayer sites comparable to those in the micas. For example, in pyrophyllite, the six nearest oxygens located around a point at (0.25, 0.25, 0.50), which is in the interlayer between adjacent hexagonal rings, are at distances of 1.979 Å (x2), 2.895 Å (x2) and 2.955 Å (x2). For a comparable site in talc, the values are 2.281 Å (x2), 2.728 Å (x2) and 2.934 Å (x2). In contrast, potassium micas have much more regular sites with average bond distances of about 2.8-2.9 Å. Therefore, pyrophyllite and talc should *not* be considered as simply an interlayer-cation-deficient analogue of the micas. However, an unusual mica-like phase, $Na_{0.84}Mg_{3.00}(Si_{3.41}Mg_{0.50}Fe_{0.09})O_{9.75}(OH)_{2.25}$, synthesized from talc, apparently maintains talc-type stacking (Drits, 1987, p. 218).

Polytypism of pyrophyllite and talc has been described by various workers. Zvyagin et al. (1969) showed that the polytypism of pyrophyllite can be represented either by vector symbols (Zvyagin, 1967) $\sigma_2\ \tau_4\ \sigma_2$ for $C1$ symmetry or $\sigma_3\ \tau_5\ \sigma_3\ \tau_1\ \sigma_3$ for Cc symmetry, where σ and τ are intralayer displacements between the octahedral and tetrahedral sheets and interlayer displacements between adjacent layers, respectively. In Ramsdell notation, these polytypes are $1Tc$ and $2M$, respectively. Recent structural determinations (see below) indicated that the $1Tc$ polytype belongs to the centric space group $C\bar{1}$. The true symmetry for the $2M$ structure is unknown. Zvyagin et al. (1969) found that natural pyrophyllite crystallizes as a $2M$ structure and a synthetic sample formed as $1Tc$, although later workers (e.g., Brindley and Wardle, 1970; Wardle

230

A

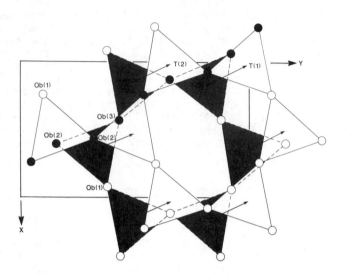

B

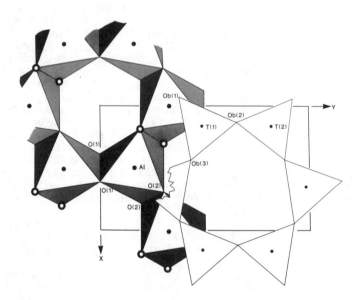

Figure 1. The pyrophyllite-1Tc structure projected down Z^*. Part (A) shows the displacement of the tetrahedral sheets *across* the interlayer region. The direction of displacement is illustrated by the arrows, which connect tetrahedral sites of adjacent tetrahedral sheets. Note that the effect is illustrated also by basal oxygens (black circles) forming lines by projecting on top of tetrahedral edges in the neighboring sheet. Part (B) illustrates portions of the tetrahedral and octahedral sheets. The tetrahedral sheet illustrated here is depicted in part (A) with shading.

A

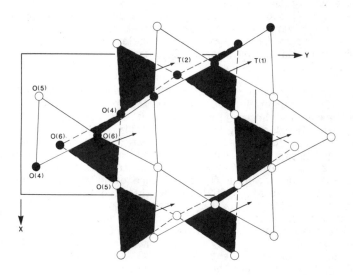

B

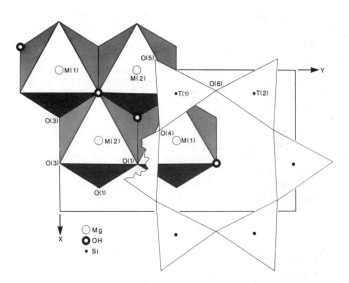

Figure 2. The talc structure projected down Z^*. Parts (A) and (B) are analogous to Figure 1.

and Brindley, 1972; Lee and Guggenheim, 1981) have found that natural samples are not restricted to the 2M form only. On the basis of synthesis work, Eberl (1979) suggested that the 1Tc structure is a high temperature form and the 2M variety is of low temperature origin. This sequence differs from muscovite in which the one-layer (1M) form is the low temperature metastable form and the 2M₁ structure is the high temperature stable form.

Although early workers (Gruner, 1934; Hendricks, 1938; Stemple and Brindley, 1960) reported that talc formed as a two-layer monoclinic polytype, Rayner and Brown (1966), Ross et al. (1968), and Akizuki and Zussman (1978) found only a one-layer triclinic form in their samples. No talcs have been identified in a two-layer form when studied in detail, and Rayner and Brown (1973) have attributed the earlier errors in polytype identification to twinning. High resolution TEM studies (Veblen and Buseck, 1980), however, have found evidence of a two-layer orthorhombic polytype. Zvyagin et al. (1969) showed that the talc stacking sequence is similar to the 1Tc (or σ_2 τ_4 σ_2) sequence of pyrophyllite. General polytype derivations for pyrophyllite and talc using the Dornberger-Schiff OD structure theory have been developed by Ďurovič and Weiss (1983) and Weiss and Ďurovič (1984). They found 22 non-equivalent ideal pyrophyllite polytypes and 10 non-equivalent theoretical talc polytypes. Calculations indicated that several of the derived polytypes have identical X-ray powder intensity profiles. Powder patterns, therefore, may be used to distinguish between fourteen pyrophyllite and seven talc polytypes, whereas single crystal methods would be necessary for further differentiation.

Pyrophyllite

An initial determination of a two-layer structure of pyrophyllite from Graves Mountain, Georgia, by Rayner and Brown (1965) could define only the subcell because of stacking disorder, which is a common feature of pyrophyllites. Structural disorder of this type produces rod-shaped streaks for $k \neq 3n$ reflections along Z^*, thereby offering little useful information about atoms which repeat at intervals other than $b/3$. Intensity maxima within the streaks suggested a two-layer stacking sequence. Therefore, the subcell is $a = 5.17$ Å, $b = b_0/3 = 2.97$ Å, $c = c_0/2 = 9.33$ Å and $\beta = 99.8°$ if $k \neq 3n$ reflections are ignored. The subcell represents an average structure of an individual layer, which was determined to be centric (space group $C2/m$). They found the tetrahedral rotation angle to be $10°$ and an undulating or corrugated basal oxygen plane resulting from a tetrahedral out-of-plane tilt of $4.1°$. Although the vacant site cannot be distinguished directly in such a subcell, the tilt is a requirement for dioctahedral structures with a larger vacant site present.

Wardle and Brindley (1972) calculated powder diffraction patterns for all possible ideal models and compared the results to powder patterns of well-crystallized material from the Coromandel region of New Zealand. A partial refinement from the powder data confirmed the general stacking sequence described by Zvyagin et al. (1969) for the 1Tc polytype, but in $C\bar{1}$ symmetry. Single crystals of well-crystallized pyrophyllite-1Tc from a hydrothermal deposit near Ibitiara, Brazil, allowed a full refinement of the structure (Lee and Guggenheim, 1981) in $C\bar{1}$ symmetry. These results (Tables 1 and 2) agree closely with those of Wardle and Brindley (1972).

Table 1. Atomic coordinates for pyrophyllite, pyrophyllite dehydroxylate, and talc.

Atom		x	y	z
		Talc (Perdakatsis and Burzlaff, 1981)		
M(1)	= Mg	0.0	0.0	0.0
M(2)	= Mg	0.50012(8)	0.83332(5)	0.9994(4)
T(1)	= Si	0.24527(7)	0.50259(4)	0.29093(3)
T(2)	= Si	0.24590(7)	0.83587(4)	0.29108(3)
O(1)		0.1991(2)	0.8344(1)	0.1176(1)
O(2)	= OH	0.6970(2)	0.6674(1)	0.1126(1)
O(3)		0.1980(2)	0.5012(1)	0.1176(1)
O(4)		0.0199(1)	0.9287(1)	0.3481(1)
O(5)		0.5202(2)	0.9109(1)	0.3494(1)
O(6)		0.2429(2)	0.6699(1)	0.3484(1)
H		0.719(4)	0.669(3)	0.203(2)
		Pyrophyllite-1Tc (Lee and Guggenheim, 1981)		
M(1)	= vac	0.0	0.0	0.0
M(2)	= Al	0.4995(2)	0.16705(9)	-0.00008(9)
T(1)	= Si	0.7449(2)	-0.00303(8)	0.29167(8)
T(2)	= Si	0.7595(2)	0.32577(8)	0.29230(8)
O(1)		0.6495(4)	0.0018(2)	0.1155(2)
O(2)		0.7314(4)	0.3079(2)	0.1158(2)
OH		0.2263(4)	0.1927(2)	0.1081(2)
Ob(1)		0.0498(4)	0.3891(2)	0.3589(2)
Ob(2)		0.7251(5)	0.1637(2)	0.3584(2)
Ob(3)		0.5452(4)	0.4426(2)	0.3325(2)
		Pyrophyllite dehydroxylate (Wardle and Brindley, 1972)		
M(1)	= vac	0.0	0.0	0.0
M(2)	= Al	0.552	0.149	0.000
T(1)	= Si	0.225	0.486	0.286
T(2)	= Si	0.749	0.312	0.286
O(1)		0.123	0.492	0.115
O(2)		0.728	0.292	0.115
Or		0.250	0.250	0.000
Ob(1)		0.037	0.378	0.355
Ob(2)		0.717	0.152	0.355
Ob(3)		0.522	0.425	0.320

Table 2. Cell parameters, derived structural parameters, and selected interatomic distances for pyrophyllite, pyrophyllite dehydroxylate, and talc.

	Pyrophyllite-1Tc (Lee and Guggenheim, 1981)	Pyrophyllite Dehydroxylate (Wardle and Brindley, 1972)	Talc (Perdakatsis and Burzlaff, 1981)
Cell parameters			
a(Å)	5.160(2)	5.191(1)	5.290(3)
b(Å)	8.966(3)	9.122(1)	9.173(5)
c(Å)	9.347(6)	9.499(1)	9.460(5)
α(°)	91.18(4)	91.17(2)	90.46(5)
β(°)	100.46(4)	100.21(2)	98.68(5)
γ(°)	89.64(3)	88.62(2)	90.09(5)
Derived structural parameters			
a_{tet} (°)	10.2	9.0	3.6
ψ_{oct} (°)	57.1	---	M(1):61.5, M(2):61.5
basal oxygen, Δz_{ave} (Å)	0.240	0.33	0.011
Interlayer separation (Å)	2.759	2.94	2.832
Sheet thickness: tetrahedral (Å)	2.153	2.13	2.176
octahedral (Å)	2.079	2.15	2.168

Selected Interatomic Distances

Pyrophyllite-1Tc — Octahedron

Al--O(1)	1.926(2)
O(1)'	1.922(2)
O(2)	1.922(2)
O(2)"	1.926(2)
OH	1.889(2)
OH"	1.888(2)
Mean	1.912

Shared edges:between Al

O(1)--O(1)'	2.415(3)
O(2)--O(2)"	2.420(3)
OH--OH"	2.338(3)
Mean	2.391

Pyrophyllite Dehydroxylate — Trigonal bipyramid

Al--O(1)'	1.82
O(1)"	1.81
O(2)'	1.84
O(2)	1.81
Or	1.80
Mean	1.82

Shared edges:between Al

O(1)"--O(1)'	2.33
O(2)--O(2)'	2.35
Mean	2.34

Talc — Octahedra

M(1)--O(1) x 2	2.082(1)	M(2)--O(1)	2.076(1)
O(2) x 2	2.052(1)	O(1)'	2.079(1)
O(3) x 2	2.080(1)	O(2)	2.057(1)
Mean	2.071	O(2)'	2.053(1)
		O(3)	2.079(1)
		O(3)'	2.078(1)
		Mean	2.070

Shared edges:

M(1) to M(2)		M(2) to M(2)	
O(1)--O(2)	2.785(1)	O(1)--O(1)'	2.819(1)
O(3)	2.817(1)	O(2)--O(2)'	2.749(1)
O(2)--O(3)	2.785(1)	O(3)--O(3)'	2.816(1)
Mean	2.796	Mean	2.795

	Pyrophyllite-1Tc (Lee and Guggenheim, 1981)	Pyrophyllite Dehydroxylate (Wardle and Brindley, 1972)	Talc (Perdakatsis and Burzlaff, 1981)

Selected Interatomic Distances

Pyrophyllite-1Tc (Lee and Guggenheim, 1981)

Octahedron

Unshared edges: between Al and vacancy

O(1)--O(2)" x 2	2.934(3)
O(1)'--OH x 2	2.853(3)
O(2)--OH" x 2	2.857(3)
Mean	2.881

triad edges

O(1)--O(2)	2.780(3)
O(1)'--OH'''	2.798(3)
O(2)"--OH"	2.761(3)
O(1)'--O(2)"	2.779(3)
O(1)--OH	2.758(3)
O(2)--OH	2.798(3)
Mean	2.779

Tetrahedra

T(1)--O(1)	1.632(2)
Ob(1)	1.615(2)
Ob(2)	1.618(2)
Ob(3)	1.602(2)
Mean	1.617
T(2)--O(2)	1.634(2)
Ob(1)	1.616(2)
Ob(2)	1.614(2)
Ob(3)	1.607(2)
Mean	1.618

Pyrophyllite Dehydroxylate (Wardle and Brindley, 1972)

Trigonal bipyramid

Unshared edges: between Al and vacancy

Or--O(1)"	3.10
O(1)'.	2.57
O(2).	2.56
O(2)'	3.11
O(1)"--O(2)"	3.18
O(1)'--O(2)	3.66
Mean	3.03

Tetrahedra

T(1)--O(1)	1.62
Ob(1)	1.63
Ob(2)	1.64
Ob(3)	1.61
Mean	1.63
T(2)--O(2)	1.62
Ob(1)	1.65
Ob(2)	1.64
Ob(3)	1.62
Mean	1.63

Talc (Perdakatsis and Burzlaff, 1981)

Octahedra

Unshared edges: triad edges, M(1)

O(1)--O(2) x 2	3.055(1)
O(3) x 2	3.063(1)
O(2)--O(3) x 2	3.053(1)
Mean	3.057

triad edges, M(2)

O(1)--O(2)	3.055(1)
O(3)	3.049(1)
O(2)--O(3)	3.062(1)
O(1)'--O(2)'	3.058(1)
O(3)'	3.056(1)
O(2)'--O(3)'	3.054(1)
Mean	3.056

Tetrahedra

T(1)--O(3)	1.621(1)
O(4)	1.623(1)
O(5)	1.623(1)
O(6)	1.625(1)
Mean	1.623
T(2)--O(1)	1.622(1)
O(4)	1.624(1)
O(5)	1.623(1)
O(6)	1.621(1)
Mean	1.623

a. Tetrahedral rotation angle is calculated from $\alpha = 0.5|120° - $ mean $O_b-O_b-O_b$ angle$|$.

b. The octahedral ψ angle, ideally 54.73°, is calculated from cos ψ = oct thickness/2 (M--O,OH).

c. Atom translations or symmetry operations are as follows: pyrophyllite ' (-x,-y,-z), " (0.5-x,0.5-y,z); pyrophyllite dehydroxylate ' (0.5-x,0.5-y,-z), " (0.5+x,0.5+y,z); talc ' (-x,-y,-z).

Figure 1 shows the major features of the dioctahedral sheet in the pyrophyllite structure. The large vacant octahedral site is located at the origin (Fig. 1B) surrounded by six equivalent in-plane Al octahedra. Adjacent Al^{3+} octahedra are linked by sharing anion edges, which are considerably shorter than the unshared anion edges of the octahedra. These shortened shared edges are a result of repulsions between adjacent Al cations and are a feature common to all dioctahedral structures. Because of the high (3^+) charge on the Al cation, these cations repel one another and try to move apart as the anions on the shared edge move closer to act as more efficient electrostatic shields. The magnitude of the adjustment of the $O(1)$-$O(1)$, $O(2)$-$O(2)$, or OH-OH pairs is limited in trioctahedral sheets because the M(1) site cation (cf. Fig. 2B) pulls (attracts) each end of the shared edge. Without this constraint in dioctahedral sheets, such sheets have severely distorted occupied octahedra. In contrast, the vacant site is very regular because the shared edges repeat at 120^0 intervals around the vacancy.

The nature of the dioctahedral sheet with both large vacant sites and small Al sites influences the attached tetrahedral sheet. In order to compensate for the size differences, the tetrahedra must tilt out of the (001) plane by 6.0^0 (Lee and Guggenheim, 1981). The tilt forces the basal oxygens to deviate from the basal plane by the amount (Δz) of 0.240 Å, which creates a corrugation effect along [110]. The angle (α) of tetrahedral rotation, which is a measure of lateral misfit between the tetrahedral and octahedral sheets, is 10.2^0. General geometric features common to all dioctahedral structures are given in more detail in Bailey (1984) and Guggenheim (1984), and considerable additional detail is given by Lee and Guggenheim (1981) in describing specific geometric features of the pyrophyllite structure. X-ray powder diffraction patterns of natural samples of pyrophyllite and talc are listed in Table 3.

The quality of the single crystal of pyrophyllite used in the refinement by Lee and Guggenheim (1981) was insufficient to determine the hydrogen coordinates. Giese (1973), however, used the atomic coordinates from Wardle and Brindley (1972) to calculate the H position by the minimum electrostatic energy technique. Repositioning the H position to conform to the refinement of Lee and Guggenheim (1981) results in coordinates of (0.153, 0.111, 0.154). The O--H vector points away from (001) by an angle of about 26^0. This angle represents the attainment of a theoretical equilibrium position for the hydrogen, which is influenced primarily by the charge of the Al cations in the two adjacent occupied octahedral sites. In contrast, dioctahedral potassium micas have a lower angle due to the effect of an occupied interlayer site. Other calculations (Giese, 1971) based on electrostatic minimization techniques for a hypothetical "muscovite" structure without the potassium interlayer cation suggested that the charge on the tetrahedral cation (Si, Al in muscovite, Si in pyrophyllite) must play an important role also in determining the O--H vector orientations.

Although van der Waals attractions are believed to be the major forces involved in interlayer bonding in pyrophyllite, interlayer bond energy calculations (Giese, 1975) indicated that there is a small electrostatic attraction between 2:1 layers, even for the ideal case of neutral layers. As these calculations had no provisions for van der Waals contributions to the layer attractions, it is not possible to establish the relative importance of each. For samples that deviate from end-member compositions, additional ionic attractions occur with tetrahedral and

Table 3A. cont.

...data of natural pyrophyllite.

Pyrophyllite-1Tc, New Zealand (Wardle and Brindley, 1972)

hkl	d(calc.),Å	d(obs.),Å	I(obs.)
001	9.197	9.20	90
002	4.598	4.60	32
110	4.419		
1̄10	4.415	4.42	100
11̄1	4.258	4.26	80
021̄	4.231		55
021	4.054	4.06	
111	3.768		
11̄2	3.749	3.765	8
111̄	3.487	3.490	6
112̄	3.450	3.454	6
022	3.181		6
003	3.178	3.178	22
093	3.066	3.068	85
112	2.973		
11̄3	2.952	2.953	17
130̄	2.740	2.741	5
201̄	2.713	2.710	8
201̄	2.575		
023̄	2.572	2.569	27
201	2.568		
211̄	2.551	2.547	27
211̄	2.539	2.532	36
202̄	2.430		
202	2.414	2.416	77
131̄	2.360		
131	2.356	2.359	6
131̄	2.341	2.341	4
132	2.321		
201	2.300	2.300	2
221̄	2.209		
131̄	2.209	2.209	5
221̄	2.193	2.195	1
221	2.175		
131̄	2.172	2.172	1
202̄	2.152	2.152	14
221	2.116	2.116	18
202	2.084	2.083	2
024̄	2.078		
024	2.070	2.070	15
131̄	2.054	2.054	16
231̄	2.026	2.026	3
231̄	1.9994	1.998	3
204̄	1.9520	1.952	4
	1.9005	1.900	4
	1.8809	1.883	

Pyrophyllite-2M, Honami, Japan (Brindley and Wardle, 1970)

hkl	d(calc.),Å	d(obs.),Å	I(obs.)
002	9.204	9.21	100
004	4.602	4.61	40
020	4.479		
110	4.428		
021̄	4.352	4.42b	60
112̄	4.248		
111	4.164		
022	4.027	4.18b	90
006	3.068	3.069	100
130̄	2.576		
202̄	2.572	2.571	15
200	2.547		
132̄	2.539	2.550*	15
026	2.531	2.534	20
132	2.426		
204̄	2.413	2.419b	45
207̄	2.352	2.352*	1
134̄	2.337	2.334	3
008	2.301	2.303	4
220	2.214	2.216b	1
134	2.168	2.166*	9
206̄	2.152	2.151*	9
222	2.082	2.086	10
136̄	2.064	2.082	
028	2.047	2.059	15

Pyrophyllite-1Tc

hkl	d(calc.),Å	d(obs.),Å	I(obs.)
133	1.8758		
222	1.8743	1.875	4
005	1.8394		
043̄	1.8234	1.841	12
134̄	1.8134	1.823	2
115̄	1.8074	1.812	3
224	1.7436		
224̄	1.7248	1.744	2
025	1.7120	1.722	1
241̄	1.6928		
025	1.6911		
311̄	1.6909	1.689	12
150	1.6898		
150	1.6886		
151̄	1.6850		
241̄	1.6827		
240	1.6802	1.677	12
240	1.6782		
151̄	1.6775		
223̄	1.6738		
310	1.6666		
310	1.6638	1.661	12
312̄	1.6628		
310̄	1.6625		
312̄	1.6583		
134̄	1.6528		
242̄	1.6500	1.650	12
115̄	1.6482		
205	1.6359		
134	1.6310	1.633	20
152	1.6268		
152̄	1.6210	1.621	2
152	1.6078		
115	1.6066	1.607	2
311̄	1.5885		
311	1.5867	1.585	4
152	1.5633	1.565	2
243̄	1.5483		
243	1.5425	1.544	2
006	1.5328		
225̄	1.5286	1.532	3
153̄	1.5149	1.514	1
153	1.5069	1.505	1
331	1.4935		
060	1.4927	1.493	30
331̄	1.4878		
224̄	1.4863		
312̄	1.4834	1.487	10
312	1.4811		
224	1.4775		
061̄	1.4759		
330	1.4730	1.472	11
330̄	1.4715		
061	1.4694		
332	1.4659		

Pyrophyllite-2M

hkl	d(calc.),Å	d(obs.),Å	I(obs.)
136	1.893	1.892	6
208	1.877	1.876*	1
0,0,10	1.841	1.842	10
206	1.811	1.814b	1
317̄	1.6928		
150	1.6901	1.6879*	14
240̄	1.6639		
240	1.6817		
310̄	1.6681	1.6662*	15
314̄	1.6617		
134	1.6458	1.6469*	20
152	1.6455		
2,0,1̄0	1.6325	1.6313*	15
1,1,10	1.6155		
048̄	1.6049	1.6104b	2
316̄	1.5832		
208	1.5766	1.5789b	4
0,0,12	1.5340	1.5348	4
225̄	1.5338		
060	1.4930	1.4935	20
332	1.4929		
314	1.4890	1.4898*	5
334̄	1.4715	1.4708	7

Table 3A. cont.

Pyrophyllite-1Tc

hkl	d(calc.),A	d(obs.),A	I(obs.)
24$\bar{4}$	1.4510	1.451	1
$\bar{1}$35	1.4451 }	1.443	2
026	1.4425 }		
153	1.4343 }		
20$\bar{6}$	1.4311 }	1.433	3
135	1.4267	1.425	3
15$\bar{4}$	1.4141	1.414	2
045	1.4093	1.408	2
243	1.3976	1.398	1
13$\bar{6}$	1.3833	1.383	17
205	1.3764 }		
22$\bar{6}$	1.3699 }	1.376	13
3$\bar{1}$3	1.3656 }		
1$\bar{3}$6	1.3632 }	1.363	13
31$\bar{5}$	1.3560 }		
06$\bar{3}$	1.3514 }	1.352	2
33$\bar{4}$	1.3454 }		
$\bar{3}$32	1.3450 }	1.345	6
332	1.3399 }		
063	1.3329 }		
3$\bar{3}$4	1.3321 }	1.333	3
007	1.3139 }		
225	1.3111 }	1.314	3

Pyrophyllite-2M

hkl	d(calc.),A	d(obs.),A	I(obs.)
156	1.4456 }		
1,3,10	1.4391 }	1.4407*b	3
24$\bar{8}$	1.4386 }		
2,0,1$\bar{2}$	1.4282 }		
332	1.4241 }	1.4260*b	4
0,4,10	1.4221 }		
33$\bar{6}$	1.4161	1.4154*b	2
1,1,12	1.3862 }		
2,0,10	1.3823 }	1.3858	15
1,3,1$\bar{2}$	1.3720 }		
316	1.3706 }	1.3697	20
334	1.3475	1.3491*	4
066	1.3425	1.3432*	1
33$\bar{8}$	1.3374	1.3365*	2
0,0,14	1.3146	1.3159	1.5

1. pyrophyllite-1Tc a=5.161(1), b=8.957(2), c=9.351(1)A, α=91.03(2), β=100.37(2), γ=89.75(2).
2. pyrophyllite-2M a=5.172, b=8.958, c=18.676A, β=100.0°, vb=very broad, b=broad, * not completely resolved.

Table 3B. X-ray powder diffraction data of natural ferripyrophyllite.

[1]ferripyrophyllite (Strassenschacht)
Chukhrov et. al. (1979)

hkl	d(obs.), A	I(obs.)
002	9.6	8
020	4.54	10
$\bar{1}$12	4.25	2
114, 006	3.17	7
$\bar{2}$02, 131	2.62	4
202, $\bar{1}$33	2.47	4
0,0,$\overline{10}$	1.89	1
150, 241, 31$\bar{1}$	1.725	3
208, 139	1.665	3
060	1.518	8

1. ferripyrophyllite-2M: a = 5.26, b = 9.10, c = 19.1 A, β = 95.5°.

Table 3C. X-ray powder diffraction data of natural talc and willemseite.

[1]Talc-1Tc (Gouverneur, New York)				[2]Willemseite (Barberton, South Africa)			
C-setting hkl	d(calc.),A	d(obs.),A	I(obs.)	hkl	d(calc.),A	d(obs.),A	I(obs.)
001	9.333	9.34	vs	002	9.36	9.40	100
002	4.667	4.68	w	004	4.68	4.68	<1
020	4.568	4.56	s	020	4.57	4.57	16
1̄10	4.531			11̄4	3.53	3.55	3
110	4.523			006	3.12	3.12	28
1̄1̄1	4.316			130	2.635	2.636	3
1̄11	4.298			13̄2	2.596	2.599	3
02̄1	4.125	4.14	vw	13̄3	2.507	2.503	23
021	4.082			008	2.339	2.338	1
11̄1	3.885	3.85	vw	13̄5	2.249	2.245	8
111	3.862			117	2.166	2.165	2
1̄1̄2	3.504	3.43	w	028	2.082	2.085	3
1̄12	3.481			137̄(?)	-	1.957	1
02̄2	3.287			0,0,10	1.871	1.873	<1
022	3.243			242̄	1.729	1.729	1
003	3.111	3.115	s	241	1.698	1.700	2
11̄2	3.061			208	1.611	1.609	2
112	3.040			0,0,12	1.559	1.560	2B
1̄1̄3	2.751			060	1.525	1.524	7
1̄13	2.733			1,3,1̄1	1.489	1.488	1
1̄30	2.632			247	1.378	1.379	3
130	2.627	2.632	wm	0,0,14	1.336	1.336	2
2̄01	2.617			260	1.318	1.318	3
200	2.606	2.598	m	262	1.290	1.289	2
1̄31	2.590			424̄	1.269	1.270	1
02̄3	2.588			425̄	1.258	1.257	2
1̄31	2.578			0,0,16	1.169	1.170	2B
023	2.555			281̄	1.051	1.051	1
1̄3̄1	2.490	2.48	s-d	0,0,18	1.039	1.039	1
131	2.472					0.9938	3
2̄02	2.442						
1̄1̄3	2.424						
201	2.415						
1̄13	2.409						
1̄32	2.382						
1̄32	2.361						
004	2.333	2.336	vw				
040	2.284	2.284	vw				
		2.219	m-d				
		2.103	m-d				
		1.994	vm				
		1.871	m				
		1.731	w				
		1.688	vw-d				
		1.558	w				
		1.527	ms				
		1.515	w				
		1.320	w				
		1.296	w				
		1.271	vw				

1. Specimen from Arnold Pit, Gouverneur, New York. Data from Ross et al., 1968. a=5.275, b=9.137, c=9.448 A, α=90.77°, β=98.92°, γ=90.0°.

2. Specimen from Barberton, South Africa. Data from deWaal and Park, 1970. a=5.316, b=9.149, c=18.994 A, β=99.96°.

Symbols: vs, very strong; s, strong; ms, medium strong; m, medium; wm, weak medium; w, weak; vw, very weak; d, diffuse, B, broad.

240

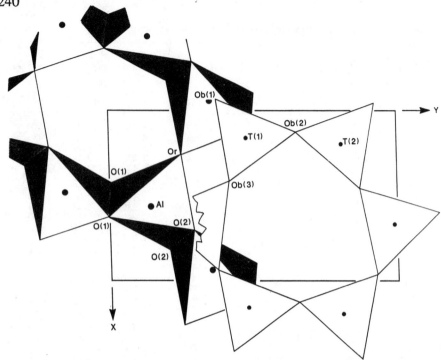

Figure 3. Portions of the pyrophyllite dehydroxylate structure projected down Z^*. Note the five-fold coordination of the Al. Other aspects of the structure, such as the displacement of the tetrahedral sheets across the interlayer region are similar to pyrophyllite (Fig. 1A).

octahedral substitutions. Based on synthesis studies, Rosenberg (1974) has suggested also that (OH) substitution for basal oxygens produces limited interlayer hydrogen bonding.

Pyrophyllite dehydroxylate

Pyrophyllite dehydroxylate (Fig. 3) is metastable and has not been found to exist in nature. It may, however, be readily synthesized by heating pyrophyllite at temperatures above the dehydroxylation reaction. Details of this reaction are given below. Dehydroxylation occurs primarily at the junction between the tetrahedral and dioctahedral sheets, and the disruption of the structure is minimal. The interlayer stacking, therefore, does not differ significantly from that of pyrophyllite (see Fig. 1A).

Wardle and Brindley (1972) showed by a structure determination from X-ray powder diffraction data (in $C\bar{1}$ symmetry) that pyrophyllite dehydroxylate differs from pyrophyllite primarily by the oxygen coordination around the aluminum cations (see Table 2, Fig. 3). In pyrophyllite dehydroxylate, the Al is in five-fold coordination. Five-fold coordinated Al sites in other structures are rare, but they have been shown to exist in the analogue muscovite dehydroxylate structure (Udagawa et al., 1974) and in andalusite (Winter and Ghose, 1979) and they are believed to exist, by analogy, in montmorillonite dehydroxylate structures (Koster van Groos and Guggenheim, 1987). In andalusite, the five-fold coordinated Al shares only edges with other five-fold coordinated Al rather than corners and edges as in pyrophyllite dehydroxylate.

The dehydroxylate structure is derived from pyrophyllite by the loss of one OH group and the H^+ associated with the neighboring OH group. The residual oxygen (Or) moves to the same z coordinate level as the Al cation plane and positions itself midway between the two closest Al cations (cf. Fig. 1 with Fig. 3). Similar to pyrophyllite, the vacant site in the dehydroxylate form is much larger than the Al sites. In contrast, however, the vacant site is considerably more distorted with point symmetry $\bar{1}$ instead of approximately $2/m$ as found in pyrophyllite. These distortions may be attributed to the nature of the dioctahedral Al sheet (see discussion above) and the positioning of the residual oxygen relative to the pair of OH groups in the low temperature form.

The structure of pyrophyllite dehydroxylate as determined by Wardle and Brindley (1972) could not be refined with the available data. Therefore, the bond lengths (Table 2) derived from their reported atomic coordinates cannot be used to discuss the details of the structure. However, the Al-Or distance is the smallest distance in the five-fold coordinated site. The analogue muscovite dehydroxylate structure has been refined by Udagawa et al. (1974) in sufficient detail to show that the Al--Or bond distance is also quite small (1.69 Å). This small interatomic distance in both structures is caused by the highly undersaturated nature of the residual oxygen (1.2 esu rather than 2.0 esu) and the electrostatic attraction of the trivalent Al to offset this charge imbalance. Guggenheim et al. (1987) have suggested that the undersaturated nature of the residual oxygen and the resulting anticipated Al movement to offset the charge imbalance plays an important role in the dehydroxylation process. This is explored in more detail below.

Talc

Rayner and Brown (1973) determined the crystal structure of talc from Weissenberg data and confirmed the stacking arrangement described by Zvyagin et al. (1969), but in $C\bar{1}$ symmetry (see discussion above). The 2:1 layer (Fig. 2) approximates monoclinic symmetry with the mirror plane extending diagonally (from the origin along the [110] direction), but this symmetry does not continue through the structure as a whole because of displacements caused by the stacking of successive layers (Fig. 2A). Perdikatsis and Burzlaff (1981) further refined the structure by X-ray single-crystal diffraction and located the hydrogen at (0.719, 0.669, 0.203), which is in close agreement with the calculated value from electrostatic minimization techniques (Giese, 1979).

Interatomic bond lengths (Table 2) for talc show that the M(1) and M(2) sites are nearly identical in size. Therefore, unlike pyrophyllite with a large vacant M(1) site, the tetrahedra do not tilt significantly out of the (001) plane and the basal plane oxygens are very nearly coplanar (Δz of 0.011 Å compared with 0.240 Å for pyrophyllite). The shapes of the octahedral sites may be described by the ψ_{oct} angle, which is the angle between a body diagonal and the vertical. For a regular octahedron ψ is 54°44', but for talc the angle is 61.5° for both M(1) and M(2), indicating that the octahedra have become thinner than the ideal.

Ideally, 2:1 layer silicate structures are limited to chemical compositions that allow the tetrahedral and octahedral sheets to articulate into a layer, as both sheets share a common junction of apical oxygens and OH groups. Octahedral thinning in talc effectively increases the lateral dimensions of the Mg rich octahedral sheet, thereby allowing better congruence with the tetrahedral sheet. Apparently, there is insufficient compensation by octahedral thinning and the tetrahedral sheet is required to reduce its lateral dimensions by the (001) in-plane counter rotation of adjacent tetrahedra of the silicate ring. This in-plane rotation (α) is 3.6^0 in talc as measured directly from the structural refinement of Perdakatsis and Burzlaff (1981), and it reduces the symmetry of the silicate ring from hexagonal to ditrigonal. It should be noted, however, that there are several factors besides the congruency between the tetrahedral and octahedral sheets that influence octahedral thinning and the magnitude of ψ_{oct}. These factors are summarized in Guggenheim and Eggleton (1987).

Radoslovich (1961) has shown that α, the tetrahedral rotation angle, may be calculated also from the relation: $\cos \alpha = b(\text{obs})/b(\text{ideal})$. The α value calculated in this way is usually an approximation as $b(\text{ideal})$ is difficult to estimate. Based on the above discussion, however, $b(\text{ideal})$ may be calculated directly for talc, as α is known to be 3.6^0 from structural considerations; $b(\text{ideal}) = 9.191$ Å, which is approximately 0.2% larger than $b(\text{obs})$. The value of 9.191 Å represents the size of the b axis after tetrahedral rotation to a value of $\alpha = 0$ and it implies that only a limited amount of a larger cation can substitute for Mg in an idealized talc structure. Therefore, as a moderate amount of ferrous iron is known to substitute for Mg in talc (see discussion above), structural accommodations are probably required for this substitution, presumably by significant octahedral site distortions. A miscibility gap between iron-rich talc and minnesotaite (see Chapter 17), if one exists, has not been determined. Alternatively, there may be an apparent polymorphic relationship (compositional overlap) between iron-rich talc and minnesotaite, although minnesotaite contains excess OH (see Guggenheim and Eggleton, 1986) and iron-rich talc may contain considerable ferric iron.

SPECTROSCOPIC STUDIES

Infrared spectra

Infrared absorption is especially sensitive to the coordination between OH groups and nearest cation neighbors. Therefore, it is of particular interest to examine systematically a solid solution series in an effort to understand variations relating to the OH group and cation configurations, and the effects on the infrared spectra. Because talc-like and pyrophyllite-like units form the basis of most layer silicates, intensive efforts have been made to take such a systematic approach. In addition, examination of spectra from OD substituted samples has proved exceedingly useful. A less satisfactory approach has been to try to use quasi-analogue or similar systems for band identification.

As with most layer silicates, sections of the infrared vibration spectra of talc and pyrophyllite may be correlated generally with the hydroxyl, Si-O bonds and the octahedral linkages (Farmer, 1974). The OH stretching frequencies occur in the 3400-3750 cm^{-1} region, and the OH bending vibrations are generally in the 600-950 cm^{-1} region. Farmer (1974) noted that the Si-O stretching vibrations in the 700-1200 cm^{-1} region only weakly affect other vibrations, whereas the Si-O bending vibrations in the 150 to 600 cm^{-1} region are strongly coupled with

Table 4. Frequencies (cm^{-1}) of infrared absorption bands of talc based on Farmer (1958), unless otherwise noted. See references for definitions of assigned infrared vibrations.

Band	Intensity	Assignment	Reference:comment
4330	vvw	ν_4 + (OH)$^-$ str.	
4200	vvw	ν_6 + (OH)$^-$ str.	
3670	w	(OH)$^-$ str.	
1919	vvw		
1866	vvw		
1818	vvw	summation bands	
1770	vvw		
1706	vvw		
1040	vs	Si-O $\nu_1\perp$	also Lazarev (1972), Rosasco and Blaha (1980)
1014	vs	Si-O ν_3‖	also Lazarev (1972), Rosasco and Blaha (1980)
890	w		
783	w		
687	m	Si-O $\nu_2\perp$	Russell et al. (1970)
669	s	Mg δ(OH)	Naumann et al. (1966), Russell et al. (1970); libration
535	m		
500	infl.		
465	vs	Si-O	Wilkins and Ito (1967)
450	vs		
444	infl.		
424	m	Si-O ν_5‖	
392	s		
384	infl.		
344	m		
311	vw		
259	m	Mg-(OH)$\perp$	Russell et al. (1970)
230	w		

intensities: w=weak; m=medium; s=strong; v=very; infl.=inflection

Table 5. Frequencies (cm^{-1}) of some infrared absorption bands of talc (from Wilkins and Ito, 1967).

Talc composition		Vibration frequency of OH close to-			
Mg - R		Mg Mg Mg	Mg Mg R	Mg R R	R R R
Mg$_{100}$		3676.5	-	-	-
Mg$_{84}$	Ni$_{16}$	3676.5	3661.9	3644.8	-
Mg$_{76}$	Ni$_{24}$	3676.5	3662.3	3645.2	3624.1
Mg$_{52}$	Ni$_{48}$	3676.5	3662.5	3645.9	3625.0
Mg$_{23}$	Ni$_{77}$	3676.7	3662.9	3646.2	3625.9
Mg$_1$	Ni$_{99}$	-	-	3646.9	3627.2
Mg$_{64}$	Co$_{46}$	3677.4	3661.4	3643.3	3622.3
Mg$_{92}$	Zn$_8$	3676.6	3664.4	3649.7	-
Mg$_{96}$	Zn$_{14}$	3676.8	3665.2	3651.6	3634.6
Mg$_{52}$	Fe$_{48}$	3678.3	3663.5	3646.0	3624.2
Mg$_{92}$	Mn$_8$	3677.4	3663.6	3650.0	-
Mg$_{55}$	Cu$_{45}$	3676.0	3669.9	3663.7	3656.3

vibrations of the octahedral cation and with the translatory vibrations of hydroxyl groups.

Talc. Farmer (1958) investigated the relationship of observed infrared absorption spectra from oriented samples and theoretical predictions based on related structures such as brucite. He identified vibration bands in which the change in dipole moment is perpendicular to (001), including those involving (OH) groups and Si-O and Mg-O bonds. Farmer noted also that excessive grinding, which produces a loss of crystallinity, affects the absorption spectra, with the Si-O vibration bands being affected before O--H stretching vibrations.

Naumann et al. (1966) pointed out that brucite was not an appropriate analogue for comparison to the talc spectra and, based on neutron inelastic scattering, suggested some modifications of the band assignments. Russell et al. (1970) re-examined talc band assignments by noting the effects of OH replacement by OD. They re-affirmed the assignment of Naumann et al. for the 669 cm^{-1} band and noted the effects of librational and translating vibrational coupling of OH and OD. Russell et al. (1970) and Lazarev (1972) further refined band assignments (see Table 4).

Vedder (1964), in a study primarily concerned with the infrared spectra of muscovite and phlogopite, noted the effect of small amounts of Fe^{2+} substitution in the Mg talc structure. Vedder (with extensions by Wilkins and Ito, 1967) determined that for any random substitution, the intensity ratios of bands near 3677 (3 Mg and OH interactions), 3662 (2 Mg + R^{2+} and OH interactions), 3645 (Mg + $2R^{2+}$ and OH interactions) and 3630 cm^{-1} ($R^{2+}+R^{2+}+R^{2+}$ and OH interactions) may be determined from

$$1:3X/Y:3(X/Y)^2:(X/Y)^3 \quad ,$$

where R^{2+} is either Ni, Co, Zn, Fe^{2+}, Mn, or Cu substitutional species in Mg talc, X is the fractional concentration of total R^{2+}, and Y is equal to (1-X), the concentration of Mg. To be valid, these ratios require (a) that the substitution is random and (b) the assumption that the molar absorption for the substituting species interacting with the OH group is identical to that of Mg. For synthetic samples containing solid solutions of each of the elements given above, Wilkins and Ito found that the observed intensity ratios closely matched the predicted values, suggesting random octahedral substitution. However, the spectra from willemseite (de Waal, 1970) suggested non-random distributions of Ni in M(1) and M(2), but no evidence for Ni-rich domain structures was detected. Wilkens and Ito also noted small but significant deviations between the two sets of values, indicating the need to modify slightly one or both of the above assumptions. Figure 4 shows the Wilkins and Ito data in the OH stretching frequency region. Note that the octahedral substitution of one R^{2+} cation species does not affect the frequency (band position) attributed to the other octahedral configurations (Table 5). Each configuration (MgMgMg, MgMgR, etc.) gives a distinct band with the intensity of the band related to the proportion of that configuration present.

Velde (1983) found a correlation in talc minerals between the average electronegativity (and mass) of the cation in the octahedral site with the OH stretch band. Also, he found that the OH stretch band frequency is determined by the electronic configuration of the nearest cation neighbors. Therefore, for nearest neighbor cations such as Mg, Ni, Co, and Fe with 3s and 3d electrons, the correlation between electronegativity and

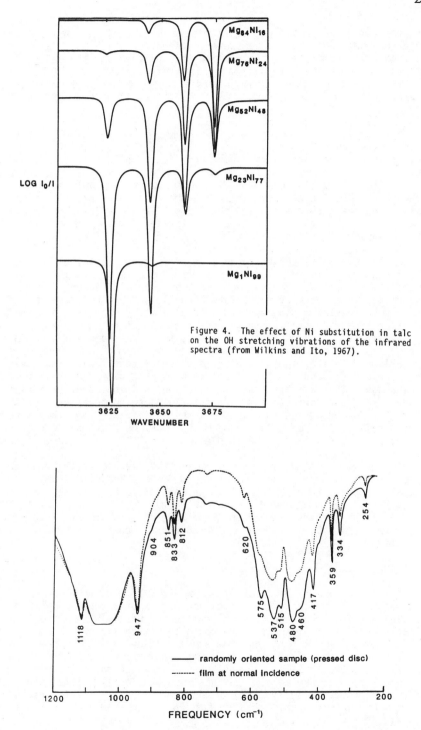

Figure 4. The effect of Ni substitution in talc on the OH stretching vibrations of the infrared spectra (from Wilkins and Ito, 1967).

Figure 5. The infrared spectra of pyrophyllite (after Russell et al., 1970).

frequency shift is readily predictable (a band shift of about 82 cm^{-1} per electronegativity unit). However, Cu and Zn, which have the 3d shell electronically complete, do not follow this pattern and the relationship between ion mass and band frequency is more complex. In Table 5, for example, note the relatively high band frequencies of 3656 cm^{-1} for RRR interactions for R = Cu, as well as similar deviations for MgRR interactions.

Pyrophyllite. Infrared band assignments are less well known for pyrophyllite than for talc, possibly because it is more difficult to synthesize and examine a large number of binary solid solution series for pyrophyllite. However, Russell et al. (1970) compared the infrared spectra (Fig. 5) of pyrophyllite to spectra of a sample treated by D$_2$O. Although the spectra were sufficiently complex to prohibit full interpretation, they were able to assign the 947 cm^{-1} band to in-plane OH librations and the 878 cm^{-1} band in the deuterated form as relating to the out-of-plane OH librations.

Rothbauer (1971) showed by neutron diffraction that the O--H vector points generally away from the two occupied M(2) sites in muscovite and toward the vacant M(1) site with a tilt of about 12^{0} above (001). This confirmed earlier work by Vedder and McDonald (1963) indicating a tilt of 16^{0}. These results may be applied generally to pyrophyllite both because vibration bands are similar (e.g., OH-stretching bands are 3675 cm^{-1} for pyrophyllite and 3660 cm^{-1} for muscovite; see Farmer and Velde, 1973) and because the thermal dehydration mechanism, which is closely related to hydroxyl orientation, appears to be identical (see below). Small variations are expected due to the effect of K in muscovite, but these similarities suggest that the presence of interlayer K is not a major influence in the O--H vector orientation. Farmer (1974) noted that the spectra of dehydrated Ca and Mg montmorillonites are significantly perturbed compared to pyrophyllite and muscovite, but it is unclear if Ca and Mg remain in the interlayer region in these samples.

A weak band at 3647 cm^{-1} in pyrophyllite has been attributed to OH groups shared between Al and Fe ions (Farmer and Russell, 1964), as its intensity appears to be related to iron content. The partial spectrum of ferripyrophyllite as given by Chukhrov et al. (1979) has peaks in this portion of the spectra at 3630 cm^{-1} and 3590 cm^{-1}, which are interpreted as analogues to the 3647 cm^{-1} and 3675 cm^{-1} peaks in pyrophyllite, respectively.

Mössbauer spectra

Talc. Blaauw et al. (1980), Coey (1980), and Heller-Kallai and Rozenson (1981) published the Mössbauer spectra of several samples of Fe-containing talc of unknown geologic origin and unreported analyses. These authors concluded that the spectra could be rationalized in terms of one doublet corresponding to high spin Fe^{2+}. These results correspond to complete ordering of Fe^{2+} into M(2) and are in contrast to the infrared studies on synthetic talcs, in which Fe^{2+} was shown to be disordered over M(1) and M(2). Blaauw et al., perhaps more cautiously, stated that "the difference between M(1) and M(2) octahedral sites in this material does not result in measurable differences in the corresponding Mössbauer doublets".

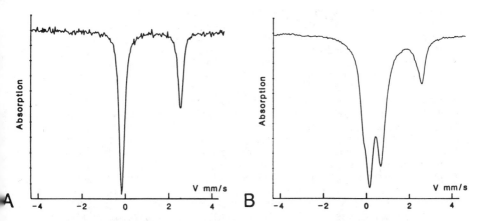

Figure 6. Mössbauer spectra after Noack et al. (1986) of iron-bearing hydrothermal talc (A) and talc from weathering profiles (B). The hydrothermal talcs give one Fe^{2+} doublet, whereas iron-bearing talcs from weathering profiles have both Fe^{2+} and Fe^{3+}. All Fe is believed to be partitioned into M(2).

The inference here is that perhaps Mössbauer analysis does not distinguish between Fe^{2+} for these sites, as suggested by Mineeva (1978) for an ideal mica structure. Dyar and Burns (·1986), however, reviewed and presented data on the effectiveness of Mössbauer to distinguish Fe^{2+} ions in M(1) and M(2) in biotites. They concluded that Mössbauer spectra can be used to distinguish between Fe^{2+} in these two sites and, presumably, this result can be extended to talc. They discussed also the possible effects of Fe^{3+} on the spectra in biotite.

Noack et al. (1986) made a very careful study to determine variations between low temperature (25-50°C) talcs from weathering profiles and higher temperature hydrothermal talcs. Talc compositions varied with up to 12 wt % FeO. The hydrothermal talcs gave results similar to earlier work with only one Fe^{2+} doublet, and they interpreted the data (Fig. 6) as indicating Fe^{2+} occupancy in M(2) only. In contrast, low temperature talc has both Fe^{2+} and Fe^{3+} in M(2). There was no evidence for tetrahedral Fe^{3+} or for the presence of Fe^{2+} or Fe^{3+} in M(1). They concluded, partly based on chemistry as well as other data, that one possible mechanism for the formation of low temperature talc involves the recrystallization of a ferrous talc, immediately followed by iron oxidation.

In summary, Mössbauer data suggest that iron in natural talc is preferentially partitioned into M(2). It is unclear if this trend continues at concentrations above 12 wt % FeO and it is uncertain also if such partitioning plays a role in the occurrence of iron-rich talc vs minnesotaite. Because the relative ease of iron to oxidize appears to be dependent on site occupancy (e.g., see the Mössbauer work of Ferrow, 1987, for annites), site-determination studies for iron-bearing talcs are a requirement for characterizing the mechanism(s) of thermal decomposition for these talcs.

Pyrophyllite. Unlike talc, which commonly has structural iron and therefore is a good candidate for Mössbauer study, pyrophyllite only rarely occurs with structural iron. However, Heller-Kallai and Rozenson (1980, 1981), assuming that most dioctahedral 2:1 layer silicates have similar dehydroxylation mechanisms, used pyrophyllite and pyrophyllite dehydroxylate as analogue structures to explain the behavior of iron in montmorillonite, muscovite, and nontronite upon heating in air for one hour. For these samples at low temperatures, iron preferred M(2) over M(1) sites. In addition, 65-95% of the iron in M(2) was found to be ferric iron, with the remaining Fe^{3+} in M(1). Upon heating, all ferrous iron oxidized and two Mössbauer doublets were observed, interpreted as corresponding to Fe^{3+} in the six-coordinated M(1) site and the five coordinated M(2) site. No cation migration was observed, and the octahedral site was found to be very distorted. Longer heating of the montmorillonite sample (at reduced temperatures) showed that the concentration of Fe^{3+} ions in the five-coordinated M(2) site decreased with a corresponding increase in Fe^{3+} concentration in M(1). Guggenheim et al. (1987) suggested that partial dehydroxylation would produce many non-equal and distorted sites which would, along with an overall reduction in crystallinity (to glass), decrease the signal of the spectrum. These effects have not been taken into account in spectroscopic studies (e.g., Mössbauer) of heat treated and partially dehydroxylated samples. Furthermore, Tsipursky et al. (1985) argued on the basis of x-ray powder and oblique texture electron diffraction data that the (Al-rich) pyrophyllite structure cannot be used to model dehydroxylation in iron-bearing layer silicates. Dehydroxylation is discussed in more detail below.

Layer silicates are useful to study two-dimensional magnetism by Mössbauer spectroscopic techniques. Coey et al. (1984) examined ferripyrophyllite and found the iron to be entirely ferric in M(2) with approximately 0.16 Fe^{3+} in tetrahedral sites (on the basis of 4.0 tetrahedral sites). Ferripyrophyllite orders antiferromagnetically at 18(2)K.

Nuclear magnetic resonance

Nuclear magnetic resonance (NMR) allows the determination of local structural environments up to about the 4th nearest neighbor, but more commonly up to nearest and next nearest neighbors. Therefore, NMR has become increasingly important to assess short range tetrahedral cation ordering, especially because it is now possible to examine solids with the development of magic-angle spinning (MAS) techniques. Silicon-29 and aluminum-27 MAS NMR spectra are of special importance for tetrahedral cation ordering studies, although the latter spectrum is more difficult to interpret because of peak broadening and splitting due to quadrupole effects. Resonance frequencies, reported as chemical shifts, involve comparison to a standard with the chemical shift given as parts-per-million differences from the standard. Chemical shifts cannot readily be calculated and hence the need exists to examine well known phases, such as pyrophyllite and talc, in order to establish a data base from which other spectra may be compared. Because paramagnetic ions (Fe, Mn, etc.) in the sample, either as substituting cations or as components of an impurity phase, cause large effects that cannot readily be interpreted, electron paramagnetic resonance (EPR) is often used as an auxiliary technique to determine the presence of magnetic impurity domains. Reviews of the basic principles and general results have been given by Smith et al. (1983), Oldfield and Kirkpatrick (1985), and Kirkpatrick et al. (1985); see also Reviews in Mineralogy (Volume 18).

Studies reporting silicon-29 spectra for pyrophyllite and talc show similar results (Lippmaa et al., 1980; Watanabe et al., 1983; Smith et al., 1983; Kinsey et al. 1985; and others) and illustrate typical spectra for what is known in notation form as Q^3 (OAl) structures (layer silicates, where 3 = number of bridging oxygens around the Si and 0 = number of ^{IV}Al next nearest neighbors). Spectra for aluminum-27 for pyrophyllite, which show all Al to be in octahedral coordination, are given in the listed references also. These results are not, in themselves, of much interest but they do help establish a data base to which other spectra may be compared.

Of more potential direct interest is the use of aluminum-27 and silicon-29 MAS NMR spectra to study the effect of temperature on the pyrophyllite structure. Frost and Barron (1984) and MacKenzie et al. (1985) presented spectra of heat-treated samples, but the former study has spectra with less detail and, consequently, little attempt was made to reconcile the data to the process of dehydroxylation. Of course, the interest here is to document also the effects of five-fold coordinated Al on the MAS NMR pattern for the dehydroxylate. A prominent feature of the aluminum-27 NMR spectra is the marked decrease in peak areas with thermal treatment without a correlated increase in intensity for a peak related to five-fold coordinated Al. Both groups of workers attributed this intensity loss to broadening caused by the very distorted Al site in the dehydroxylate.

McKenzie et al. (1985) concluded from their study that pyrophyllite dehydroxylates in a single-step process and they suggested that the pyrophyllite material that they studied had considerable amounts of interlayer water. Pyrophyllite, however, has no interlayer water (see the section on the crystal structure), and Guggenheim et al. (1987) re-interpreted the NMR data on the basis of a two-step dehydroxylation mechanism and electrostatic considerations. Site distortions in an octahedral sheet undergoing dehydroxylation may be predicted also by this model, which explains further the loss of peak area in the aluminum-27 spectrum. Details of the dehydroxylation process based on Guggenheim et al. (1987) are given below.

Other spectroscopic studies

Vibrational bands at frequencies other than the infrared have been investigated in order to understand long range interactions (OH--OH, Si sheet--Si sheet, etc.) as in the case of Raman spectra, or the silicate sheet network as in the case of the far infrared. Loh (1973) presented Raman spectra for talc from 100-1100 cm^{-1} and interpreted these spectra on the basis of weakly coupled (isolated) Si tetrahedra or octahedral molecules. Steger (1975) concluded that such an approach is too simple and does not adequately explain the spectra. A Raman microprobe was used by Blaha and Rosasco (1978) and Rosasco and Blaha (1979) to produce spectra (100-3800 cm^{-1}) on particles of microscopic size. They found no difference in the spectra of fibrous and non-fibrous talc forms and were able to make additional vibrational mode assignments. Such fibrous forms were found by Stemple and Brindley (1960) to be intergrowths resulting from a structurally-controlled transformation from tremolite to talc. A preliminary Raman spectrum of pyrophyllite presented by Wiewiora et al. (1979) showed only one band at 3668 cm^{-1}.

Ishii et al. (1967) presented far infrared spectra (60-1200 cm^{-1}) for synthetic talc and synthetic pyrophyllite samples, although these were not of end-member composition. Synthetic pyrophyllites, including a deuterated sample and end-member compositions, were reported in the 50-400 cm^{-1}

250

range by Velde and Couty (1985). Both studies discuss vibrational band assignments.

Optical spectra ($\sim$8,000-30,000 cm^{-1}) for talc containing Fe^{2+} vs Fe^{2+}, Fe^{3+} are compared by Noack et al. (1986). These talcs were examined also by Mössbauer spectroscopy (see above).

THERMAL DECOMPOSITION

The thermal decomposition of layer silicates, including talc and pyrophyllite, has been reviewed periodically since the 1960's (see Dent Glasser et al., 1962; Brindley, 1963; Brett et al., 1970; Brindley, 1970; Brindley, 1975; Wesolowski, 1984). The recent review on the thermal decomposition of talc (Wesolowski, 1984) is of special interest to Western readers because the emphasis is on the published literature from the USSR and Eastern European countries.

The earlier reviews (e.g., Brindley, 1975) summarized the nature of the poorly crystallized, metastable phases developed prior to the formation of the high-temperature stable phases. These transitional forms are topotactic. The dehydroxylate phase of pyrophyllite retains the basic structure of pyrophyllite, minus one H^+ and an OH group, with the Al in five-fold coordination (see structure section). The unit cells of both the pyrophyllite and its dehydroxylate are analogous. On the other hand, platy talc undergoes simultaneous dehydration and recrystallization to enstatite but, again, with minimal atomic adjustments. The description of the mechanisms by which these topotactic reactions occur is tradition-ally developed from the characterization of the parent and the product materials and their relative structural orientations. Although it would be more desirable to measure dynamic structural changes directly, des-cribing a rapidly changing and disordered phase is difficult. As in the earlier reviews, the nature of these reactions is stressed in this section also but, in contrast, crystal chemical models of the process of dehydroxylation are given wherever possible to unify existing diffraction, spectroscopic, and thermal analysis data. Equilibrium reaction data involving talc and pyrophyllite are discussed elsewhere in this chapter.

Taylor (1962) described dehydroxylation in terms of two categories. The mechanism is considered "homogeneous" for cases where elements of water are lost uniformly throughout the crystal. "Inhomogeneous" reactions involve sections of the crystal acting somewhat differently. For these reactions, "no oxygen is lost from those regions of the crystal where the topotactic change occurs. Instead, cation migrations occur, and donor and acceptor regions come into existence..." For example, in brucite, Mg^{2+} ions migrate to the acceptor region with a counter migration of protons. Brett et al. (1970) noted that although the porous nature of the brucite products may be explained by such a process, it is difficult to reconcile the counter-migration of two positively charged ions. Although "homogeneous" and "inhomogeneous" are generally accepted, these terms are misleading because they suggest no other alternatives. Alter-natives are certainly possible, however, and suggested reaction models have been called "modified" inhomogeneous reactions (e.g., Brindley and Hayami, 1964, for chrysotile serpentine), "proton tunnelling" mechanisms (e.g., Freund, 1965, for brucite), and others. Freund and Nièpce (1988) recently reviewed the dehydration mechanism of brucite and portlandite.

Pyrophyllite

The aluminous dioctahedral 2:1 layer silicates, pyrophyllite, muscovite, and Na,K montomorillonite appear to dehydroxylate in similar fashions, although margarite and Mg-exchanged montmorillonites probably do not. For the former cases, the shapes of the thermal analysis peaks and the changes in the IR patterns from the low temperature and high temperature forms are similar. In addition, structural refinements from X-ray data have shown that the primary difference between the pyrophyllite low and high temperature forms vs the muscovite low and high temperature forms is the variation in stacking and not variations in the 2:1 layer, where the dehydroxylation reaction occurs. X-ray powder data and thermal analysis (Koster van Groos and Guggenheim, 1987) suggest an identical relationship for Na,K montmorillonites. Therefore, although this review emphasizes pyrophyllite dehydroxylation, data from muscovite and Na,K montmorillonite are introduced where necessary for the discussion or when such data are not available for pyrophyllite.

Guggenheim et al. (1987) suggested a dehydroxylation process for pyrophyllite and muscovite based on electrostatic considerations. Since both the pyrophyllite and the pyrophyllite dehydroxylate structures (Wardle and Brindley, 1972 and as given above) are well-described, it is possible to predict from Pauling's electrostatic valency principle how bond distances change during the dehydroxylation process. The prediction of bond distances is important to explain the process because, in ionic substances, the bond distance is inversely proportional to bond strength. How bond strengths change during the process will determine the mechanism of dehydroxylation.

The electrostatic valency principle describes relative stability for ionic crystals. For any anion, a stable structure has the sum of the bond strengths that reach the anion from all neighboring cations equal to the formal charge on the anion, but opposite in sign. Pauling bond strengths are defined as the cation charge divided by cation coordination number and are expressed in electrostatic valency units, esu. If the pyrophyllite dioctahedral sheet is examined (Fig. 7A), each oxygen has its formal charge of -2.0 satisfied by its cation neighbors.. For example, the oxygen in the OH group has two octahedral aluminum neighbors with Pauling bond strengths (PBS) of 3/6 each and a H^+ (PBS = 1/1). Therefore, that oxygen is fully satisfied (3/6 + 3/6 + 1/1 = 2.0) as are all the other oxygen ions in pyrophyllite.

Pyrophyllite undergoing heat treatment transforms to the dehydroxylate form (Fig. 7B) by the reaction of 2(OH) -> $H_2O(\uparrow)$ + O, where the residual oxygen is greatly undersaturated at 1.2 esu (two ^VAl bonds at PBS = 3/5 each). The undersaturated nature of the residual oxygen and the four oversaturated oxygens around the aluminum account for the metastable nature of the dehydroxylate form. This pattern may be used also to predict qualitatively the movement of Al and the resulting Al--O bond distances. The Al, because of its high positive (3^+) charge, will move away from certain anions (note arrows in Fig. 7B).

Bond distances for a pyrophyllite structure undergoing dehydroxylation may be predicted also at any stage of partial dehydroxylation. For example, Figure 7C shows a pair of residual OH groups in an otherwise dehydroxylated structure. Of interest is the *pattern* of oversaturated oxygens in the Al--O,OH octahedron. Note that the five-coordinated Al

252

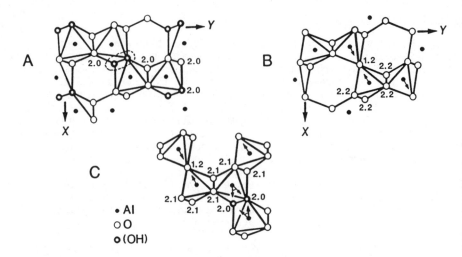

Figure 7. The Al polyhedra for pyrophyllite (A), pyrophyllite dehydroxylate (B), and pyrophyllite in transition (C). Numbers next to anions are Pauling bond strength (PBS) summations. Arrows emanating from Al cations indicate anticipated movement toward anions with lesser PBS summations and away from oversaturated anions. The circle in (A) indicates two OH groups which dehydroxylate to form H_2O and a residual oxygen; note (B). See text for discussion (after Guggenheim et al., 1987).

neighbors of the Al--O,OH octahedron produces an asymmetric distribution of four oxygen ions (each at 2.1 esu), thereby resulting in four longer Al--O bonds for that octahedron. Consequently, Al--OH bonds must shorten (become stronger). The anticipated movement (note arrows in Fig. 7C) of Al ions toward the OH groups oversaturates slightly the oxygen of the group, which, in turn, lengthens (weakens) the O--H bond. This latter slight oversaturation is not depicted numerically in the figure. The model, as presented thus far, considers an extreme case where a pair of Al--O,OH octahedra are linked only to five-coordinated Al. Depending on the degree of dehydroxylation, however, Al neighbors may be in either octahedra exclusively or combinations of five- and six-fold coordination. Therefore, Figures 7A and 7C represent the two configurational extremes for a pair of OH groups in a partially dehydroxylated structure.

Furthermore, the undersaturated nature of the residual oxygen explains why re-hydroxylation of a partially dehydroxylated sample of pyrophyllite (Heller et al., 1962) or muscovite (e.g., Vedder and Wilkins, 1969) is relatively easy. No bonds need to be broken upon re-hydroxylation and the residual oxygen is better satisfied electrostatically as an (OH) group. Re-hydroxylation occurs most readily at (a) high water fugacities (at high water pressures) and (b) if the coordination about the aluminum site is distorted, as a highly distorted configuration is unstable. Maximum distortions occur when the five-fold sites involved in re-hydroxylation have both five-fold and six-fold Al neighbors. When all neighboring Al sites are in five-fold coordination, the driving force for re-hydroxylation is greatly reduced.

Early thermal analysis studies (e.g., Bradley and Grim, 1951; Anton, 1969) showed dehydroxylation to take place over a temperature range of 150-250°C. The temperature at which dehydroxylation started varied also but commonly occurred around 450°C. The details of the thermal gravimetric curves for the dehydroxylation interval apparently depended on crystallite size and shape and on the vapor pressure (Brindley, 1975). In most cases, however, the differential thermal analysis (D.T.A.) peak is asymmetric with the greater part of dehydroxylation occurring at the higher temperature portions. Early D.T.A. studies suffered from the use of equipment requiring large samples (1-10 g) often heated in air. Temperature gradients in the sample were high and, since the material evolved water vapor, water fugacity was influenced by both sample and experimental conditions. Modern commercial equipment reduces these problems by the use of very small sample size in inert atmosphere chambers, although evolved water vapor may still influence results (e.g., Koster van Groos and Guggenheim, 1987). Thermal gravimetric analytical (T.G.A.) and differential thermal gravimetric (D.T.G.) curves obtained by MacKenzie et al. (1985) resolved two dehydration events (Fig. 8).

The model presented in Figure 7 may be used to explain the observed thermal analysis data (Guggenheim et al., 1987). Dehydroxylation occurs relatively rapidly when OH groups have first and second Al neighbors in six-coordination (Fig. 7A). When second Al nearest neighbors become five-coordinated after partial dehydroxylation, the pattern of over-saturated oxygens requires that the Al--OH bond strengthens, thereby slowing the dehydroxylation process. A point will be reached in the experiment where all second nearest Al neighbors are in five-fold coor-dination (Fig. 7C). At this point, dehydroxylation becomes rapid as applied thermal energy increases. The variation in Al--O,OH bond strengths during partial dehydroxylation explains the large temperature interval and the bimodal nature of the reaction.

Both Heller et al. (1962) and Anton (1969) presented pyrophyllite and pyrophyllite dehydroxylate infrared data. Both data sets are similar, with the OH-related bands at 947 cm^{-1} (in-plane OH librations) and 3675 cm^{-1} (OH stretching) being absent in the dehydroxylate. The 878 cm^{-1} band (out-of-plane OH librations) was absent from the spectra presented by Anton for both pyrophyllite and the dehydroxylate, and from the spectrum of the dehydroxylate phase only, as presented by Heller et al. Unassigned bands at 575, 540, and 520 cm^{-1} were present in pyrophyllite but not in the dehydroxylate, with the former two bands appearing to be replaced by a broad band at 566 cm^{-1} in the dehydroxylate.

Infrared data for muscovite and its dehydroxylate are similar also to the pyrophyllite series. In addition, Gaines and Vedder (1964) noted a small amount of peak broadening with decreasing intensity and a small wave number shift for the OH stretching band from 3628 cm^{-1} to 3600 cm^{-1} with dehydroxylation. In situ heating of muscovite in a spectrophotometer (Aines and Rossman, 1985) showed a similar wave number shift, although perhaps to a smaller extent. Guggenheim et al. (1987) attributed this effect during dehydroxylation to the movement of the Al cation toward the remaining OH groups. This movement slightly oversaturates the oxygen of the group, which weakens the O--H bond and causes a downward shift for the 3628 cm^{-1} band. Also, it is expected that the O--H orientation would be affected slightly by the Al movement. In contrast, there is no apparent wave number shift in pyrophyllite, which may indicate that these effects are smaller in pyrophyllite. The model is supported by the thermal analysis data of Guggenheim et al. (1987), which indicated that muscovite retains OH to about 1000°C whereas pyrophyllite is fully dehydroxylated by 850°C, although it is unclear how interlayer cations may affect the data. The change in orientation and the weakening of the O--H bond in

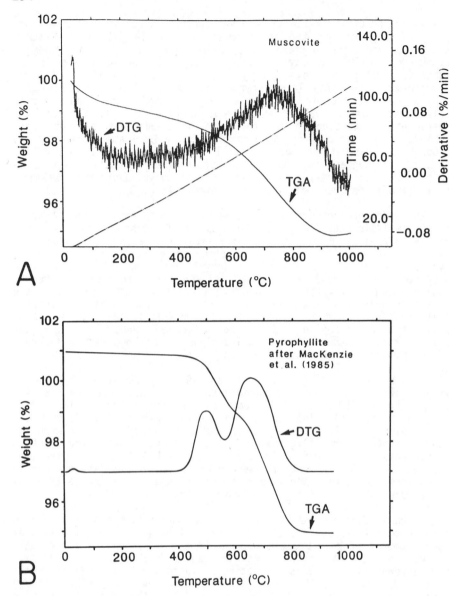

Figure 8. Thermal analysis curves for muscovite (A) compared to pyrophyllite (B). Details of the run conditions are given in Guggenheim et al. (1987).

muscovite may be related to the proposed proton "delocalization" process of Fripiat et al. (1965). However, the latter process was derived from experiments in which the water fugacity was not adequately controlled, thereby perhaps causing the reaction to appear "smeared out".

The above discussion considers kinetic aspects from a structural view based on a model derived from Pauling bond strength considerations. Direct experimental data are necessary to confirm the model, although the model's apparent success in unifying the available data is encouraging. Such a model is qualitative, however, in that it cannot easily be used, for example, to estimate diffusion coefficients or rate equations. Quantitative determinations for pyrophyllite were attempted by Holt et al. (1963), but they did not do the experiments under either a controlled atmosphere or a constant water fugacity, nor did they adequately determine grain size uniformity and other sample features.

The dehydroxylation model as presented here differs considerably from the traditional concepts of homogeneous vs inhomogeneous reactions. Clearly, the model is not homogeneous because different sections of the crystal respond to temperature effects differently. These responses are predicted from Pauling's rules. However, the model does not conform to an inhomogeneous reaction either as defined by Taylor (1962), as large scale cation migrations are not a necessary part of the model.

Talc

Talc thermal decomposition has received less extensive study than pyrophyllite, presumably because the mechanism of decomposition involves a rapid transformation to enstatite without transitional phases. Wardle and Brindley (1972) noted the importance of the octahedral vacant site in pyrophyllite to allow readjustment of the anions following dehydroxylation. In contrast, the trioctahedral arrangement in talc prevents such readjustment and a more extensive recrystallization is required upon de-protonization. This mechanism is generally recognized as inhomogeneous.

The recrystallization reaction may be represented by the equation:

$$Mg_3Si_4O_{10}(OH)_2 \rightarrow 3\ MgSiO_3 + SiO_2 + H_2O$$
$$\text{talc} \rightarrow 3\ \text{enstatite} + \text{silica} + \text{water}$$

The transformation of a structure with sheets of tetrahedra to one with chains may be readily achieved by the isolation of tetrahedral chains along the X direction of the tetrahedral sheet. For cases where the tetrahedral rotation angle, α, is small and the sheet is nearly hexagonal, regular chains may be derived either along X_1 or the pseudohexagonal X_2, X_3 axes.

Nakahira and Kato (1964), in an electron diffraction study of talc heat-treated in air, found that enstatite (e) and cristobalite (c) transformed from talc (t) with the following parameter/cell relationships:

$$a_t(5.3\ \text{Å}) \parallel c_e(5.2\ \text{Å}); \qquad a_t(5.3\ \text{Å}) \parallel (110)_c$$
$$b_t(9.1\ \text{Å}) \parallel b_e(8.8\ \text{Å})$$
$$d(001)_t(9.35\ \text{Å} \times 2) \parallel a_e(18.2\ \text{Å}); \qquad (001)_t \parallel (111)_c \quad.$$

Nakahira and Kato (1964) were uncertain if the enstatite transformation was topotactic relative to the true (a, b) cell repeats or if the pseudo-axes were involved, because twinning was a common feature of their talc samples. Daw et al. (1972) observed the topotactic relation to the true cell, but Bapst and Eberhart (1970) found the three orientations of enstatite for their samples. Likewise, the cristobalite orientation of Nakahira and Kato (1964) should be considered tentative because of the

256

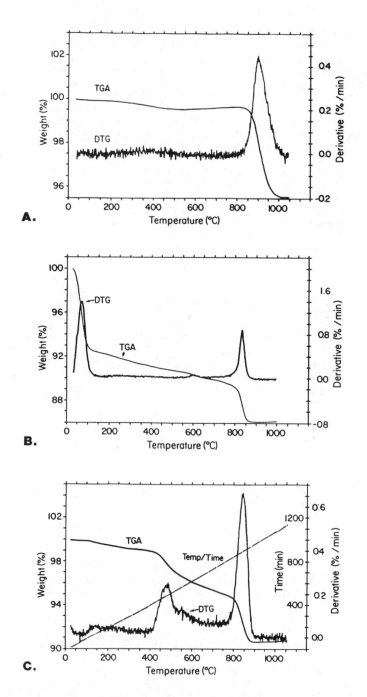

Figure 9. Thermal analysis data for (A) well-crystallized talc, Mt. Windara, Australia, (B) kerolite, Carter's mine, Madison Co., North Carolina, USA, and (C) the 10 Å phase of Bauer and Sclar (1981). These curves illustrate the thermal loss of water from talc and talc-like phases. Runs were made at 10°C/min on a Dupont 1090 Analyzer in N_2 atmosphere using approximately 25 mg samples. P. van Krieken and T. Zalm, analysts.

small number of cristobalite reflections and substantial reflection overlap between the diffraction sets of talc, enstatite, and cristobalite. Daw et al. (1972) could not confirm the presence of cristobalite in their electron diffraction study of in situ heated talc. It is likely that amorphous silica is present in cases where a crystalline silica phase is not found.

Nakahira and Kato (1964) described the talc to enstatite transformation as involving the loss of eight protons in talc and the gain of 4 Mg per unit cell of enstatite. The oxygen framework remains largely undisturbed with only small distortions occurring. The Mg ions migrate to different coordination sites to compensate for the charge imbalance caused by proton loss. Presumably, silica is produced in regions from which Mg cations migrate to the enstatite.

Daw et al. (1972) found that voids, usually associated with dislocations (see also Amelinckx and Delavignette, 1961), formed early in the transformation process and acted as nucleation sites for the enstatite, suggesting that deprotonization and recrystallization are not simultaneous. Kinetics of the transformation depend on heating rate, etc., and the two steps may not always be discernable. Ward (1975), however, determined that the reaction follows first order kinetics and it is consistent with the breaking of the Mg--OH bond followed by ·the migration of Mg. In early crystallite formation, Daw et al. found that the enstatite cell was only quasi-oriented with respect to the talc cell, although the topotactic relationship increased with crystallite size and the voids eventually disappeared. Stemple and Brindley (1960) found a topotactic relationship between tremolite and a fibrous variety of talc resulting from the transformation of tremolite to talc.

Representative thermal analysis curves of talc and kerolite are given in Figure 9. No thermal effects for talc are observed until about 900°C. An apparent single but broad peak occurs between about 900-1050°C with the maximum peak height at about 1000°C.

PETROLOGY AND PHASE EQUILIBRIA

Introduction

Pyrophyllite and talc are rock-forming minerals that occur primarily in a limited range of metamorphic rock-types and certain hydrothermal deposits. Minnesotaite, although now recognized as a modulated 2:1 layer silicate (see Chapter 17), has structural and petrologic affinities to talc, and is therefore also treated in detail here. The simple compositions of these minerals place them somewhat external to the compositional space occupied by most common lithologies, with the result that their occurrence tends to be limited to rocks with large amounts of Al, Mg, and Fe, respectively, relative to other components. They react out with advancing grade of metamorphism: pyrophyllite and minnesotaite under greenschist facies conditions, and talc under amphibolite facies conditions.

The three minerals are best identified by X-ray diffraction and/or by microprobe analysis, although talc is routinely described from ultramafic rocks on the basis of its optics alone. The distinction between Fe-talc and minnesotaite requires X-ray diffraction. Pyrophyllite may be

258

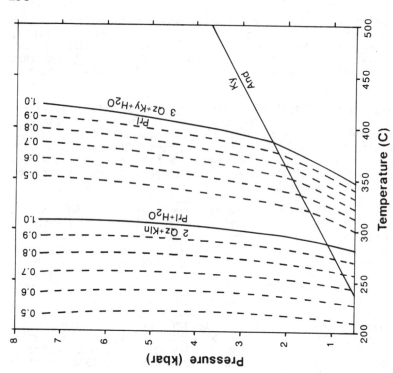

Figure 10 (left). P-T phase diagram for pyrophyllite in the system Al$_2$O$_3$-SiO$_2$-H$_2$O. Prl = pyrophyllite, Dsp = diaspore, Kln = kaolinite, Qz = quartz, Co = corundum. 1: Kln + 2 Qz = Prl + H$_2$O; 2: 2 Kln = Prl + 2 Dsp + 2 H$_2$O; 3: Prl + 6 Dsp = 4 Ky + 4 H$_2$O; 4: Prl = Ky + 3 Qz + H$_2$O.

Figure 11 (right). Pyrophyllite terminal reactions as a function of the activity of H$_2$O. Abbreviations as in Figure 10.

recognized from basal reflections at 9.20 Å, 4.60 Å and 3.07 Å, and distinguished from talc by $d(060)$ at 1.49 Å (Table 3). Minnesotaite has strong reflections at 9.54 Å (cf. talc 9.34 Å), 3.18 Å, and 2.53 Å (see Chapter 17, Appendix). Optical properties are summarized in Deer et al. (1962) and Troeger (1979).

Parageneses of pyrophyllite

Pyrophyllite occurs in metamorphic rocks containing more Al than can be accomodated in other aluminous minerals such as muscovite, paragonite, chlorite, chloritoid, and lawsonite. Thus, it is most frequently found in very aluminous metapelites, including metabauxites, aluminous meta-quartzites, and in rocks relatively enriched in Al by base-leaching during hydrothermal alteration. It is also precipitated in hydrothermal veins. In contrast to kaolinite, its occurrence in sediments is very minor.

With regard to phase equilibria, almost all workers have assumed that pyrophyllite is pure $Al_2Si_4O_{10}(OH)_2$. This seems justified by most of the available chemical analyses of natural pyrophyllite (Kodama, 1958; Deer et al., 1962; Heller et al., 1962; Swindale and Hughes, 1968; Loughnan and Steggles, 1976; Fransolet and Bourguinon, 1978; Papezik and Keats, 1976; Ianovici et al., 1981; Feenstra, 1985; Phillips, 1987). The occurrence of submicroscopic, regularly or randomly interstratified phases, such as montmorillonite, kaolinite, and sericite (Kodama, 1958; Gradusov and Zotov, 1976; Page, 1980), in addition to accidental inclusions, requires the use of TEM procedures for proper characterization and accurate interpretation of chemical data. The presence of small amounts of tetrahedral Al is characteristic of pyrophyllite crystallized metastably in the laboratory (e.g., Rosenberg and Cliff, 1980) and apparently also in nature, although it remains to be seen if this substitution occurs in well crystallized natural material with the appropriate basal spacing (see above). Unlike talc, only rarely has F been determined (Sykes and Moody, 1978) and little attention has been paid to the possible influence of F on the stability of pyrophyllite.

In the system Al_2O_3-SiO_2-H_2O, pyrophyllite + H_2O is stable over a temperature range varying in width from 70 to 130°C, and is limited in the P-T plane by reactions involving kaolinite, quartz, kyanite, andalusite, diaspore, and H_2O (Fig. 10). Figure 10, and all other calculated phase diagrams in this chapter, was calculated using a version of PTX-SYSTEM (GEO-CALC) written by Perkins et al. (1986) and the optimized thermodynamic data base of Berman (1988) for 67 rock-forming minerals. The pyrophyllite equilibria as depicted in Figure 10 are consistent, within the limits of experimental error, with experiments by Fyfe and Hollander (1964), Kerrick (1968), Haas (1972), Haas and Holdaway (1973), Hemley et al. (1980), Chatterjee et al. (1984), and D.C. McPhail (unpublished). Additional discussion of the experimental work may be found in Day (1976), Helgeson et al. (1978), Perkins et al. (1979), Halbach and Chatterjee (1982), and Hewitt and Wones (1984). The phase diagram (Fig. 10) precludes the stable association of kaolinite with corundum or aluminosilicate, and of pyrophyllite with corundum. At near surface conditions, including the hot-spring environment of 100°C and 1 atm, pyrophyllite is less stable than kaolinite + alpha-quartz and kaolinite + alpha-cristobalite, but apparently more stable than kaolinite + amorphous silica (or with H_2O saturated in amorphous silica). Figure 10 has the same topology as that recommended by Day (1976), is similar to one of two models (model 4)

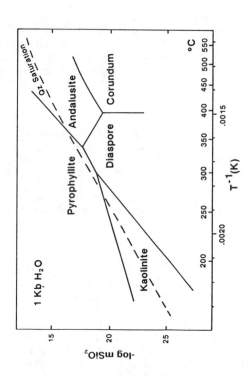

Figure 13 (right). Pyrophyllite phase relations in the system Al₂O₃-SiO₂-H₂O at 1 kbar H₂O in terms of aqueous SiO₂ and inverse temperature, after Hemley et al. (1980). Note that the thermodynamic properties of phases in this diagram are slightly different from those in all other diagrams in this chapter.

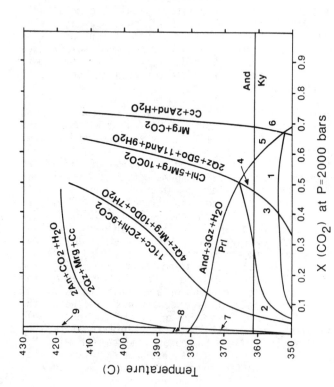

X (CO₂) at P=2000 bars

Figure 12 (left). T-X(CO₂) phase diagram for pyrophyllite at 2000 bars in the system CaO-MgO-Al₂O₃-SiO₂-H₂O-CO₂. Calculated using GEO-CALC, data in Berman (1988), and the MRK expression for non-ideal mixing of CO₂ and H₂O from Kerrick and Jacobs (1981). Mrg = margarite, Chl = chlorite, An = anorthite, And = andalusite, Ky = kyanite, Cc = calcite, Do = dolomite. See Bucher-Nurminen et al. (1983) for a discussion of margarite equilibria in progressive metamorphism. 1: 2 Prl + Cc = Mrg + 6 Qz + H₂O + CO₂; 2: 11 Prl + 5 Do = 5 Mgr + Chl + 31 Qz + 2 H₂O + 10 CO₂; 3: Prl = And + 3 Qz + H₂O; 4: 2 Prl + 31 Ky + 15 Do + 25 H₂O + 15 Mrg + 3 Chl + 30 CO₂; 5: 2 Prl + 31 And + 15 Do + 25 H₂O = 15 Mrg + 3 Chl + 30 CO₂; 6: Prl = Ky + 3 Qz + H₂O; 7: 2 Ky + Cc + H₂O = Mrg + CO₂; 8: 3 Mrg + 5 Cc + 6 Qz = 4 Zo + H₂O + CO₂; 9: 2 Zo + Mrg + 2 Qz = 5 An + 2 H₂O; 10: 3 An + Cc + H₂O = 2 Zo + CO₂.

advocated by Perkins et al. (1979), and corresponds to one of three solutions (solution 29) of Stout (1985). The phase diagram calculated by Chatterjee et al. (1984, Fig. 3) from internally consistent data in the system $CaO-Al_2O_3-SiO_2-H_2O$ differs seriously from Figure 10 only in having a quartz-absent invariant point in the field of andalusite (at 1.4 kbar and $333^{\circ}C$), resulting from the intersection of curves 2 and 3. This invariant point limits the paragenesis pyrophyllite + diaspore to higher pressures and permits the stable coexistence of kaolinite and andalusite at low pressure. Based on the common occurrence of pyrophyllite + diaspore, even in areas of low pressure hydrothermal alteration, and the suspect nature of reported kaolinite + andalusite parageneses, Burt (1978) and Day (1978) agreed that invariant point (Qz) is either at very low pressure or not present at all.

The optimized thermodynamic data indicate an upper pressure limit for pyrophyllite provided by the H_2O-conserved reaction: pyrophyllite = 2 diaspore + 4 quartz at 10-13 kbar (Fig. 10) according to Berman (1988) and 15 to 18 kbar according to Chatterjee et al. (1984). The location of this reaction is sensitive to small uncertainties in the thermodynamic data, and may eventually be best fixed by field data. The high-pressure paragenesis diaspore + quartz has been reported in lawsonite-blueschist facies rocks in the Celebes (DeRoever, 1947) and in the French Alps (Ellenberger, 1958; Goffe, 1982). This type of occurrence must be carefully distinguished from diaspore that has formed metastably with quartz in the near-surface environments of weathering, diagenesis or hydrothermal alteration (Berg, 1937; Chandler et al., 1967; Day, 1976). Interestingly, for all varieties of SiO_2 other than alpha-quartz, the thermodynamic data indicate that diaspore + SiO_2 is everywhere more stable than pyrophyllite.

When H_2O becomes progressively diluted by other fluid species (e.g., CH_4, CO_2), pyrophyllite occupies a slightly broader temperature stability field which is displaced to lower temperatures (Fig. 11). In binary mixtures of H_2O and CO_2, pyrophyllite + calcite are stable over a wide range of fluid compositions (Fig. 12) consistent with the occurrence of pyrophyllite in a variety of calcareous rocks, as described for example by Frey (1987a). Low values of $X(CO_2)$, however, render pyrophyllite + calcite less stable than margarite + quartz or zoisite (Fig. 12), or, at higher pressures, lawsonite + quartz (Nitsch, 1972). To understand paths of metasomatic alteration in mineralizing environments, mineral reactions in the system $Al_2O_3-SiO_2-H_2O$ can be written in terms of H_2O and aqueous silica (e.g., pyrophyllite + 5 H_2O = andalusite + 3 H_4SiO_4), and conveniently viewed in plots of $\log mSiO_2(aq)$ versus $1/T$ (Fig. 13). Hemley et al. (1980) discussed the geological implications of changes in solution composition and mineral assemblage following various paths across this diagram, for example: prograde and retrograde buffered paths, isothermal/isobaric metasomatic paths, externally-controlled-solution paths, and kinetically controlled paths. For instance a kinetic control (sluggish crystallization of quartz) explains the crystallization of pyrophyllite instead of stable kaolinite at very low temperatures in situations of vigorous acid attack on silicate rocks (e.g., hydrothermal alteration in volcanic systems), which produces a fluid supersaturated in quartz (Hemley et al., 1980). The system $K_2O-Na_2O-Al_2O_3-SiO_2-H_2O$ (Montoya and Hemley, 1975; Bowers et al., 1984, p.106f.) is relevant to the growth of pyrophyllite from silicic igneous rocks and sediments through hydrolysis reactions such as:

$$2 \text{ muscovite} + 6 \text{ quartz} + 2 \text{ H}^+ = 3 \text{ pyrophyllite} + 2 \text{ K}^+ \quad .$$

The phase diagrams provided by Bowers et al. (1984) illuminate the relationships between pyrophyllite and numerous other minerals in other systems in terms of the activities of species in an aqueous fluid and various saturation conditions.

The succession of metamorphic assemblages indicated by the phase diagram for Al_2O_3-SiO_2-H_2O is consistent with prograde parageneses found in the field: namely, kaolinite -> pyrophyllite -> kyanite in quartz-bearing aluminous pelites (e.g., Frey, 1978), and with assemblages found in quartz-free metabauxites. Pyrophyllite is described in metamorphic rocks that also contain quartz, kaolinite, illite, muscovite, paragonite, mixed layer muscovite-paragonite, andalusite, kyanite, diaspore, chlorite, sudoite, cookeite, dolomite, calcite, hematite, chloritoid, margarite, lawsonite, and carpholite. Coexistence of pyrophyllite with albite and K-feldspar is generally not found in metamorphic rocks (although there are exceptions, Frey, 1987b), since paragonite and muscovite, respectively, are more stable alternatives. Pyrophyllite has optical properties (colorless, high birefringence, 2V(-) around 50-60 degrees, straight extinction) very similar to those of muscovite, and thus may well have been overlooked in many rocks.

An isograd marking the first appearance of pyrophyllite, by the reaction of kaolinite with quartz (reaction 1, Fig. 10), has been mapped over a distance of 200 km in shales, marls, impure limestones, and laterites in the Helvetic nappes and Prealps of Switzerland (Frey, 1987a). P-T conditions for the isograd were estimated to be 1-2 kbar and 240-260°C, based on data from H_2O-CH_4 fluid inclusions and vitrinite reflectance. Equilibrium between kaolinite, pyrophyllite, alpha-quartz, and H_2O under these P-T conditions would require an H_2O-activity of 0.6 to 0.8 (Fig. 11), a range of values that is in agreement with the compositions of fluid inclusions in quartz in the area (Mullis, 1979; Frey, 1987a). The same isograd was mapped over a distance of 15 km in metasediments in the Eastern Alps of Austria (Schramm, 1978). Franceschelli et al. (1986) recognized four metamorphic zones: (1) kaolinite, (2) kaolinite + pyrophyllite, (3) pyrophyllite, and (4) kyanite in the low-grade Verrucano metasediments of the Northern Apennines. The coexistence of kaolinite, pyrophyllite, and quartz over a range of conditions (Henderson, 1971; Paradis et al., 1983; Franceschelli et al., 1986; Juster, 1987) can be explained by disequilibrium (metastable persistence or retrograde growth of kaolinite), internal fluid buffering, or domainal equilibrium (possibly caused by local variation in activity of SiO_2 or by the physical separation of previously generated immiscible H_2O-rich and CH_4-rich fluids; Juster, 1987). The growth of pyrophyllite in preference to kaolinite in deeply buried sediments (e.g., Chennaux et al., 1970) may be related to a number of factors: hydrostatic pressure less than lithostatic pressure, substantial amounts of CH_4 or salts in the pore fluid, or metastable growth from fluids supersaturated in SiO_2 due to dissolution of metastable detrital material.

In most cases, the final disappearance of pyrophyllite during prograde metamorphism is not affected by its terminal reaction 4 (Fig. 10), but by reactions such as:

$$6 \text{ pyrophyllite} + 2 \text{ chlorite} = 9 \text{ chloritoid} + 20 \text{ quartz} + 5 \text{ H}_2O$$
(e.g., Zen, 1960; Frey, 1978; Paradis et al., 1983, Baltatzis, 1980),

2 pyrophyllite + calcite = margarite + 6 quartz + H_2O + CO_2
(Frey, 1978; Bucher-Nurminen et al., 1983; Fig. 12 here), and

11 pyrophyllite + 5 dolomite = 5 margarite + chlorite + 31 quartz +
$$2 H_2O + 10 CO_2$$
(Frey, 1978; Fig. 12 here).

However, with the high dP/dT gradients in lawsonite-blueschist terranes, the coexistence of pyrophyllite and chlorite ends as a result of a prograde *hydration* reaction:

4 pyrophyllite + chlorite + 2 H_2O = 5 MgFe-carpholite + 9 quartz

(e.g., Viswanathan and Seidel, 1979; Goffé, 1982; Chopin and Schreyer, 1983). The *retrograde* formation of pyrophyllite at the expense of chlorite, chloritoid, and diaspore in hydrothermal veins in metapelites was described in the French Western Alps by Goffé et al. (1987).

Hydrolysis reactions are clearly responsible for the metasomatism that brings many natural rocks into the composition space of the simple system in which pyrophyllite, kaolinite, and diaspore can crystallize. This includes the environment of advanced argillic alteration of silicic and intermediate silicate rocks associated with acidic hydrothermal systems in hot springs, fumaroles, and shallow volcanic/plutonic areas (e.g., Hildebrand, 1961; Swindale and Hughes, 1968; Iwao, 1970; Gradusov and Zotov, 1975) and in association with porphyry coppers (e.g., Meyer and Hemley, 1967; Meyer at al., 1968; Gustafson and Hunt, 1975; Hemley et al., 1980; Beane and Titley, 1981). Depending on T, P, and fluid properties, pyrophyllite in these occurrences is accompanied by: quartz, sericite, dickite, kaolinite, diaspore, montmorillonite, andalusite, topaz, fluorite, zunyite, alunite, jarosite, and sulfide minerals of high sulfidation state (pyrite, covellite, digenite). Gradusov and Zotov (1975) noted that there appeared to be two types of pyrophyllite occurrence: a high temperature type (250-300°C) associated with quartz in metamorphic rocks, and a low temperature type (up to 100°C) associated with opal in solfatara fields. In near-surface and surface environments, kaolinite + alpha-quartz is more stable than pyrophyllite; however, the data of Berman (also those of Helgeson et al., 1978) indicate that pyrophyllite is more stable than kaolinite + amorphous silica at 100°C and 1 bar, a result that is at variance with the widespread occurrence of kaolinite in acid sulfate hot-springs (Allen and Day, 1935), where the waters are saturated in amorphous silica (Ghiorso et al., 1988).

The pyrophyllite deposits of North Carolina (Zen, 1961; Stuckey, 1967; Sykes and Moody, 1978) were produced from silicic and intermediate volcanic rocks during an extended period of overlapping hydrothermal events (argillic or greisen-like) and low-grade regional metamorphism and deformation. Other examples of similar occurrences are found elsewhere in the northern and southern Appalachians (e.g., Papezik and Keats, 1976). Similar base-leaching processes are involved, of course, in the development of bauxitic soils, which recrystallize to pyrophyllite-bearing assemblages during low-grade metamorphism (e.g., Robert, 1971; Jansen and Schuiling, 1976; Goffé and Saliot, 1977; Sharma, 1979; Goffé, 1982; Feenstra, 1985).

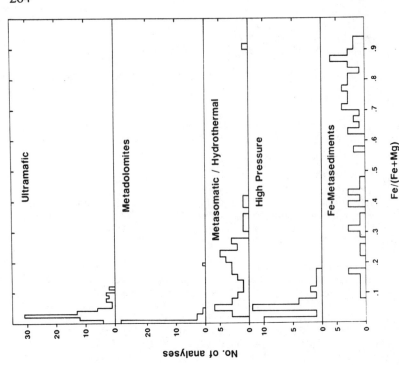

Figure 14 (left). P-T phase diagram relevant to talc stability in the system MgO-SiO₂-H₂O. Calculated using GEO-CALC and thermodynamic data of Berman (1988). Tc = talc, Atg = antigorite, Chr = chrysotile, Fo = forsterite, Br = brucite, En = enstatite, Ath = anthophyllite, Qz = quartz. 1: 9 Tc + 4 Fo = 5 Ath + 4 H₂O; 2: Ath + Fo = 9 En + H₂O; 3: Atg + 14 Tc = 90 En + 45 H₂O; 4: Atg = 14 Fo + 20 En + 31 H₂O.

Figure 15 (right). Compositional range of members of the talc-minnesotaite series from various rock-types and environments. Analytical data from published papers and theses. Note scale and interval changes on y and x axes, respectively.

Phillips (1988) has presented geochemical, petrological, and structural evidence for a remarkable, regional generation of pyrophyllite and chloritoid in pelites (formerly chlorite), metaconglomerites, and metabasalts, due to extraction of SiO_2, MgO, FeO, and CaO by an infiltrating, synmetamatic fluid in the Witwatersrand goldfields.

Parageneses of talc

Talc is more enriched in Mg relative to Fe than any coexisting silicate, oxide, or carbonate mineral. As a result, talc preferentially occurs in highly magnesian rocks such as metamorphosed ultramafics and siliceous dolomites. Talc may not occur in these rocks, however, in the presence of significant amounts of Al_2O_3, CaO, and K_2O, which give rise instead to chlorite, tremolite, and phlogopite, respectively. Talc breaks down on its own composition at $800^{\circ}C$ or less (Fig. 14), and consequently does not occur as a magmatic mineral in ordinary mafic or ultramafic igneous rocks. On the other hand, at 298 K and 1 bar, talc is more stable than serpentine + quartz (Bricker et al., 1973; Hemley et al., 1977a; Berman et al., 1986) and sepiolite (Jones, 1986). Thus, talc has an occurrence in surface environments, although many additional factors govern its occurrence vis-a-vis alterative hydrous magnesian silicates such as sepiolite, palygorskite, kerolite, and stevensite (Jones, 1986; Chapter 16 this volume). Talc has been reported occasionally in evaporites (Kosolov et al., 1969; Braitsch, 1970; Holser, 1979; Scrivenor and Sanderson, 1982), dolomitic (Raymond, 1962; Isphording, 1984) and non-dolomitic limestones, recent beach sands, and very rarely in sandstones and shales (Friedman, 1965) and sea-floor sediments (Hathaway, 1979; see "Hydrothermal Talc" below). Because of equilibrium on a local (mm) scale, talc can also occur in rocks whose bulk composition is outside the usual limited range for talc -- for example, as an alteration product of olivine or orthopyroxene formed during incipient retrograde metamorphism or hydrothermal alteration of gabbro. As was the case for pyrophyllite, the combination of high water/rock ratio and suitable fluid compositions has in certain instances led to large-scale mass-transfer and the stabilization of talc (in many cases a ferroan variety) in hydrothermal environments. Finally, at very high metamorphic pressures, talc is found in a greater compositional range of rock-types, owing to the instability of chlorite. Ferroan talc occurs in some iron-formations and is discussed with minnesotaite below.

Talc in ultramafic rocks. The most abundant occurrence of talc is in metamorphosed ultramafic rocks. In isochemically metamorphosed peridotites and pyroxenites (usually large rock masses), talc coexists with olivine or serpentine minerals respectively. Small masses of ultramafic rock have generally been exposed to SiO_2-saturated, frequently CO_2-bearing, aqueous pore fluids during metamorphism, resulting in virtually monomineralic bodies of talc, or rocks rich in talc, but accompanied by such minerals as tremolite, anthophyllite, chlorite, magnesite, dolomite, magnetite, and quartz (e.g., Chidester, 1962; Sanford, 1982). An increase in modal talc during carbonation results, of course, simply from the increase in Si/Mg ratio of the silicate fraction of the rock due to the conversion of Mg-silicate to Mg-carbonate.

Wet chemical and microprobe analyses of talc from metamorphic ultramafic rocks (e.g., Trommsdorff and Evans, 1972; Evans and Trommsdorff, 1974a) generally reveal only significant amounts of FeO in addition to the end-member components. In isochemically metamorphosed ultramafic

266

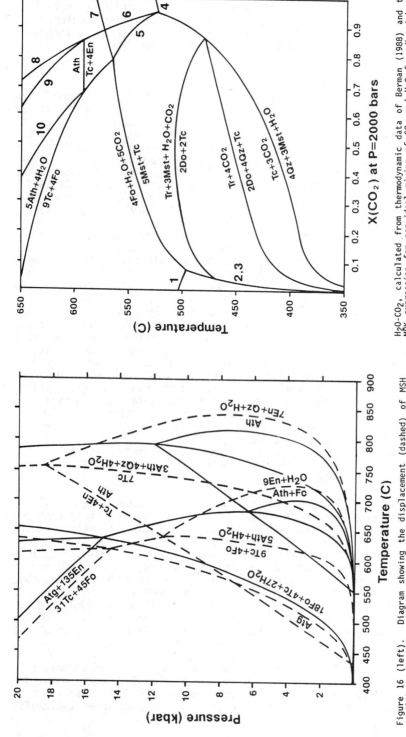

X(CO₂) at P=2000 bars

Temperature (C)

Pressure (kbar)

Temperature (C)

Figure 16 (left). Diagram showing the displacement (dashed) of MSH equilibria (full lines) caused by the presence of Fe in amounts typical of metamorphic ultramafic rocks. Compositions (X_{Mg}) are taken to be: olivine 0.89, orthopyroxene 0.90, talc 0.97, anthophyllite 0.88, antigorite 0.95. Unlabelled curves can be read off Figure 14.

Figure 17 (right). T-X(CO₂) phase diagram showing stability of talc + dolomite and talc + magnesite at 2000 bars in the system CaO-MgO-SiO₂-H₂O-CO₂, calculated from thermodynamic data of Berman (1988) and the MRK expression for non-ideal mixing of CO₂ and H₂O from Kerrick and Jacobs (1981). Mineral abbreviations as in Figure 12; Do = dolomite, Mst = magnesite, Tr = tremolite. Some stable reactions not involving talc have been omitted. 1: Atg = 4 Tc + 18 Fo + 27 H₂O; 2: 15 Tr + 2 Atg + 60 CO₂ = 30 Do + 47 Tc + 30 H₂O; 3: 2 Atg + 45 CO₂ = 17 Tc + 45 Mst + 45 H₂O; 4: Mst + Qz = En + CO₂; 5: Mst + Tc = 4 En + H₂O + CO₂; 6: Tc = 3 En + Qz + H₂O; 7: Mst + En = Fo + CO₂; 8: Ath = 7 En + Qz + H₂O; 9: 7 Tc = 3 Ath + 4 Qz + 4 H₂O; 10: Ath + Fo = 9 En + H₂O.

rocks, most talc compositions fall in the range 1-4 mol % Fe-talc end-member (Fig. 15). Talc in meta-serpentinite rich in magnetite may, however, contain less than 1 mol % Fe-end-member (Vance and Dungan, 1977). Conversely, talc in high-variance metasomatic bodies and veins tends to be more Fe-rich (e.g., Sanford, 1982). Al_2O_3 and NiO (cf. talc in other lithologies) are generally detectable by microprobe analysis, whereas Cr_2O_3, TiO_2, MnO, CaO, Na_2O, K_2O (cf. minnesotaite), and F (cf. talc in siliceous dolomites) are not. Because the talc in metamorphic ultramafic rocks is so close to end-member composition, its maximum stability limits are well represented by the phase diagram for the simple system $MgO-SiO_2-H_2O$ (MSH, Fig. 14); FeO reduces the stability of talc parageneses by varying amounts (see below). Talc alone ($+H_2O$ pore-fluid) is stable over almost the entire metamorphic range of PT-conditions. The stability field for talc + forsterite (relevant to metaperidotites) declines in width from a maximum of $150^\circ C$ at low metamorphic pressures to zero at about 15 kbar, beyond which point calculations indicate meta-peridotites should be represented by antigorite + forsterite or enstatite + forsterite (Berman et al., 1986). The high-pressure coexistence of antigorite + enstatite (in SiO_2-enriched metapyroxenite or metaperidotite, for example) has yet to be observed in the field in eclogite-facies terrains. The talc + olivine field persists undiminished under very low P conditions, precluding the stable coexistence of anthophyllite with serpentine minerals. Contact aureoles developed in ultramafic rock at low pressures around major plutons (e.g., Trommsdorff and Evans, 1972; Springer, 1974; Frost, 1975; Arai, 1975; Matthes and Knauer, 1981) typically have an intermediate zone characterized by talc + olivine roughly one kilometer in width, consistent with the phase diagram (Fig. 14) and the likely thermal gradients. A common feature of the fabric of talc-olivine rocks is the tendency for olivine to be large and elongate (Evans and Trommsdorff, 1974b), producing "jackstraw" texture (Hietanen, 1977; Snoke and Calk, 1978), which superficially resembles the spinifex texture of komatiites (e.g., Flinn and Moffat, 1986, cf. Nesbitt and Hartman, 1986).

The paragenesis talc + enstatite is the high pressure or low temperature equivalent of anthophyllite (Fig. 14). Although the PT-location of this H_2O-conserved reaction in the system MSH is now quite well defined (Berman et al., 1986, p. 1359), its use as a PT-indicator is still limited by the large but imperfectly known effect of Fe. Figure 16 compares the diagram for MSH with the locations of equilibria for an average metamorphic peridotite calculated assuming Raoult's Law behavior for all the minerals. For such Mg-rich compositions, more sophisticated solution models (e.g., Engi, 1986) give (one-site) activity coefficients in the range 1.000-1.005 at $700^\circ C$. Compared to the MSH system, the equilibrium: talc + 4 enstatite = anthophyllite is displaced 6-9 kbar higher (or $120-140^\circ C$ lower), and dehydration reactions are shifted, some upwards and some downwards, by as much as $50^\circ C$. Use of an ideal Mg-Fe site-preference model for anthophyllite (Seifert and Virgo, 1974) would slightly increase the calculated shifts. Figure 16 also suggests that it is possible for anthophyllite and antigorite to occur stably at the same pressure and temperature; they would require very different H_2O-activities, however.

Metaperidotites are not uncommonly carbonate bearing, and, except for serpentine and brucite, their constituent minerals are stable in CO_2-rich fluids. Talc alone is stable in a binary H_2O-CO_2 fluid up to an $X(CO_2)$ of about 0.95 (Fig. 17). In the presence of magnesite, dolomite, and calcite, respectively, the stability of talc is progressively reduced to

268

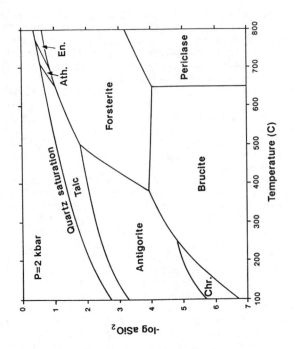

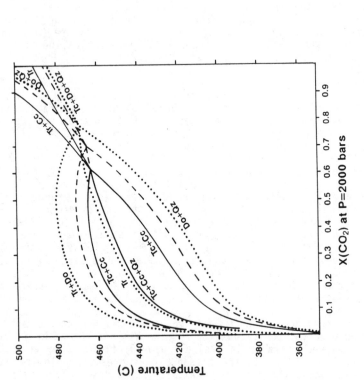

Figure 18 (left). T-X(CO$_2$) phase diagram showing stability of talc + calcite and the influence of F substitution for OH at 2000 bars in the system CaO-MgOSiO$_2$-H$_2$O-CO$_2$-F. Data sources as in Figure 17. Full lines: X$_{OH}$ in talc = 0.0, dashed lines: X$_{OH}$ in talc = 0.2, dotted lines: X$_{OH}$ in talc = 0.4. OH/F partitioning data from Abercrombie et al. (1987).

Figure 19 (right). Stability of talc and related minerals in the system

more restricted portions of the T-X(CO$_2$) diagram (Figs. 17, 18). Talc + enstatite replaces anthophyllite below 600°C at 2 kbar in MSH when X(CO$_2$) exceeds 0.67. The paragenesis talc + enstatite + magnesite, sometimes found in "sagvandites" (Evans and Trommsdorff, 1974a; Ohnmacht, 1974), is stable only at high values of X(CO$_2$), between 0.78 and 0.95 at 2 kbar for example (Fig. 17).

The growth of talc as a result of the metasomatic alteration of metaperidotite has been described and discussed, qualitatively and quantitatively, by a number of authors (e.g., Bricker et al., 1973; Brady, 1977; Hemley et al., 1977b; Nesbit and Bricker, 1978; Frantz and Mao, 1979; Sanford, 1982). Alteration assemblages resulting from gradients in silica activity can be read from diagrams like Figure 19. This diagram (cf. Fig. 13; Hemley et al., 1977a) was calculated from the mineral data of Berman (1988) and the aqueous species data of Helgeson et al. (1981), using the program PTA (Berman et al., 1988). The relationships among activities of other aqueous species, talc, and alternative minerals in more complex systems are shown by Bowers et al. (1984).

Talc in siliceous dolomites. Talc also occurs in small amounts in metamorphosed siliceous dolomitic limestones, in both contact (e.g., Tilley, 1948; Cooper, 1957; Guitard, 1966; Moore and Kerrick, 1976; Hover-Granath et al., 1983) and regional environments (e.g., Trommsdorff, 1966, 1972; Puhan and Hoffer, 1973, cf. Puhan, 1988; Jansen et al., 1978; Hoinkes, 1983; Hoisch, 1987). In some cases, a talc + calcite zone has been defined in the field, representing the beginning stage of metamorphism. Talc has also been described as having formed by retrograde replacement of tremolite and forsterite in marbles. Indeed, many of the economic deposits of talc associated with dolomitic rocks in the U.S. formed by retrograde alteration of pre-existing silicates (Brown, 1973).

Except for F (see below) and Al$_2$O$_3$ (mostly less than 1.0 wt %), and for some unusual parageneses (Schreyer et al., 1980, for example, described talc with up to 36% Na-phlogopite in solid solution in an evaporite-related metadolomite), the composition of talc in siliceous dolomites is well represented by the end-member formula; the proportion of Fe-talc component is even smaller than in ultramafic talc (Fig. 15). Microprobe analyses provide little support for an excess of the brucite component in either metadolomite or ultramafic talc (cf. Smolin et al., 1974). In the mixed-volatile system CaO-MgO-SiO$_2$-CO$_2$-H$_2$O, the stability field of talc + calcite is defined by divariant reactions involving dolomite, calcite, quartz, tremolite, talc, and fluid. In the T-X(CO$_2$) plane at 2000 bars (Fig. 18), the talc + calcite field has a temperature range of less than 50°C and an upper limit of X(CO$_2$) provided by an isobaric invariant point (Skippen, 1974; Slaughter et al., 1975). Because of the acute angle of intersection of isobaric univariant curves at the invariant point, the upper limit of X(CO$_2$) for talc + calcite is difficult to determine accurately, and in fact there has been disagreement in the literature regarding the effect of pressure on the size of the talc + calcite field. The most recent experiments (Eggert and Kerrick, 1981) indicate a contraction with increasing pressure, a conclusion supported by internally consistent thermodynamic data sets (Gottschalk and Metz, 1984; Berman et al., 1985; Berman, 1988; Fig. 20 here). Mirror-image relationships are seen in the isothermal P-X(CO$_2$) plane, drawn at 400°C in Figure 21 (see also Franz and Spear, 1983). An occurrence of talc + aragonite + quartz (Evans and Misch, 1976), suggesting a stability field for talc + CaCO$_3$ at about 350°C and 8 kbar, is in agreement with the

270

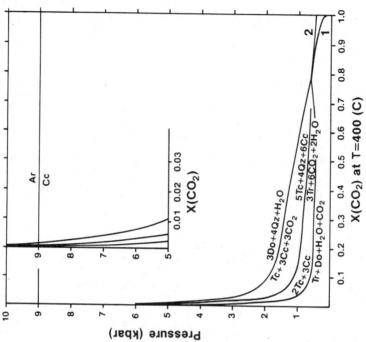

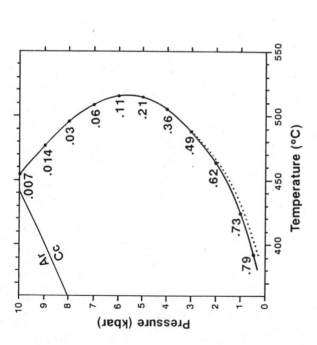

Figure 20 (left). P-T location of the univariant assemblage talc +
tremolite + quartz + dolomite + calcite + H_2O-CO_2 fluid (full line).
Attached numbers are values of $X(CO_2)$. Dotted line: maximum T for the
assemblage talc + calcite (above 3 kbar it coincides with the univariant
assemblage). Data sources as in Figure 17. Aragonite = calcite equilibrium from Johannes and Puhan (1971).

Figure 21 (right). P-X(CO_2) phase diagram showing stability of talc +
calcite at 400°C. Data sources as in Figures 17 and 20. 1: Tr + 3 Cc +
7 CO_2 = 5 Do + 8 Qz + H_2O; 2: 2 Do + 4 Qz + Tc = Tr + 4 CO_2.

calculated phase diagram (Figs. 20, 21), although $X(CO_2)$ must be small. Flowers and Helgeson (1983) modelled the metamorphism of siliceous dolomites using polybaric T-$X(CO_2)$ diagrams reflecting low and medium pressure metamorphic gradients, and computed the relative abundances of talc, tremolite, etc. as a function of temperature for different whole-rock compositions.

The limited field of stability for talc + calcite in the end-member phase diagram (Fig. 18) accounts, in part, for why talc is much less common than tremolite in low-grade metamorphosed siliceous dolomites. In addition, if metamorphism is accompanied by buffering of the composition of the pore fluid (apparently the case for carbonate rocks, at least in the outer parts of contact aureoles), only small quantities of talc will be produced before it reacts out to form tremolite at the isobaric invariant point (Greenwood, 1975). Alternatively, in siliceous dolomites with clay impurity, it has been argued (Gordon and Greenwood, 1970; Trommsdorff, 1972; Skippen, 1974; Rice, 1977) that the T-$X(CO_2)$ reaction path misses the field of talc + calcite, because of buffering along the isobaric univariant reaction: dolomite + K-feldspar + H_2O = calcite + phlogopite + CO_2. The direct reaction of dolomite + quartz to tremolite + calcite, commonly observed in natural impure dolomitic marbles (e.g., Hutcheon and Moore, 1973), can also be explained by the destabilizing effect on talc + calcite of Fe, whence actinolite forms from ferroan dolomite, even in the presence of an H_2O-rich pore fluid. Conversely, the substitution of F for OH in talc and tremolite can greatly stabilize talc + calcite, since of course there is no corresponding uptake of F by dolomite. Abercrombie et al. (1987), using analyses of coexisting talc and tremolite in metamorphosed siliceous carbonate rocks from the Grenville Province of Canada, showed that the K_D for F/OH exchange between tremolite and talc, while about 2.0 (i.e., F favors tremolite) for small contents of F (Allen, 1978; Mercolli, 1980; Walther, 1983), is in fact dependent on the F/OH ratio and passes through unity for large F-contents (i.e., X_F = 0.5 in both minerals). The non-ideality is attributed to talc (see also Duffy and Greenwood, 1979), since OH-F partitioning data for tremolite and phlogopite suggest ideal behavior. The effect of F in talc and tremolite on the phase relations is shown in Figure 18 for values of X_F = 0.2 and 0.4 in talc and of 0.3 and 0.45 in tremolite in exchange equilibrium with talc (Abercrombie et al., 1987). Fluorine extends the stability field of talc + calcite to both lower and higher temperatures (into the field of sillimanite, for example) and to higher values of $X(CO_2)$ in the fluid. Since F has not routinely been analyzed in talc (but see Ross et al., 1968, Table 4) or other minerals in metamorphosed siliceous dolomites, it is not known to what extent the occurrence of talc has been influenced by elevated f_{HF} in the pore fluid. Walther (1983), for example, suggested that talc formed from tremolite in the veins in dolomites at Campolungo, Ticino, Switzerland, due to local enrichment of HF in the fluid, possibly through dissolution of nearby fluorite.

A theoretical framework for the analysis of metasomatism in siliceous dolomites was presented by Walther and Helgeson (1980), who provided diagrams relating the activities of species in the fluid to P, T, and $X(CO_2)$. The stability of talc + quartz at elevated temperature and pressure in aqueous solutions containing $MgCl_2$ was determined by Frantz and Popp (1979). Recent work on fluid inclusions in calcareous meta-sediments (Sisson et al., 1981; Trommsdorff et al., 1985) has revealed the presence of brines and largely immiscible brine + CO_2 fluids at ambient metamorphic temperatures. If typical of the pore fluid in metamorphic siliceous dolomites, such immiscibility must be taken into account in the topology of isobaric relationships in $T-X(CO_2)$ and $P-X(CO_2)$ diagrams such as Figures 18 and 21. The consequences of NaCl in the fluid were explored by Jacobs and Kerrick (1981) and Bowers and Helgeson (1983).

Hydrothermal and other low-temperature parageneses of talc. In recent years, talc has been recognized along with iron, copper, and zinc sulfides as a direct precipitate in modern submarine hydrothermal environments and in ancient analogues of such environments. In some cases talc is the most abundant silicate phase. It is distinctly more Fe-rich than talc in ultramafics and siliceous dolomites (Fig. 15). Talc precipitates as a result of the rapid mixing of hot silica-saturated hydrothermal fluid with cold Mg-rich bottom waters or pore fluids. Conditions have been reproduced experimentally (e.g., Seyfried and Bischoff, 1977) and theoretically (Janecky and Seyfried, 1984; Bowers et al., 1985). Important controls on the formation of talc are the temperature and water/rock ratio of the hydrothermal fluid and the degree of mixing with seawater. Kinetic factors may also be significant, since the equilibrium models predict more talc than is generally the case.

Siever and Kastner (1967) reported minor amounts of talc with sepiolite and zeolites in sediments on the crest of the Mid-Atlantic Ridge, but the provenance of the talc was unclear (an ultramafic source seems likely, M. Kastner, Univ. Cal.-San Diego, pers. comm., 1988). Lonsdale et al. (1980) described terraces and ledges of massive ferroan talc (19 to 25% of the Fe-end-member) together with pyrrhotite and smectite on the ocean floor along a segment of a spreading ridge in the Guaymas Basin of the Gulf of California. Isotopic data suggested a formation temperature of about 280°C. Koski et al. (1985) described the formation of talc (7.5% Fe-end-member) at approximately 270°C in a pyrrhotite-rich hydrothermal mound in the southern trough of the Guaymas Basin. Small amounts of talc have also been described together with anhydrite in mounds and in the walls of polymetallic (pyrite, chalcopyrite, sphalerite, wurtzite) sulfide chimneys along the East Pacific Rise (Spiess et al., 1980; Haymon and Kastner, 1981; Styrt et al., 1981; Goldfarb et al., 1983). These chimneys are discharging hydrothermal fluid at maximum temperatures of 350-380°C. Zierenberg and Shanks (1983) described cross-cutting veins of talc and anhydrite, together with Mg-saponite, chlorite, and sulfides, in association with recent metalliferous sediments in the Atlantis II Deep of the Red Sea Rift. The veins are thought to have formed at high temperature (390°C) in the conduits of discharging brine solutions.

Direct precipitation of talc from a Permian sea-floor hydrothermal system was inferred by Aggarwal and Nesbitt (1984) for the Chu Chua massive chalcopyrite-pyrite deposit, British Columbia. The hydrothermal system developed above alkalic basalt and was characterized by a high water-rock ratio and rapid fluid ascent. Increasing degrees of mixing of a 300°C (or higher) hydrothermal fluid with seawater were responsible for the precipitation in lenticular masses first of ferroan talc, then of ferroan talc + magnetite, and finally the metal sulfides on top. An association of concordant lenses of talc-actinolite rock (±chlorite, stilpnomelane, sulfide, magnetite) with a Cu-Zn volcanogenic massive sulfide deposit was described in a detailed geochemical study by Costa et al. (1983) of the Mattagami Lake deposit in the Archean Abitibi Greenstone Belt in N. W. Quebec. Footwall rhyolite underwent Mg, Fe, Mn, and S addition and Si, Na, Ca, and K depletion. In response to the evolving fluid composition, the talc precipitated ranged in composition from 5 to 34% of the Fe-end-member, with the smallest Fe values at the stratigraphic base of the deposit. Isotopic and fluid inclusion study suggested deposition between 250°C and 300°C. The talc is believed to have precipitated (cf. Roberts and Reardon, 1978) along with the sulfides from a dynamic seafloor brine pool fed by discharging hydrothermal solutions. Late-stage sulfide-silicate replacement veins containing talc cut the chloritic footwall rhyolite. A similar association of talc-rock (±carbonate and chlorite), metarhyolite, and massive Cu-Zn sulfide mineralization occurs in Silurian rocks at the Penobscot mine, Harborside, Maine (Bouley and Hodder, 1984). Talc, along with chlorite, quartz, phengite, phlogopite, albite, and dolomite, was reported in stratified alteration zones in association with volcanogenic massive sulfides in the Ambler district of Alaska (Schmidt, 1986). Notwithstanding these examples, talc is not a typical silicate mineral for associated massive base-metal sulfide deposits. In the Salton Sea geothermal system, talc was identified in the pores of sandstone in the biotite zone (325-360°C) of alteration (McDowell and Elders, 1980).

Although some deposits of talc described previously as epigenetic are more likely of seafloor syngenetic origin, there nevertheless remain some cases where the evidence for a replacement origin is good. These include talc deposits associated with metasomatic chlorite in Cambrian dolostones near Winterboro, Alabama (Blount and Vassiliou, 1980), and those near Dillon, Montana, which formed during Proterozoic alteration of Archean dolomitic marbles and, to a lesser extent, of quartzo-feldspathic gneisses (Anderson et al., 1988). In the latter case, a shallow hydrothermal system removed Ca and added massive quantities of Mg and Si, resulting in the replacement of serpentine, tremolite, and dolomite by talc, and the growth of talc in open fractures. The geochemistry of a further example of metasomatic talc and chlorite-rich rocks, derived in the French Pyrenees from dolomitic marbles and other lithologies, was discussed in detail by Moine (1982) and Moine et al. (1982). Although waters issuing from ultramafic rocks at the surface are supersaturated in talc and other Mg-silicates, precipitation of talc is rare (Pfeifer, 1977).

It is interesting to note that while the formation of talc from pre-existing olivine and orthopyroxene is ordinarily viewed as a hydration reaction, the effect of heated (550°C) meteoric waters on olivine in gabbro, Isle of Skye, Scotland, was described by Ferry (1985) in terms of a *hydrolysis* reaction, releasing hydrogen and H_2O:

$$4\ Mg_{1.5}Fe_{0.5}SiO_4 + 6.24\ H^+ = Mg_{2.88}Fe_{0.12}Si_4O_{10}(OH)_2 + 0.627\ Fe_3O_4 + 1.493\ H_2O + 0.627\ H_2 + 3.12\ Mg^{++}\ .$$

High temperature hydrothermal activity following crystallization of the Skaergaard Intrusion produced talc (X_{Fe} = up to 0.24) in veins and altered wallrock, accompanied by hornblende, cummingtonite, chlorite, and biotite (Bird et al., 1986).

Talc is not a common mineral constituent of soils. However, nickel- (e.g., 19.3 wt % NiO) and ferric iron-rich varieties occur in the saprolite zone of laterites developed on serpentinized ultramafic rock (Maynard, 1983; Noack et al., 1986). The talc of talc deposits appears able to survive intense tropical weathering (Akpanika et al., 1987). At its type locality in South Africa, willemseite, the Ni-analog of talc, occurs

274

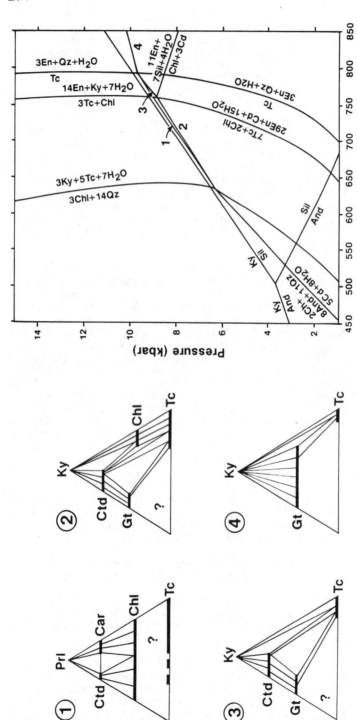

Figure 23. P-T phase diagram illustrating talc parageneses in the system MgO-Al₂O₃-SiO₂-H₂O. Thermodynamic data from Berman (1988). Chl = clinochlore, Ky = kyanite, And = andalusite, Sil = sillimanite, Cd = hydrous cordierite, En = aluminous enstatite, after Gasparik and Newton (1984). Calculated with talc activity = 1.0 (see text). 1: 2 Tc + 6 Sil + Qz = 3 Cd + 2 H₂O; 2: 11 Tc + 29 Sil = 14 Cd + Chl + 7 H₂O + 6 + 2 Sil = 14 Cd + En + H₂O; 4: 2 Sil + 2 En + 4 H₂O = Cd.

Figure 22. Phase relations involving talc in the high-pressure metamorph-ism of pelites, after Goffé and Chopin (1986). Car = carpholite, Gt = garnet, Ctd = chloritoid. P and T increase from (1) to (4).

in a nickel-rich ore deposit by reaction at the contact between quartzite and ultramafic rock (de Waal, 1970).

Talc in high-pressure rocks. As a result of the relative instability of chlorite and biotite, talc becomes a significant phase in the high-pressure (eclogite facies) metamorphism of alumina-rich rocks at medium to high temperatures. Critical parageneses produced in quartz-bearing rocks are talc + phengite, talc + garnet, talc + chloritoid, and talc + kyanite. First discovered by Abraham and Schreyer (1976), the pair talc + phengite (compositionally equivalent to biotite + kyanite + quartz + H_2O) has since been shown to be widespread in high-grade pelitic blueschists in the Western Alps (Chopin, 1981; Chopin and Monié, 1984). In the progressive metamorphism of quartz + phengite bearing pelitic rocks under high P-T gradients (Fig. 22), the gradual breakdown of chlorite (starting at the Fe-end) leads first to the assemblage chloritoid + talc + garnet (or chloritoid + talc + chlorite in Mg-rich samples); and then to chloritoid + talc + kyanite (not even Mg-rich chlorite stable); and finally the disappearance of Mg-chloritoid leads to the assemblage kyanite + pyropic garnet + talc (+ coesite) in rocks of high Mg/Fe ratio (Chopin, 1981; 1984). These assemblages show regional zonation in high-pressure metamorphosed crystalline basement and ophiolitic cover rocks in the 200 km long arc of the Western Alps (Goffé and Chopin, 1986). The Fe-content of talc in these parageneses ranges from zero in highly oxidized metasediments to at least 15% of the iron end-member (Fig. 15). Modest amounts of Al_2O_3 may be coupled with a Na_2O substitution (Raheim and Green, 1974; Abraham and Schreyer, 1976) or represent a minor pyrophyllite component (Chopin, 1981; cf. McKie, 1959). High Al is not a necessary property of naturally occurring high-pressure talc.

The association of talc with kyanite has been reported not only in pelitic schists but also in metagabbros (Chinner and Dixon, 1973; Abraham et al., 1974; Miller, 1974; Cooper, 1980; Barnicoat and Fry, 1986), evaporite-related metasediments (Schreyer and Abraham, 1977) and other rocks of debatable parentage (McKie, 1959; Vrana and Barr, 1972; Kulke and Schreyer, 1973; Raheim and Green, 1974; Vrana, 1975; Udovkina et al., 1980). The pressure and temperature significance of this "whiteschist" association has been discussed by Schreyer (1973, 1977, 1988). Compositionally equivalent parageneses (Fig. 23) are (i) chlorite + quartz at low temperature, (ii) cordierite at low pressure (+ chlorite or corundum if SiO_2-undersaturated), (iii) kyanite/sillimanite + enstatite/gedrite at high temperature, (iv) Mg-carpholite + quartz at very high pressure and low temperature (Chopin and Schreyer, 1983), and (v) pyrope + coesite at very high pressure and high temperature (Schreyer, 1985, 1988). Another alternative is yoderite + quartz (not considered in the MASH system depicted in Fig. 23), although this has not been reported other than in Tanzania (McKie, 1959; Basu and Mruma, 1985). Under conditions of $P(H_2O)$ = P(total) the minimum pressure for talc + kyanite is just above 6 kbar (i.e., is indistinguishable from that implied by kyanite itself), or even less under lower activities of H_2O, as noted by Schreyer (1977). Figure 23 suggests that a very narrow stability field for talc + sillimanite is possible. The calculated position of the reaction: 3 Chl + 14 Qz = 3 Ky + 5 Tc + 7 H_2O (Fig. 23) differs from the experimental brackets of Massonne et al. (1981), perhaps because of Al in the high-pressure synthetic talc (Newton, 1972; Berman, 1988). When talc and kyanite are accompanied by chloritoid, pressures of 15 to 20 kbar are indicated, according to Goffé and Chopin (1986). Grew and Sandiford (1984) described an occurrence of the rare paragenesis talc + staurolite from Antarctica, and attributed it to quartz undersaturation, temperatures near 700°C, and pressures of 8 kbar or more. The same mineral pair, together with pyrope, chlorite, and kyanite, were also reported by Chopin (1987). In quartzite at the Mautia Hill locality, Tanzania, Basu and Mruma (1985) described the assemblage Al-talc + viridine + piemontite + yoderite + phlogopite + tremolitic hornblende. In this case, the talc has both tetrahedral and octahedral Al.

Parageneses of minnesotaite

Compositional and stability constraints on minnesotaite limit its occurrence to very-low and low-grade Fe-rich metasediments (iron-formations and ironstones) and a few hydrothermal deposits.

By assigning Al where necessary to both tetrahedral and octahedral sites, classical and microprobe analyses of minnesotaite can generally be recast into a formula that has the same T/O ratio as talc (4/3). Further-more, the existence of a minor "greenalite" component in minnesotaite (Floran and Papike, 1975) is not supported by the recent structural determination (Guggenheim and Eggleton, 1986; Chapter 17). Many of the apparently low anhydrous totals of microprobe analyses of minnesotaite now become acceptable when structural H_2O is calculated from either the P-cell [$(Fe,Mg)_3Si_4O_{9.6}(OH)_{2.8}$] or C-cell [$(Fe,Mg)_3Si_4O_{9.55}(OH)_{2.89}$] formula determined by Guggenheim and Eggleton (1986) rather than, as formerly, from a talc formula (Chapter 17). The few quality wet chemical analyses of minnesotaite in fact reported appropriate amounts of H_2O. Published analyses of minnesotaite, ferroan talc, and talc in Fe-rich metasediments (mainly iron-formations) cover almost the entire possible range in X_{Fe} (Fig. 15); it remains to be seen if the scarcity of compositions in the range X_{Fe} = 0.45 to 0.6 has any significance with regard to talc-minnesotaite phase relations. Except for material altered by surface waters, the iron is almost entirely ferrous. Among the other minor constituents, MnO is generally less than 0.5 wt % in minnesotaite from Precambrian banded iron-formations, which are low in whole-rock MnO, but it is high in some of the other occurrences (see below). Al_2O_3 is typically present in the 0.1 to 1.5 wt % range, K_2O varies up to 0.6 wt %, and Na_2O is detectable but generally less than 0.1 wt %. The K is believed to be located in interlayer regions, as described by Guggenheim and Eggleton (1986; Chapter 17).

Until recently, published experimental studies (e.g., Smith, 1957; Flaschen and Osborne, 1957: Forbes, 1969, 1971; Grubb, 1971) have not succeeded in determining the P-T stability limits of minnesotaite. Results of a study of the metastable equilibrium:

$$10 \text{ minnesotaite} = 15 \text{ olivine} + 25 \text{ quartz} + 14 \text{ } H_2O$$

in the system $MgO-FeO-SiO_2-H_2O$ (balanced using the P-cell formula of minnesotaite) have been reported in an abstract (Evans et al., 1982) and a thesis (Engi, 1986). In this work, the synthesis of minnesotaite was not attempted, but instead natural minnesotaite was reacted with quartz, H_2O, and synthetic olivines to form product olivines with compositions bracketing the olivine limb of the $T-X_{Mg}$ phase loop for the above reaction at 2 kbar (Fig. 24). Extrapolation to pure fayalite indicated equilibrium in the MgO-free system at about 455°C and 2 kbar. Calculation from this result of the stable breakdown reaction of minnesotaite:

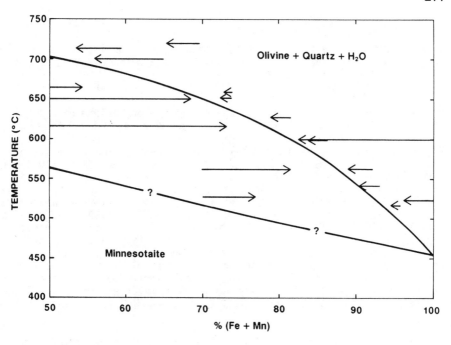

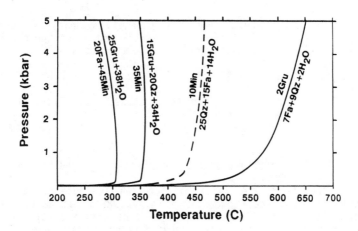

Figure 24 (top). Experimental brackets on the olivine limb of the T-X phase loop at 2 kbar for the metastable breakdown of minnesotaite to olivine + quartz + H_2O (Engi, 1986). Arrows show amount and direction of change of olivine in run products measured by microprobe. Starting materials in different runs: Fa100, Fa70, Fa50, or no olivine, 2 natural minnesotaites, synthetic minnesotaite (Fe50), quartz, and H_2O.

Figure 25 (bottom). Calculated P-T phase diagram (see text) showing stable (full lines) and metastable (dashed line) phase relations of minnesotaite, grunerite, fayalite, and quartz in the system $FeO-SiO_2-H_2O$. Data sources: fayalite and quartz, Berman (1988); Cp for minnesotaite ($Fe_3Si_4O_{9.6}(OH)_{2.8}$) and grunerite, Berman and Brown (1985, Table 4); minnesotaite, V = 153.4 cm^3, S(298K) = 378 J/K, H_f(298K) = -4941200. J.; grunerite, V = 278.7 cm^3, S(298K) = 744.25 J/K (Anovitz et al., 1988), H_f(298K) = -9609500. J.

$$35 \text{ minnesotaite} = 15 \text{ grunerite} + 20 \text{ quartz} + 34 \text{ } H_2O$$

is made uncertain by the lack of agreement on the stability of grunerite (Miyano and Klein, 1986, and references therein) with respect to fayalite:

$$2 \text{ grunerite} = 7 \text{ fayalite} + 9 \text{ quartz} + 2 \text{ } H_2O \text{ ,}$$

and the possible error in estimating the entropy and volume of end-member minnesotaite. Figure 25 was calculated assuming an equilibrium temperature of $600^{\circ}C$ at 2 kbar for the breakdown reaction of end-member grunerite. This temperature is approximately midway between the curves calculated from the data of Fonarev (1981) and those of Miyano and Klein (1986); a natural grunerite with only 3.8 mol % Mg + Ca end-members was found to react to olivine, quartz, and H_2O at $624^{\circ}C$ and 2 kbar (Engi, 1986). The calculated upper stability limit of minnesotaite is somewhat higher in temperature than suggested by Miyano (1978a), Mel'nik (1982), and Miyano and Klein (1983). The breakdown curve changes from positive to negative dP/dT slope around 2 kbar; the revision in the formula of minnesotaite (1.4 H_2O instead of 1.0 H_2O) has changed the volume of reaction and led to an increase in the estimated entropy and volume of minnesotaite. These revised thermodynamic data suggest that there may not be a pressure maximum for minnesotaite, corresponding to an invariant point involving greenalite, minnesotaite, grunerite, quartz, and H_2O (cf. Evans et al., 1982).

The applicability of the FSH diagram (Fig. 25) to natural Mg-bearing minnesotaite is hard to assess, since few data exist on the compositions of minnesotaite coexisting with grunerite (Floran and Papike, 1978, Tables 1 and 5) or with any ferromagnesian mineral. Towards more Mg-rich compositions, Fe-Mg partitioning is strongly composition dependent, however, and one notes that the Mg-end-member analog of the minnesotaite breakdown reaction (with different crystal structures on both sides):

$$7 \text{ talc} = 3 \text{ anthophyllite} + 4 \text{ quartz} + 4 \text{ } H_2O$$

is in equilibrium at a much higher temperature, e.g., $710^{\circ}C$ at 2 kbar.

The lower limit of stability of minnesotaite corresponds to the reaction:

$$\text{greenalite} + \text{quartz} = \text{minnesotaite} + H_2O$$

(greenalite reactions are left unbalanced here because of its non-stoichiometric formula, see Guggenheim et al., 1982). It is unlikely that this stable greenalite reaction will ever be experimentally bracketed; higher temperature, metastable greenalite reactions (e.g., producing fayalite + quartz + H_2O) would be better targets for experimental reversal (Engi, 1986). Field estimates place the entry of minnesotaite in the 1-2 kbar range at about $200^{\circ}C$ (French 1973), $150^{\circ}C$ (Perry et al., 1973; Klein and Fink, 1976), $130^{\circ}C$ (Miyano, 1978a), and $100^{\circ}C$ (Klein, 1983).

Unlike talc, minnesotaite is not stable in the presence of a CO_2-rich aqueous fluid, siderite + quartz being the stable assemblage when $X(CO_2)$ exceeds a certain value (Frost, 1979). Haase (1982b), for example, suggested a maximum value of 0.2 at 2 kbar, corresponding to an isobaric invariant point at $360^{\circ}C$ containing the minerals minnesotaite, grunerite, quartz, and siderite. Anovitz et al. (1988) calculated a maximum $X(CO_2)$

of less than approximately 0.01 for minnesotaite. The stability of minnesotaite with respect to oxygen fugacity has not been accurately defined experimentally. Burt (1972) presumed hematite and minnesotaite to be incompatible. Miyano (1978b) compiled analytical data to show that, with rare exceptions, only the Mg-rich members of the talc-minnesotaite series (X_{Fe} less than 0.34) occur in Precambrian iron-formations along with hematite + magnetite, and those with higher X_{Fe} occur exclusively with magnetite. The T-f(O_2) diagrams of Frost (1979) incorporate similar reasoning.

By far the most important occurrence of minnesotaite is in Precambrian banded iron-formations, in North America particularly the Sokoman, Gunflint, Biwabik, and Negaunee Iron Formations in Labrador, Quebec, Ontario, Minnesota, and Michigan. In many of the world's iron formations, ferroan talc is more common than minnesotaite (Klein, 1983), although, as discussed earlier, the distinction has usually been made on a compositional rather than structural basis. Field data suggest that minnesotaite forms as a late diagenetic phase (Klein and Bricker, 1977), although it is most abundant as a metamorphic mineral forming a minnesotaite-bearing zone at higher grades than greenalite + quartz (e.g., Floran and Papike, 1978), and before the appearance of grunerite (Klein, 1982: 1983). Minnesotaite is usually described as a colorless to pale green acicular mineral, typically forming sprays, or bow-tie and rosette structures, having formed at the expense of greenalite granules, stilpnomelane, siderite and ankerite (French, 1973; Klein, 1974; Floran and Papike, 1975, 1978; Klein and Fink, 1976; Gole, 1980; Klein and Gole, 1981; Haase, 1982a). The prograde reactions are:

$$\text{greenalite} + \text{quartz} = \text{minnesotaite} + H_2O \quad ,$$

and

$$15 \text{ siderite} + 20 \text{ quartz} + 7 \ H_2O = 5 \text{ minnesotaite} + 15 \ CO_2 \quad .$$

A reaction forming magnetite as well as minnesotaite from greenalite is most realistically written with oxygen conserved (cf. Floran and Papike, 1978):

$$\text{greenalite} = \text{minnesotaite} + \text{magnetite} + H_2 + H_2O \quad .$$

Minnesotaite is also reported to have formed by retrograde alteration from fayalite and grunerite in higher grade iron-formation hornfels (Floran and Papike, 1978). Minerals typically accompanying minnesotaite are quartz, greenalite, stilpnomelane, magnetite, siderite, ankerite, calcite, riebeckite, chlorite, pyrite, pyrrhotite, and graphite.

The first occurrence of grunerite in progressive metamorphism overlaps partially with the occurrence of minnesotaite (Klein, 1982; 1983). In Barrovian-style terranes the grunerite isograd is found close to the biotite isograd (Klein, 1978; Haase, 1982a), and temperatures for the grunerite isograd have been estimated at 350-400°C (French, 1968), 275-345°C (James and Clayton, 1962), 300-400°C (Mukhopadhyay et al., 1980), 310-360°C (Gole, 1981), and 300-350°C (Haase, 1982b; Klein, 1983). Grunerite typically forms from siderite or ankerite and quartz and, although minnesotaite quickly diminishes in amount above the isograd, textural evidence for the direct conversion of minnesotaite to grunerite has not been widely reported (but see Gair, 1975, Gole, 1981, and Klein

280

and Gole, 1981). The above temperature estimates are in some cases slightly lower than expected from Figure 25.

Mn-rich minnesotaite has been reported as a retrograde mineral in Phanerozoic blueschist-facies ironstones (e.g., Wood, 1982), in one case by replacement of deerite (Helper, 1986). On the other hand, ferroan talc (up to 23% Fe-end-member) probably crystallized in high-pressure equilibrium with deerite in a quartz-magnetite-garnet ironstone from the Western Alps, according to Chopin (1985).

Mn-rich minnesotaite has also been reported, along with Fe-talc, greenalite, siderite, magnetite, Pb-Zn-Fe sulfides, and quartz/opal/chalcedony in a rhyodacite-related, Tertiary iron-rich hydrothermal ore deposit developed in limestone at Emilia-San Valentin, Sierra de Cartagena, SE Spain (Oen et al., 1975; Kager and Oen, 1983). A similar epigenetic Tertiary hydrothermal iron-rich Pb-Zn sulfide deposit in Cambrian marble at the Bluebell Mine, Kootenay Lake, British Columbia, produced Mn-rich minnesotaite, greenalite, knebelite, quartz, and various carbonates along with the sulfides (Ohmoto and Rye, 1970). The broad miscibility gap (X_{Fe} = 0.1 to 0.7) between talc and minnesotaite proposed by Kager and Oen (1983) is hard to reconcile with the compositions of these minerals in hydrothermal deposits and metamorphosed iron-formations (Fig. 15), the majority of which formed in the temperature range 200-350°C.

ACKNOWLEDGMENTS

We thank Drs. L. Heller-Kallai, Hebrew University of Jerusalem, C. Chopin, Laboratoire de Géologie, ENS, Paris, R.G. Berman, University of British Columbia, Vancouver, C. Klein, University of New Mexico, Albuquerque, L.M. Anovitz, University of Arizona, Tucson, T. Miyano, University of Tsukuba, Japan, and C. Owen, Smithsonian Institution, Washington, for critically reviewing portions of the manuscript and Dr. S.W. Bailey for editorial comments. Support for this work was provided by the Petroleum Research Fund of the American Chemical Society under grant 17263-AC2 and the National Science Foundation under grants EAR87-04681, EAR86-08838, and EAR85-07757. We thank P. van Krieken and T. Zalm of the University of Utrecht for the thermal analytical results and Dr. D.L. Bish, Los Alamos National Laboratory, and C.B. Sclar, Lehigh University, for samples.

REFERENCES

Abercrombie, H.J., Skippen, G.B. and Marshall, D.D. (1987) F-OH substitution in natural tremolite, talc, and phlogopite. Contrib. Mineral. Petrol. 97, 305-312.

Abraham, K. and Schreyer, W. (1976) A talc-phengite assemblage in piemontite schist from Brezovica, Serbia, Yugoslavia. J. Petrol. 17, 421-439.

_____, Hormann, P.K. and Raith, M.(1974) Progressive metamorphism of basic rocks from the southern Hohe Tauern area, Tyrol (Austria). N. Jahrb. Mineral. Abh. 122, 1-35.

Aggarwal, P.K. and Nesbitt, B.E. (1984) Geology and geochemistry of the Chu Chua massive sulfide deposit, British Columbia. Econ. Geol. 79, 815-825.

Aines, R.D. and Rossman, G.R. (1985) The high temperature behavior of trace hydrous components in silicate minerals. Am. Mineral. 70, 1169-1179.

Akizuki, M. and Zussman, J. (1978) The unit cell of talc. Mineral. Mag. 42, 107-110.

Akpanika, O.I., Ukpong, E.E. and Olade, M.A. (1987) Mineralogy and geochemical dispersion in tropical residual soils overlying a talc deposit in southwestern Nigeria. Cem. Geol. 63, 109-119.

Alietti, A. (1956) Il minerale a strati misti saponite-talco di Monte Chiara (Val di Taro. Appennino Emiliano). Rend. Accad. Naz. Lincei VIII-21, 201-207.

Allen, E.T. and Day, A.L. (1935) Hot Springs of the Yellowstone National Park. Carnegie Inst. Wash. Publ. 466, 525p.

Allen, J.M. (1978) F-OH distribution among minerals in marbles and implications for the stability of talc + calcite (abs). Geol. Soc. Am. Abstr. with Prog. 10, p. 358.

Amelinckx, S. and Delavignette, P. (1961) Electron microscope observation of dislocations in talc. J. Appl. Phys. 32, 341-351.

Anderson, D.L., Mogk, D.W. and Childs, J.F. (1988) Petrogenesis and timing of talc formation in the Ruby Range, S.W. Montana. Econ. Geol. (in review).

Anovitz, L.M., Essene, E.J., Hemingway, B.S., Komada, N.L. and Westrum, E.F. (1988) The heat capacities of grunerite and deerite: phase equilibria in the system Fe-Si-C-O-H and implications for the metamorphism of banded iron formations. Trans. Am. Geophys. Union, EOS 69, 515.

Anton, O. (1969) Study on the thermal transformation of pyrophyllite. Rev. Roum. Geol., Geophys., Geogr.-Serie de Geologie 13, 29-38.

Arai, S. (1975) Contact metamorphosed dunite-harzburgite complex in the Chigoku District, Western Japan. Contrib. Mineral. Petrol. 52, 1-16.

Ayres, D.E. (1972) Genesis of iron-bearing minerals in banded iron-formation mesobands in the Dales Gorge Member, Hamersley Group, Western Australia. Econ. Geol. 67, 1214-1233.

Bailey, S.W. (1980) Structures of layer silicates. Ch. 1 in Crystal Structures of Clay Minerals and their X-ray Identification, G.W. Brindley and G. Brown, eds., Mineral. Soc. Great Britain Monograph No. 5, London.

_____ (1984) Crystal chemistry of the true micas. Ch. 2 in Micas, S.W. Bailey, ed., Rev. in Mineral. 13, 13-60.

Baltatzis, E. (1980) Chloritoid-forming reaction in the eastern Scottish Dalradian: a possibility. N. Jahrb. Mineral. Mh., H. 7, 306-313.

Bapst, G. and Eberhart, J.P. (1970) Transformation of talc to $MgSiO_2$. Bull. Group Fr. Argiles 22, 17-23.

Barnicoat, A.C. and Fry, N.(1986) High-pressure metamorphism of the Zermatt-Saas ophiolite zone, Switzerland. J. Geol. Soc. Lond. 143, 607-618.

Basu, N.K. and Mruma, A.H. (1985) Mineral chemistry and stability relations of talc-piemontite viridine bearing quartzite of Mautia Hill, Mpwapwa district, Tanzania. Indian J. Earth Sci. 12(4), 223-230.

Bauer, J.F. and Sclar, C.B.(1981) The "10 Å phase"in the system $MgO-SiO_2-H_2O$. Am. Mineral. 66, 576-585.

Beane, R.E. and Titley, S.R. (1981) Porphyry copper deposits. Part II. Hydrothermal alteration and mineralization. Econ. Geol., 75th Anniv. Vol., 235-269.

Berg, E.L. (1937) An occurrence of diaspore in quartzite. Am. Mineral. 22, 997-999.

Berman, R.B. (1988) Internally-consistent thermodynamic data for minerals in the system $Na_2O-K_2O-CaO-MgO-FeO-Fe_2O_3-Al_2O_3-SiO_2-TiO_2-H_2O-CO_2$.J.Petrol. 29,in press.

_____ and Brown, T.H.(1985)Heat capacity of minerals in the system $Na_2O-K_2O-CaO-MgO-FeO-Fe_2O_3-Al_2O_3-SiO_2-TiO_2-H_2O-CO_2$: representation, estimation, and high temperature extrapolation. Contrib. Mineral. Petrol. 89, 168-183.

_____, Brown, T.H. and Greenwood, H.J. (1985) An internally consistent thermodynamic data base for minerals in the system $Na_2O-K_2O-CaO-MgO-FeO-Fe_2O_3-Al_2O_3-SiO_2-TiO_2-H_2O-CO_2$. Atomic Energy of Canada Ltd. Tech. Rep. 377, 62p.

_____, Brown, T.H. and Perkins, E.H. (1988) GE0-CALC update: software for calculation of pressure-temperature-$X(CO_2)$-activity phase diagrams. Geol. Assoc. Can. Program with Abstr. 13, A9.

_____, Engi, M., Greenwood, H.J. and Brown, T.H. (1986) Derivation of internally consistent thermodynamic data by the technique of mathematical programming: a review with application to the system $MgO-SiO_2-H_2O$. J. Petrol., 27, 1331-1364.

282

Bird, D.K., Roger, R.D. and Manning, C.E. (1986) Mineralized fracture systems of the Skaergaard intrusion, East Greenland. Medd. om Grønland. Geosci. 16, 1-68.

Blaauw, C., Stroink, G. and Leiper, W. (1980) Mössbauer analysis of talc and chlorite, J. de Phys. C1, 411-412.

Blaha, J.J. and Rosasco, G.J. (1978) Raman microprobe spectra of individual microcrystals and fibers of talc, tremolite, and related silicate minerals. Anal. Chem. 50, 892-896.

Blount, A.M. and Vassiliou, A.H. (1980) The mineralogy and origin of the talc deposits near Winterboro, Alabama. Econ. Geol. 75, 107-116.

Bouley, B.A. and Hodder, R.W. (1984) Strata-bound massive sulfide deposits in Silurian-Devonian volcanic rocks at Harborside, Maine. Econ. Geol. 79, 1693-1702.

Bowers, T.S. and Helgeson, H.C. (1983) Calculation of the thermodynamic and geochemical consequences of nonideal mixing in the system H_2O - CO_2 - NaCl on phase relations in geological systems. Metamorphic equilibria at high pressures and temperatures. Am. Mineral., 68, 1059-1075.

_____, Jackson, K.J. and Helgeson, H.C. (1984) Equilibrium activity diagrams for coexisting minerals and aqueous solutions at pressures and temperatures to 5 kb and 600°C. Springer-Verlag, New York, 397 p.

_____, von Damm, K.L. and Edmond, J.M. (1985) Chemical evolution of mid-ocean ridge hot springs. Geochim. Cosmochim. Acta 49, 2239-2252.

Bradley, W.F. and Grim, R.E. (1951) High temperature thermal effects of clay and related materials. Am. Mineral. 36, 182-201.

Brady, J.B. (1977) Metasomatic zones in metamorphic rocks. Geochim. Cosmochim. Acta 41, 113-125.

Braitsch, O. (1970) Salt deposits, their origin and composition. Springer-Verlag, Berlin, 297 p.

Brett, N.H., MacKenzie, K.J.D. and Sharp, J.H. (1970) The thermal decomposition of hydrous layer silicates and their related hydroxides. Chem. Soc. London. Quart. Rev. 24, 185-207.

Bricker, O.P., Nesbitt, H.W. and Gunter, W.D. (1973) The stability of talc. Am. Mineral. 58, 64-72.

Brindley, G.W. (1963) Crystallographic aspects of some decomposition and recrystallization reactions. In Progress in Ceramic Science v. 3 J.E. Burke, ed., 1-56. MacMillan Co., New York.

_____ (1970) Reorganization of dioctahedral hydrous layer silicates by dehydroxylation. Mineral. Soc. Japan Spec. Paper 1, 70-73.

_____ (1975) Thermal transformations of clays and layer silicates. Proc. Int'l. Clay Conf. 1975, 119-129.

_____, Bish, D.L. and Wan, H-M. (1977) The nature of kerolite, its relation to talc and stevensite. Mineral. Mag. 41, 443-452.

_____ and Hayami, R. (1964) Kinetics and mechanisms of dehydration and recrystallization of serpentine - I. Clays & Clay Minerals 19, 35-47.

_____ and Thi Hang, Pham (1973) The nature of garnierites-I. Structures, chemical compositions and color characteristics. Clays & Clay Minerals 21, 27-40.

_____ and Wardle, R. (1970) Monoclinic and triclinic forms of pyrophyllite and pyrophyllite anhydride. Am. Mineral. 55, 1259-1272.

Brown, C.E. (1973) Talc. U.S. Geol. Surv. Prof. Paper 820, 619-626.

Bucher-Nurminen, K., Frank, E. and Frey, M. (1983) A model for the progressive regional metamorphism of margarite-bearing rocks in the Central Alps. Am. J. Sci. 283A, 370-395.

Burt, D.M. (1972) The system Fe-Si-C-O-H: a model for metamorphosed iron formations. Carnegie Inst. Wash. Yrbk. 71, 435-443.

_____ (1978) A working model of some equilibria in the system alumina-silica-water. Discussion. Am. J. Sci. 278, 244-249.

Chandler, F.W., Young, G.M. and Wood, J. (1967) Diaspore in early Proterozoic quartzite of Ontario. Canad. J. Earth Sci. 6, 337-340.

Chatterjee, N.D., Johannes, W. and Leistner, H.(1984) The system $CaO-Al_2O_3-SiO_2-HO2$:new phase equilibria data, some calculated phase relations, and their petrological applications. Contrib. Mineral. Petrol. 88, 1-13.

Chennaux, G., Dunoyer de Segonzac, G. and Petracco, F. (1970) Genèse de la pyrophyllite dans le Paléozoique du Sahara occidental. C.R. Acad. Sci. Paris, t. 270, Série D, 2405-2406.

Chidester, A.H. (1962) Petrology and geochemistry of selected talc-bearing ultramafic rocks and adjacent country rocks in North-central Vermont. U.S. Geol. Surv. Prof. Pap. 345, 207p.

Chinner, G.A. and Dixon, J.E. (1973) Some high-pressure parageneses of the Allalin gabbro, Valais, Switzerland. J. Petrol. 14, 185-202.

Chopin, C. (1981) Talc-phengite: a widespread assemblage in high grade pelitic blueschists of the Western Alps. J. Petrol. 22, 628-650.

_____ (1984) Coesite and pure pyrope in high-grade pelitic blueschists of the Western Alps: a first record and some consequences. Contrib. Mineral. Petrol. 86, 107-118.

_____ (1987)Very-high-pressure metamorphism in the Western Alps:implications for subduction of continental crust. Phil. Trans. Royal Soc. London A321, 183-197.

_____ and Schreyer, W. (1983) Magnesiocarpholite and magnesiochloritoid: two index minerals of pelitic blueschists and their preliminary phase relations in the model system MgO-Al_2O_3-SiO_2-H_2O. Am. J. Sci. 283A, 72-96.

_____ and Monié, P. (1984) A unique magnesiochloritoid-bearing high-pressure assemblage from the Monte Rosa, Western Alps: petrologic and ^{40}Ar-^{39}Ar radiometric study. Contrib. Mineral. Petrol. 87, 388-398.

Chukhrov, F.V., Zvyagin, B.B., Drits, V.A., Gorshkov, A.I., Ermilova, L.P., Goilo, E.A. and Rudnitskaya, E.S. (1979) The ferric analogue of pyrophyllite and related phases. In M.M. Mortland and V.C. Farmer, eds., Proc. Int'l. Clay Conf., Oxford, 1978, Elsevier, Amsterdam, 55-64.

Coey, J.M.D. (1980) Clay minerals and their transformations studied with nuclear techniques: The contribution of Mössbauer spectroscopy. Atom. Energy Rev. 18, 73-124.

_____, Chukhrov, F.V. and Zvyagin, B.B. (1984) Cation distribution, Mössbauer spectra, and magnetic properties of ferripyrophyllite. Clays & Clay Minerals 32, 198-204.

Cooper, A.F. (1980) Retrograde alteration of chromian kyanite in metachert and amphibolite whiteschist from the Southern Alps, New Zealand, with implications for uplift on the Alpine Fault. Contrib. Mineral. Petrol. 75, 153-164.

Cooper, J.R. (1957) Metamorphism and volume losses in carbonate rocks near Johnson Camp, Cochise County, Arizona. Geol. Soc. Am. Bull., 68, 577-610.

Costa, U.R., Barnett, R.L. and Kerrich, R. (1983) The Mattagami Lake mine Archean Zn-Cu sulfide deposit, Quebec: hydrothermal coprecipitation of talc and sulfides in a sea-floor brine pool - evidence from geochemistry, ^{18}0/^{16}0, and mineral chemistry. Econ. Geol. 78, 1144-1203.

Daw, J.D., Nicholson, P.S. and Embury, J.D. (1972) Inhomogeneous dehydration of talc. J. Am. Ceram. Soc. 55, 149-151.

Day, H.W. (1976) A working model of some equilibria in the system alumina-silica-water. Am. J. Sci. 276, 1254-1285.

_____ (1978) A working model of some equilibria in the system alumina-silica-water. Reply. Am. J. Sci. 278, 250-253.

_____, Chernosky, J.V. and Kumin, H.J. (1985) Equilibria in the system MgO-SiO_2-H_2O: a thermodynamic analysis. Am. Mineral. 70, 237-248.

Deer, W.A., Howie, R.A. and Zussman, J. (1962) The Rock Forming Minerals. Vol. 3, Sheet Silicates. Wiley, New York, 270 p.

Dent Glasser, L.S., Glasser, F.P. and Taylor, H.F.W. (1962) Topotactic reactions in inorganic oxy-compounds. Quart. Rev. 16, 343-360.

DeRoever, W.P. (1947) Igneous and metamorphic rocks in the Central Celebes, in Geological explorations of the Island of Celebes. H.A. Brouwer: Amsterdam, North Holland Publishing Co., p. 65-73.

de Waal, S.A. (1970) Nickel minerals from Barberton, South Africa: III. Willemseite, A nickel-rich talc. Am. Mineral. 55, 31-42.

Drits, V.A. (1987) Electron diffraction and high-resolution electron microscopy of mineral structures. Springer-Verlag, Berlin.

Duffy, C.J. and Greenwood, H.J. (1979) Phase equilibria in the system MgO-MgF2-SiO_2-H_2O. Am. Mineral. 64, 1156-1174.

284

Ďurovič, S. and Weiss, Z. (1983) Polytypism of pyrophyllite and talc. Part I. OD interpretation and MDO polytypes. Silikáty 27, 1-18.

D'yakonov, Yu. S. (1963a) Results from an X-ray study of ceriolites. Doklady Akad. Nauk SSSR 148, Earth Science Sections (English Transl., Am. Geol. Inst.), 107-109.

_____ (1963b) X-ray investigations of ceriolite. Mineral. Sbornik Vses. Nauchn.-Issled. Geol. Inst. 3, 203-212 (in Russian).

Dyar, M.D. and Burns, R.G. (1986) Mössbauer spectral study of ferruginous one-layer trioctahedral micas. Am. Mineral. 71, 955-965.

Eberl, D. (1979) Synthesis of pyrophyllite polytypes and mixed layers. Am. Mineral. 64, 1091-- 1096.

Eggert, R.G.and Kerrick, D.M.(1981) Metamorphic equilibria in the siliceous dolomite system: 6 kbar experimental data and geologic implications. Geochim. Cosmochim. Acta 45, 1039-1049.

Ellenberger, F. (1958) Etude Géologique du pays de Vanoise (Savoie). Mem. Carte géol. de la France.

Engi, M.(1986)Towards a thermodynamic data base for geochemistry.Consistency and optimal representation of the stability relations in mineral systems. Habilitation thesis, E.T.H. Zuerich, 110 p.

Evans, B.W. and Misch, P. (1976) A quartz-aragonite-talc schist from the lower Skagit Valley, Washington. Am. Mineral. 61, 1005-1008.

_____ and Trommsdorff, V. (1974a) Stability of enstatite + talc and CO_2-metasomatism of metaperidotite. Am. J. Sci., 274, 274-296.

_____ and _____ (1974b) On elongate olivine of metamorphic origin. Geology 2, 131-132.

_____, Engi, M. and Owen, C. (1982) Stability of minnesotaite. Geol. Soc. Am. Abstr. Programs 14, 485.

Farmer, V.C. (1958) The infrared spectra of talc, saponite and hectorite. Mineral. Mag. 31, 829-845.

_____ (1974) The layer silicates. In V.C. Farmer (ed.), Infrared Spectra of Minerals. Mineral. Soc. Great Britain Mono. 331-363.

_____ and Russell, J.D. (1964) The infrared spectra of layer silicates. Spectrochim. Acta 20, 1149-1173.

_____ and Velde, B. (1973) Effects of structural order and disorder on the infrared spectra of brittle mica. Mineral. Mag. 39, 282-288.

Fawcett, J.J. (1962) The alumina content of talc. Carnegie Inst. Wash. Yrbk. 62, 139-140.

_____ and Yoder, H. S. (1966) Phase relationships in the system $MgO-Al_2O_3-SiO_2-H_2O$. Am. Mineral. 51, 353-380.

Feenstra, A. (1985) Metamorphism of bauxites on Naxos, Greece. Geologica Ultraiectina 39, 1-206.

Ferrow, E. (1987) Mössbauer and x-ray studies on the oxidation of annite and ferriannite. Phys. Chem. Minerals 14, 270-275.

Ferry, J.M. (1985) Hydrothermal alteration of Tertiary igneous rocks from the Isle of Skye, northwest Scotland. I. Gabbros. Contrib. Mineral. Petrol. 91, 264-282.

Flaschen, S.S. and Osborn, E.F. (1957) Studies in the system iron oxide - silica - water at low oxygen partial pressures. Econ. Geol. 52, 923-943.

Flinn, D. and Moffat, D.T. (1986) A reply to R.W. Nesbitt and L.A. Hartman. Geol. J. 21, 207-209.

Floran, R.J. and Papike, J.J. (1975) Petrology of the low-grade rocks of the Gunflint Iron formation, Ontario-Minnesota. Geol. Soc. Am. Bull. 86, 1169-1190.

_____ and _____ (1978) Mineralogy and petrology of the Gunflint Iron formation, Minnesota--Ontario: Correlation of compositional and assemblage variations at low to moderate grade. J. Petrol. 19, 215-288.

Flowers, G.C. and Helgeson, H.C. (1983) Equilibrium and mass transfer during progressive metamorphism of siliceous dolomites. Am. J. Sci. 283, 230-286.

Forbes, W.C. (1969) Unit-cell parameters and optical properties of talc on the join $Mg_3Si_4O_{10}$-$(OH)_2$ - $Fe_3Si_4O_{10}(OH)_2$. Am. Mineral. 54, 1399-1400.

_____ (1971) Iron content of talc in the system $Mg_3Si_4O_{10}(OH)_2$ - $Fe_3Si_4O_{10}(OH)_2$. J. Geol. 79, 63-74.

285

Franceschelli, M., Leoni, L., Memmi, I. and Puxeddu, M. (1986) Regional distribution of Al-silicates and metamorphic zonation in the low-grade Verrucano metasediments from the Northern Apennines, Italy. J. Metam. Geol. 4, 309-321.

Fransolet, A.M. and Bourguignon, P. (1978) Pyrophyllite, dickite et kaolinite dans les filons de quartz du Massif de Stavelot. Bull. Soc. Royale Sci. de Liège, 47e année, 5-8, 213-221.

Frantz, J.D. and Mao, H.K. (1979) Bimetasomatism resulting from intergranular diffusion: II. Prediction of multimineralic zone sequences. Am. J. Sci. 279, 302-323.

_____ and Popp, R.K.(1979)Mineral-solution equilibria. I. An experimental study of complex-ing and thermodynamic properties of aqueous $MgCl_2$ in the system $MgO-SiO_2-H_2O-HCl$. Geochim. Cosmochim. Acta 43, 1223-1239.

Franz, G. and Spear, F.S. (1983) High pressure metamorphism of siliceous dolomites from the Central Tauern Window, Austria. Am. J. Sci., 283A, 396-413.

French, B.M.(1968) Progressive contact metamorphism of the Biwabik Iron Formation, Mesabi Range, Minn. Minnesota Geol. Surv. Bull. 45.

_____ (1973) Mineral assemblages in diagenetic and low-grade metamorphic iron formation. Econ. Geol. 68, 1063-1074.

Freund, F. (1965) Dehydration of hydroxides. Angew. Chem. Int'l. Ed. Eng. 4, 445.

_____ and Nièpce, J.C.(1988) Protons in simple ionic hydroxides: The dehydration mechanism. In Adv. Inorgan. Chem., C.R.A. Catlow, ed., v. 32, Academic Press, in press.

Frey, M. (1978) Progressive low-grade metamorphism of a black shale formation, Central Swiss Alps, with special reference to pyrophyllite and margarite bearing assemblages. J. Petrol. 19, 95-135.

_____ (1987a)The reaction-isograd kaolinite + quartz = pyrophyllite + H_2O. Schweiz. Mineral. Petrogr. Mitt. 67, 1-11.

_____ (1987b) Very low-grade metamorphism of clastic sedimentary rocks. In Frey, M., ed., Low temperature metamorphism: Blackie, Glasgow, 9-58.

Friedman, G.M. (1965) Occurrence of talc as a clay mineral in sedimentary rocks. Nature 207, 283-284.

Fripiat, J.J., Rouxhet, P. and Jacobs, H. (1965) Proton delocalization in micas. Am. Mineral. 50, 1937-1958.

Frost, B.R. (1975) Contact metamorphism of serpentinite, chloritic blackwall, and rodingite at Paddy-go-easy Pass, Central Cascades, Washington. J. Petrol. 16, 272-313.

_____ (1979)Metamorphism of iron-formation parageneses in the system Fe-Si-C-O-H. Econ. Geol. 74, 775-785.

Frost, R.L. and Barron, P.F. (1984) Solid-state silicon-29 and aluminum-27 nuclear magnetic resonance investigation of the dehydroxylation of pyrophyllite. J. Phys. Chem., 88, 6206-6209.

Fyfe, W.S. and Hollander, M.A.(1964)Equilibrium dehydration of diaspore at low temperatures. Am. J. Sci. 262, 709-712.

Gaines, G.L., Jr. and Vedder, W. (1964) Dehydroxylation of muscovite. Nature 201, 495.

Gair, J.E. (1975) Bedrock geology and ore deposits of the Palmer quadrangle, Marquette County, Michigan. U.S. Geol. Surv. Prof. Paper 769, 159p.

Gasparik, T. and Newton, R.C. (1984) The reversed alumina contents of orthopyroxene in equilibrium with spinel and forsterite in the system $MgO-Al_2O_3-SiO_2$. Contrib. Mineral. Petrol. 85, 186-196.

Ghiorso, M.S., Carmichael, I.S.E. and Wells, J.T. (1988) Mineral solution equilibria in active sulfate hot springs. Geochim. Cosmochim. Acta (in review).

Giese, R.F. (1971) Hydroxyl orientation in muscovite as indicated by electrostatic energy calculations. Science 172, 263-264.

_____ (1973) Hydroxyl orientation in pyrophyllite. Nature, Phys. Sci. 241, 151.

_____ (1975) Interlayer bonding in talc and pyrophyllite. Clays & Clay Minerals 23, 165-166.

_____ (1979) Hydroxyl orientations in 2:1 phyllosilicates. Clays & Clay Mineral. 27, 213-223.

Goffé, B. (1979) La lawsonite et les associations à pyrophyllite-calcite dans les sediments alumineux du Briançonnais. Premières occurrences. C.R. Acad. Sci., Paris, D 289, 813-816.

_____ (1982) Définition du faciès à FeMg carpholite-choritoide, un marqueur de métamorph-isme de HP-BT dans les metasédiments alumineux. Thèse d'Etat, Univ. P. et M. Curie, Paris, 233 p.

286

_____ and Chopin, C. (1986) High-pressure metamorphism in the Western Alps: zoneography of metapelites, chronology and consequences. Schweiz. Mineral. Petrogr. Mitt., 66, 41-52.

_____ and Saliot, P. (1977) Les associations minéralogiques des roches hyperalumineuses du Dogger de Vanoise - Leur signification dans le métamorphisme régional. Soc. France Minéral. Crist. Bull. 100, 302-309.

_____ and Velde, B. (1984) Contrasted metamorphic evolutions in thrusted cover units of the Brionconnais zone (French Alps): a model for the conservation of HP-LT-metamorphic mineral assemblages: Earth Planet. Sci. Letters 68, 351-360.

_____, Murphy, W.W. and Lagache, M. (1987) Experimental transport of Si, Al and Mg in hydrothermal solution: application to vein mineralization during high-pressure, low-temperature metamorphism in the French Alps. Contrib. Mineral. Petrol. 97, 438-450.

Goldfarb, M.S., Converse, D.R., Holland, H.D. and Edmond, J.M. (1983) The genesis of hot spring deposits on the East Pacific Rise, 21° N. Econ. Geol., Monograph 5, 184-197.

Gole, M.J. (1980) Mineralogy and petrology of very-low-metamorphic grade Archean banded iron-formations, Weld Range, Western Australia. Am. Mineral. 65, 8-25.

Gordon, T.M. and Greenwood, H.J. (1970) The reaction: dolomite + quartz + water = talc + calcite + carbon dioxide. Am. J. Sci., 268, 225-242.

Gottschalk, M. and Metz, P. (1984) Konsistente Gleichgewichtsdaten fur Reaktionen in metamorphen kieseligen Dolomiten. Fortschr. Mineral. 62, 78.

Gradusov, B.P.and Zotov, A.V.(1975)Low-temperature pyrophyllite in surface deposits around acid springs on Kunashir Island. Doklady, Acad. Sci. USSR, Earth Sci. Sect. 224, 141-143.

Greenwood, H.J. (1975) Buffering of pore fluids by metamorphic reactions. Am. J. Science, 275, 573-593.

Grew, E.S. and Sandiford, M. (1984) A staurolite-talc assemblage in tourmaline-phlogopite-chlorite schist from northern Victoria Land, Antarctica, and its petrogenetic significance. Contrib. Mineral. Petrol. 87, 337-350.

Grubb, P.L.C. (1971) Silicates and their parageneses in the Brockman Iron Formation of Wittenoom Gorge, Western Australia. Econ. Geol. 66, 281-292.

Gruner, J.W. (1934) The crystal structures of talc and pyrophyllite. Z. Kristallogr. Kristallgeom. 88, 412-419.

_____ (1944) The composition and structure of minnesotaite, a common iron silicate in iron formations. Am. Mineral. 29, 363-372.

Guitard, M.G. (1966) Le métamorphisme et les faciès mineralogiques des marbres dérivant des dolomies silicieuse dans les massifs du Canigou et de la Caranca (Pyrénées-Orientales). Acad. Sci. Paris C.R. 262, 245-247.

Guggenheim, S. (1984) The brittle micas. Ch. 3 in Micas, S.W. Bailey, ed., Rev. in Mineral. 13, 61-104.

_____ and Bailey, S.W. (1982) The superlattice of minnesotaite. Can. Mineral. 20, 579-584.

_____ and Eggleton, R.A. (1986) Structural modulations in iron-rich and magnesium-rich minnesotaite. Can. Mineral. 24, 479-497.

_____, Chang, Y-H. and Koster van Groos, A.F., (1987) Muscovite dehdyroxylation: High temperature studies. Am. Mineral. 72, 537-550.

_____, Bailey, S.W., Eggleton, R.A. and Wilkes, P. (1982) Structural aspect of greenalite and related minerals. Can. Mineral. 20, 1-18.

Gundersen, J.N. and Schwartz, G.M (1962) The geology of the metamorphosed Biwablik Iron-formation, East Mesabi District, Minnesota. Minn. Geol. Surv. Bull. 43, 135 p.

Gustafson, L.B. and Hunt, J.P. (1975) The porphyry copper at El Salvador, Chile. Econ. Geol. 70, 857-912.

Haas, H. (1972) Diaspore-corundum equilibrium determined by epitaxis of diaspore on corundum. Am. Mineral. 57, 1375-1385.

_____ and Holdaway, M.J. (1973) Equilibrium in the system Al_2O_3-SiO_2-H_2O involving the stability limits of pyrophyllite and thermodynamic data of pyrophyllite. Am. J. Sci. 273, 449-464.

Haase, C.S. (1982a)Metamorphic petrology of the Negaunee iron formation, Marquette District Northern Michigan:Mineralogy, metamorphic reactions, and phase equilibria. Econ. Geol 77, 60-81.

_____ (1982b) Phase equilibria in metamorphosed iron-formations: qualitative T-X(CO_2) petrogenetic grids. Am. J. Sci. 282, 1623-1654.

Halbach, H. and Chatterjee, N.D. (1982) The use of linear parametric programming for determining internally consistent thermodynamic data for minerals.In,High-pressure Researches in Geosciences, W. Schreyer, ed., 475-491.E.Schweizerbart'sche Verlagsbuchhandlung, Stuttgart.

Hathaway, J.C. (1979) Clay minerals, Ch. 5 in R.G. Burns, ed., Rev. in Mineral. 6, 123-150.

Haymon, R.M. and Kastner, M. (1981) Hot spring deposits on the E.P.R. at 21^0 N: preliminary description of minerals and genesis. Earth Planet. Sci. Letters 53, 363-381.

Helgeson, H.C., Delaney, J.M., Nesbit, H.W. and Bird, D.K.(1978)Summary and critique of the thermodynamic properties of rock-forming minerals. Am. J. Sci. 278A, 225 p.

_____, Kirkham, D.H. and Flowers, G.C. (1981) Theoretical prediction of the thermodynamic behavior of aqueous electrolytes at high pressures and temperatures: IV. Calculation of activity coefficients, osmotic coefficients, and apparent molal and standard and relative partial molal properties to 600^0C and 5 kb. Am. J. Sci. 281, 1249-1516.

Heller, L., Farmer, V.C., Mackenzie, R.C., Mitchell, B.D. and Taylor, H.F.W. (1962) The dehydroxylation and rehydroxylation of trimorphic dioctahedral clay minerals. Clay Mineral. Bull. 5, 56-72.

Heller-Kallai, L. and Rozenson, I.(1980)Dehydroxylation of dioctahedral phyllosilicates. Clays & Clay Minerals 28, 355-368.

_____ and ___ (1981) The use of Mössbauer spectroscopy of iron in clay mineralogy. Phys. Chem. Minerals 7, 223-238.

Helper, M.A. (1986) Deformation and high P/T metamorphism in the central part of the Condrey Mountain Window, North-central Klamath Mountains, California and Oregon. In B.W. Evans and E.H. Brown, ed., Geol. Soc. Am. Memoir 164, 125-142.

Hemley, J.J., Montoya, J.W., Christ, C.L. and Hostetler, P.B. (1977a) Mineral equilibria in the $MgOSiO_2$-H_2O system: Talc-chrysotile- forsterite-brucite stability relations. Am. J. Sci. 277, 322-351.

_____, ___, Marinenko, J.W. and Luce, R.W. (1980) Equilibria in the system Al_2O_3-SiO_2-H_2O and some general implications for alteration/mineralization processes. Econ. Geol. 75, 210-228.

_____, ___, Shaw, D.R. and Luce, R.W. (1977b) Mineral equilibria in the MgO-SiO_2-H_2O system: II. Talc-antigorite-forsterite-enstatite stability relations and some geologic implications in the system. Am. J. Sci. 277, 353-383.

Henderson, G.V. (1971) The origin of pyrophyllite-rectorite in shales of north-central Utah. Utah Geol. and Mineral. Surv. Spec. Studies no. 34, p. 46.

Hendricks, S.B. (1938) On the crystal structure of talc and pyrophyllite. Z. Kristallogr. Kristallgeom. 99, 264-274.

Hewitt, D.A. and Wones, D.R. (1984) Experimental phase relations of the micas. Rev. in Mineral. 13, 201-256.

Hietanen, A. (1977) Blades of olivine in ultramafic rock from Northern Sierra Nevada, California. J. Res. U.S. Geol. Surv. 5, n. 2, 217-219.

Hildebrand, F.A. (1961) Hydrothermally altered rocks in eastern Puerto Rico. U.S. Geol. Surv. Prof. Paper 424b, 219-221.

Hoinkes, G. (1983) Cretaceous metamorphism of metacarbonates in the Austroalpine Schneeberg complex. Tirol. Schweiz. Mineral. Petrogr. Mitt. 63, 95-114.

Hoisch, T. (1987) Heat transport by fluids during Late Cretaceous regional metamorphism in the Big Maria Mountains, southeastern California. Geol. Soc. Am. Bull. 98, 549-553.

Holser, W.T. (1979) Mineralogy of evaporites, Ch. 8 in R.G. Burns, ed., Marine Minerals, Rev. in Mineral. v. 6, 211-294.

Holt, J.B., Cutler, I.B. and Wadsworth, M.E. (1964) Kinetics of the thermal dehydration of hydrous silicates. Clays & Clay Minerals 12, 55-67.

Hover-Granath, V.C., Papike, J.J. and Labotka, T.C. (1983) The Notch Peak contact metamorphic aureole, Utah: Petrology of the Big Horse Limestone Member of the Orr Formation. Geol. Soc. Am. Bull. 94, 889-906.

Hutcheon, I. and Moore, J.M. (1973) The tremolite isograd near Marble Lake, Ontario. Canad. J. Earth Sci. 10, 936-947.

288

Ianovici, V., Neacsu, G. and Neacsu, V. (1981) Pyrophyllite occurrences and their genetic relations with the kaolin minerals in Romania. Bull. Mineral. 104, 768-775.
Ishii, M., Shimanouchi, T. and Nakahira, M. (1967) Far infra-red absorption spectra of layer silicates. Inorg. Chim. Acta 1, 387-392.
Isphording, W. (1984) The clays of Yucatan, Mexico: a contrast in genesis. In Palygorskite-sepiolite: occurrences, genesis, and uses. Developments in Sedimentology 37, 59-73, Elsevier, New York.
Iwao, S. (1970) Clay and silica deposits of volcanic affinity in Japan. In T. Tatsumi, ed., Volcanism and Ore Genesis, p. 267-283. Univ. Tokyo Press, Tokyo.
_____ and Udagawa, S. (1969) Pyrophyllite and "Roseki" clays. Part II, Section II in The clays of Japan, Editorial Subcommittee, eds., 1969 Int'l. Clay Conf., p. 71-87.
Jacobs, G.K. and Kerrick, D.M.(1981) Devolatilization equilibria in H_2O-CO_2 and H_2O-CO_2-NaCl fluids:an experimental and thermodynamic evaluation at elevated temperatures and pressures. Am. Mineral. 66, 1135-1153.
James, H.L. and Clayton, R.N. (1962) Oxygen isotope fractionation in metamorphosed iron-formation of the Lake Superior region and in other iron-rich rocks, in Engel, A.E.J., Jones, H.C. and Leonard,H.F.,ed., Petrologic Studies: a volume to honor A.F. Buddington. Geol. Soc. Am., 217-239.
Janecky, D.R. and Seyfried, W.E. (1984) Formation of massive sulfide deposits on oceanic ridge crests: incremental reaction models for mixing between hydrothermal solutions and seawater. Geochim. Cosmochim. Acta 48, 2723-2738.
Jansen, J.B.H. and Schuiling, R.D. (1976) Metamorphism on Naxos: petrology and geothermal gradient. Am. J. Sci. 276, 1225-1253.
_____, van de Kraats, A.H., van der Rijst, H. and Schuiling, R.D. (1978) Metamorphism of siliceous dolomites at Naxos, Greece. Contrib. Mineral. Petrol. 67, 279-288.
Johannes, W. and Puhan, D. (1971) The calcite-aragonite transition, reinvestigated. Contrib. Mineral. Petrol. 31, 28-38.
Jones, B.F. (1986) Clay mineral diagenesis in lacustrine sediments. In Studies in Diagenesis, F.A. Mumpton, ed., U.S. Geol. Surv. Bull. 1578, 291-300.
Juster, T.C. (1987)Mineralogic domains in very low grade pelitic rocks. Geology 15,1010-1013.
Kager, P.C.A. and Oen, I.S. (1983) Iron-rich talc-opal-minnesotaite spherulites and crystal-lochemical relations of talc and minnesotaite. Mineral. Mag. 47, 229-231.
Kerrick, D.M. (1968) Experiments on the upper stability limit of pyrophyllite at 1.8 and 3.9 kb pressure. Am. J. Sci. 266, 204-214.
_____ and Jacobs, G.K. (1981)A modified Redlich-Kwong equation for H_2O,CO_2, and H_2O-CO_2 mixtures at elevated pressures and temperatures. Am. J. Sci., 281, 735-767.
Khodyrev, O.Yu. and Agoshkov, V.M. (1986)Phase transitions in serpentine in the $MgO-SiO_2$-H_2O system at 40-80 kbar. Geochem. Int'l. 23, 47-52.
Kinsey, R.A., Kirkpatrick, R.J., Hower, J., Smith, K.A. and Oldfield, E.(1985)High resolution alumina-27 and silicon-29 nuclear magnetic resonance spectroscopic study of layer silicates, including clay minerals. Am. Mineral. 70, 537-548.
Kirkpatrick, R.J., Smith, K.A., Schramm, S., Turner, G. and Yang, W-H. (1985) Solid-state nuclear magnetic resonance spectroscopy of minerals. Ann. Rev. Earth Planet. Sci. 13, 29-47.
Klein, C. (1974) Greenalite, stilpnomelane, minnesotaite, crocidolite, and carbonates in a very low-grade metamorphic Precambrian iron-formation. Canad. Mineral. 12, 475-498.
_____ (1982) Amphiboles in iron-formations. In D.R. Veblen and P.H. Ribbe, eds., Rev. in Mineral. 9B, 88-98.
_____ (1983) Diagenesis and metamorphism of Precambrian banded iron-formations. In A.F. Trendall and R.C. Morris, eds., Iron-formations: Facts and Problems. New York, Elsevier, 417-469.
_____ and Bricker, O.P. (1977) Some aspects of the sedimentary and diagenetic environment of Proterozoic banded iron-formation. Econ. Geol. 72, 1457-1470.
_____ and Fink, R.P. (1976) Petrology of the Sokoman Iron Formation in the Howells River area, at the western edge of the Labrador Trough. Econ. Geol. 71, 453-487.
_____ and Gole, M.J. (1981) Mineralogy and petrology of parts of the Marra Mamba Iron formation, Hamersley Basin, Western Australia. Am. Mineral. 66, 507-525.

Kodama, H. (1958) Mineralogical study on some pyrophyllites in Japan. Mineral. J. (Japan) 2, 236-244.

Kosolov, A.S., Pustyl'nikov, A.M. Moshkina, I.A. and Mel'nikova, Z.M. (1969) Talc in Cambrian salts of the Kantaseyevka Depression. Akad. Nauk SSSR Proc. Earth Sci. Sect. 185, 127-129.

Koski, R.A., Lonsdale, P.F., Shanks, W.C., Berndt, M.E. and Howe, S.S. (1985) Mineralogy and geochemistry of a sediment hosted sulfide deposit from the southern trough of Guaymas Basin, Gulf of California. J. Geophys. Res. 90, B8, 6695-6707.

Koster van Groos, A.F. and Guggenheim, S. (1987) High pressure differential thermal analysis (HP-DTA) of the dehydroxylation of sodium montmorillonite and K-exchanged montmorillonite. Am. Mineral. 72, 1162-1167.

Kulke, H.G. and Schreyer, W. (1973) Kyanite-talc schist from Sar e Sang, Afghanistan. Earth Planet. Sci. Letters 18, 324-328.

Lazarev, A.N. (1972) Vibrational spectra and structure of silicates. Consultants Bureau, New York, p. 116-125.

Lee, J.H. and Guggenheim, S. (1979) Single crystal X-ray refinement of pyrophyllite-1$\underline{Tc}$. Am. Mineral. 66, 350-357.

Lesher, C.M. (1978) Mineralogy and petrology of the Sokoman Iron Formation near Ardua Lake, Quebec. Can. J. Earth Sci. 15, 480-500.

Lippmaa, E., Magi, M., Samoson, A., Englehardt, G. and Grimmer, A.-R. (1980) Structural studies of silicates by solid-state high-resolution ^{29}Si NMR. J. Am. Chem. Soc. 102, 4889-4893.

Loh, E. (1973) Optical vibrations in sheet silicates. J. Phys. C: Solid State Phys. 6, 1091-1104.

Lonsdale, P.F., Bischoff, J.L., Burns, V.M., Kastner, M. and Sweeney, R.E. (1980) A high temperature hydrothermal deposit on the seabed at a Gulf of California spreading center. Earth Planet. Sci. Letters 49, 8-20.

Loughnan, F.C. and Steggles, K.R. (1976) Cookeite and diaspore in the Back Creek pyrophyllite deposit near Pambula, N.S.W. Mineral. Mag. 40, 765-772.

MacKenzie, K.J.D., Brown, J.W.M., Meinhold, R.H. and Bowden, M.E. (1985) Thermal reactions of pyrophyllite studied by high-resolution solid-state ^{27}Al and ^{29}Si nuclear magnetic resonance spectroscopy. J. Am. Ceram. Soc. 68, 266-272.

Maksimović, Z. (1966) β-kerolite-pimelite series from Goles Mountain, Yugoslavia. Proc. Int'l. Clay Conf., Jerusalem, Israel 1, 97-105.

Massonne, H-J., Mirwald, P.W. and Schreyer, W. (1981) Experimentelle Uberprufung der Reaktionskurve Chlorit + Quartz = Talk + Disthen im System MgO-Al$_2$O$_3$-SiO$_2$-H$_2$O. Fortschr. Mineral. 59, 122-123.

Matthes, S. and Knauer, E. (1981) The phase petrology of the contact metamorphic serpentinites near Erbendorf, Oberpfalz, Bavaria. N. Jahrb. Mineral. Abh. 141, 59-89.

Maynard, J.B. (1983) Geochemistry of sedimentary ore deposits. Springer-Verlag, New York, 305 p.

McDowell, S.D. and Elders, W.A. (1980) Authigenic layer silicate minerals in Borehole Elmore 1, Salton Sea Geothermal Field, California, USA, Contrib. Mineral. Petrol. 74, 293-310.

McKie, D. (1959) Yoderite, a new hydrous magnesium iron aluminosilicate from Mautia Hill, Tanganyika. Mineral. Mag. 32, 282-307.

Mel'nik, Y.P. (1982) Precambrian Banded Iron-Formations: Physicochemical Conditions of Formation. Elsevier, Amsterdam, 310 p.

Mercolli, I. (1980) Fluor-Verteilung in Tremolit und Talk in den metamorphen Dolomiten des Campolungo (Tessin) und ihre phasen petrologische Bedeutung. Schweiz. Mineral. Petrogr. Mitt. 60, 31-44.

Meyer, C. and Hemley, J.J. (1967) Wall rock alteration, in Barnes, H.L., ed., Geochemistry of hydrothermal ore deposits. Holt, Rinehart and Winston, New York, 670 p.

_____, Shea, E.P. and Goddard, C.C. (1968) Ore deposits at Butte, Montana, in Ridge, J.D., ed., Ore deposits of the United States, 1933-1967 (Graton-Sales vol.). Am. Inst. Mining Metall. Petrol. Engineers 2, 1373-1416.

Miller, C. (1974) On the metamorphism of the eclogites and high-grade blueschists from the Penninic Terrane of the Tauern Window, Austria. Schweiz. Mineral. Petrogr. Mitt. 54, 371-384.

Mineeva, R.M. (1978) Relationship between Mössbauer spectra and defect structure in biotite from electric field gradient calculations. Phys. Chem. Minerals 2, 267-277.

Miyano, T. (1978a)Phase relations in the system Fe-Mg-Si-O-H and environments during low-grade metamorphism of some Precambrian iron formations.J.Geol.Soc. Japan 84, 679-690.

_____ (1978b) Effects of CO_2 on mineralogical differences in some low-grade metamorphic iron-formations. Geochem. J. 12, 201-211.

_____ and Klein, C.(1983) Phase relations of orthopyroxene, olivine, and grunerite in high-grade metamorphic iron-formation. Am. Mineral. 68, 699-716.

_____ and Klein, C. (1986) Fluid behavior and phase relations in the system Fe-Mg-Si-C-O-H: application to high grade metamorphism of iron-formations. Am. J. Sci. 286, 540-575.

Moine, B. (1982) Géochimie des transformations métasomatiques à l'origine du gisement de talc et chlorite de Trimouns (Luzenac, Ariège, France). II. Approche des conditions physico-chimiques de la métasomatose. Bull. Mineral. 105,76-88.

_____, Gavoille, B. and Thiebaut, J. (1982) Géochimie des transformations métasomatiques à l'origine du gisement de talc et chlorite de Trimouns (Luzenac, Ariège, France). I. Mobilité des éléments et zonalités. Bull. Minéral. 105, 62-75.

Montoya, J.W. and Hemley, J.J. (1975) Activity relations and stabilities in alkali feldspar and mica alteration reactions. Econ. Geol. 70, 577-583.

Moore, J.N. and Kerrick, D.M. (1976) Equilibria in siliceous dolomites of the Alta aureole, Utah. Am. J. Sci. 276, 502-524.

Mukhopadhyay, D., Baral, M.C. and Neogi, R.K. (1980) Mineralogy of the banded iron-formation in the southeastern Bababudan Hills, Karnataka, India. N. Jahrb. Mineral. Abh. 139, 303-327.

Mullis, J. (1979) The system methane-water as a geological thermometer and barometer from the external part of the Central Alps. Bull. Minéral. 102, 526-536.

Nakahira, M. and Kato, T. (1964) Thermal transformations of pyrophyllite and talc as revealed by X-ray and electron diffraction studies. Clays & Clay Minerals 12, 21-27.

Naumann, A.W., Safford, G.J. and Mumpton, F.A. (1966) Low frequency (OH)⁻ motions in layer silicate minerals. Clays & Clay Minerals 14, 367-383.

Nesbit, H.W. and Bricker, O.P.(1978)Low temperature alteration processes affecting ultramafic bodies. Geochim. Cosmochim. Acta 42, 403-409.

Nesbitt, R.W. and Hartman, L.A. (1986) Comments on "A peridotitic komatiite from the Dalradian of Shetland" by D. Flinn and D.T. Moffat. Geol. J. 21, 201-205.

Newman, A. C. D. and Brown, G. (1987) The chemical constitution of clays. Ch. 1 in Chemistry of Clays and Clay Minerals, A. C. D. Newman, ed., Mineral. Soc. Great Britain Monograph 6, London.

Newton, R.C. (1972) An experimental determination of the high pressure stability limits of magnesian cordierite under wet and dry conditions. J. Geol. 80, 398-420.

Nisio, P. and Lardeaux, J.-M. (1987) Retromorphic Fe-rich talc in low-temperature eclogites: example from Monviso (Italian Western Alps). Bull. Minéral. 110, 427-437.

Nitsch, K.-H.(1972)The P-T-X(CO₂)-stability field of lawsonite. Contrib. Mineral. Petrol. 34, 116-134.

Noack, Y., Decarreau, A. and Manceau, A. (1986) Spectroscopic and oxygen isotopic evidence for low and high temperature origin of talc. Bull. Minéral. 109, 253-263.

Oen, I.S., Fernández, J.C. and Manteca, J.I. (1975) The lead-zinc and associated ores of La Union Sierra de Cartagena, Spain. Econ. Geol. 70, 1259-1278.

Ohmoto, H. and Rye, R.O.(1970)The Bluebell Mine, British Columbia. I. Mineralogy, paragen-eses, fluid inclusions, and the isotopes of hydrogen, oxygen, and carbon. Econ. Geol. 65, 417-437.

Ohnmacht, W. (1974) Petrogenesis of carbonate-orthopyroxenites (sagvandites) and related rocks from Troms, Northern Norway. J. Petrol. 15, 303-324.

Oldfield, E. and Kirkpatrick, R.J. (1985) High-resolution nuclear magnetic resonance of in-organic solids. Science 227, 1537-1544.

Page, R.H.(1980)Partial interlayers in phyllosilicates studied by TEM. Contrib. Mineral. Petrol. 75, 309-314.

Papezik, V.S. and Keats, H.G. (1976) Diaspore in a pyrophyllite deposit on the Avalon Peninsula, Newfoundland. Can. Mineral. 14, 442-449.

Paradis, S., Velde, B. and Nicot, E. (1983) Chloritoid-pyrophyllite-rectorite facies rocks from Brittany, France. Contrib. Mineral. Petrol. 83, 342-347.

Perdikatsis, B. and Burzlaff, H. (1981) Strukturverfeinerung am Talk $Mg_3[(OH)_2Si_4O_{10}]$. Z. Kristallogr. 156, 177-186.

Perkins, D. III, Essene, E.J., Westrum, E.F. and Wall, V.J. (1979) New thermodynamic data for diaspore and their application to the system Al_2O_3-SiO_2-H_2O. Am.Mineral.64,1080-1090.

Perkins, E.H., Brown, T.H. and Berman, R.G. (1986) PTX-SYSTEM: three programs for calculation of pressure-temperature-composition phase diagrams. Computers & Geosci. 12, 749-755.

Perry, E.C., Tan, F.C. and Morey, G.B. (1973) Geology and stable isotope geochemistry of the Biwabik Iron Formation, northern Minnesota. Econ. Geol. 68, 1110-1126.

Pfeifer, H-R. (1977) A model for fluids in metamorphosed ultramafic rocks. Schweiz. Mineral. Petrogr. Mitt. 57, 361-396.

Phillips, G.N. (1987) Metamorphism of the Witwatersrand gold fields: conditions during peak metamorphism. J. Metam. Geol. 5, 307-322.

_____ (1988) Widespread fluid infiltration during metamorphism of the Witwaterrand gold-fields: generation of chloritoid and pyrophyllite. J. Metam. Geol. 6, 311-332.

Puhan, D. (1988) Reverse age relations of talc and tremolite deduced from reaction textures in metamorphosed siliceous dolomites of the southern Damara Orogen (Namibia). Contrib. Mineral. Petrol. 98, 24-27.

_____ and Hoffer, E. (1973) Phase relations of talc and tremolite in metamorphic calcite-dolomite sediments in the southern portion of the Damara Belt (South West Africa). Contrib. Mineral. Petrol. 40, 207-214.

Raheim, A. and Green, D.H. (1974) Talc-garnet-kyanite-quartz schist from an eclogite-bearing terrane, western Tasmania. Contrib. Mineral. Petrol. 43, 223-236.

Raymond, L.R. (1962) The petrology of the Lower Magnesian Limestone of north-east Yorkshire and south-east Durham. Quart. J. Geol. Soc. Lond. 118, 39-64.

Rayner, J.H. and Brown, G. (1965) Structure of pyrophyllite. Clays & Clay Minerals 13, 73-84.

_____ and _____ (1966) Triclinic form of talc. Nature 212, 1352-1353.

_____ and _____ (1973) The crystal structure of talc. Clays & Clay Minerals. 21, 103-114.

Reed, B.L. and Hemley, J.J. (1966) Occurrences of pyrophyllite in the Kekiktuk Conglomerate, Brooks Range, northeast Alaska. U.S. Geol. Surv. Prof. Paper 550C, 162-166.

Rice, J.M. (1977) Progressive metamorphism or impure dolomite limestone in the Marysville aureole, Montana. Am. J. Sci. 277, 1-24.

Robert, P. (1971) Sur un gisement de bauxite de l'ile d'Eubee (Grece). C. R. Acad. Sc. Paris, t. 272, no. 26, 3228-3230.

Roberts, R.G. and Reardon, E.J. (1978) Alteration and ore-forming processes at Mattagami Lake Mine, Quebec. Can. J. Earth Sci. 15, 1-21.

Robinson, G.W. and Chamberlain, S.C.(1984)The Sterling mine, Antwerp, New York. Mineral. Record 15, 199-216.

Rosasco, G.J. and Blaha, J.J. (1980) Raman microprobe spectra and vibrational mode assignments of talc. Applied Spectr. 34, 140-144.

Rosenberg, P.E. (1974) Pyrophyllite solid solutions in the system Al_2O_3-SiO_2-H_2O. Am. Mineral. 59, 254-260.

_____ and Cliff, G. (1980) The formation of pyrophyllite solid solutions. Am. Mineral. 65, 1217-1219.

Ross, M., Smith, W.L. and Ashton, W.H. (1968) Triclinic talc and associated amphiboles from Gouverneur mining district; New York. Am. Mineral. 53, 751-769.

Russell, J.D., Farmer, V.C. and Velde, B. (1970) Replacement of OH by OD in layer silicates, and identification of the vibrations of these groups in infra-red spectra. Mineral. Mag. 37, 869-879.

Saksena, B.D. (1961) Infrared absorption studies of some silicate structures. Trans. Faraday Soc. 57, 242-258.

Sanford, R.F. (1982) Growth of ultramafic reaction zones in greenschist to amphibolite facies metamorphism. Am. J. Sci. 282, 543-616.

Schmidt, J.M. (1986) Stratigraphic setting and mineralogy of the Arctic volcanogenic massive sulfide prospect, Ambler District, Alaska. Econ. Geol. 81, 1619-1643.

292

Schramm, J.M. (1978) Anchimetamorphes Permoskyth an der Basis des Kaisergebirges (Sudrand der Nordlichen Kalkalpen zwischen Worgl und St. Johann in Tirol, Osterreich). Geol. Palaeont. Mitt. Innsbruck 8, 101-111.

Schreyer, W. (1973) Whiteschists: a high pressure rock and its geological significance. J. Geol. 81, 735-739.

_____ (1977) Whiteschists: their compositions and pressure-temperature regimes based on experimental, field, and petrographic evidence. Tectonophys. 43, 127-144.

_____ (1985) Metamorphism of crustal rocks at mantle depths: high pressure minerals and mineral assemblages in metapelites. Fortschr. Mineral. 63, 227-261.

_____ (1988) Experimental studies on metamorphism of crustal rocks under mantle pressures. Mineral. Mag. 52, 1-26.

_____ and Abraham, K. (1976) Three-stage metamorphic history of a whiteschist from Sar e Sang, Afghanistan, as part of a former evaporite deposit. Contrib. Mineral. Petrol. 59, 111-130.

_____, _____ and Kulke, W. (1980) Natural sodium phlogopite coexisting with potassium phlogopite and sodian aluminian talc in a metamorphic evaporate sequence from Derrag, Tell Atlas, Algeria. Contrib. Mineral. Petrol. 74, 223-233.

_____, Medenbach, O., Abraham, K., Gebert, W. and Muller, W. F. (1982) Kulkeite, a new metamorphic phyllosilicate mineral: ordered 1:1 chlorite/talc mixed-layer. Contrib. Mineral. Petrol. 80, 103-109.

Sclar, C.B., Carrison, L.C. and Schwartz, C.M. (1965) High pressure synthesis and stability of a new hydronium-bearing layer silicate in the system MgO-SiO_2-H_2O. Trans. Am. Geophys. Union, EOS 46, 184.

Scrivenor, R.C. and Sanderson, R.W. (1982) Talc and aragonite from the Triassic halite deposits of the Burton Row borehole, Brent Knoll, Somerset. Report, U.K. Instit. Geol. Sci. 82-1, 58-60.

Seifert, F. and Virgo, D. (1974) Temperature dependence of intracrystalline Fe^{2+}-Mg distribution in a natural anthophyllite. Carnegie Inst. Wash. Yrbk. 73, 405-411.

Seyfried, W. and Bischoff, J.L. (1977) Hydrothermal transport of heavy metals by seawater: the role of seawater/basalt ratio. Earth Planet. Sci. Lett. 34, 71-77.

Sharma, R.P. (1979) Origin of the pyrophyllite-diaspore deposits of the Bundelkhand Complex, central India. Mineral. Deposita 14, 343-352.

Siever, R.A. and Kastner, M. (1967) Mineralogy and petrology of some Mid-Atlantic Ridge sediments. J. Marine Res. 25, 263-278.

Sisson, V.B., Crawford, M.L. and Thompson, P.H. (1981) CO_2 - brine immiscibility at high temperatures, evidence from calcareous metasedimentary rocks. Contrib. Mineral. Petrol., 78, 371-378.

Skippen, G.B. (1974) An experimental model for low pressure metamorphism of siliceous dolomitic marble. Am. J. Sci. 274, 487-509.

Slaughter, J., Kerrick, D.M. and Wall, V.J. (1975) Experimental and thermodynamic study of equilibria in the system CaO-MgO-SiO_2-CO_2-H_2O. Am. J. Sci. 275, 143-171.

Smith, J.R. (1957) Reconnaissance in the system FeO-Fe_2O_3-SiO_2-H_2O. Ann. Report Director Geophys. Lab. Carnegie Inst. Wash. Yrbk. 56, 230-231.

Smith, K.A., Kirkpatrick, R.J., Oldfield, E. and Henderson, D.M. (1983) High-resolution silicon-29 nuclear magnetic resonance spectroscopic study of rock-forming silicates. Am. Mineral. 68, 1206-1215.

Smolin, P.P., Zvyagin, B.B., Drits, V.A., Sidorenko, O.V. and Alexandrova, V.A. (1974) Structural identification of natural varieties of talc and variation in the ordering of their structures. Akad. Nauk. USSR Doklady 218, 120-123.

Snoke, A.W. and Calk, L.C. (1978) Jackstraw-textured talc-olivine rocks, Preston Peak area, Klamath Mountains, California. Geol. Soc. Am. Bull. 89, 223-230.

Spiess, F.N. and others (Rise Project Group) (1980) East Pacific Rise: hot springs and geophysical results. Science 207, 1421-1433.

Springer, R.K. (1974) Contact metamorphosed ultramafic rocks in the Western Sierra Nevada foothills, California. J. Petrol. 15, 160-195.

Steger, E. (1975) Comment on Optical vibrations in sheet silicates. (letter to editor) J. Phys. C: Solid State Phys. 8, L261-L263.

Stemple, I.S. and Brindley, G.W. (1960) A structural study of talc and talc-tremolite relations. J. Am. Ceram. Soc. 43, 34-42.

Stoessell, R.K. (1988) 25°C and 1 atm dissolution experiments of sepiolite and kerolite. Geochim. Cosmochim. Acta 52, 365-374.

_____ and Hay, R.L. (1978) The geochemical origin of sepiolite and kerolite at Amboseli, Kenya. Contrib. Mineral. Petrol. 65, 255-267.

Stout, J.H. (1985) A general chemographic approach to the construction of ternary phase diagrams, with application to the system Al_2O_3-SiO_2-H_2O. Am. J. Sci. 285, 385-408.

Stubican, V. and Roy, R. (1961) Isomorphous substitution and infrared spectra of the layer lattice silicates. Am. Mineral. 46, 32-51.

Stuckey, J.L.(1967)Pyrophyllite deposits in North Carolina.North Carolina Dept.Conservation, Div. Mineral Resources Bull. 80, 50 p.

Styrt, M.M., Brackman, A.J., Holland, H.D., Clark, B.C., Pisutha-Arnond, V., Eldridge, C.S. and Ohmoto, H. (1981) The mineralogy and the isotopic composition of sulfur in hydrothermal sulfide/sulfate deposits on the East Pacific Rise, $21°$ N latitude, East Pacific Rise. Earth Planet. Sci. Letters 53, 382-390.

Swindale, L.D. and Hughes, I.R. (1968) Hydrothermal association of pyrophyllite, kaolinite, diaspore, dickite, and quartz in the Coromandel area, New Zealand. New Zealand J. Geol. Geophys. 11, 1163-1183.

Sykes, M.L. and Moody, J.B. (1978) Pyrophyllite and metamorphism in the Carolina slate belt. Am. Mineral. 63, 96-108.

Taylor, H.F.W. (1962) Homogeneous and inhomogeneous mechanisms in the dehydroxylation of minerals. Clay Mineral. Bull. 5, 45-55.

Tilley, C.E. (1948) Earlier stages in the metamorphism of siliceous dolomites. Mineral. Mag., 28, 272-276.

Troeger, W.E. (1979) Optical Determination of Rock-Forming Minerals, Part I, Determinative Tables. Schweizerbart'sche Verlagsbuchhandlungen, Stuttgart.

Trommsdorff, V.(1966)Progressive metamorphose Kieseliger Karbonatgesteine in den Zentralalpen zwischen Bernina und Simplon. Schweiz. Mineral. Petrogr. Mitt. 46, 431-460.

_____ (1972) Change in T-X during metamorphism of siliceous dolomitic rocks of the central Alps. Schweiz. Min. Petr. Mitt., 52, 1-4.

_____ and Evans, B.W. (1972) Progressive metamorphism of antigorite schist in the Bergell tonalite aureole (Italy). Am. J. Sci. 272, 423-437.

_____, Skippen, G. and Ulmer, P. (1985) Halite and sylvite as solid inclusions in high-grade metamorphic rocks. Contrib. Mineral. Petrol., 89, 24-29.

Tsipursky, S.T., Kameneva, M.Y. and Drits, V.A.(1985)Structural transformations of Fe^{3+}-containing 2:1 dioctahedral phyllosilicates in the course of dehydration. 5th Meeting European Clay Groups, Prague, 569-577.

Udagawa, S., Urabe, K. and Hasu, H. (1974) The crystal structure of muscovite dehydroxylate. Japan. Assoc. Mineral., Petrol., Econ. Geol. 69, 381-389.

Udovkina, N.G., Muravitskaya, G.N. and Laputina, I.P.(1980)Talc-garnet-kyanite rocks of the Kokchetav block, Northern Kazakhstan. Doklady Earth Sci. Sect. 237, 202-205.

Vance, J.A. and Dungan, M.A. (1977) Formation of peridotites by deserpentinization in the Darrington and Sultan areas, Cascade Mountains, Washington. Geol. Soc. Am. Bull. 88, 1497-1508.

Veblen, D.R. and Buseck, P.R.(1980)Microstructures and reaction mechanisms in biopyriboles. Am. Mineral. 65, 599-623.

Vedder, W. (1964) Correlations between infrared spectrum and chemical composition of mica. Am. Mineral. 49, 736-768.

_____ and McDonald, R.S. (1963) Vibrations of the OH ions in muscovite. J. Chem. Phys. 38, 1583-1590.

_____ and Wilkins, R.W.T. (1969) Dehydroxylation and rehydroxylation, oxidation and reduction of micas. Am. Mineral. 54, 482-509.

Velde, B. (1983) Infrared OH-stretch bands in potassic micas, talcs and saponites; influence of electronic configuration and site charge compensation. Am. Mineral. 68, 1169-1173.

_____ and Couty, R. (1985) Far infrared spectra of hydrous layer silicates. Phys. Chem. Minerals 12, 347-352.

294

_____ and Kornprobst, J. (1969) Stabilite des silicates d'alumine hydrates. Contrib. Mineral. Petrol. 21, 63-74.

Veniale, F. and van der Marel, H.W. (1969) Identification of some 1:1 regular interstratified trioctahedral clay minerals. Proc. Internat. Clay Conf., Tokyo, Japan 1, 233-244.

Viswanathan, K. and Seidel, E.(1979)Crystal chemistry of Fe-Mg carpholites.Contrib. Mineral. Petrol. 70, 41-48.

Vrana, S. (1975) Magnesian-aluminous rocks, the associated ore mineralization and the problem of magnesium-iron metasomatism, Krystalinikum 11, 101-114.

_____ and Barr, M.W.C. (1972) Talc-kyanite-quartz schists and other high-pressure assemblages from Zambia. Mineral. Mag. 299, 837-846.

Walther, J.V. (1983) Description and interpretation of metasomatic phase relations at high pressures and temperatures: 2. Metasomatic reactions between quartz and dolomite at Campolungo, Switzerland. Am. J. Sci. 283A, 459-485.

_____ and Helgeson, H.C.(1980)Description and interpretation of metasomatic phase relations at high pressures and temperatures. I. Equilibrium activities of ionic species in nonideal mixtures of CO_2 and H_2O. Am. J. Sci. 280, 575-606.

Ward, J.R. (1975) Kinetics of talc dehydration. Thermchim. Acta 13, 7-14.

Wardle, R. and Brindley, G.W. (1972) The crystal structures of pyrophyllite, 1Tc, and of its dehydroxylate. Am. Mineral. 57, 732-750.

Watanabe, K. and Usui, A. (1977) Mineral paragenesis of pyrophyllite deposits in the Shin-yo Mine, Nagano Prefecture. J. Mineral. Soc. Japan 13, Special Issue, 217-225 (in Japanese).

Watanabe, T., Shimizu, H., Masuda, A. and Saito, H. (1983) Studies of [29]Si spin-lattice relaxation times and paramagnetic impurities in clay minerals by magic-angle spinning ^{29}Si-- NMR and EPR. Chem. Letters 1983, 1293-1296.

Weiss, Z. and Ďurovič, S. (1984) Polytypism of pyrophyllite and talc. Silikáty 28, 289-309.

Wesolowski, M.(1984)Thermal decomposition of talc.A review.Thermochem. Acta 78, 395-421.

Wiewiora, A., Wieckowski, T. and Sokolowska, A. (1979) The Raman spectra of kaolinite sub-group minerals and of pyrophyllite. Archiwum Mineralogiezne 35, 5-11.

Wilkins, R.W.T. and Ito, J. (1967) Infrared spectra of some synthetic talcs. Am. Mineral. 52, 1649-1661.

Winter, J.K. and Ghose, S. (1979) Thermal expansion and high temperature crystal chemistry of the Al_2SiO_5 polymorphs. Am. Mineral 64, 573-586.

Wood, R.M. (1982) The Laytonville Quarry (Mendocino County, California) exotic block: iron-rich blueschist-facies subduction-zone metamorphism. Mineral. Mag. 45, 87-99.

Yamamoto, K. and Akimoto, S-I. (1977) The system $MgO-SiO_2-H_2O$ at high pressures and temperatures--Stability field for hydroxyl-chondrodite, hydroxyl-clinohumite and 10 Å phase. Am. J. Sci. 277, 288-312.

Yoder, H.S.Jr.(1952) The $MgO-Al_2O_3-SiO_2-H_2O$ system and related metamorphic facies. Am. J. Sci. 250, 569-627.

Zajac, I.S. (1974) The stratigraphy and mineralogy of the Sokoman Formation in the Knob Lake area, Quebec and Newfoundland. Geol. Surv. Can. Bull. 220.

Zen, E-an (1960) Metamorphism of Lower Paleozoic rocks in the vicinity of the Taconic range in west-central Vermont. Am. Mineral. 45, 129-175.

_____ (1961) Mineralogy and petrology of the system $Al_2O_3-SiO_2-H_2O$ in some pyrophyllite deposits of North Carolina. Am. Mineral. 46, 52-66.

Zierenberg, R.A. and Shanks, W.C. III (1983) Mineralogy and geochemistry of epigenetic features in metalliferous sediment, Atlantis II Deep, Red Sea. Econ. Geol. 78, 57-72.

Zvyagin, B.B. (1967) Electron-diffraction analysis of clay mineral structures. Plenum Press, New York.

_____, Mishchenko, K.S. and Soboleva, S.V. (1969) Structure of pyrophyllite and talc in relation to the polytypes of mica-type minerals. Soviet Phys. Cryst. 13, 511-515.

Chapter 9 J.V. Chernosky, Jr., R.G. Berman & L.T. Bryndzia

STABILITY, PHASE RELATIONS, AND THERMODYNAMIC PROPERTIES OF CHLORITE AND SERPENTINE GROUP MINERALS

INTRODUCTION

Chlorite group minerals are found as authigenic phases in sedimentary rocks, are ubiquitous in low to medium grade metamorphic rocks, and occur in zones of hydrothermal alteration. Serpentine group minerals are formed by the alteration of ultrabasic and some carbonate rocks under pressure-temperature conditions ranging from those found at the earth's surface to those corresponding to the low eclogite or high greenschist facies of regional metamorphism. The chemistry and perhaps structure (polytype) of serpentine and chlorite minerals undoubtedly record information about the intensive conditions under which they have formed.

Our objective in this chapter is to summarize the progress experimentalists have made toward elucidating the phase relations of the serpentine and chlorite group minerals over the past four decades. A considerable body of experimental phase equilibrium data is available over a wide range in temperature for chrysotile, antigorite, lizardite, clinochlore, and Fe,Mg-chlorite for $P(H_2O)=P(Total)$ and for clinochlore and chrysotile in supercritical H_2O-CO_2 fluids. Our approach to this review differs fundamentally from the approach taken by reviewers of experimentally determined phase relations in previous MSA Short Courses. Rather than presenting phase equilibrium data as depicted on published phase diagrams, we have elected to synthesize and evaluate critically the data within the context of an internally consistent thermodynamic data base. The advantage of this approach is that it provides a rational basis for comparison of experimental data pertaining to different equilibria and affords the opportunity to assess which of the experimental data are in agreement and which are not; the disadvantage is that inevitably choices have to be made among incompatible data sets. Although we have justified the relatively few choices we have had to make, we do not claim infallibly. We view this synthesis as a point of departure for future work rather than as a culmination of work already done. The major challenge facing experimentalists is to provide data that will constrain equation of state parameters which completely describe the behavior of serpentine and chlorite minerals as a function of intensive variables such as T, $P(H_2O)$, $f(O_2)$, $f(S_2)$, and, most importantly, compositional variables describing solid solution in these minerals.

In this work, we make use of the thermodynamic data base of Berman (1988) because it is more up to date and the equilibrium curves calculated with it fit the experimental data more closely than curves calculated using the data bases of Helgeson et al. (1978) and Holland and Powell (1985). A brief description of the Berman (1988) data base is presented in the next section. Agreement among the calculated and experimentally located positions for serpentine and Mg-chlorite equilibria is remarkable when one considers that the experiments were performed by numerous investigators over a period of time when experimental techniques were continually being refined. First order estimates of the thermodynamic properties of the Fe-chlorite endmember, chamosite, have been retrieved from limited experimental data for which the composition of the Fe,Mg-chlorite has been determined.

The focus of this review will be on information that experimental work provides about the physicochemical conditions under which serpentine and chlorite minerals form in nature. Experimental phase equilibrium and thermodynamic data have allowed petrologists to construct petrogenetic grids which delineate the P-T stabilities of various Fe,Mg-chlorite- and serpentine-bearing assemblages in the system $Mg-Fe-Si-Al-O_2-H_2O-CO_2$. The high degree of consistency among the phase equilibrium and thermodynamic data discussed in this review supports the validity of these petrogenetic grids which incorporate experimentally located equilibria extrapolated to conditions beyond the experiments and equilibria located solely on the basis of the thermodynamic data.

296

Table 1. Phase Names, Abbreviations, and Chemical Formulae

Phase	Abbreviation	Formula
Almandine	Alm	$Fe_3Al_2Si_3O_{12}$
Amesite	Am	$Mg_4Al_4Si_2O_{10}(OH)_8$
Andalusite	And	Al_2SiO_5
Antigorite	Atg	$Mg_{48}Si_{34}O_{85}(OH)_{62}$
Brucite	Br	$Mg(OH)_2$
Calcite	Cc	$CaCO_3$
Chamosite (daphnite)	Ch	$Fe_5Al(AlSi_3)O_{10}(OH)_8$
Chrysotile	Chr	$Mg_3Si_2O_5(OH)_4$
Clinochlore	Cln	$Mg_5Al(AlSi_3)O_{10}(OH)_8$
Cordierite	Cd	$(Mg,Fe)_2Al_4Si_5O_{18}$
Corundum	Co	Al_2O_3
Crocidolite	Cr	$Na_2(Mg,Fe)_5Si_8O_{22}(OH)_2$
Diopside	Di	$CaMgSi_2O_6$
Dolomite	Do	$CaMg(CO_3)_2$
Enstatite (ortho)	En	$MgSiO_3$
Forsterite	Fo	Mg_2SiO_4
Garnierite	Ga	$Ni_3Si_2O_5(OH)_4$
Greenalite	Grn	$(Fe^{2+},Fe^{3+})_{5-6}Si_4O_{10}(OH)_8$
Grossular	Gr	$Ca_3Al_2Si_3O_{12}$
Hercynite	Hr	$FeAl_2O_4$
Kaolinite	Kln	$Al_2Si_2O_5(OH)_4$
Kyanite	Ky	Al_2SiO_5
Lizardite	Lz(x)	$(Mg_{6-x}Al_x)(Si_{4-x}Al_x)O_{10}(OH)_8$
Magnesite	Mst	$MgCO_3$
Magnetite	Mt	Fe_3O_4
Minnesotaite	Mi	$Fe_3Si_4O_{10}(OH)_2$
Muscovite	Ms	$KAl_3Si_3O_{10}(OH)_2$
Phlogopite	Phl	$KMg_3(AlSi_{13}O_{10})(OH)_2$
Pyrope	Py	$Mg_3Al_2Si_3O_{12}$
Pyrophyllite	Prl	$Al_2Si_4O_{10}(OH)_2$
Quartz	Qz	SiO_2
Sapphirine	Sa	$Mg_7Al_{18}Si_3O_{40}$
Silica (aqueous)	S	$SiO_2(aq)$
Sillimanite	Si	Al_2SiO_5
Spinel	Sp	$MgAl_2O_4$
Sudoite	Su	$Mg_2Al_4Si_3O_{10}(OH)_8$
Talc	Tc	$Mg_3Si_4O_{10}(OH)$
Tremolite	Tr	$Ca_2Mg_5Si_8O_{22}(OH)_2$
Carbon dioxide	CO_2	CO_2
Water	W	H_2O

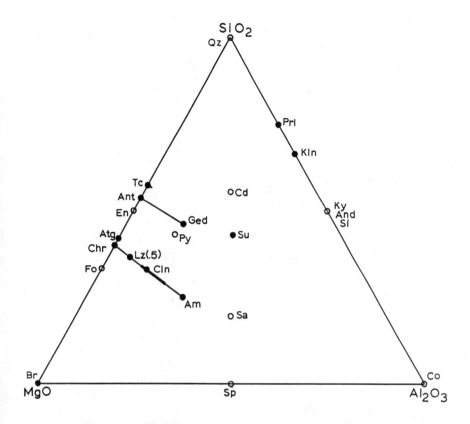

Figure 1. Crystalline phases in the system MgO-Al₂O₃-SiO₂- H₂O projected, in mole percent, from the H₂O apex to the anhydrous base of a tetrahedron. Light lines show various crystalline solutions; heavy solid line shows approximate compositional range of the high temperature, II$\underline{b}$, Mg,Al-chlorites; open circles refer to anhydrous phases; solid dots refer to hydrous phases. Abbreviations are given in Table 1.

Because they have fairly simple chemistries and are generally the same polytype (II$\underline{b}$), chlorites in low to medium grade metamorphic rocks also have enormous and largely untapped potential for geothermometry; in fact, their potential utility for geothermometry is not matched by any other phase in these rocks. Unfortunately, our knowledge of activity-composition relations for Fe,Mg-chlorites is limited, restricting their usefulness in geothermometry at present. It is unlikely that the serpentine minerals, with the possible exception of lizardite, will prove useful in geothermometry principally because they exhibit limited chemical variation. However, polytypism may one day prove useful as an indicator of metamorphic grade.

Following the AIPEA nomenclature committee's recommendation (Bailey, 1980), we shall refer to the Fe²⁺ dominant trioctahedral chlorite as chamosite rather than daphnite. The aluminous 7 Å serpentines will be referred to as lizardite rather than septechlorite or 7 Å chlorite. Table 1 contains a list of abbreviations and chemical formulae for the phases considered in this review. Phases in the MgO-Al₂O₃-SiO₂-H₂O (MASH) system are plotted on Figure 1. When referring to crystalline solutions we shall append (ss) to mineral names. Table 2 contains a list of experimentally investigated equilibria involving serpentine and chlorite minerals which are considered in this review.

298

Table 2. Experimentally Investigated Equilibria Involving Serpentine and Chlorite Group Minerals

<u>Chlorites</u>

*1.	5Cln = 10 Fo + 3Sp + 1Cd + 20W	Yoder (1952); Nelson & Roy (1958)
		Fawcett and Yoder (1966)
		Chernosky (1974); McPhail (1985)
*2.	6Cln + 29Qz = 8Tc + 3Cd + 16W	Fawcett & Yoder (1966)
		Velde (1973); Fleming & Fawcett (1976)
		Chernosky (1978)
		Massone (unpublished)
*3.	2Cln + 8Ky + 11Qz = 5Cd + 8W	Seifert and Schreyer (1970)
*4.	2Cln + 8And + 11Qz = 5Cd + 8W	Seifert and Schreyer (1970)
*5.	2Cln + 8Si + 11Qz = 5Cd + 8W	Seifert and Schreyer (1970)
*6.	6Cln + 13And + 11Prl = 15Cd + 35W	Seifert and Schreyer (1970)
*7.	2Cln + 8Prl = 5Cd + 13Qz + 16W	Seifert and Schreyer (1970)
*8.	2Cln + 19And = 5Cd + 11Co + 8W	Seifert (1973)
*9.	2Cln + 19Si = 5Cd + 11Co + 8W	Seifert (1973)
*10.	2Cln + 19Ky = 5Cd + 11Co + 8W	Seifert (1973)
*11.	3Cln + 11Ky = 5Tc + 14Co + 7W	Massone (unpublished)
*12.	3Cln + 14Qz = 5Tc + 3Ky + 7W	Massone et al. (1981)
		Massone (unpublished)
*13.	2Cln + 6En = 7Fo + Cd + 8W	Fawcett and Yoder (1966)
		Chernosky (1976)
		Jenkins and Chernosky (1986)
*14.	Cln = Fo + 2En + Sp + 4W	Fawcett and Yoder (1966)
		Staudigel and Schreyer (1977)
		Jenkins (1981)
		Jenkins and Chernosky (1986)
15.	Cln + 2Sp + 6Co = Sa + 4W	Seifert (1974)
		Doroshev & Malinovskii (1974)
		Ackermand et al. (1975)
16.	29Cln + 154Co = 6Cd + 19Sa + 116W	Seifert (1974)
17.	5Cln + 10Sa = 9Cd + 77Sp + 20W	Seifert (1974)
18.	5Cln + 2Cd = 22En + Sa + 20W	Seifert (1974)
19.	5Cln + Sa = 18En + 14Sp + 20W	Seifert (1974)
*20.	Ms + Cln + 2Qz = Cd + Phl + 4W	Seifert (1970)
		Bird and Fawcett (1973)
*21.	3Cln + 5Ms = 5Phl + 8Ky + Qz + 12W	Bird and Fawcett (1973)
		Massone (unpublished)
*22.	3Cln+2Cc = 5Fo+2Di+3Sp+12W+2CO₂	Chernosky and Berman (in press)
*23.	Cln+2Do = 3Fo+2Cc+Sp+4W+2CO₂	Widmark (1980)
		Chernosky and Berman (in press)
*24.	Cln + 2Mst = 3Fo + Sp + 4W + 2CO₂	Chernosky and Berman(unpublished)
*25.	Cln + 2Co = 2Sp + Py + 4W	Ackermand et al. (1975)
*26.	2Cln = Sp + Py + 3Fo + 8W	Staudigel and Schreyer (1977)
*27.	5Cln + Cd = 7Sp + 20En + 20W	Seifert (1974)
		Jenkins and Chernosky (1986)
*28.	5Cln + 20Co = 3Cd + 19Sp + 20W	Seifert (1974)
29.	Gr + Cln = 3Di + 2Sp + 4W	Wang and Greenwood (1988)
30.	5Su = 2Cln + 8And + Qz + 12W	Fransolet and Schreyer (1984)
31.	5Su = 2Cln + 8Ky + Qz + 12W	Fransolet and Schreyer (1984)

Table 2. (Continued)

32.	$15Su = 6Cln + 23And + Prl + 35W$	Fransolet and Schreyer (1984)
33.	$15Su = 6Cln + 23Ky + Prl + 35W$	Fransolet and Schreyer (1984)
34.	$5Su = 2Cln + 7Ky + Kln + 10W$	Fransolet and Schreyer (1984)
35.	Fe-Cln=Fe-Cd+Fa+Mt(ss)+Hr(ss)+fluid	Turnock (1960)
36.	Fe-Cln=Mt(ss)+Hr(ss)+Fe-Cd+Qz+fluid	Turnock (1960)
37.	Fe-Cln=Hematite+Prl+Mullite+fluid	Turnock (1960)
38.	Fe-Cln=Hematite+Co+Qz+fluid	Turnock (1960)
39.	$Fe,Mg-Cln+Ms+Qz=CoBiotite+Al_2SiO_5+W$	Hirschberg and Winkler (1969)
40.	$Fe,Mg-Cln+Ms+Qz=Alm(ss)+Biotite+Al_2SiO_5+W$	Hirschberg and Winkler (1968)
41.	Fe-Cln + Qz + fluid = Alm + fluid	Hsu (1968)
42.	Qz+Fe-Cln+Mt+fluid=Alm+fluid	Hsu (1968)
43.	Fe-Cln+Qz+Mt=Mt(ss)+Hr(ss)+Qz+fluid	Hsu (1968)
44.	Fe,Mg-Cln+Ms=Staurolite+Biotite+Qz+fluid	Hoschek (1969)
45.	Fe-Cln=Fe-Cd+Mt(ss)+Qz+fluid	James et al. (1976)
46.	Fe-Cln+Qz=Fe-gedrite+Mt(ss)+fluid	James et al. (1976)
47.	Fe,Mg-Cln=olivine+Cd(ss)+Sp(ss)+fluid	McOnie et al. (1975)
48.	Fe,Mg-Cln=olivine+Cd(ss)+Hr+fluid	McOnie et al. (1975)
49.	$Fe,Mg-Cln+O_2=Cd(ss)+Mt(ss)+orthoamphibole(ss)+fluid$	McOnie et al. (1975)
50.	$Fe,Mg-Cln+O_2=Cd(ss)+Qz+Mt(ss)+fluid$	McOnie et al. (1975)
51.	$Fe,Mg-Cln+O_2=olivine+Co+Mt+W$	McOnie et al. (1975)
52.	$Fe,Mg-Cln+Actinolite+Epidote+$ Albite+Qz$\pm$Sphene=Al-amphibole Ilmenite+W	Apted and Liou (1983)
53.	$Fe,Mg-Cln+Ms=Cd(ss)+Biot(ss)+Al_2SiO_5+W$	Burnell and Rutherford (1984)
54.	Fe,Mg-Cln+Ms=Biotite+Co+W	Burnell and Rutherford (1984)
55.	$Fe,Mg-Cln+O_2=Al_2SiO_5+Mt+Qz+W$	Bryndzia and Scott (1987a)
56.	$Fe,Mg-Cln+S_2=Al_2SiO_5+Pyrrhotite+Qz+O_2+W$	Bryndzia and Scott (1987a)

Serpentines

*57.	$5Chr = 6Fo + Tc + 9W$	Bowen and Tuttle (1949); Pistorius (1963) Scarfe and Wyllie (1976) Kitahara et al. (1966) Chernosky (1982) Yamamoto and Akimoto (1977)
*58.	$Chr + Br = 2Fo + 3W$	Bowen and Tuttle (1949) Kitahara et al. (1966) Pistorius (1963); Johannes (1968)
*59.	$Atg = 18Fo + 4Tc + 27W$	Evans et al. (1976)
*60.	$2Chr + 3CO_2 = Tc + 3Mst + 3W$	Johannes (1969)
*61.	$Chr + Mst = 2Fo + 2W + CO_2$	Johannes (1969)
62.	$Atg + 30S = 16Tc + 15W$	Hemley et al. (1977b)
63.	$Chr + 2S = Tc + W$	Hemley et al. (1977a)
64.	$5Lz(0.2)=9Fo+Cln+1.6Tc+14.4W$	Chernosky (1973)
65.	$10Lz(0.5)=12Fo+3Tc+5Cln+18W$	Caruso and Chernosky (1979)

Designates equilibria used in consistency analysis by Berman (1988).

Equilibria 35 through 56 are not balanced because many of the participating phases have variable compositions.

EVALUATION OF PHASE EQUILIBRIUM DATA AND THERMODYNAMIC ANALYSIS

Evaluation and assessment of the relative quality of phase equilibrium data requires consideration of many factors, the most important being the care with which (a) reversal is demonstrated, (b) starting materials have been prepared and characterized, (c) experimental apparatii have been calibrated prior to and controlled during an experiment, and (d) experimental products have been characterized and their compositions determined.

A major difference exists between synthesis experiments and reversal experiments. Many of the earlier experiments were of the synthesis type wherein a mineral assemblage was produced from starting materials such as glasses, gels, or mixtures of oxides. The potential for nucleating metastable phases when using such starting materials is great due to the high free energy of these materials (Fyfe et al., 1958); once nucleated, metastable phases tend to persist because the free energy difference between metastable and stable assemblages is often relatively small. Equilibria located using synthesis experiments often lie at too high a temperature because significant overstepping of an equilibrium curve is usually required to induce nucleation and growth of the stable assemblage.

We shall be concerned primarily with reversal type experiments in this review. In such experiments, one mineral assemblage is produced from starting materials consisting of another chemically equivalent mineral assemblage. Experimentalists typically use a starting material containing all of the phases which participate in an equilibrium; when all phases are present, reaction rates are governed by relative solubilities of the phases rather than by ease of nucleation. The balanced chemical equilibrium relating two assemblages is said to be "bracketed" or "reversed" when the high temperature assemblage is converted to the low temperature assemblage and vice versa. Because the driving force for reaction, $\Delta_r G$, diminishes as an equilibrium is approached, reaction rates close to the equilibrium are sluggish and complete conversion of one assemblage to another is often not realized on a laboratory time scale. In this situation, reaction direction must be determined by comparing modal amounts of phases with a petrographic microscope or by comparing relative peak intensities on an XRD trace of the starting material with that of the experimental product. Because considerable change (30-50%) in relative XRD peak intensities must be observed in order to overcome the effect of preferred grain orientation on peak intensities, reversals based on small changes (<20%) must be viewed with a measure of skepticism. The occurrence in either the starting material or experimental products of minerals other than those appearing in the balanced chemical reaction also complicates interpretation of the results because the actual reaction taking place may be difficult to ascertain.

When preparing starting material to be used for determining the location of an equilibrium, each of the phases participating in the equilibrium may be synthesized separately and then combined, or an assemblage of phases may be synthesized directly from a single mix. Neither method is entirely satisfactory. It is easier to characterize phases synthesized individually; hence changes in composition or in structural state which occur during hydrothermal treatment are more readily evaluated. However, there is no guarantee that an equilibrium distribution of components among all phases will be achieved during hydrothermal treatment. On the other hand, when an assemblage is synthesized from a single bulk composition, an equilibrium distribution of components may be more easily achieved; however, the initial compositions of the phases and the redistribution of components as a function of P, T, and length of hydrothermal treatment are more difficult to evaluate. Internal consistency is more difficult to demonstrate when the products of a previous experiment containing phases which may vary in composition are used as starting materials for a subsequent experiment carried out at a different pressure, temperature, or oxygen fugacity.

True equilibrium is impossible to demonstrate in experimental studies. Even though an equilibrium is reversed in the laboratory, a third assemblage having a lower free energy, hence more stable, than either of the assemblages participating in the equilibrium in question may exist; this is of particular importance when considering the stabilities of serpentine minerals. Nevertheless, experimental reversals of metastable equilibria are useful for the derivation of thermodynamic properties of participating phases. The presence of additional phases or components in a rock may prevent reactions which occur in pure synthetic systems from ever being realized in rocks. Various solution models must be evaluated before using experimental data obtained for end-member phases to deduce the conditions of formation of rocks containing minerals which are often non-ideal crystalline solutions. It is imperative to ascertain the reliability of experimental results by comparing them to compatibility relations observed in rocks and to a comprehensive thermodynamic data base.

In all phase equilibrium experiments the information gained is that the experimental products are more stable than the starting materials, and, in reversal experiments, each experiment has the potential to yield one half-bracket for an equilibrium. Because the thermodynamic information conveyed is that the Gibbs free energy of the product assemblage is lower than that of the starting assemblage, reversal experiments can by analyzed most rigorously by using the mathematical techniques of linear programming (Gordon, 1973) which cast the results of reversal experiments in terms of inequalities in the Gibbs free energies of reaction. Berman et al. (1986) advocate use of the technique of mathematical programming for derivation of thermodynamic data because it utilizes and takes into account the mathematical nature of both phase equilibrium data (inequality constraints which are characterized by a step function rather than by a normal distribution of errors around a "best" value) and direct measurements (calorimetric and volumetric) of individual phase properties which are characterized by a normal distribution. Detailed discussion of the use of this technique as well as a comparison with the use of regression techniques is given by Berman et al. (1986) and will not be repeated here.

The data base of Berman (1988) gives standard state properties for 67 minerals in the system $Na_2O-K_2O-CaO-MgO-FeO-Fe_2O_3-Al_2O_3-SiO_2-TiO_2-H_2O-CO_2$. These properties were derived from and are consistent in large measure with some 1200 phase equilibrium brackets involving 180 different equilibria, as well as available calorimetric and volumetric measurements on individual phase properties. Use was made of the equations of state of Haar et al. (1984) for water, Kerrick and Jacobs (1981) for carbon dioxide and water-carbon dioxide mixtures, and Helgeson et al. (1981) for aqueous species. Unless otherwise stated, all equilibria shown in the figures in this chapter were computed with these equations of state and the Berman (1988) mineral data.

PHASE RELATIONS OF SERPENTINE GROUP MINERALS

The importance of understanding serpentine phase relations has grown with the realization that they might prove useful in deciphering the physicochemical conditions prevailing during metamorphism of alpine peridotites, ophiolites, oceanic ultrabasic rocks, and some carbonate rocks. In this section we review and evaluate published phase equilibrium and thermodynamic data pertaining to chrysotile, antigorite, and lizardite. Although serpentine phase equilibria have only been investigated in relatively simple chemical systems, the experimental data provide useful constraints on the pressures and maximum temperatures attending serpentinization and allow calculation of internally consistent thermodynamic properties for chrysotile, lizardite, and antigorite. Two factors have plagued experimental studies aimed at elucidating serpentine phase relations. First, antigorite nucleates very sluggishly, allowing the metastable growth and persistence of chrysotile and low-Al lizardite. Second, Al readily substitutes into lizardite and dramatically affects this mineral's stability.

Antigorite

Observations of antigorite textural relations and distribution in contact metamorphic aureoles (Trommsdorff and Evans, 1972 and 1977) leave little doubt that this mineral has a true stability field. Although antigorite is a common mineral in serpentinites, experimentalists have found it a very difficult phase to synthesize hydrothermally because chrysotile and lizardite tend to nucleate rapidly and persist metastably even in experiments of long duration. Although Jasmund and Sylla (1971) first reported the successful synthesis of Mg and Ni endmember antigorite, additional work (Jasmund and Sylla, 1972) indicated that the synthetic product was a 6-layer lizardite. Korytkova and Makanova (1971) and Kortkova et al. (1972) synthesized antigorite by reacting olivine with alkaline, silica-bearing fluids at 250°-300°C and $P(H_2O) = 0.5$-0.9 kbar. Iishi and Saito (1973) reportedly synthesized antigorite together with lizardite and chrysotile at temperatures of 450°-550°C at water pressures from 517 to 1724 bars in five days and concluded that a Mg:Si ratio below the ideal ratio in chrysotile is necessary for antigorite synthesis. Evans et al. (1976) synthesized antigorite in a piston-cylinder apparatus but do not provide any information about the synthesis conditions or the synthetic product.

Although Yoder (1967) was able to react natural antigorite to forsterite and talc, he was unable to reverse the reaction in experiments of relatively short duration (< 200 h). Evans et al. (1976) successfully reversed the reaction Atg = 18Fo + 4Tc + 27W at water pressures from 2 to 15 kbar (Fig. 2c). The starting material for their experiments consisted of 10% by weight of natural antigorite from Piz Lunghin containing 0.7 wt % Al$_2$O$_3$ and 1.23 wt % FeO (total Fe expressed as FeO) mixed with synthetic talc and forsterite. The low pressure experiments (2, 4, and 6 kbar) were performed in cold-seal vessels and the high pressure experiments (10 and 15 kbar) were performed in a piston-cylinder apparatus. Although the oxygen fugacity was not buffered, the authors considered the experimental results reliable because two experiments at P(H$_2$O) = 10 kbar using an all synthetic starting material produced the same results as the starting material containing natural antigorite.

Hemley et al. (1977b) determined the solubilities of natural antigorite (containing 1.3 wt % Al$_2$O$_3$ and 3.7 wt % FeO; total Fe as FeO) and talc at high temperature by investigating the equilibria Atg + 30S = 16Tc + 15W (Fig. 3c) and 2Tc = 3Fo + 5S + 2W (Fig. 3a) at P(H$_2$O) = 1 kbar. Dissolution reactions tend to occur more rapidly than thermal decomposition reactions, and the composition of the solution in which the solids are dissolving is a very sensitive indicator of the direction and extent of reaction. Combination of these two equilibria results in the equilibrium Atg = 18Fo + 4Tc + 27W, which was reversed by Evans et al. (1976). The Atg + 30S = 16Tc + 15W curve calculated by Berman (1988) is inconsistent (Fig. 3c) with the solubility data of Hemley et al. (1977b) who mention that "reversals are difficult to demonstrate for talc-antigorite, and more data points at long run times would be desirable".

Because Helgeson et al. (1978) based their thermodynamic data for antigorite on the mineral-solution equilibria of Hemley et al. (1977b), their thermodynamic data for this mineral differ substantially from those of Berman (1988) which were derived from the experimental data of Evans et al. (1976). Our reasons for preferring the experimental data of Evans et al. (1976) will be detailed after the stability of chrysotile is discussed.

Chrysotile

Clinochrysotile (2M_{cl}) is readily synthesized hydrothermally in several days from a mixture of periclase (MgO), SiO$_2$ glass and water at temperatures 10-20°C below its decomposition to Fo + Tc + W (Fig. 2a, b). Fluor-hydroxyl chrysotile (Ushio and Saito, 1970) and Ni chrysotile (Roy and Roy, 1954) have been synthesized but their stabilities have not been investigated. Nelson and Roy (1958) and Chernosky (1975) observed that chrysotile synthesized from starting material containing Al$_2$O$_3$ incorporated a minor amount of this element. Both Page (1966) and Moody (1976) synthesized chrysotile from a starting material containing iron; Page was unable to determine whether the synthetic product contained iron whereas Moody suggested that it was iron-free.

Bowen and Tuttle (1949) first investigated the stability relations of Mg endmember chrysotile at water pressures to 2800 bars. Although Bowen and Tuttle's results were duplicated by Yoder (1952) and by Roy and Roy (1955) at a water pressure of 2000 bars, reversibility was not demonstrated in any of these studies. Johannes (1968) reversed the equilibrium Chr + Br = 2Fo + 3W at water pressures to 7 kbar (Fig. 2d). Although Johannes did not identify the serpentine mineral used in his experiments, the reported positions and intensities of XRD reflections from the synthetic product coincide with those given by Whittaker and Zussman (1956) for chrysotile (2M_{cl}).

The equilibrium 5Chr = 6Fo + Tc + 9W has been reversed (Fig. 2a) at low pressures in cold-seal vessels by Scarfe and Wyllie (1967), Kalinin and Zubkov (1981), and Chernosky (1982). Scarfe and Wyllie (1967) used a novel technique based on topotaxy to determine that the back reaction, Fo + Ta + W = Chr, had occurred. They first synthesized the high temperature assemblage, then lowered the temperature for a period of time before opening the capsule to determine if the textural relations in the charge suggested that it had reacted to serpentine. Scarfe and Wyllie did not identify the serpentine mineral present in their experiments.

303

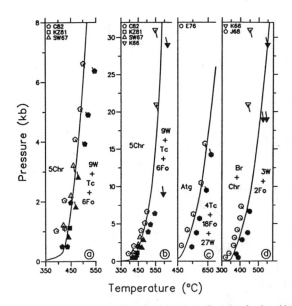

Figure 2. Comparison of computed phase equilibria involving chrysotile and antigorite with experimental data in the system MgO-SiO$_2$-H$_2$O: C82 - Chernosky (1982); E76 - Evans et al. (1976); J68 - Johannes (1968); K66 - Kitahara et al. (1966); SW67 - Scarfe and Wyllie (1967); KZ81 - Kalinin and Zubkov (1981). Symbols show the locations of experimental data after adjustment of nominal data (end of lines connected to symbol centers) for experimental uncertainties. Open and closed symbols depict half-brackets on opposite sides of the equilibrium.

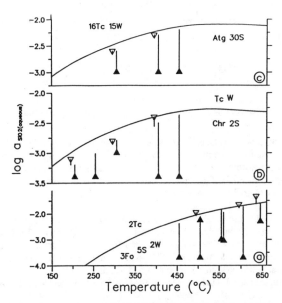

Figure 3. Comparison of computed aqueous silica (S) activities (referenced to equilibria written with one SiO$_2$) with 1 kbar experimental data of Hemley et al. (1977a and b) in the system MgO-SiO$_2$-H$_2$O. Filled and open symbols show experiments approached from undersaturation and supersaturation, respectively, while connected lines indicate final silica concentration.

304

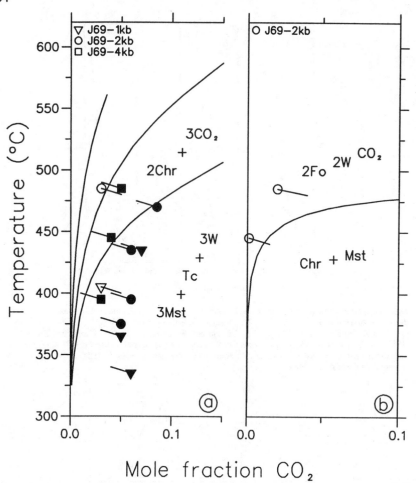

Figure 4. Comparison of computed phase equilibria with experimental data of Johannes (1969) for (a) 3Mst + Tc + 3W = 2Chr + 3CO₂ and (b) Chr + Mst = 2Fo + 2W + CO₂. Symbols as in Figure 2.

Several investigators have attempted to locate the equilibria Chr + Br = 2Fo + 3W and 5Chr = 6Fo + Tc + 9W at high pressures. Pistorius (1963) attempted to locate these equilibria using a piston-anvil apparatus at water pressures from 5 to 60 kbar. His data are unreliable because mixtures of oxides were used as the sole starting materials. Furthermore, data obtained using a piston anvil device are highly suspect due to high pressure gradients across the sample.

Kitahara et al. (1966) used a piston-cylinder apparatus to reverse both equilibria at water pressures from 6 to 30 kbar (Figs. 2b and 2d). Thermodynamic analysis (Berman, 1988) shows that their experimental data for 5Chr = 6Fo + Tc + 9W and for Chr + Br = 2Fo + 3W are in good agreement with the low pressure data. Yamamoto and Akimoto (1977) used a tetrahedral-anvil apparatus in an attempt to locate both equilibria at water pressures between 29 and 77 kbar. Although their results agree, within experimental error, with the data of Kitahara et al. (1966) at 30 kbar, the reactions were not reversed.

Hemley et al. (1977a) determined the solubilities of a natural chrysotile (Fig. 3b) containing 0.4 wt % Al₂O₃ and 0.78 wt % FeO (total Fe expressed as FeO) and a synthetic chrysotile at P(H₂O) = 1 kbar according to the reaction Chr + 2S = Tc + W; the experimental results for both the synthetic and natural chrysotiles were very similar.

The stability of natural chrysotile containing 1.92 wt % FeO has been experimentally investigated by Johannes (1969) in a fluid phase containing H_2O and CO_2. Although the chrysotile polytype used in the starting material was not mentioned, Helgeson et al. (1978, p. 96) cite a written communication from Otterdoom suggesting that it was orthochrysotile. Johannes (1969) loosely bracketed the equilibrium 2Chr + 3CO_2 = 3Mst + Tc + 3W at fluid pressures of 1, 2, and 4 kbar (Fig. 4a) and obtained a reversed half-bracket on the high temperature side of the equilibrium Chr + Mst = 2Fo + 2W + CO_2 at a fluid pressure of 2 kbar (Fig. 4b). The equilibrium curves calculated by Berman (1988) are consistent with all but one of Johannes' (1969) reversals of the equilibrium 2Chr + 3CO_2 = 3Mst + Tc + 3W at 1 kbar.

Moody (1976) investigated the decomposition of natural olivine containing 7 mol % FeO to Chr + Br + Mt at oxygen fugacities defined by the iron-magnetite buffer. Her results suggest that an olivine having this iron content decomposes at a temperature 15°C lower than pure forsterite. Although chrysotile and lizardite were both present in the experimental products, Moody suggested that lizardite was metastable because the amount of chrysotile apparently increased with respect to lizardite as the length of hydrothermal treatment increased. Moody was unable to determine the compositions of the serpentine phases present in the experimental products but suggested that chrysotile remained iron-free. In light of the observation that natural chrysotile incorporates iron, it is unlikely that an equilibrium distribution of iron and magnesium was attained among the phases in Moody's experiments.

Although phase equilibrium data for chrysotile were obtained within the stability field of antigorite they constrain the thermodynamic properties of chrysotile and allow us to compute the phase relations of this mineral at low temperatures. Berman et al. (1986) derived thermodynamic properties for chrysotile and antigorite from the phase equilibrium brackets of Chernosky (1982), Scarfe and Wyllie (1967), and Kalinin and Zubkov (1981) rather than from the solubility data of Hemley et al. (1977a, b) for two reasons. First, the former three sets of phase equilibrium data for chrysotile are internally consistent (Figs. 2a, b). Second, thermodynamic properties of chrysotile and antigorite derived solely from the solubility data (Delany and Helgeson, 1978) render chrysotile metastable with respect to antigorite and brucite at all temperatures and pressures; this is contradicted by the observation that chrysotile has formed directly from antigorite (Evans et al., 1976, p. 84 and references therein). The thermodynamic properties derived for chrysotile and antigorite by Berman (1988) are very similiar to those derived by Berman et al. (1986) but incorporate the effects of thermal expansion and compressibility on molar volumes. The new thermodynamic data for chrysotile and antigorite are consistent with all phase equilibrium brackets except the 0.5 kbar and 2 kbar half-brackets of Chernosky (1982) for 5Chr = 6Fo + Tc + 9W which lie at 10° and 5°C lower, respectively, than the calculated curve (Fig. 2a).

Lizardite

Compositional differences among the serpentine minerals are subtle (Whittaker and Wicks, 1970) and control the stabilities of these minerals because to a large extent they control the manner in which the octahedral and tetrahedral sheets are articulated. Lizardite phase relations are complex because its decomposition temperature increases dramatically with increasing aluminum content. Increasing the aluminum content via the coupled Tschermak's substitution Mg + Si = 2Al results in an expansion of the lateral dimensions of the tetrahedral sheet and a contraction in the lateral dimensions of the octahedral sheet. Lizardite's decomposition temperature is maximized at the composition for which the lateral dimensions of the octahedral and tetrahedral sheets become equivalent. Radoslovich and Norrish (1962) calculated that the unit cell parameter, b, for both sheets will be identical for the composition Lz(0.75) whereas Chernosky (1975) calculated that the areas of both sheets will be identical for Lz(0.6).

Nelson and Roy (1958) first demonstrated that lizardites readily crystallize from all aluminous compositions along the chrysotile-amesite join at temperatures below about 500°C and suggested that lizardites more aluminous than Lz(0.5) slowly recrystallize to chlorite at higher temperatures. Gillery (1959) synthesized a 1-layer lizardite polytype from starting materials ranging in composition from x = 0.25 to x = 0.65. Velde (1973) investigated the limits of lizardite and chlorite solid solutions for 26 compositions in the MASH system and noted the formation of trioctahedral 15 Å and 17 Å expanding phases.

As discussed earlier, the difficulty with which antigorite nucleates permits us to synthesize chrysotile ($2M_{Cl}$) and determine its phase relations under conditions where it is thermodynmically metastable. Similarly, Al-free lizardite, Lz(O), has been synthesized (Chernosky, 1973 and 1975) in the system $MgO\text{-}SiO_2\text{-}H_2O$ (MSH) under conditions where antigorite is the stable serpentine phase. Synthesis experiments at temperatures below about 420°C produce the metastable assemblage Chr (polytype unknown) + Lz(O). Recrystallization of the Chr+Lz(O) assemblage at 438°C and 2 kbar results in chrysotile ($2M_{Cl}$) whereas recrystallization at 418°C results in Lz(O), suggesting that Lz(O) has a lower thermal stability than chrysotile at $P(H_2O)=2$ kbar.

Evidently, in the MSH system at $P(H_2O) = 2$ kbar, partial curling of the tetrahedral and octahedral sheets accompanied by tetrahedral inversion as in antigorite is the preferred mechanism of misfit relief for temperatures ranging from about 500° down to about 250°C where chrysotile becomes more stable than antigorite + brucite (see Fig. 7 below). Curling of the octahedral and tetrahedral sheets as in chrysotile is the preferred mechanism for achieving misfit relief below about 250°C. Alternate clockwise and anticlockwise twist of tetrahedra as in lizardite is the least preferred mechanism of misfit relief; if Lz(0) has a true stability field, it must lie at temperatures below 250°C at $P(H_2O) = 2$ kbar.

As aluminum is added to the starting material, the two phase assemblage Lz+Chr is produced for bulk compositions ranging from x=0.05 to 0.25 (Nelson and Roy, 1958; Gillery, 1959 and Chernosky, 1975); the Lz+Chr assemblage has also been synthesized from iron-bearing starting materials (Page, 1966 and Moody, 1976). Chernosky (1975) observed that chrysotile tubes coexisting with lizardite become shorter, thicker and less abundant as the aluminum content of the starting material increases to x=0.25 and inferred that for a given bulk composition lizardite was always more aluminous than the coexisting chrysotile. Although the experimental data were obtained at pressure-temperature conditions within the stability field of antigorite, they suggest that Lz+Chr can coexist at lower temperatures and are, therefore, consistent with the observation that this assemblage is widespread in rocks. Synthesis experiments using compositions more aluminous than x=0.25 result in a single phase, lizardite.

Two equilibria involving synthetic lizardites having the compositions Lz(0.2) or 3.7 wt % Al_2O_3, and Lz(0.5) or 9.25 wt % Al_2O_3, have been reversed (Fig. 5) at water pressures below 10 kbar: 5Lz(0.2) = 9.6Fo + Cln + 1.6 Ta + 14.4W (Chernosky, 1973) and 10Lz(0.5) = 12Fo + 5Cln + 2Tc + 18W (Caruso and Chernosky, 1979). The experimentally located position of the equilibrium involving Lz(0.2) is approximate because with increasing temperature lizardite becomes more aluminous than Lz(0.2) as it decomposes. Furthermore, the lizardite present in the starting material probably was more aluminous than Lz(0.2) because an aluminous chrysotile was also present. The location of the equilibrium involving Lz(0.5) is more reliable because the starting material did not contain impurities and the composition of the lizardite used in the experiments was close to the composition for which the lateral dimensions of the octahedral and tetrahedral sheets are equivalent. The experimental results suggest that lizardite stability increases remarkably with aluminum content (Fig. 5) and it quickly becomes more stable than chrysotile and antigorite.

Petrogenetic grids involving serpentine minerals

A theoretical multisystem analysis (Evans et al., 1976) involving the phases Atg, Chr, Br, Tc, Fo, and W in the system $MgO\text{-}SiO_2\text{-}H_2O$ (Fig.6) suggests that the stability fields for Chr and for Chr+Br do not extend to as high a temperature as suggested by experimental work on the equilibria Chr+Br=2Fo+3W and 5Chr=6Fo+Tc+9W (Figs. 2a, 2b, and 2d). Unfortunately, phase equilibrium data alone do not permit discrimination between the two theoretically possible nets shown in Figure 6. Data from natural occurrences, however, indicate that the assemblage Atg+Br is common at higher metamorphic grades than Chr (Dietrich and Peters, 1971; Trommsdorff and Evans, 1974) and that antigorite commonly forms from Chr (Hess et al., 1952; Wilkonson, 1953; Francis, 1956; Coats, 1968; Springer, 1974) and vica versa (Evans et al., 1976). These obervations led Evans et al. (1976) to conclude that only the topology shown in Figure 6a is correct. This requires that the vapor conservative reactions 15Chr+Tc=Atg and Atg+3Br also appear on the diagram and that the reactions 5Chr=6Fo+Tc+9W and Chr+Br=2Fo+3W be metastable with respect to the analagous reactions involving antigorite: Atg = 18Fo+4Tc+27W and Atg + 20Br = 34Fo + 51W.

307

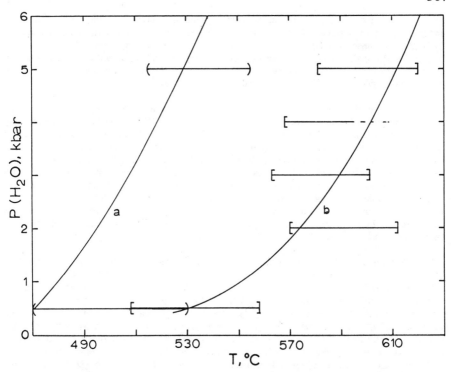

Figure 5. Experimental data for equilibria a) 5Lz(0.2) = 9.6Fo + 1.6Tc + Cln + 14.4W (Chernosky, 1973) and b) 10Lz(0.5) = 2Tc + 12Fo + 5Cln + 18W (Caruso and Chernosky, 1979). Brackets have been expanded by 5°C from the nominal temperature of each experiment to account for calibration uncertainties. Curves computed with the thermodynamic data derived in this chapter.

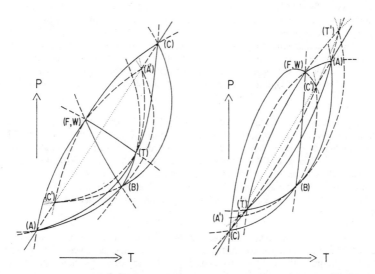

Figure 6. Two theoretically possible nets for the MSH system involving the phases antigorite (A), brucite (B), chrysotile (C), forsterite (F), talc (T), and water (W). Continuous lines are stable, dashed and dotted lines metastable. Primes indicate metastable invariant points. After Evans et al. (1976).

A revised phase diagram depicting the thermodynamically consistent stability fields of chrysotile and antigorite in the system MSH (Fig. 7) with P(Total) = P(H$_2$O) has been calculated with the program GEO-CALC (Berman et al., 1987) and the thermodynamic data base of Berman (1988). The vapor conservative reaction 17Chr = Atg + Br has a negative slope and is displaced toward lower temperatures by about 50°C relative to the position calculated by Evans et al. (1976); the equilibrium 15Chr + Tc = Atg does not appear at temperatures above 25°C for water pressures below 10 kbar.

We can explore the phase relations among all three serpentine minerals by adding Al$_2$O$_3$ to the MSH system and including the phases Lz(ss) and Cln. Chernosky (1973) and Caruso and Chernosky (1979) reversed equilibria involving Lz(0.2) and Lz(0.5) (Fig. 5) and discussed two alternate topologies for antigorite-lizardite phase relations. In one (Caruso and Chernosky, 1979, Fig. 4a), Lz(0.5) is stable at low pressure, whereas in the other (Caruso and Chernosky, 1979, Fig. 4b) Lz(0.5) is stabilized by high pressure. As noted by Caruso and Chernosky (1979), the experimental data clearly support the former topology. This topology is consistent with most observed assemblages in metamorphosed serpentinites. However, it is not compatible with the observation that Atg + Cln, rather than the chemically equivalent assemblage Lz + Tc, is stable throughout many ultramafic rocks in the Italian Alps regionally metamorphosed at water pressures on the order of 7 kbar, as well as in the Bergell aureole where fluid pressures probably did not exceed 2 kbar (Trommsdorff and Evans, 1972).

In order to clarify this apparent discrepancy, we have retrieved thermodynamic properties for Lz(0.5) (Table 3) from the phase equilibrium data of Caruso and Chernosky (1979) for 10Lz(0.5)=12Fo + 3Tc + 5Cln + 18W using the thermodynamic data of Berman (1988) for minerals other than lizardite. Although the width of the experimental brackets permits the entropy of Lz(0.5) to range from 225 to 413 Jmol^{-1}K^{-1}, the range of calculated phase topologies that results is consistent with Figure 4a of Caruso and Chernosky (1979). Figure 8 shows the phase relationships computed with our preferred entropy for Lz(0.5) (Table 3). The invariant point at 34 kbar and 750°C limits the stability of the assemblage Cln + Atg to higher pressures and temperatures, which is clearly incompatible with field observations. As the entropy of Lz(0.5) decreases, the invariant point at 34 kbar and 750°C moves to lower pressures approaching 10 kbar and 610°C for the minimum entropy, but this much lower entropy is physically unreasonable.

The only way to reconcile the occurrence of Atg + Cln in nature with the experimental data is with solid solution effects. The aluminum content of lizardites in metamorphosed serpentinites does not exceed x = 0.3 (5.5 wt % Al$_2$O$_3$) and is generally much lower. Chernosky (1973) reversed the equilibrium 5Lz(0.2) = 9.6Fo + Cln + 1.6Tc + 14.4W at P(H$_2$O) = 0.5 kbar between 476 and 525°C. Calculations using the thermodynamic properties for Lz(0.2) (3.7 wt % Al$_2$O$_3$) obtained from the midpoint of this bracket locate the brucite absent invariant point, [Br], at approximately 13 kbar and 625°C. Assuming that the equilibrium lies close to the low temperature half-bracket shifts the position of this invariant point to the more geologically reasonable conditions of 5 kbar and 560°C. This pressure and temperature will be further lowered by decreasing the aluminum content of the lizardite and/or by shifting the Atg = 18Fo + 4Tc + 27W equilibrium to higher temperatures by incorporating Al or Cr in antigorite (Trommsdorff and Evans, 1972).

To the extent that the aluminum content of natural lizardite and antigorite serve to lower the [Br] invariant point in Figure 8 to geologically reasonable pressures, Figures 7 and 8 are compatible with the following assemblages generated during prograde metamorphism: Chr + Br (Evans et al., 1976), Chr + Tc (Evans et al., 1976), Chr + Lz (Muller, 1959; Dietrich, 1969), Chr + Atg (Frost, 1975; Evans, 1977), Atg+Fo (Evans, 1974; Frost, 1975; Coleman, 1971). The most common prograde sequence Lz --> Atg --> Fo is allowed over a wide pressure interval. Amesite as well as the 15 Å and 17 Å phases described by Velde (1973) are omitted from Figure 8 because we are not convinced that these phases are stable at elevated temperatures.

In order to compute phase relationships among these minerals more accurately, it will be necessary to derive solution properties, particularly for both lizardite and chlorite. The limited amount of experimental data presently available makes it impossible to derive a unique set of solution and endmember properties for lizardite. For this reason, phase relations in the Al-free system are displayed schematically in Figure 9. For the purpose of this figure, we have assumed that Lz(0) has a stability field at elevated pressure. Although the lack of an accurate entropy value for Lz(0) precludes calculation of the

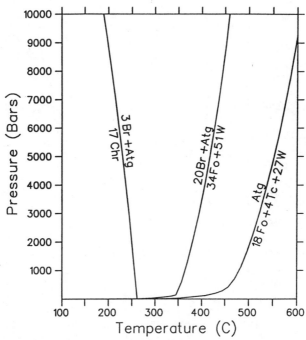

Figure 7. Equilibria in the system MSH computed for a(H₂O) = 1 with the thermodynamic data base of Berman (1988).

Table 3. Standard state thermodynamic properties (Joules) of Lz(0.5) at 1 bar, 298.15K

Property	Value	Source
$\Delta_f H$	-8,860,268	Caruso & Chernosky (1979) phase equilibrium data (see text)
$S^{\bullet}$	405.3	Estimated from: S(Cln) + S(En) -1/2 S(Co)
V	21.255	Caruso & Chernosky (1979, Table 3)
$Cp = K_0 + K_1 T^{-1/2} + K_2 T^{-2} + K_3 T^{-3}$		
K_0	1209.6	Estimated from: Cp(Cln) + 1/2[Cp(SiO₂) + Cp(MgO) -Cp(Al₂O₃)] (Berman and Brown, 1985)
K_1	-11,123.	
K_2	379,080.	
K_3	-1,338,102,000.	
$V^{P,T}/V^{1 bar, 298} = 1 + V_1(P-1) + V_2(P-1)^2 + V_3(T-298) + V_4(T-298)^2$		
V_1	-1.819×10^{-6}	Same as clinochlore
V_2	0.0	
V_3	26.452×10^{-6}	
V_4	0.0	

310

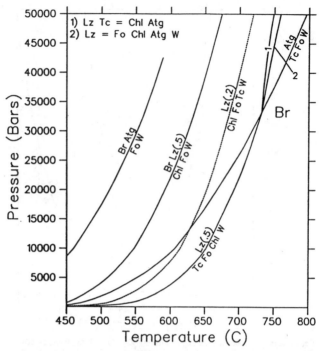

Figure 8. Computed phase equilibria for 10Lz(0.5) = 2Tc + 12Fo + 5Cln + 18W using thermodynamic properties of Lz(0.5) (Table 3) derived from the experimental data of Caruso and Chernosky (1979). Dotted line shows computed equilibrium for 5Lz(0.2) = 9.6Fo + 1.6Tc + Cln + 14.4W using thermodynamic data for Lz(0.2) derived from the experimental data of Chernosky (1973).

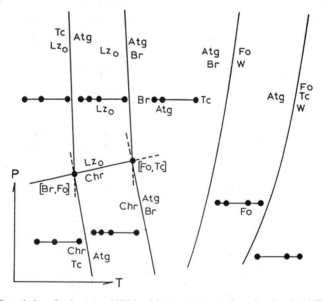

Figure 9. Theoretical net for the system MSH involving the phases aluminum-free lizardite,Lz(0), Chr, Atg, Br, Fo, Tc, and W.

slope of the polymorphic transition Lz(0) = Chr, we have assigned it a shallow positive slope, consistent with experimental data (Chernosky, 1975) which indicate that Lz(0) reacts to Chr with increasing temperature and that increasing pressure favors Lz(0) because it has a smaller molar volume than Chr. Observations that chrysotile forms authigenically by dissolution of low Al lizardite and reprecipitation as chrysotile (Mumpton and Thompson, 1975; Craw et al., 1987) further suggests that the Lz(0) = Chr phase boundary, if it is stable at 1 atm, must lie at elevated temperatures. It should be pointed out that because chrysotile can accommodate a minor amount of aluminum, the polymorphic transition is not restricted to the Al-free system.

The phase relations depicted on Figures 7, 8 and 9 will change as the activity of H_2O is lowered by diluting the fluid phase with an additional component such as CO_2 or by decreasing $P(H_2O)$ relative to P(Total) such as might occur during the metamorphism of an ultramafic rock which has not been completely penetrated by water. In both instances the dehydration equilibria will migrate toward lower temperatures relative to the vapor-conservative equilibria whose positions remain unaffected by the presence or absence of a fluid phase. The main effect of these changes is to move the position of the [Br] invariant point to a higher pressure. The consequences of metamorphism in a water deficient regime are explored in greater detail by Wicks and O'Hanley (Chapter 5, this volume).

Chrysotile is the only serpentine mineral whose stability in supercritical H_2O-CO_2 fluids has been experimentally investigated. If the arguments presented above concerning the metastability of the equilibria Chr + Br = 2Fo + 3W and 5Chr = 6Fo + Tc + 9W with respect to analogous equilibria involving antigorite are accepted, then the equilibria 2Chr + $3CO_2$ = 3Mst + Tc + 3W and Chr + Mst = 2Fo + 2W + CO_2 reversed by Johannes (1967) in H_2O-CO_2 fluids (Fig. 4a and b) must also be metastable with respect to analogous equilibria involving antigorite.

Johannes' (1967) experimental data show that chrysotile is only stable in very H_2O-rich fluids (XCO_2<0.07) at a total pressure of 2 kbar. However, Trommsdorff and Evans (1977) argued that the higher thermal stability of antigorite relative to chrysotile would permit antigorite to coexist stably with fluids having a higher CO_2 content [X(CO_2)=0.2] at total pressure of 2 kbar. A T-X(CO_2) diagram at a total pressure of 2 kbar (Fig. 10 and Table 4) calculated with the program GEO-CALC (Berman et al., 1987) shows the phase relations among Atg, En, Fo, Tc, Mgs, Di, Tr, Do, and Cc in the system CMSHCO2. This 10 phase multisystem consists of 462 possible equilibria, only 18 of which were computed to be stable at a total pressure of 2 kbar. Chrysotile, which is metastable at temperatures above 250°C, was not included in the multisystem analysis. Enstatite does not participate in any of the stable equilibria. The topology of the calculated phase diagram is consistent with the most commonly found assemblages in ophicarbonate rocks (Trommsdorff and Evans, 1977), Atg+Fo+Tr+Cc, Atg+Tr+Cc+Do, and Atg+Di+Tr+Cc, and reveals that antigorite is only stable in H_2O-rich fluids [X(CO_2) < 0.08].

Garnierite and greenalite

Noll and Kirchner (1952) and Roy and Roy (1954) synthesized garnierite from compositions in the system NiO-SiO_2-H_2O. Roy and Roy found that if Na^+ or Cl^- were present in the starting material chrysotile formed, whereas if these ions were absent a platy phase, presumably lizardite, formed. Synthesis experiments indicated that the platy phase having the composition $Ni_3Si_2O_5(OH)_4$ decomposed to Ni-talc + Ni-olivine at 530 $\pm$ 15°C, $P(H_2O)$ = 689 bars. Roy and Roy were unable to synthesize serpentine from compositions in which Mg had been replaced by Mn^{2+}, Zn^{2+}, or Co^{2+}, and concluded that Ni^{2+} was the largest divalent cation which could replace Mg^{2+} completely in the serpentine structure. Roy and Roy (1954) also synthesized a magnesia-germania serpentine from a starting material containing $3MgO$ + $2GeO_2$ at 450°C, $P(H_2O)$ = 1278 bars; the synthetic product consisted of hexagonal plates which decomposed at 520$\pm$15°C at 689 bars to $Mg_3Ge_4O_{10}(OH)_4$ (germanium talc) and Mg_2GeO_4. The composition $3NiO\cdot2GeO_2$ yielded a platy, serpentine-like 1:1 structure with a basal spacing of 7.50 Å. Jasmund and Sylla (1971) reported the synthesis of Ni-antigorite but further work showed that they had synthesized 6-layer lizardite (Jasmund and Sylla, 1972).

312

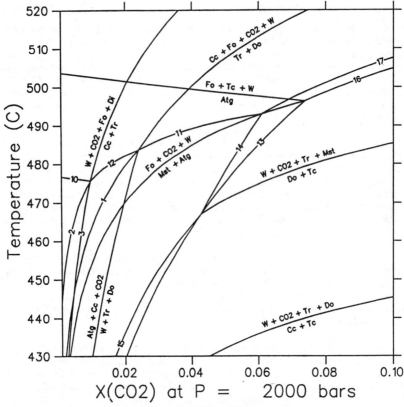

Figure 10. T-X(CO$_2$) diagram for the system CMSHCO$_2$ at P(fluid) = 2 kbar involving the phases Atg, Fo, Tc, Mgs, Di, Tr, Do, and Cc computed with the thermodynamic data base of Berman (1988).

Table 4. Equilibria shown on Figure 10.

1.	31 W + 20 CO$_2$ + 34 Fo + 20 Cc = Atg + 20 Do
2.	93 W + 20 CO$_2$ + 62 Fo + 20 Di = 3 Atg + 20 Cc
3.	45 CO$_2$ + 107 Di + Atg = 45 Cc + 31 Tr
4.	4 Atg + 141 Cc + 73 CO$_2$ = 107 W + 17 Tr + 107 Do
5.	13 Cc + 8 Fo + 9 CO$_2$ + W = Tr + 11 Do
6.	3 W + 5 CO$_2$ + 2 Fo + 11 Di = 5 Cc + 3 Tr
7.	W + CO$_2$ + Tr + Do = 3 Cc + 2 Tc
8.	31 W + 20 CO$_2$ + 34 Fo = Atg + 20 Mst
9.	18 Fo + 4 Tc + 27 W = Atg
10.	27 W + 4 Tr + 18 Fo = Atg + 8 Di
11.	383 W + 80 CO$_2$ + 20 Tr + 282 Fo = 13 Atg + 40 Do
12.	321 W + 40 CO$_2$ + 20 Tr + 214 Fo = 11 Atg + 40 Cc
13.	2 Atg + 45 CO$_2$ = 45 W + 17 Tc + 45 Mst
14.	4 Atg + 34 Do + 73 CO$_2$ = 107 W + 17 Tr + 141 Mst
15.	2 Atg + 15 Tr + 60 CO$_2$ = 30 W + 47 Tc + 30 Do
16.	4 Fo + 5 CO$_2$ + W = Tc + 5 Mst
17.	2 Do + 8 Fo + 9 CO$_2$ + W = Tr + 13 Mst
18.	W + CO$_2$ + Tr + 3 Mst = 2 Do + 2 Tc

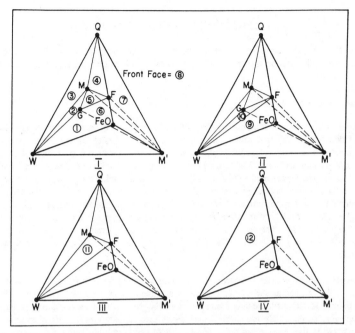

Figure 11. Sequence of stable low temperature assemblages in the system FeO-Fe_3O_4-SiO_2-H_2O in four temperature ranges: I = temperatures less than 250°C, II = 250° - 470°C, III = 470° - 480°C, and IV = temperatures greater than 480°C. Circled numbers refer to the different three and four condensed phase assemblages in the system. Abbreviations: M = magnetite, Q = quartz, F = fayalite, M' = minnesotaite, and G = greenalite. After Flaschen and Osborn (1957).

In an experimental study of the system iron oxide-silica-water, Flaschen and Osborn (1957) succeeded in synthesizing greenalite at low but unbuffered oxygen fugacities. They found that fayalite decomposed to greenalite + magnetite at temperatures below 250°C and water pressures ranging from 172 to 1,207 bars. Greenalite was found to decompose to fayalite + minnesotaite + H_2O at approximately 470°C. The sequence of stable assemblages deduced by Flaschen and Osborn are schematically illustrated in Figure 11. It must be emphasized that the oxygen fugacity was not buffered and that the Fe^{3+}/Fe^{2+} ratios of greenalite and minnesotaite probably varied.

In an attempt to duplicate experimentally some of the paragenetic changes found in the Brockman Iron Formation in Western Australia, Grubb (1971) succeeded in synthesizing greenalite and minnesotaite using gel starting materials having various amphibole compositions in the system Na_2O-MgO-FeO-Al_2O_3-SiO_2-H_2O-O_2 at temperatures between 150-350°C, and water pressures from 100 to 1000 bars. Although Grubb did not determine the stability of greenalite, he concluded that:

(1)   Presence of Al_2O_3 in the starting material suppressed crocidolite growth in favor of greenalite.

(2)   When MgO is added to the starting material in any significant quantity, greenalite becomes the predominant end-product.

(3)   Minnesotaite is frequently associated with greenalite at 150°C but upon long hydrothermal treatment (3 years) minnesotaite becomes the predominant phase.

STABILITY OF ALUMINOUS LIZARDITE

In the previous section we considered the synthesis and stability of lizardites having aluminum contents $x < 0.8$; in this section we shall consider the stability of lizardites having aluminum contents in the range $0.8 < x < 2.0$.

Synthesis of Lz(ss) with $x > 0.8$

Several investigators (Nelson and Roy, 1958; Gillery, 1959; Shirozu and Momoi, 1972; Velde, 1973; and Chernosky, 1975) have shown that lizardites can be synthesized from all compositions along the chrysotile-amesite join. Gillery (1959) synthesized a 1-layer lizardite polytype from starting materials ranging in composition from $x = 0.25$ to 0.65 and a 6-layer polytype from compositions more aluminous than $x = 1.5$ at water pressures between 1.03 and 3.9 kbar and temperatures below about 450°C; intermediate compositions contained mixtures of both polytypes. Gillery concluded that composition was the predominant factor in determining the lizardite polytype produced. Shirozu and Momoi (1972) synthesized a lizardite 6(3) polytype (6-layer approximating a 3-layer structure) at temperatures between 300-400°C and a 6(2) polytype (6-layer approximating a 2-layer structure) at 600°C for compositions ranging from about $x = 1.25$ to $x = 2.0$ and concluded that the 6(2) polytype is favored by increasing temperatures.

As explained previously, the decomposition temperature of lizardite is maximized when the lateral dimensions of the octahedral and tetrahedral sheets are identical ($x = 0.6$ to 0.75). Once the optimum aluminum content is exceeded, the lateral dimensions of the tetrahedral sheet begin to grow larger than the lateral dimensions of the octahedral sheet. Progressively greater distortion is required in order to achieve articulation between the sheets, resulting in a concomitant decrease in the stability of Lz(ss). Attempts to synthesize a lizardite more aluminous than Lz(2.0) (Nelson and Roy, 1958; Chernosky, unpublished data) have failed; it is likely that octahedral sheets with higher charges than those in amesite are unstable (Bailey, written communication).

Stability of lizardite ($x > 0.8$)

Numerous investigators (Yoder, 1952; Roy and Roy, 1955; Nelson and Roy, 1958; Segnit, 1963; and Velde, 1973) have attempted to locate the Lz(1.0) = Cln polymorphic transition in P-T space experimentally but the results are ambiguous and the interpretations of the experimental data conflicting. The Lz(1.0) = Cln phase boundary has been placed at temperatures ranging from 375°C (Velde, 1973) to 510°C (Yoder, 1952) principally because it was established solely on the basis of synthesis experiments. Although all of these investigators were able to transform lizardite to chlorite experimentally at elevated temperatures, none were able to reverse the reaction, i.e. to transform clinochlore to lizardite.

Interpretations of the experimental data on the equilibrium Lz(ss) = Mg, Al-chlorite differ substantially. Yoder (1952) suggested that Lz(1.0) is metastable with respect to clinochlore at water pressures ranging from 0.14 to 2.1 kbar and temperatures above 520°C; Nelson and Roy (1958) suggested that lizardites with aluminum contents greater than $x = 0.4$ were metastable with respect to chlorites at all temperatures. On the other hand, Velde (1973) postulated that all aluminous lizardites were stable with respect to chlorites at temperatures below about 380°C at $P(H_2O) = 2$ kbar. Velde's (1973, p. 307) interpretation was not based upon experimental evidence but on the observation that "sedimentary chlorites found on ocean and estuary bottoms are 7 Å whereas those in sedimentary rocks are usually 14 Å". It is evident that experimental data do not provide reliable information on the stability of chlorite relative to aluminous lizardite with $0.8 < x < 1.5$.

More recently, Cho and Fawcett (1986) investigated the rate at which Lz(1.0) transforms to Cln in the temperature range 600-700°C at $P(H_2O) = 2$ kbar and found that Lz(1.0) persists metastably for over 72 days at 600°C before finally transforming to clinochlore (Fig. 12). Although the uncertainties in

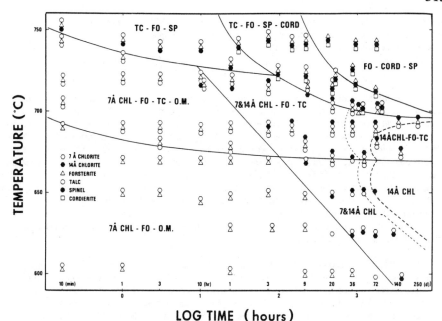

Figure 12. A time-temperature-transformation diagram for the conversion of oxide mix (O.M.) to various assemblages. Long-dashed curve is the calculated line for the complete conversion of Lz(1.0) to chlorite and short-dashed curve for 75% conversion. After Cho and Fawcett (1986).

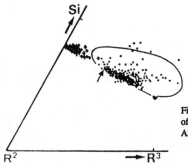

Figure 13. Natural chlorite and serpentine analyses plotted as a function of R^2-R^3-Si content. Dots are 14 Å chlorites, crosses are serpentines. After Velde (1973).

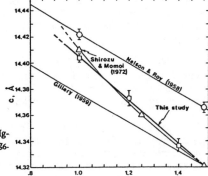

Figure 14. Variation in c as a function of composition of synthetic Mg-chlorite. x refers to the number of Al cations in the formula $(Mg_{6-x})(Si_{4-x}Al_x)O_{10}(OH)_8$. After Jenkins and Chernosky (1986).

determining the percent conversion of Lz(1.0) to Cln were large, Cho and Fawcett were able to calculate the rate at which the transformation occurred as a function of temperature. Because the rate constant k is related to an activation energy by the Arrhenius equation:

$$k = A \exp(-H_a/RT)$$

where A is the frequency factor, H_a is the activation energy, R is the gas constant and T is absolute temperature, Cho and Fawcett were able to calculate an activation energy for the Lz(1.0) to Cln transformation. The calculated activation energy, 90 $\pm$ 35 kcal mol^{-1} at temperatures below 670°C, was interpreted to be consistent with Roy and Roy's (1955) suggestion that the polymorphic transition involves first the breaking of Si-O bonds in every other tetrahedral sheet of lizardite followed by formation of new Si-O bonds with the tetrahedra pointing in the opposite direction. Cho and Fawcett (1986) suggested that such a rearrangement of Si-O bonds might be the rate-determining process in the Lz(1.0) = Cln transformation. Extreme sluggishness of this transformation at 600°C suggests that it would take unrealistically long experiments to reverse at temperatures below 300°C even if it were a stable equilibrium. It appears that the only way to assess experimentally the stability of Lz(ss) relative to chlorite is to extract thermodynamic properties for Lz(ss) from phase equilibrium data and then calculate this mineral's phase relations at low temperatures.

Because natural trioctahedral Mg-chlorites with aluminum contents x > 1.5 do not exist, it is possible that lizardites having aluminum contents ranging from 1.5 < x < 2.0 are stable.Analyses of natural lizardites reveal few examples with compositions between 1.5 < x <1.9; however, there is a cluster of natural lizardites having the amesite, Lz(2.0), composition. Experimental information for this composition is based on synthesis experiments which are inconsistent. For example, Nelson and Roy (1958) suggest that 7 Å amesite reacts to 14 Å amesite at 480°C and that the 14 Å phase is stable to 650°C at water pressures of about 1 kbar; on the other hand, Velde (1973, p. 303) suggested that 7 Å amesite is metastable above 500°C. The 14 Å amesite referred to by Roy and Roy (1954) was undoubtedly a chlorite having a composition near x = 1.5. Chernosky (unpublished data) has observed that an initially pure 7 Å amesite progressively reacts to a less aluminous Lz + Sp + an unidentified phase at temperatures above 450°C at P(H$_2$O) = 2 kbar.

STABILITY OF MG-CHLORITE

Although Nelson and Roy (1958), Velde (1973), and Fleming and Fawcett (1976) report synthesis of chlorites having alumina contents in the range 0.5 < x < 2.0, detailed experimental work (Shirozu and Momoi, 1972; Chernosky, unpublished data) suggests that the IIb trioctahedral chlorites lying on the chrysotile-amesite join have compositions restricted to 0.8 < x <1.5. Synthesis experiments using a starting material with x < 0.8 result in a more aluminous chlorite + Fo + Tc whereas experiments using a starting material with x > 1.5 result in a less aluminous chlorite + Sp. These observations are consistent with the rather limited solid solution range of natural trioctahedral chlorites (presumably IIb) reported by Shirozu (1960) and Velde (1973). It is only in the silicon-rich chlorites, which form via the substitution 0.5(MgAl^{vi}) = Si$_{0.5}$ (a vacancy on an octahedral site compensates for the increased charge resulting from the introduction of silicon on a tetrahedral site), that Velde (1973) reports aluminum contents higher than x = 1.5 (Fig. 13). However, Fransolet and Schreyer (1984) suggest that the silicon-rich chlorites might be regarded as solid solutions between trioctahedral and dioctahedral chlorites which form via the substitution 2Al = 3Mg.

Shirozu and Momoi (1972) report larger basal spacings and higher vibration frequencies in the hydroxyl-stretching region on infrared absorption spectra for synthetic Mg, Al-chlorites relative to the corresponding natural chlorites. They attribute these differences to aluminum preference for the octahedral sites in the hydroxide interlayer relative to the 2:1 layer of natural chlorites.

All synthetic chlorites which have been characterized as to structural type are "high temperature" IIb polytypes as defined by Bailey and Brown (1962); synthetic chlorites which have not been characterized as to structural type have been synthesized under conditions which suggest that they also are IIb polytypes. We have no information on the compositional and temperature ranges of the other structural types. Extensive criteria for the identification of serpentine and chlorite polytypes by X-ray powder diffraction are given by Bailey (1988).

Thirty-one equilibria involving clinochlore or a Mg-chlorite somewhat more aluminous than clinochlore have been reversed in the laboratory. Of these, 28 involve H_2O and were investigated under conditions of $P(H_2O) = P(total)$ while three involve supercritical, $H_2O + CO_2$ mixtures and were investigated under conditions of $P(H_2O + CO_2) = P(total)$.

Mg-chlorite-bearing equilibria investigated experimentally are listed in Table 2 and are plotted on Figures 15-24; equilibria considered in the consistency analysis of Berman (1988) are identified in Table 2 with an asterisk. Five of the equilibria in Table 2 involve sudoite and another five involve sapphirine; the absence of thermodynamic data for these phases combined with the variable composition of sapphirine, precludes calculation of these ten equilibrium curves.

Due to space limitations it is not feasible to discuss all 31 Mg-chlorite bearing equilibria in detail. Rather, we shall discuss some generic problems common to these equilibria and then focus on equilibria whose experimentally located positions are in dispute as well as on equilibria for which there is a discrepancy between the calculated and experimentally determined position.

Inspection of Figures 15-24 reveals a high degree of consistency among most of the experimentally located equilibria and the equilibrium curves calculated using the thermodynamic data base of Berman (1988). This degree of consistency was not achieved by Helgeson et al. (1978) largely because their estimated third law entropy of clinochlore (465.0 $Jmol^{-1}K^{-1}$) is considerably higher than the value derived (435.15) by Berman (1988) and the value determined calorimetrically (397.6) by Henderson et al. (1983). Comparison of Berman's (1988) entropy with the calorimetric value is indicative of almost complete disorder on tetrahedral and octahedral sites in the synthetic clinochlores used in phase equilibrium studies (maximum disorder = 41.4 $Jmol^{-1}K^{-1}$). The analysis of phase equilibrium data by Holland and Powell (1985) also resulted in a high value for the entropy of clinochlore (462 $Jmol^{-1}K^{-1}$), because equilibria involving both cordierite and clinochlore were not considered. Thermodynamic properties for two natural chlorites measured by Hemingway et al. (1984) and derived from solubility data on four natural chlorites by Kittrick (1982) are difficult to compare to those obtained by Berman (1988) because they apply to compositions far removed from the clinochlore endmember.

Several lines of evidence suggest that the most stable Mg-chlorite has a composition of x = 1.2 at its decomposition temperature. Fawcett and Yoder (1966) reported that the composition of Mg-chlorite in experiments conducted at water pressures above 3.5 kbar became more aluminous (x = 1.2) than clinochlore (x = 1.0) with increasing pressure and attributed this progressive increase in Al content to increasing pressure; Fleming and Fawcett (1976) confirmed this report. These observations are consistent with analyses of natural Mg-chlorites from ultramafic rocks (Frost, 1975) and dolomitic limestones (Rice, 1977; Bucher-Nurminen, 1981 and 1982) which indicate that at the highest grades the Mg-chlorite has a composition near x = 1.2.

Bird and Fawcett (1973), Chernosky (1974), and Staudigel and Schreyer (1977) were unable to detect that the Mg-chlorite in their experiments became more aluminous, although Bird and Fawcett and Staudigel and Schreyer used an aluminous chlorite (x = 1.2) in their starting materials. Chernosky used clinochlore in the starting material, but the XRD technique he used to monitor compositional variation in the chlorites apparently was not sensitive enough to detect the difference in Al content between clinochlore nucleii present in the starting material and more aluminous chlorite rimming the nucleii in the experimental products.

Jenkins and Chernosky (1986) resolved this discrepancy experimentally by demonstrating that Mg-chlorites synthesized from bulk compositions with varying Al contents at $P(H_2O) = 3$ and 14 kbar converged on the composition x = 1.2 with increasing temperature rather than increasing pressure; they also calibrated the variation in the c unit cell parameter of Mg-chlorite as a function of Al content (Fig.14). Jenkins and Chernosky (1986) observed that both synthesis and bracketing experiments conducted at temperatures above about 680°C at $P(H_2O) = 3$ kbar produced a chlorite more aluminous that clinochlore.

The data of Jenkins and Chernosky (1986) suggest that one must be very cautious in retrieving thermodynamic data from phase equilibrium data involving Mg-chlorite. Departures from clinochlore composition must be accounted for by reducing the activity of this endmember in the Mg-chlorite crystalline solution. Inspection of the synthesis conditions and the unit cell parameter, c, of Mg-chlorites used in starting materials by Bird and Fawcett (1973), Velde (1973), McOnie et al.(1975), Fleming and Fawcett (1976), and Staudigel and Schreyer (1977) reveals that the chlorites were more aluminous than clinochlore. Even if an equilibrium is reversed at temperatures below 680°C, the use of aluminous chlorite in the starting material may displace the position of an equilibrium toward higher temperatures relative to its position with stoichiometric clinochlore.

Berman (1988) found good agreement between experimentally determined and calculated positions for a number of high temperature Mg-chlorite bearing equilibria using an $a_{clinochlore}$ = 0.7 to account for this nonstoichiometry. The displaced equilibria calculated using this reduced activity for clinochlore are shown as dotted lines on Figures 15-20.

Phase Relations of Mg-chlorite with P(H₂O) = P(Total)

The upper thermal stability of Mg-chlorite is bounded by the equilibrium 5Cln = 10Fo + Cd + 3Sp + 20W for water pressures below about 3 kbar, by the equilibrium Cln = Fo + 2En + Sp + 4W for water pressures between about 3 and 16 kbar and by the equilibrium 2Cln = 3Fo + Py + Sp + 8W for water pressures above 16 kbar (Fig. 15).

The equilibrium 5Cln = 10Fo + Cd + 3Sp + 20W has been reversed experimentally by Fawcett and Yoder (1966), Chernosky (1978) and McPhail (1985); Figure 15 illustrates that the calculated equilibrium passes through the experimental brackets obtained in all of these studies. Earlier experimental work (Yoder, 1952 and Roy and Roy, 1955) on this equilibrium was based on synthesis rather than on reversed experiments and suggested a location for the equilibrium tens of degrees higher than shown on Figure 15. In calculating the position of this equilibrium, Berman (1988) explicitly accounted for the variable water content in cordierite using a model (McPhail, 1985) similiar to one used by Helgeson et al. (1978) and Newton and Wood (1979).

The equilibrium Cln = Fo + Sp + 2En + 4W has been reversed experimentally at water pressures from 3 to 18 kbar. Earlier, unreversed experimental work (Segnit, 1963) suggested that clinochlore decomposed by the equilibrium 5Cln = 7Fo + 2Tc + 5Sp + 18W at water pressures to 20 kbar. However, Fawcett and Yoder (1966) showed that the assemblage Tc + Sp was metastable. Figure 15 illustrates that the calculated equilibrium curve is consistent with the experimental data of Fawcett and Yoder (1966), Staudigel and Schreyer (1977), Jenkins (1981), and Jenkins and Chernosky (1986). In computing the position of this equilibrium, Berman (1988) used the ideal mixing model of Gasparik and Newton (1984) to model the variable solubility of Al in enstatite in equilibrium with forsterite + spinel.

The equilibrium 2Cln = 3Fo + Py + Sp + 8W has been experimentally reversed by Staudigel and Schreyer (1977) at water pressures ranging from 20 to 35 kbar (Fig. 15). The experimental brackets for this equilibrium are only consistent with the displaced equilibrium (Fig. 15, dotted line) calculated assuming that the $a_{clinochlore}$ = 0.7. The ideal mixing model of Gasparik and Newton (1984) was adopted to model the aluminum content of enstatite in equilibrium with pyrope.

The upper stability of the assemblage Cln + Qz is governed by the equilibrium 6Cln + 29Qz = 8Tc + 3Cd + 16W at water pressures below about 6 kbar and the equilibrium 3Cln + 14Qz = 5Tc + aluminosilicate + 7W at higher pressures (Fig. 16). The former equilibrium has been investigated by Fawcett and Yoder (1966), Velde (1973), Fleming and Fawcett (1976), Chernosky (1978), and Massone (unpublished data). The calculated equilibrium curve is consistent with Massone's unpublished results, all of Chernosky's (1978) brackets except the high temperature reversal at P(H₂O) = 3 kbar which lies just below the calculated curve, and the very wide brackets established using natural cordierite and

319

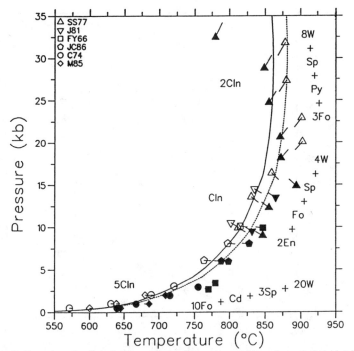

Figure 15. Comparison of computed phase equilibria with experimental data on the upper thermal stability of synthetic Mg-chlorite: FY66 - Fawcett and Yoder (1966); JC86 - Jenkins and Chernosky (1986); C74 - Chernosky (1974); M85 - McPhail (1985); J81 - Jenkins (1981); SS77 - Staudigel and Schreyer (1977). Dotted lines were calculated with $a_{clinochlore}$ = 0.70 (see text). Symbols as in Figure 2.

chlorite by Fawcett and Yoder (1966). Velde (1973) located the equilibrium about 25°C higher at 1 kbar, and Fleming and Fawcett (1976) located it 50°C higher at 2 kbar. The Mg-chlorites used as starting materials by Velde and Fleming and Fawcett were more aluminous (x = 1.2) than clinochlore and were synthesized from bulk compositions richer in SiO_2 than clinochlore. These compositional differences lead to a reduced activity for clinochlore that is appropriate to account for the 25°C displacement of Velde's data but not the 50°C displacement of Fleming and Fawcett's data.

Massone's (1981 and unpublished data) brackets above water pressures of 6 kbar for 3Cln + 14Qz = 3Ky + 5Tc + W (Fig. 16) lie at lower temperatures compared to the calculated curve with pure phases. Berman (1988), mindful of Fawcett and Yoder's (1966) observation that talc incorporates up to about 4 wt % Al_2O_3 at P(H_2O) = 10 kbar, suggested that the discrepancy illustrated on Figure 16 could be accounted for by the Al content of talc increasing at elevated pressures.

The experiments of Jenkins and Chernosky (1986) and Seifert (1974) for the equilibrium 5Cln + Cd = 20En + 7Sp + 20W (Fig. 17a) are slightly inconsistent, with the computed equilibrium curve supporting Jenkins and Chernosky's brackets. The discrepancy could be caused by a combination of much shorter experiment durations (by a factor of 5 compared to Jenkins and Chernosky) and the presence of substantial lizardite in the starting materials. Seifert's (1974) 1 kbar bracket for the equilibrium 5Cln + 20Co = 3Cd + 19Sp + 20W (Fig. 17b) lies about 60°C higher than the calculated curve, although the discrepancy diminishes with increasing temperature and vanishes at 4 kbar. Inasmuch as Seifert observed nearly complete conversion of the high temperature assemblage to Cln + Co in experiments of long duration, we can offer no explanation for the discrepancy.

The unpublished experimental brackets of Massone at water pressures of 2 and 6 kbar lie about 35°C lower than the higher pressure experimental brackets of Bird and Fawcett (1973) for the equilibrium 3Cln + 5Ms = 8Ky + 5Phl + Qz + 12W (Fig. 18b). The agreement between the calculated curve and the

320

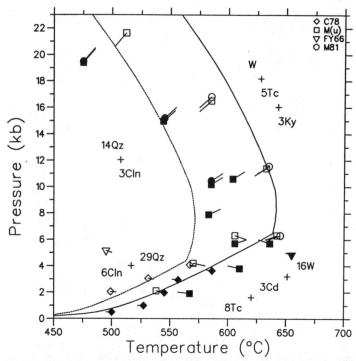

Figure 16. Comparison of computed phase equilibria with experimental data for upper stability of synthetic Mg-chlorite + quartz; C78 - Chernosky (1978); FY66- Fawcett and Yoder (1966); M81 - Massone et al. (1981); M(u) - (unpublished data). Dotted curves were calculated with $a_{talc} = 0.60$ (see text). Symbols as in Figure 2.

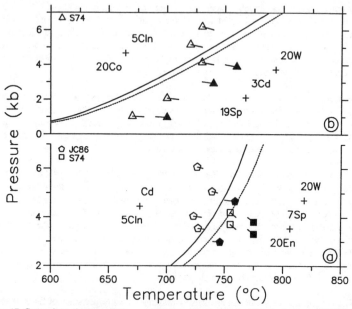

Figure 17. Comparison of computed phase equilibrium with experimental data in the system MASH: S74 - Seifert (1974); JC86 - Jenkins and Chernosky (1986). Dotted line was calculated with $a_{clinochlore} = 0.70$ (see text). Symbols as in Figure 2.

experimental data (Fig. 18b) of Bird and Fawcett (1973) must be viewed with skepticism and may be the result of cancellation of several opposing effects. First, Bird and Fawcett used corundum + quartz rather than kyanite in their starting material; use of this metastable assemblage will shift the position of the reaction by approximately 25°C to higher temperature. Second, the chlorite synthesized at 750°C by Bird and Fawcett was more aluminous than clinochlore ($x = 1.25$) and this will also shift the equilibrium to higher temperatures by about 15-20°C. Lastly, Bird and Fawcett observed that phlogopites in equilibrium with kyanite + quartz changed their compositions significantly towards eastonite. It seems likely that this latter factor accounts for the lower temperature of Massone's brackets as well as for the coincidental agreement between Bird and Fawcett's experimental data and the position of the equilibrium calculated for stoichiometric minerals by Berman (1988).

Experimental data for the ten equilibria shown in Figures 18a, 19, and 20 are thermodynamically consistent and require no further elaboration. Also in agreement are Seifert and Schreyer's (1970) half-brackets for the pyrophyllite-bearing equilibria, 6Cln + 13And + 11Prl = 15Cd + 35W and 2Cln + 8Prl = 5Cd + 13Qz + 16W, and Wang and Greenwood's (1988) data for the equilibrium Gr + Cln = 3Di + 2Sp + 4W. Five equilibria (Table 2, equilibria 30-34) bearing on the stability of sudoite have been experimentally investigated by Fransolet and Schreyer (1984); these equilibria all involve clinochlore. Due to very sluggish reaction rates, the only experimental brackets obtained by these investigators are very broad and do not constrain meaningfully the thermodynamic parameters of sudoite.

Five equilibria involving Mg-chlorite and sapphirine are listed in Table 2 and plotted on Figures 21 and 22. The reaction coefficients for these equilibria (Table 2, equilibria 15-19) were calculated assuming the 7:9:3 stoichiometry for sapphirine (Table 1), although a range in compositions has been observed in experimental products. Attempts to derive a consistent set of thermodynamic properties for sapphirine from the experimental data failed. All sapphirine bearing equilibrium curves (Figs. 21 and 22) were therefore drawn by eye.

Stability of Mg-chlorite in H_2O-CO_2 fluids

Three equilibria involving Mg-chlorite have been bracketed experimentally in supercritical, H_2O-CO_2 fluids (Fig. 23) and are thermodynamically consistent with the experimental data summarized above. The phase equilibrium data of Chernosky and Berman (1988) for the equilibrium Cln + 2Do = 3Fo + 2Cc + Sp + 4W + 2CO_2 (Fig. 23b) are not in agreement with the experimental data of Widmark (1980). The 5°C discrepancy at 1 kbar can be accounted for by Widmark's use of an aluminous Mg-chlorite ($x = 1.2$) rather than clinochlore, however the 10°C discrepancy at 2 kbar cannot be ascribed solely to differences in the composition of the Mg-chlorites used in the starting materials.

STABILITY OF FE, MG-CHLORITE

There have been numerous experimental phase equilibrium studies of equilibria involving Fe and Fe, Mg-chlorite (Table 2). Although attempts have been made to extract thermodynamic data for chamosite, the Fe-endmember chlorite, from the phase equilibrium data, there is no agreement among the derived thermodynamic values (Table 5).

Published experimental data on the stability of Fe-chlorite in the system FASH are limited to the work of Turnock (1960), Hsu (1968), and James et al. (1976). The stability of Fe, Mg-chlorites with compositions on the join chamosite-clinochlore has been investigated by McOnie et al. (1975); this study was later expanded by Fleming and Fawcett (1976) to include quartz and chlorites with compositions more aluminous than those on the join chamosite-clinochlore. Bryndzia and Scott (1987a) investigated redox equilibria involving the sulfidation and oxidation of intermediate Fe, Mg-chlorite. The latter three studies were restricted to the system FMASH.

322

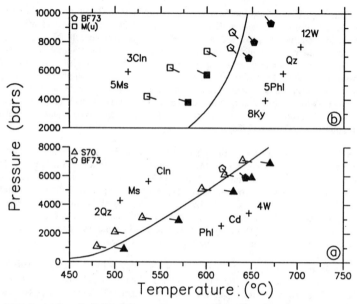

Figure 18. Comparison of computed phase equilibria with experimental data in the system KMASH: S70 - Seifert (1970); BF73 - Bird and Fawcett (1973); M(u) - Massone (unpublished data). Symbols as in Figure 2.

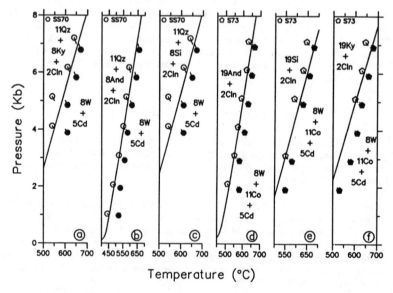

Figure 19. Comparison of computed phase equilibria with experimental data in the system MASH: SS70 - Seifert and Schreyer (1970); S73 - Seifert (1973). Symbols as in Figure 2.

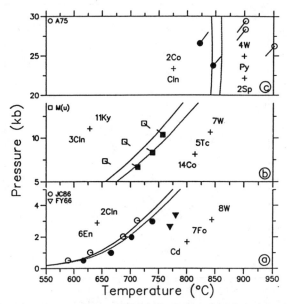

Figure 20. Comparison of computed phase equilibria with experimental data in the system MASH: FY66 - Fawcett and Yoder (1966); A75 - Ackermand et al. (1975); JC86 - Jenkins and Chernosky (1986); M(u) - Massone (unpublished data). Dotted curves were calculated with $a_{clinochlore}$ = 0.70. Symbols as in Figure 2.

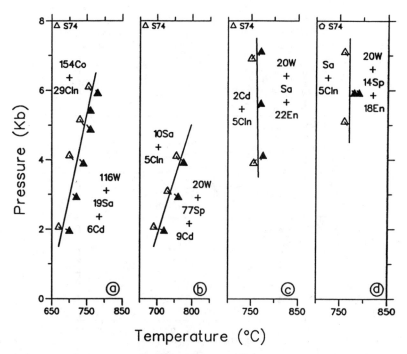

Figure 21. Experimental data bearing on the stability of sapphirine in the system MASH: S74 - Seifert (1974). Equilibria drawn by eye through the brackets. Symbols as in Figure 2.

324

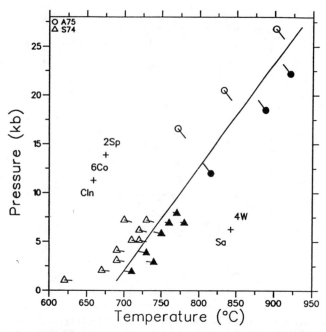

Figure 22. Experimental data bearing on the stability of sapphirine in the system MASH: S74 - Seifert (1974); A75 - Ackermand et al. (1975). Equilibria drawn by eye through the brackets. Symbols as in Figure 2.

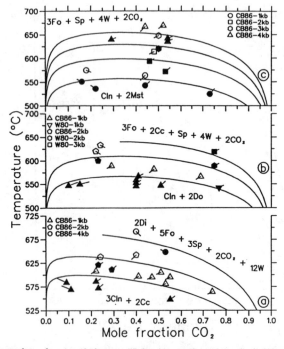

Figure 23. Comparison of computed phase equilibria with experimental data in the system MASHCO$_2$: W80 - Widmark (1980); CB86 - Chernosky and Berman (1986a in Fig. 23a and 23b; 1986b in Fig. 23c). Symbols as in Figure 2.

Table 5. Summary of thermodynamic parameters for chamosite, $Fe_5Al_2Si_3O_{10}(OH)_8$, at 298.15 K.

Formation from the elements			Reference
ΔH^o_f	ΔG^o_f	S^o	
(kJ mol^{-1})		J·mol^{-1}·K^{-1}	
-7223.05±46	-6697.33±27	594.96±54.4	Bryndzia and Scott (1987a)
-7072.99		596.22	Walshe (1986)
-7080.83[1]			" "
-7110.64[2]			" "
	-6486.46		Sverjensky (1985)
	-6560.51		Anovitz and Essene (1982)

[1]Estimated using the method of Tardy and Garrels (1974).
[2]Derived from the low-temperature, 2.0 kbar bracket of James et al. (1976).

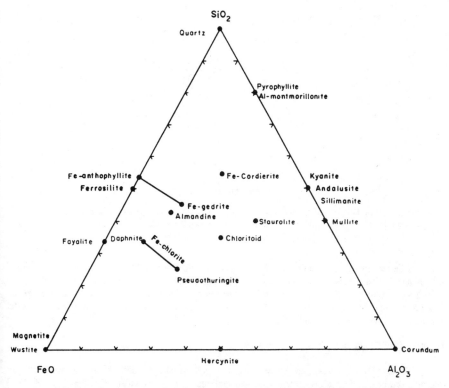

Figure 24. Compositions of phases (in mole %) in the system $FeO-Al_2O_3-SiO_2-H_2O-O_2$ projected onto the $FeO-Al_2O_3-SiO_2$ plane. Note that daphnite is chemically equivalent to chamosite. From James et al. (1976).

326

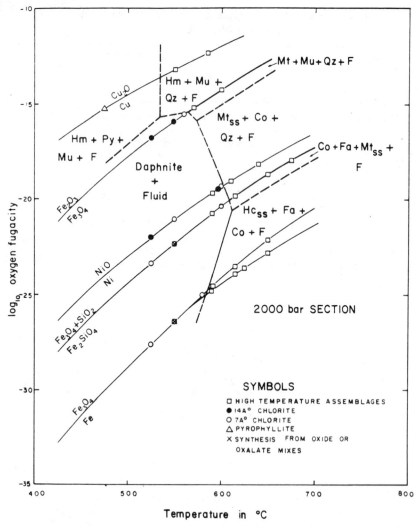

Figure 25. Log f(O₂)-T plot of chamosite ("daphnite") stability relations at P(H₂O) = 2.07 kbar. From James et al. (1976), modified after Turnock (1960), as discussed in the text.

The reaction of intermediate Fe, Mg-chlorite to cordierite in model pelitic systems, containing K_2O as an additional component, was investigated experimentally by Hirschberg and Winkler (1968) and Burnell and Rutherford (1984); Hoschek (1969) investigated the reaction of Fe, Mg-chlorite to staurolite. Apted and Liou (1983) investigated the reaction of a metastable chloritic phase to amphibole in a model basaltic system containing CaO and TiO_2.

The stability of Fe-chlorite

Turnock (1960) investigated the stability of Fe-chlorite as a function of temperature and oxygen fugacity. The Fe-chlorites in his experiments were synthesized from chamosite [$Fe_5Al_2Si_3O_{10}(OH)_8$] and "pseudothuringite" [$Fe_4Al_2Si_2O_{10}(OH)_8$] bulk compositions (Fig. 24). He reported that Fe-lizardite, which he called 7 Å Fe-chlorite, often nucleated before 14 Å Fe-chlorite and

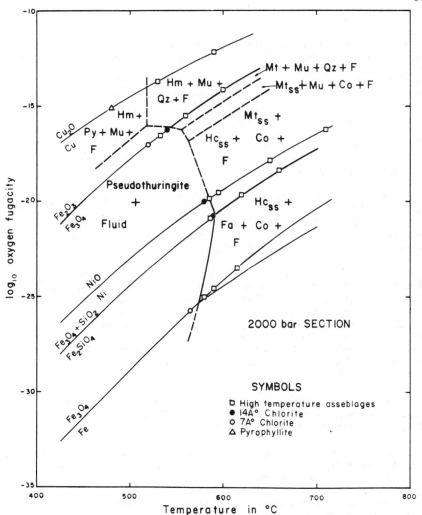

Figure 26. Log $f(O_2)$-T plot of "thuringite" stability relations at $P(H_2O)$ = 2.07 kbar. From James et al. (1976), after Turnock (1960), as discussed in the text.

persisted; he assumed that the Fe-Lz + Fe-Cln assemblage was stable. Turnock also reported that experiments conducted using starting materials of chamosite and "pseudothuringite" bulk composition often resulted in Mt + Qz in addition to Fe-chlorite (Figs. 25 and 26). At $P(H_2O)$ = 2.0 kbar, Cd + fayalite + Mt(ss) + Hr(ss) + Qz + fluid were the reported decomposition products of Fe-chlorite at oxygen fugacity conditions bounded by the iron-magnetite (IM) and magnetite-hematite (MH) oxygen fugacity buffers. At the higher oxygen fugacity conditions defined by the Cu-Cu$_2$O buffer and at temperatures less than approximately 550°C, Fe-chlorite decomposed to an assemblage of Prl + mullite(?) + hematite + fluid; above about 550°C Fe-chlorite decomposed to Co + Qz + hematite + fluid. According to James et al. (1976), the decomposition products of Fe-chlorite in Turnock's experiments at the Cu-Cu$_2$O buffer are metastable; they suggest that the stable low-temperature (<550°C) assemblage should be hematite + And + mullite, and that the stable high-temperature assemblage should be hematite + mullite + Qz. Turnock (1960) did demonstrate, however, that Fe-chlorite synthesized from a chamosite bulk composition (Fig. 25) was stable to a higher temperature than Fe-chlorite synthesized from a "pseudothuringite" bulk composition (Fig. 26). At $P(H_2O)$ = 2.0 kbar, Fe-chlorite synthesized from the chamosite bulk composition decomposed at ~610°C while Fe-chlorite synthesized from the "pseudothuringite" bulk composition decomposed at ~590°C. For both bulk compositions, Fe-chlorite was thermally most stable at oxygen fugacity conditions close to the fayalite-magnetite-quartz (FMQ) oxygen fugacity buffer (Figs. 25 and 26).

Hsu (1968) experimentally reversed the following three equilibria, with oxygen fugacities buffered by iron-quartz-fayalite (IQF), IM, FMQ and nickel-bunsenite (NNO):

$$\text{IQF: Fe-Cln} + \text{Qz} + \text{fluid} = \text{Alm} + \text{fluid} \tag{1}$$

$$\text{IM and FMQ: Qz} + \text{Fe-Cln} + \text{Mt} + \text{fluid} = \text{Alm} + \text{fluid} \tag{2}$$

$$\text{NNO: Fe-Cln} + \text{Qz} + \text{Mt} = \text{Mt(ss)} + \text{Hr(ss)} + \text{Qz} + \text{fluid} \tag{3}$$

Hsu (1968) estimated that the Fe-chlorite synthesized in his experiments had a composition lying midway between chamosite and "pseudothuringite" (i.e., $FeO:Al_2O_3:SiO_2$ between 5:1:3 and 4:2:2, see Fig. 24). In addition to synthesizing Fe-chlorite, he reported that minor amounts of metastable Fe-lizardite also formed. Metastable phases such as Fe-cordierite and fayalite were also produced in the low-pressure and low-temperature region of the almandine stability field.

The stability of Fe-chlorite determined by Hsu (1968) at the IQF and IM buffers most likely represents metastable equilibrium; an interpretation consistent with the numerous metastable phases reported in these experiments by Hsu (1968). In all probability, the IQF buffered experiments failed due to the unreliable nature of the IQF buffer which has a limited capacity to buffer oxygen and is very short lived (G. B. Skippen, pers. comm., 1985). Anovitz and Essene (1982) also concluded that the IQF datum at 528°C and 1 kbar determined by Hsu (1968) did not represent stable equilibrium. In contrast, experiments conducted by Hsu (1968) at the FMQ buffer do not appear to be plagued by metastability because the FMQ buffer is long lived, it has an excellent buffering capacity for oxygen, and the experiments were conducted at much higher temperatures.

For $P(H_2O) = 0.5$ to 3.0 kbar, $T = 517°$-610°C, and oxygen fugacities ranging from FMQ to IQF, Hsu's experiments suggest that the assemblage Alm + Qz + Fe-Cln is stable. However, according to Hsu (1968), almandine is not stable at oxidation states defined by the NNO buffer (equilibrium 3).

James et al. (1976) re-determined the stability of an Fe-chlorite intermediate in composition between chamosite and "pseudothuringite" (Fig. 24) in experiments whose oxygen fugacity was buffered by NNO. Attempted synthesis of chamosite (i.e. $FeO:Al_2O_3:SiO_2 = 5:1:3$) yielded minor quartz, fayalite, magnetite, and hercynite, in addition to Fe-chlorite. Using XRD, James et al. (1976) estimated the composition of the Fe-chlorite to be $[Fe^{2+}_{4.00}(Al,Fe^{3+})_{1.87}][Si_{2.4}]O_{10}(OH)_8$; based on the structural formula for chlorite proposed by Foster (1962).

The low-pressure (<4 kbar) and high-pressure (>4 kbar) decomposition of Fe-chlorite at $P(H_2O)$ = P(Total) and $f(O_2)$ = NNO reported by James et al. (1976) is governed by equilibria (4) and (5), respectively:

$$\text{Fe-Cln} = \text{Fe-Cd} + \text{Mt(ss)} + \text{Qz} + \text{fluid} \tag{4}$$

and

$$\text{Fe-Cln} + \text{Qz} = \text{Fe-gedrite} + \text{Mt(ss)} + \text{fluid} \tag{5}$$

The reported decomposition temperature for Fe-chlorite in equilibrium (3) is approximately 60°C below the maximum (metastable) decomposition temperature for Fe-chlorite reported by Turnock (1960) (Figs. 25 and 26). Above $P(H_2O) = 4$ kbar, Fe-Cln + Qz reacts to a gedrite-bearing assemblage with increasing temperature. Equilibrium (5) has been bracketed by James et al. (1976) at 630±10°C, 8.5±0.14 kbar and 615±15°C, 7.2±0.14 kbar with a half-bracket at 585±10°C and 4.8±0.1 kbar. Equilibria (1), (2), (3) and (4) have been plotted on Figure 27.

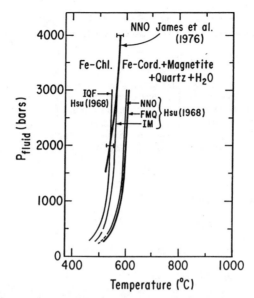

Figure 27. Stability of Fe-chlorite based on the experimental studies of James et al. (1976) and Hsu (1968). IQF, IM, FMQ, and NNO refer to the solid oxygen fugacity buffers used in the experiments. The curves labelled IQF, IM, FMQ, and NNO from Hsu (1968) refer to equilibria (1), (2), and (3) as discussed in the text; almandine formed at the expense of Fe-chlorite $\pm$ Mt $\pm$ Qz + fluid in experiments performed with the IQF, IM, and FMQ buffers. The curve labelled NNO from James et al.(1976) refers to equilibrium (4), as discussed in the text. The heavy solid lines indicate Fe-chlorite equilibria that have been reliably located. Light lines indicate metastable phase relations, as discussed in the text.

The curve for equilibrium (3) plots ~5°C below the curve defining the stability of Fe-chlorite established by Hsu (1968) for oxidation conditions defined by the FMQ buffer, and approximately 50°C higher than the temperature at which an Fe-chlorite of similar bulk composition, and under the same experimental conditions, was observed to decompose by James et al. (1976). They argued that the products of equilibrium (3) are metastable and should be Fe-Cd + Mt + Qz. This is in accord with the results of Fleming and Fawcett (1976) who showed that at $P(H_2O) = 2.0$ kbar and $f(O_2) =$ NNO, Fe-rich chlorite + Qz decomposed to a cordierite-bearing assemblage at $590\pm10°C$.

The Fe^{3+} content of the Fe-chlorites synthesized by Hsu (1968) and James et al. (1976) is unknown. Preliminary Mössbauer measurements reported by Bryndzia and Scott (1987a) on synthetic chlorite annealed at oxygen fugacities ranging from NNO to HM indicate that Fe-chlorite only contains detectable Fe^{3+} at the highest oxygen fugacity conditions (<5% at 600°C, $f(O_2) =$ HM). This suggests that the Fe-chlorite produced in the experiments of Hsu (1968) and James et al. (1976) probably contained only minor Fe^{3+}.

Bryndzia and Scott (1987b) have shown that oxygen fugacity has a profound effect on the stability of Fe, Mg-chlorite. The higher oxygen fugacity imposed by the NNO buffer should *lower* the thermal stability of Fe, Mg-chlorite relative to the FMQ buffer. In this context, the stabilities of Fe-chlorite (of similar bulk composition) determined by Hsu (1968) and James et al. (1976) at the NNO buffer (Fig. 27) do plot at lower temperatures relative to the FMQ buffer, as would be predicted.

The disposition of the two univariant curves defining the stability of Fe-chlorite with approximately the same bulk composition, but at two different oxygen fugacities (NNO; James et al., 1976 and FMQ; Hsu, 1968) suggests that almandine may not be stable relative to Fe-cordierite at the oxygen fugacity conditions of the NNO buffer. If this is true, it has important implications for the formation of both cordierite and almandine in natural parageneses. For example, in sediments of appropriate bulk composition metamorphosed at low pressures ($P(H_2O) < 3$ kbar) and moderate temperatures (550°-600°C) such as might occur in a contact aureole, the presence or absence of graphite may very well be the critical factor which determines whether almandine or cordierite will form at the expense of Fe, Mg-chlorite.

The stability of intermediate Fe, Mg-chlorite

McOnie et al. (1975) investigated the stability of intermediate Fe, Mg-chlorites in the system FMASH, with compositions on the join chamosite-clinochlore (Fig. 28). Their experiments were conducted at $P(H_2O)$ = 2.07 kbar with oxygen fugacities buffered by NNO and without excess quartz. They used stoichiometric mixes of oxides as starting materials, and in some cases, the crystalline products of prior experiments. McOnie et al. (1975) reported that the Fe, Mg-chlorites decomposed to the following five assemblages (Fig. 29), in order of increasing Fe/(Mg+Fe) of the starting material:

(a) Fe, Mg-Cln = olivine + Cd + Sp + W,

(b) Fe, Mg-Cln = olivine + Co + Hr + W,

(c) Fe, Mg-Cln + O_2 = olivine + Co + Mt + W,

(d) Fe, Mg-Cln + O_2 = Co + orthoamphibole + Mt + W, and

(e) Fe, Mg-Cln + O_2 = Co + Qz + Mt + W.

Two important observations may be made regarding Figure 29. First, in this system, quartz is not a reactant, and the thermal stability of Fe, Mg-chlorite is a function only of the Fe/(Fe+Mg) ratio in the chlorite, with Mg-rich chlorite being stable at the highest temperatures. Second, cordierite, and not almandine, was the observed decomposition product of the Fe-rich chlorite close to the chamosite bulk composition, consistent with the results of Turnock (1960) and James et al. (1976).

Fleming and Fawcett (1976) determined experimentally the thermal stability of Mg and Fe, Mg-Cln + Qz bearing assemblages in the system FMASH, at $P(H_2O)$ = 2.07 kbar and $f(O_2)$ = NNO, using oxide mixes and crystalline run products as starting materials. Bulk compositions used by these investigators lie on the anthophyllite-gedrite-ferroanthophyllite-ferrogedrite plane (Fig. 28), and intersect Fe, Mg-Cln + Qz tie lines over a wide range of aluminum contents and Fe/(Fe+Mg) ratios. Fleming and Fawcett's (1976) results are summarized in Figures 30, 31, and 32. Figure 30 shows the stability of Mg-Cln + Qz in the iron-free system; Figure 31 shows the stability of Fe, Mg-Cln + Qz for bulk compositions having a constant Fe/(Fe+Mg) ratio of 0.5 and a variable Al/(Al+Si) ratio. Figure 32 shows the stability of Fe, Mg-Cln + Qz for bulk compositions having variable Fe/(Fe+Mg) ratios and a constant Al/Si ratio.

Fleming and Fawcett (1976) report that in the iron-free system (Fig. 30), the stable assemblages below 590±10°C are, with increasing alumina content, Tc + Cln + Qz, Cln + Qz, Cd + Cln + Qz; above this temperature the stable assemblage is Cd + Tc + Cln (i.e. quartz is absent). For bulk compositions with Fe/(Fe+Mg) = 0.5 (Fig. 31), the assemblages stable below 595±10°C are, with increasing alumina content, Tc + Mt + Fe, Mg-Cln + Qz, Mt + Fe, Mg-Cln + Qz, Fe, Mg-Cln + Qz, Cd + Fe, Mg-Cln + Qz; above about 600°C, the assemblage orthoamphibole + Cd + Tc + Mt is stable and chlorite is absent. Their experiments show that the thermal stability of chlorite + quartz is independent of both the alumina content (Fig. 31) and the Fe/(Fe+Mg) ratio (Fig. 32) of the starting material.

It is interesting to compare the results for chlorite stability in the quartz-free system as determined by McOnie et al. (1975) with the results of Fleming and Fawcett (1976) in the quartz-bearing system. The most striking difference is that at $P(H_2O)$ = 2.07 kbar the maximum thermal stability of pure Mg-chlorite is reduced by approximately 175°C by the presence of quartz (compare Figs. 15 and 16 with Figs. 29 and 32), while Turnock (1960), Hsu (1968), and James et al. (1976) showed that Fe-chlorite can coexist with quartz at its upper limit of stability. Fe-chlorite reacts to produce a quartz-saturated assemblage (e.g. equilibrium 4), while Mg-chlorite reacts to produce assemblages undersaturated with respect to quartz.

331

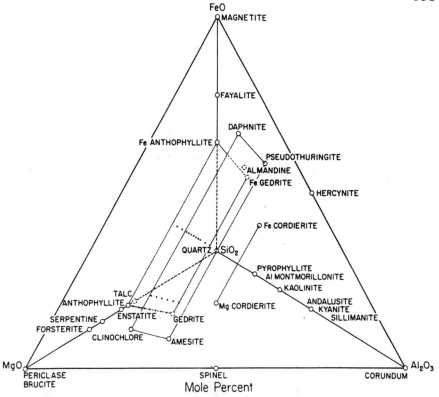

Figure 28. Compositions of phases in the system MgO-FeO-Al₂O₃-SiO₂-H₂O projected onto the H₂O-free tetrahedron. The line for cordierite and planes for orthorhombic amphibole and chlorite represent solid-solution ranges. Dots represent compositions of starting materials in the experiments of Fleming and Fawcett (1976). Note that daphnite is equivalent to chamosite. From Fleming and Fawcett (1976).

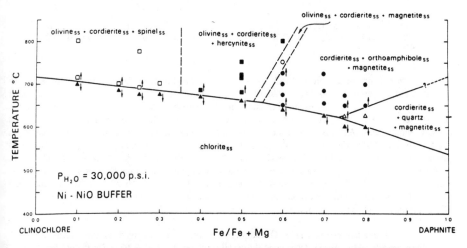

Figure 29. T-X section at P(H₂O) = 2.07 kbar, f(O₂) = NNO depicting the stability of Fe, Mg-chlorite on the clinochlore-chamosite join, $Mg_5Al_2Si_3O_{10}(OH)_8$-$Fe_5Al_2Si_3O_{10}(OH)_8$. Arrows indicate reaction direction. From McOnie et al. (1975).

332

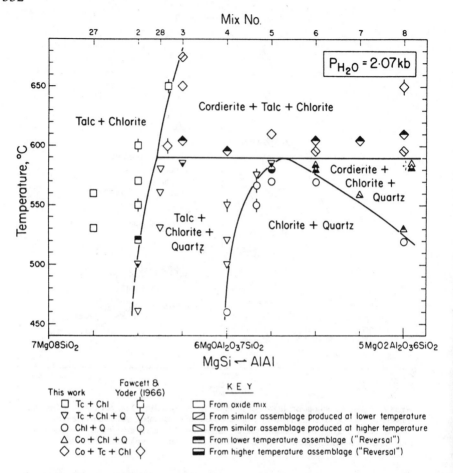

Figure 30. T-X section at P(H₂O) = 2.07 kbar, f(O₂) = NNO depicting the stability of Mg-chlorite + quartz along the 7MgO·8SiO₂·2H₂O - 5MgO·2Al₂O₃·6SiO₂·2H₂O join (see Fig. 28). From Fleming and Fawcett (1976).

The experimental results of McOnie et al. (1975) and Fleming and Fawcett (1976) may be used to constrain the temperatures of contact metamorphism of rocks of appropriate bulk composition. The almost monomineralic chloritic rocks common beneath some of the base metal sulfide deposits that occur in the Rouyn-Noranda area of Quebec have bulk compositions similar to those in the above mentioned experimental studies. As a result of isochemical contact metamorphism, some of the Fe-chlorite-rich rocks have been converted to Cd + anthophyllite + Fe-chlorite + Mt ± quartz assemblages. McOnie et al. (1975) showed that this assemblage had approximately the same bulk composition as their assemblage (d), discussed previously. They inferred that temperatures of metamorphism of the chloritic rocks were in the range 620° to 660°C, if f(O₂) conditions were approximately those of the NNO buffer. Fleming and Fawcett (1976) suggest that contact metamorphic temperatures were closer to 600°C, since Fe-chlorite + Qz occur together in some of the rocks.

The additional component K₂O introduces biotite and muscovite to the system FMASH. Hirschberg and Winkler (1968) determined the stability of Fe, Mg-Cln + Ms + Qz under conditions of P(H₂O) = P(Total), using natural minerals as starting materials. The Fe/(Fe+Mg) ratio of the starting material was obtained by mixing two natural chlorites, one with Fe/(Fe + Mg) = 0.8 and the other nearly

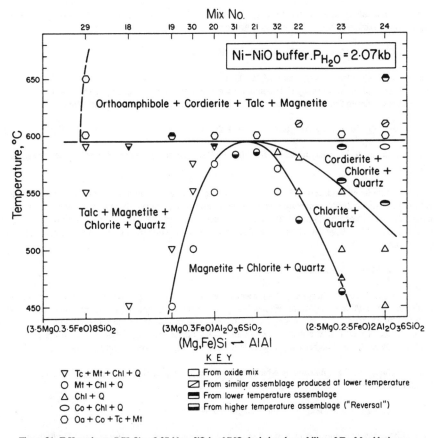

Figure 31. T-X section at P(H₂O) = 2.07 kbar, f(O₂) = NNO depicting the stability of Fe, Mg-chlorite + quartz along the join 3.5MgO·3.5FeO·8SiO₂·2H₂O - 2.5MgO·2.5FeO·2Al₂O₃·6SiO₂·2H₂O (see Fig. 28). The chlorites have fixed Fe/Fe+Mg = 0.5 but variable Al/Si. From Fleming and Fawcett (1976).

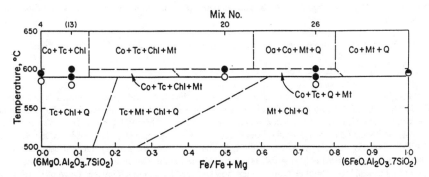

Figure 32. Schematic T-X section at P(H₂O) = 2.07 kbar, f(O₂) = NNO depicting the stability of Fe, Mg-chlorite + quartz along the 6MgO·Al₂O₃·7SiO₂ - 6FeO·Al₂O₃·7SiO₂ join which connects orthorhombic amphiboles of intermediate aluminum content (see Fig. 28). The chlorites have variable Fe/Fe+Mg but constant Al/Si ratios. Circles represent assemblages of Fe, Mg-chlorite + quartz interpreted as being stable; dots represent assemblages in which this assemblage was interpreted as being unstable. From Fleming and Fawcett (1976).

pure Mg-clinochlore. The oxygen fugacity was not specified but was probably buffered by the pressure vessels (approximately NNO). For the equilibrium

$$\text{Chlorite} + \text{Ms} + \text{Qz} = \text{Cd} + \text{biotite} + \text{Al}_2\text{SiO}_5 + \text{W} \tag{6}$$

brackets for both chlorites were obtained at $505\pm10°C$, 0.5 kbar; $513\pm10°C$, 1.0 kbar; $527\pm10°C$, 4.0 kbar. At 6.0 kbar, with a starting material containing the iron-rich chlorite (i.e. Fe/(Fe+Mg) = 0.8), almandine rather than cordierite formed according to the following equilibrium which was not reversed:

$$\text{Fe, Mg-Cln} + \text{Ms} + \text{Qz} = \text{biotite} + \text{Alm} + \text{Si} + \text{W}. \tag{7}$$

For the bulk composition having an Fe/(Fe+Mg) = 0.4, almandine did not crystallize at water pressures to 7 kbar.

Burnell and Rutherford (1984) also re-investigated equilibria involving chlorite + muscovite using natural quartz, andalusite, sillimanite, and synthetic muscovite, chlorite, biotite, and cordierite as starting materials. Their experiments were conducted under conditions of $P(H_2O) = P(Total)$ with oxygen fugacity buffered by FMQ. Mg-rich chlorite breaks down in metapelitic rocks according to the continuous equilibrium:

$$\text{Fe, Mg-Cln} + \text{Ms} = \text{biotite} + \text{Cd} + \text{W}. \tag{8}$$

Burnell and Rutherford (1984) reversed equilibrium (8) by growing chlorite from biotite + cordierite. The Fe/(Fe + Mg) ratios of co-existing biotite and cordierite in the experimental products were obtained by electron probe microanalysis, but the co-existing chlorites were too fine-grained to analyze. Their results showed that at a given pressure and $f(O_2)$, intermediate Fe, Mg-chlorite, in the presence of muscovite, was stable to higher temperatures than either the Mg or Fe-rich chlorites (Fig. 33).

The terminal equilibrium by which chlorite disappears in metapelitic rocks is (Burnell and Rutherford , 1984):

$$\text{Fe, Mg-Cln} + \text{Ms} = \text{Cd} + \text{biotite} + \text{Al}_2\text{SiO}_5 + \text{W}. \tag{9}$$

To facilitate reaction and demonstrate reversibility for equilibrium (9), Burnell and Rutherford (1984) seeded their experiments with either sillimanite or andalusite and noted that identical run products were obtained regardless of which Al_2SiO_5 phase was used as a seed. They report reversals for equilibrium (9) at $640\pm10°C$, 4.0 kbar; $632\pm7°C$, 3.0 kbar and at $612\pm12°C$, 2.0 kbar (Fig. 33). Their results show that at 2.0 and 4.0 kbar, the 4-phase assemblage Fe, Mg-Cln + biotite + Co + Al_2SiO_5 is stable approximately 85° to 95°C *higher* than the analogous quartz-bearing assemblage (equilibrium 6) reported by Hirschberg and Winkler (1968).

Hoschek (1969) determined the stability of staurolite by reversing the following equilibrium:

$$\text{Fe, Mg-Cln} + \text{Ms} = \text{staurolite} + \text{biotite} + \text{Qz} + \text{W} \tag{10}$$

with $P(H_2O) = P(Total)$ and oxygen fugacity buffered by FMQ. Natural minerals, mixed in proportions which yielded a bulk compostion with Fe/(Fe+Mg) = 0.6, were used as starting materials. To achieve this bulk composition Hoschek (1969) mixed a natural Fe-rich chlorite having Fe/(Fe+Mg) = 0.82 with a Mg-rich chlorite having Fe/(Fe+Mg) = 0.41. Equilibrium (10) was bracketed at 7 kbar, $565\pm15°C$ and 4.0 kbar, $540\pm15°C$ (see Fig 35 below). Although the bulk composition of the starting material used to reverse equilibrium (10) was Fe/(Fe+Mg) = 0.6, no compositional data are given for the Fe, Mg-chlorite in the equilibrium assemblage. Hoschek (1969) notes, however, that since a much broader range of Fe/(Fe+Mg) values prevails in rocks of pelitic composition, the assemblage Fe, Mg-Cln + Ms + staurolite + biotite + Qz should be stable over a small temperature interval.

Apted and Liou (1983) investigated the stability of a metastable intermediate Fe, Mg-chloritic phase in a model basaltic system containing CaO and TiO_2. They established limiting conditions for the

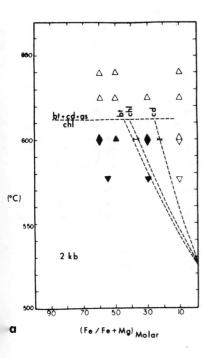

a

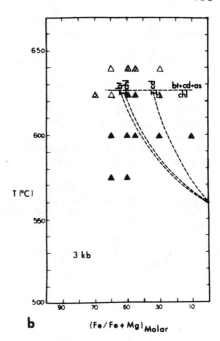

b

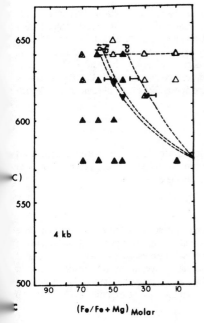

c

Figure 33. T-X sections at $P(H_2O)$ = 2, 3, and 4 kbar, $f(O_2)$ = FMQ depicting the continuous equilibrium Cln + Ms = Biotite + Cd. Triangles indicate direction from which equilibrium was approached. Triangles pointing up contained chlorite in the starting material whereas those pointing down did not contain chlorite. Open triangles indicate that chlorite was not observed while solid triangles indicate that chlorite was observed in the experimental products. The positions of the loops are based on analyzed biotite and/or cordierite compositions. At each pressure, the loop is terminated at high temperature by the terminal equilibrium Cln + Ms = Biotite + Cd + Aluminosilicate. From Burnell and Rutherford (1984).

greenschist to amphibolite facies transition, which, in nature, is marked by the decomposition of chlorite to amphibole via the continuous reaction:

Fe, Mg-Cln + actinolite + epidote + albite + Qz + sphene =

aluminous amphibole + ilmenite + W. (11)

They used a natural basalt glass as a starting material and conducted experiments at P(Fluid) = P(Total) with oxygen fugacities buffered by either FMQ, NNO, or HM. The chloritic phase synthesized by Apted and Liou (1983) could not be properly characterized. They noted that it may have been metastable, relative to a 14 Å chlorite, but due to its metastable persistence they used it to model chlorite stability in their experiments. Reaction of the chloritic phase in their experiments was monitored using X-ray diffraction.

Apted and Liou (1983) observed that with increasing oxygen fugacity the chloritic phase synthesized from the natural basaltic glass breaks down at progressively lower temperatures; for example, at 7.0 kbar, the chloritic phase decomposed at about 535°C for MH, at 550°C for NNO, and at 560°C for FMQ.

Bryndzia and Scott (1987a) studied the following oxidation and sulfidation reactions involving the chamosite component of chlorite (written in terms of a single iron atom in the chlorite formula):

$$FeAl_{2/5}Si_{3/5}O_2(OH)_{8/5} \text{ (in Fe, Mg-chlorite)} + 1/6O_2 =$$

$$1/5Al_2SiO_5 + 1/3Fe_3O_4 + 2/5SiO_2 + 4/5H_2O \qquad (12)$$

and

$$FeAl_{2/5}Si_{3/5}O_2(OH)_{8/5} \text{ (in Fe, Mg-chlorite)} + 1/2S_2 =$$

$$FeS \text{ (in pyrrhotite)} + 1/5Al_2SiO_5 + 2/5SiO_2 + 4/5H_2O + 1/2O_2. \qquad (13)$$

The experiments were conducted using well characterized natural silicates and synthetic oxide and sulfide phases as starting materials. Oxygen and sulfur fugacities were buffered by the assemblages pyrite-pyrrhotite-magnetite (PPM) and pyrrhotite-magnetite (PM). The experiments were seeded with an Al_2SiO_5 polymorph and employed a triple layer arrangement in double sealed gold capsules (Bryndzia and Scott, 1987a; Fig. 1). Reversals for equilibria (12) and (13) were obtained for $P(H_2O) = P(Total) = 2.07$-$6.0$ kbar, and 575-625°C; with $Fe/(Mg^{2+} + Fe^{2+} + Mn^{2+})$ in Fe, Mg-chlorite ranging from 0 to 0.4. Problems with metastability in their experiments were minor and limited to the appearance of talc in some of the Mg-rich bulk compositions (Bryndzia and Scott, 1987a, Table 1).

Thermodynamic analysis

Based on electron microprobe analyses of their chlorite experimental products, Bryndzia and Scott (1987a) were able to evaluate activity-composition relationships for chlorite. They derived a one-site chlorite solution model for calculating chamosite activity, given by the expression:

$$a_{cham} = \{36/5[Fe^{2+}/6][Al(Oct)/6]\}^{1/5}$$

with the factor 36/5 arising in order that $a_{cham} = 1$ for stoichiometric chamosite. Iron and aluminum were assumed to mix randomly over available octahedral sites and mixing of silicon and aluminum on tetrahedral sites was ignored because it is coupled by Tschermack's exchange with Al^{vi}. Activities of chamosite calculated from the model are equivalent, within analytical error, to the mole fraction of iron in the chlorite formula, suggesting that solution of the chamosite component in chlorite may be taken to be ideal (i.e. $a = X_{cham}$) for the range of chlorite compositions encompassed by the experiments ($Fe/ \Sigma R^{2+} = 0$ to 0.4).

The equilibrium constant for reaction (12) is given by

$$\log K(12) = 4/5 \log f(H_2O) - 1/6\log f(O_2) - \log a_{cham} \qquad (14)$$

Table 6 contains values of log K(12,1bar) calculated from the experimental results of Bryndzia and Scott (1987a, Table 3) and the expression:

$$\log K(1 \text{ bar},T) = \log K(P,T) + \Delta V_s{}^{\bullet}(P-1)/2.303 \text{ RT} \qquad (15)$$

where $\Delta V_s{}^{\bullet}$ (= -8.757 J bar^{-1}) is the molar volume of the solid phases in the equilibrium at 298.15 K and 1 bar taken from Hutcheon (1979, Table 2); R is the gas constant, and log K(P,T) is the equilibrium constant for the reaction at the pressures and temperatures given in Table 6. The following equilibrium constants for reaction (12) were determined from the data in Table 6:

$10^3/T(K)$	log K(12;T,1bar)
1.113	5.903
1.145	5.862 $\pm$ 0.042 ($\pm$1s, n = 6)
1.179	5.768

These equilibrium constants have been plotted in Figure 34.

Figure 34 shows that Bryndzia and Scott's (1978a) data plot in a linear array, consistent with equilibrium being established in their experiments. Most of their experiments were conducted at 600°C with a half bracket at 625°C and a bracket at 575°C; these data allow a reliable estimate of the equilibrium constant. The heavy solid line in Figure 34 is a least squares fit to the data of Bryndzia and Scott (1987a, Figure 6); the regression equation is given by:

$$\log K(12) = 40.505 - 9.870 \text{ x } 10^3/T \qquad (16)$$

Bryndzia and Scott (1987a) estimated the thermodynamic parameters of chamosite (Table 5) based on the least squares fit illustrated in Figure 34. Walshe's (1986) estimate of the absolute standard state entropy of chamosite (596.22 J mol^{-1}K^{-1}) is virtually identical to the estimate of Bryndzia and Scott (1987a) (594.97$\pm$54.4 J mol^{-1}K^{-1}). The thermodynamic data given in Table 5 indicate that there is no agreement among the published values for the enthalpy of formation of chamosite. This may be due partly to the fact that some of the published estimates were based on chlorite with an assumed, but unknown composition. For example, the low-temperature, 2 kbar bracket of James et al. (1976) and the estimated chlorite composition was used by Walshe (1986) to estimate the enthalpy of formation of chamosite (-7110.64 kJ mol^{-1}); this value differs significantly from the value estimated by Bryndzia and Scott (1987a) (-7223.05 kJ mol^{-1}) using equation (16). The composition of the Fe, Mg-chlorite in the experiments of James et al. (1976) is not well known and can lie anywhere along the join chamosite-"pseudothuringite" (Fig. 24).

At the present time, the reported discrepancies in the free energy and enthalpy data for chamosite obtained from reversed experiments (Bryndzia and Scott, 1987a) and from various approximation procedures (Walshe, 1986; Sverjensky, 1985) remain unresolved.

Geological application of chlorite redox equilibria

Fe, Mg-chlorite is very common in many sulfide deposits, occurring in rocks of widespread bulk composition and metamorphic grade. Using the experimental data in Figure 34 and equations (12), (14), and (16), it is possible to estimate peak metamorphic pressures and temperatures for compatible chlorite-quartz-aluminosilicate-sulfide-oxide assemblages, similar to those which occur in amphibolite-grade rocks associated with the metamorphosed Cu-Zn sulfide ore bodies at the Stall Lake and Anderson Lake mines of the Snow Lake area, northern Manitoba. The geology and metamorphism of the Snow Lake area has been

338

Table 6. Summary of Fe-Mg chlorite equilibrium from Bryndzia and Scott (1987a).

T(^{0}C)	P$_{Total}$(bars)	(1) log f$_{H_2O}$	(2) log fo$_2$	(3) log a$_{cham}$	(4) $\dfrac{\Delta V_s{}^{o}(P-1)}{2.303\ RT}$	(5) log K$_{(1bar)}$
600	2070	3.03	-17.74	-0.592	-0.113	5.860
600	2070	3.03	-17.90	-0.572	-0.113	5.866
600	2070	3.03	-16.44	-0.863	-0.113	5.914
600	6000	3.60	-16.41	-0.539	-0.327	5.827
600	6000	3.60	-17.05	-0.400	-0.327	5.795
575	4500	3.37	-17.23	-0.453	-0.253	5.768
600	5000	3.47	-16.48	-0.660	-0.273	5.910
625	5500	3.57	-15.80	-0.706	-0.292	5.903

[1] Burnham, Holloway, and Davis (1969).
[2] Bryndzia and Scott (1987a, table 3).
[3] a=X (Bryndzia and Scott, 1987a; table 3).
[4] $\Delta V_s{}^{o}$ = -8.757 j·bar^{-1}, obtained from molar volume data discussed in the text.
[5] Equilibrium constant at 1bar and T, for chamosite with a single iron atom in its formula.

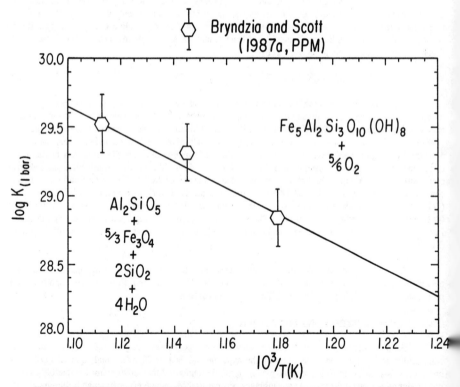

Figure 34. Log K(12,1bar) vs. 1000/T(K) diagram summarizing the Fe-chlorite experiments of Bryndzia and Scott (1987a) written in terms of equilibrium (12), as discussed in the text. Error bars in log K are based on the 600°C datum of Bryndzia and Scott (1987a). PPM refers to the solid oxygen fugacity buffer used in the experiments. The heavy solid line is a least squares fit to the data of Bryndzia and Scott (1987a), as explained in the text.

discussed elsewhere (Bryndzia and Scott, 1987b) and only details pertaining to Fe, Mg-chlorite will be given here.

The "biotite-sillimanite" isograd (Froese and Gasparrini, 1975) is based on the equilibrium:

$$Fe, Mg\text{-}Cln + staurolite + Ms + Qz = biotite + Si + W \qquad (17)$$

The isograd assemblage chlorite-muscovite-biotite-staurolite-Al_2SiO_5 is common and is usually associated with pyrite + pyrrhotite $\pm$ magnetite (PPM) (Hutcheon, 1979; Froese and Moore, 1980).

Fe, Mg-chlorite co-existing with quartz and aluminosilicate has a unique composition under conditions of fixed pressure [P(H_2O) = P(Total)] and temperature if the oxygen fugacity is buffered (Bryndzia and Scott, 1987b). Under such conditions, chlorite stability may be represented by a redox equilibrium such as (12), for which the equilibrium constant is given by 1/5 x equation (16). Chlorite isopleths may therefore be plotted on a pressure-temperature diagram (Fig. 35) by solving equation (14) for the activity of chamosite at a given pressure, temperature, and oxygen fugacity.

Figure 35 shows the calculated chlorite and measured sphalerite isopleths as a function of pressure and temperature. Isopleths of the iron content of sphalerite buffered by pyrite and pyrrhotite were determined from equation (1) of Hutchinson and Scott (1981).

Chlorite compositions from the Stall Lake and Anderson Lake mines are summarized in Bryndzia and Scott (1987b, Table 1) and were obtained from assemblages which contained quartz, aluminosilicate, and pyrite + pyrrhotite + magnetite. The mole fractions of iron in chlorites from the two deposits have very small deviations from their average values suggesting that both oxygen and sulfur fugacities were indeed buffered by the PPM assemblage (Anderson Lake, x = 0.41 $\pm$ 0.03, n = 3; Stall Lake, x = 0.35 $\pm$ 0.02, n = 8).

Based on the presence of kyanite in the footwall rocks of the Anderson Lake deposit and sillimanite in the Stall Lake deposit, the metamorphic pressure and temperature may be determined from the intersection of the shaded region in Figure 35 with Holdaway's (1971) kyanite-sillimanite phase boundary. The preferred values for the peak metamorphic temperature and pressure are 580°C and 5.6 kbar. Chlorite isopleths in Figure 35 are very steep and, therefore, sensitive to variations in temperature.

Also plotted for reference in Figure 35 is the experimentally determined stability of staurolite from Hoschek (1969). This equilibrium establishes a useful lower pressure-temperature boundary for metamorphic conditions in the Snow Lake area since both staurolite and biotite are ubiquitous in the chlorite-biotite-staurolite and biotite-staurolite-sillimanite zones in which the two deposits occur. As noted by Froese and Gasparrini (1975), the staurolite-biotite isograd reaction for the Snow Lake area:

$$Fe, Mg\text{-}Cln + Alm + Ms = biotite + staurolite + Qz + W \qquad (18)$$

must lie to the right of Hoschek's staurolite forming reaction in Figure 35.

To illustrate quantitatively the effect of oxygen fugacity on the stability of Fe, Mg-chlorite, the position of the 0.35 isopleth has been calculated for oxygen fugacity conditions defined by the HM and FMQ buffers which lie at higher and lower oxygen fugacities, respectively, relative to the PPM buffer (Fig. 35). It is apparent from Figure 35 that the oxygen fugacity (at fixed P and T) has a pronounced effect on the stability of Fe, Mg-chlorite, as indicated by the relative position of the 0.35-chlorite isopleths. Relative to the PPM buffer, higher oxygen fugacities such as those at the HM buffer decrease the stability of Fe, Mg-chlorite by approximately 55°C, whereas lower oxygen fugacities such as those defined by the FMQ buffer increase the relative stability of Fe, Mg-chlorite by a similar amount. These results are in accord with the experimental data of Ganguly and Newton (1968) and Apted and Liou (1983) for redox equilibria involving chloritoid and a chloritic phase, respectively.

340

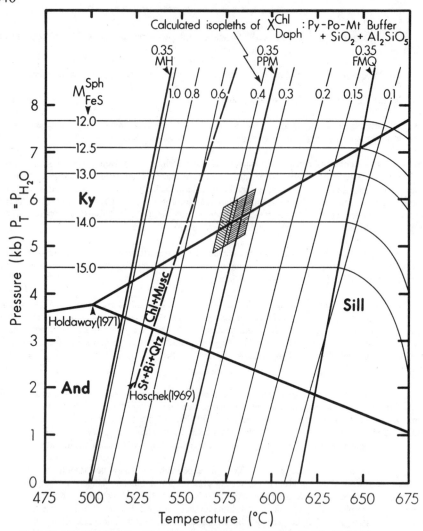

Figure 35. Calculated isopleths of the chamosite (daphnite) component in Fe, Mg-chlorite ($X_{chamosite}$ = Fe/ΣR^{2+}) and iron content of sphalerite in mole percent FeS (M_{FeS}) in equilibrium with pyrite-pyrrhotite-magnetite (PPM) as a function of temperature and pressure with P(H_2O) = P(Total). Aluminosilicate phase relations are from Holdaway (1971). Staurolite stability is from Hoschek (1969), as discussed in the text. MH, PPM, and FMQ refer to the position of the 0.35-chlorite isopleth with oxygen fugacities buffered by magnetite-hematite, pyrite-pyrrhotite-magnetite, and fayalite-magnetite-quartz, respectively. The shaded area represents conditions of metamorphism for chlorite + quartz + aluminosilicate + PPM assemblages from the Anderson and Stall Lakes mines.

In addition to the sensitivity of $X_{chamosite}$ in Fe, Mg-chlorite to f(O_2) and f(S_2), the composition of the fluid phase must also be considered. For example, with P(H_2O) = 0.5P(Total), chlorite isopleths would be lowered by approximately 35°C on Figure 35, i.e. in the same direction as an increase in the oxygen fugacity (Bryndzia and Scott, 1987a,b). In studies of natural parageneses, therefore, it is important to recognize that the stability of Fe, Mg-chlorite may be influenced by the presence of graphite or carbonate which indicate that the a(H_2O) in the fluid phase was less than unity. The present example serves to demonstrate the utility of P - T - $X_{chamosite}$ diagrams for geothermometry and geobarometry, particularly when combined with other experimentally reversed equilibria defining isograd reactions on a petrogenetic grid.

SUGGESTIONS FOR FUTURE WORK

A number of the unanswered questions raised in this review can be addressed by additional phase equilibrium work. Although equilibria involving serpentine minerals are plagued by very sluggish reaction rates and equilibria involving Fe-Mg chlorite require careful monitoring of compositional changes attending hydrothermal treatment, the following experiments should prove fruitful:

1. Experimentally reverse the Chr = Atg + Br equilibrium.

2. Experimentally investigate the stabilities of the chrysotile polytypes $2M_{cl}$ and $2Or$.

3. Perform additional experimental work on aluminous lizardite and chlorite crystalline solutions over a range of bulk compositions with the aim of deducing activity-composition relations. Attempt to locate experimentally the equilibrium Lz(ss) = Cln + Fo + Atg + W. These data will enable calculation of refined thermodynamic parameters for lizardite which will, in turn, permit quantitative assessment of its phase relations.

4. Extend the limited range of chlorite activity-composition relationships measured by Bryndzia and Scott (1987a) and investigate the effects of the Tschermak coupled substitution on Fe, Mg-chlorite.

5. Experimentally calibrate the chlorite-biotite geothermometer, or alternatively, experimentally calibrate activity-composition relations for chlorite and biotite crystalline solutions and calculate the geothermometric relations.

ACKNOWLEDGMENTS

Extensive discussions concerning serpentine phase relations with D.S. O'Hanley and Fe, Mg-chlorite phase relations with R.C. Newton are gratefully acknowledged. Reviews of portions of the manuscript by S.W. Bailey, Bear McPhail, R.C. Newton, and D.S. O'Hanley led to many improvements and are appreciated. LTB acknowledges partial support from NSF Grant CHE-8700937 awarded to O.J. Kleppa and use of the Central Facilities of the University of Chicago Materials Research Laboratory.

REFERENCES

Ackermand, D., Seifert, F. and Schreyer, W. (1975) Instability of sapphirine at high pressures. Contrib. Mineral. Petrol. 50, 79-92.

Anovitz, L.M. and Essene, E.J. (1982) Phase equilibria in the system Fe-Al-Si-O-H. Geol. Soc. Am. Abstr. Programs 14, 434.

Apted, M.J. and Liou, J.C. (1983) Phase relations among greenschist, epidote-amphibolite, and amphibolite in a basaltic system. Am. J. Sci. 283-A, 328-354.

Bailey, S.W. (1980) Summary of recommendations of the AIPEA nomenclature committee. Can. Mineral. 18, 143-150.

_____ (1988) X-ray diffraction identification of the polytypes of mica, serpentine, and chlorite. Clays & Clay Minerals 36, 193-213.

_____ and Brown, B.E. (1962) Chlorite polytypism: I. Regular and semi-random one-layer structures. Am. Mineral. 47, 819-850.

Berman, R.G. (1988) Internally-consistent thermodynamic data for minerals in the system $Na_2O-K_2O-CaO-MgO-FeO-Fe_2O_3-Al_2O_3-SiO_2-TiO_2-H_2O-CO_2$. J. Petrol. 29, 445-522.

_____ and Brown, T.H. (1985) The heat capacity of minerals in the system $K_2O-Na_2O-CaO-MgO-FeO-Fe_2O_3-Al_2O_3-SiO_2-TiO_2-H_2O-CO_2$: representation, estimation, and high temperature extrapolation. Contrib. Mineral. Petrol. 89, 168-183.

_____, _____ and Perkins, E.H. (1987) GEO-CALC: software for calculation and display of pressure-temperature-composition phase diagrams. Am. Mineral. 72, 861-862.

342

_____, Engi, M., Greenwood, H.J. and Brown, T.H. (1986) Derivation of internally-consistent thermodynamic data by the technique of mathematical programming, a review with application to the system $MgO-SiO_2-H_2O$. J. Petrol. 27, 1331-1364.

Bird, G.W. and Fawcett, J.J. (1973) Stability relations of Mg-chlorite, muscovite and quartz between 5 and 10 kb water pressure. J. Petrol. 14, 415-428.

Bowen, N.L. and Tuttle, O.F. (1949) The system $MgO-SiO_2-H_2O$. Geol. Soc. Am. Bull. 80, 1947-1960.

Bryndzia, L.T. and Scott, S.D. (1987a) The composition of chlorite as a function of sulfur and oxygen fugacity: An experimental study. Am. J. Sci. 287, 50-76.

_____ and _____ (1987b) Application of chlorite-sulfide-oxide equilibria to metamorphosed sulfide ores, Snow Lake area, Manitoba. Econ. Geol. 82, 963-970.

Bucher-Nurminen, K. (1981) Petrology of chlorite-spinel marbles from NW Spitsbergen (Svalbard). Lithos 14, 203-213.

_____ (1982) On the mechanism of contact aureole formation in dolomitic country rock by the Adamello intrusion (northern Italy). Am. Mineral. 67, 1101-1117.

Burnell, J.R. and Rutherford, M.J. (1984) An experimental investigation of the chlorite terminal equilibrium in pelitic rocks. Am. Mineral. 69, 1015-1024.

Burnham, C.W., Holloway, J.R. and Davis, N.F. (1969) Thermodynamic properties of water to 1000°C and 10,000 bars. Geol. Soc. Am. Special Paper 132, 96 p.

Caruso, L.J. and Chernosky, J.V., Jr. (1979) The stability of lizardite. Can. Mineral. 17, 757-769.

Chernosky, J.V., Jr. (1973) *An Experimental Investigation of the Serpentine and Chlorite Group Minerals in the System $MgO-Al_2O_3-SiO_2-H_2O$*. Ph.D. Thesis, Massachusetts Institute of Technology, Cambridge, MA.

_____ (1974) The upper stability of clinochlore at low pressure and the free energy of formation of Mg-cordierite. Am. Mineral. 59, 496-507.

_____ (1975) Aggregate refractive indices and unit cell parameters of synthetic serpentine in the system $MgO-Al_2O_3-SiO_2-H_2O$. Am. Mineral. 60, 200-208.

_____ (1976) The stability of anthophyllite - a reevaluation based on new experimental data. Am. Mineral. 61, 1145-1155.

_____ (1978) The stability of clinochlore + quartz at low pressure. Am. Mineral. 63, 73-82.

_____ (1982) The stability of clinochrysotile. Can. Mineral. 20, 19-27.

_____ and Berman, R.G. (1986a) The stability of clinochlore in mixed volatile, H_2O-CO_2 fluids. Trans. Am. Geophys. Union 67, 407.

_____ and _____ (1986b) Experimental reversal of the equilibrium: clinochlore + 2 magnesite = 3 forsterite + spinel + 2 CO_2 + 4 H_2O. Trans. Amer. Geophys. Union 67, 1279.

_____ and _____ (1988) The stability of Mg-chlorite in supercritical H_2O-CO_2 fluids. Am. J. Sci., Wones Vol., in press.

Cho, M. and Fawcett, J.J. (1986) A kinetic study of clinochlore and its high temperature equivalent forsterite-cordierite-spinel at 2 kbar water pressure. Am. Mineral. 71, 68-77.

Coats, C.J.A. (1968) Serpentine minerals from Manitoba. Can. Mineral. 9, 321-347.

Coleman, R.G. (1971) Petrologic and geophysical nature of serpentinites. Geol. Soc. Am. Bull. 82, 897-918.

Craw, D., Landis, C.A. and Kelsey, P.I. (1987) Authigenic chrysotile formation in the matrix of Quaternary debris flows, Northern Southland, New Zealand. Clays & Clay Minerals 35, 43-52.

Dietrich, V. and Peters, T. (1971) Regionale Verteilung der Mg-Phyllosilikate in dem Serpentiniten des Oberhalbsteins. SMPM 51, 329-348.

Delany, J.M. and Helgeson, H.C. (1978) Calculation of the thermodynamic consequences of dehydration in subducting oceanic crust to 100 kb and 800°C. Am. J. Sci. 278, 638-686.

Doroshev, A.M. and Malinovskii, I. (1974) Upper pressure boundary of sapphirine stability. Dokl. Akad. Nauk. SSSR 219, 959-961.

Evans, B.W. (1977) Metamorphism of alpine peridotite and serpentinite. Annual Rev. Earth Planet. Sci. 5, 397-447.

_____ and Trommsdorff, V. (1970) Regional metamorphism of ultramafic rocks in the central Alps: Parageneses in the system $CaO-MgO-SiO_2-H_2O$. Schweiz. Mineral. Petrogr. Mitt. 50, 481-492.

_____ Johannes, W., Oterdoom, H. and Trommsdorff, V. (1976) Stability of chrysotile and antigorite in the serpentine multisystem. Schweiz. Mineral. Petrogr. Mitt. 56, 79-93.

Fawcett, J.J. and Yoder, H.S., Jr. (1966) Phase relations of the chlorites in the system $MgO-Al_2O_3-SiO_2-H_2O$. Am. Mineral. 51, 353-380.

343

Flaschen, S.S. and Osborn, E.F. (1957) Studies of the system iron oxide-silica-water at low oxygen partial pressures. Econ. Geol. 52, 923-943.

Fleming, P.D. and Fawcett, J.J. (1976) Upper stability of chlorite + quartz in the system $MgO-FeO-Al_2O_3-SiO_2-H_2O$ at 2 kbar water pressure. Am. Mineral. 61, 1175-1193.

Foster, M.D. (1962) Interpretation of the composition and a classification for the chlorites. U.S. Geol. Surv. Prof. Paper 414-A, 33 p.

Francis, G.H. (1956) The serpentine mass in Glen Urquhart, Inverness-shire Scotland. Am. J. Sci. 254, 201-226.

Fransolet, A.-M. and Schreyer, W. (1984) Sudoite, di/trioctahedral chlorite: a stable low-temperature phase in the system $MgO-Al_2O_3-SiO_2-H_2O$. Contrib. Mineral. Petrol. 86, 409-417.

Froese, E. and Gasparrini, E. (1975) Metamorphic zones in the Snow lake area, Manitoba. Can. Mineral. 13, 162-167.

_____ and Moore, J.M. (1980) Metamorphism in the Snow Lake area, Manitoba. Can. Geol. Surv. Paper 78-27, 16 p.

Frost, B.R. (1975) Contact metamorphism of serpentinite, chloritic blackwall and rodingite at Paddy-Go-Easy Pass, Central Cascades, Washington. J. Petrol. 16, 272-213.

Fyfe, W.S., Turner, F.J. and Verhoogen, J. (1958) Metamorphic facies and metamorphic reactions: Geol. Soc. Am. Memoir 73, 259 p.

Ganguly, J. and Newton, R.C. (1968) Thermal stability of chloritoid at high pressure and relatively high oxygen fugacity. J. Petrol. 9, 444-466.

Gasparik, T. and Newton, R.C. (1984) The reversed alumina contents of orthopyroxene in equilibrium with spinel and forsterite in the system $MgO-Al_2O_3-SiO_2$. Contrib. Mineral. Petrol. 85, 186-196.

Gillery, G.H. (1959) X-ray study of synthetic Mg-Al serpentine and chlorites. Am. Mineral. 44, 143-152.

Gordon, T.M. (1973) Determination of internally consistent thermodynamic data from phase equilibrium experiments. J. Geol. 81, 199-208.

Greenwood, H.J. (1967) Mineral equilibria in the system $MgO-SiO_2-H_2O-CO_2$, in Abelson, P.H., Ed., Researches in Geochemistry: Vol. II, John Wiley and Sons, New York, 663 p.

Grubb, P.L.C. (1971) Silicates and their paragenesis in the Brockman Iron Formation of Wittenoom Gorge, Western Australia. Econ. Geol. 66, 281-292.

Haar, C., Gallagher, J.S. and Kell, G.S. (1984) NBS/NRC Steam Tables. Thermodynamic and transport properties and computer programs for vapor and liquid states of water in SI units. Hemisphere Publishing Co., Washington, D.C.

Helgeson, H.C., Delany, J.M., Nesbitt, H.W. and Bird, D.K. (1978) Summary and critique of the thermodynamic properties of rock-forming minerals. Am. J. Sci. 278A, 229 p.

_____, Kirkham, D.H. and Flowers, G.C. (1981) Theoretical prediction of the thermodynamic behavior of aqueous electrolytes at high pressures and temperatures: IV. Calculation of activity coefficients, osmotic coefficients, and apparent molal and standard and relative partial molal properties to $600°C$ and 5kb. Am. J. Sci. 281, 1249-1516.

Hemingway, B.S., Robie, R.A., Kittrick, J.A., Grew, E.S., Nelen, J.A. and London, D. (1984) The heat capacities of osumilite from 298.15 K to 1000 K, the thermodynamic properties of two chlorites to 500 K, and the thermodynamic properties of petalite to 1800 K. Am. Mineral. 69, 701-710.

Hemley, J.J., Montoya, J.W., Christ, C.L. and Hosteller, P.B. (1977a) Mineral equilibria in the $MgO-SiO_2-H_2O$ system: I: Talc-chrysotile-forsterite-brucite stability relations. Am. J. Sci. 277, 322-351.

_____, _____, Shaw, D.R. and Luce, R.W. (1977b) Mineral equilibria in the $MgO-SiO_2-H_2O$ system: II: Talc-antigorite-forsterite-anthophyllite-enstatite stability relations and some geologic implications in the system. Am. J. Sci. 277, 353-383.

Henderson, C.E., Essene, E.J., Anovitz, L.M., Westrum, E.F.,Jr., Hemingway, B.S. and Bowman, J.R. (1983) Thermodynamics and phase equilibria of clinochlore, $(Mg_5Al)(Si_3Al)O_{10}(OH)_8$. Trans. Am. Geophys. Union 64, 466.

Hess, H.H., Smith, R.J. and Dengo, G. (1952) Antigorite from the vicinity of Caracas, Venezuela. Am. Mineral. 37, 68-75.

Hirschberg, A. and Winkler, H.G.F. (1968) Stabilitatsbezeichnungen zwischen chlorit, cordierit und almandin bei der metamorphose. Contrib. Mineral. Petrol. 18, 17-42.

Holdaway, M.J. (1971) Stability of andalusite and the aluminum silicate phase diagram. Am. J. Sci. 271, 97-131.

_____ and Lee, S.M. (1977) Fe-Mg cordierite stability in high-grade pelitic rocks based on experimental, theoretical, and natural observations. Contrib. Mineral. Petrol. 63, 175-198.

344

Holland, T.J.B. and Powell, R. (1985) An internally consistent thermodynamic dataset with uncertainties and correlations: 2. Data and results. J. Metamorphic Geol. 3, 343-370.

Hoschek, G. (1969) The stability of staurolite and chloritoid and their significance in metamorphism of pelitic rocks. Contrib. Mineral. Petrol. 22, 207-232.

Hostetler, P.B. and Christ, C.L. (1968) Studies in the system $MgO-SiO_2-CO_2-H_2O$ (I): The activity-product constant of chrysotile. Geochim. Cosmochim. Acta 32, 485-497.

Hsu, L.C. (1968) Selected phase relationships in the system Al-Mn-Fe-Si-O-H: A model for garnet equilibrium. J. Petrol. 9, 40-83.

Huebner, J.S. (1971) Buffering techniques for hydrostatic systems at elevated pressures. In: Ulmer, G.C., Ed., Research techniques for high pressure and high temperature. Springer-Verlag, Berlin. 367 p.

Hutcheon, I. (1977) The metamorphism of sulfide-bearing pelitic rocks from Snow Lake, Manitoba. Ph.D. Thesis, Carleton University, Ottawa, 205 p.

Hutchison, M.N. and Scott, S.D. (1981) Sphalerite geobarometry in the Cu-Fe-Zn system. Econ. Geol. 76, 143-153.

Iishi, K., and Saito, M. (1973) Synthesis of antigorite. Am. Mineral. 58, 915-919.

James, R.S., Turnock, A.C. and Fawcett, J.J. (1976) The stability and phase relations of iron chlorite below 8.5 kb P_{H2O}. Contrib. Mineral. Petrol. 56, 1-25.

Jasmund, K. and Sylla, H.M. (1971) Synthesis of Mg- and Ni-antigorite. Contrib. Mineral. Petrol. 34, 84-86.

Jasmund, K. and Sylla, H.M. (1972) Synthesis of Mg and Ni antigorite: A correction. Contr. Mineral. Petrol. 34, 346.

Jenkins, D.M. (1981) Experimental phase relations of hydrous peridotites modelled in the system $H_2O-CaO-MgO-Al_2O_3-SiO_2$. Contrib. Mineral. Petrol. 77, 166-176.

_____ and Chernosky, J.V., Jr. (1986) Phase equilibria and crystallochemical properties of Mg-chlorite. Am. Mineral. 71, 924-936.

Johannes, W. (1967) Zur Bildung und Stabilitat von Forsterit, Talk, Serpentin, Quarz und Magnesit in system $MgO-SiO_2-H_2O-CO_2$. Contrib. Mineral. Petrol. 15, 233-250.

_____ (1968) Experimental investigation of the reaction forsterite + H_2O = serpentine + brucite. Contrib. Mineral. Petrol. 19, 309-315.

_____ (1969) An experimental investigation of the system $MgO-SiO_2-H_2O-CO_2$. Am. J. Sci. 267, 1083-1104.

_____ (1975) Zur Synthese und thermischen Stabilitat von Antigorit. Fortschr. Mineral. 53, 36.

Kalinin, D.V. and Zubkov, M.Y. (1981) Kinetic investigation of the $MgO-SiO_2-H_2O$ system, reaction: serpentine = forsterite + talc + water. Soviet Geol. Geophysics. 22-9, 61-68.

Kerrick, D.M. (1974) Review of metamorphic mixed-volatile (H_2O-CO_2) equilibria. Am. Mineral. 59, 729-762.

_____ and Jacobs, G.K. (1981) A modified Redlich-Kwong equation for H_2O, CO_2, and H_2O-CO_2 mixtures at elevated pressures and temperatures. Am. J. Sci. 281,735-767.

King, E.G., Barany, R. Weller, W.W. and Pankratz, L.B. (1967) Thermodynamic properties of forsterite and serpentine. U.S. Bureau of Mines, Rept. Inv. 6962.

Kitahara, S., Takenouchi, S. and Kennedy, G.C. (1966) Phase relations in the system $MgO-SiO_2-H_2O$ at high temperatures and pressures. Am. J. Sci. 264, 223-233.

Kittrick, J.A. (1982) Solubility of two high-Mg and two high-Fe chlorites using multiple equilibria. Clays & Clay Minerals 30, 167-179.

Korytkova, E.N. and Makarova, I.A. (1971) Experimental study of the serpentinization of olivine. Doklady Akad. Nauk. SSSR 196, 144-145.

_____ and _____ (1972) Experimental investigation of the hydrothermal alteration of olivine in connection with the formation of asbestos. Geochem. Int'l. 9, 957-961.

Massone, H.-J., Mirwald, P.W. and Schreyer, W. (1981) Experimentelle Uberprufung der Reaktionskurve Chlorit + Quartz = Talk + Disthen im System $MgO-Al_2O_3-SiO_2-H_2O$. Fortschr. Mineral. 59, 122-123.

McPhail, D.C. (1985) The stability of Mg-chlorite. M.Sc. Thesis, Univ. British Columbia, Vancouver, B.C.

McOnie, A.W., Fawcett, J.J. and James, R.S. (1975) The stability of intermediate chlorites of the clinochlore-daphnite series at 2 Kb PH_2O. Am. Mineral. 60, 1047-1062.

Moody, J.B. (1976) An experimental study on the serpentinization of iron-bearing olivines. Can. Mineral. 14, 462-478.

Mueller, P. (1959) Vesuvianfuhrende Gesteine vom Piz Lunghin, Graubunden. Hamburger Beitr. 2. angew. Min. Kristallopysik Petrogenese 2, 136-163.

Mumpton, F.A. and Thompson, C.S. (1975) Mineralogy and origin of Coalinga asbestos deposit. Clays & Clay Minerals 23, 131-143.

Nelson, B.W. and Roy, R. (1958) Synthesis of the chlorites and their structural and chemical compositions. Am. Mineral. 43, 707-725.

Newton, R.C. and Wood, B.J. (1979) Thermodynamics of water in cordierite and some petrologic consequences of cordierite as a hydrous phase. Contrib. Mineral. Petrol. 68, 391-405.

Noll, W. and H. Kircher (1952) Synthese des Garnierites. Naturwiss. 10, 233-234.

Page, N.J. (1966) Mineralogy and chemistry of the serpentine group minerals and the serpentinization process. Ph.D. Thesis, University of California, Berkeley.

Pistorius, G.W.F.T. (1963) Some phase relations in the system $MgO-SiO_2-H_2O$ to high pressures and temperatures. N. Jahrb. Mineral. Monatsh. 11, 283-293.

Radoslovich, E.W. and Norrish, K. (1962) The cell dimensions and symmetry of layer lattice silicates. I. Some structural considerations. Am. Mineral. 47, 599-616.

Rice, J.M. (1977) Contact metamorphism of impure dolomitic limestone in the Boulder Aureole, Montana. Contrib. Mineral. Petrol. 59, 237-259.

Robie, R.A., Hemingway, B.S. and Fisher, J.R. (1979) Thermodynamic properties of minerals and related substances at 298.15 K and 1 bar (10^5 pascals) pressure and at higher temperatures. U.S. Geol. Surv. Bull. 1452, 455 p.

Rost, F. (1949) Das Serpentinit-Gabbro-Vorkommen von Wurlitz und seine Mineralien. Heidelberger Beitrage Mineral. Petrog. 1, 626-688.

Roy, D.M. and Roy, R. (1954) An experimental study of the formation and properties of synthetic serpentines and related layer silicate minerals. Am. Mineral. 39, 957-975.

_____ and _____ (1955) Synthesis and stability of minerals in the system $MgO-Al_2O_3-SiO_2-H_2O$. Am. Mineral. 40, 147-178.

Rucklidge, J.C. and Zussman, J. (1965) The crystal structure of the serpentine mineral, lizardite $Mg_3Si_2O_5(OH)_4$. Acta Crystallogr. 19, 381-389.

Scarfe, C.M. and Wyllie, P.J. (1967) Serpentine dehydration curves and their bearing on serpentinite deformation in orogenesis. Nature 215, 945-946.

Segnit, R.E. (1963) Synthesis of clinochlore at high pressures. Am. Mineral. 48, 1080-1089.

Seifert, F. (1970) Low temperature compatibility relations of cordierite in haplopelites of the system $K_2O-MgO-Al_2O_3-SiO_2-H_2O$. J. Petrol. 11, 73-99.

_____ (1973) Stability of the assemblage cordierite-corundum in the system $MgO-Al_2O_3-SiO_2-H_2O$. Contrib. Mineral. Petrol. 41, 171-178.

_____ (1974) Stability of sapphirine: a study of the aluminous part of the system $MgO-Al_2O_3-SiO_2-H_2O$. J. Geol. 82, 173-204.

_____ and Schreyer, W. (1970) Lower temperature stability limit of Mg cordierite in the range 1-7 kb water pressure: a redetermination. Contrib. Mineral. Petrol. 27, 225-238.

Shirozu, H. (1958) X-ray powder patterns and cell dimensions of some chlorites in Japan. Mineral. J. (Japan) 2, 209-223.

_____ (1960) Ionic substitution in iron-magnesium chlorites. Mem. Faculty Sci., Kyushu Univ., D, Geol. 9, 183-186.

_____ and Momoi, H. (1972) Synthetic Mg-chlorite in relation to natural chlorite. Mineral. J. (Japan) 6, 464-476.

Springer, R.K. (1974) Contact metamorphosed ultramafic rocks in the Western Sierra Nevada Foothills, California. J. Petrol. 15, 160-195.

Staudigel, H. and Schreyer, W. (1977) The upper thermal stability of clinochlore, $Mg_5Al[AlSi_3O_{10}](OH)_8$, at 10-35 kb PH_2O. Contrib. Mineral. Petrol. 61, 187-198.

Sverjensky, D.A. (1985) The distribution of divalent trace elements between sulfides, oxides, silicates and hydrothermal solutions: I. Thermodynamic basis. Geochim. Cosmochim. Acta 49, 853-864.

Tardy, Y. and Garrels, R.M. (1974) A method of estimating the Gibbs energies of formation of layer silicates. Geochim. Cosmochim. Acta 38, 1101-1116.

Trommsdorff, V. and Evans, B.W. (1972) Progressive metamorphism of antigorite schist in the Bergell tonalite aureole (Italy). Am. J. Sci. 272, 423-437.

_____ and _____ (1974) Alpine metamorphism of peridotitic rocks. Schweiz. Mineral. Petrogr. Mitt. 54, 333-352.

346

_____ and _____ (1977) Antigorite-ophicarbonates: phase relations in a portion of the system CaO-MgO-SiO$_2$-H$_2$O-CO$_2$. Contrib. Mineral. Petrol. 60, 39-56.

Turnock, A.C. (1960) The stability of iron chlorites. Carnegie Inst. Wash. Yrbk. 59, 98-103.

Ushio, M. and Saito, H. (1970) Hydrothermal experiments on materials corresponding to fluor-hydroxyl-chrysotile [Mg$_6$Si$_4$O$_{10}$F$_x$(OH)$_{8-x}$]. Yogyo-Kyokai-Shi 78, 359-364.

Velde, B. (1973) Phase equilibria in the system MgO-Al$_2$O$_3$-SiO$_2$-H$_2$O: chlorites and associated minerals. Mineral. Mag. 39, 293-312.

Walshe, J.L. (1986) A six-component chlorite solid solution model and the conditions of chlorite formation in hydrothermal and geothermal systems. Econ. Geol. 81, 681-703.

Wang, X. and Greenwood, H.J. (1988) An experimental study of the equilibrium: grossular + clinochlore = 3 diopside + 2 spinel + 4 H$_2$O. Can. Mineral. 26, 269-281.

Widmark, E.T. (1980) The reaction chlorite + dolomite = spinel + forsterite + calcite + carbon dioxide + water. Contrib. Mineral. Petrol. 72, 175-179.

Wilkinson, J.F.G. (1953) Some aspects of the alpine-type serpentinites of Queensland. Geol. Mag. 90, 305-321.

Whittaker, E.J.W. and Wicks, F.J. (1970) Chemical differences among the serpentine "polymorphs": a discussion. Am. Mineral. 55, 1025-1047.

_____ and Zussman, J. (1956) The characterization of serpentine minerals by x-ray diffraction. Mineral. Mag. 31, 107-126.

Wicks, F.J. and Whittaker, E.J.W. (1975) A reappraisal of the structures of the serpentine minerals. Can. Mineral. 13, 227-243.

Yamamoto, K. and Akimoto S. (1977) The system MgO-SiO$_2$-H$_2$O at high pressures and temperatures - stability field for hydroxyl-chondrodite, hydroxyl-clinohumite and 10Å -phase. Am. J. Sci. 277, 288-312.

Yoder, H.S. (1952) The MgO-Al$_2$O$_3$-SiO$_2$-H$_2$O system and related metamorphic facies. Am. J. Sci., Bowen Vol., 569-627.

_____ (1967) Spilites and serpentinites. Carnegie Inst. Wash. Yrbk. 65, 269-280.

Chapter 10   S.W. Bailey

CHLORITES: STRUCTURES AND CRYSTAL CHEMISTRY

INTRODUCTION

The chlorite group of minerals derives its name from the green color of most varieties. The color is known to vary widely in different specimens, however, and to include different shades of green, black, brown, orange, red, pink, purple, blue, yellow, grey, and even white. The morphology is equally diverse. Pseudohexagonal platelets parallel to the basal pinacoid (001) occur in the best crystallized varieties. These plates range in width from less than a millimeter up to several inches. Occasionally, as in the chlorite in serpentine-chromite deposits from Erzincan, Turkey, and in Swiss Alpine veins, small prismatic and pyramidal faces may be developed as well. Scaly flakes, wedge-shaped aggregates of crystals, spherules, rosettes, and fine-grained earthy masses are of much more common occurrence than the well-formed crystals.

Although the crystal morphology may approximate rhombohedral or hexagonal geometry, the true symmetry is usually triclinic, or occasionally monoclinic. There is a perfect basal cleavage parallel to (001). The cleavage flakes are flexible but inelastic, with a luster varying from pearly or vitreous to dull and earthy. The hardness on the cleavage is about 2.5. The density varies between 2.6 and 3.3, depending on composition.

Chemically, the chlorites are hydrous silicates incorporating medium-sized octahedral cations, primarily Mg, Al, and Fe, but occasionally Cr, Mn, Ni, V, Cu, Zn, and Li. There is a continuous solid solution series between the Mg and Fe species. Most chlorites are trioctahedral and belong to the Mg-Fe series. Dioctahedral chlorites are rare, as are intermediate forms that combine one trioctahedral sheet with one dioctahedral sheet. Al substitutes for Si between the approximate limits 0.4 to 1.8 atoms per 4 tetrahedral positions. Occasionally Fe^{3+}, B^{3+}, Zn^{2+}, or Be^{2+} may substitute for Si.

Chlorite is a common accessory mineral in low-to-medium grade regional metamorphic rocks and may be the most abundant mineral in metamorphic rocks of the chlorite zone. It occasionally occurs in igneous rocks, usually forming secondarily by deuteric or hydro-thermal alteration of primary ferromagnesian minerals, such as mica, pyroxene, amphibole, garnet, and olivine. Chlorite is found in pegmatites and fissure vein deposits. It is a common constituent of altered basic rocks and of hydrothermal alteration zones around ore bodies. In sedimentary rocks, chlorite is a common, but usually minor, component. Occasionally, chlorite makes up the bulk of the clay mineral fraction of sedimentary rocks. Some chlorite is of detrital origin in sediments, but some chlorite forms during diagenesis. Certain low-temperature structural polytypes have been found to be characteristic of the diagenetic, oolitic chlorite in iron-formation rocks and of the chloritic cement in porous sandstones and conglomerates. Similarly in soils, chlorite is a common, but usually minor, component. In most cases, soil chlorite is probably inherited from the parent material.

NOMENCLATURE

Because of the variety of chemical substitutions that are possible in the chlorite structure, many species names and classification schemes have been proposed. The classification schemes have varied over the years in accordance with our understanding of the compositions, properties, and structures of the chlorites. A summary of the early classifications of Tschermak, Clarke, Dalmer, Gossner, Orcel, Winchell, Hallimond, Hey, Lapham, Foster, and Phillips may be found in Bailey (1975).

In this chapter the simplified classification scheme recommended by the Nomenclature Committee of AIPEA is followed (see Bailey, 1980). The chlorite group is subdivided into four sub-groups-- trioctahedral chlorite, dioctahedral chlorite, di,trioctahedral chlorite, and tri,dioctahedral chlorite. *Trioctahedral chlorite*, the common form, is trioctahedral in both the 2:1 layer and the interlayer sheet. *Dioctahedral chlorite* is dioctahedral in both octahedral sheets. An example is *donbassite*. *Di,trioctahedral* chlorite is dioctahedral in the 2:1 layer but trioctahedral in the interlayer. *Cookeite* and *sudoite* are examples, with cookeite being Al- and Li-rich and sudoite being Al-rich but Li-poor. The only known example of a structure with a trioctahedral 2:1 layer but a dioctahedral interlayer is *franklinfurnaceite*, which may be classified as structurally intermediate between a chlorite and a brittle mica because of the presence of Ca between each 2:1 layer and the interlayer.

Trioctahedral chlorite species are named based on the suggestions of Bayliss (1975) in which the dominant divalent octahedral cation present determines the species name. Recommended species names are *clinochlore* for Mg-dominant [end-member = $(Mg_5Al)(Si_3Al)O_{10}(OH)_8$], *chamosite* for Fe^{2+}-dominant [end-member = $(Fe_5^{2+}Al)(Si_3Al)O_{10}(OH)_8$], *pennantite* for Mn^{2+}-dominant [end-member = $(Mn_5^{2+}Al)(Si_3Al)O_{10}(OH)_8$], *nimite* for Ni-dominant [end-member = $(Ni_5Al)(Si_3Al)O_{10}(OH)_8$], and *baileychlore* for Zn-dominant [end-member = $(Zn_5Al)(Si_3Al)O_{10}(OH)_8)$]. All other species and varietal names should be discarded because arbitrary subdivisions according to octahedral and tetrahedral compositions have been shown to have little structural significance. Tetrahedral compositions and trivalent octahedral cations are not considered in the recommended species names, nor is the distribution of octahedral cations between the 2:1 layer and the interlayer. Adjectival modifiers, such as those of Schaller (1930), may be used to indicate either important octahedral cations other than the dominant cation or unusual tetrahedral compositions. Bayliss (1975) gave modifiers appropriate for many of the chlorite species listed in other nomenclature systems.

The chlorite 14 Å unit structure described above consists of a 2:1 layer plus an interlayer hydroxide sheet, instead of consisting of 2:1:1 or 2:2 type layers. This usage emphasizes the similarity of chlorite to other 2:1 type phyllosilicates containing interlayer materials. The terms "talc layer" and "brucite sheet" are not suitable for describing the component parts of the trioctahedral chlorite structure because the minerals talc and brucite admit very little substitution of Mg by Al, which is an essential feature of trioctahedral chlorite (Brindley et al., 1968). However, "talc-like" and "brucite-like" are suitable terms for these structural components if an emphasis on the analogy to these minerals is desired.

CHEMICAL COMPOSITION

The trioctahedral chlorite structure consists of 2:1 layers ideally of composition $(R^{2+},R^{3+})_3(Si_{4-x}Al_x)O_{10}(OH)_2$ that alternate in the structure with octahedral interlayer sheets ideally of composition $(R^{2+},R^{3+})_3(OH)_6$. The tetrahedral portion of each 2:1 layer has a negative charge x owing to substitution of x ions of Al^{3+}, or occasionally of Fe^{3+}, B^{3+}, Zn^{2+}, or Be^{2+} for Si^{4+}. The interlayer sheet has a positive charge (except in franklinfurnaceite) due to substitution of R^{3+} ions for R^{2+} and serves to neutralize the negative charge on the 2:1 silicate layer. In most cases, it is not possible to determine if the tetrahedral charge is compensated entirely within the interlayer sheet or whether the octahedral portion of the 2:1 layer also acquires a positive charge. There is some evidence from structural refinements and infrared spectra that the octahedral sheet within the 2:1 layer often does have a net charge. The charge tends to be positive if the net tetrahedral charge exceeds -1.0 or negative if the net tetrahedral charge is less than -1.0. This implies that a charge of -1.0 for the entire 2:1 layer and of +1.0 for the interlayer is the stable state. The main constituents of the two octahedral sheets are Mg, Fe^{2+}, Al, and Fe^{3+}, but with important substitutions of Cr, Ni, Mn, V, Cu, Zn, or Li in certain varieties. Any medium-sized cation will fit in the octahedral sites. It is unlikely that the larger Na and Ca cations can occupy either octahedral sites or sites located between the 2:1 layer and the interlayer sheet in the structure of common rock-forming chlorites. See the later section on franklinfurnaceite for details.

Foster (1962) showed that very few chlorites have total octahedral Al approximately equal to tetrahedral Al. In most cases, octahedral Al is low, and other trivalent cations, such as Fe^{3+} or Cr^{3+}, are present to proxy for Al in balancing the negative charge on the tetrahedral sheet. If the total number of trivalent octahedral cations is approximately equal to tetrahedral Al, the octahedral occupancy is close to 6.00 atoms. If the total number of trivalent octahedral cations is greater than tetrahedral Al, Foster showed that the total octahedral occupancy is less than 6.00 atoms by an amount equal to one-half the excess of octahedral trivalent cations over tetrahedral Al. This relationship, illustrated in Figure 1, indicates that the excess trivalent octahedral cations replace divalent cations in the ratio 2:3. Such chlorites have vacancies in the octahedral sites and are called leptochlorites in earlier classifications. A generalized composition can be written as $(R^{2+}_{6-x-3y}R^{3+}_{x+2y}\square_y)(Si_{4-x}R^{3+}_x)O_{10}(OH)_8$, where $\square$ represents a vacancy.

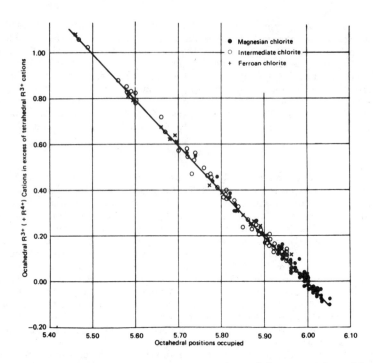

Figure 1. Linear relationship between octahedral cation positions occupied per 6.0 sites and octahedral trivalent cations in excess of tetrahedral trivalent cations. From Foster (1962).

Foster found two series of ionic replacements in the chlorites, replacement of Mg by Fe^{2+} and replacement of tetrahedral and octahedral Al by Si and Mg, respectively. Replacement of Mg by Fe^{2+} in the octahedral sheets is ion for ion and does not cause any change in the layer charge. A complete replacement series of this type can be demonstrated. The amount of Fe^{3+} present bears no necessary relation to the amount of Fe^{2+} in a particular specimen, but there is a general trend for Fe^{2+}-rich chlorites to contain more Fe^{3+} than do the Fe^{2+}-poor chlorites. In most cases this Fe^{3+} must be considered primary, unless there are definite indications of oxidation.

The replacement of tetrahedral and octahedral Al by Si and Mg occurs in accordance with the equation

$$Si^{4+}(IV) + Mg^{2+}(VI) = Al^{3+}(IV) + Al^{3+}(VI) \qquad (1)$$

Si varies between 2.34 and 3.45 atomic positions in the samples of Foster's study. This replacement is especially well shown by chlorites of low Fe^{2+} content. A complete replacement of Mg by Fe takes place between Si contents of 2.5 and 2.7, and most of the ferrous chlorites contain more than 4 wt.% Fe_2O_3.

In a later paper, Foster (1964) considered the water content of chlorites. Of 110 reliable analyses, 42% give near-theoretical (OH) atomic values (8.00 ± 0.20 atoms), 44.4% give deficient (OH) values

(< 7.80), and 13.6% give high (OH) values (> 8.20). For those chlorites having (OH) contents of 8.00 ± 0.20, there is a systematic decrease in H_2O by analysis with increase in FeO content. Thus, the reported H_2O content may vary by nearly 3% and still correspond to the theoretical (OH) content, depending on the Mg-Fe content of the sample. There is also a tendency for (OH)-deficient chlorites to be iron-rich. There is little correlation between ferric iron and excess oxygen in these analyses, however, and internal oxidation and dehydration will only explain (OH) deficiency in some of the Fe-chlorites. It cannot explain (OH) deficiency in other Fe-chlorites or in iron-poor chlorites. Of the (OH)-deficient chlorites, 44% also have abnormally high values for octahedral occupancy. The number of octahedral cations obtained by allocation from a chemical analysis depends on the conversion factor, which is determined in part by the H_2O present by analysis. But the summations are too high to be a result solely of inaccurate analysis. A possible cause of both the (OH) deficiencies and of the abnormally high octahedral cations totals would be a small percentage of (OH) vacancies. There is no verification of the existence of such vacancies and, pending proof, Foster considers such an explanation highly speculative. It is more likely that interstratification with less hydrous phyllosilicates is involved.

Bailey and Brown (1962) obtained the compositions of over 100 chlorites using the calibrated X-ray spacing graphs of Shirozu (1958). Their results for the compositional limits of chlorite parallel those of Foster. Si was found to vary between limits of 2.30 and 3.39, with an average of 2.69 atoms. The Fe/(Fe + Mg) ratio varies from 0.01 to 0.96. There is a broad trend for increasing tetrahedral Al substitution to be accompanied by a corresponding increase in octahedral Fe (Fig. 2) within the most common structural form (IIb) of chlorite. This trend was also noted by Shirozu (1960) for his specimens and explained as a requirement to maintain a certain degree of fit between the lateral dimensions of the tetrahedral and octahedral sheets. The scatter of the data plus the existence of other chlorite structural types in which high octahedral Fe or Zn are not accompanied by high Al^{IV} (see Fig. 13) suggest that other factors are involved.

For trioctahedral chlorites with high tetrahedral sheet charges (> -1.0), the excess of charge above unity most likely is compensated by substitutions within the octahedral sheet of the 2:1 layer and the remainder within the interlayer sheet. Such substitutions exist in the only high tetrahedral charge chlorite for which the structure has been determined in detail (Rule and Bailey, 1987). In the less well refined structure of a chlorite with a low tetrahedral charge, Rule and Radke (1988) suggested the tetrahedral charge is supplemented by a negative charge on the 2:1 octahedral sheet created by octahedral vacancies. These two examples suggest that total 2:1 layer charge approaches -1.0, but further study is needed for verification. Infrared absorption spectra support the idea of a positive charge on the 2:1 octahedral sheets of chlorites that have high negative tetrahedral charges. Shirozu (1980) found that considerable Al tends to substitute in the 2:1 octahedral sheet as well as its ubiquitous presence in the interlayer sheet. For two chlorites from the Furotobe and Ashio mines, Japan, the IR spectra are interpreted to

352

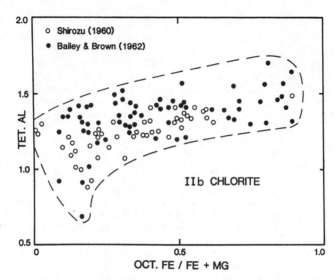

Figure 2. Graph of octahedral Fe/(Fe + Mg) vs. tetrahedral Al per 4.0 sites for chlorite of IIb structural type. Composition determined from spacing graphs of Shirozu (1958). Modified from Bailey and Brown (1962).

indicate that a 2:1 octahedral charge of +0.5 in both specimens reduces tetrahedral charges of -1.7 and -1.6 to resultant layer charges of -1.2 and -1.1.

Schreyer et al. (1986) used observed intergrowths of chlorites from the manganese deposit of the Lienne Valley, Belgium, to indicate a miscibility gap in trioctahedral Mn-Mg-Fe chlorites that are also Al-rich. The gap begins along the binary Mg-Mn series (approximately $Mn^{2+}/(Mn^{2+} + Mg) = 0.25$ to 0.50) and extends into the ternary system about 30% of the way to the Fe end member. It is uncertain whether the miscibility gap closes to form a solvus at higher temperature.

The dioctahedral chlorite *donbassite*, reported from certain sediments, soils, bauxites, hydrothermal alteration zones, and fissure veins, has an octahedral cation total slightly greater than 4.0 per formula unit, primarily Al. Substitution of Al for Si in the tetrahedral sheets of the best crystallized specimens ranges from 0.6 to 1.3 atoms per 4.0 positions. The octahedral cation total ranges from about 4.2 to 4.5. This indicates that the positive charge on the interlayer sheet arises primarily from the presence of Al cations in excess of 2.0 in this sheet, in accord with a theoretical formula of $Al_{4+x/3}(Si_{4-x}Al_x)O_{10}(OH)_8$ (Eggleton and Bailey, 1967). Some positive charge also may result from anion vacancies or from the presence of extra H protons. Small amounts of octahedral Li are often present (Rozinova and Dubik, 1983).

Cookeite, a mineral known since 1862, is an example of a di,trioctahedral chlorite with approximately 5.0 octahedral cations per formula unit. The ideal composition is $(LiAl_4)(Si_3Al)O_{10}(OH)_8$. Černý (1970) studied 13 cookeites occurring primarily in Li-rich pegmatites to show that Si tends to remain nearly constant at 3.0 per 4.0 tetrahedral positions. Tetrahedral Al occasionally may be substituted by B or Be. The exact relationship of cookeite to *manandonite*, believed to be a di,trioctahedral chlorite with substantial tetrahedral B, is uncertain. The total octahedral occupancy in cookeite is close to 5.0, but may be as high as 5.35. The main octahedral substitution is $3(Li,alk)^{1+}$ for Al^{3+}. Total Li was found to vary from 0.75 to 1.5 atoms per 6.0 octahedral positions, with a possible extension to even lower Li contents. There is a small octahedral substitution of divalent cations in some specimens. Fluorine substitutes frequently for OH.

Sudoite is another di,trioctahedral chlorite that has been identified in hydrothermal alteration zones, sediments, fissure veins, and low-grade metamorphic rocks from many locations. The ideal composition is $Al_2(Si_3Al)O_{10}(OH)_2$ for the dioctahedral 2:1 layer and $Mg_2Al(OH)_6$ for the trioctahedral interlayer. Chemical analyses suggest tetrahedral Al can range from about 0.6 to 1.1 atoms per formula unit, octahedral Al from 2.7 to 3.2, Mg from 1.2 to 2.3, with smaller amounts of Fe, Mn, and Li, but some of the data come from wet chemical analyses of fine grained, impure specimens (Eggleton and Bailey, 1967).

Table 1 lists analyses and structural formulas for representative samples of the chlorite group.

STRUCTURAL TYPES

Layer-interlayer units

The trioctahedral chlorite structure consists of negatively charged 2:1 layers of ideal composition $(R^{2+},R^{3+})_3(Si_{4-x}Al_x)O_{10}(OH)_2$ that alternate regularly with positively charged interlayer sheets of ideal composition $(R^{2+},R^{3+})_3(OH)_6$. In addition to the electrostatic interaction between the charges on the layer and interlayer, the hydroxyl surfaces of each interlayer sheet have been found to maintain long hydrogen bond contacts with the basal oxygen atoms of the 2:1 layers above and below (about 2.9 Å between the centers of the OH and O atoms). This general outline of the chlorite structure was determined by Pauling (1930) with further details supplied by McMurchy (1934), Garrido (1949), Robinson and Brindley (1949), Brindley et al. (1950), and Steinfink (1958a,b). Bailey and Brown (1962) showed that these earlier workers had studied the same structural type of chlorite but that other structural arrangements, which differ in the orientation and position of the interlayer sheet relative to the adjacent 2:1 layer, also occur in nature in lesser abundance. The relationship of the interlayer sheet to the adjacent 2:1 layers is the single most important factor in determining the resulting chlorite structures and their stabilities.

If the octahedral cations within the first 2:1 layer of a chlorite structure are in the set of positions labeled I (see Fig. 1 of Chapter 2), then a stagger of $a/3$ along $-X_1$ occurs between the upper tetrahedral sheet relative to the lower tetrahedral sheet, as illustrated in side view in Figure 3. Single crystal X-ray studies show that most chlorites have either regular-stacking 1-layer structures or what has been termed semi-random stacking. In either case adjacent 14 Å units must be the same (identical for 1-layer

Table 1. Chemical analyses and structural formulas of representative chlorites

Wt %	1	2	3	4	5	6	7	8	9	10	11	12	13	14	15
Al_2O_3	15.23	15.37	10.0	22.23	20.35	17.64	20.21	18.60	18.8	15.21	44.28	48.4	46.03	35.40	12.4
SiO_2	32.20	31.58	31.4	25.50	22.81	20.82	24.35	22.64	22.0	27.27	38.26	34.7	34.65	32.90	32.0
Cr_2O_3	3.16	3.37	9.3	0.03	-	-	-	-	-	<0.01	-	-	-	-	-
MgO	35.56	34.93	35.1	18.58	6.11	4.15	5.57	1.48	0.8	10.13	-	0.05	1.58	14.73	4.6
TiO_2	0.03	0.04	-	-	-	Tr.	0.04	-	-	-	-	-	-	0.06	n.a.
MnO	0.00	0.00	-	-	0.56	-	0.48	38.93	43.4	0.06	-	-	-	0.01	0.15
FeO	1.19	1.42	1.4	20.41	34.87	37.96	36.27	0.0	0.0	2.78	0.48	-	-	2.69	12.9
Fe_2O_3	n.a.	n.a.	0.5	n.a.	4.84	8.70	2.13	4.43	4.1	4.35	1.31	-	0.55	-	n.a.
NiO	0.24	0.20	-	-	-	-	-	-	-	29.49	-	-	-	-	-
CoO	-	-	-	-	-	-	-	-	-	0.38	-	-	-	-	-
ZnO	-	-	-	-	-	-	-	-	-	-	-	-	-	-	30.5
Li_2O	-	-	-	-	-	-	-	-	-	-	2.00	2.45	-	-	n.a
CaO	-	-	-	-	0.10	-	0.10	-	-	-	-	0.12	1.82	0.04	1.0
Na_2O	-	-	-	-	-	-	-	-	0.3	-	-	-	-	0.00	-
BaO	-	-	-	-	-	-	-	1.33	-	-	-	0.01	1.08	0.01	-
H_2O+	-	-	12.3	12.25	10.80	10.31	10.46	9.40	9.0	10.48	13.00	14.1	13.96	-	-

1. Day Book Body, North Carolina, IIb-4 clinochlore, $(Mg_{2.97}Al_{0.03})(Si_{3.02}Al_{0.98})O_{10}(OH)_2 \cdot (Mg_{1.98}Al_{0.69}Cr_{0.23}Fe^{3+}_{0.04}Fe^{2+}_{0.04}Ni_{0.02})(OH)_6$ Phillips et al. (1980)

2. Siskiyou Co., California, IIb-4 clinochlore, $(Mg_{2.95}Al_{0.05})(Si_{2.99}Al_{1.01})O_{10}(OH)_2 \cdot (Mg_{1.97}Al_{0.66}Cr_{0.25}Fe^{3+}_{0.06}Fe^{2+}_{0.06})(OH)_6$ Phillips et al. (1980)

3. Erzincan, Turkey, Ia-4 clinochlore, $(Mg_{3.0})(Si_{3.0}Al_{1.0})O_{10}(OH)_2 \cdot (Mg_{2.0}Cr_{0.7}Al_{0.2}Fe_{0.1})(OH)_6$ Brown & Bailey (1963)

4. Washington, D.C., IIb-2 clinochlore, $(Mg_{1.64}Fe_{0.96}Al_{0.40})(Si_{2.64}Al_{1.36})O_{10}(OH)_2 \cdot (Mg_{0.69}Fe_{0.31}Al_{1.0})(OH)_6$ Rule & Bailey (1987)

5. Sahama mine, Japan, Ibb($\beta = 90°$) chamosite, $(Si_{2.54}Al_{1.46})(Fe^{2+}_{3.24}Fe^{3+}_{0.41}Mg_{1.01}Al_{1.20}Mn_{0.05}Ca_{0.01})O_{10}(OH)_8$ Shirozu (1958)

6. Schmiedefeld, Thuringia, Germany, IIb chamosite, $(Si_{2.34}Al_{1.66})(Fe^{2+}_{3.69}Fe^{3+}_{0.76}Mg_{0.72}Al_{0.83})O_{10}(OH)_8$ Jung & Kohler (1930)

7. Tolgus mine, Cornwall, Germany, IIb chamosite, $(Si_{2.71}Al_{1.29})(Fe^{2+}_{3.38}Fe^{3+}_{0.18}Mg_{0.92}Al_{1.36}Mn_{0.05})O_{10}(OH)_8$ Hallimond (1939)

8. Benallt mine, Wales, IIb pennantite (2-layer), $(Si_{2.70}Al_{1.30})(Mn^{2+}_{3.92}Fe^{3+}_{0.39}Mg_{0.26}Al_{1.32}Ba_{0.08})O_{10}(OH)_8$ Smith et al. (1946)

Table 1 (continued)

9. Benallt mine, Wales, Ia pennantite (grovesite), $(Si_{2.57}Al_{1.43})(Mn_{4.29}Fe_{0.36}Mg_{0.14}Al_{1.15}Ba_{0.01})O_{10}(OH)_8$
   Bannister et al. (1955)

10. Barberton, S. Africa, IIb nimite, $(Si_{3.01}Al_{0.99})(Ni_{2.62}Mg_{1.67}Fe^{2+}_{0.26}Fe^{3+}_{0.36}Al_{0.98}Mn_{0.01}Co_{0.03}Ca_{0.04})O_{10}(OH)_8$
    de Waal (1970)

11. Djalair, Middle Asia, Ia-6 cookeite, $(Al_{1.86}Fe^{2+}_{0.04}Fe^{3+}_{0.09})(Si_{3.38}Al_{0.62})O_{10}(OH)_2 \cdot (Al_{2.1}Li_{0.7})(OH)_6$
    Vrublevskaya et al. (1975)

12. Jeffrey Quarry, N. Little Rock, Ark., Ia cookeite (2-layer "r"),
    $(Al_{2.0})(Si_{3.03}Al_{0.97})O_{10}(OH)_2 \cdot (Li_{0.86}Al_{1.05}Ca_{0.02}Mg_{0.01})(OH)_6$ Miser & Milton (1964)

13. Nagol Tarasovski, U.S.S.R., Ia donbassite (2-layer "r"), $(Al_{2.0})(Si_{3.12}Al_{0.88})O_{10}(OH)_2 \cdot (Al_{2.0}Mg_{0.24}Ca_{0.2})(OH)_6$
    Drits & Lazarenko (1967)

14. Ottre, Belgium, IIb sudoite (2-layer "s"), $(Al_{2.01})(Si_{3.00}Al_{1.00})O_{10}(OH)_2 \cdot (Mg_{2.00}Al_{0.82}Fe^{3+}_{0.18}Li_{0.01})(OH)_6$
    Lin & Bailey (1985)

15. Chillagoe, Australia, $Iba(\beta - 97°)$ baileychlore,
    $(Zn_{2.50}Al_{0.14}O_{0.36})(Si_{3.55}Al_{0.45})O_{10}(OH)_2 \cdot (Fe_{1.20}Al_{1.03}Mg_{0.76}Mn_{0.01})(OH)_6$ Rule & Radke (1988)

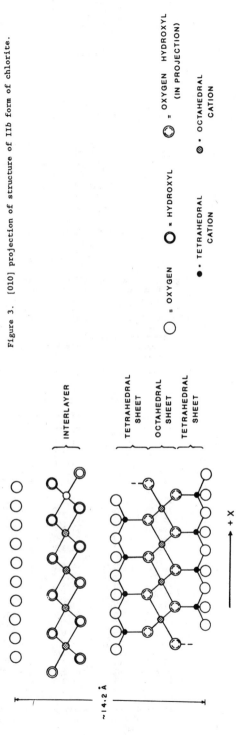

Figure 3. [010] projection of structure of IIb form of chlorite.

355

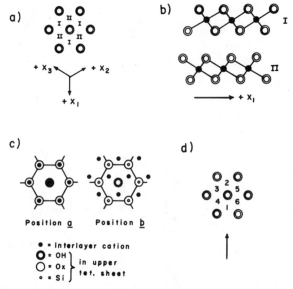

Figure 4. (a) Definition of I and II sets of octahedral sites that can be occupied by cations in the interlayer sheet, relative to a fixed set of pseudohexagonal axes. (b) Occupation of I or II cation sets give different slants to the interlayer sheet, relative to fixed axes. (c) Positions *a* and *b* of interlayer sheet give different patterns of superposition of interlayer cations onto hexagonal rings of 2:1 layer below. (d) Six positions for center of an hexagonal ring at base of upper 2:1 layer in projection onto the upper hydroxyl plane of the interlayer sheet. The central OH of the figure must lie on the mirror plane of the 2:1 layer below. Arrow indicates direction of *a*/3 intralayer shift within the first 2:1 layer.

Table 2. Summary of one-layer chlorite polytypes

Symbol	Space Group	ΔX,ΔY	Unique Angle	Equivalent Structure	Enantiomorphic Structure	Zvyagin Symbol
Iab-1	Cm	-1/3,0	β = 97°	Iba-2	-	
Ia-2*	C2/m	-1/3,0	β = 97°	-	-	$|\sigma_3'|\tau_0|\sigma_3'|$
Iab-3	C$\bar{1}$	-1/3,0	β = 97°	Iba-6	Ia-5	
Ia-4*	C1	-1/3,0	β = 97°	Ia-6	-	$|\sigma_1'|\tau_+|\sigma_1'|$
Iab-5	C$\bar{1}$	-1/3,0	β = 97°	Iba-4	Ia-3	
Ia-6	C1	-1/3,0	β = 97°	Ia-4	-	
Ib-1*	C2/m	0,0	β = 90°	-	-	$\sigma_3'\tau_6\sigma_3'$,
Iba-2*	Cm	-1/3,0	β = 97°	Iab-1	-	$|\sigma_6\tau_6|\sigma_6$
Ib-3*	C1	0,-1/3	β = 102°	Ib-5	-	$\sigma_4\tau_5\sigma_4$,
Iba-4*	C$\bar{1}$	-1/3,0	β = 97°	Iab-5	Iba-6	$|\sigma_2\tau_4|\sigma_2$
Ib-5	C1	0,-1/3	α = 102°	Ib-3	-	
Iba-6	C1	-1/3,0	β = 97°	Iab-3	Iba-4	
IIa-1*	C2/m	-1/3,0	β = 97°	-	-	$|\sigma_3|\tau_0|\sigma_3|$
IIab-2*	Cm	0,0	β = 90°	IIba-1	-	$|\sigma_3|\tau_0\sigma_3|$
IIa-3*	C1	-1/3,0	β = 97°	IIa-5	-	$|\sigma_1|\tau_+|\sigma_1|$
IIab-4*	C$\bar{1}$	0,-1/3	α = 102°	IIba-5	IIab-6	$|\sigma_4|\tau_5\sigma_4|$
IIa-5	C1	-1/3,0	β = 97°	IIa-3	-	
IIab-6	C1	0,-1/3	α = 102°	IIba-3	IIab-4	
IIba-1	Cm	0,0	β = 90°	IIab-2	-	
IIb-2*	C2/m	-1/3,0	β = 97°	-	-	$\sigma_6\tau_6\sigma_6$
IIba-3	C$\bar{1}$	0,-1/3	α = 102°	IIab-6	IIba-5	
IIb-4*	C1	-1/3,0	β = 97°	IIb-6	-	$\sigma_2\tau_4\sigma_2$
IIba-5	C$\bar{1}$	0,-1/3	α = 102°	IIab-4	IIba-3	
IIb-6	C1	-1/3,0	β = 97°	IIb-4	-	

*Selected among 12 unique polytypes. Only one member of an enantiomorphic pair included in unique polytypes. Equivalent structures are related by 180° rotation about Y axis.

types and on average for semi-random types). Therefore the structural variations permissible within a 2:1 layer, which are well known for the micas but applicable here only for the relatively rare multi-layer chlorite structures, may be neglected at the outset.

By restricting the discussion to regular 1-layer structures or to irregular layer sequences, only one orientation of the initial 2:1 layer need be considered. The interlayer sheet will be designated I or II, depending on which set of octahedral cation positions (Fig. 4a) is occupied in the interlayer. For easy reference, the interlayer orientation I may be depicted with its octahedral slant in the same direction as that within the 2:1 layer, whereas for orientation II the two octahedral slants are opposed (Fig. 4b). However, this description of the two directions of octahedral slant cannot be used in the terminology of regular-stacking multi-layer chlorites, to be discussed later.

A given interlayer, oriented as I or II, may be placed on the initial 2:1 layer in six ways to provide the necessary hydrogen bond contacts between the adjacent oxygen and hydroxyl surfaces. These six positions can be divided into two sets with three positions per set. Because the interlayer sheet by itself has trigonal symmetry and all atoms in it repeat at intervals of $b/3$, it suffices to select one position from each set and to label the sets differently, say a and b. The three a positions are equivalent to one another, as are the three b positions. In the a positions one of the three interlayer cations projects onto the center of a hexagonal ring of apical oxygens in the nearest anion plane of the adjacent 2:1 layer and the other two cations project onto tetrahedral cations of this ring (Fig. 4c). In the three b positions the interlayer sheet is shifted by $a/3$ so that the interlayer cations form triads symmetrically disposed in projection to the center of the hexagonal ring below. It is convenient to designate these four structural arrangements Ia, Ib, IIa, and IIb. They represent four different layer-interlayer assemblages or unit structures.

Regular-stacking polytypes

The regular 1-layer structures or polytypes result from different ways of positioning the repeating 2:1 layer upon the interlayer sheet to obtain long hydrogen bonds between the adjacent oxygen and hydroxyl surfaces. As with the superposition of the interlayer sheet on the 2:1 layer below, there are six ways (Fig. 4d) to position the upper 2:1 layer on top of the interlayer to provide hydrogen bonds. These also can be grouped into two sets of positions, which also could be called a and b but will be distinguished here by "even" or "odd" numbers instead. The numbered positions represent the 6 possible positions of the center of an hexagonal ring in the base of the upper 2:1 layer as superimposed on the uppermost OH surface of the interlayer. The central OH of the figure must be selected to lie on the symmetry plane of the 2:1 layer below.

Bailey and Brown (1962) showed that theoretically there can be 12 different regular 1-layer chlorite polytypes, designated according to the orientation of the interlayer sheet (I or II), the position (*a* or *b*) of the interlayer sheet on the initial 2:1 layer, and the position (1 through 6) of the repeating 2:1 layer on the interlayer sheet. These combinations give rise to 24 regular polytypes, I*a*-1 through I*a*-6, I*b*-1 through I*b*-6, and so forth (Table 2). Only 12 of these 24 polytypes prove to be truly different; the others are either enantiomorphic or equivalent by 180° rotation about *Y*. Two of the 12 unique polytypes are based on a triclinic-shaped unit cell with $\alpha \cong$ 102° and have triclinic symmetry. Two are based on an orthorhombic-shaped unit cell with $\alpha = \beta = \gamma = 90°$ and have monoclinic symmetry. The remaining eight polytypes are based on a monoclinic-shaped unit cell with $\beta \cong 97°$ and include one monoclinic symmetry and one triclinic symmetry representative of each of the four structural unit types (I*a*, I*b*, II*a*, II*b*). The symbol itself (e.g., I*a*-4) does not specify the directions of the resulting axes or the crystallographic angles of the structures, but the displacement of the resultant *Z* axis in (001) projection is listed in terms of its *a*/3 and *b*/3 components along resultant *X* and *Y* in Table 2 under the column heading $\Delta x, \Delta y$ [e.g., -1/3,0 = $-a_1/3$ displacement and an ideal β angle of 97° from the relation β = arccos (-*a*/3*c*)]. Superposition symbols 1 and 2 indicate the lower tetrahedral rings of the second layer are centered on the symmetry plane of the first layer (Fig. 4d) to give symmetric structures of monoclinic symmetry. Symbols 3-6 indicate asymmetric positions of the rings relative to the symmetry plane and resultant triclinic symmetry.

The polytype designations used thus far in this chapter are arbitrary and somewhat misleading because they imply that the 2:1 layers are positioned symmetrically relative to the interlayer, i.e., in position *a* on both sides or *b* on both sides. In fact, for 4 of the 12 unique 1-layer polytypes a I or II interlayer sheet has an *a* position relative to the 2:1 layer on one side but a *b* position relative to the layer on the other side. For these asymmetric structures a mixed symbol, such as I*ba* or II*ab*, can be used. This is a modification of the original Bailey and Brown terminology. Because these asymmetric structures give rise to unit cells of different shape (orthorhombic or triclinic vs. monoclinic) than for the symmetric structures, they have been distinguished in the earlier literature by the magnitude of the β angle. For example, I*b*(β=90°) is symmetric with *b* positions on both sides of the interlayer, whereas I*b*(β=97°) = I*ba* is asymmetric with the interlayer having a *b* relationship to the 2:1 layer below but an *a* relationship to the 2:1 layer above. I*ba* is equivalent to I*ab* by 180° rotation about *Y*.

Only with the interlayer in the *a* orientation relative to the 2:1 layers both above and below, and then only with specific positions of those layers, namely I*a*-2 and II*a*-1, do adjacent 2:1 layers adopt the mica configuration with their six-fold rings superimposed (but separated by the interlayer sheet). In all other regular 1-layer chlorites the rings in the nearest surfaces of adjacent layers are offset either by $\pm b_1/3$, by $+ a_3/3$ ($= a_1/3 + b_1/3$), or by $- a_2/3$ ($= - a_1/3 + b_1/3$).

Zvyagin (1963) and Zvyagin and Mishchenko (1966) confirmed the regular 1-layer chlorite structures of Bailey and Brown (1962) by independent derivations. They used a more complex but more precise analytical scheme of symbols to describe the resulting structures.

A 14 Å structural unit is designated by the symbol σ. If the octahedra in the interlayer and the 2:1 layer are parallel, i.e., slant in the same direction, a prime sign is added (σ'). Absence of a prime sign implies that the two octahedral sheets are antiparallel. If the interlayer cations project onto the upper tetrahedral cations of the 2:1 layer below, a vertical line is added to the right of the symbol $(\sigma|)$. If there is a similar superposition of interlayer cations relative to tetrahedral cations at the base of the 2:1 layer above, a vertical line is added to the left of the symbol $(|\sigma)$.

In the case of regular stacking sequences, a subscript i is added to the layer symbol (σ_i) to define a vector within the 2:1 layer from an origin in the octahedral sheet to an origin in the tetrahedral sheet immediately above. Because the 2:1 layer is usually centrosymmetric, the same vector and symbol would hold between the octahedral sheet and the lower tetrahedral sheet. The octahedral origin is either a vacant octahedral site, if dioctahedral, or a *trans* octahedral site on the symmetry plane of the layer, if trioctahedral. The tetrahedral origin is the center of a tetrahedral 6-fold ring. The subscript i can take values from 1 to 6 at 60° increments around $Z*$, the normal to the layers, as shown below. A second symbol τ_j represents the relative displacement in (001) projection across the interlayer between the origins of the nearest tetrahedral sheets of adjacent 2:1 layers. The subscript j defines the components of the displacement along the resultant X and Y axes of the structures. The subscript can take values of $j = 0, +, -$ for interlayer shifts of $0, + b/3$, and $- b/3$, respectively, or of $j = 1$ through 6 for shifts of $\pm a/3$ along the three pseudohexagonal X axes. The table for assigning the subscript values i and j is given below, where the vectors between the tetrahedral and octahedral origins involved must be measured as multiples of $a/3$ and $b/3$ along the resultant X and Y axes. Note that a displacement with components $1\bar{1} = - a_1/3 + b_1/3$ is equivalent to $+ a_2/3$ so that all displacements i,j = 1-6 correspond to $\pm a/3$ along one or another of the three pseudohexagonal X axes and that the reversed signs between σ_i and τ_j in the table insure that $2\sigma_i = \tau_i$ [e.g., because $2\sigma_2$ (or $+ 2a_2/3$) $= \tau_2$ (or $a_1/3 - b_1/3 = - a_2/3$)]. The correlation of these symbols with those of Bailey and Brown is shown in Table 2 for the 12 unique 1-layer polytypes.

i or j	σ_i	τ_j
1	$1\bar{1}$	$\bar{1}1$
2	$\bar{1}1$	$1\bar{1}$
3	10	$\bar{1}0$
4	$\bar{1}\bar{1}$	11
5	$1\bar{1}$	$\bar{1}1$
6	$\bar{1}0$	10
0	-	00
+	-	$0\bar{1}$
-	-	01

In regular 2-layer structures, the second 2:1 layer may not adopt the same orientation as that of the initial layer, and the structural variations possible within each 2:1 layer no longer can

360

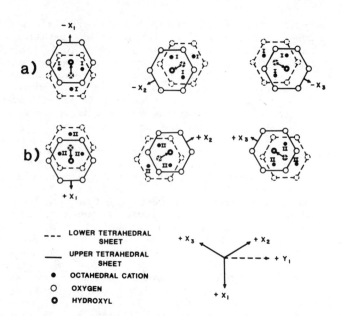

Figure 5. Upper tetrahedral sheet of 2:1 layer can be shifted relative to the lower tetrahedral sheet by $a/3$ along any of the three pseudohexagonal X axes. (a) Shift is in negative directions if octahedral sites I are occupied, or (b) in positive directions if sites II are occupied.

be ignored. Within each 2:1 layer, the upper tetrahedral sheet can be shifted by $a/3$ relative to the lower tetrahedral sheet along $-X_1$, $-X_2$, $-X_3$ or $+X_1$, $+X_2$, $+X_3$ of a set of pseudohexagonal axes depending on occupation of the I or II octahedral cation sites, respectively, in the 2:1 layer. These shifts are exactly the same as in the micas (Fig. 5), and simply represent different slant directions of the enclosed octahedral sheet. In addition to these six ways of constructing the 2:1 layer, there are four arrangements for the interlayer sheet (Ia, Ib, IIa, IIb) and six different positions for the second 2:1 layer. For a fixed orientation, therefore, there are 6 x 4 x 6 = 144 different one-layer types that can be combined with one another to form a regular two-layer chlorite. The number of theoretical two-layer structures is given by $(n^2 - n)/2 = 10,296$, where n = 144.

Lister and Bailey (1967) with the aid of stacking models showed that many of these theoretical two-layer structures are equivalent after ± 60° or 180° rotation about the normal to (001) or 180° rotation about the Y axis. The probable number of different two-layer chlorite structures is 1,134. Of these, 1,009 have monoclinic-shaped unit cells, and 125 have orthorhombic-shaped cells. They use six symbols to describe a two-layer structure analytically, three for each 14 Å unit. For each 14 Å unit, the first symbol represents the direction of $a/3$ stagger within the 2:1 layer relative to an arbitrary fixed set of pseudohexagonal axes (X_1, X_2, X_3, $\bar{X}_1$, $\bar{X}_2$, $\bar{X}_3$). Because of equivalence relationships it is always possible to specify this direction as $\bar{X}_1$. The second symbol indicates the orientation and position of the interlayer sheet relative to the 2:1 layer below. The symbols Ia, Ib, IIa, IIb are used but modified here to include *ab* or *ba* for asymmetric positions of the interlayer sheet. The third symbol describes the position of the upper 2:1 layer relative to the interlayer sheet (1 through 6). For the second 2:1 layer it is important to specify the direction of its *a*/3 stagger relative to the same set of axes used for the first layer. Negative shift directions indicate the octahedral cations in the second 2:1 layer are in the I set of positions relative to the fixed axes of the first layer, positive directions indicate the II set of positions. The orientation and position of the interlayer sheet are again indicated by the symbols I*a*, I*b*, II*a*, or II*b*, modified as necessary for asymmetric positions. The position (1 through 6) of the lower tetrahedral sheet of the third 2:1 layer is given again by Figure 4, where the figure must be oriented so that its -*a*/3 stagger is parallel to that of the first 2:1 layer of the structure. It is usually true that the numbers 1 through 6 will not be distributed in projection above a N-S mirror plane in the 2:1 layer below, as in Figure 4, but instead about a mirror plane intersecting either sites 4-5 or 3-6 corresponding to $\pm X_2$ or $\pm X_3$ staggers in the second layer. An unfortunate consequence of using the fixed axes of the first layer to describe the second layer is that layer symbols may be derived that do not correspond to the unique 1-layer polytypes listed in Table 2. For example, a symbol $\overline{X}_2$-I*a*-2 may be derived for a second layer that would be designated $\overline{X}_1$-I*a*-6 (or simply I*a*-6) if it occurred as a 1-layer structure. The layer type symbol may change also, as from I*a* to II*a*.

Drits and Karavan (1969) also derived the possible two-layer chlorite polytypes and their different characteristics. They recognized only 148 different trioctahedral structures, in contrast to the results of Lister and Bailey cited above. The structures are described by the analytical symbols developed by Zvyagin, and the symbols do not suffer the ambiguities mentioned above for the Bailey-Brown and Lister-Bailey systems.

Dioctahedral and di,trioctahedral chlorites, which are often two-layer in nature, are included in the Drits and Karavan derivation. For both types it is assumed that the vacant octahedral site in the 2:1 layer is at the center of symmetry in the *trans* M(1) position, so that its position need not be specified in the symbol. But for dioctahedral interlayers the symbol τ_j must be replaced by $\tau_j = \tau_{j1} + \tau_{j2}$. The symbol τ_{j1} represents a vector from an origin in the upper tetrahedral sheet of a 2:1 layer to the vacant site in the interlayer above, and τ_{j2} represents a vector from the same vacant site to an origin in the lower tetrahedral sheet of the 2:1 layer above. For example, a symbol τ_+ representing a + b_1/3 displacement of the origins of the nearest tetrahedral sheets of adjacent 2:1 layers for a trioctahedral chlorite might be described as $\tau_0 \tau_+$ for a dioctahedral chlorite in order to state the location of the vacant site in the interlayer.

Any of these nomenclature systems can be extended to regular-stacking chlorites with more than two layers in the repeat along Z. Fortunately, multi-layer chlorites are rare in nature.

Semi-random stacking

The observation that most chlorites give X-ray reflections that are sharp for index $k = 3n$ but are streaked for index $k \neq 3n$ indicates that the layers and interlayers, although irregular as to relative positions in adjacent 14 Å units, still maintain hydrogen bond contacts and only adopt those positions that are related to one another by shifts of $\pm b/3$ along the three pseudohexagonal Y axes. This has been termed semi-random stacking because it is not completely random. In the terminology of Bailey and Brown (1962), semi-random stacking consists of irregular 14 Å unit sequences along $Z*$ in which each interlayer sheet adopts at random one of the three a positions or one of the three b positions on top of the 2:1 layer and the next 2:1 layer adopts at random one of the three "even" or the three "odd" positions on top of the interlayer sheet. Thus there is no regular repeat in the Z direction for some atoms. In each of these four kinds of irregular sequences (a, b, "odd", "even") the three individual positions within each set are related to one another by $\pm b/3$ along the three pseudohexagonal Y axes (Figs. 4,6), in accord with the diffraction evidence. It is not permissible to mix the a with the b sets or the "even" with the "odd" sets in a given crystal, because this would involve relative shifts of $\pm a/3$ --for which there is usually no diffraction evidence. Shirozu and Bailey (1965) suggest that it is likely that the stacking randomness involves semi-random articulation of the interlayer sheet with the 2:1 layers on both sides, and not just on one side, although there would be little resultant difference in diffraction intensities because of the trigonal symmetry of the interlayer sheet.

Each of the four structural layer-interlayer chlorite units can be involved in two different semi-random stacking sequences, which could be termed the "even" and "odd" possibilities (equivalent to the a and b possibilities). The eight resultant structures are reduced to six different structures because of equivalence relationships after 180° rotation about Y. Four of the six structures are based on a monoclinic-shaped unit cell with $\beta \cong 97°$, one for each different layer-interlayer type, and the other two structures are based on an orthorhombic-shaped unit cell and Ibb or IIab units (Table 3). The I$b(\beta=97°)$ and II$a(\beta=90°)$ structures are asymmetric (=Iba and IIab). The symmetry of all of the semi-random stacking sequences is probably triclinic, being a statistical average of one monoclinic and two triclinic arrangements if the three stacking positions in each set are adopted equally.

IDENTIFICATION

X-ray study

Oriented aggregates. Identification of well-crystallized chlorites by means of oriented slides in a powder diffractometer is relatively simple and straightforward. Such chlorites are characterized by a sharp, integral sequence of 00ℓ lines based on a repeat of $d = c \sin \beta \cong 14.2$ Å. The layers do not expand on solvation with glycerol or ethylene glycol. The layers shrink only a few tenths of an Ångstrom on heating up to 700°C.

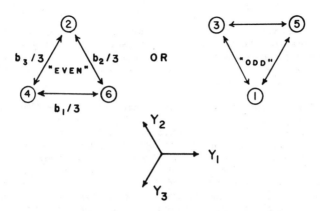

Figure 6. Semi-random adoption by 2:1 layers of the three even or three odd positions, as defined in Figure 4, above the interlayer sheet means layers are related to one another by random shifts of $\pm b/3$ along the three pseudohexagonal Y axes (normal to the three X axes).

Table 3. Chlorite semi-random stacking structures

$\beta \cong 97°$	$\beta = 90°$
Ia = Ia-2,4,6	--
Iba = Iba-2,4,6 = Iab-1,3,5	Ib = Ib-1,3,5
IIa = IIa-1,3,5	IIab = IIab-2,4,6 = IIba-1,3,5
IIb = IIb-2,4,6	--

On heating chlorite to about 500° to 650°C, a characteristic change in 00ℓ intensities is observed. The 14 Å reflection increases in intensity by a factor of 2 to 5 times, and at the same time many of the higher-order 00ℓ reflections decrease in intensity. Brindley and Ali (1950) showed that the great increase in intensity of the 14 Å peak is due to reducing the amount of electron density at the mid-plane of the interlayer sheet. This is accomplished by dehydroxylation of the interlayer in which two-thirds or more of the interlayer OH groups leave the structure at that temperature in the form of water. The interlayer cations then migrate toward the positions vacated by the OH groups in order to satisfy their coordination needs. This leaves only a small amount of scattering power (cations) at the middle of the interlayer, which is an asymmetry of distribution between the two octahedral sheets that causes an enhancement of the 14 Å peak at the expense of many of the other 00ℓ peaks. This unique feature has been used extensively in the identification of chlorite present in mixtures. The other layer silicates that give reflections in the 14 Å region (vermiculite, smectite, and interstratified minerals) will shift to lower spacings at this temperature, and the increased intensity of the 001 chlorite line can usually be seen even for small percentages of chlorite. The test is not completely diagnostic, however, because Nelson and Roy (1954) and Brindley and Wan (1978) showed that many of the 7 Å layer silicates develop a line between 13.2 and 13.8 Å on heating at this temperature. Caillère and Hénin (1960) also pointed out that the temperature of complete dehydroxylation depends on the composition

of the interlayer. A well-crystallized Mg-rich interlayer
dehydroxylates at about 640°C, an Al-hydroxy interlayer at 500°C,
an Fe^{2+}-hydroxy interlayer at 450°C, and an Fe^{3+}-hydroxy interlayer
at 250°C. Degree of crystallinity and grain size also affect the
temperature of dehydroxylation.

In mixtures of chlorite with kaolinite, the 00ℓ kaolinite
reflections may exactly or nearly superimpose on the even orders of
the chlorite 00ℓ reflections. Chlorites with appreciable amounts of
tetrahedral Al have slightly smaller d values than kaolinite for
these reflections. Biscaye (1964) showed for such specimens that the
doubled peaks can be resolved into their components by use of a
diffractometer with slow scan and fine collimation. For unresolved
peaks, Bradley (1954) demonstrated that heating in the range 400°
to 600°C may displace the components of the peaks sufficiently for
resolution and that identification can be made from the observed
directions of shift and from the relative decrease in intensity
with temperature for each component. Another technique that may
be used to give resolution is intersalation of the kaolinite with
various salts. Intersalation with potassium acetate followed by
washing with ammonium nitrate shifts the kaolinite 001 peak from
7.16 Å up to 11.6 Å. It is also possible to remove either the
chlorite or the kaolinite selectively from the mixture.
Trioctahedral chlorite is soluble in warm dilute hydrochloric acid,
whereas the dioctahedral species and kaolinite are relatively
insoluble. Kaolinite (and serpentine) become amorphous at 500°C, and
then can be preferentially dissolved by 0.5 N sodium hydroxide.

Chlorite in soils and sediments may have incomplete interlayer
sheets or form a randomly interstratified system with an expansible
component. In these cases, in contrast to well-crystallized species,
expansion may take place as a result of solvation, and shrinking as a
result of heating. The chloritic nature usually can be recognized by
the fact that shrinking is incomplete and that some intensification
of the 001 reflection occurs on heating. The name *swelling chlorite*
has been given to specimens that expand on solvation, but do not
collapse appreciably on heating. Martin Vivaldi and MacEwan (1960)
suggested that these specimens are chlorites with imperfect interlayer sheets. Regular and random interstratifications involving
chlorite are discussed in Chapter 15.

Random power mounts

A random orientation of particles must be achieved, either in
a powder diffractometer mount or in a conventional film camera, to
examine $h0\ell$ spacings and intensities in order to identify the type of
chlorite structural unit present. Table 4 lists the $h0\ell$ structure
amplitudes (F values) for the six possible structural types, calculated for a clinochlore composition and ideal atomic parameters. The
six semi-random stacking structures give different $k = 3n$ X-ray
intensities and thus can be identified by X-ray powder patterns. The
d values for the four structural types based on a monoclinic-shaped
unit cell with $\beta = 97°$ are similar, although the intensities
(proportional to F^2) are different. The d values change as β changes
and thus are quite different for the orthorhombic-shaped unit cell.

365

Table 4. Calculated and observed h0ℓ structure amplitudes of chlorite polytypes[a]

β = 97° Polytypes

d(Å)	h0ℓ	F(calc.) Ia	Iba	IIa	IIb	F(obs.) Ia	IIb
2.66	20$\bar{1}$	44	41	60	7	44	29
2.65	200	86	52	52	15	116	6
2.59	202	68	91	6	96	79	98
2.55	201	58	58	134	117	36	148
2.44	203	1	192	96	151	26	134
2.39	202	218	20	117	98	228	105
2.26	204	44	99	151	89	64	94
2.20	203	26	61	80	39	33	28
2.07	205	53	135	46	38	52	44
2.01	204	147	41	55	153	168	176
1.88	206	38	89	112	87	62	98
1.83	205	19	32	44	104	29	105
1.72	207	60	182	100	73	73	41
1.67	206	217	54	189	84	185	80
1.57	208	120	111	124	194	124	168
1.52	207	46	136	39	27	33	40
1.435	209	65	78	91	32	55	27
1.398	208	47	131	82	192	49	197
1.335	401	81	36	91	4	78	10
1.332	402	94	93	76	57	98	35
1.325	400	66	119	30	138	47	128
1.320	2,0,10	209	67	162	30	184	33
1.318	403	98	40	141	85	74	82
1.305	401	38	147	17	111	19	92
1.294	404	137	70	79	129	143	114
1.288	209	79	160	131	84	71	63
1.275	402	94	61	157	15	95	33
1.261	405	75	82	12	11	68	11
1.237	403	14	72	46	13	8	11
1.221	406	91	17	35	68	89	76
1.195	404	13	73	33	101	14	95
1.192	2,0,10	4	62	54	14	0	12

β = 90° Polytypes

d(Å)	h0ℓ	F(calc.) Ib	IIab	F(obs.) Ib
2.67	200	8	25	75
2.62 {	201	32	78	41
	201	58	55	39
2.50 {	202	208	142	316
	202	49	97	53
2.33 {	203	22	112	43
	203	44	115	50
2.13 {	204	164	57	210
	204	68	93	127
1.95 {	205	58	58	49
	205	7	121	16
1.77 {	206	194	88	218
	206	28	102	88
1.62 {	207	62	176	61
	207	35	91	15
1.478 {	208	134	75	127
	208	190	102	182
1.358 {	209	60	68	53
	209	71	169	56
1.335	400	10	42	49
1.329 {	401	96	78	74
	401	71	110	71
1.312 {	402	1	51	64
	402	185	135	209
1.285 {	403	87	132	74
	403	20	43	16
1.254 {	2,0,10	33	110	16
	2,0,10	150	69	124
1.250 {	404	63	64	103
	404	59	17	92
1.208 {	405	45	62	30
	405	23	60	5

[a]Calculated for clinochlore composition, assuming ideal hexagonal nets and no cation ordering with $a = 5.34$ Å, $c \sin \beta = 14.20$ Å. F(obs.) values for the Ia, IIb, and Ib structures are taken from Bailey (1986), Phillips et al. (1980), and Shirozu and Bailey (1965), respectively.

Some deviations of observed intensities from the ideal F values are to be expected as a result of structural distortion, isomorphous substitution, and cation ordering, as well as reflection multiplicity, crystal perfection, and the large effect of the Lorentz-polarization factor at low θ values. These deviations are not great enough, however, to prevent identification of the structural types.

Identification of the Ibb-orthohexagonal ($\beta=90°$) chlorite must be made with caution in the case of mixtures. The pattern is very similar, with the major exception of the 14 Å line, to those of the hexagonal forms of lizardite (Chapter 3) and berthierine (Chapter 6). X-ray data may not be definitive without heating and chemical treatments for mixtures of this chlorite with 7 Å serpentine or berthierine. The orthohexagonal Ibb-lines are also similar to the more intense lines in the powder pattern given by the monoclinic-cell Iba assemblage. Although the weaker lines in the latter pattern may be used to identify the cell shape and to differentiate the two chlorite types, these lines fade as crystallinity decreases. The patterns then differ only by variations in spacings in the observable 20ℓ lines of about 0.05 Å for the ideal composition, a difference that might be partly or wholly compensated by compositional variation and by distortion of the ideal cell shape. Careful indexing on the basis of both cells is advisable if the lines tend to be diffuse and the spacings intermediate between the ideal values given in Table 4 for the orthorhombic- and monoclinic-shaped Ib cells.

Trioctahedral chlorite of the Ia type gives a powder pattern very similar to that of trioctahedral vermiculite, which also is based on a Ia type structural unit, but the vermiculite pattern has a considerably more intense 14 Å reflection. The monoclinic-cell IIa and IIb chlorite patterns also should be somewhat similar to one another, according to the ideal F values of Table 4, and this may be one reason the IIa structure has not been reported for a true chlorite. It is seen below that the latter (IIa) is also the least stable chlorite.

The $k \neq 3n$ reflections that identify the regular-stacking chlorite polytypes are so weak that they often do not show up on X-ray powder patterns. Grinding will destroy the regularity of stacking and, in some cases, may even change the layer-interlayer relationships (which affect the strong powder lines). For example, Shirozu (1963) showed that grinding a I$b(\beta=90°)$ orthohexagonal chlorite transformed it to the monoclinic cell I$ba(\beta=97°)$ or to the Ia structure. These transformations only require linear shifts of the interlayers and layers relative to one another. Transformation to a IIa or IIb form is structurally more difficult and would require reorganization of the interlayer sheet (changing octahedral cation sets) in addition to shifts. If possible, it is best to avoid grinding of chlorites for powder analysis and to use dicing or comminution by a steel file instead. A Gandolfi camera may also be used if only a crystal or a solid specimen is available. Identification of the regular-spacing monoclinic and triclinic polytypes within each chlorite group usually cannot be made from powder patterns, even when the $k \neq 3n$ reflections are visible. This is due to the small differences in spacings and intensities between the $k \neq 3n$ reflections that are diagnostic of each polytype (Table 5). Single crystals should be used for such identification.

Representative powder patterns are illustrated in Figure 7 and the data tabulated for reference in Tables 5 and 6. Only four of the six possible layer-interlayer assemblages have been recognized in nature. No IIa or IIab structures have been reported to date (excluding the hybrid species franklinfurnaceite).

Single crystals

It may be impossible to distinguish semi-random from regular layer sequences from a powder pattern because of the weakness of the diagnostic $k \neq 3n$ reflections. These reflections are best studied by single-crystal methods. Photographs taken during rotation about the cleavage normal, as illustrated in Figure 8, will show discrete $k \neq 3n$ spots for regular stacking polytypes. The spots degenerate to streaks for semi-random stacking.

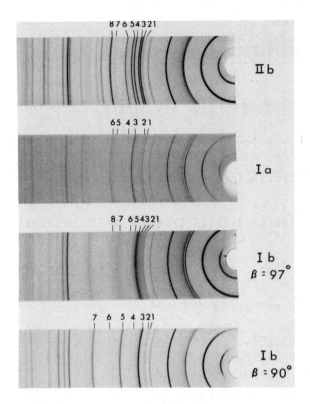

Figure 7. X-ray powder patterns of chlorites (graphite-monochromatized FeKα radiation, 114.6 mm camera diameter). Lines diagnostic of the four structural types are numbered and listed in accompanying Table 6.

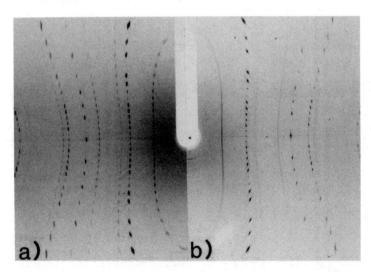

Figure 8. Single crystal X-ray photographs taken by rotation about the normal to the (001) cleavage. (a) Regular stacking of layers gives discrete spots for reflections of all possible indices, whereas (b) semi-random stacking of layers gives continuous streaks for all $k \neq 3n$ indices but discrete spots for all $k = 3n$ indices. From Bailey (1975).

Table 5 X-ray Powder Patterns of Selected Chlorites

A. Triclinic IIb-4 clinochlore

hkl	I	d(obs.)	d(calc.)
001	90	14.2 Å	14.21 Å
002	100	7.1	7.10
003	50	4.75	4.736
020	25	4.58	4.613
110			4.581
1̄10			4.577
021	10	4.39	4.394
111			4.382
02̄1	2	4.22	4.226
1̄11	7	4.06	4.224
111			4.063
1̄12			4.050
112	2	3.86	3.860
022	2	3.67	3.672
1̄12			3.667
004	70	3.554	3.552
02̄3	2	3.31	3.313
023	1	3.13	3.128
113			3.122
114	1	2.97	2.964
005	20	2.840	2.842
130	2	2.655	2.658
201			2.656
1̄31			2.654
131	25	2.580	2.583
202			2.581
1̄32			2.581
132	50	2.540	2.547
2̄01			2.537
1̄32			2.537
203	40	2.438	2.440
132			2.438
1̄33			2.436
133	20	2.380	2.388
2̄02			2.377
133			2.376
1̄34	18	2.257	2.260
134			2.258
2̄03			2.254
1̄34	2	2.195	2.201
205			2.190
134			2.189
007	2	2.065	2.069
1̄35			2.068
135			2.062
204	20	2.030	2.030
1̄35	30	2.010	2.011
206	18	2.000	2.000
135			2.000
1̄36	5	1.884	1.885
136			1.884
205	5	1.875	1.879
136	15	1.831	1.831
205	5	1.823	1.822
150			1.822
	5	1.760	1.742
			1.760

B. Monoclinic IIb-2 clinochlore

hkl	I	d(obs.)	d(calc.)
001	90	14.2 Å	14.14 Å
002	100	7.1	7.07
003	40	4.72	4.712
020	5	4.62	4.632
111	3	4.52	4.524
1̄11	3	4.25	4.243
112	3	4.05	4.065
022	3	3.88	3.874
1̄12	3	3.69	3.678
004	90	3.535	3.534
023	1	3.30	3.303
113	1	3.14	3.127
114	1	2.95	2.963
005	20	2.83	2.827
201			2.668
130	10	2.665	2.669
114			2.667
202			2.593
131	55	2.592	2.593
201			2.549
131	60	2.550	2.550
132			2.446
203	55	2.445	2.446
132			2.388
202	45	2.387	2.388
133			2.301
220	3	2.31	2.301
221			2.312
204			2.262
133	30	2.262	2.262
203	1	2.20	2.197
116			2.197
205	7	2.065	2.068
134			2.068
007	10	2.020	2.020
204			2.004
135	40	2.005	2.005
206			1.883
135	20	1.882	1.882
205			1.824
136	15	1.824	1.824
226			1.744
151	5	1.745	1.745
240			1.744
207			1.714
136	3	1.714	1.713
206			1.661
137	20	1.662	1.662
208			1.564
137	40	1.564	1.564
060	65	1.543	1.544
331			1.543
062	30	1.508	1.508
331			1.508
063	10	1.468	1.467
332			1.467
0,0,10			1.414
064	20	1.415	1.415
208	40	1.392	1.392
139			1.393
065	5	1.354	1.355
334			1.355
400	20	1.325	1.326
262			1.326
403	5	1.320	1.320
261			1.321
401	3	1.305	1.305
263			1.306
404	15	1.297	1.296
262			1.296
209			1.283
1,3,10	7	1.283	1.283
422			1.283
171			1.282

A. Specimen from Mcito Andei, Mochako district, Kenya, Harvard Museum #97703. a = 5.317, b = 9.227, c = 14.324 Å, α = 90.3°, β = 97.3°, γ = 89.9°.

B. Specimen from Waterworks Tunnel, Washington, D.C., USNM #45875. a = 5.347, b = 9.263, c = 14.250 Å, β = 97.2°.

Both patterns taken in Gandolfi cameras, 114.6 mm diameter, with graphite-monochromated $FeK\alpha_1$ radiation. Indices assigned by comparison with single crystal intensities. Powder intensities estimated visually. Data from Bailey (unpublished).

C. Orthohexagonal Ibb (β = 90°) chamosite

hkℓ	I	d(obs.)	d(calc.)
001	90	14.20 Å	14.180 Å
002	100	7.10	7.090
003	30	4.73	4.727
02;11	30 B	4.67	4.671
004	80	3.543	3.545
005	30	2.835	2.836
200,130	30	2.696	2.697
201,201̄,131,131̄	20	2.649	2.650
202,132̄	90	2.520	2.521
203,203̄,133,133̄	10	2.341	2.343
204,204̄,134,134̄	60	2.149	2.147
007	5	2.023	2.026
205,135̄	5	1.953	1.954
206,136̄	40	1.778	1.778
15;24,31̄	1	1.77	1.766
207,137̄	5	1.620	1.620
060,330	55	1.558	1.557
062,332,332̄	40	1.520	1.521
208,208̄,138,138̄	30	1.480	1.481
064,334,334̄	25	1.425	1.426
0,0,10	10	1.418	1.418
209,209̄,139,139̄	5	1.361	1.360
400,260	10 B	1.344	1.349 / 1.343
402,262	30	1.325	1.325
066,336,336̄	10	1.300	1.300
404,404̄,264,264̄	10	1.261	1.260
2,0,10;1,3,10	10	1.255	1.255
0,0,12	10	1.183	1.182

C. Specimen from Arakawa mine, Japan.
Specimen has semi-random stacking.
a = 5.394, b = 9.342, c = 14.180 Å,
β = 90°. Conditions as for A.

D. Baileychlore-Iba

hkℓ	I	d(obs.)	d(calc.)
001	95	14.3 Å	14.290 Å
002	100	7.14	7.145
003	5	4.76	4.763
			4.628
02;11	30	4.600	4.602
004	30	3.573	3.572
005	10	2.860	2.858
201,130	50	2.660	2.668 / 2.667
202,131	5 B	2.59	2.594 / 2.593
132,203	35 B	2.450	2.450 / 2.449
205	2 B	2.27	2.268
204,133	15	2.080	2.078 / 2.076
134	2 B	2.040	2.041
007	2 B	1.89	1.892
206,135	20	1.745	1.748 / 1.737
207,136	10	1.722	1.724
060,331	60	1.542	1.543
062,333,331	25	1.508	1.508
209,138	10	1.44	1.443
00,10	3	1.428	1.429
064,335,333	15	1.415	1.416
065,336,334	1	1.36	1.358
261,402,260	20 B	1.335	1.336 / 1.334
262	20	1.295	1.297 / 1.296
20,10;13,11	2	1.198	1.198

Specimen from Red Dome deposit near
Chillagoe, Queensland, Australia.
a = 5.346, b = 9.257, c = 14.401 Å,
β = 97.12°. Graphite-monochromatized
FeKα radiation. Camera diameter 114.6
mm. Intensities estimated visually.
From Rule and Radke (1988).

E. Pennantite-Ia

hkℓ	I	d(obs.)	d(calc.)
001	50	14.3 Å	14.29 Å
002	100	7.15	7.143
003	15	4.77	4.762
020			4.683
110	5	4.66	4.651
112	2	3.71	3.719
005	70	3.57	3.572
004	20	2.86	2.857
131,200	60	2.681	2.680
132,201	15	2.577	2.577 / 2.576
133	95	2.413	2.413 / 2.412
202	5	2.33	2.325
220,115,133	23	2.285	2.286 / 2.284
204,134	15	2.090	2.090 / 2.089
205	35	2.027	2.026
135,204	18	1.903	1.903 / 1.901
206,151	5	1.763	1.764 / 1.763 / 1.762
240,226			
136,207	2	1.732	1.732 / 1.731
206,137	35	1.679	1.679
137,208	25	1.580	1.581 / 1.579
060,331	35	1.560	1.561 / 1.560
062,331,333	25	1.524	1.525 / 1.524
332,334,063	2	1.482	1.483 / 1.482
138,209	10	1.446	1.448 / 1.447
064,333	15	1.429	1.430 / 1.429
335;00,10,139,208	5	1.407	1.429 / 1.407
065,334,336	7	1.368	1.370 / 1.369
401,402	10	1.349	1.350 / 1.348
139,20,10	20	1.332	1.332 / 1.331
262,404	15	1.310	1.310 / 1.309
066,335,337	10	1.304	1.305 / 1.304

Specimen from Franklin, New Jersey.
a = 5.401, b = 9.365, c = 14.398 Å.
β = 97.2°. Camera diameter 114.6 mm.
Graphite-monochromatized FeKα radiation.
Data from Bailey (unpublished).

F. Ninite-IIb

hkℓ	I	d(obs.)	d(calc.)
001	25	14.2	14.2
002	100	7.10	7.10
003	16	4.74	4.73
020	1	4.60	4.61
004	45	3.55	3.55
005	7	2.841	2.838
200	3	2.644	2.639
202	2	2.582	2.580
201	2	2.540	2.539
203	2	2.438	2.436
202	2	2.379	2.380
204	V 1	2.258	2.255
205	2	2.062	2.064
007	3	2.028	2.027
204	V 1	2.003	2.003
206	2	1.8253	1.8812
205			1.8239
206	2	1.6620	1.6625
208	1	1.5661	1.5647
060	1B	1.5349	1.5359
062	1B	1.5012	1.5007
0,0,10	1B	1.4188	1.4192
064	V 1	1.4101	1.4093
208	1B	1.3948	1.3948
400	1B	1.3190	1.3200
401	1B	1.3001	1.2995
404	1B	1.2901	1.2904
406	1B	1.2179	1.2184
0,0,12		1.1821	1.1826

a = 5.320, b = 9.214, c = 14.302 Å
β = 97.10°. Specimen from
Barberton, South Africa.
(de Waal, 1970)

G. Cookeite-Ia

hkl	I	d(obs.)	d(calc.)
002	90	14.06 Å	14.077 Å
004	35	7.03	7.039
006	60	4.70	4.692
111̄	20	4.47	4.459 / 4.443
110			
112̄	10	4.35	4.366 / 4.321
111			
111̄	10 B	4.17	4.184 / 4.118
112			
023		4.03	4.031
112̄			
113	5 B	3.87	3.866
024̄		3.77	3.770
115̄	1	3.66	3.674
008	60	3.518	3.519
116̄	3	3.410	3.404
116	3	3.060	3.073
0.0.10	45	2.815	2.815
200	60	2.560	2.561 / 2.558
132̄			
132	80	2.505	2.505 / 2.503
204̄			
202̄	3	2.465	2.469 / 2.464
134̄			
134	2	2.370	2.372 / 2.368
206̄			
204̄	100	2.316	2.321 / 2.314
136̄			
0.2.11	1 B	2.220	2.220 / 2.202
136			
208̄	5	2.200	2.197
206̄	3	2.140	2.144 / 2.138
138̄			
0.2.12̄	1	2.070	2.077
138			
2.0.10̄	20	2.109	2.021 / 2.016
208̄	45 B	1.960	1.964 / 1.958
1.3.10̄	1	1.790	1.792 / 1.787
1.3.12			
0.0.16	5	1.760	1.760
1.3.12̄	20 B	1.686 / 1.680	1.685 / 1.681
2.0.14̄			
2.0.12̄	35 B	1.638 / 1.630	1.637 / 1.632
1.3.14̄			
1.3.14	20	1.542	1.541
2.0.16̄	15	1.538	1.537
33̄2	75	1.487	1.489 / 1.488
060			
331			1.487
332	15	1.456	1.457 / 1.456
064			
336̄			1.455
334	15	1.419	1.420 / 1.418
338̄			
066̄			1.418
1.3.16̄	10	1.412	1.414 / 1.411
2.0.18̄			
2.0.16̄	7	1.374	1.377 / 1.373 B
1.3.18̄			
336	7	1.371	1.373 / 1.370
068			
3.3.10̄			1.369

Specimen from N. Little Rock, Arkansas, has 2-layer "r" structure with Iα layers. Cell dimensions: a = 5.158, b = 8.927, c = 28.351 Å, β = 96.8°. Conditions as for A.

H. Cookeite-IIb

hkl	I	d(obs.)	d(calc.)
002	80	14.10 Å	14.100 Å
004	60	7.05	7.050
006	80	4.70	4.700
111̄	65	4.45	4.463 / 4.445
110			
112̄	5	4.35	4.373 / 4.321
111			
112	5	4.13	4.117
024̄	5 B	3.73	3.775 / 3.684
115			
008	40	3.52	3.525
026	1	3.22	3.239
118̄	1	2.92	2.915
00.10	30	2.820	2.820
119	5	2.705	2.700
118	1	2.62	2.630
200,132̄	25 B	2.560	2.561
204,132	15	2.505	2.507
202,134̄	100	2.468	2.468
206,134	80	2.373	2.373
204,136̄	20	2.319	2.319
208,136	20	2.203	2.203
206,138̄	1	2.140	2.143
20.10,138	15	2.021	2.022
208.13.10	75	1.962	1.962
20.12,13.10̄	12	1.848	1.847
20.10,13.12	40	1.792	1.791
00.16	5	1.762	1.762
20.14,13.12̄	15	1.686	1.686
20.12,13.14	3	1.638	1.636
20.16,13.14̄	55	1.542	1.542
060,332	90	1.490	1.490
064,332,336̄	15	1.458	1.458
066,334,338̄	20	1.420	1.420
20.16,13.18̄	40 B	1.375	1.372 / 1.376
068,336,33.10̄			
06.10,338,33.12̄	8	1.319	1.317

Specimen from Norway, Maine. Cell dimensions: a = 5.161, b = 8.938, c = 28.41 Å, β = 97°. Conditions as for A.

J. Orthohexagonal 2-layer clinochlore (β = 90°)

hkl	I	d(obs.)	d(calc.)
002	90	14.46	14.36 Å
004	100	7.206	7.183
006	80	4.809	4.789
110,020,111,021	80	4.619	4.590
112,022	20	4.412	4.398
113,023	30	4.175	4.161
114,024	20	3.899	3.885
115,025,008	90	3.600	3.600
116,026	10	3.334	3.325
117,027	5	3.076	3.068
0.0.10,118,028	40	2.878	2.873
200,130,201,131	30	2.664	2.661
119,029,202,132	80	2.626	2.626
203,133	40	2.575	2.569
204,134	80	2.504	2.500
205,135,0.0.12	100	2.424	2.429
206,136	70	2.340	2.330
223,043,207,137	10	2.249	2.245
225,045	10	2.143	2.143
0.0.14,209,139	60	2.050	2.050
2.0.10,1.3.10	50	1.958	1.955
2.0.12,1.3.12	40	1.786	1.781
310,240,150	15	1.749	1.746
312,242,152	15	1.738	1.733
2.0.13,1.3.13	30	1.704	1.702
2.0.14,1.3.14	45	1.629	1.626
2.0.15,1.3.15	30	1.558	1.556
330,060	90	1.541	1.540
1.1.18,0.2.18,334,064	45	1.507	1.507
2.2.15,0.4.15,336,066	20	1.468	1.470
0.0.20	10	1.437	1.436
2.0.17,1.3.17	10	1.429	1.427
068,338	20	1.415	1.415

Specimen from Errincan, Turkey. $a = 5.335$, $b = 9.240$, $c = 28.735$ Å, $\beta = 90°$. Two-layer chlorite in which the individual layers are of the Ibb type. Modified from Lister and Bailey (1967).

---

hkl	I	d(obs.)	d(calc.)	hkl	d(obs.)	d(calc.)
002	40	14.2 Å	14.16 Å	2,1,3,1̄2	1.820	1.820
004	80	7.1	7.08	2,0,1̄0	1.803	1.805
006	100	4.72	4.721	1,3,1̄2		1.804
020	4	4.50	4.547	0,0,16	1.770	1.770
021			4.496	312		1.717
11̄1			4.483	311	1.711	1.713
111			4.387	0,4,11		1.712
022	10	4.35	4.384	1,3,1̄2		1.706
1̄12			4.340	153̄	1.705	1.705
022			4.318	314		1.705
1̄14	10	4.01	4.024	153		1.675
1̄13			4.004	245̄	1.673	1.674
113	5	3.87	3.915	154		1.650
023			3.905	246	1.650	1.650
025	40	3.535	3.561	2,0,1̄2		1.647
008			3.540	1,3,1̄4		1.646
025			3.531	245	1.600	1.600
1̄15	1	3.36	3.361	1,3,1̄4		1.600
115			3.350	2,0,1̄6	1.560	1.559
026	2	3.28	3.289	1,3,1̄4	1.549	1.559
1̄16			3.260	332̄		1.549
116	2	3.10	3.102	332		1.516
0,0,10			3.092	060	1.515	1.516
202̄	25	2.835	2.832	332̄		1.514
202			2.619	332		1.480
130	10	2.615	2.618	336̄	1.481	1.480
13̄2			2.610	332		1.480
132			2.550	064		1.479
204̄	20	2.545	2.549	338̄		1.447
132			2.544	066	1.422 B	1.447
20̄2	90	2.502	2.503	334̄		1.441
13̄4			2.502	338		1.441
134			2.412	334		1.441
206̄	35	2.408	2.412	066	1.419	1.439
134			2.412	0,6,0,20		1.416
136			2.402	068̄	1.398	1.398
0,0,12̄	15	2.358	2.364	336		1.389
068			2.360	068	1.384	1.389
136			2.357	1,5,1̄3		1.383
204			2.348			
136	10	2.345	2.347			
204			2.346			
1̄,1,11	20	2.237	2.370			
208			2.269			
206	5	2.165	2.165			
138			2.164			
0,0,14	1	2.050	2.051			
0,2,1̄0	10	2.020	2.050			
1,3,1̄0	20	1.995	2.023			
208	18	1.978	1.996			
1,3,1̄0			1.979			
2,0,1̄0	5	1.870 B	1.871			
1,3,1̄0			1.858			

Specimen from Ottre, Belgium. $a = 5.247$, $b = 9.094$, $c = 28.557$ Å. $\alpha = 90.5°$, $\beta = 97.3°$, and $\gamma = 89.9°$. Structure is of the g-type with IIb layers. Conditions as for A.

Table 6. Diagnostic X-ray diffraction lines of chlorite

	Ia			Ibα (97°)			Ib (90°)	
hkl	d	Int.	hkl	d	Int.	hkl	d	Int.
1. 200	2.65	1 1/2	1. 201̄	2.68	2 1/2	1. 200	2.69	2
2. 202	2.59	6	2. 202	2.60	1 1/2	2. 201	2.65	2 1/2
3. 202	2.39	6	3. 201	2.55	1/2	3. 202	2.51	10
4. 204	2.27	1	4. 203	2.47	7B	4. 203	2.34	1
5. 205	2.07	1/2	5. 202	2.40	1/2	5. 204 –	2.15	4
6. 204	2.01	3	6. 204	2.30	1/2	6. 205	1.96	1
			7. 205	2.11	2B	7. 206,008	1.78	3
			8. 204	2.01	1			

$a = 5.34$ Å $\beta = 97.1°$
$b = 9.25$
$c = 14.43$

$a = 5.36$ Å $\beta = 97.8°$
$b = 9.29$
$c = 14.45$

$a = 5.39$Å $\beta = 90°$
$b = 9.34$
$c = 14.14$

	IIb	
hkl	d	Int.
1. 201̄	2.66	1 1/2
2. 202	2.59	6
3. 201	2.55	5
4. 203	2.45	7
5. 202	2.39	4
6. 204	2.26	4
7. 205	2.07	1
8. 204	2.01	6

$a = 5.37$ Å $\beta = 97.0°$
$b = 9.30$
$c = 14.19$

372

The same general principles given in Chapter 2 for identification of 1:1 layer structures can be applied to chlorites, because in both types of structure there is an alternation of tetrahedral and octahedral sheets. The octahedral cations, which repeat at $b/3$ intervals in the chlorite structure, contribute only to X-ray reflections of index $k = 3n$ and not to reflections of index $k \neq 3n$. The octahedral anions also repeat ideally at $b/3$ intervals and contribute strongly to reflections of index $k = 3n$, but because of distortion from the ideal $b/3$ repeat pattern they may also contribute a small amount to $k \neq 3n$ reflections. The tetrahedral cations and basal oxygen atoms contribute to all reflections, but their contributions are most noticeable for the $k \neq 3n$ reflections. The different structural types of chlorite are easily identified, as a result, by $h0\ell$ precession photographs. Examples of observed and calculated patterns are illustrated in Figure 9.

There are two unique regular stacking polytypes that are related to, and give the same $h0\ell$ intensities as, each of the six possible semi-random stacking sequences. One of these regular polytypes has monoclinic symmetry, and the other has triclinic symmetry (Table 2). For this reason, comparison of $hk\ell$ and $\overline{hk\ell}$ intensities on precession or Weissenberg films serves to differentiate between the possible regular polytypes after the layer-interlayer type has been determined from the $h0\ell$ intensities [F($hk\ell$) = F($\overline{hk\ell}$) for monoclinic symmetry but not for triclinic]. An example of this application using precession photographs is shown in Figure 10, in which the symmetry left and right across Y^* and up and down across Z^* distinguishes the monoclinic IIb-2 polytype from the triclinic IIb-4. The polytypes IIab-4 and Ibb-3 are based on a triclinic-shaped unit cell with crystallographic $\alpha = 102°$ ($\alpha^* = 78°$), although crystallographic $\beta = 90°$. Neither polytype has been reported to date, but the calculated $0k\ell$ precession pattern is shown in Figure 10.

Optical properties

Graphs relating the variation in optical properties to composition within the chlorite series have been presented by many investigators. Figure 11 is taken from Albee (1962). Because a sequence of two 1:1 layers in a serpentine structure gives the same succession of tetrahedral and octahedral sheets as found in a chlorite 2:1 layer plus an interlayer sheet, optical properties by themselves do not distinguish well between serpentines and chlorites of similar compositions. It is best to determine first by X-ray study whether a 7 Å or a 14 Å structural unit is involved before using Figure 11 or comparable graphs.

Because chlorite in almost all cases is triclinic but nearly monoclinic, the optic plane is nearly parallel to (010), and the acute bisectrix is nearly normal to (001). The sign of elongation is always opposite that of the optic sign. Refractive indices (Fig. 11) and birefringence vary primarily as a function of the octahedral heavy atoms. At a β index of 1.630 ± 0.003, corresponding to a (Fe + Mn + Cr)/(Fe + Mn + Cr + Mg) content of 52%, the mineral is nearly isotropic for most specimens, and the optic sign changes from negative on the Fe-rich side to positive on the Mg-rich side.

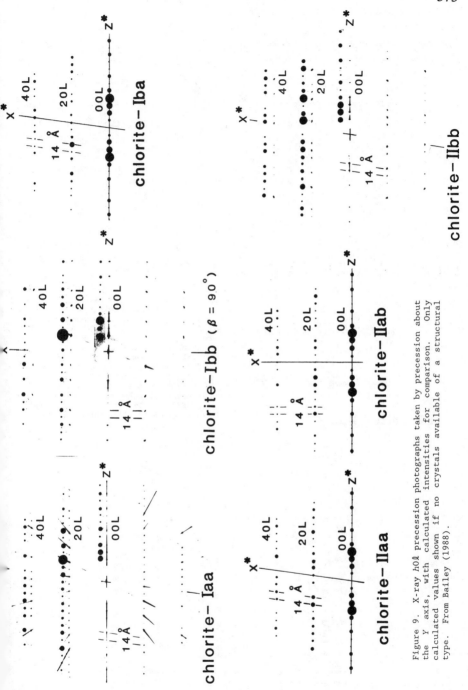

Figure 9. X-ray h0ℓ precession photographs taken by precession about the Y axis, with calculated intensities for comparison. Only calculated values shown if no crystals available of a structural type. From Bailey (1988).

374

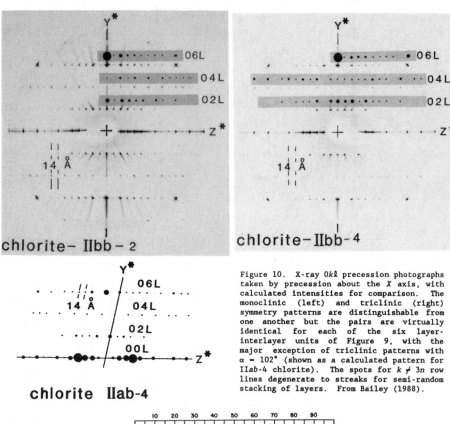

chlorite— IIbb-2

chlorite—IIbb-4

chlorite IIab-4

Figure 10. X-ray $0k\ell$ precession photographs taken by precession about the X axis, with calculated intensities for comparison. The monoclinic (left) and triclinic (right) symmetry patterns are distinguishable from one another but the pairs are virtually identical for each of the six layer-interlayer units of Figure 9, with the major exception of triclinic patterns with $\alpha = 102°$ (shown as a calculated pattern for IIab-4 chlorite). The spots for $k \neq 3n$ row lines degenerate to streaks for semi-random stacking of layers. From Bailey (1988).

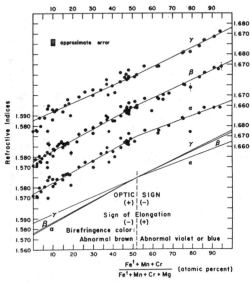

Figure 11. Graph of Albee (1962) relating refractive indices to total octahedral heavy atom content.

Abnormal interference colors occur near the sign change due to the mineral becoming isotropic for part of the spectrum and anisotropic for another part. The abnormal colors are blue to violet on the Fe-rich side and brown on the Mg-rich side. The birefringence increases in both directions away from the sign change.

The refractive indices and optic sign can be used in conjunction with the presence or absence of abnormal interference colors to give a reasonable estimate of the heavy atom content. But the scatter of points in Figure 11 suggests that extreme accuracy cannot be expected at the present time. Optical scatter is to be expected because of the multiple substitutions, vacant sites, oxidation, and varying O/OH ratios that are possible in chlorite as well as the experimental difficulties arising from the platy habit and oftentimes fine-grain size, dark color, and abnormal interference colors. No study has been made of the variation of optical properties with structural unit types. However, Albee (1962) observed that chlorites from amygdules and veins in mafic igneous rocks do not fit Figure 11 because of anomalous optic signs and considerably lower indices. Bailey and Brown (1962) showed that the less common structural types are likely to be found in low-temperature environments such as these, so that these deviations may be related to the structures. Craw and Jamieson (1984) also reported anomalous optical properties from low-temperature chlorites and suggested the anomalies are a result of metastable stacking sequences.

Most phyllosilicates are optically negative with two high refractive indices for vibration directions essentially in the plane of the layer and a lower index for a vibration direction essentially normal to the layer. The lower index is due to weaker bonding in that direction and consequent lesser secondary polarization involving the electric dipoles. For the chlorites this trend is followed in Fe-rich compositions, whereas specimens at Al-rich compositions are optically positive (Fig. 11). One possible explanation for the reversal of this trend in Al-rich chlorites is interaction involving the highly polarizable OH anions present in the chlorite interlayer and the increasing number of small Al^{3+} cations of high field strength in the interlayer as the Al end-member is approached.

Kepezhinskas (1965) made a statistical study of the variations of chlorite optical properties with composition, as expressed in the following equations for the β refractive index and the birefringence $(\gamma - \alpha)$.

$$\beta = 1.5757 + 0.00023Si + 0.004Al^{VI} + 0.0164Fe^{3+}$$

$$+ 0.0218Fe^{2+} - 0.0010Mg \quad . \tag{2}$$

$$(\gamma - \alpha) = 0.0646 - 0.0097Si - 0.0035^{IV} - 0.01335Fe^{3+}$$

$$- 0.0094Fe^{2+} - 0.0048Mg \quad . \tag{3}$$

For determinative purposes, Kepezhinskas recommends using a combination of optical properties and the specific gravity D to obtain the ferruginosity (F = atomic percentage of heavy elements).

$$F = 592.346\beta + 50.761D - 1065.282 \qquad (4)$$

or

$$F = 368.56n_c' + 74.42D - 1086.11(\gamma - \alpha) - 769.05 \quad , \quad (5)$$

where n_c' is the refractive index having its vibration direction closest to the Z crystallographic axis.

Infrared absorption

Tuddenham and Lyon (1959) state that the amount of substitution of Al for Si and the total iron content of natural chlorites can be determined from the infrared absorption bands in the 9.0 to 10.5 μm region (Si-O stretching band). The amount of tetrahedral substitution is estimated from the number and the shapes of the bands in this region. The total Fe content is measured from the position of the maximum absorption band in the same region. The 7 Å aluminian serpentines of similar composition can be recognized by the presence of absorption bands in the 2.7 to 2.8 μm region (O-H stretching band) that are either very weak or not present in true chlorites.

Stubican and Roy (1961) made an infrared study of Mg-Al synthetic chlorites and aluminian serpentines. They noted that a band at 1080 cm^{-1} found in the spectra of some natural chlorites is not present in the synthetic products, perhaps due to Si,Al disorder in the latter. The spectra of the 14 Å and 7 Å forms are similar, but with some distinctions that permit identification. There is generally no resolved band at 910 cm^{-1} in true chlorites, and the two bands in the 750 to 850 cm^{-1} frequency region are not as well resolved as in the 7 Å minerals. The authors state that the position of the strongest Si-O band in the 665 to 685 cm^{-1} region can be used for quantitative determination of octahedral and tetrahedral Al in both the chlorites and 7 Å serpentines.

Hayashi and Oinuma (1965, 1967) studied the infrared absorption spectra in the range 450 to 1200 and 3200 to 3800 cm^{-1} for several analyzed natural chlorites. They stated that dioctahedral chlorite (probably having mixed di,trioctahedral sheets) is recognizable by strong absorptions at 475 to 555, 692, 3340, 3520, and 3620 cm^{-1} and by lack of noticeable absorption near 760 cm^{-1}. The latter absorption is noticeable for Mg-rich trioctahedral chlorites and strong for Fe-rich species. They found a shift in the wave number of the absorption band at 540 to 560 cm^{-1} with increasing amount of octahedral Al, but a decrease in wave number for increasing Mg and Fe. They modified the results of Stubican and Roy (1961) by finding that Mg and Fe influence the position of the absorption band in the region 620 to 692 cm^{-1}, in addition to the effect of tetrahedral and octahedral Al. The wave number decreases with a decrease in octahedral Al and with increase in Mg and Fe. The b parameter of the unit cell increases in a regular manner with decrease in the wave numbers of the four absorption bands in the 540 to 560, 620 to 692, 744 to 765, and 970 to 1005 cm^{-1} regions. Hayashi and Oinuma (1967) also noted a decrease in frequency of vibration for the two OH absorption bands in the regions 3400 to 3436 and 3560 to 3586 cm^{-1}

with increase in Fe content.   This was interpreted as a smaller O-OH interlayer distance for Fe-rich chlorites.

Shirozu (1980) studied the IR stretching band of OH in more detail.   Two broad bands near 3560 cm^{-1} and 3420 cm^{-1} (bands 2 and 3, respectively) are ascribed to the interlayer OH.   These may be accompanied by a small band at 3680-3620 cm^{-1} (band 1) ascribed to the inner OH of the 2:1 layer.   The vibration frequency of the latter should be controlled primarily by the 2:1 octahedral cations and secondarily by the tetrahedral cations, but because of its diffuse nature and overlap with band 2 little information has been derived. Bands 2 and 3 are due to two different O--OH distances across the 2:1 layer--interlayer junction, and they are interpreted as caused by different electrostatic forces on the basal oxygen atoms arising from their linkage to different tetrahedral cation pairs.   Band 2 is ascribed to (SiSi)O--OH and band 3 to (SiAl)O--OH vibrations.   The intensities of these two absorption bands best fit a disordered distribution of tetrahedral Si,Al, in which Al--Al neighbors also are permitted.   Although the intensities of the two bands are dependent only on the tetrahedral composition and its distribution, the vibration frequencies are due to the compositions of all of the tetrahedral and octahedral sheets.   O--OH distances between layer and interlayer calculated from the stretching frequencies and intensities of bands 2 and 3 are in fairly good agreement with those obtained by one-dimensional refinement of z atomic positions using 00ℓ X-ray intensities, and it is concluded that each O--OH distance is shortened by the surplus negative charge on the basal oxygens caused by the linked tetrahedral Al.

In di,trioctahedral sudoite three broad OH stretching bands are noted at 3620, 3520, and 3340 cm^{-1}, in contrast to the two bands for trioctahedral species (Oinuma and Hayashi, 1966; Hayashi and Oinuma, 1967; Shirozu and Hayashi, 1976).   The intensification of the band at 3620 cm^{-1} ascribed to the inner OH is believed due to overlap with a third interlayer OH stretching band.   This implies three different O-OH distances, which may be due to an expected corrugation of the basal oxygen surface of the dioctahedral 2:1 layer (Shirozu and Hayashi, 1976).

Shirozu and Ishida (1982) interpreted the details of IR patterns in the 800-700 cm^{-1} region of untreated and of partially deuterated trioctahedral chlorites as due to overlap of lattice vibrations with libration (bending) of the interlayer OH, expressed by the (SiAl)O--OH band near 800 cm^{-1} and the (SiSi)O--OH band near 720 cm^{-1}, and of the inner OH of the 2:1 layer (near 650 cm^{-1}).   The librational frequencies increase with stronger layer to interlayer bonding and thus correlate with the OH stretching frequencies cited above in a reciprocal relationship.   For the di,trioctahedral species sudoite and cookeite, in which the O--H vectors of the inner OH are believed to be nearly parallel to (001), the inner OH libration band is shifted to about 930 cm^{-1}.   The interlayer OH libration bands of sudoite (Mg,Al-rich) are similar to those of trioctahedral chlorites but those of cookeite (Al,Li-rich) are different.

Shirozu (1985) summarized the effects on the IR absorption spectra in the 4000-400 cm^{-1} range of the tetrahedral and octahedral

378

compositions for trioctahedral chlorites along the Mg-Fe join. The bands in the 3620-3400 cm^{-1} range, 1100-900 cm^{-1} range, and near 750 cm^{-1} can be used to give approximate tetrahedral compositions. The relative intensities of the two OH stretching bands cited previously as bands 2 and 3 can be related to tetrahedral compositions by the equation

$$E_{3420\sim}/(E_{3560\sim} + E_{3420\sim}) = 0.078 + 0.3x \qquad (6)$$

where x is the number of tetrahedral Al cations per four tetrahedral sites. The octahedral composition can be estimated by the nature of the bands at 3680-3615 cm^{-1}, in the 700-600 cm^{-1} range, and at 470-410 cm^{-1}. The frequency of the OH libration band (overlapping with two lattice vibration bands for Mg-rich compositions) varies progressively with Mg,Fe content, and was combined with equation (6) into a determinative graph for semi-quantitative estimation of the composition by Shirozu (1985). The graph is not suitable for dioctahedral chlorites or for trioctahedral chlorites with appreciable octahedral vacancies. No spectral differences due to different chlorite polytypes were noted.

Mössbauer analysis

Mössbauer spectra have proven useful in micas for determining the $Fe^{2+}:Fe^{3+}$ ratio, the amount of tetrahedrally coordinated Fe^{3+}, and the amounts of Fe^{2+} and Fe^{3+} in the octahedral trans M(1) and cis M(2) sites of the 2:1 layer. For the more complex structure of the chlorites, however, the application has been less successful. The $Fe^{2+}:Fe^{3+}$ ratio can be determined, and Goodman and Bain (1979) reported the presence of tetrahedral Fe^{3+} in several chlorites. Although Fe^{3+} in one octahedral environment and Fe^{2+} in two different octahedral environments were reported by Kodama et al. (1982) and Townsend et al. (1986) for three chlorites, it did not prove possible to assign the observed spectra to individual octahedral sites in the 2:1 layer or the interlayer. The Mössbauer parameters are reported as similar to those in biotite, where assignments can be made in cis and trans octahedra, but the possibility of Fe in somewhat similar octahedral sites in the interlayer must be considered also. Although cis and trans octahedra do not exist in the interlayer, the M(4) site on a center of symmetry is different than the two noncentrosymmetric M(3) sites. Structural refinement indicates R^{3+} cations tend to concentrate in M(4) and R^{2+} in M(3) in the most stable (ordered) state, but that such ordering is not always complete.

DETERMINATION OF COMPOSITION

Direct analysis

The usual direct method for complete analysis of chlorite is wet chemical analysis. The absolute quantities of the elements present in the unit cell can be obtained if the density and the volume of the unit cell are available in addition to the chemical analysis. In most cases this is not possible, and the allocation to a structural formula must be made on the basis of some assumption about the formula unit. For chlorites, one of the following assumptions is usually made: (1) total (O + OH) = 18; (2) total O = 14, excluding

H_2O; (3) total cation valence = 28, excluding H; (4) total number of cations = 10, excluding H. The latter assumption (4) is not valid in terms of our modern knowledge of chlorite compositions, because octahedral vacancies are common. For direct analysis by electron microprobe, where H_2O is not measured, either assumption (2) or (3) must be used and the valencies of the transition metals present must be determined by other means (e.g., Mössbauer or optical spectra).

Indirect analysis

Graphs of the (001) and (060) spacings and of 00ℓ intensities versus composition have been used to determine the tetrahedral and octahedral cation populations in chlorite and the distribution of heavy atoms between the two octahedral sheets. Increasing substitution of Al for Si in the tetrahedral sites is said to decrease the layer thickness, measured by $d(001) = c \sin \beta$, because of the increased layer charge and greater bond strength between layer and interlayer (Shirozu, 1958). Increasing substitution of the larger Fe^{2+} ion for Mg or Al in the octahedral sites tends to increase the lateral dimensions of the sheets, as measured by the $d(060)$ value. The two octahedral sheets are separated along Z by $c/2$. For this reason, the scattered contributions from the two sheets are exactly in phase for the even orders of 00ℓ, so that these F values vary as a function of total heavy atom content, irrespective of distribution between sheets. If the heavy atoms are distributed symmetrically between the two octahedral sheets, the odd-order F values are independent of octahedral composition due to exact cancellation of the scattered contributions from the two sheets. The odd-order F values then are entirely due to the contributions from the tetrahedral sheets. For an asymmetric distribution of heavy atoms, the odd-order F values have an added contribution that is a function of the difference in scattering power between the two octahedral sheets. Experimental determination of the ratios of structure amplitudes for different 00ℓ reflections theoretically should permit determination of total heavy atom content as well as degree of asymmetry. There are pitfalls in the use of all of these indirect methods, but useful semi-quantitative information can be obtained for many specimens.

Bailey (1975) attempted to test the validity of the various graphs and regression equations available at that time by comparing their predictions with data found experimentally for four chlorites whose structures had been determined in detail. He concluded that both tetrahedral Al and total octahedral heavy atoms can be estimated by X-ray spacing methods with an average error of about 10%, although individual determinations may be in error up to 20%. A 10% error for these specimens corresponds approximately to 0.1 atoms of Al^{IV} or Fe^{2+} per formula unit.

Tetrahedral Al was found to be estimated best by equation (7) from Brindley (1961)

$$d(001) = 14.55 \text{ Å} - 0.29x \qquad (7)$$

where $x = Al^{IV}$ per 4.0 tetrahedral positions, or from equation (8) from Kepezhinskas (1965)

$$d(001) = 14.648 \text{ Å} - 0.378x \tag{8}$$

Some caution must be used in such estimates because structural factors and octahedral composition may affect $d(001)$ quite independently of tetrahedral Al content. For example, the four test chlorites do not show a linear variation of $d(001)$ with Al^{IV}, primarily as a result of the closer approach of the 2:1 layer and the interlayer with increasing heavy atom content. All equations cited above were derived for trioctahedral chlorites and should not be applied to dioctahedral species. Eggleton and Bailey (1967) and Shirozu and Higashi (1976) found that trioctahedral $d(001)$ graphs predicted Al^{IV} values that were too high by 0.3 to 0.6 atoms for several analyzed dioctahedral chlorites. Lin and Bailey (1985) suggested the smaller $d(001)$ values for dioctahedral and di,trioctahedral species are due primarily to a thinner dioctahedral sheet in the 2:1 layer plus possibly a slightly thinner total interlayer space.

Total octahedral heavy atoms were found to be estimated best by b cell parameter equation (9) derived from the data of von Engelhardt (1942), by equations (10) and (11) from the data of Shirozu (1958), or by regression equation (12) from Kepezhinskas (1965).

$$b = 9.22 \text{ Å} + 0.028Fe^{2+} \tag{9}$$

$$b = 9.210 \text{ Å} + 0.039(Fe^{2+},Mn) \tag{10}$$

$$b = 9.210 \text{ Å} + 0.037(Fe^{2+},Mn) \tag{11}$$

$$F = 527.025b - 39.461 \ d(001) - 4283.797 \tag{12}$$

where F is ferruginosity or total heavy atom content. The equations above were calibrated primarily by common IIb chlorites and only yield Al^{IV} and $(Fe, Mn)^{VI}$ values. In order to derive the total composition it is usually assumed that tetrahedral Al equals octahedral Al and that no octahedral vacancies are present. Whittle (1986) showed by TEM/EDAX analyses that the latter two assumptions are not valid for sedimentary diagenetic chlorites of the $Ib(\beta=90°)$ structural type, for which he found that octahedral Al usually exceeds tetrahedral Al by more than 20% and that the total cations per formula unit are correspondingly reduced, presumably by compensating octahedral vacancies. He also found that equation (7) overestimates Al^{IV} and is closer to Al^{VI} for $Ib(\beta=90°)$ chlorites and that equation (11) overestimates octahedral Mg + Fe. The equations have not been tested by modern analytical tools for the more common IIb chlorites.

Methods using the relative intensities (I) or structure amplitudes (F) of the basal reflections are not as accurate as the spacing methods because of (1) variation in the z parameters of the atoms independent of heavy atom content and (2) the extreme sensitivity of F and I to heavy atom content. Such methods are, at present, the only means short of structural analysis, however, for determining any asymmetry in the distribution of heavy atoms between the 2:1 octahedral sheet and the interlayer. Best results for asymmetry were obtained using the tables of Brindley and Gillery

(1956) or the curve of Petruk (1964). The latter curve plots asymmetry as the number of heavy atoms in the 2:1 layer minus those in the interlayer. Walker (1987) noted that the significance of this difference is unambiguous only for a value of zero, for which there is no asymmetry of distribution. For all other values the significance of the difference depends on the total number of heavy atoms present, however, and a better function to use is the ratio of the number of heavy atoms in the octahedral sheet of the 2:1 layer to the total number of heavy atoms present (the mole fraction). This distribution index varies from 0 to 1 for the extreme cases of no heavy atoms in the 2:1 layer to no heavy atoms in the interlayer, respectively, whereas a value of 0.5 indicates a symmetric distribution. This index allows easier comparison of samples having different total heavy atom contents.

Extreme asymmetry in the distribution of heavy atoms can even lead to misidentification of chlorite as serpentine. A Ni,Fe-rich specimen from Western Australia, identified originally as serpentine because of lack of an observable 14 Å powder line, was shown later to be a chlorite having about 20 more electrons per formula unit in the interlayer than in the 2:1 octahedral sheet (Bailey and Riley, 1977). This asymmetry reduces the 001 intensity drastically, but a 14 Å periodicity is still evident from other 00l spacings (e.g., from strong 003 and 007 reflections).

CRYSTAL CHEMISTRY

Structural stabilities

Little is known about the thermodynamic stability of the different structural types of chlorite because the IIb form is the only one that has been produced synthetically and is also by far the most abundant in nature. IIb chlorite thus is assumed to be the most stable form and all other structural types are assumed to be metastable--probably low temperature forms that follow Ostwald's step-crystallization rule. At least a qualitative idea of their relative stabilities can be gained by looking at the details of each structure.

The different layer-interlayer-layer sequences that are possible in the chlorite structure create varying amounts of cation-cation repulsion and cation-anion attraction as a result of the different superposition of sheets. Bailey and Brown (1962) and Shirozu and Bailey (1965) attempted to explain the observed relative abundances of the chlorite structural types according to the relative stabilities indicated by these interatomic forces. Repulsion between the superimposed interlayer and tetrahedral cations in the Ia and IIa structural units is considered the most important single factor in reducing the stability of these two unit types relative to Ib and IIb units. Three other factors considered are:

(1) repulsion between the superposed interlayer and 2:1 layer octahedral cations, possible only in the Ib unit structure;

(2) attraction between the superposed 2:1 layer octahedral cations and interlayer hydroxyls found in the Ia and IIb units; and

(3) relative distances of the attractions from the surface basal oxygens to the octahedral cations within the 2:1 layer and to the interlayer cations, due to tetrahedral rotation of the basal oxygens toward the nearest interlayer hydroxyl groups.

The effects of all the repulsive and attractive forces mentioned are summarized for the four layer-interlayer structural types in Table 7, where a plus or minus sign indicates that the stability is either enhanced or decreased by the effect.

Table 7 applies only to structures in which adjacent 2:1 layers are positioned symmetrically relative to the interlayer, i.e., position a on both sides or b on both sides. For asymmetric Iba and IIab structures the difference in positions of adjacent layers must be factored into Table 7. Side views of all of the layer-interlayer assemblages are shown in Figure 12. Weighting of the several interatomic forces according to the distances over which they must operate leads to a qualitative ranking of structural stabilities

$$II b > I b (\beta=90°) > I ba (\beta=97°) > I a > II ab (\beta=90°) > II a (\beta=97°)$$

143   37   13   10   0   0

Listed under each structural type above is the number of examples of each found in a survey of 303 different chlorite specimens by Bailey and Brown (1962), and the ranking of structural stabilities is in agreement with the abundances observed in their study. The IIb structure accounts for 80 per cent of the specimens examined. This is the structural type present in the specimens mentioned previously as having been examined by Pauling, McMurchy, Garrido, Robinson and Brindley, Brindley et al., and Steinfink in their studies. It must be considered the most stable structural arrangement. The ortho-hexagonal Ib structure (Ib-odd, $\beta=90°$) had been recognized as having a different powder pattern with $\beta = 90°$ by von Engelhardt (1942) and by Shirozu (1955). The Ia and Iba structures with monoclinic-shaped unit cells had not been recognized prior to the Bailey and Brown study. Neither of the two possible IIa and IIab stacking sequences have been reported to date, except in the chlorite-brittle mica hybrid franklinfurnaceite structure (see below).

Table 7 shows that the structural factors enhancing stability are favorable for the Ia structure, except for the repulsions between interlayer cations and tetrahedral cations. The various layer-interlayer arrangements mentioned as possible for chlorite in this chapter are possible also for vermiculite, which differs from chlorite primarily in having an incomplete interlayer with only a few exchangeable interlayer cations. This reduces the cation-cation repulsion factor drastically, thereby stabilizing vermiculite in the Ia configuration.

Table 7. Estimated structural stabilities of four layer-interlayer units

Atoms Concerned	Distance	IIb	Ib	Ia	IIa	Type of Force
T-M$_{Int}$	4.4Å	++	++	- -	- -	} cation-cation repulsion
M$_{2:1}$-M$_{Int}$	7.1	+	-	+	+	
M$_{2:1}$-(OH)$_{Int}$	6.1	+		+		cation-anion attraction
M$_{2:1}$-O$_S$	3.4	+	-	+	-	} difference of attraction due
M$_{Int}$-O$_S$	3.9	-		-		} to rotation of tetrahedra

T - tetrahedral cation; M$_{2:1}$ - octahedral cation in 2:1 layer; M$_{Int}$ - interlayer cation; O$_S$ - surface oxygen in 2:1 layer; (OH)$_{Int}$ - interlayer hydroxyl.

Structural factors that enhance stability are marked + or ++. Factors that diminish stability are marked by - or --. Resultant relative stabilities are IIb > Ib > Ia > IIa.

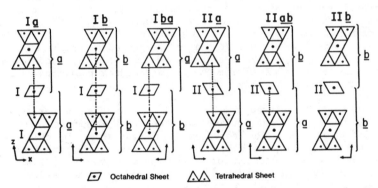

Figure 12. [010] diagrammatic views of six different layer-interlayer assemblages of chlorite to show differing amounts and kinds of cation-cation repulsion.

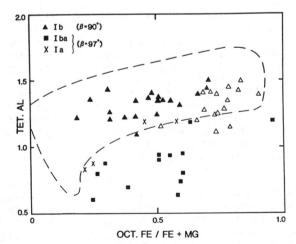

Figure 13. Graph of octahedral Fe/Fe+Mg vs. tetrahedral Al per 4.0 sites for metastable, lower temperature chlorites. Compositions determined from spacing graphs of Shirozu (1958). Dashed line encloses composition field of stable IIb chlorite, as determined in Figure 2. Modified from Bailey and Brown (1962).

Environmental significance of polytypes

There appears to be a correlation between the type of chlorite structure adopted and the energy available in the environment of formation. Bailey and Brown (1962) noted that the stable chlorite in the chlorite zone of regional metamorphism and in medium-to-high temperature ore deposits is almost exclusively the IIb form. The other structural types are found in lower-temperature assemblages, including oolitic primary chlorite in iron formations, diagenetic chloritic cement in porous sandstones, chloritic alteration products of pyroclastics, low-rank chloritic metamorphic products, chlorite in low-temperature hydrothermal veins, and chloritic deuteric alteration products of ferromagnesian minerals. It was discussed above that there is a correlation between the frequency of occurrence of the structural types and their relative stabilities, as judged by structural energy comparisons. Bailey and Brown suggested that when sufficient energy is available in the environment of formation, the most stable, lowest structural energy chlorite will crystallize. This is the IIb form. At lower temperatures, one of the higher structural energy units will form metastably. This structure may remain indefinitely or, if energy is added to the system through metamorphism, may invert to the stable IIb type. They also suggested, based on the use of the indirect spacing graphs of Shirozu (1958), that the $Iba(\beta=97°)$ structure in their study tends to have lesser substitution of Al for Si in tetrahedral positions than the other structural types (Fig. 13).

Hayes (1970) viewed the presence of the high-temperature IIb chlorite in sediments as evidence of a detrital origin in most cases. He pointed out that the occurrence of clay-size IIb chlorite in unmetamorphosed, eugeosynclinal sediments is to be expected from the weathering and erosion of metamorphic terranes. The lower-temperature Ib chlorite found in non-eugeosynclinal sediments is authigenic in most cases, according to the evidence of thin-section petrography. It often occurs as euhedral plates. It may also occur as clay pellets and may be similar to glauconite in color, morphology, and association with organisms. Hayes recognized a sequence of transformations in the structural form of diagenetic chlorite in sediments as a function of depth of burial. The lowest temperature form nearest the surface is a type I chlorite with disordered stacking. This transforms sequentially to $Iba(\beta=97°)$ and $Ib(\beta=90°)$ with greater burial depth and greater temperature. Thermal energy approaching that of low-grade metamorphism is usually necessary to cause $Ib(\beta=90°)$ chlorite to convert to the more stable IIb structure. This sequence is analogous to the transformations $1M_d$ → $1M$ → $2M_1$ in muscovite. The position of the Ia chlorite structure in these sedimentary sequences was uncertain in Hayes' study, but was thought to be related to the uplift and weathering of Ib chlorite so that Ia chlorite is intermediate in evolution toward Ia vermiculite.

Weaver et al. (1984) found Ia chlorite porphyroblasts in high grade diagenetic rocks having incipient slaty cleavage (T_{max} = 250°C) in a shale-slate transition sequence in the southern Appalachians, whereas the IIb polytype occurs in the higher grade rocks with slaty cleavage in the lower epizone (T = 250° to 330°C). The data of Bailey and Brown (1962) indicate that other Ia chlorites can be of hydrothermal origin or form by recrystallization of I*b* chlorite near quartz veins. Many chlorites of dioctahedral nature or of unusual compositions containing Li, Mn, Ni, and Cr are likely to be of the I*a* type.

Walker (1987) summarized occurrences of all type I chlorites. He noted that type I tends to form preferentially in lithologies that afford open space in which the chlorite can crystallize, as indicated by vug and vein fillings and by its preference for sandstones and granular limestones (as opposed to pelites and micritic limestones). Type I also may form preferentially in an environment of high pH, as indicated by its frequent occurrence in limestones, and in oxidizing conditions associated with iron formations.

Variation of composition with temperature and genesis

Figure 2 shows a broad trend for increasing tetrahedral Al to be correlated with an increasing Fe/(Fe + Mg) ratio for a random selection of the common II*b* chlorites. Bailey and Brown (1962) noted that in some, but not all, of these specimens increasing octahedral Fe and tetrahedral Al contents seem to be correlated with increasing temperature of formation. Likewise, Cathelineau and Nieva (1985) found that tetrahedral charge correlates positively with iron content and negatively with octahedral vacancy for a suite of hydrothermal chlorites in altered andesites from the Los Azufres geothermal area of Mexico. They also found that Al^{IV} and Fe^{VI} are positively correlated with temperature, whereas octahedral vacancy is negatively correlated with temperature. They suggested using a plot of Al^{IV} versus temperature determined from bore hole measurements and liquid inclusions for their specimens as a chlorite geothermometer (Fig. 14). Mg and the Fe/(Fe + Mg) ratio are very poorly correlated with temperature. The Fe/(Fe + Mg) ratio, however, is dependent on the bulk composition of the host rock. They found a 1:1 trend in 11 different areas between Fe/(Fe + Mg) in the host rock and in the chlorite. Shikazono and Kawahata (1987) likewise found a 1:1 ratio for the MgO/FeO ratio in fresh rock and in chlorite altered from the rock in many areas. But two significant deviations from this linearity occur for hydrothermal chlorites in Kuroko type ore deposits and in Neogene vein deposits from Japan. Chlorites in Kuroko deposits are extremely Mg-rich as a consequence of the ore fluids being dominated by Mg-rich seawater. Chlorites in Neogene Cu-Pb-Zn vein deposits contain much larger, but variable, amounts of Fe. Isotope studies indicate the latter ore fluids are dominated by meteoric waters.

Structural details of trioctahedral II*b* chlorite

Triclinic symmetry. The triclinic II*b*-4 polytype of ideal symmetry $C\bar{1}$ is the most abundant regular-stacking one-layer chlorite that occurs in nature, and the crystal structures of five specimens have been refined by Steinfink (1958b), Phillips et al. (1980), Joswig et al. (1980), Joswig and Fuess (1989), and Zheng and Bailey (1989). The results of these refinements are similar in that no ordering of tetrahedral or octahedral cations was found within the 2:1 layers, even when refined in subgroup symmetry, but partial to

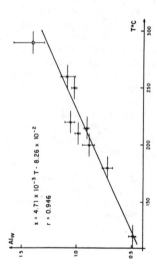

$x = 4.71 \times 10^{-3} \, T - 8.26 \times 10^{-2}$

$r = 0.946$

Figure 14. Plot of tetrahedral Al versus temperature for chlorites from the Los Azufres (Mexico) geothermal system. From Cathelineau and Nieva (1985).

Table 8. Details of Refined Chlorite Structures

Reference	Species	Composition	α_{tet}	τ_{tet}	ψ_{oct}
1. Joswig et al. (1980)	clinochlore IIb4 (Zillerthal, Tirol)	$(Mg_{2.40}Al_{0.30}Fe_{0.30})(Si_{3.16}Al_{0.84})O_{10}(OH)_2 \cdot$ $(Mg_{2.50}Al_{0.50})(OH)_6$	6.5	110.6,110.7	58.7,58.7x2 / 60.8x2,60.4
2. Phillips et al. (1980)	clinochlore IIb-4 (Siskiyou, CA)	$(Mg_{2.95}Al_{0.05})(Si_{2.99}Al_{1.01})O_{10}(OH)_2 \cdot$ $(Mg_{1.97}Al_{0.66}Cr_{0.25}Fe^{3+}_{0.06}Fe^{2+}_{0.06})(OH)_6$	7.2	110.8,111.0	58.8,58.8x2 / 61.4x2,59.6
3. Phillips et al. (1980)	clinochlore IIb-4 (Day Book Body, NC)	$(Mg_{2.97}Al_{0.03})(Si_{3.02}Al_{0.98})O_{10}(OH)_2 \cdot$ $(Mg_{1.98}Al_{0.69}Cr_{0.23}Fe^{3+}_{0.04}Fe^{2+}_{0.04}Ni_{0.02})(OH)_6$	7.2	110.7,110.8	58.8,58.8x2 / 61.4x2,59.6
4. Zheng & Bailey (1989)	clinochlore Ib-4 (Mochako, Kenya)	$(Mg_{2.73}Fe^{3+}_{0.06}Cr_{0.13})(Si_{2.96}Al_{1.04})O_{10}(OH)_2 \cdot$ $(Mg_{2.0}Al_{0.54}Fe^{2+}_{0.09}Fe^{3+}_{0.17}Ni_{0.01})(OH)_6$	6.9	110.8,110.8	58.6,58.6x2 / 61.3x2,59.4

	Reference	Mineral / Locality	Formula				
5.	Joswig & Fuess (1989)	clinochlore Ib-4 (Achmatov, USSR)	$(Mg_{2.70}Fe^{2+}_{0.28})(Si_{2.85}Al_{1.15})O_{10}(OH)_2 \cdot$ $(Mg_{1.85}Al_{0.97}Fe^{3+}_{0.18})(OH)_6$	7.4	110.8,110.8	58.8, 58.7x2	61.2x2, 60.1
6.	Rule & Bailey (1987)	clinochlore IIb-2 (Washington, DC)	$(Mg_{1.68}Fe_{0.52}Al_{0.23}Fe^{3+}_{0.14})(Si_{2.62}Al_{1.38})O_{10}(OH)_2 \cdot$ $(Mg_{1.28}Fe^{2+}_{0.64}Al_{1.05})(OH)_6$	8.5	111.1	58.8, 58.7x2	61.4x2, 58.3
7.	Zheng & Bailey (1989)	clinochlore IIb-2 (Mochako, Kenya)	$(Mg_{2.73}Fe^{3+}_{0.06}Cr_{0.13})(Si_{2.96}Al_{1.04})O_{10}(OH)_2 \cdot$ $(Mg_{2.0}Al_{0.54}Fe^{2+}_{0.09}Fe^{3+}_{0.17}Ni_{0.01})(OH)_6$	6.8	110.9	58.7, 58.6x2	61.0x2, 59.8
8.	Joswig et al. (1989)	clinochlore IIb-2 (Achmatov, USSR)	$(Mg_{2.70}Fe^{2+}_{0.28})(Si_{2.85}Al_{1.85}Al_{1.15})O_{10}(OH)_2 \cdot$ $(Mg_{1.85}Al_{0.97}Fe^{3+}_{0.18})(OH)_6$	7.5	110.7	58.7, 58.7x2	61.0x2, 60.1
9.	Bailey (1986)	clinochlore Ia-4 (Erzincan, Turkey)	$Mg_{3.0}(Si_{3.0}Al_{1.0})O_{10}(OH)_2 \cdot$ $(Mg_{2.0}Cr_{0.7}Al_{0.2}Fe_{0.1})(OH)_6$	5.5	111.0,110.8	58.7, 58.7x2	60.6x2, 59.6
10.	Shirozu & Bailey (1965)	chamosite Ibb(odd) (Tazawa, Japan)	$(Fe_{2.0}Mg_{0.8}Al_{0.2})(Si_{2.7}Al_{1.7}Al_{1.3})O_{10}(OH)_2 \cdot$ $(Fe_{1.4}Mg_{0.5}Al_{1.1})(OH)_6$	5.0	110.6	58.7, 58.7x2	61.2x2, 61.2
11.	Aleksandrova et al. (1972)	donbassite Ia-2 (Novaya Zemlya, USSR)	$Al_{2.00}(Si_{3.14}Al_{0.86})O_{10}(OH)_2 \cdot$ $(Al_{2.10}Fe^{3+}_{0.04}Fe^{2+}_{0.01}Mg_{0.08}Li_{0.26})(OH)_6$	13.5	110.3,110.3	62.2(□), 56.4, 56.7	58.0, 58.7, 62.6
12.	Peacor et al. (1988)	franklinfurnaceite IIaa-1 (Franklin, NJ)	$(Mn^{2+}_{1.50}Mn^{3+}_{1.00}Mg_{0.35}Zn_{0.15})Si_{1.95}Zn_{2.05})O_{10}(OH)_2 \cdot$ $Ca_2(Mn^{2+}_{0.55}Fe^{3+}_{0.75}Al_{0.25}Mg_{0.30}Zn_{0.15})(OH)_6$	23.5	109.0,110.4	59.6, 57.9, 59.9	54.5, 56.8, 59.6 (□)

388

Table 8 (continued)

	Sheet Thicknesses (Å) Tet.	Oct.	Int.	Separation 2:1-Int. (Å)	Mean Bond Lengths (Å) T--O	M--O,OH	Int. O--OH	Intralayer Shift in 2:1	Shift Adj. 2:1 Layers	Resultant Shift
1.	2.242	2.161	1.995	2.823	1.651, 1.650	2.081, 2.078x2 / 2.045x2, 2.022	2.915, 2.907, 2.915	+0.335a_2	+0.337a_3	-0.337a_1 / -0.000b_1
2.	2.251	2.151	1.985	2.807	1.656, 1.661	2.078, 2.078x2 / 2.076x2, 1.960	2.914, 2.913, 2.852	+0.362a_2	+0.361a_3	-0.348a_1 / -0.014b_1
3.	2.247	2.147	1.976	2.810	1.654, 1.652	2.074, 2.075x2 / 2.068x2, 1.962	2.916, 2.908, 2.864	+0.357a_2	+0.357a_3	-0.345a_1 / -0.012b_1
4.	2.241	2.171	1.994	2.802	1.652, 1.652	2.082, 2.082x2 / 2.079x2, 1.957	2.906, 2.857, 2.906	+0.36a_2	+0.360a_3	-0.34a_1 / -0.01a_1
5.	2.252	2.157	1.985	2.797	1.654, 1.657	2.079, 2.077x2 / 2.057x2, 1.991	2.890, 2.894, 2.862	+0.351a_2	+0.35a_3	-0.346a_1 / -0.009b_1
6.	2.265	2.166	2.024	2.730	1.668	2.092, 2.84x2 / 2.117x2, 1.929	2.774, 2.831x2	+0.335a_1	+0.370a_1	-0.295a_1
7.	2.243	2.170	2.005	2.805	1.655	2.086, 2.082x2 / 2.067x2, 1.993	2.875, 2.903x2	+0.334a_1	+0.348a_1	-0.317a_1
8.	2.248	2.157	1.990	2.792	1.654	2.078, 2.077x2 / 2.055x2, 1.997	2.882, 2.860x2	+0.334a_1	+0.347a_1	-0.319a_1
9.	2.240	2.167	2.031	2.827	1.650, 1.652	2.083, 1.081x2 / 2.068x2, 2.007	2.912, 2.912, 2.916	-0.334a_3	+0.333b_2	-0.334a_1
10.	2.252	2.182	1.975	1.752	1.670	2.102x3 / 2.050x3	2.848x2, 2.856	disordered stacking		0
11.	2.231	2.118	2.011	2.768	1.675, 1.617	2.270(□), 1.913, 1.928 / 1.935, 2.181, 1.898	2.663, 3.061, 2.801	-0.375a_1	0	-0.375a_1
12.	2.428	2.178	2.361	2.502	1.971, 1.648	2.154, 2.051, 2.174 / 2.032, 2.155, 2.330(□)	---	-0.320a_1	-0.004a_1	-0.324a_1

complete ordering of octahedral R^{3+} cations relative to R^{2+} was found in the interlayer sheet.

All five refined specimens would be termed clinochlore in the nomenclature of Bayliss (1975) with small substitutions of octahedral Cr^{3+} in three specimens. There is no evidence for tetrahedral Cr^{3+} in any chlorite. According to optical absorption spectra, Cr^{3+} bonded to six OH groups in the interlayer sheet of chlorite gives it a purple color whereas Cr^{3+} bonded primarily to oxygen in the octahedral sheet of the 2:1 layer imparts a green color, as in the mica fuchsite (Neuhaus, 1960). Phillips et al. (1980) found by 1-dimensional electron density maps for 10 purple chromian clinochlores that Cr^{3+} tends to concentrate preferentially in the interlayer sheet, and for two triclinic specimens refined in detail Cr+Al orders into site M(4) on an inversion center in the interlayer (Table 8).

Location of a trivalent element in site M(4) leads to repulsion of the tetrahedral sheets above and below the interlayer, to which they are only weakly bound by long hydrogen bonds, so that the distance from M(4) to the nearest tetrahedral cations is increased and the crystallographic α angle deviates from 90° by 0.4 to 0.5°. Cr^{3+} gains additional crystal field stabilization energy by locating in the less distorted M(4) site. A similar ordering pattern and triclinic slewing of the cell was found by Zheng and Bailey (1989) for a IIb-4 specimen from Kenya originally refined incompletely by Steinfink (1958b, 1961). Although the same interlayer ordering pattern was found also by neutron diffraction for a IIb-4 specimen from Zillertal, Tyrol, by Joswig et al. (1980), the ordering in that case proved to be incomplete and the unit cell was not distorted to triclinic geometry. A clinochlore IIb-4 crystal from the Achmatov mine, Ural Mtns., USSR, refined with X-ray data by Joswig and Fuess (1989) also showed partial ordering of Al in interlayer site M(4). The degree of ordering found was greater than that in the Zillerthal specimen but less than that in the three triclinic specimens refined by Phillips et al. (1980) and by Zheng and Bailey (1989), and the α angle was accordingly intermediate in value in the Achmatov specimen.

Bish and Giese (1981) calculated by the method of minimization of electrostatic energy that ordering of a trivalent element into site M(4) of the IIb-4 structure substantially increases its interlayer bonding energy and resultant stability. They also calculated that a chlorite structure with neutral layers is theoretically possible due to the strength of the long hydrogen bonds, but a charged interlayer dramatically increases the interlayer bonding energy. All recent refinements have located the hydrogen protons associated with the OH groups, but the reported positions can be considered accurate only for those studies using neutron diffraction. For the interlayer OH groups the H^{+} protons participate in slightly bent hydrogen bonds of approximate length 2.9 Å to the adjacent basal oxygen surfaces.

Details of these structures, as well as those of other chlorites described in this section are listed in Table 8.

Monoclinic symmetry. The crystal structure of a regular-stacking 1-layer II*b*-2 ferroan clinochlore of monoclinic symmetry from Washington, D.C., was refined partially by Steinfink (1958a, 1961) and in more detail by Rule and Bailey (1987). The original wet chemical analysis of this specimen indicated the Fe present to be primarily ferrous, whereas a later analysis obtained by Steinfink (1958a) indicated it to be primarily ferric. As a result, the specimen was described as having oxidized some time after its original collection. This is not the case, however, as Mössbauer study of the original specimen has shown most of the Fe still to be present in the ferrous state (Heller-Kallai, 1982 and 1985, personal communications to the author). The latter finding is confirmed by the octahedral bond lengths in the refinement by Rule and Bailey (1987), which are consistent with the larger cation size of Fe^{2+} and not with the smaller Fe^{3+}.

Although the partial refinement of this specimen by Steinfink (1958a, 1961) suggested ordering of tetrahedral Si and Al in subgroup *C*2, the more detailed refinement by Rule and Bailey (1987) indicated complete disorder of the tetrahedral cations with no indication of symmetry lower than *C*2/*m*. The trivalent Al in the interlayer sheet is ordered into site M(4) on the inversion center, however, as also found by Steinfink. There is a small ordering of Mg, Fe^{2+}, and Fe^{3+} within the 2:1 layer so that the *trans* octahedral site M(1) is slightly larger than the *cis* site M(2). An important observation is that the excess of negative charge above unity due to tetrahedral substitution of Al for Si (1.378 atoms Al^{IV}) is compensated entirely within the octahedral sheet of the 2:1 layer. It is suggested that an interlayer charge of +1.0 is the most stable state for the chlorite structure.

Zheng and Bailey (1989) refined the structure of another II*b*-2 polytype found intergrown laterally along (001) with the II*b*-4 polytype in crystals from the Kenya locality discussed in the previous section. Both the triclinic and monoclinic structures were refined. The results of the refinement of the monoclinic component are very similar to those from the Washington, D.C., specimen, except that the tetrahedral charge is only -0.98. Although the intergrown monoclinic and triclinic polytypes are of similar compositions, the monoclinic polytype is significantly less ordered in the interlayer than the triclinic polytype [M(4)--0,OH = 1.993 Å vs. 1.957 Å, respectively]. Because the crystallization conditions for the two intergrown crystals must be identical, the authors cite the relative degrees of ordering found as evidence that the monoclinic structure is inherently less stable than the triclinic structure (in accord with its lesser abundance in nature). See Table 8 for the structural details.

Joswig et al. (1989) found co-existing triclinic and monoclinic II*b* polytypes in the Achmatov mine chlorite discussed above. The results of refinement of the triclinic II*b*-4 polytype using X-ray data already have been described. The monoclinic II*b*-2 polytype was refined with neutron data and, similar to the results of Zheng and Bailey (1989), the true space group was determined to be *C*2/*m* with no ordering of tetrahedral cations. Interlayer Al is partially ordered into site M(4). In contrast to the laterally intergrown polytypes of

Zheng and Bailey (1989) the co-existing polytypes (not stated to be intergrown) of the Achmatov mine chlorite, however, differ by only a small amount in the degree of ordering of interlayer Al. The observed bond lengths are M(4)--OH = 1.991 and 1.997 Å for the triclinic and monoclinic polytypes, respectively. The laterally intergrown polytypes of Zheng and Bailey (1989) have observed M(4)--OH values of 1.957 and 1.993 Å for the triclinic and monoclinic polytypes, where the smaller values are a result of greater ordering of Al into M(4).

General observations. On the basis of the structural results obtained up to that time, Rule and Bailey (1987) predicted that ordering of a trivalent cation into octahedron M(4) in the interlayer should be universal for all stable trioctahedral chlorites. Factors controlling the ordering are (1) energy minimization by the location of the source of the positive charge on the interlayer sheet in one octahedron rather than in two; this avoids disorder of divalent and trivalent interlayer cations over the M(3) sites or intermixed domainal structures; (2) the possibility of lessening of cation-cation repulsion between M(4) and tetrahedral cations by tetrahedral sheet offsets if a trivalent cation is in site M(4); (3) higher crystal-field stabilization energy by location of certain trivalent transition metals, such as Cr, in the less distorted octahedron M(4); and (4) local charge balance by a trivalent cation in M(4) positioned exactly between ordered Al-rich tetrahedra in the sheets immediately above and below the interlayer for *a* positions of the interlayer sheet relative to the 2:1 layers. The latter situation is known to be realized to date only in the structure of franklinfurnaceite (see below).

Rule and Bailey (1987) concluded further that most chlorites of the II*b* and I*bb*(β=90°) types should have a disordered tetrahedral cation distribution. This is because the *b* position of the interlayer sheet relative to the 2:1 layers above and below provides no preferential driving force for concentration of Si or Al in any tetrahedron as a consequence of the expected interlayer cation ordering. Interlayer site M(4) is equidistant from the two adjacent tetrahedral sites into which Si and Al might order. All five regular-stacking trioctahedral chlorites studied in detail to date have disorder of the tetrahedral cations but ordering of the inter-layer cations. The direction of tetrahedral rotation is always to move the acceptor basal oxygens closer to the donor OH groups to shorten hydrogen bond contacts.

Structural details of trioctahedral I*a* chlorite

No regular-stacking examples of the monoclinic I*a*-2 polytype have been reported to date for trioctahedral chlorites. A purple chromian clinochlore from Erzincan, Turkey, crystallizes as the triclinic I*a*-4 polytype of ideal symmetry *C*1. Although the original refinement of the I*a*-4 structure by 2-dimensional film methods (Brown and Bailey, 1963) suggested ordering of both tetrahedral and interlayer cations in a pattern that provided local charge balance, refinement of the same crystal by modern techniques (Bailey, 1986) showed that only the interlayer cations are ordered. The interlayer R^{3+} cations ($Cr_{0.7}Al_{0.2}$) are preferentially ordered into site M(4) on the inversion center, as indicated both by occupancy refinement and M--OH bond lengths (Table 8). This is similar to the ordering pattern found in the triclinic IIb-4 structure, but the a position of the interlayer sheet relative to the adjacent 2:1 layers does not allow the distortion to triclinic geometry observed for IIb-4. Slightly bent hydrogen bonds of length 2.9 Å exist between layer and interlayer.

Structural details of the trioctahedral Ib chlorite

Orthorhombic-shaped cell. No regular-stacking examples of the $Ibb(\beta=90°)$ chlorite have been reported. Single crystal photographs always show continuous streaking of the $k \neq 3n$ reflections along $Z*$ similar to that illustrated in Figure 8. Shirozu and Bailey (1965) interpreted the intensities of the observed continuous streaking as requiring not only random linear displacements of the layers by $\pm b/3$ along the three pseudohexagonal Y axes, but also random $\pm$ 120° rotations about the cleavage normal. Random linear displacements are equivalent to random adoption of superposition sites 1, 3, and 5 of Figure 4 for this structure, and random 120° rotations are equivalent to random shifts of $a/3$ within the 2:1 layers at the octahedral junction (Fig. 5). Refinement of a specimen of chamosite from the Tazawa mine, Japan, by means of only the sharp $k = 3n$ reflections gave the details of an average layer of symmetry $C2/m$ (Shirozu and Bailey, 1965). The average structure does not permit determination of any Si,Al ordering that may be present. A 5.0° tetrahedral rotation of basal oxygens toward the nearest interlayer OH positions was observed, but the H^+ proton positions were not determined. Electron density peak heights indicated that 60% of the iron is concentrated in the octahedral sheet of the 2:1 layer. Smaller octahedral M--OH distances in the interlayer (2.05 Å versus 2.10 Å) and a thinner octahedral sheet in the interlayer (1.98 Å versus 2.18 Å) indicate that most of the octahedral Al is in the interlayer, as it must be in order to have a positively charged interlayer.

Monoclinic-shaped cell. No crystals of sufficient perfection to allow refinement of the structure of the $Iba(\beta=97°)$ chlorite have been reported.

Multi-layer trioctahedral chlorites

The regular layer sequences within multi-layer trioctahedral chlorites have been determined for only a few specimens, and no detailed structural refinements have been reported. The multi-layer specimens are listed below, and their layer sequences are given in both the terminology of Lister and Bailey (1967) and Drits and Karavan (1969). In some cases the latter symbols have been modified to apply to a different unit cell shape than used by the authors.

L-B symbol	D-K symbol

Brindley et al. (1950)
chlorite "C" $\bar{X}_1$-IIb-2: $\bar{X}_3$-IIb-2 $\sigma_2\tau_2\sigma_6\tau_2\sigma_2$
(β-97°)

chlorite "D" $\bar{X}_1$-IIb-2:$\bar{X}_3$-IIb-4: X_3-IIb-2 $\sigma_3\tau_3\sigma_1\tau_5\sigma_1\tau_3\sigma_3$
(α-94°)

Ogniben & Quareni (1964)
clinochlore $\bar{X}_1$-IIb-6: $\bar{X}_2$-IIb-6 $\sigma_2\tau_6\sigma_4\tau_6\sigma_2$
(β-97°)

Lister & Bailey (1967)
"kämmererite" $\bar{X}_1$-Iab-3: X_2-IIab-6 $\sigma_5|\tau_1\ \sigma_4|\tau_2\ \sigma_5|$

The chromian chlorite of Lister and Bailey (1967) is unique in that the octahedral cations alternate regularly between the I and II sites in successive 14 Å units in both the 2:1 layer and the inter-layer sheet. Both regular and semi-random stacking variants have been recognized. For the latter, the two-layer periodicity shows up in the k = 3n reflections even though the $k \neq$ 3n reflections degenerate to streaks (Figs. 8, 15). Because the k = 3n reflections are strong, the different nature of this chlorite is also evident in its powder pattern (Table 5).

Di,trioctahedral chlorite

Cookeite. Vrublevskaja et al. (1975) described a triclinic 1-layer cookeite in metamorphosed bauxite from Djalair in Middle Asia, which was identified by the authors by means of high voltage electron diffraction (oblique texture) as the Ia-6 polytype of Bailey and Brown (1962) or $|\sigma_5|\tau_-|\sigma_5|=|\sigma_5|\tau_+\tau_+|\sigma_5|$ in the terminology of Zvyagin. This cookeite with regular-stacking of layers appears to be a well crystallized specimen that bears some structural similarity to the more abundant but less regular form of cookeite. Lister (1966) examined single crystals of seven less well crystallized cookeites of the latter type and concluded that they are composed of Ia layers that simulate a 1-layer stacking sequence. Diffuse 0$k\ell$ reflections of the $k \neq$ 3n type, however, suggest, that these are really imperfect 2-layer structures similar to the "s" structure derived by Mathieson and Walker (1954) in a study of vermiculite. The Lister and Bailey symbol for this arrangement is $\bar{X}_1$-Ia-4:$\bar{X}_1$-Ia-6, and the Drits and Karavan symbol is $|\sigma_3|\tau_+|\sigma_3|\tau_-|\sigma_3|$ for a cell with β = 97°. Lister (1966) also identified a 2-layer cookeite from two localities that is well crystallized and that can be correlated with the monoclinic "r" or "q" structures of Mathieson and Walker. These have symbols "r" = $\bar{X}_1$-Ia-4:$\bar{X}_3$-Ia-4 = $|\sigma_6|t_-|\sigma_4|\tau_+|\sigma_6|$ and "q" = $\bar{X}_1$-Ia-6:$\bar{X}_3$-Ia-2 = $|\sigma_1|\tau_-|\sigma_5|\tau_+|\sigma_1|$ for cells with β = 97° and are geometrically equivalent. Bailey (1975) summarized results of an incomplete structural analysis of one of the "r" or "q" cookeite crystals from North Little Rock, Arkansas, U.S.A. The data suggest ordering of tetrahedral Si,Al and an asymmetry of distribution of Si and Al between the two tetrahedral sheets of the 2:1 layer similar to that found in the brittle mica margarite-2M_1 by Guggenheim and Bailey

394

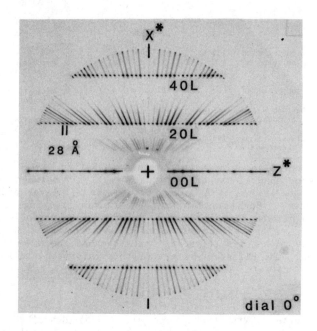

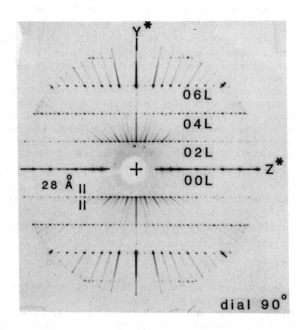

Figure 15. Precession photographs of a 2-layer chlorite in which octahedral cations in the 2:1 layer and the interlayer alternate regularly between the I and II sets of positions in adjacent layers. The 2-layer periodicity remains in *h0ℓ* patterns even for semi-random stacking of layers (Fig. 8b). From Bailey (1988).

(1975), but further refinement is needed for verification of both features. Local charge balance also was noted in this structure. The low-charge Li^{1+} in the interlayer sheet is located on a line directly between tetrahedral cations in the 2:1 layers above and below (only possible in *a* positions of the interlayer relative to the adjacent 2:1 layers). The higher-charged Al^{3+} cations in the interlayer sheet have a tetrahedral cation on only one side, however, with the center of a six-fold ring on the other side.

Although the great majority of cookeites are based on I*a* structural units, Cerný et al. (1971) described cookeite-II*b* from a pegmatite at Dobra Voda, Czechoslovakia, and the present writer (unpublished) has identified cookeite-II*b* from a pegmatite from Norway, Maine, U.S.A. (Table 5).

Sudoite. Eggleton and Bailey (1967) and Shirozu and Higashi (1976) attempted a partial refinement of the structure of a fine-grained sudoite from X-ray powder data. Their results confirmed that the sudoite specimens studied are composed of II*b* units, that the 2:1 layer is dioctahedral, and that the interlayer is trioctahedral. Drits and Lazarenko (1967) and Lin and Bailey (1985) used single crystals to show that specimens from the Urals and Belgium, respectively, have 2-layer structures of the "*s*" type. The stacking symbols in this case become "*s*" = $\bar{X}_1$-II*b*-4:$\bar{X}_1$-II*b*-6 = $\sigma_6\tau_2\sigma_6\tau_4\sigma_6$, where σ indicates a di,trioctahedral structure. One-dimensional electron density projections by Lin and Bailey showed that the smaller $d(001)$ value of sudoite relative to trioctahedral species is due primarily to its thinner dioctahedral 2:1 layer.

Dioctahedral donbassite

Drits and Lazarenko (1967) identified a specimen from the Donbass region, U.S.S.R., as having a 2-layer structure like that of the equivalent "*r*" or "*q*" structures and consisting of I*a* units. The structure was not refined. Alesksandrova et al. (1972) refined the crystal structure of a low-Li donbassite from Novaya Zemlya, U.S.S.R. They used 380 single crystal X-ray reflections in several cycles of least-squares refinement to attain a final R value of 9.9%. The structure is that of a regular 1-layer monoclinic form of space group $C2$. It can be described as a distorted I*a*-2 = $|\sigma_6'|\tau_-\tau_+|\sigma_6'|$ structure in which the vacant octahedral site in the 2:1 layer lies on the symmetry plane of the layer, but the partially vacant octahedral site in the interlayer lies off the projection of the mirror plane (thereby eliminating the mirror for the crystal as a whole). Adjacent layers adopt the mica configuration. The octahedra in the 2:1 layer are rotated by 9.3° but those in the interlayer by only 6.5° because of the counter attraction of the partly filled third octahedral site in the interlayer. A tetrahedral rotation of 13.5° was found in a direction that tends to shorten the hydrogen bonds from the basal oxygens to the adjacent OH of the interlayer. The mean T--O distances of 1.68 Å and 1.62 Å suggest that all of the tetrahedral Al is ordered in one tetrahedron. The authors postulated that the charge differences resulting from the ordering are responsible for the observed flattening of the larger (Si,Al) tetrahedron and extension of the smaller Si tetrahedron, although both cations lie at the center of mass of their individual tetrahedra.

At the present time this donbassite is the only chlorite for which ordering of tetrahedral Si,Al is claimed. There are differences of 0.12 Å and 0.08 Å between the largest and smallest individual T--O bonds within the two tetrahedra of this structure, however, so that a more accurate refinement is desired for verification of the ordering. Although X-ray refinement by film techniques of two trioctahedral chlorites originally indicated tetrahedral cation ordering, later refinement by modern techniques proved both chlorites to be disordered. Film refinements that use only a few hundred reflections and that give residual R values of 10% or greater must be considered suspect.

Octahedral compositions outside the Mg-Fe-Al system

Pennantite. Smith et al. (1946) defined pennantite as the Mn^{2+}-dominant member of the trioctahedral chlorite series. Type material from the Benallt mine in Wales occurs as small, orange-brown flakes in manganese ore and has a composition $(Si_{2.70}Al_{1.30})$ $(Mn^{2+}_{3.92}Al_{1.32}Fe^{3+}_{0.39}Mg_{0.26})O_{10}(OH)_8$. Single crystal X-ray study identified the chlorite as a 2-layer type. Because the powder pattern was stated to be similar to that of "thuringite" from Schmiedefeld, Thuringia (Table 1), the structure can be inferred to be of the IIb type. No further structural details are known.

The name "grovesite" was given by Bannister et al. (1955) to a mineral forming dark brown rosettes on manganese ore in the Benallt mine. A structural formula allocated by the present writer on the basis of 28 positive charges is $(Si_{2.57}Al_{1.43})(Mn^{2+}_{4.29}Al_{1.15}Fe^{3+}_{0.36}$ $Mg_{0.14})O_{10}(OH)_8$, which is only slightly different from that of type pennantite from the same locality. The X-ray powder patterns of "grovesite" and pennantite are different, however, and the authors believed "grovesite" to be the Mn-analogue of berthierine (based on a 1:1 layer), despite the presence of a 14 Å line in the pattern. Peacor et al. (1974) found "grovesite" in manganese ores at Bald Knob, North Carolina, and identified it as a one-layer chlorite with $\beta = 97°$. Bayliss (1975) identified the powder pattern listed by Peacor et al. as belonging to chlorite of the Ia structural type. Thus, pennantite and "grovesite" are polytypes, and pennantite has priority as the species name.

Nimite. De Waal (1970) defined nimite as the Ni-dominant member of the trioctahedral chlorite series. The mineral occurs as irregular, yellow-green veins in talc and opaque nickel ore in the Bon Accord area near Barberton, South Africa. The composition was given as $(Si_{3.01}Al_{0.99})(Ni_{2.62}Mg_{1.67}Al_{0.98}Fe^{2+}_{0.26}Fe^{3+}_{0.36}Mn_{0.01}Co_{0.03}$ $Ca_{0.04})O_{10}(OH)_8$. The X-ray powder pattern (Table 5) indicates the structure to be of the IIb type. No other structural details are known.

The Barberton nimite contains 29.49 wt.% NiO. Specimens from other localities with lesser amounts of NiO have often been called "schuchardite" in the literature. The NiO content ranges from 1 to 15% in these specimens, and Fleischer (1969) noted that in all cases Ni < Mg. The correct name is nickeloan clinochlore, therefore, and the name "schuchardite" should be relegated to the synonymy. The

X-ray pattern of many of these Ni-bearing clinochlores is that of
the Ia structural type. Brindley and de Souza (1975) pointed out
that many of these specimens are partly or wholly altered to Ia
vermiculite. Bailey and Riley (1977) showed that a dark green
nickeloan Ia clinochlore (NiO = 14.6 wt.%) from Woodline Well in
Western Australia has a 2-layer "s" structure and weathers readily to
golden yellow vermiculite, for which the "s" structure is believed to
be the most stable arrangement (Shirozu and Bailey, 1966; Calle et
al. 1976). The Woodline Well chlorite is unusual in that the 14 Å
line on its X-ray powder pattern is extremely weak due to
concentration of the heavy elements Ni and Fe in the interlayer
sheets.

Baileychlore. Rule and Radke (1988) gave the name baileychlore
to trioctahedral chlorites in which Zn is the dominant divalent
octahedral cation present. The only occurrence reported to date is
as dark green rims on colloform calcite veins within a strongly
oxidized collapse karst-breccia at the Red Dome deposit near
Chillagoe, Queensland, Australia. The chlorite formed by
remobilization of primary Zn-minerals by ground water. The powder
pattern (Table 5) is that of the I$ba(\beta=97°)$ structural type, in
accord with the low temperature origin.

The highest reported ZnO content in a baileychlore is 30.5 wt.%
and corresponds to a structural formula of $(Zn_{2.50}Fe_{1.20}Al_{1.17}Mg_{0.76}$
$Mn_{0.01}\square_{0.36})$ $(Si_{3.55}Al_{0.45})O_{10}(OH)_8$. A one-dimensional electron
density projection and an asymmetry value of +1.29 suggest a
composition $(Zn_{2.50}Al_{0.14}\square_{0.36})^{-0.58}$ for the octahedral sheet within
the 2:1 layer and $(Fe_{1.20}Al_{1.03}Mg_{0.76}Mn_{0.01})^{+1.03}$ for the interlayer
sheet. This is an example of a chlorite in which the low negative
charge on the tetrahedral sheets (-0.45) is supplemented by a net
negative charge on the octahedral sheet (-0.58) to give an overall
net negative charge on the layer that approximates unity.

Related structures

Franklinfurnaceite. The name franklinfurnaceite was given by
Dunn et al. (1987) to a dark brown platy mineral found in the
Franklin mine at Franklin, New Jersey. The overall composition is
$Ca_{1.875}(Mn^{2+}_{2.235}Mn^{3+}_{1.120}Fe^{3+}_{0.970}Al_{0.065}Zn_{0.555}Mg_{0.125})_{5.07}(Si_{2.085}$
$Zn_{1.915})O_{10.215}(OH)_{7.785}$, which is that of a chlorite plus extra Ca.
The mineral is unique in many ways. It is the only known example of
a tri,dioctahedral chlorite with a trioctahedral 2:1 layer but a
dioctahedral interlayer. The chlorite has a regular-stacking one-
layer structure of the IIaa-1 type and is thus the only known example
of a IIa chlorite. The IIaa-1 structure has adjacent 2:1 layers with
their ditrigonal rings opposed in the mica configuration. The extra
Ca lie in the ditrigonal rings on each side of the interlayer. The
large tetrahedral Zn cation creates a lateral misfit that causes a
large tetrahedral rotation angle of 23.5° This rotation causes a
nearly ideal three-dimensional closest-packing of the anions, which
creates an octahedral site for Ca and establishes the structure type
as IIa. The Ca is coordinated on one side by a triad of basal
oxygens that have rotated toward the center of the ditrigonal ring
and on the other side by a triad of OH groups of the interlayer. The
Ca causes the H^+ protons of the OH group to point into the interlayer

398

sheet so that there is no hydrogen bonding system. The dioctahedral interlayer has a net negative charge and the Ca is required to bond together the negatively charged 2:1 layer with the negatively charged interlayer. Franklinfurnaceite is thus a hybrid mineral that is intermediate in its structural details between a chlorite and a brittle mica.

The structural refinement of franklinfurnaceite by Peacor et al. (1988) found complete ordering of the Si_2Zn_2 tetrahedral cations. Due to ordering of the interlayer cations and the a position of the interlayer sheet, optimum local charge balance is achieved by positioning of interlayer Fe^{3+} on a straight line between tetrahedral Zn^{2+} cations above and below, interlayer Mn^{2+} between tetrahedral Si^{4+}, and interlayer vacancies between Ca sites. The trioctahedral sheet within the 2:1 layer has ordered Mn^{3+} and divalent cations. The cation ordering patterns reduce the symmetry from the ideal $C2/m$ to subgroup $C2$. The authors state that Ca substitution in common rock-forming chlorites is unlikely and may occur only if the structure type is IIa, the interlayer is dioctahedral, and if substitutions of substantial Al and/or Fe^{3+} occur in tetrahedral sheets and in both octahedral sheets.

Gonyerite. Frondel (1955) described gonyerite as a trioctahedral Al-poor Mn-chlorite that occurs in hydrothermal veinlets cutting skarn at Langban, Sweden. The composition was given as $(Si_{3.75}Fe^{3+}_{0.17}Al_{0.08})(Mn_{3.25}Mg_{1.95}Fe^{3+}_{0.64}Zn_{0.04}Pb_{0.02}Ca_{0.01})O_{10.2}(OH)_{7.8}$. The X-ray powder pattern, however, cannot be indexed on the basis of a conventional chlorite structure. Later study by Guggenheim and Eggleton (1987) showed gonyerite to have a modulated chlorite structure, as discussed in Chapter 17.

ACKNOWLEDGMENTS

Previously unpublished research in this chapter was sponsored in part by grant EAR-8614868 from the National Science Foundation and in part by grant 17966-AC2-C from the Petroleum Research Fund, administered by the American Chemical Society. Dr. W. Joswig kindly provided details of two structural refinements prior to publication. Dr. S. Guggenheim provided constructive criticism of the text.

REFERENCES

Albee, A.L. (1962) Relationships between the mineral association, chemical composition and physical properties of the chlorite series. Am. Mineral. 47, 851-870.
Aleksandrova, V.A., Drits, V.A. and Sokolova, G.V. (1972) Structural features of dioctahedral one-packet chlorite. Soviet Phys.-Crystallogr. 17, 456-461.
Bailey, S.W. (1975) Chlorites. Ch. 7 in: Soil Components, Vol. 2 Inorganic Components, J.E. Gieseking (Ed.). New York, Springer-Verlag, 191-263.
___ (1980) Structures of layer silicates. Ch. 1 in: Crystal Structures of Clay Minerals and their X-ray Identification, G.W. Brindley and G. Brown (Eds.). London, Mineralogical Soc. Mono. 5, 1-123.

399

___ (1986) Re-evaluation of ordering and local charge balance of Ia chlorite. Can. Mineral. 24, 649-654.

___ and Brown, B.E. (1962) Chlorite polytypism: I. Regular and semi random one-layer structures. Am. Mineral. 47, 819-850.

___ and Riley, J.F. (1977) An unusual chlorite from Western Australia. Mineral. Mag. 41, 541-544.

___ (1988) X-ray identification of the polytypes of mica, serpentine, and chlorite. Clays & Clay Minerals 36, 193-213.

Bannister, F.A., Hey, M.H. and Smith, W.C. (1955) Grovesite, the manganese-rich analogue of berthierine. Mineral. Mag. 30, 645-647.

Bayliss, P. (1975) Nomenclature of the trioctahedral chlorites. Canadian Mineral. 13, 178-180.

Biscaye, P.E. (1964) Distinction between kaolinite and chlorite in recent sediments by x-ray diffraction. Am. Mineral. 49, 1281-1289.

Bish, D.L. and Giese, R.F., Jr. (1981) Interlayer bonding in IIb chlorite. Am. Mineral. 66, 1216-1220.

Bradley, W.F. (1954) X-ray diffraction criteria for the characterization of chloritic material in sediments. Clays & Clay Minerals 2, 324-334.

Brindley, G.W. and Ali, S.Z. (1950) X-ray study of thermal transforamations in some magnesian chlorite minerals. Acta Crystallogr. 3, 25-30.

___, Bailey, S.W., Faust, G.T., Forman, S.A. and Rich, C. I. (1968) Report of the Nomenclature Committee (1966-67) of the Clay Minerals Society. Clays & Clay Minerals 16, 322-324.

___ and de Souza, J.V. (1975) Nickel-containing montmorillonites and chlorites from Brazil, with remarks on schuchardite. Mineral. Mag. 40, 141-152.

___ and Gillery, F.H. (1956) X-ray identification of chlorite species. Am. Mineral. 41, 169-186.

___, Oughton, B.M. and Robinson, K. (1950) Polymorphism of the chlorites. I. Ordered structures. Acta Crystallogr. 3, 408-416.

___ and Wan, H-M. (1978) The 14 Å phase developed in heated dickites. Clay Minerals 13, 17-24.

Brown, B.E. and Bailey, S.W. (1963) Chlorite polytypism: II. Crystal structure of a one-layer Cr-chlorite. Am. Mineral. 48, 42-61.

Caillère, S. and Henin, S. (1960) Relation entre la constitution cristallochimique des phyllites et leur temperature de deshydratation application au cas des chlorites. Bull. Soc. Fr. Cer. 48, 63-67.

Calle, C. de la (1976) Crystal structure of two-layer Mg-vermiculites and Na,Ca-vermiculites. Proc. Int. Clay Conf. 1975, Mexico City, S.W. Bailey (Ed.) 201-209.

Cathelineau, M. and Nieva, D. (1985) A chlorite solid solution geothermometer: the Los Azufres (Mexico) geothermal system. Contrib. Mineral. Petrol. 91, 235-244.

Cerny, P. (1970) Compositional variations in cookeite. Can. Mineral. 10, 636-647.

___, Povondra, P. and Stanek, J. (1971) Two cookeites from Czechoslovakia: a boron-rich variety and a IIb polytype. Lithos 4, 7-15.

400

Craw, D. and Jamieson, R. A. (1984) Anomalous optics in low-grade chlorite from Atlantic Canada. Can. Mineral. 22, 269-280.

de Waal, S.A. (1970) Nickel minerals from Barberton, South Africa: II. Nimite, a nickel-rich chlorite. Am. Mineral. 55, 18-30.

Drits, V.A. and Lazarenko, E.K. (1967) Structural and mineralogical characteristics of donbassites. Mineralog. Sbornik, L'vov Geol. Obshch. 21, 40-48 (in Russian).

___ and Karavan Yu. V. (1969) Polytypes of the two-packet chlorites. Acta Crystallogr. B25, 632-639.

Dunn, P.J., Peacor, D.R., Ramik, R.A., Su. S.-C. and Rouse, R. C. (1987) Franklinfurnaceite, a Ca-Fe(III)-Mn(III)-Mn(II) zincosilicate isotypic with chlorite, from Franklin, New Jersey. Am. Mineral. 72, 812-815.

Eggleton, R.A. and Bailey, S.W. (1967) Structural aspects of dioctahedral chlorite. Am. Mineral. 52, 673-689.

Fleischer, M. (1969) New mineral names. Nimite. Am. Mineral 54, 1740.

Foster, M.D. (1962) Interpretation of the composition and a classification of the chlorites. U.S. Geol. Survey Prof. Paper 414-A, 1-33.

___ (1964) Water content of micas and chlorites. U.S. Geol. Survey Prof. Paper 474-F, 1-15.

Frondel, C. (1955) Two chlorites: gonyerite and melanolite. Am. Mineral. 40, 1090-1094.

Garrido, J. (1949) Structure cristalline d'une chlorite chromifère. Bull. Soc. franc. Minéral. Cristallogr. 87, 321-355.

Guggenheim, S. and Bailey, S.W. (1975) Refinement of the margarite structure in subgroup symmetry. Am. Mineral. 60, 1023-1029.

___ & Eggleton, R. A. (1987) Modulated 2:1 layer silicates: Review, systematics, and predictions. Am. Mineral. 72, 724-738.

Goodman, B.A. and Bain, D.C. (1979) Mössbauer spectra of chlorites and their decomposition products. Proc. 6th Internat. Clay Conf. 1978, Oxford, 65-74. Elsevier, Amsterdam.

Hallimond, A.F. (1939) On the relation of chamosite and daphnite to the chlorite group. Mineral. Mag. 25, 441-465.

Hayashi, H. and Oinuma, K. (1965) Relationship between infrared absorption spectra in the region of 450-900 cm^{-1} and chemical composition of chlorite. Am. Mineral. 50, 476-483.

___ & ___ (1967) Si-0 absorption band near 1000 cm^{-1} and OH absorption bands of chlorite. Am. Mineral. 52, 1206-1210.

Hayes, J.B. (1970) Polytypism of chlorite in sedimentary rocks. Clays & Clay Minerals 18, 285-306.

Joswig, W. and Fuess, H. (1989) Refinement of a one-layer triclinic chlorite. Clays & Clay Minerals 37 (in press).

___, ___ and Mason, S.A. (1989) Neutron diffraction study of a one layer monoclinic chlorite. Clays & Clay Minerals 37, (in press).

___, ___ and Rothbauer, R. (1980) A neutron diffraction study of a one-layer triclinic chlorite (penninite). Am. Mineral. 65, 349-352.

Jung, H. and Köhler, E. (1930) Untersuchungen uber den Thuringit von Schmiedefeld, in Thüringen. Chemie Erde 5, 182-200.

Kepezhinskas, K.B. (1965) Composition of chlorites as determined from their physical properties. Dokl. Akad. Nauk. S.S.S.R., Earth Sci. Sect. 164, 126-129 (Engl. transl.).

Kodama, H., Longworth, G. and Townsend, M.G. (1982) A Mössbauer investigation of some chlorites and their oxidation products. Can. Mineral. 20, 585-592.

Lin, C-y. and Bailey, S.W. (1965) Structural data for sudoite. Clays & Clay Minerals 33, 410-414.

Lister, J. (1966) The crystal structure of two cholorites. Ph.D. thesis, Univ. Wisconsin, Madison, WI, U.S.A.

_____ and Bailey, S.W. (1967) Chlorite polytypism: IV. Regular two-layer structures. Am. Mineral. 52, 1614-1631.

Martin-Vivaldi, J.L. and MacEwan, D.M.C. (1960) Corrensite and swelling chlorite. Clay Minerals Bull. 4, 173-181.

Mathieson, A. McL. and Walker, G.F. (1954) Crystal structure of magnesium-vermiculite. Am. Mineral. 39, 231-255.

McMurchy, R.C. (1934) The crystal structure of the chlorite minerals.Z. Kristallogr. 88, 420-432.

Miser, H.D. and Milton, C. (1964) Quartz, rectorite, and cookeite from the Jeffrey Quarry, near North Little Rock, Pulaski County, Arkansas. Bull. Arkanses Geol. Commission 21, 1-29.

Nelson, B.W. and Roy, R. (1954) New data on the composition and identification of chlorites. Clays & Clay Minerals 2, 335-348.

Neuhaus, A. (1960) Über die Ionenfarben der Kristalle und Minerale am Beispiel der Chromfärbungen. Z. Kristallogr. 113, 195-233.

Ogniben, G. and Quareni, S. (1964) Studi sul clinocloro di Val Devero. Ricerca Scientifica 34, 469-476.

Pauling, L. (1930) The structure of the chlorites. Proc. Natl. Acad. Sci. 16, 578-582.

Peacor, D.R., Essene, E.J., Simmons, W.B., Jr. and Bigelow, W.C. (1974) Kellyite, a new Mn-Al member of the serpentine group from Bald Knob, North Carolina, and new data on grovesite. Am. Mineral. 39, 957-975.

_____, Rouse, R.C. and Bailey, S.W. (1988) The crystal structure of franklinfurnaceite: A tri-dioctahedral zincosilicate intermediate between chlorite and mica. Am. Mineral. 73 (in press).

Petruk, W. (1964) Determination of the heavy atom content in chlorite by means of the X-ray diffractometer. Am. Mineral. 49, 61-71.

Phillips, T.L., Loveless, J.K. and Bailey, S.W. (1980) Cr^{3+} coordination in chlorites: a structural study of ten chromian chlorites. Am. Mineral. 65, 112-122.

Robinson, K. and Brindley, G.W. (1949) A note on the crystal structure of the chlorite minerals. Proc. Leeds Phil. Soc. 5, 102-108.

Rozinova, E.L. and Dubik, O. Yu. (1983) Dioctahedral chlorites. Mineralog. Zhurnal 3, 14-31 (in Russian).

Rule, A.C. and Bailey, S.W. (1987) Refinement of the crystal structure of a monoclinic ferroan clinochlore. Clays & Clay Minerals 35, 129-138.

_____ and Radke, F. (1988) Baileychlore, the Zn end member of the trioctahedral chlorite series. Am. Mineral. 73, 135-144.

Schaller, W.T. (1930) Adjectival ending of chemical elements used as modifiers to mineral names. Am. Mineral. 15, 566-574.

Schreyer, W., Fransolet, A.-M. and Abraham, K. (1986) A miscibility gap in trioctahedral Mn-Mg-Fe chlorites: Evidence from the Lienne Valley manganese deposit, Ardennes, Belgium. Contrib. Mineral. Petrol. 94, 333-342.

Shikazono, N. and Kawahata, H. (1987) Compositional differences in chlorite from hydrothermally altered rocks and hydrothermal ore deposits. Can. Mineral. 25, 465-474.

Shirozu, H. (1955) Iron-rich chlorite from Shogase, Kochi Preference, Japan. Mineral. J. (Japan) 1, 224-232.

___ (1958) X-ray powder patterns and cell dimensions of some chlorites in Japan, with a note on their interference colors. Mineral. J. (Japan) 2, 209-223.

___ (1960) Ionic substitution in iron-magnesium chlorites. Mem. Faculty Sci., Kyushu Univ., D, Geol. 9, 183-186.

___ (1963) Structural changes of some chlorites by grinding. Mineral. J. (Japan) 4, 1-11.

___ (1980) Cation distribution, sheet thickness, and O--OH space in trioctahedral chlorites--an X-ray and infrared study. Mineral. J. (Japan) 10, 14-34.

___ (1985) Infrared spectra of trioctahedral chlorites in relation to chemical composition. Clay Science 6, 167-176.

___ and Bailey, S.W. (1965) Chlorite polytypism. III. Crystal structure of an orthohexagonal iron chlorite. Am Mineral. 50, 868-885.

___ and ___ (1966) Crystal structure of a two-layer Mg-vermiculite. Am. Mineral. 51, 1124-1143.

___ and Higashi (1976) Structural investigations of sudoite and regularly interstratified sericite/sudoite. Mineral. J. (Japan) 8, 158-170.

___ and Ishida, K. (1982) Infrared study of some 7 Å and 14 Å layer silicates by deuteration. Mineral. J. (Japan) 11, 161-171.

Smith, W.C., Bannister, F.A. and Hey, M.H. (1946) Pennantite, a new manganese-rich chlorite from Benallt mine, Rhiw, Carnarvonshire. Mineral. Mag. 27, 217-220.

Steinfink, H. (1958a) The crystal structure of chlorite. I. A monoclinic polymorph. Acta Crystallogr. 11, 191-195.

___ (1958b) The crystal structure of chlorite. II. A triclinic polymorph. Acta Crystallogr. 11, 195-198.

___ (1961) Accuracy in structure analysis of layer silicates: some further comments on the structure of prochlorite. Acta Crystallogr. 14, 198-199.

Stubican, V. and Roy, R. (1961) Isomorphous substitution and infrared spectra of the layer lattice silicates. Am. Mineral. 46, 32-51.

Townsend, M.G., Longworth, G. and Kodama, H. (1986) Magnetic interaction at low temperature in chlorite and its products of oxidation: A Mössbauer investigation. Can. Mineral. 24, 105-115.

Tuddenham, W.M. and Lyon, R.J.P. (1959) Relation of infrared spectra and chemical analysis for some chlorites and related minerals. Anal. Chem. 31, 377-380.

Von Engelhardt, W. (1942) Die Strukturen von Thuringit, Bavalit und Chamosit und ihre Stellung in der Chloritgruppe. Z. Kristallogr. 104, 142-159.

Vrublevskaja, Z.V., Delitsin, I.S., Zvyagin, B.B. and Soboleva, S.V. (1975) Cookeite with a perfect regular structure, formed by bauxite alteration. Am. Mineral. 60, 1041-1046.

Walker, J.R. (1987) Structural and compositional aspects of low-grade metamorphic chlorite. Ph.D. thesis, Dartmouth College, Dartmouth, New Hampshire, U.S.A. 99 pp.

Weaver, C.E., Highsmith, P.B. and Wampler, J.M. (1984) Chlorite. In Weaver, C. E. and Associates, Shale-Slate Metamorphism in the Southern Appalachians. Elsevier, Amsterdam, pp. 99-139.

Whittle, C.K. (1986) Comparison of sedimentary chlorite compositions by X -ray diffraction and analytical TEM. Clay Minerals 21, 937-947.

Zheng, Hong and Bailey, S.W. (1989) The structures of intergrown triclinic and monoclinic IIb chlorites from Kenya. Clays & Clay Minerals 37, (in press).

Zvyagin, B.B. (1963) Theory of the polymorphism of chlorites. Soviet Phys.-Crystallogr. 8, 23-27 (Engl. transl.).

___ and Mishchenko, K.S. (1966) Identification of single-packet semidisordered polymorphic modifications of chlorite. Soviet Phys.-Crystallogr. 10, 463-465. (Engl. transl.).

Chapter 11 J. Laird

CHLORITES: METAMORPHIC PETROLOGY

"Minerals are the ultimate geological units." *M.J. Buerger (1948)*.

INTRODUCTION

Chlorite is a common mineral in metamorphosed pelitic, felsic/intermediate, mafic, aluminous calc-silicate, and ultramafic rocks. It also occurs in metamorphosed Mn-rich rocks and is rare in metamorphosed iron formations. Chlorite occurs over a wide range of temperature, pressure, and activity of H_2O space, from diagenesis to granulite and to eclogite facies. Therefore, the mineral assemblages in which chlorite occurs are useful in determining metamorphic grade, facies series, and field gradient. Provided one has control on bulk rock composition, change in chlorite composition along the Tschermak substitution ($Al_2Mg_{-1}Si_{-1}$) and along $FeMg_{-1}$ is helpful in estimating metamorphic grade changes. Several exchange equilibria involving $FeMg_{-1}$ [chlorite] have been proposed as geothermometers and geobarometers.

After an introduction to petrography, problems, and compositional variation, petrologic data on chlorite in each of the rock types noted above are summarized. The manuscript concludes with a discussion of pertinent geothermometers and geobarometers. Earlier summaries of chlorite consider wet chemical analyses (Albee, 1962; Foster, 1962). Unless noted otherwise, chemical analyses summarized herein are electron microprobe analyses and are normalized to 14 oxygens assuming all iron is ferrous (see below for a discussion of these normalization assumptions). Chlorite analyses summarized in an earlier short course (Laird, 1982), and referred to herein, are normalized to total cations minus (Na+K+Ca) = 10, and Fe^3/Fe^2 is estimated from stoichiometry. These data were renormalized for the figures in this manuscript (except Fig. 29).

Exchange vector notation, procedure for writing net-transfer reactions, and directions for reaction space are summarized by J.B. Thompson Jr. (1981, 1982a,b). Mineral abbreviations suggested by Kretz (1983) are adopted. Abbreviations for exchange vectors are: TK (the combined Al^{VI}, Cr, and Ti Tschermak substitutions), FM ($FeMg_{-1}$) and AM (Al_2Mg_{-3}, dioctahedral substitution).

Many petrologic papers use the nomenclature of Hey (1954) and less commonly Foster (1962), in part no doubt because the former nomenclature is presented in many mineralogy textbooks. However, the boundaries between "varieties" have little structural meaning (Brown and Bailey, 1962). Chlorite composition discussed below is between clinochlore and chamosite using the nomenclature adopted by Bailey (1980).

In addition to mineral composition diagrams from other papers, analyses from different rock types are compared on three diagrams (atomic units): (1) Al vs. Fe vs. Mg in rectangular coordinates allows $FeMg_{-1}$ and total Al to be visualized, (2) $Al^{VI} + 2Ti + Cr - 1$ vs. $Al^{IV} - 1$ illustrates the Tschermak substitutions ($Al^{VI}Al^{IV}Mg_{-1}Si_{-1}$, $CrAl^{IV}Mg_{-1}Si_{-1}$, and $TiAl^{IV}_2Mg_{-1}Si_{-2}$) with clinochlore-chamosite, $(Mg,Fe)_5AlSi_3AlO_{10}(OH)_8$, as the origin. Analyses plotting above a 1:1 line from the origin show dioctahedral substitution and have octahedral vacancies, and (3) $Al^{IV} - 1$ vs. $Mg/(Mg+Fe)$ and $Al^{VI} + 2Ti + Cr - 1$ vs. $Mg/(Mg+Fe)$ allow tetrahedral and octahedral substitutions, respectively, to be examined relative to FM.

Petrography

Much is to be learned about the petrology of chlorite-bearing metamorphic rocks with the petrographic microscope. Chlorite is easily recognized by its "grass green" pleochroism, anomalous birefringence, and basal cleavage. In low grade rocks chlorite might be confused with other green pleochroic minerals such as biotite and actinolite, but these minerals are usually easy to distinguish in crossed polarized light (the anomalous birefringence of chlorite gives it away).

One can estimate (Fe+Mn+Cr)/(Fe+Mg+Mn+Cr) in chlorite from routine petrographic observation; the color of anomalous birefringence and sign of elongation are generally all that need be determined. Figure 1 shows the relationship between petrographic data and chlorite composition (wet chemical). Albee's (1962) prediction that the change in optic sign and sign of elongation occurs at about 52 atom % $(Fe_{total}+Mn+Cr)$ /$(Fe_{total}+Mn+Cr+Mg)$ is confirmed by subsequent electron microprobe data from pelitic schist (Black, 1975, Table 1; Hollacher, 1981, p. 101) and mafic schist (Table 1). Although both very Mg-rich and very Fe-rich compositions are pale green in crossed polarized light, the sign of elongation is different.

As with many easy "things," however, there are potential problems. Craw and Jamieson (1984) reported that the change in optic sign is not the same for different chlorite polytypes. Figure 1 is constructed for chlorite polytype IIb. While this polytype is most commonly found in metamorphic rocks (Brown and Bailey, 1962; Curtis et al., 1985), deformation can result in recrystallization to another polytype, especially at low temperatures (Craw and Jamieson, 1984). Ductile deformation is very apparent in the "normal" metamorphic chlorites summarized in Table 1.

Interpreting the textures of chlorite is not an exact science. However, sometimes nature is kind. Crystal form often tells one what mineral was pseudomorphed by chlorite; garnet is a good example. Mineral reactions may be inferred from textural relationships; good examples in calc silicate rocks are given by Ferry (1976) and Bucher-Nurminen (1982b). Cross-folial chlorite is probably "late." How late is not distinguishable from petrographic data alone. However, examination of the structural geology and geologic setting helps distinguish the process by which the "late" chlorite formed.

Textures in complex polymetamorphic terranes may be more controversial. Is chlorite in or around a porphyroblast retrograde or prograde? With what mineral compositions is the chlorite analyzed in equilibrium? Field data, petrologic relationships, and geologic common sense often help one narrow down the possible explanations.

Interpretation of chlorite compositions

There are several problems with interpreting chemical data for chlorite. Chlorite may be interstratified or intergrown with another sheet silicate, and the composition obtained does not represent chlorite completely. When intergrown chlorite-muscovite or chlorite-biotite are seen petrographically, analysis of these grains may be avoided. However, chlorite and other sheet silicates may be interstratified and intergrown on a larger scale.

Interstratified chlorite-smectite is apparent from X-ray diffraction data from diagenetic and low grade metamorphic rocks (e.g., McDowell and Elders, 1983; Coombs et al. 1976; Frey 1978; Robinson and Bevins, 1986; and Wang et al. 1986). Transmission electron microscopy (TEM) shows that interstratified chlorite and illite forms in diagenetic volcanogenic sediments (Ahn et al., 1988) and that chlorite retrograded from biotite may contain interstratified chlorite and biotite (Banos and Amouric, 1984, and references therein).

Interstratified chlorite-muscovite occurs in chlorite and biotite grade pelitic schist (Franceschelli et al., 1986). Backscattered scanning electron microscopy shows intergrowths of chlorite and phengite in prehnite-pumpellyite facies metagraywackes and in chlorite to garnet grade schists from New Zealand (White et al., 1985). Intergrowths

407

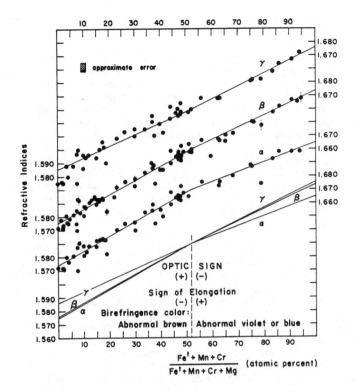

Figure 1. Correlation between petrographic data and chemical composition of chlorite form Albee (1962, Fig. 4). See Table 1 for further data on anomalous birefringence.

Table 1. Relationship between chemistry, birefringence and elongation. Chlorite from mafic schist, Vermont.*

$\dfrac{Fe + Mn + Cr}{Fe + Mn + Cr + Mg}$	Anomalous birefringence	Length fast/slow
0.25 → 0.35	light green	fast
0.35 → 0.45	green-brown	fast
0.45 → 0.51	brown-purple	fast
0.54 → 0.57	purple-blue	slow

*Composition is in atom %. Fe is total iron.

Table 2. Electron microprobe analyses (in wt %) from: (1) well, (2) sort of, and (3) very poorly polished chlorite.*

	(1)	(2)	(3)
SiO_2	25.89	25.34	25.78
Al_2O_3	21.85	21.86	21.78
TiO_2	0.05	0.06	0.08
Cr_2O_3	0.02	0.02	0.05
FeO_{total}	21.36	20.89	20.95
MnO	0.23	0.21	0.27
MgO	18.39	18.77	18.64
H_2O^*	12.22	12.85	12.44

*Obtained with CAMECA model MBX electron microprobe and converted to oxide wt % using the Bence-Albee (1968) technique and Albee-Ray (1970) correction factors. 15KV, 12 na, 16 micron wide raster. Grains are from mafic schist and are adjacent to each other.
FeO_{total} = total iron as FeO.
H_2O^* by difference from 100%.

increase in width with metamorphic grade but may still be less than two microns (White et al., 1985, p. 416).

Even higher grade rocks are not "immune." Ferry et al. (1987) reported that below 400°C interstratified smectite-chlorite occurs in contact metamorphosed alkali olivine basalts. Between 400 and 500°C chlorite forms as a replacement of smectite pseudomorphs of olivine and as individual grains. Veblen (1983) used high resolution TEM to show that intergrown chlorite and other sheet silicates occurred at staurolite grade. Furthermore, chlorite may have excess or missing brucite-like layers, so that chlorite analyses that do not seem to make sense stoichiometrically might not be stoichiometric.

Many of us rely on the electron microprobe and do not do X-ray diffraction or TEM studies or worry about what polytype is being analyzed. Notable exceptions are the papers referenced above, experimental studies, and petrologic studies on fine-grained rocks. Fortunately, it is now routine at many electron microprobe laboratories to examine an area with a backscatter detector before analyzing it; therefore intergrown sheet silicates may be more readily identified and can be avoided.

One should treat with caution an analysis with appreciable K_2O, Na_2O, or CaO. In her compilation of chlorite analyses, Foster (1962) excluded any analysis with more than 0.5 wt % Na_2O, K_2O, or CaO. Ernst (1983, p. 317) found that K, Na, and Ca appear to be covariant in chlorite analyses from Taiwan and proposed that some analyses represent submicroscopic intergrowths of chlorite, phengite, paragonite, and margarite.

Another potential problem with interpreting electron microprobe analyses of chlorite is that sheet silicates are not easy to polish, especially if one is also worried about polishing other minerals with different hardnesses in a sample. A *very brief* test of this problem shows that adjacent grains of chlorite with good to very poor polish have very similar electron microprobe compositions using a raster of about 16 microns (Table 2) .

A major problem with interpreting electron microprobe analyses of chlorite (and other common rock-forming minerals such as amphibole, biotite, or muscovite) is deciding how to normalize the data to a formula. If one normalizes to anhydrous oxygens, what does one use for Fe^2/Fe_{total}? In general, petrologists seem to normalize electron microprobe analyses of chlorite to 14 anhydrous oxygens and assume all iron is ferrous. Yet wet chemical analyses compiled by Foster (1962) show up to 4 wt % Fe_2O_3 in Mg-rich chlorite (zero to 0.4 $Fe^2/(Fe^{2+}+Mg+Mn)$) and up to 20 wt % Fe_2O_3 in more Fe-rich chlorite (see also Albee, 1962, Fig. 3). How much of this Fe_2O_3 is due to late-stage oxidation of FeO or to impurities in the mineral separate?

Another method used for normalizing chlorite analyses is to assume no octahedral vacancies and normalize total cations to 10. Then one can estimate Fe^2/Fe_{total} from stoichiometry. Yet, Foster (1962) reported that octahedral vacancies are common, up to 0.6 per six octahedral cations. We could all go back to wet chemical analyses, but obtaining a chlorite separate that is 100% pure is not only time consuming but easier said than done.

All estimates of TK, AM, and FM substitutions are dependent upon errors in the mineral formula estimated. Examples of the magnitude of this problem are given in Table 3 and discussed below. McDowell and Elders (1980, Tables 3 and 4) presented electron microprobe analyses of chlorite from the Salton Sea Geothermal Field. If data are normalized to 10 cations (excluding Na, K, and Ca), total positive charge is over 28 even if all the iron is ferrous. If the analyses are normalized to 14 oxygens assuming all iron is ferrous, octahedral vacancies range from 0.01 to 0.22. If the analyses are normalized to 14 oxygens and Fe^3/Fe_{total} is 10 atomic percent, vacancies range from 0.05 to 0.46! One can create or destroy roughly the range of octahedral vacancies in a series of chlorite analyses just by normalizing one analysis in different ways (Table 3). TK also varies significantly

Table 3. Effects of normalization assumptions on chlorite formulas.

Sample and normalization.*	$Al^{VI} + 2Ti + Cr + Fe^3_{-1}$	Al^{IV}_{-1}	$\dfrac{Mg}{Mg + Fe^2}$	Vacancy
Salton Sea Geothermal Field (McDowell & Elders, 1983, 675 m)				
10 cations, all Fe^2	0.58	-0.10	0.51	0
14 oxygens, all Fe^2	0.45	-0.03	0.51	0.23
14 oxygens, 10% Fe^3	0.62	-0.01	0.54	0.30
Blueschist facies metasediment, New Zealand (Black, 1975, #10741)				
10 cations, analyzed Fe^3	0.54	0.10	0.52	0
14 oxygens, all Fe^2	0.22	0.12	0.48	0.05
14 oxygens, analyzed Fe^3	0.45	0.15	0.52	0.15
Kyanite zone pelitic schist, Vermont (Albee, 1965a, LA1OK)				
10 cations, analyzed Fe^3	0.64	0.41	0.64	0
14 oxygens, all Fe^2	0.47	0.41	0.62	0.03
14 oxygens, analyzed Fe^3	0.63	0.42	0.64	0.09
Greenschist facies mafic schist, Vermont (Laird & Albee, 1981b, LA435A)				
10 cations	0.23	0.23	0.77	0
14 oxygens, all Fe^2	0.10	0.22	0.75	-0.06

*Normalization to 10 cations is exclusive of Na, K and Ca

for the three normalizations, but variation along AM (represented by vacancies) is the greatest.

Different normalizations result in a large range of estimated $Al^{VI} + 2Ti + Cr + Fe^3$ for blueschist facies chlorite and a smaller, but distinct, variation in octahedral occupancy for chlorite from kyanite grade pelitic schist and greenschist (Table 3). Assuming either 10 cations or 14 oxygens and using ferrous/ferric estimates from wet chemical data results in a smaller difference in site occupancy than assuming all iron is ferrous.

In general, if chlorite analyses from biotite to staurolite grade mafic schist in Vermont are normalized to 10 cations (less Na+K+Ca), minor Fe^3 (0.15 formula proportion maximum; Laird and Albee, 1981b, p. 135) is required to maintain charge balance. If the analyses are normalized to 14 oxygens assuming no ferric iron, the octahedral site may be greater than six.

Because of the problems noted above, it seems prudent not to place too much emphasis on small changes in chlorite composition. It does seem reasonable to propose and test generalizations that make sense in several geographic areas where rocks of similar composition, temperature, pressure, and aH_2O occur.

BULK ROCK COMPOSITION

Albee (1962) concluded that chlorite composition depends on rock composition as well as metamorphic grade. He reported that chlorite composition varies in the following order of increasing $(Fe_{total}+Mn+Cr)/(Fe_{total}+Mn+Cr+Mg)$: chlorite within metamorphosed carbonate rocks, serpentinite, mafic schist, pelitic schist, and metamorphosed iron-formations. This relationship makes geologic sense and is supported by more recent data (Fig. 2). In addition, chlorite from calcareous meta-argillite is distinctly more Fe-rich than chlorite from impure metadolomites and metarodingite. Very Mg-rich and intermediate FM compositions are reported for chlorite from Mn-rich rocks.

Chlorite analyses from metamorphosed felsic rocks overlap data for pelitic and Fe-rich mafic rocks (Fig. 2). Elimination of the three Al-poor analyses, which may include

410

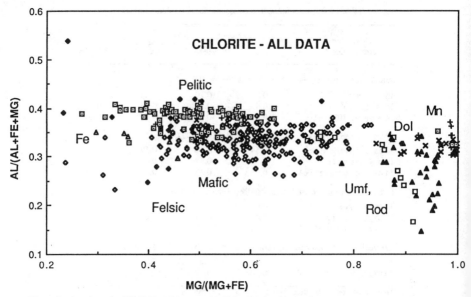

Figure 2. Atomic ratio Al/(Al+Fe+Mg) versus Mg/(Mg+Fe) for chlorite from metamorphosed pelitic (square with dot), mafic (diamond), felsic (filled diamond), calc-silicate (x), and ultramafic (filled triangle, Umf) rocks and from rodingite (square, Rod), iron formation (triangle), and Mn-rich rocks (plus). Mg-rich chlorite in calc-silicate is from impure meta-dolomite (Dol). See Appendix 1 for data references.

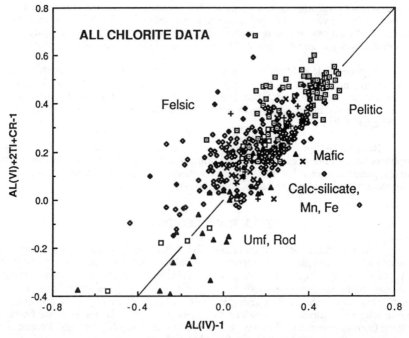

Figure 3. Chlorite composition (formula proportion $Al^{VI} + 2Ti + Cr - 1$ versus $Al^{IV} - 1$) for the same samples plotted on Figure 2, using the same symbols. The origin is end member clinochlore/chamosite. Compositional variation along the 1:1 line represents TK.

interstratified phyllosilicates (see below) however, shows that chlorite from felsic rocks tends to have Al compositions similar to chlorite in pelitic rocks and to be more Al-rich than chlorite in mafic rocks (Fig. 2). Ernst (1983, p. 317) reported that chlorite from felsic and other metaclastic schists from Taiwan tends to be more Al- and Fe-rich than chlorite in metavolcanic rocks.

Figure 2 shows that chlorite from pelitic and mafic rocks spans a wide range in FM, between clinochlore and chamosite. Within individual samples chlorite does not usually show a wide variation in composition except at very low metamorphic grades or in samples where prograde and retrograde chlorite occur (but see Kurata and Banno, 1974, who described heterogeneous chlorites from pelitic schists). In diagenetic and low metamorphic grade samples, chlorite composition is strongly dependent upon the allogenic parent (McDowell and Elders, 1983).

It is not surprising that Figure 2 shows chlorite from metapelitic rocks to be more Al-rich than chlorite from other rock types. Chlorite from ultramafic rocks and meta-rodingite shows the least Al. Chlorite from all rock types shows compositional variation along TK (Fig. 3), with greater advancement from clinochlore-chamosite shown by chlorite from pelitic rocks than chlorite from other bulk rock compositions. If one believes the normalization assumptions, chlorite from ultramafic rocks, rodingite, and some mafic and felsic rocks may have more than 3/4ths of the tetrahedral sites filled by Si (i.e., $Al^{IV} - 1$ is less than zero, Fig. 3).

Most analyses plot close to the one-to-one line on Figure 3 and do not show much AM substitution. Variations are discussed in the appropriate sections below and may represent AM substitution within the chlorite structure *or* interstratified/intergrown phyllosilicates.

Are more Fe-rich chlorite compositions more aluminous as has been suggested, for example, for chlorite from metamorphosed silicic volcanics and pelitic rocks in Maine (Stewart and Flohr, 1984; D.B. Stewart, pers. comm.) and for pumpellyite-actinolite facies metagraywacke in Switzerland (Coombs et al., 1976, Fig. 7)? Such a relationship seems possible from Figure 2 and implies that TK plus or minus AM are coupled with FM. This hypothesis will be addressed in more detail below where rock types are treated separately.

PELITIC ROCKS

Mineralogical questions

Figures 4 and 5 test the hypothesis that more aluminous chlorite in pelitic schist is more Fe-rich. One can make reasonable sense out of the data if samples are distinguished by facies series. Medium-pressure facies series and regional low-pressure facies series chlorite show a positive correlation between increasing $Al^{VI} + 2Ti + Cr$ (representing TK and AM) and FM as does high-pressure facies series chlorite. Less of a correlation is seen when Al^{IV} (presumably representing TK) is plotted against FM. Mineralogically, why should TK and FM be correlated?

The extent of AM does not appear to be related simply to pressure or temperature (Fig. 6). Chlorite in contact metamorphosed, chlorite to garnet grade pelitic schist generally has more AM than chlorite in medium-pressure facies series (weakly metamorphosed to kyanite grade) pelitic schist. Labotka (1981, Fig. 7B; 1980, Fig. 7) found more compositional variation and AM in low-pressure facies series regional metamorphic chlorite than medium-pressure chlorite. However, garnet and higher grade regional low-pressure metamorphic chlorites from Maine are hard to distinguish compositionally from medium-pressure facies series chlorite. Furthermore, high-pressure facies series chlorite may also have significant AM. Because of the large effect of mineral normalization on the chlorite

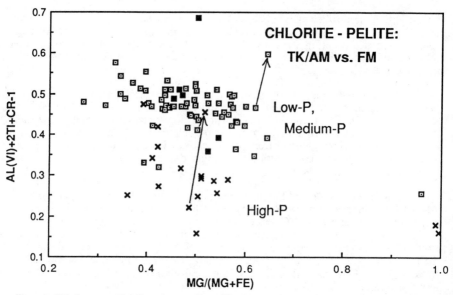

Figure 4. Chlorite composition (formula proportion $Al^{VI} + 2Ti + Cr - 1$ versus atomic ratio $Mg/(Mg+Fe)$ from metamorphosed pelitic rocks. See Appendix for data sources. Vertical axis combines the TK and AM substitutions. Contact metamorphism (filled square), regional medium- and low-P metamorphism (square), high-P metamorphism (x). Arrows show effects of changing normalization from 14 cations, all ferrous iron to 14 cations, analyzed Fe^2/Fe^3. In the latter case Fe^3 is added to the vertical axis, and the horizontal axis involves Fe^2 but not Fe^3.

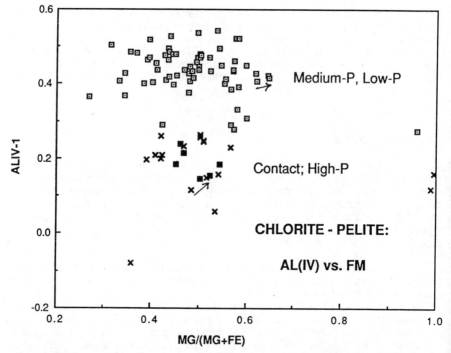

Figure 5. Chlorite composition (formula proportion $Al^{IV} - 1$ versus atomic ratio $Mg/(Mg+Fe)$) from metamorphosed pelitic rocks. Same data and symbols as Figure 4. Filled symbols are for contact metamorphic rocks; low- and medium-pressure regional samples have similar compositions. Effect of mineral normalization is minor.

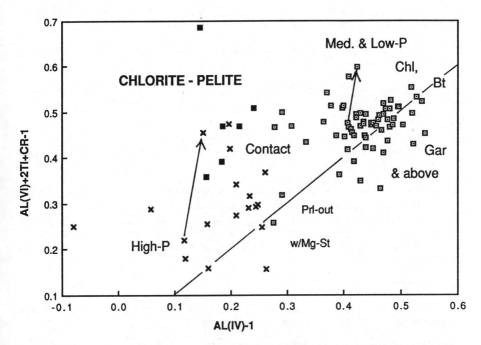

Figure 6. Chlorite composition on formula proportion diagram showing TK along 1:1 line and relative AM (distance above line) in pelitic schist. Same data and symbols as Figures 4 and 5. Arrows show change in mineral normalization as per Figures 4 and 5. See text for discussion of metamorphic grade.

formula (Table 3, Fig. 6), one wonders if one is attempting to make mineralogical sense out of graphical nonsense. Could some analyses be compositions of interstratified/intergrown sheet silicates and thus complicate the interpretation? Were different polytypes analyzed?

Returning to the rocks

The lowest metamorphic grade mapped by Barrow (1912) in pelitic rocks from Scotland is the chlorite zone. However, chlorite occurs over a wide range of pelitic rocks, from those that are diagenetic to high grade rocks in the kyanite, sillimanite, or andalusite and cordierite zones.

From an earlier compilation of electron microprobe analyses, Velde and Rumble (1977) concluded that chlorite with muscovite is generally $(Fe,Mg)_{4.5}Al_{1.5}$ - $Si_{2.5}Al_{1.5}O_{10}(OH)_8$, similar in composition to the cluster of garnet to staurolite, andalusite, or kyanite zone analyses on Figure 6. Velde and Rumble (1977) reported that low grade chlorite showed a greater variation in composition as is also seen on Figure 6.

Diagenesis to chlorite grade

In the marginal basin of Wales, Robinson and Bevins (1986) reported interstratified chlorite-smectite in diagenetic pelitic rocks and chlorite in anchizone pelitic rocks. X-ray diffraction data show that interstratified chlorite-montmorillonite occurs in unmetamorphosed claystones from the Central Alps, Switzerland (Frey, 1978). Chlorite occurs in the anchizone (above the reaction kaolinite + quartz = pyrophyllite + H_2O, about 236 to 267°C

414

at 1.3 to 2.1 kbar based on coal rank and fluid inclusion data, Frey, 1987) and at higher grade but is not interstratified with another mineral.

Modal chlorite and TK should increase with progressive metamorphism by the following reactions (J.B. Thompson, 1979):

$$6Al_2Si_4O_{10}(OH)_2 = Fe_5AlSi_3AlO_{10}(OH)_8 + 5Al_2Fe_{-1}Si_{-1} + 26SiO_2 + 2H_2O \qquad (1)$$
$$[Prl] \qquad\qquad [Chl] \qquad\qquad\qquad\qquad [Qtz]$$

and

$$6Al_2Si_2O_5(OH)_4 = Fe_5AlSi_3AlO_{10}(OH)_8 + 5Al_2Fe_{-1}Si_{-1} + 14SiO_2 + 8H_2O \ . \qquad (2)$$
$$[Kln] \qquad\qquad [Chl] \qquad\qquad\qquad\qquad [Qtz]$$

Frey (1978) estimated from X-ray diffraction data that chlorite becomes slightly more aluminous from unmetamorphosed rocks to the anchizone (Al^{IV} = 1.3 - 1.1 and Al^{VI} = 1.2 - 1.1 in unmetamorphosed rocks; Al^{IV} = 1.3 and Al^{VI} = 1.3 in the anchizone). Except for chlorite in very Mg-rich staurolite grade pelitic schist, chlorite from pyrophyllite-out to chloritoid-in pelitic schist shows the least TK than any medium-pressure facies series pelitic sample on Figure 6.

Biotite isograd

Three of the reactions written by J.B. Thompson (1979, reactions 11) for the biotite isograd involve chlorite and can be written as follows:

$$6KAl_2Si_3AlO_{10}(OH)_2 =$$
$$[Ms]$$

$$6KAlSi_3O_8 + Fe_5AlSi_3AlO_{10}(OH)_8 + 5Al_2Fe_{-1}Si_{-1} + 2SiO_2 + 2H_2O \ , \qquad (3)$$
$$[Kfs] \qquad\quad [Chl] \qquad\qquad\qquad [Ms,Chl,Bt] \qquad [Qtz]$$

$$3KAl_2Si_3AlO_{10}(OH)_2 + Fe_5AlSi_3AlO_{10}(OH)_8 =$$
$$[Ms] \qquad\qquad\qquad [Chl]$$

$$3KFe_3Si_3AlO_{10}(OH)_2 + 7SiO_2 + 4Al_2Fe_{-1}Si_{-1} + 4H_2O \ , \qquad (4)$$
$$[Bt] \qquad\qquad\quad [Qtz] \qquad [Ms,Chl,Bt]$$

and

$$2KAlSi_3O_8 + Fe_5AlSi_3AlO_{10}(OH)_8 =$$
$$[Kfs] \qquad\quad [Chl]$$

$$2KFe_3Si_3AlO_{10}(OH)_2 + Al_2Fe_{-1}Si_{-1} + 4SiO_2 + 2H_2O \ . \qquad (5)$$
$$[Bt] \qquad\qquad\quad [Ms,Chl,Bt] \quad [Qtz]$$

Modal chlorite increases by (3) but decreases by (4) and (5). These reactions are equivalent to reactions 11 (with m = 6), 13, and 11 (with m = 2), respectively, written by Miyashiro and Shido (1985) for the biotite isograd. Reaction 4 is equivalent to the biotite isograd written by Wang et al. (1986) for low-pressure facies series metamorphism in Japan. Pattison (1987, he chose Mg-end members) noted that (4) is equivalent to reactions given earlier for the biotite isograd in the Scottish Dalradian.

Each reaction (3 to 5) predicts that chlorite and/or biotite should become more Tschermak-rich with increasing dehydration (and presumably grade) and/or that muscovite becomes less phengitic. While data scatter makes generalization difficult, Miyashiro and Shido (1985, p. 461) concluded that TK in chlorite decreases from the chlorite to biotite zone, medium-pressure facies series. Stewart and Flohr (1984 and written comm., 1987) reached a similar conclusion for chlorite in low-pressure facies series silicic volcanic rocks and pelites from Maine (see discussion under FELSIC ROCKS). Chlorite and biotite zone chlorite from Idaho (Lang and Rice, 1985a) show more TK than higher grade samples from the same and other areas (Fig. 6).

Garnet isograd

For Fe-rich compositions chlorite decreases in mode at the almandine isograd by:

$$Cld + Chl + 2Qtz = 2Alm + 5H_2O \qquad (6)$$

and

$$2Fe_5AlSi_3AlO_{10}(OH)_8 + Al_2Fe_{-1}Si_{-1} + 4SiO_2 =$$
$$\text{[Chl]} \qquad\qquad \text{[Ms,Chl,Bt]} \quad \text{[Qtz]}$$

$$3Fe_3Al_2Si_3O_{12} + 8H_2O . \qquad (7)$$
$$\text{[Grt]}$$

Equation 7 is equivalent to equation 27 written by Miyashiro and Shido (1985). For more Mg-rich compositions garnet may "come in" by a continuous reaction describing "movement" of the garnet + chlorite + biotite field toward Mg-richer compositions with grade (e.g., by a reaction such as number 15 below) as observed by Lang and Rice (1985a) and predicted by Loomis (1986).

Continuous reaction (7) should decrease TK with dehydration (and presumably grade) as is predicted by J.B. Thompson (1979) and seen in the data compiled on Figure 6.

More Tschermak substitution

The breakdown of Fe-rich chlorite to more Al-rich minerals (chloritoid, staurolite, Al_2SiO_5) should decrease TK with dehydration as predicted by J.B. Thompson (1979). Breakdown of chlorite to cordierite ($\pm$ other phases as pertinent) may or may not form Tschermak with increasing temperature (Fig. 7).

Chlorite, biotite, and muscovite can change composition along this substitution. All minerals change composition sympathetically in regional and contact metamorphosed pelitic rocks from Scotland (Pattison, 1987). In their data compilation Miyashiro and Shido (1985, Fig. 5) identified a greater change in muscovite composition than in chlorite composition. A consistent change in TK is not seen on Figure 6 among chlorite from garnet and higher grade pelitic schist, perhaps because change in TK is small at medium and higher metamorphic grades, and small variations in bulk composition or errors in mineral formulas can mask a consistent variation (if present). Furthermore, some pertinent reactions increase TK while others decrease TK (Karabinos, 1985).

Petrogenetic grid

In the last 20 years or so, several petrogenetic grids for garnet and higher grade pelitic rocks have been constructed using the method of Schreinemakers (e.g., see Zen, 1966) and data from rocks and experiments (e.g., Albee, 1965b; Hess, 1969; A.B. Thompson, 1976b; Harte and Hudson, 1979; Labotka, 1981; Spear and Cheney, 1986). Yardley et al. (1980) pointed out that the reactions on such petrogenetic grids are discontinuous and concluded that the reactions actually taking place in pelitic rocks from Connemara are continuous. Nevertheless, these grids provide a framework for interpreting pelitic assemblages, and the continuous reactions can be readily included within this framework as is illustrated by A.B. Thompson et al. (1977b), Harte and Hudson (1979), and Pattison (1987), for example.

416

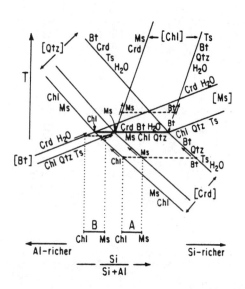

Figure 7. Schematic temperature versus XAl-Si grid (constant pressure) for reactions involving the phases Ms, Bt, Chl, Qtz, Crd, H_2O, and the Tschermak exchange (Ts) from Pattison (1987, Fig. 11). Two isopleths are shown for each reaction involving Ts. For initial mineral compositions A, Ts increases in Chl, Ms, and Bt with grade; but for initial mineral compositions B, Ts decreases in Chl and Ms with grade. For both initial compositions A and B, T remains constant at the reaction Ms + Chl + Qtz = Crd + Bt + H_2O until one reactant is consumed. If Chl is consumed (as shown) Ts decreases in Ms and Bt as temperature increases.

While there are differences in the grids published, they predict many of the same changes in chlorite-bearing assemblages (+quartz, muscovite) as a function of pressure, temperature, aH_2O, and X_{Mg}. Most show that chorite is stable with garnet + quartz + muscovite to higher temperature the greater the pressure of metamorphism (Fig. 8), but see Loomis (1986). The grids are applicable to a wide range of geologic localities, a few of which are discussed below. Provided one keeps in mind the possible effects of changing mineral composition and fluid composition (see discussion of <u>Fluid phase</u> below), one can use these petrogenetic grids to estimate the general field gradient across a metamorphic terrane from petrographic data. For example, Clark et al. (1987) and Klaper and Bucher-Nurminen (1987) have used the grids plus field and structural data to understand P-T paths.

Medium-pressure facies series. For pelitic compositions near the garnet-chlorite join, medium-pressure facies series metamorphism results in the sequence (all assemblages with quartz, muscovite):

garnet + biotite + chlorite / chloritoid + garnet + chlorite,

garnet + biotite + chlorite / staurolite + garnet + chlorite,

staurolite + garnet + biotite / staurolite + chlorite + biotite,

and staurolite + garnet + biotite / staurolite + kyanite +biotite / kyanite + biotite + chlorite (e.g., Albee, 1968, Fig. 25-2; middle arrow Fig. 8).

This sequence of mineral assemblages (or part of it) is identified, for example, in the "typical" Barrovian terrane, eastern Scotland (Harte and Hudson (1979, sequence A), in Death Valley, California (Labotka, 1980), in the Nufenenpass area, Switzerland (Klaper and Bucher-Nurminen, 1987), in South Australia (Clark et al., 1987), and in the

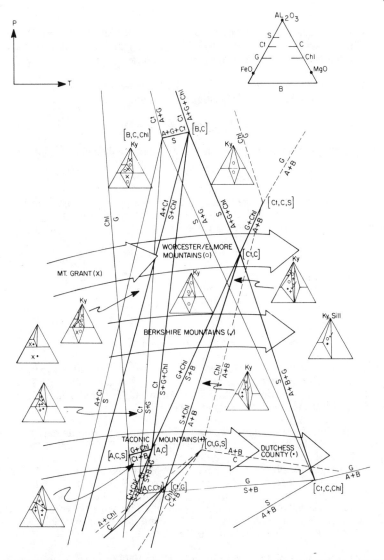

Figure 8. Schematic petrogenetic grid for pelitic schist from the western northern Appalachians compiled by Laird (1988, Fig. 3) using the grid of Harte and Hudson (1979, Fig. 2). Arrows represent field gradients inferred from reported assemblages. Mineral assemblages are shown on a J.B. Thompson (1957) projection from Ms and Qtz. G (garnet), Ct (Chloritoid), S (staurolite), A (kyanite, sillimanite, or andalusite), C (cordierite), B (biotite). Bold lines (KFMASH), dashed lines (KMASH), light lines (KFASH reactions). (Qtz and Ms are eliminated from reactions for clarity.)

Appalachians (e.g., Fig. 8, middle arrow). Loomis (1986, Fig. 5) predicted the same sequence of assemblages (above chloritoid out) for average pelitic schist at 5 kbar, 500 to 550°C.

That the sequence of assemblages summarized above actually forms during progressive metamorphism is nicely protected in garnet porphyroblasts from the Gassetts schist, Vermont. A.B. Thompson et al. (1977b) deduced the sequence of mineral assemblages illustrated in Figure 8 (middle arrow) from the *petrographic* and mineral chemistry data below ($X_{Fe} = 100Fe/(Fe+Mg)$):

(a) Inclusions within the garnet core ($X_{Fe} = 0$) are chlorite ($X_{Fe} = 37$) and chloritoid ($X_{Fe} = 77$). Because biotite is not included within the garnet core, the effective bulk composition must not allow the assemblage garnet + chlorite + biotite.

(b) Closer to the garnet rim ($X_{Fe} = 81$), staurolite ($X_{Fe} = 77$), chlorite ($X_{Fe} = 38$), and chloritoid ($X_{Fe} = 77$) occur, suggesting the reaction

$$Cld + Qtz = St + Grt + H_2O \qquad (8)$$

(c) Chloritoid is not present at the garnet rim nor in the matrix, indicating

$$Cld + Qtz = Chl + Grt + St \ . \qquad (9)$$

(d) Kyanite is found near the edge of garnet porphyroblasts and in the matrix. Because biotite is not present at the edge of garnet porphyroblasts, garnet must have reached its maximum size before the reaction

$$Grt + Chl + Ms = St + Bt \ ; \qquad (10)$$

and this reaction plus the reactions

$$St + Chl + Ms = Bt + Ky \qquad (11)$$

and

$$St + Qtz = Grt + Ky + H_2O \qquad (12)$$

occurred in close succession. This prediction is in agreement with estimates for the temperatures at which these reactions take place (e.g., see Lang and Rice, 1985b, Fig. 14; Harte and Hudson, 1979, Fig. 3). At 5 kbar Loomis (1986) predicted reaction (10) takes place between 500 and 530°C and (11) at 530 to 550°C.

Low-pressure facies series metamorphism. Field and petrographic data from Scotland, Death Valley, and the Appalachians (Harte and Hudson, 1979; Labotka, 1980, 1981; data compiled by Laird, 1988, respectively) indicate that compared to medium pressure facies series metamorphism, progressive metamorphism at low pressure results in the early demise of chlorite plus garnet (Figs. 8 (bottom arrow), 9, and 10). Chlorite breaks down over a large range in X_{Fe} with a small change in grade. Cordierite is identified more commonly. Hudson (1980) reported that chlorite becomes unstable between the cordierite and andalusite isograds in the Buchan zone of Scotland, and Hudson (1985) estimated that these isograds occur at 430±15°C and 490±15°C, respectively.

The cordierite + biotite + chlorite three-phase field (+quartz, muscovite) occupies a narrow range in Fe-Mg (or in T? Fig. 9, Panamint Mountains), and thus is not common in the field as reported, for example, by Hudson (1980) for Buchan metamorphism in northeastern Scotland. Shiba (1988) reported that the region of cordierite + chlorite is too narrow to be mapped in metapelites from the Hidaka, low-pressure facies series terrane in Japan.

Nevertheless, much discussion has ensued concerning whether chlorite projects to the Fe- or Mg-side of the cordierite-biotite tie line on J.B. Thompson's (1957) AFM projection from muscovite and quartz. Guidotti et al. (1975) recognized that for staurolite and sillimanite grade, low-pressure facies series metamorphism in Maine, chlorite is more Fe-rich than coexisting cordierite and biotite, and thus the breakdown of chlorite in Mg-rich pelitic rocks is

$$Chl + Ms + Qtz = Al\text{-silicate} + Crd + Bt + H_2O \ . \qquad (13)$$

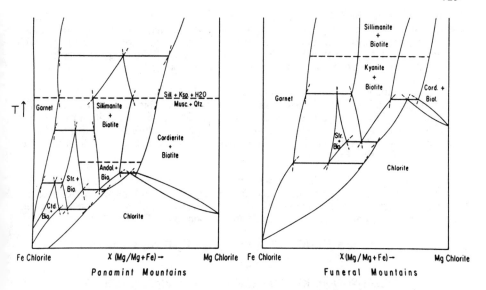

Figure 9. Schematic temperature-XMg diagrams for pelitic schist from Death Valley, California, metamorphosed to low-pressure facies series (Panamint Mountains) and medium-pressure facies series (Funeral Mountains) from Labotka (1981, Fig. 10). X-axis is along the garnet-chlorite join in the J.B. Thompson (1957) projection form quartz and muscovite. Garnet + chlorite is stable through a range of T for medium-pressure metamorphism but is unstable at low-pressure metamorphism. Bio (biotite), Ctd (chloritoid), Cord (cordierite), Str (staurolite), Andal. (andalusite), Musc (muscovite), Sill (sillimanite), Ksp (K feldspar).

Labotka (1981) also found this reaction applicable to low-pressure facies series metapelites from Death Valley, California (Fig. 10) as did Labotka et al. (1981) in contact metamorphosed argillite and graywacke from Minnesota and van Bosse and Williams-Jones (1988) in pelitic hornfels from Quebec. Shiba (1988) reported that cordierite, chlorite, and biotite are nearly colinear and that the reaction between the two lower grade zones is

$$Chl + Ms + Qtz = Crd + Bt + H_2O \ . \tag{14}$$

From regional and contact metamorphosed pelitic schists from Scotland, Pattison (1987) "divided" this reaction into the continuous reactions involving TK (Fig. 7).

Hudson (1980) found that chlorite projects just on the Mg side of the cordierite-biotite tie line in the Thompson (1957) AFM projection from quartz and muscovite and suggested that in low-pressure facies series metapelites the position of chlorite with respect to the cordierite-biotite tie line might "flip" with increasing temperature from the Mg side to the Fe side, perhaps because of different rates in $MgFe_{-1}$ between chlorite and cordierite. Pattison (1987) reported that the three phase field cordierite + chlorite + biotite moves to more Fe-rich compositions with increasing temperature.

In contrast chlorite is distinctly more Mg rich than biotite in medium-pressure facies series pelitic rocks (e.g., Lang and Rice, 1985b, Fig. 2; Labotka, 1980) and in regional low-pressure facies series rocks from Maine with andalusite but without cordierite (Holdaway et al. 1988). The chlorite-biotite Fe/Mg distribution is discussed further in the section below on GEOTHERMOMETERS AND GEOBAROMETERS.

$FeMg_{-1}$ substitution

$T-X_{Mg}$ diagrams constructed from chemical data (e.g., Albee, 1972, Fig.13; A.B.

Thompson, 1976a, Fig.2; Lang and Rice, 1985b, Fig. 14; Fig. 9 herein) show that chlorite associated with garnet becomes more Mg-rich with increasing metamorphic temperature. Further support comes from analytical and theoretical studies of chlorite compositions with garnet ± biotite as presented by A.B. Thompson et al. (1977b, Fig. 5), Clarke et al. (1987), Klaper and Bucher-Nurminen (1987), and Kurata and Banno (1974).

Consequently, if pressure and aH_2O are relatively constant, one can estimate the relative change in temperature across a terrane where chlorite + biotite + garnet + muscovite + quartz occur from petrographic data, i.e., estimate the chlorite composition from Figure 1. This can be particularly useful in an area where garnet + chlorite + biotite + quartz + muscovite is stable over a large range in temperatures (as is clearly possible at medium-pressure facies series, Figs. 8 and 9).

That $MgFe_{-1}$ should increase with increasing temperature makes perfectly good sense from the following continuous reaction written among geologically reasonable and conviently chosen mineral compositions. (One can easily change an additive component composition by adding or subtracting the desired amount of $MgFe_{-1}$ or TK.)

$$3Mg_3Fe_2AlSi_3AlO_{10}(OH)_8 + KAl_2Si_3AlO_{10}(OH)_2 + 3SiO_2 =$$
$$[Chl] \qquad\qquad [Ms] \qquad\qquad [Qtz]$$

$$4Fe_3Al_2Si_3O_{12} + KFe_2MgSi_3AlO_{10}(OH)_2 + 8MgFe_{-1} + 12H_2O . \qquad (15)$$
$$[Grt] \qquad\qquad [Bt] \qquad\qquad [Chl,Ms,Bt,Grt]$$

Fluid phase

Because dehydration reactions are dependent on aH_2O as well as pressure (P) and temperature (T), change in aH_2O will change the position (in P-T space) of a dehydration reaction, and different assemblages may be formed. Lang and Rice (1985a, Fig. 7) showed $T-XH_2O$ grids for the reactions (several of which include chlorite) above and below the [Chloritoid, Cordierite] invariant point on Figure 8. Reactions between the staurolite zone and staurolite-kyanite zone can be explained by increasing temperature and XH_2O (Fig. 11B). Pressure need not be increased as would be expected from the P-T diagram. These results have obvious implications when it comes to using mineral assemblages to interpret P-T paths. Is the rock recording a clockwise loop, a counter-clockwise loop, or is XH_2O changing?

Furthermore, with increasing temperature modal chlorite may decrease, increase, and then decrease (following the dotted arrow on Fig. 11B), perhaps making the textural relationships somewhat less than easy to understand. For example, chlorite surrounding garnet need not imply retrogradation.

Lang and Rice (1985a) concluded that the metamorphic reactions are controlling XH_2O. Dickenson (1988) came to a similar interpretation for two outcrops of low-pressure facies series, staurolite-andalusite grade metapelites in northwestern New Hampshire. He showed that the common problem of interpreting coexisting (apparently) chlorite with staurolite, biotite, Al silicate, quartz, and muscovite need not be a problem. While there are crossing tie lines on an AFM projection, tie lines do not cross on an AFM plus H_2O diagram (Fig. 12). His data show that the activity of H_2O is not constant over each outcrop. Is XH_2O internally controlled or could it be externally controlled and change? If samples are taken at various positions with respect to quartz veins or structural weakness, could different crossing tie lines result due to variation in externally controlled XH_2O?

High-pressure facies series

Using the normalization assumptions that total cations = 14 and all iron is ferrous, chlorite in high-pressure facies series pelitic schist shows less TK than in medium-pressure

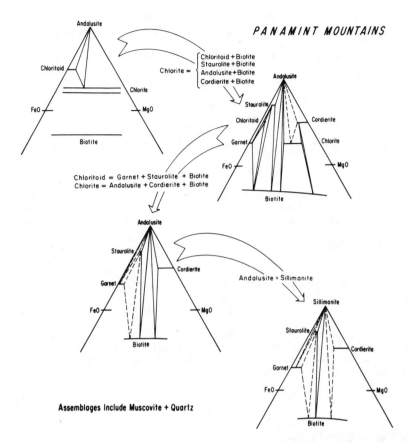

Figure 10. Low-pressure facies series metamorphism in Death Valley, California, from Labotka (1981, Fig. 8) illustrated on J.B. Thompson (1957) projections. Dashed tie lines are inferred; the assemblages are not identified in these pelitic rocks (from the Panamint Mountains, Fig. 9). Metamorphic grade increases from top to bottom, roughly along the bottom arrow in Figure 8 (but Labotka's grid is a bit different from Fig. 8).

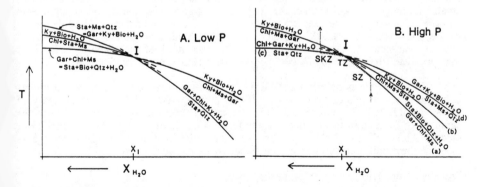

Figure 11. Schematic T-XH₂O diagrams below (A) and above (B) the chloritoid, cordierite-absent invariant point on Figure 8 from Lang and Rice (1985a, Fig. 7). Dotted arrow in (B) is the T-XH₂O path inferred by Lang and Rice for progressive metamorphism from the staurolite zone (SZ) to the staurolite-kyanite zone (SKZ), by way of a transition zone (TZ). Bio (biotite), Gar (garnet), Sta (staurolite).

422

facies series pelitic schist (Figs. 4-6). More Al-rich compositions are more Fe-rich. If analyzed Fe^3/Fe^2 are used in mineral normalization, the same relationship is seen (Fig. 6).

Blueschist facies assemblages for pelitic and mafic schists are summarized for a variety of high-pressure facies series terranes in Figure 13. In aluminous compositions chlorite is stable with paragonite and chloritoid or paragonite and albite below about 10 to 15 kbar. The reaction

$$Ab + Chl = Gln + Pg \qquad (16)$$

allows glaucophane to become stable in aluminous compositions, and chlorite occurs with glaucophane and paragonite. [Chlorite + paragonite + chloritoid are still stable.] The reaction

$$Pg + Chl = Cld + Gln \qquad (17)$$

stabilizes the assemblage chloritoid + glaucophane + chlorite which is common in high-pressure facies series pelitic schist from Greece and the Western Alps and was recently identified in western France (Guiraud et al., 1987) and Vermont (Bothner and Laird, 1987). This assemblage is not common in New Caledonia (Ghent et al., 1987) or Japan as summarized on Figure 13.

Chopin (1981, 1986) and Chopin and Schreyer (1983) presented petrogenetic grids for high-pressure facies series pelitic rocks. Chlorite is finally consumed, with increasing grade, by the reaction

$$Qtz + Chl = Tlc + Ky + H_2O . \qquad (18)$$

This reaction is among those used by Newton (1986, Fig. 11) to distinguish different P-T conditions for eclogites and may explain why chlorite is uncommon in eclogite facies metaclastic rocks from southwestern Japan, western California, and the western Alps (see summary diagrams by Ernst (1977, Figs. 3-5).

FELSIC ROCKS

Diagenesis to zeolite facies

Allogenic and authigenic chlorite occur in sandstone at the Salton Sea Geothermal Field (McDowell and Elders, 1980, 1983). Variation in composition of authigenic chlorite is dependent on the allogenic parent (alkali feldspar, plagioclase, biotite, white mica, or chlorite). Increasing temperature (depth in a borehole) from 150 to 350°C resulted in a small but regular compositional change in both fine-grained and coarse-grained authigenic chlorite. Octahedral Al decreases while both (Mg+Fe) and total octahedral population increase, consistent with less AM as seen on Figure 14. Sedimentary chlorite (below 200°C) has more vacancies than metamorphic chlorite (Curtis et al., 1985) and is polytype Ib rather than polytype IIb (as is found in metamorphic rocks (Curtis et al., 1985, Fig. 8).

Zeolite facies burial metamorphism of volcanogenic sandstones in New Zealand resulted in interstratified chlorite-smectite (Boles and Coombs, 1977). Chemical analyses show less TK than is seen in chlorite from the Salton Sea, from pumpellyite facies samples, or from chlorite and biotite grade felsic volcanic rocks (Fig. 14). Recent TEM studies (Ahn et al., 1988) showed interstratified chlorite-illite in volcanogenic sediments from the same unit southeast of the area studied by Boles and Coombs (1977).

Prehnite-pumpellyite to greenschist facies

In metagraywacke from New Zealand chlorite (with intergrown phengite, White et al., 1985) first forms in the prehnite-pumpellyite facies (Bishop, 1972). With increasing metamorphism into the pumpellyite-actinolite and greenschist facies, modal pumpellyite and chlorite decrease perhaps by the reaction

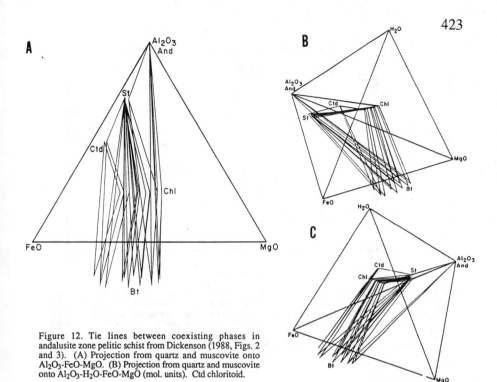

Figure 12. Tie lines between coexisting phases in andalusite zone pelitic schist from Dickenson (1988, Figs. 2 and 3). (A) Projection from quartz and muscovite onto Al_2O_3-FeO-MgO. (B) Projection from quartz and muscovite onto Al_2O_3-H_2O-FeO-MgO (mol. units). Ctd chloritoid.

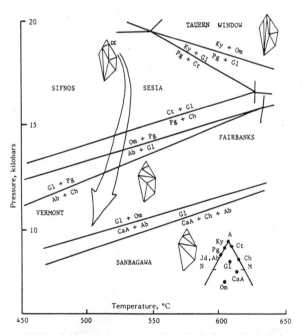

Figure 13. P-T grid for high-P facies series metamorphism illustrating parageneses for several terranes. Diagram modified to show P-T path implied by later data from Vermont (Bothner and Laird, 1987). Composition diagram is an epidote + garnet + quartz + H_2O projection onto Al + Fe^3(A), Na(N), and Mg(M). Ct (chloritoid), Ch (chlorite), CaA (calcic amphibole), Gl (glaucophane). From Brown and Forbes (1986, Fig. 14); data sources therein.

424

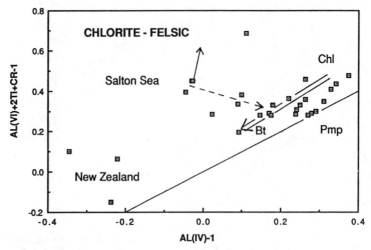

Figure 14. Chlorite analyses from metamorphosed felsic rocks. See Appendix for data sources. 1:1 line represents TK; origin is chlinochlore/chamosite. Arrow shows effect of mineral normalization (from 14 oxygens, all Fe^2 to 14 oxygens, 10% Fe^3/Fe_{total}). Double arrow indicates progressive metamorphism from chlorite to biotite grade (Stewart and Flohr, written comm., 1987). Dashed arrow shows change in chlorite composition with increasing depth in the Salton Sea Geothermal Field. Analyses labeled New Zealand are from Boles and Coombs (1977). Analyses labeled Pmp vary from zeolite to pumpellyite-actinolite facies.

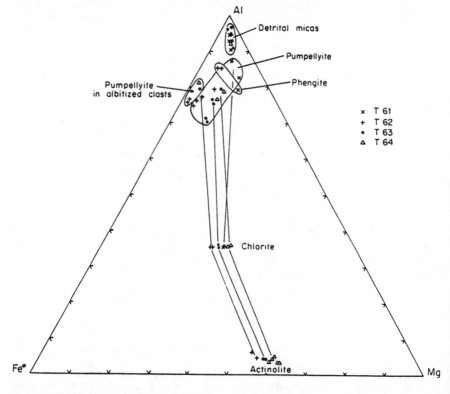

Figure 15. Coexisting compositions of Chl and other minerals in pumpellyite-actinolite facies metagraywacke, Switzerland (from Coombs et al., 1976, Fig. 7). Atom units Al, Mg, Fe_{total}.

$$Pmp + Chl + Qtz = Act + Ep + H_2O \ . \qquad (19)$$

However, in the Valais, Switzerland, all five minerals coexist in volcanogenic metagraywacke because: Mg and Fe cannot be treated as one component, CO_2 is not in excess, or the variance is low (Coombs et al., 1976). The Fe-Mg substitution is generally consistent among amphibole, chlorite, and pumpellyite, with chlorite more Fe-rich than associated amphibole (Fig. 15; see also Baltatzis and Katagas, 1984, Fig. 8, for similar compositional relationships in mafic and intermediate metavolcanics). The conditions of metamorphism are estimated to be 250 to 370°C at pressures of several kilobars.

Felsic volcanic rocks

The biotite-in reaction identified in felsic volcanics and pelites from coastal Maine is reaction (4) above (Stewart and Flohr (1984 and pers. comm.). Between the chlorite and biotite zones, chlorite becomes less Tschermak-rich (Fig. 14) and more Mg-rich, whereas muscovite becomes more Tschermak-rich and more Fe-rich. These results indicate that the exchange reactions between the two phases "favor" TK [Ms] and FM [Ms] with increasing temperature. TK changes more in muscovite than in chlorite or biotite. $FeMg_{-1}$ varies extensivley in chlorite and biotite due to variation in bulk rock composition and metamorphic grade. The Tschermak and $FeMg_{-1}$ substitutions are coupled in chlorite and biotite; more Fe-rich compositons are richer in both Al^{IV} and Al^{VI} (Fig. 16).

MAFIC ROCKS

Chlorite provides much of the green in greenschist but decreases modally with metamorphism into the amphibolite facies. Because chlorite is ubiquitous in low grade metamorphosed mafic rocks, it provides a convenient phase from which to project mineral assemblages (Liou et al., 1985, Fig. 6). Fe/(Fe+Mg) is variable (Fig. 2) and sensitive to rock composition. Maruyama et al. (1986, p. 13) suggested that this ratio in chlorite can be used to monitor bulk rock Fe/(Fe+Mg).

Figure 17 is a petrogenetic grid for mafic rocks based on experimental data, assemblages in rocks, and thermodynamic calculations. The reactions in Na_2O-CaO-MgO-Al_2O_3-SiO_2-H_2O that are used for the facies boundaries are (from Liou et al., 1985):

Zeolite to prehnite-pumpellyite facies

$$Lmt + Pmp = Czo + Chl + Qtz + H_2O \qquad (19)$$

Prehnite-pumpellyite facies to pumpellyite-actinolite facies

$$Prh + Chl + Qtz = Pmp + Tr + H_2O \qquad (20)$$

Prehnite-actinolite to greenschist facies

$$Prh + Chl + Qtz = Czo + Tr + H_2O \qquad (21)$$

Pumpellyite-actinolite to greenschist facies

$$Pmp + Chl + Qtz = Czo + Tr + H_2O \qquad (22)$$

Blueschist to greenschist facies

$$Gln + Czo + Qtz + H_2O = Chl + Tr + Ab \qquad (23)$$

426

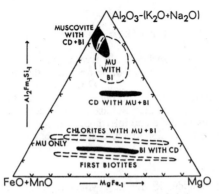

Figure 16. Range of chlorite, biotite, cordierite, and muscovite compositions from silicic volcanics and pelites (Stewart and Flohr, 1984; written comm., 1987) showing correlation of TK and FM. Diagram is in mol % units and projected from alkali feldspar.

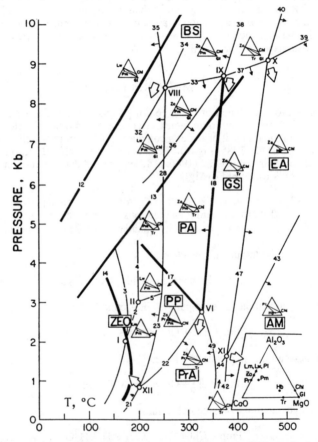

Figure 17. Petrogenetic grid for metamorphosed mafic rocks from Liou et al. (1985, Fig. 2). Bold lines are reactions studied experimentally. Facies: ZEO (zeolite), PP (prehnite-pumpellyite), PrA (prehnite-actinolite), PA (pumpellyite-actinolite), GS (greenschist), EA (epidote-amphibolite), AM (amphibolite), BS (blueschist). Reactions may be inferred from mineral composition diagrams. Reactions (19) to (23) in text are reactions (23), (17), (49), (18), and (37), respectively, on figure. Large arrows indicate movement of invariant point with increasing Fe_2O_3, and small arrows indicate movement of univariant lines with increasing Fe_2O_3. Gl (glaucophane), Hb (hornblende), lm (laumontite), lw (lawsonite), Pm (pumpellyite), Pr (prehnite).

Cho and Liou (1987) presented a grid that includes calcite (Fig. 18). The calcite + chlorite field increases in P-T space with increasing XCO_2. Zen (1974) presented petrogenetic nets for prehnite and pumpellyite assemblages assuming both that CO_2 is and is not a boundary value component. Harte and Graham (1975, Fig. 7) constructed a T-XCO_2 grid involving the phases Act, Ank, Chl, Cal, Ep, H_2O, and CO_2.

Diagenesis to greenschist facies

In general analyses of chlorite from zeolite and pumpellyite facies mafic schist plot above the 1:1 line on Figure 19 and indicate AM substitution. The analyses span all but the very Mg-rich chlorite compositions seen in mafic schist from high-pressure and low-pressure facies series terranes (Fig. 2).

In zeolite facies metabasalts from the Horokanai ophiolite, Japan (Ishizuka, 1985) and East Taiwan ophiolite (Liou, 1979), chlorite composition is a function of the "protomineral" or protolith. For example, chlorite formed from relict clinopyroxene is not the same composition as chlorite formed from devitrified glass. Therefore, chlorite composition in some samples may vary considerably at these temperatures (e.g., 150-250°C at 0.6 to 1.6 km in eastern Taiwan). Chlorite-smectite is reported but so too is chlorite.

The smectite to chlorite transition occurs between the zeolite and greenschist facies in the Point Sal ophiolite, northern California (Bettison and Schiffman, 1988). Upper lavas have randomly interstratified chlorite-smectite and smectite; only minor chlorite occurs. With depth randomly interstratified chlorite-smectite and smectite decrease while chlorite increases in mode. Chlorite is the most abundant phase at the bottom of the section. Si and Ca decrease from interstratified chlorite-smectite to chlorite. Similar relationships are seen in the Del Puerto ophiolite, central California (Evarts and Schiffman, 1983). The chlorite-smectite to chlorite transition occurs just above (not as deep as) the transition from pumpellyite to epidote facies, the latter is at about 225°C.

X-ray diffraction data show that the transition from interstratified chlorite and chlorite sensu stricto is between about 600 and 700 meters below the top of a mafic to intermediate volcanic pile in northern Ecuador, between pumpellyite-prehnite-epidote assemblages and actinolite-epidote-chlorite assemblages (Aguirre and Atherton, 1987). In geothermal wells from Iceland, the interstratified chlorite-smectite to chlorite transition is reported at between 200 and 240°C (Kristmannsdottir, 1975, 1979, and Franzson et al., 1986, as referenced by Bettison and Schiffman, 1988, p. 74).

Burial metamorphism of mafic volcanics, Hamersley basin, Western Australia resulted in little change in chlorite composition through 3 km depth (Smith et al., 1982). What small change in chlorite composition noted in chlorite from prehnite-pumpellyite and prehnite-actinolite facies metabasite from North Wales is sympathetically related to rock composition (Bevins and Merriman, 1988, Fig. 7) as is chlorite composition from prehnite-pumpellyite facies mafic volcanics from southern British Columbia (Beddoe-Stephens, 1981). Baltatzis and Katagas (1984) reported that in basic and intermediate metavolcanics of about the same X_{Fe}, chlorite is more Fe-rich in pumpellyite-actinolite facies rocks (X_{Fe}=0.52) than in greenschist facies rocks (X_{Fe} = 0.40). However, the data scatter is rather large.

Other papers presenting petrographic descriptions of pumpellyite facies metavolcanics and metavolcanic sediments include Coombs et al. (1970), Zen (1974), and Kuniyoshi and Liou (1976a). A symposium on low-grade metamorphism (*Canadian Mineralogist*, 1974, 12, p. 437-552) adds further to our petrographic understanding of chlorite assemblages in low grade rocks.

Cho et al. (1986) and Kuniyoshi and Liou (1976b) described low-grade contact metamorphism in mafic volcanics from Vancouver Island. Chlorite occurs in both zeolite

428

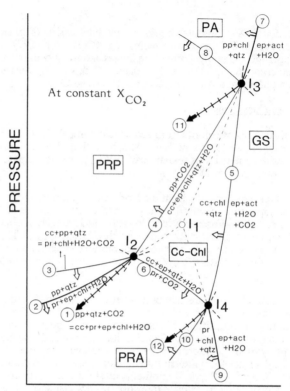

Figure 18. P-T diagram for reactions within prehnite-actinolite (PRA), prehnite-pumpellyite (PRP), and green-schist facies (GS) mafic rocks from Cho and Liou (1987, Fig. 9). Addition of calcite to Figure 17 results in a calcite-chlorite (Cc-Chl) field that increases with increasing XCO_2. Arrows show change of univariant liens with increasing Fe^3. Pumpellyite (pp), prehnite (pr).

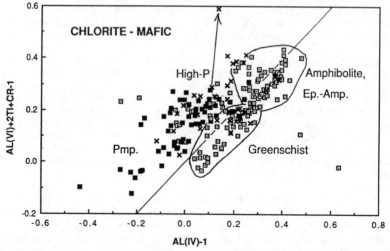

Figure 19. Chlorite analyses from metamorphosed mafic rocks. (See Appendix for data sources). Analyses above 1:1 line show AM; TK parallels this line; the origin is clinochlore/chamosite. Filled squares are analyses from zeolite to pumpellyite-actinolite facies. Envelopes surround analyses from greenschist facies samples and epidote-amphibolite to amphibolite facies samples (medium- and low-pressure facies series). X = high-pressure facies series. Effect of data normalization shown by arrow from 14 oxygens all Fe^2 to 14 oxygens, analyzed Fe^3/Fe^2.

and prehnite-pumpellyite samples. While zeolite facies chlorite shows a larger range in TK than prehnite-pumpellyite facies chlorite, there is no consistent variation between TK and metamorphic grade (Cho et al, 1986, Fig. 10). Although significant overlap occurs Fe/(Fe+Mg) is generally somewhat smaller at higher metamorphic grade.

With increasing metamorphism into the greenschist facies, Cho and Liou (1987) reported no consistent change in chlorite composition, but Maruyama et al. (1983, Fig. 3) showed an increase in Al. The main variation in X_{Fe} is probably related to initial bulk composition. Matrix chlorite is generally more Fe rich than chlorite in amygdules. Reactions between the prehnite-pumpellyite and greenschist facies initially produce modal chlorite:

$$Pmp + Qtz + CO_2 = Cal + Prh + Ep + Chl + H_2O \qquad (24)$$

at low XCO_2; at higher temperature modal chlorite was consumed:

$$Cal + Chl + Qtz = Ep + Act + H_2O + CO_2 . \qquad (25)$$

[Reactions 24 and 25 are numbers 1 and 5 on Fig. 18, respectively.]

Greenschist to amphibolite facies

In an earlier short course on amphiboles, Laird (1982) included a discussion of chlorite in metamorphosed mafic rocks. Some of the main points (with respect to chlorite) are summarized below. (The conclusions do not change if chlorite analyses are normalized to 14 oxygens rather than 10 cations, as was done by Laird, 1982.) References and documentation are in the earlier short course notes. Other recent discussions of this transition include Moody et al. (1983), Apted and Liou (1983), and Maruyama et al. (1983, 1986).

(1) Chlorite is a common mineral in greenschist to low grade amphibolite along with amphibole, epidote, plagioclase, and quartz plus a Ti phase (titanite, ilmenite, or rutile) and Fe oxide (magnetite or hematite), and plus or minus a K-mica (white mica or biotite) and a carbonate (calcite, ankerite, or dolomite).

(2) Within this common assemblage modal chlorite decreases with metamorphic grade.

(3) TK increases from greenschist to amphibolite as is shown on Figure 19 herein. Amphibole shows a greater increase in TK between the greenschist and amphibolite facies than chlorite does (Laird, 1982a, Fig. 64).

(4) Of the three controlling net-transfer reactions pertinent between greenschist and low grade amphibolite, two involve chlorite:

$$22Ab + 2Clinochlore = 5Tr + 15TK + 22NaSiCa_{-1}Al_{-1} + 6Czo + 7Qtz , \qquad (26)$$
$$\text{[Amp,Chl]} \qquad \qquad \text{[Amp,Pl]}$$

and

$$7Clinochlore + 12Czo + 14Qtz = 12Tr + 25TK + 22H_2O . \qquad (27)$$
$$\text{[Amp,Chl]}$$

These reactions explain (2) and (3) and define the Beta and Gamma axes, respectively, on Figure 20. Reaction (26) is the greenschist-blueschist reaction (similar to #37 from Liou et al., 1985) and is shown on Figure 17. TK increases from greenschist to blueschist; the majority is included within amphibole to make glaucophane.

(5)   Amphibole is more Mg-rich than coexisting chlorite in greenschist (with the *common* assemblage), but the two minerals have about the same $Mg/(Mg+Fe)$ in epidote-amphibolite and low-grade amphibolite (Laird, 1982, Figs. 55 and 63).

Maruyama et al. (1983) reported that chlorite in low-pressure facies series basaltic rocks shows an increase in Al_2O_3 associated with a decrease in SiO_2 between greenschist and epidote amphibolite facies. This relationship makes sense because the variation in TK for medium- and for low-pressure facies series metamorphism is controlled by reaction (27) as seen in Figure 20. Changes in TK and $FeMg_{-1}$ appear to be correlated based on the data of Maruyama et al. (1983). However, Beddoe-Stephens (1981) found no correlation in mafic volcanics from British Columbia. A correlation is not very obvious when the data are taken as a whole (Figs. 21, 22). Variation between low-, medium-, and high-pressure metamorphism is not obvious. Temperature appears to be more important. The apparent correlation between increasing Al with increasing Mg is controlled by analyses from zeolite to pumpellyite facies. Why should chlorite in pelitic and felsic rocks show a positive correlation between Al and Fe while chlorite in mafic rocks do not?

Apted and Liou (1983, Fig. 5) compared their experimental data on the chlorite-out and epidote-out reactions with those of other whole rock experimental studies on mafic compositions. While the reaction boundaries depend on bulk rock composition and fO_2, the following generalization is valid (Fig. 20C): At moderate and higher pressure chlorite is consumed before epidote is; whereas at low pressure epidote may be consumed before chlorite is. This generalization is very nicely illustrated in reaction space (Fig. 20A,B) as first shown by Thompson et al. (1982).

Figure 20 can be used at least qualitatively to estimate pressure and temperature. Regardless of which bulk composition composes the reaction space, or if other average basaltic compositions are used, the 2 kbar data of Liou et al. (1974) plot distinctly above the 5 and 7 kbar data of Apted and Liou (1983), which plot above the sample of glaucophane schist. Temperature increases away from the origin. Empirical data from mafic schist support these relationships (Laird et al., 1984, Fig. 3).

<u>Fe-rich mafic compositions</u>

Laird (1980, Fig. 4) showed that Fe-rich greenschist from Vermont is composed of stilpnomelane + actinolite + chlorite while Fe-rich epidote-amphibolite and low-grade amphibolite is composed of garnet + hornblende + chlorite, both with the other minerals in the *common* assemblage. Chlorite in these assemblages is more Fe-rich than chlorite in mafic schist without stilpnomelane or garnet. Chlorite with stilpnomelane (or garnet) but without amphibole is more aluminous than chlorite with both stilpnomelane (or garnet) and amphibole. (See Fig. 55 in *Rev. in Mineralogy*, Vol.. 9B, p. 121).

In the Shuksan Suite blueschist, chlorite associated with stilpnomelane, Na-amphibole, and Na-pyroxene is more Fe rich than chlorite associated with lawsonite, Ca-amphibole, Na-amphibole, and pumpellyite (Brown, 1986, Fig. 8). Similarly, chlorite in stilpnomelane-bearing metabasite from the Franciscan (lawsonite and pumpellyite zones) is Fe-rich compared to stilpnomelane absent assemblages (Maruyama and Liou, 1988, Fig 12). At low pressure facies series, chlorite and garnet do not commonly coexist in amphibole-bearing assemblages (Labotka, 1987).

<u>Contact metamorphism</u>

Low grade contact metamorphism of mafic rocks is described above (data by Cho e al. 1986 and Cho and Liou, 1987). Chlorite in contact metamorphosed alkali olivine basalts from the Isle of Skye, Scotland is confined to the amphibole zone where it occurs as individual grains and as replacement of smectite pseudomorphs after olivine (Ferry et al. 1987). At temperatures below the amphibole zone (i.e., less than 400°C) smectite occurs

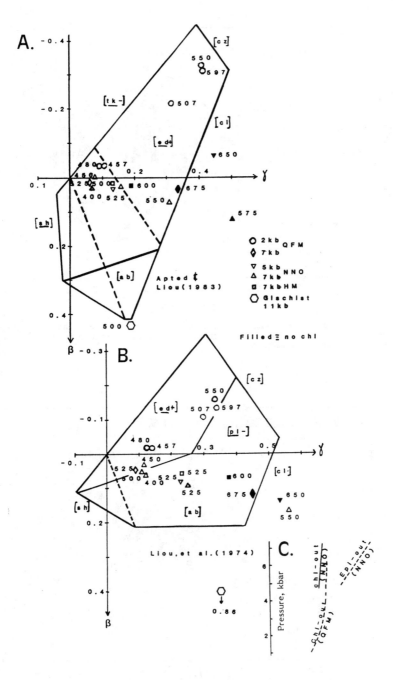

Figure 20. Reaction space for mafic schist (A, B) constructed by the method of J.B. Thompson et al. (1982) and pressure-temperature diagram (C) for chlorite-out and epidote-out reactions from Apted and Liou (1983). Beta and Gamma are equivalent to reactions (26) and (27) in text, respectively. Bulk composition for (A) and (B) are from Apted and Liou (1983) and Liou et al. (1974), respectively. Experimental data from both papers are shown on both (A) and (B) as is a sample of glaucophane schist from Laird and Albee (1981a). Numbers indicate temperature of experimental run. 5 and 7 kbar data at QFM, NNO, and HM buffers are distinguishable more by temperature than pressure. Chlorite was not present in experimental runs indicated by filled symbols.

432

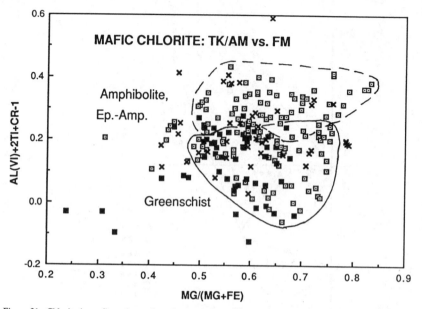

Figure 21. Chlorite in mafic rocks on formula proportion Al^{VI} + 2Ti + Cr – 1 versus atom ratio Mg/(Mg+Fe) diagram. Data and symbols are the same as in Figure 19. Envelopes distinguish greenschist facies from epidote-amphibolite plus amphibolite facies.

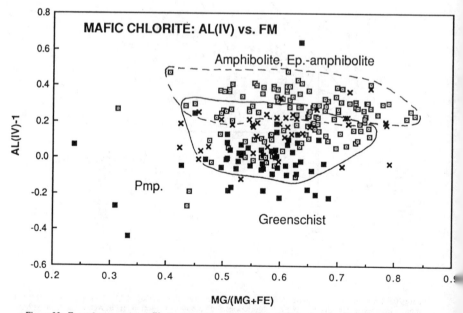

Figure 22. Formula proportion Al^{IV} – 1 versus atom ratio Mg/(Mg+Fe) for chlorite from mafic schist. Data and symbols are the same as for Figures 19 and 21. Envelopes distinguish chlorite from greenschist and from epidote-amphibolite plus amphibolite facies samples.

but chlorite does not. Chlorite is not identified in the orthopyroxene + olivine zone (T = about 900–1030°C).

Blueschist and eclogite

Chlorite in high-pressure facies series mafic schist seems to show AM substitution, but this substitution is very dependent on mineral normalization (Fig. 19). Maruyama and Liou (1988, p. 18) reported that progressive metamorphism results in a slight increase in TK from chlorite in Franciscan metabasite.

Within the pumpellyite facies chlorite can coexist with aegerine-jadeite, aragonite, and paragonite (Takayama, 1986, Fig. 5) and with pumpellyite + aegerine-jadeite (Maruyama and Liou, 1988 ; Sakakibara, 1986) or pumpellyite + aragonite (Brown and Ghent, 1983). Chlorite is consumed in several reactions that form glaucophane (reactions 14-16, 18, 19 as summarized by Laird 1982), but it forms in high-pressure facies series metamorphism of metagabbro by the breakdown of plagioclase to jadeite, lawsonite, chlorite, and quartz (Pognante and Kienast, 1987) or the breakdown of paragonite (reaction 20, Laird, 1982).

Chlorite is stable at least to the epidote zone and 550°C and is reported to be consumed between the epidote and omphacite zones (see for example summaries by Yokoyama et al., 1986, Fig. 5; Laird, 1982, Figs. 71, 72; and Ernst, 1977, Figs. 3-5). However, chlorite in eclogites may be present due to retrogradation (e.g., Moore, 1984), and it should be stable in less Na-rich compositions at the same conditions of meta-morphism (Fig. 13; see also Fig. 16, Laird and Albee, 1981a).

Garbenschiefer

Amphibole garbenschiefer are very pretty rocks with fascicles of amphibole, generally with white mica(s), Fe-rich to very Mg-rich chlorite, and (if one is lucky) Al-rich phases such as chloritoid, staurolite, or kyanite. Selverstone et al. (1984) showed that the assemblages with staurolite and kyanite are not due simply to bulk composition. Rather, the breakdown of chlorite + epidote + plagioclase ± garnet allows amphibole to coexist with staurolite, kyanite, or paragonite (Fig. 23). Similar relationships are shown by Helms et al. (1987, Fig. 5). P.H. Thompson and Leclair (1987, Fig. 8) constructed petrogenetic grids to explain these assemblages and proposed that hornblende + chloritoid is stable at lower temperature than staurolite + hornblende, and that kyanite + hornblende forms at higher P.

Selverstone et al. (1984, p. 529) suggested that the assemblage hornblende + kyanite + staurolite ± paragonite may indicate high-pressure facies series metamorphism. Pressures of about 8 kbar are indicated for garbenschiefer from the Austrian and Swiss Alps and Blue Ridge (Selverstone et al., 1984; Klaper and Bucher-Nurminen, 1987; Helms et al., 1987). However, medium-pressure facies series hornblende + kyanite garbenschiefer occurs locally in southeastern Vermont (Hepburn et al., 1984). The garbenschiefer texture also occurs in medium- and low-pressure facies series. [See Boxwell and Laird, 1987, and Laird and Albee, 1981b for a discussion of the petrology in these samples, including data on chlorite.]

ALUMINOUS CALC-SILICATE ROCK

Variation in chlorite composition from aluminous calc-silicate rocks is primarily along TK and FM. While more data would be helpful, chlorite from impure dolomites appears to show less TK than chlorite from calcareous argillaceous rocks (Fig. 24). Chlorite from impure dolomites is distinctly more Mg-rich (Fig. 2).

434

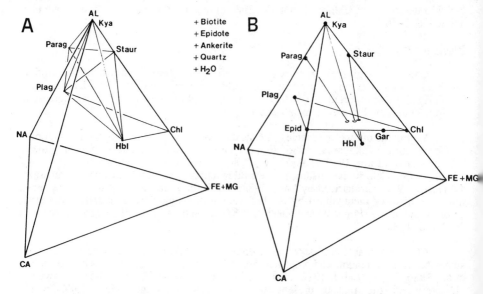

Figure 23. Minerals within garbenschiefer illustrated on Al-Na-Ca-(Fe+Mg) tetrahedron (atom units) from Selverstone et al. (1984, Figs. 6 and 8). (A) Projected from biotite, epidote, ankerite, quartz, and H$_2$O. (B) Similar to (A) but not projected from epidote. With progressive metamorphism chlorite + epidote (Epid) + plagioclase (Plag) become unstable so that hornblende (Hbl) can occur with aluminous minerals such as kyanite (Kya), staurolite (Staur), and paragonite (Parag). Gar = garnet.

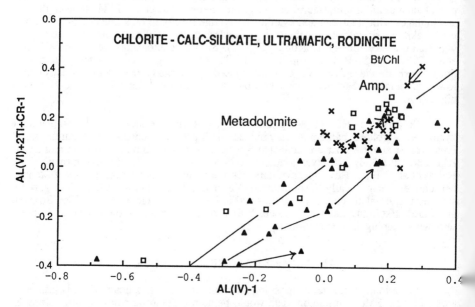

Figure 24. Formula proportion Al^{VI} + 2Ti + Cr − 1 versus Al^{IV} − 1 for chlorite from metamorphosed ultramafic rocks (filled triangle), calc-silicate rocks(x), and rodingite (square). Data sources in Appendix. Origin is clinochlore/chamosite. The 1:1 line represents TK. Single arrows show increasing contact metamorphism of ultramafic rocks. Dashed arrow shows increasing contact metamorphism of impure dolomite. Double arrow shows increasing low-pressure facies series, regional metamorphism of argillaceous calc-silicate.

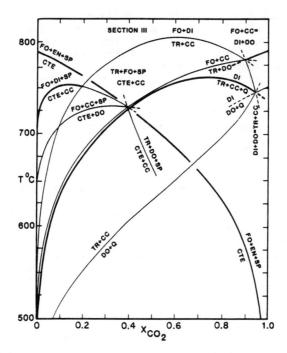

Figure 25. Temperature-XCO_2 diagram for impure marble and metamorphosed peridotite from Lieberman and Rice (1986, Fig. 13). Chlorite breaks down with increasing T and XCO_2 to forsterite + enstatite + spinel and is stable to higher temperature with decreasing XCO_2. CC (calcite), CTE (chlorite), DO (dolomite), SP (spinel), Q (quartz).

Chlorite + calcite + quartz occurs in unmetamorphosed calc-silicate rocks (e.g., Bucher-Nurminen, 1982a) and is inferred to form during retrogradation of calc-silicate gneisses in Scotland (Winchester, 1972) and retrograde metamorphism of eclogite facies calc-silicate in the Alps (Franz and Spear, 1983). This assemblage becomes unstable with progressive metamorphism at both medium- and low-pressure facies series.

Medium-pressure facies series

Comparison of mineral assemblages in intercalated carbonate-rich, pelitic, mafic, and ultramafic rocks from Naxos, Greece (Jansen and Schuiling, 1976) shows that chlorite is stable in carbonate-rich rocks until the staurolite-kyanite zone, low-amphibolite facies, and until forsterite + anthophyllite occur in ultramafic rocks. Similarly, chlorite is stable in aluminous calc-silicate rocks from east-central Alaska in the biotite and garnet zones of intercalated pelitic rocks and greenschist and epidote amphibolite facies of intercalated mafic rocks (Laird et al., 1986). In the staurolite-kyanite zone chlorite becomes unstable (in aluminous calc-silicate rocks) in favor of amphibole, plagioclase, and epidote.

However, chlorite + calcite + clinopyroxene + olivine + spinel is identified in upper amphibolite facies marble from northern California (Lieberman and Rice, 1986). Based on Figure 25 this assemblage implies XCO_2 less than 0.2 using the inferred temperature (760 to 900°C).

Low-pressure facies series regional metamorphism

The Vassalboro Formation in south-central Maine provides a good natural laboratory for studying low-pressure facies series metamorphism in argillaceous carbonate rocks.

While not abundant chlorite is a common mineral in ankerite, biotite, and amphibole zones. Ferry (1976, 1983a,b) identified the reactions responsible for increasing modal chlorite between the ankerite and biotite zones and for decreasing chlorite from the biotite to amphibole to zoisite zones (in order of progressive metamorphism). Based on petrographic relationships, mineral composition data, and pertinent mineral composition diagrams, "ideal" reactions produce and then consume chlorite with increasing grade (Ferry, 1976):

Biotite-chlorite isograd

$$Ms + Qtz + Ank + CO_2 = Cal + Chl + Bt + H_2O . \tag{28}$$

Amphibole-anorthite isograd

$$Chl + Cal + Qtz + \text{intermediate plagioclase} =$$
$$\text{calcic amphibole} + \text{calcic plagioclase} + H_2O + CO_2 . \tag{29}$$

Further data on mode, rock composition, and the fluid phase give a more detailed understanding of these reactions and their proportion in whole rock reactions (Ferry, 1983a). Reaction (28) becomes:

$$2.668Ms + 0.190Ilm + 10.061Ank + 1.803 Qtz + 3.990H_2O + 0.079HCl =$$
$$Chl + 2.697Bt + 10.061Cal + 10.061CO_2 + 0.79NaCl . \tag{30}$$

Reaction (29) is:

$$Chl + 0.029Ttn + 2.722Cal + 5.602Qtz + 0.158Ab =$$
$$0.983Amp + 0.844An + 3.017H_2O + 2.722CO_2 \tag{31}$$

(Ferry, 1983b). (Addition of net-transfer reactions involving exchange vectors does not affect reactions controlling modal chlorite because mafic phases do not change composition appreciably at these grades.) Neither reaction is particularly important to the whole rock reactions. Advancement along (30) is 0.029 moles/1000 cm^3 rock and along (31) is 0.032 moles/1000 cm^3 rock (Ferry, 1983b; similar numbers are given by Ferry, 1983a).

Contact metamorphism

Rice (1977a,b) identified chlorite in low- and high-grade contact metamorphosed dolomitic limestone from Montana. With increasing metamorphism (decreasing proximity to the Boulder batholith) chlorite coexisting with calcite and dolomite increases in Al^{IV} and (Al^{VI} + 2Ti) along TK (Fig. 24). The greatest change in composition is before spinel or forsterite is stable. Chlorite coexists with forsterite and spinel but is finally consumed by the reaction:

$$Chl + 2Dol = 3Fo + 2Cal + Spl + 2CO_2 + 4H_2O \tag{32}$$

(as seen on Fig. 25).

Metasomatism, including rodingite

Mineral assemblages in dolomitic rock contact metamorphosed by granite in northern Italy can be explained by progressive metamorphism (Bucher-Nurminen, 1982a, Fig. 6). However, there is a large increase in modal silicates at the forsterite isograd (perhaps ten-fold, Bucher-Nurminen, 1982a, p. 1113), suggesting infiltration of SiO$_2$ and Al$_2$O$_3$ rich fluids from the granitic intrusion into the dolomitic country rock. Chlorite is inferred to increase in mode by metasomatism. Chlorite changes very little in composition between about 450 and 600°C, and what compositional variation is present is not consistent with respect to temperature. Further data on metasomatic reactions in impure dolomite are presented by Bucher-Nurminen (1982b).

Rodingites are calc-silicate rocks formed by metasomatism of mafic to felsic rocks associated with serpentinized untramafic rocks. Chlorite analyses from metamorphosed rodingites show a compositional variation similar to that of metamorphosed ultramafic rocks (Fig. 24). Compositional variation is primarily along TK and due to a combination of differences in metamorphic grade, bulk rock composition, and "protomineral (s)" (Lan and Liou, 1981, p. 371).

Rice (1983) presented an excellent review of metarodingites and several P-T-X diagrams. These diagrams are applicable to metamorphosed rodingites and to silica-undersaturated aluminous limestones. Lieberman and Rice (1986) showed that such diagrams are very sensitive to the thermodynamic data and calculation procedures used.

ULTRAMAFIC ROCKS

Chlorite is the Al-phase in metamorphosed Alpine-type peridotite from the prehnite-pumpellyite facies into the amphibolite and hornblende hornfels facies (Evans, 1977 and 1982, Table 6). Tracy et al. (1984) reported chlorite in hornblendite metamorphosed at 650 to 700°C and 6±1 kbar. Abbott and Raymond (1984) reported chlorite + tremolite + olivine up to upper amphibolite or granulite facies with temperature about 700°C and pressure greater than 7 kbar. Chlorite breaks down in the sillimanite zone (of associated pelitic rocks) to forsterite + enstatite + green spinel + H_2O (Evans, 1977; see Fig. 25 herein). In contrast to chlorite in other rock types, except for metamorphosed rodingites, chlorite in low grade ultramafic rocks (below enstatite in) shows less TK than clinochlore (Fig. 24). Cr and Ni also appear to be significant in metamorphosed ultramafic rocks, but comparison with other rock types is precluded by the fact that Cr is often not reported in chlorite from other rock types and Ni rarely is reported.

Contact metamorphism

Springer (1974) listed chlorite as one of the first minerals to form by contact metamorphism of serpentinite from the Sierra Nevada Mountains. In Washington state and British Columbia, respectively, Frost (1975) and Pinsent and Hirst (1977) showed that chlorite within contact metamorphosed peridotite and blackwall becomes more Tschermak-rich with increasing metamorphic grade/distance toward the igneous heat source (Fig. 24).

Metasomatism

Chlorite blackwall occurs at the contact of ultramafic rocks and their country rocks. Sanford (1982) showed that in northern Appalachian occurrences $Al_2Mg_{-1}Si_{-1}$ and $FeMg_{-1}$ diffused into the country rock and that this metasomatism is manifested in the chlorite composition (Fig. 26). Al and Fe are distinctly greater in chlorite from the country rock; and Mg , Cr, and Ni are less in chlorite from the country rock. These observations are consistent with data presented by Zhong et al. (1985) who showed increasing Al and Fe in chlorite with increasing metasomatism of ultramafic rocks.

IRON FORMATIONS

Chlorite is rare in iron formations (e.g., Floran and Papike, 1978, p. 228) because the bulk rock composition is generally very low in Al. Where chlorite appears to be present petrographically, it may not be possible to test the hypothesis with X-ray diffraction data (i.e., distinguish 7 Å and 14 Å phases) because it is so fine-grained and not very abundant (see Klein, 1978). Chamosite (7 Å), chlorite, and interstratified 7 Å and 14 Å phases are reported in banded iron formation from western Australia (Gole, 1980).

438

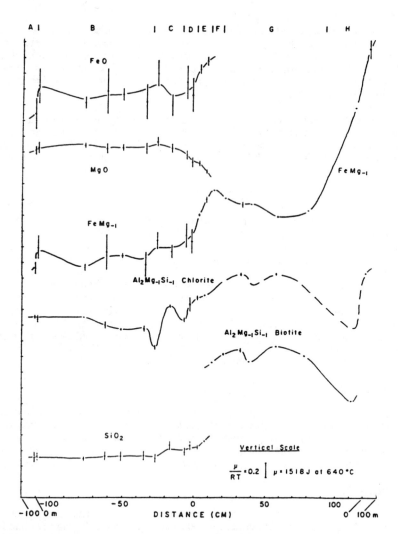

Figure 26. Variation in chemical potential within rocks, chlorite, and biotite across an ultramafic/mafic country rock contact from Sanford (1982, Fig. 20). The ultramafic (A) is separated from the country rock by metasomatic zones: Tlc + carbonate + Chl + Mag (B), Tlc + Chl + Mag (c), Tlc + Amp + Chl (D), and Chl (E). The country rock is Hbl + Pl + Bt + Chl (G) and Pl + Qtz + Bt + Chl + Ms + Fe-Ti oxide (H).

Chlorite is most common at low metamorphic grades but persists into the staurolite-kyanite zone (Klein, 1978, Fig. 4; 1982, Fig. 45) and into the sillimanite zone (Haase, 1982a). Low grade (about 150°C or less) chlorite occurs along the west edge of the Labrador Trough (Klein and Fink, 1976) and is more Fe-rich in quartzite below the iron formation than in chert. Chlorite from northern Michigan iron formations show a slight increase in Al with increasing metamorphic grade (chlorite to sillimanite zone) and perhaps an increase in X_{Fe}. However, there is a large range in X_{Fe} due to differences in bulk composition (Haase, 1982a, p. 71). Figure 2 suggests more Al-rich chlorite in iron formations is more Fe-rich.

Haase (1982a,b) constructed a petrogenetic grid for iron formations in the system $CaO-FeO-Al_2O_3-SiO_2-CO_2-H_2O$ (Fig. 27). For Al-rich bulk compositions, chlorite should

439

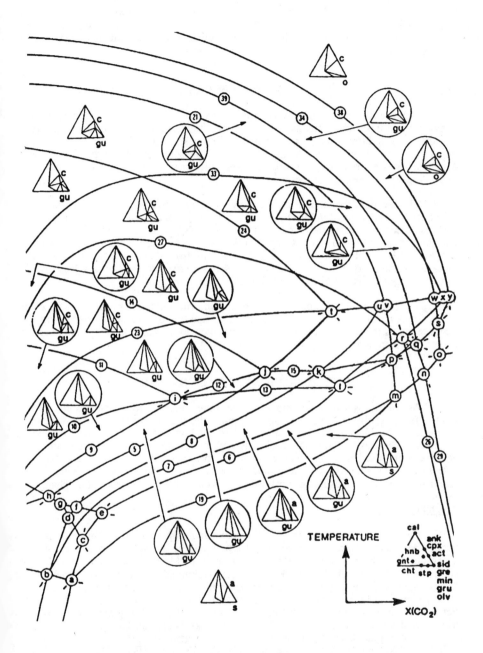

Figure 27. T-XCO₂ diagram for iron formations from Haase (1982b, Fig. 4). Triangle shows mineral assemblages in CaO-Al₂O₃-FeO (XFe = 0.7) space. Quartz is omnipresent; magnetite may or may not be present. gnt (garnet), gre (greenalite), hnb (hornblende), min (minnesotaite), olv (olivine), sid (siderite). Reaction numbers are from Haase (Table 4).

be stable over a wide range of temperature and XCO_2 conditions, (with garnet + calcite at low T and garnet + hornblende at higher T). Chlorite + siderite occurs at lower temperature than chlorite + stilpnomelane. With increasing temperature stilpnomelane breaks down; chlorite + grunerite are stable at lower XCO_2 than chlorite + olivine.

Mn-RICH ROCKS

Chlorite is identified in high-pressure facies series, Mn-rich cherts (Mottana, 1986) and in piemontite-bearing, greenschist to amphibolite facies schists (Kawachi et al., 1983; Abraham and Schreyer, 1976) and quartzite (Reinecke,1986). Mottana (1986, p. 284) noted that while chlorite is commonly described in blueschist facies manganiferous cherts, chemical analyses are often not given and X-ray diffraction data required to distinguish 14 and 7 Angstom phases are rarely presented.

The few data shown on Figure 2 show that very Mg-rich chlorite and chlorite with about 50 atom percent Mg/(Mg+Fe) occur. Compositional variation along TK is minor (Fig. 3). Schreyer et al. (1986) reported a miscibility gap exists in trioctahedral Mn-Mg-Fe chorite from phyllite and slate. Copper, cobalt, and nickel are reported by Abraham and Schreyer (1976). Mn is somewhat richer in chlorite from these rock types, between 0.1 and 0.2 formula proportion (normalized to 14 oxygens). Mn is generally less than 0.1 formula proportion in chlorite from other rocks.

The slope of tie lines between coexisting chlorite and white mica varies for green-schist facies metamorphism in New Zealand as a function of rock composition (Fig. 28). Chlorite varies more in composition than white mica does and is more Mg-rich in assemblages with piemontite.

GEOTHERMOMETRY AND GEOBAROMETRY

As chlorite is such a common mineral in metamorphic rocks, calibrated geother-mometers and geobarometers involving this mineral can be very helpful in understanding the metamorphic history of an area. Most calibrations involve the $FeMg_{-1}$ exchange and are empirical. Besides the "normal" problems with bulk composition, change in intensive variables not considered, and accuracy of the independent calibrations used for reference, assessing the accuracy of proposed geothermometers and geobarometers involving chlorite requires examination of estimates for Fe^2/Fe_{total} in chlorite and other coexisting phases.

Amphibole-chlorite

While chlorite is more Fe-rich than coexisting amphibole in prehnite-pumpellyite facies (Bevins and Merriman, 1988, Fig. 6), pumpellyite-actinolite facies (Fig. 14), and greenschist facies (e.g., Laird, 1980, Fig. 4) mafic rocks, chlorite and amphibole have about the same values of Mg/(Mg+Fe2) at epidote-amphibolite and low-grade amphibolite facies. K_d [(Fe2+Mn)/Mg]Chl/Amp changes from 1.9 to 1.0 (Laird, 1982, Fig. 63). (In these papers Laird normalized chlorite analyses to total cations less Na, K, and Ca to 10. Ferric and ferrous iron were estimated from stoichiometry.) This general relationship is true even if all iron is estimated as ferrous, although Harte and Graham (1975, Fig. 3) showed that amphibole in epidote-amphibolite and low grade amphibolite facies is more Fe-rich than coexisting chlorite. Ernst et al. (1981, Fig. 13) found (Fe/Mg)Chl/(Fe/Mg)Act = 1.41 in greenschist from Taiwan and (Fe/Mg)Chl/(Fe/Mg)Hbl = 1.06 for amphibolite.

A $ln K_d$ versus $1/T$ diagram for the data presented in the amphibole short course (Laird, 1982, Fig. 63) shows a "reasonably" linear relationship (Fig. 29), but the scatter of points at each metamorphic grade is considerable due to errors in estimating the formulas of amphibole (especially) and chlorite, to polymetamorphism, and to errors in independent

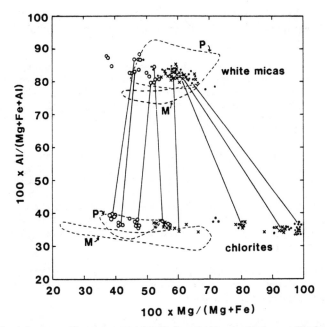

Figure 28. Coexisting compositions (atom units) for chlorite and white mica from greenschist facies samples, New Zealand (Kawachi et al., 1983, Fig. 7). Piemontite quartzose schist (filled dot), quartzose schist (x), and quartzofeldspathic schist (o). Envelopes delimit data compiled by Kawachi et al. (1983) from previous papers: Pelitic and quartzofeldspathic rocks (P), mafic rocks (M).

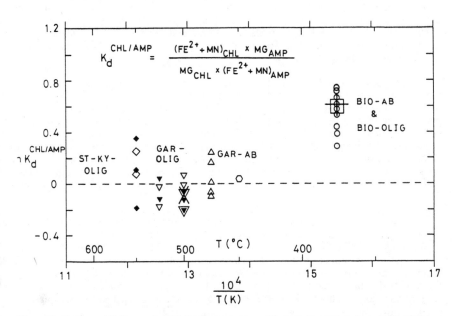

Figure 29. $Ln\ K_d$ vs. $1/T$ diagram for FM distribution between chlorite and amphibole from mafic schist in Vermont. Data are from Laird (1982, Fig. 63). Metamorphic grade is indicated by associated pelitic rocks; plagioclase composition is for the mafic rocks. + and X (low-pressure facies series), hexagon and square (high-pressure facies series), other symbols (medium-pressure facies series). Filled = samples with garnet.

estimates of temperature. While a temperature variation in this distribution seems real, it may not be large and cannot be quantified without a means of determining Fe^2/Fe_{total} more accurately.

No variation in K_d between medium-pressure facies series and low-pressure facies series is seen among the scatter on Figure 29. A more detailed study of the pressure dependence of $FeMg_{-1}$ between actinolite and chlorite by Moonsup Cho (pers. comm.), however, shows that K_d between actinolite and chlorite is dependent on pressure and may provide a geobarometer for low grade metabasites, with P (kbar) = $10.5 \ln K_d + 0.5$, where $K_d = (Mg/Fe)Act/(Mg/Fe)Chl$.

An attempt to quantifiy the Tschermak exchange between amphibole and chlorite was mde by Laird (1982, Fig. 64). A small change in K_d is identified between greenschist and epidote-amphibolite facies mafic schist, and K_d TK[Chl/Amp] appears to decrease by an order of magnitude with increasing metamorphism into the amphibolite facies (Laird, 1982, Fig. 64). However, error in estimates of Tschermak substitution increase as metamophic grade (and Al in amphibole and chlorite) increases. Again, accurate means to estimate Fe^2/Fe_{total} and mineral formulas precludes this from being a useful geothermometer at the present.

Rice (1977a) also showed that the distribution of Al^{IV} between amphibole and chlorite changes abruptly in contact metamorphosed impure dolomites near the final breakdown of chlorite (by reaction 32 above).

<u>Garnet-chlorite</u>

Empirical geothermometers based on the Fe-Mg exchange between garnet and chlorite have been proposed by Kurata and Banno (1974) and, more recently, Dickenson and Hewitt (1986) and Ghent et al. (1987). Dickenson and Hewitt (1986) based their geothermometer on the garnet-biotite geothermometer of Ferry and Spear (1978). At greenschist and low amphibolite facies (400 to 520°C, 5.7 to 11.3 kbar), their geothermometer (as modified, Dickenson, pers. comm.), 55841 − 10.76 T (K) + 0.212 P (bars) + 15RT $\ln K_d$ = 0 (with R in calories) gives temperatures within 15°C of garnet-biotite temperatures from samples not used in their calibration. (K_d is defined as $(Mg/Fe)Gar/(Mg/Fe)Chl.$)

Holdaway et al. (1988, Table 5) showed that in garnet-grade, low pressure facies series pelitic schist the garnet-chlorite geothermometer of Dickenson and Hewitt (1986, modified as given above) gives temperatures from 7° below to 1°C above the garnet-biotite geothermometer of Ferry and Spear (1978). In the staurolite zone the garnet-chlorite geothermometer gives temperatures 15°C below to 7°C above the garnet-biotite geothermometer. The larger variation at higher temperature may be because staurolite grade chlorite is retrograde. That staurolite grade, garnet-biotite and garnet-chlorite temperatures are so close may be because the biotite-chlorite Fe/Mg K_d is the same at both grades and chlorite formed from biotite in the staurolite zone (Holdaway et al., 1988, Fig. 5 and p. 31).

Combining the Ferry-Spear (1978) garnet-biotite geothermometer and the biotite-chlorite equilibrium data presented by Lang and Rice (1985b), Ghent et al. (1987) obtained: T = 2109.92 + 0.00608P/0.6867 − $\ln$ K for the garnet-chlorite Fe-Mg exchange equilibrium. Assuming both chlorite and garnet are ideal solutions, Ghent et al. (1987, p. 249) reported temperatures of 378 to 506°C for omphacite zone pelitic schist. However, using the nonideal solution model for garnet given by Ganguly and Saxena (1984) and ideal solution in chlorite, Ghent et al. (1987) obtained 391 to 536°C for the same samples.

Ghent et al. (1987, p. 249) concluded that the range of temperatures (nearly 150°C) obtained for the same metamorphic grade may represent "incomplete equilibration between garnet and chlorite during cooling." Because these temperaturess are below (50°C and

15°C, respectively) those estimated from oxygen isotope thermometry and Fe-Mg exchange between garnet and clinopyroxene, they concluded that: "Garnet-chlorite Fe-Mg exchange thermometry does not appear to yield reliable estimates of peak metamorphic temperatures for" the omphacite zone metasediments studied. Better agreement is obtained for lower temperature samples, 350-385°C assuming ideal solution for garnet and chlorite and 420-465°C assuming nonideal solution for garnet with ideal solution for chlorite.

Chloritoid-chlorite

Ghent et al. (1987) summarized data for Fe-Mg between chloritoid and chlorite (Fig. 30). $K_d = (Fe/Mg)Cld/(Fe/Mg)Chl$ appears to decrease with increasing temperature (see also Ghent et al., 1987, Fig. 3). However, the variation in K_d is not large, about 10 for greenschist facies rocks, 4.2 to 6.4 for high pressure metasediments from the epidote zone, and 3.9 to 5.5 for the omphacite zone (Ghent et al., 1987; see also Ashworth and Evirgen, 1984). Therefore, this exchange is a relative temperature indicator but not a quantitative geothermometer.

Biotite-chlorite

The Fe-Mg exchange between biotite and chlorite is a poor geothermometer as the exchange varies little with metamorphic grade. Lang and Rice (1985b) reported little change in K_d (Mg/Fe)Bt/(Mg/Fe)Chl (= ~0.91) between chlorite-biotite grade pelitic rocks and above the staurolite isograd, medium-pressure facies series. Holdaway et al. (1988, Fig. 5) showed the biotite-chlorite exchange to be constant ($K_d = 0.87$) in garnet and staurolite grade, low-pressure facies series pelitic schist. Nearly this same ratio ($K_d = (Fe/Mg)Chl/(Fe/Mg)Bt = 0.89$) was given by Ernst et al. (1981, Fig. 13) for mafic schist in Taiwan. Ghent et al. (1987, p. 248, Fig. 3) also showed little fractionation of Fe and Mg between chlorite and biotite from blueschist and eclogite facies pelitic rocks.

From data on natural assemblages metamorphosed at 3.5 to 5 kbar and their garnet-chlorite geothermometer, combined with the garnet-biotite geothermometer of Ferry and Spear (1978), Dickenson and Hewitt (1986) predicted K_d (as defined above) = 0.96 to 0.86 between 400 and 550°C. They gave the relationship (Dickenson, pers. comm.) 6429 - 12.547T(K) + 0.073(bars) + 15RT ln (Mg/Fe)Chl/(Mg/Fe)Bt, with R in calories, and noted that this small fractionation of Mg and Fe between chlorite and biotite allows one to test whether chlorite and biotite are in equilibrium.

Using experimental data from Velde (1965) on the reaction muscovite (20% celadonite, 80% muscovite) = biotite + orthoclase + quartz + H_2O and thermodynamic calculations, Powell and Evans (1983) proposed a geobarometer for biotite + muscovite + chlorite + quartz assemblages. Bucher-Nurminen (1987) recalibrated this geobarometer including further experimental data. His results illustrated in Figure 31 allow one to estimate pressure if temperature is known. For pressures less than 6 kbar and temperatures greater than 450°C, Bucher-Nurminen (1987, Table 2), the two empirical geobarometers give similar pressures (±1 kbar). At higher pressure, the Powell and Evans geobarometer underestimates pressure.

CONCLUDING REMARKS

Petrography remains a very important tool in identifying chlorite (and associated minerals) and estimating the composition of chlorite along the FM exchange. Our understanding of mineral assemblages including chlorite is quite well known. Electron microprobe data on chlorite composition support the general variations identified by Albee (1962). Hypotheses presented for variation in composition as a function of metamorphic grade and facies series need further testing and examination from a mineralogical/crystallographic point of view. Further TEM and X-ray diffraction studies will be important in

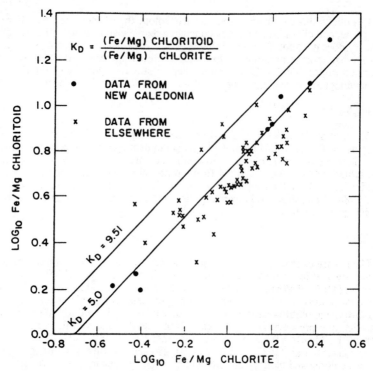

Figure 30. *Ln K_d* diagram compiled by Ghent et al. (1987, Fig. 4) for the FM exchange between chloritoid and chlorite. Lines of constant K_d are for reference only.

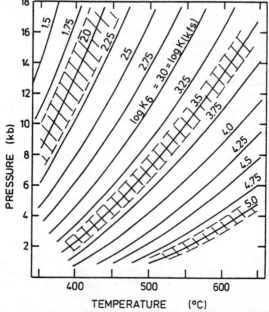

Figure 31. P-T diagram for biotite-chlorite geobarometer from Bucher-Nurminen (1987, Fig. 3). Contours are for *ln* K of: 3celadonite + clinochlore = muscovite + 3phlogopite + 7quartz + 4H$_2$O. Ruled areas represent error bands for the 2.0, 3.5, and 5.0 contours.

assessing interstratification, intergrowth, and polytype. Any advancement in our understanding of Fe^2/Fe^3 and stuctural vacancies in chlorite will be welcomed.

ACKNOWLEDGMENTS

Electron microprobe data on chlorite from mafic schist in Vermont were obtained and assessed petrologically with the financial assistance of NSF Grants DES 69-0064 and EAR 75-03416 to Arden L. Albee and EAR-83-19383 to Jo Laird and W.A. Bothner. The typing and computer expertise of N. Simoneau, T.J. Koch and W.A. Bothner were invaluable. Reviews by Mel Dickenson, Wally Bothner, S.W. Bailey, and P.H. Ribbe are sincerely appreciated. Apologies to those of you whose papers I missed or misinterpreted.

APPENDIX

Data sources for Figures 2-6, 14, 19, 21, 22, 24, 29. Rocks listed by composition.

I. Pelitic Rocks

   A. Medium-pressure facies series

      Albee (1965a)   Clark et al. (1987)
      Duebendorfer (1988)   Enami & Zang (1988)
      Hollocher (1981)   Karabinos (1985)
      Klaper & Bucher-Nurminen (1987)   Lang & Rice (1985a)
      Phillips (1987)   A.B. Thompson et al. (1977a)

   B. Regional low-pressure facies series

      Dickenson (1988)   Guidotti (1974)
      Holdaway et al. (1988)   Novak & Holdaway (1981)

   C. Contact metamorphism

      Fletcher & Greenwood (1979)   Labotka et al. (1981)
      Pattison (1987)   Shiba (1988)

   D. High-pressure facies series

      Black (1975)   Chopin (1986)
      Kurata & Banno (1974)   Laird & Albee (1981a)
      Laird (unpublished)

II. Felsic Rocks

      Boles & Coombs (1977)   Coombs et al. (1976)
      Evarts & Schiffman (1983)   McDowell & Elders (1980, 1983)
      Sanford (1982)   Stewart & Flohr (written comm., 1987)

III. Mafic Rocks

   A. Zeolite, prehnite-pumpellyite, and prehnite-actinolite facies

      Aguirre & Atherton (1987)   Bettison & Schiffman (1988)
      Bevins & Merriman (1988)   Cho & Liou (1987)
      Cho et al. (1986)   Evarts & Schiffman (1983)

Ishizuka (1985)   Maruyama & Liou (1985, 1988)

B.   Greenschist facies

Bettison & Schiffman (1988)   Ishizuka (1985)
Laird & Albee (1981b)   Laird (unpublished data)
Maruyama & Liou (1985)   Morten et al. (1987)

C.   Epidote amphibolite and low-grade amphibolite facies

Baroz et al. (1987)   Ishizuka (1985)
Labotka (1987)   Laird & Albee (1981b)
Laird (unpublished data)   Maruyama et al. (1983)
Spear (1982)

D.   High-pressure facies series, garbenschiefer

Black (1975)   Brown & Ghent (1983)
Cotkin (1987)   Helms et al. (1987)
Guiraud et al. (1987)   Laird & Albee (1981a)
Laird (unpublished)   Moore (1984)
Pognante & Kienast (1987)   Selverstone et al. (1984)
Thompson & Leclair(1987)

IV.   Aluminous calc-silicate rocks

A.   Calcareous argillite

Ferry (1976)

B.   Impure dolomite

Bucher-Nurminen (1982, 1987)   Lieberman & Rice (1986)
Rice (1977a,b)

C.   Rodingite

Lan & Liou (1981)

V.   Ultramafic rocks

Frost (1975)   Pinsent & Hirst (1977)
Sanford (1982)   Tracy et al. (1984)

VI.   Fe-formation

Haase (1982b)
Klein & Fink (1976)
Klein & Gole (1981)

VII.   Mn-deposits

Abraham & Schreyer (1976)
Kawachi et al. (1983)
Mottana (1986)

REFERENCES

Abbott, R.N., Jr. and Raymond, L.A. (1984) The Ashe Metamorphic Suite, northwest North Carolina: metamorphism and observations on geologic history. Am. J. Sci. 284, 350-375.

Abraham, K. and Schreyer, W. (1976) A talc-phengite assemblage in piemontite schist from Brezovica, Serbia, Yugoslavia. J. Petrol. 17, 421-439.

Aguirre, L. and Atherton, M.P. (1987) Low-grade metamorphism and geotectonic setting of the Macuchi Formation, western Cordillera of Ecuador. J. Metamorphic Geol. 5, 473-494.

Ahn, J.H., Peacor, D.R., and Coombs, D.S. (1988) Formation mechanisms of illite, chlorite and mixed-layer illite-chlorite in Triassic volcanogenic sediments from the Southland syncline, New Zealand. Contrib. Mineral. Petrol. 99, 82-89.

Albee, A.L. (1962) Relationships between the mineral association, chemical composition and physical properties of the chlorite series. Am. Mineral. 47, 851-870.

Albee, A.L. (1965a) Phase equilibria in three assemblages of kyanite-zone pelitic schists, Lincoln Mountain quadrangle, central Vermont. J. Petrol. 6, 246-301.

Albee, A.L. (1965b) A petrogenetic grid for the Fe-Mg silicates of pelitic schists. Am. J. Sci. 263, 512-536.

Albee, A.L. (1968) Metamorphic zones in northern Vermont. In E-an Zen, W.S. White, J.B. Hadley, and J.B. Thompson, Jr., Eds., Studies of Appalachian Geology: Northern and Maritime. Interscience, 329-341.

Albee, A.L. (1972) Metamorphism of pelitic schists: reaction relations of chloritoid and staurolite. Geol. Soc. Am. Bull. 83, 3249-3268.

Albee, A.L. and Ray, L. (1970) Correction factors for electron probe microanalysis of silicates, oxides, carbonates, phosphates, and sulfates. Analyt. Chem. 42, 1408-1414.

Apted, M.J. and Liou, J.G. (1983) Phase relations among greenschist, epidote-amphibolite, and amphibolite in a basaltic system. In H.J. Greenwood, Ed., Studies in Metamorphism and Metasomatism (Orville, vol.). Am. J. Sci. 283-A, 328-354.

Ashworth, J.R. and Evirgen, M.M. (1984) Mineral chemistry of regional chloritoid assemblages in the chlorite zone, Lycian nappes, southwest Turkey. Mineral. Mag. 48, 159-165.

Bailey, S.W. (1980) Summary of recommendations of AIPEA nomenclature committee on clay minerals. Am. Mineral. 65, 1-7.

Baltatzis, E.G. and Katagas, C.G. (1984) The pumpellyite-actinolite and contiguous facies in part of the Phyllite-Quartzite Series, central Northern Peloponnesus, Greece. J. Metamorphic Geol. 2, 349-363.

Banos, J.O. and Amouric, M. (1984) Biotite chloritization by interlayer brucitization as seen by HRTEM. Am. Mineral. 69, 869-871.

Baroz, F., Bebien, J., and Ikenne, M. (1987) An example of high-pressure low-temperature metamorphic rocks from an island-arc: the Paikon series (Innermost Hellenides, Greece). J. Metamorphic Geol. 5, 509-527.

Barrow, G. (1912) On the geology of lower Deeside and the southern Highland Border. Proc. Geol. Assoc. 23, 274-290.

Beddoe-Stephens, B. (1981) Metamorphism of the Rossland volcanic rocks, southern British Columbia. Canadian Mineral. 19, 631-641.

Bence, A.E. and Albee, A.L. (1968) Empirical correction factors for the electron microanalysis of silicates and oxides. J. Geol. 76, 382-403.

Bettison, L.A. and Schiffman, P. (1988) Compositional and structural variations of phyllosilicates from the Point Sal ophiolite, California. Am. Mineral. 73, 62-76.

Bevins, R.E. and Merriman, R.J. (1988) Compositional controls on coexisting prehnite-actinolite and prehnite-pumpellyite facies assemblages in the Tal y Fan metabasite intrusion, North Wales: implications for Caledonian metamorphic field gradients. J. Metamorphic Geol. 6, 17-39.

Bishop, D.G. (1972) Progressive metamorphism from prehnite-pumpellyite to greenschist facies in the Dansey Pass area, Otago, New Zealand. Geol. Soc. Am. Bull. 83, 3177-3198.

Black, P.M. (1975) Mineralogy of New Caledonia metamorphic rocks IV Sheet silicates from the Ouégoa district. Contrib. Mineral. Petrol. 49, 269-284.

Boles, J.R. and Coombs, D.S. (1977) Zeolite facies alteration of sandstones in the southland syncline, New Zealand. Am. J. Sci, 277, 982-1012.

Bothner, W.A. and Laird, J.(1987) Structure and metamorphism at Tillotson Peak, north-central Vermont. In D.S. Westerman, Ed., Guidebook for Field Trips in Vermont, Vol. 2. NEIGC, Norwich University, 383-405.

Boxwell, M. and Laird, J. (1987) Metamorphic and deformational history of the Standing Pond and Putney Volcanics in southeastern Vermont. In D.S. Westerman, Ed., Guidebook for Field Trips in Vermont, Vol. 2. NEIGC, Norwich University, 1-20.

Brown, B.E. and Bailey, S.W. (1962) Chlorite polytypism: I. Regular and semi-random one-layer structures. Am. Mineral. 47, 819-850.

448

Brown, E.H. (1986) Geology of the Shuksan Suite, North Cascades, Washington, U.S.A. In B.W. Evans and E.H. Brown, Eds., Blueschists and Eclogites. Geol. Soc. Am. Memoir 164, 143-154.

Brown, E.H. and Forbes, R.B. (1986) Phase petrology of eclogitic rocks in the Fairbanks district, Alaska. In B.W. Evans and E.H. Brown, Eds., Blueschists and Eclogites. Geol. Soc. Am. Memoir 164, 155-167.

Brown, E.H. and Ghent, E.D. (1983) Mineralogy and phase relations in the blueschist facies of the Black Butte and Ball Rock areas, northern California Coast Ranges. Am. Mineral. 68, 365-372.

Bucher-Nurminen, K. (1982a) On the mechanism of contact aureole formation in dolomitic country rock by the Adamello intrusion (northern Italy). Am. Mineral. 67, 1101-1117.

Bucher-Nurminen, K. (1982b) Mechanism of mineral reactions inferred from textures of impure dolomitic marbles from East Greenland. J. Petrol. 23, 325-343.

Bucher-Nurminen, K. (1987) A recalibration of the chlorite-biotite-muscovite geobarometer. Contrib. Mineral. Petrol. 96, 519-522.

Buerger, M.J. (1948) The role of temperature in mineralogy. Am. Mineral. 33, 101-121.

Cho, M. and Liou, J.G. (1987) Prehnite-pumpellyite to greenschist facies transition in the Karmutsen metabasites, Vancouver Island, B.C. J. Petrol. 28, 417-443.

Cho, M., Liou, J.G., and Maruyama, S. (1986) Transition from the zeolite to prehnite-pumpellyite facies in the Karmutsen metabasites, Vancouver Island, British Columbia. J. Petrol. 27, 467-494.

Chopin, C. (1981) Talc-phengite: a widespread assemblage in high-grade pelitic blueschists of the western Alps. J. Petrol. 22, 628-650.

Chopin, C. (1986) Phase relations of ellenbergite, a new high-presure Mg-Al-Ti-silicate in pyrope-coesite-quartzite from the Western Alps. In B.W. Evans, and E.H. Brown, Eds., Blueschists and Eclogites. Geol. Soc. Am. Memoir 164, 31-42.

Chopin, C. and Schreyer, W.(1983) Magnesiocarpholite and magnesiochloritoid: two index minerals of pelitic blueschists and their preliminary phase relations in the model system $MgO-Al_2O_3-SiO_2-H_2O$. In H.J. Greenwood, Ed., Studies in Metamorphism and Metasomatism (Orville vol.). Am. J. Sci. 283-A, 72-96.

Clark, G.L., Guiraud, M., Powell, R., and Burg, J.P. (1987) Metamorphism in the Olary Block, South Australia: compression with cooling in a Proterozoic fold belt. J. Metamorphic Geol. 5, 291-306.

Coombs, D.S., Nakamura, Y., and Vuagnat, M. (1976) Pumpellyite-actinolite facies schists of the Taveyanne Formation near Loèche, Valais, Switzerland. J. Petrol. 17, 440-471.

Coombs, D.S., Horodyski, R.J., and Naylor, R.S. (1970) Occurrence of prehnite-pumpellyite facies metamorphism in northern Maine. Am. J. Sci. 268, 142-156.

Cotkin, S.J. (1987) Mineralogy and conditions of metamorphism in an Early Paleozoic blueschist, Schist Skookum gulch, northern California. Contrib. Mineral. Petrol. 96, 192-200.

Craw, D. and Jamieson, R.A. (1984) Anomalous optics in low-grade chlorite from Atlantic Canada. Canadian Mineral. 22, 269-280.

Curtis, C.D., Hughes, C.R., Whiteman, J.A., and Whittle, C.K. (1985) Compositional variation within some sedimentary chlorites and some comments on their origin. Mineral. Mag. 49, 375-386.

Dickenson, M.P., III (1988) Local and regional differences in the chemical potential of water in amphibolite facies pelitic schists. J. Metamorphic Geol. 6, 365-381.

Dickenson, M.P. and Hewitt, D.A. (1986) A garnet-chlorite geothermometer. Geol. Soc. Am., Abstr. Programs 18, 584.

Duebendorfer, E.M. (1988) Evidence for an inverted metamorphic gradient associated with a Precambrian suture, southern Wyoming. J. Metamorphic Geol. 6, 41-63.

Enami, M. and Zang, Q. (1988) Magnesium staurolite in garnet-corundum rocks and ecoglite from the Donghai district, Jiangsu province, east China. Am. Mineral. 73, 48-56.

Ernst, W.G. (1977) Tectonics and prograde versus retrograde P-T trajectories of high-pressure metamorphic belts. Rend. Soc. Ital. Mineral. Petrol. 33, 191-220.

Ernst, W.G. (1983) Mineral parageneses in metamorphic rocks exposed along Tailuko Gorge, Central Mountain Range, Taiwan. J. Metamorphic Geol. 1, 305-329.

Ernst, W.G., Liou, J.G., and Moore, D.E. (1981) Multiple metamorphic events recorded in Tailuko amphibolites and associated rocks of the Suao-Nanao area, Taiwan. Geol. Soc. China Memoir 4, 391-441.

Evans, B.W. (1977) Metamorphism of Alpine peridotite and serpentinites. In F.A. Donath, Ed., Annual Review of Earth and Planetary Sciences 5, 397-447.

Evans, B.W. (1982) Amphiboles in metamorphosed ultramafic rocks. In D.R. Veblen and P.H. Ribbe, Eds., Amphiboles: Petrology and Experimental Phase Relations. Rev. Mineral. 9B, 98-113.

Evarts, R.C. and Schiffman, P. (1983) Submarine hydrothermal metamorphism of the Del Puerto ophiolite, California. Am. J. Sci. 283, 289-340.

Ferry, J.M. (1976) Metamorphism of calcareous sediments in the Waterville-Vassalboro area, south-central Maine: mineral reactions and graphical analysis. Am. J. Sci. 276, 841-882.

Ferry, J.M. (1983a) Mineral reactions and element migration during metamorphism of calcareous sediments from the Vassalboro Formation, south-central Maine. Am. Mineral. 68, 334-354.

Ferry, J.M. (1983b) Applications of the reaction progress variable in metamorphic petrology. J. Petrol. 24, 343-376.

Ferry, J.M. and Spear, F.S. (1978) Experimental calibration of the partitioning of Fe and Mg between biotite and garnet. Contrib. Mineral. Petrol. 66, 113-117.

Ferry, J.M., Mutti, L.J., and Zuccala, G.J. (1987) Contact metamorphism/hydrothermal alteration of Tertiary basalts from the Isle of Skye, northwest Scotland. Contrib. Mineral. Petrol. 95, 166-181.

Fletcher, C.J.N. and Greenwood, H.J. (1979) Metamorphism and structure of Penfold Creek area, near Quesnel Lake, British Columbia. J. Petrol. 20, 743-794.

Floran, R.J. and Papike, J.J. (1978) Mineralogy and petrology of the Gunflint iron formation, Minnesota-Ontario: correlation of compositional and assemblage variations at low to moderate grade. J. Petrol. 19, 215-288.

Foster, M.D. (1962) Interpretation of the composition and a classification of the chlorites. U.S. Geol. Surv. Prof. Paper 414-A, 22 pp.

Franceschelli, M., Mellini, M., Memmi, I., and Ricci, C.A. (1986) Fine-scale chlorite-muscovite association in low-grade metapelites from Nurra (NW Sardinia), and the possible misidentification of metamorphic vermiculite. Contrib. Mineral. Petrol. 93, 137-143.

Franz, G. and Spear, F.S. (1983) High pressure metamorphism of siliceous dolomites from the central Tauern Window, Austria. In H.J. Greenwood, Ed., Studies in Metamorphism and Metasomatism (Orville vol.). Am. J. Sci. 283-A, 396-413.

Frey, M. (1978) Progressive low-grade metamorphism of a black shale formation, central Swiss Alps, with special reference to pyrophyllite and margarite bearing assemblages. J. Petrol. 19, 93-135.

Frey, M. (1987) The reaction-isograd kaolinite + quartz = pyrophyllite + H_2O, Helvetic Alps, Switzerland. Schweiz. mineral. petrogr. Mitt. 67, 1-11.

Frost, B.R. (1975) Contact metamorphism of serpentinite, chloritic blackwall and rodingite at Paddy-Go-Easy Pass, central Cascades, Washington. J. Petrol. 16, 272-313.

Ganguly, J. and Saxena, S.K. (1984) Mixing properties of aluminosilicate garnets: constraints from natural and experimental data and applications to geothermobarometry. Am. Mineral. 69, 88-97.

Ghent, E.D., Stout, M.Z., Black, P.M., and Brothers, R.N. (1987) Chloritoid-bearing rocks associated with blueschists and eclogites, northern New Caledonia. J. Metamorphic Geol. 5, 239-254.

Gole, M.J. (1980) Mineralogy and petrology of very-low-metamorphic grade Archaean banded iron-formations, Weld Range, Western Australia. Am. Mineral. 65, 8-25.

Guidotti, C.V. (1974) Transition from staurolite to sillimanite zone, Rangeley quadrangle, Maine. Geol. Soc. Am. Bull. 85, 475-490.

Guidotti, C.V., Cheney, J.T., and Conatore, P.D. (1975) Coexisting cordierite + biotite + chlorite from the Rumford quadrangle, Maine. Geology 3, 147-148.

Guiraud, M., Burg, J.P., and Powell, R. (1987) Evidence for a Variscan suture zone in the Vendée, France: a petrological study of blueschist facies rocks from the Bois de Cené. J. Metamorphic Geol. 5, 225-237.

Haase, C.S. (1982a) Metamorphic petrology of the Negaunee Iron Formation, Marquette district, northern Michigan: Mineralogy, metamorphic reactions, and phase equilibria. Econ. Geol. 77, 60-81.

Haase, C.S. (1982b) Phase equilibria in metamorphosed iron-formations: qualitative T-X(CO2) petrogenetic grids. Am. J. Sci. 282, 1623-1654.

Harte, B. and Graham, C.M. (1975) The graphical analysis of greenschist to amphibolite facies mineral assemblages in metabasites. J. Petrol. 16, 347-370.

Harte, B. and Hudson, N.F.C. (1979) Pelite facies series and the temperatures and pressures of Dalradian metamorphism in E Scotland. In A.L. Harris, C.H. Holland, and B.E. Leake, Eds., The Caledonides of the British Isles—Reviewed. Scottish Academic Press, 323-337.

Helms, T.S., McSween, H.Y., Jr., Labotka, T.C., and Jarosewich, E. (1987) Petrology of a Georgia Blue Ridge amphibolite unit with hornblende + gedrite + kyanite + staurolite. Am. Mineral. 72, 1086-1096.

Hepburn, J.C., Trask, N.J., Rosenfeld, J.L., and Thompson, J.B., Jr. (1984) Bedrock geology of the Brattleboro quadrangle, Vermont-New Hampshire. Vermont Geol. Surv. Bull. 32, 162 pp.

Hess, P.C. (1969) The metamorphic paragenesis of cordierite in pelitic rocks. Contrib. Mineral. Petrol. 24, 191-207.

Hey, M.H. (1954) A new review of the chlorites. Mineral. Mag. 30, 277-292.

Holdaway, M.J., Dutrow, B.L., and Hinton, R.W. (1988) Devonian and Carboniferous metamorphism in west-central Maine: the muscovite-almandine geobarometer and the staurolite problem revisited. Am. Mineral. 73, 20-47.

Hollocher, K.T. (1981) Retrograde metamorphism of the Lower Devonian Littleton Formation in the New Salem area, west-central Massachusetts. Contrib. No. 37, Dept. of Geology and Geography, Univ. Mass., 269 pp.

450

Hudson, N.F.C. (1980) Regional metamorphism of some Dalradian pelites in the Buchan area, N.E. Scotland. Contrib. Mineral. Petrol. 73, 39-51.

Hudson, N.F.C. (1985) Conditions of Dalradian metamorphism in the Buchan area, NE Scotland. J. Geol. Soc. London 142, 63-76.

Ishizuka, H. (1985) Prograde metamorphism of the Horokanai ophiolite in the Kamuikotan Zone, Hokkaido, Japan. J. Petrol. 26, 391-417.

Jansen, J.B.H. and Schuiling, R.D. (1976) Metamorphism of Naxos: petrology and geothermal gradients. Am. J. Sci. 276, 1225-1253.

Karabinos, P. (1985) Garnet and staurolite producing reactions in a chlorite-chloritoid schist. Contrib. Mineral. Petrol. 90, 262-275.

Kawachi, Y., Grapes, R.H., Coombs, D.S., and Dowse, M. (1983) Mineralogy and petrology of a piemontite-bearing schist, western Otago, New Zealand. J. Metamorphic Geol. 1, 353-372.

Klaper, E.M. and Bucher-Nurminen, K. (1987) Alpine metamorphism of pelitic schists in the Nufenen Pass area, Lepontine Alps. J. Metamorphic Geol. 5, 175-194.

Klein, C. (1978) Regional metamorphism of Proterozoic iron-formation, Labrador Trough, Canada. Am. Mineral. 63, 898-912.

Klein, C. (1982) Amphiboles in iron-formations. In D.R. Veblen and P.H. Ribbe, Eds., Amphiboles: Petrology and Experimental Phase Relations. Rev. Mineral. 9B, 88-98.

Klein, C. and Gole, M.J. (1981) Mineralogy and petrology of parts of the Marra Mamba Iron Formation, Hamersley basin, Western Australia. Am. Mineral. 66, 507-525.

Klein, C. and Fink, R.P. (1976) Petrology of the Sokoman Iron Formation in the Howells River area, at the western edge of the Labrador Trough. Econ. Geol. 71, 453-487.

Kretz, R. (1983) Symbols for rock-forming minerals. Am. Mineral. 68, 277-279.

Kuniyoshi, S. and Liou, J.G. (1976a) Burial metamorphism of the Karmutsen volcanic rocks northeastern Vancouver Island, British Columbia. Am. J. Sci. 276, 1096-1119.

Kuniyoshi, S. and Liou, J.G. (1976b) Contact metamorphism of the Karmutsen Volcanics, Vancouver Island, British Columbia. J. Petrol. 17, 73-99.

Kurata, H. and Banno, S. (1974) Low-grade progressive metamorphism of pelitic schists of the Sazare area, Sanbagawa metamorphic terrain in central Sikoku, Japan. J. Petrol. 15, 361-382.

Labotka, T.C. (1980) Petrology of a medium-pressure regional metamorphic terrane, Funeral Mountains, California. Am. Mineral. 65, 670-689.

Labotka, T.C. (1981) Petrology of an andalusite-type regional metamorphic terrane, Panamint Mountains, California. J. Petrol. 22, 261-296.

Labotka, T.C. (1987) The garnet + hornblende isograd in calcic schists from an andalusite-type regional metamorphic terrain, Panamint Mountains, California. J. Petrol. 28, 323-354.

Labotka, T.C., Papike, J.J., Vaniman, D.T., and Morey, G.B., (1981) Petrology of contact metamorphosed argillite from the Rove Formation, Gunflint Trail, Minnesota. Am. Mineral. 66, 70-86.

Laird, J. (1980) Phase equilibria in mafic schist from Vermont. J. Petrol. 21, 1-37.

Laird, J. (1982) Amphiboles in metamorphosed basaltic rocks: greenschist to amphibolite facies and blueschist-greenschist-eclogite relations. In D.R. Veblen and P.H. Ribbe Eds., Amphiboles: Petrology and Experimental Phase Relations. Rev. Mineral. 9B, 113-159.

Laird, J. (1988) Arenig to Wenlock age metamorphism in the Appalachians. In A.L. Harris, and D.J. Fettes, Eds., The Caledonian-Appalachian Orogen, 311-345.

Laird, J. and Albee, A.L. (1981a) High-pressure metamorphism in mafic schist from northern Vermont. Am. J. Sci. 281, 97-126.

Laird, J. and Albee, A.L. (1981b) Pressure, temperature, and time indicators in mafic schist: their application to reconstructing the polymetamorphic history of Vermont. Am. J. Sci. 281, 127-175.

Laird, J., Foster, H.L., and Biggs, D.L. (1986) Petrologic evidence for a terrane boundary in south-central Circle quadrangle, Alaska. Geophys. Res. Letts. 13, 1035-1038.

Laird, J., Lanphere, M.A., and Albee, A.L. (1984) Distribution of Ordovician and Devonian metamorphism in mafic and pelitic schists from northern Vermont. Am. J. Sci. 284, 376-413.

Lan, C.-Y. and Liou, J.G. (1981) Occurrence, petrology and tectonics of serpentinites and associated rodingites in the Central Range, Taiwan. Geol. Soc. China Memoir 4, 343-389.

Lang, H.M. and Rice, J.M. (1985a) Regression modelling of metamorphic reactions in metapelites, Snow Peak, northern Idaho. J. Petrol. 26, 857-887.

Lang, H.M. and Rice, J.M. (1985b) Geothermometry, geobarometry and T-X (Fe-Mg) relations in metapelites, Snow Peak, northern Idaho. J. Petrol. 26, 889-924.

Lieberman, J.E. and Rice, J.M. (1986) Petrology of marble and peridotite in the Seiad ultramafic complex, northern California, USA. J. Metamorphic Geol. 4, 179-199.

Liou, J.G. (1979) Zeolite facies metamorphism of basaltic rocks from the East Taiwan ophiolite. Am. Mineral. 64, 1-14.

Liou, J.G. and Maruyama, S. (1987) Paragenesis and compositions of amphiboles from Franciscan jadeite-glaucophane type facies series metabasites at Cazadero, California. J. Metamorphic Geol. 5, 371-395.

451

Liou, J.G., Kuniyoshi, S., and Ito, K. (1974) Experimental studies of the phase relations between green-schist and amphibolite in a basaltic system. Am. J. Sci. 274, 613-632.

Liou, J.G., Maruyama, S., and Cho, M. (1985) Phase equilibria and mineral paragenesis of metabasites in low-grade metamorphism. Mineral. Mag. 49, 321-333.

Loomis, T.P. (1986) Metamorphism of metapelites: calculations of equilibrium assemblages and numerical simulations of the crystalization of garnet. J. Metamorphic Geol. 4, 201-229.

Maruyama, S. and Liou, J.G. (1985) The stability of Ca-Na pyroxene in low-grade metabasites of high-pressure intermediate facies series. Am. Mineral. 70, 16-29.

Maruyama, S. and Liou, J.G. (1987) Clinopyroxene—a mineral telescoped through the processes of blueschist facies metamorphism. J. Metamorphic Geol. 5, 529-552.

Maruyama, S. and Liou, J.G. (1988) Petrology of Franciscan metabasites along the jadeite-glaucophane type facies series, Cazadero, California. J. Petrol. 29, 1-37.

Maruyama, S., Cho, M., and Liou, J.G. (1986) Experimental investigations of blueschist-greenschist transition equilibria: pressure dependence of Al_2O_3 contents in sodic amphiboles—A new geobarometer. In B.W. Evans and E.H. Brown, Eds., Blueschists and Eclogites. Geol. Soc. Am. Memoir 164, 1-16.

Maruyama, S., Suzuki, K., and Liou, J.G. (1983) Greenschist-amphibolite transition at low pressures. J. Petrol. 24, 583-604.

McDowell, S.D. and Elders, W.A. (1980) Authigenic layer silicate minerals in borehole Elmore #1, Salton Sea Geothermal Field, California, USA. Contrib. Mineral. Petrol. 74, 293-310.

McDowell, S.D. and Elders, W.A. (1983) Allogenic layer silicate minerals in borehole Elmore #1, Salton Sea Geothermal Field, California. Am. Mineral. 68, 1146-1159.

Miyashiro, A. and Shido, F. (1985) Tschermak substitution in low- and middle-grade pelitic schists. J. Petrol. 26, 449-487.

Moody, J.B., Meyer, D., and Jenkins, J.E. (1983) Experimental characterization of the greenschist/amphibolite boundary in mafic systems. Am. J. Sci. 283, 48-92.

Moore, D.E. (1984) Metamorphic history of a high-grade blueschist exotic block from the Franciscan complex, California. J. Petrol. 25, 126-150.

Morten, L., Bargossi, G.M., Martinez Martinez, J.M., Puga, E., and Diaz de Federico, A. (1987) Metagabbro and associated eclogites in the Lubrin area, Nevado-Filabride Complex, Spain. J. Metamorphic Geol. 5, 155-174.

Mottana, A. (1986) Blueschist-facies metamorphism of manganiferous cherts: a review of the alpine occurrences. In B.W. Evans and E.H. Brown, Eds., Blueschists and Eclogites. Geol. Soc. Am. Memoir 164, 267-299.

Newton, R. C. (1986) Metamorphic temperatures and pressures of Group B and C eclogites. In B.W. Evans and E.H. Brown, Eds., Blueschists and Eclogites. Geol. Soc. Am. Memoir 164, 17-30.

Novak, J.M. and Holdaway, M.J. (1981) Metamorphic petrology, mineral equilibria, and polymetamorphism in the Augusta quadrangle, south-central Maine. Am. Mineral. 66, 51-69.

Pattison, D.R.M. (1987) Variations in Mg/(Mg+Fe), F, and (Fe, Mg)Si = 2Al in pelitic minerals in the Ballachulish thermal aureole, Scotland. Am. Mineral. 72, 255-272.

Phillips, G.N. (1987) Metamorphism of the Witwaterstrand gold fields: conditions during peak metamorphism. J. Metamorphic Geol. 5, 307-322.

Pinsent, R.H. and Hirst, D.M. (1977) The metamorphism of the Blue River ultramafic body, Cassiar, British Columbia, Canada. J. Petrol. 18, 567-594.

Pognante, U. and Kienast, J.-R. (1987) Blueschist and eclogite transformations in Fe-Ti gabbros: a case from the Western Alps ophiolites. J. Petrol. 28, 271-292.

Pognante, U., Talarico, F., Rastelli, N., and Ferrati, N. (1987) High pressure metamorphism in the nappes of the Valle dell' Orco traverse (Western Alps collisional belt). J. Metamorphic Geol. 5, 397-414.

Powell, R. and Evans, J.A. (1983) A new geobarometer for the assemblage biotite-muscovite-chlorite-quartz. J. Metamorphic Geol. 1, 331-336.

Reinecke, T. (1986) Phase relationships of sursassite and other Mn-silicate in highly oxidized low-grade, high-pressure mafic rocks from Evvia and Andros Islands, Greece. Contrib. Mineral. Petrol. 94, 110-126.

Rice, J.M. (1977a) Contact metamorphism of impure dolomitic limestone in the Boulder aureole, Montana. Contrib. Mineral. Petrol. 59, 237-259.

Rice, J.M. (1977b) Progressive metamorphism of impure dolomitic limestone in the Marysville aureole, Montana. Am. J. Sci. 277, 1-24.

Rice, J.M. (1983) Metamorphism of rodingites: Part I. Phase relations in a portion of the system CaO-MgO-Al_2O_3-SiO_2-CO_2-H_2O. In H.J. Greenwood, Ed., Studies in Metamorphism and Metasomatism (Orville vol.). Am. J. Sci. 283-A, 121-150.

Robinson, D. and Bevins, R.E. (1986) Incipient metamorphism in the Lower Paleozoic marginal basin of Wales. J. Metamorphic Geol. 4, 101-113.

Sakakibara, M. (1986) A newly discovered high-pressure terrane in eastern Hokkaido, Japan. J. Metamorphic Geol. 4, 401-408.

452

Sanford, R.F. (1982) Growth of ultramafic reaction zones in greenschist to amphibolite facies metamorphism. Am. J. Sci. 282, 543-616.

Schreyer, W., Fransolet, A.M., and Abraham, K. (1986) A miscibility gap in trioctahedral Mn-Mg-Fe chlorites: evidence from the Lienne Valley Manganese deposit, Ardennes, Belgium. Contrib. Mineral. Petrol. 94, 333-342.

Selverstone, J., Spear, F.S., Franz, G., and Morteani, G. (1984) High-pressure metamorphism in the SW Tauern Window, Austria: P-T paths from hornblende-kyanite-staurolite schists. J. Petrol. 25, 501-531.

Shiba, M. (1988) Metamorphic evolution of the southern part of the Hidaka belt, Hokkaido, Japan. J. Metamorphic Geol. 6, 273-296.

Smith, R.E., Perdrix, J.L., and Parks, T.C. (1982) Burial metamorphism in the Hamersley Basin. J. Petrol. 23, 75-102.

Spear, F.S. (1982) Phase equilibria of amphibolites from the Post Pond Volcanics, Mt. Cube quadrangle, Vermont. J. Petrol. 23, 383-426.

Spear, F.S. and Cheney, J.T. (1986) Yet another petrogenetic grid for pelitic schists. Geol. Soc. Am., Abstr. Programs 18, 758.

Springer, R.K. (1974) Contact metamorphosed ultramafic rocks in the western Sierra Nevada foothills, California. J. Petrol. 15, 160-195.

Stewart, D.B. and Flohr, M.K. (1984) Systematic compositional variation in layer silicates from chlorite to cordierite grade, coastal volcanic belt, Maine. Geol. Soc. Am., Abstr. Programs 16, 618.

Takayama, M. (1986) Mode of occurrence and significance of jadeite in the Kamuikotan metamorphic rocks, Hokkaido, Japan. J. Metamorphic Geol. 4, 445-454.

Thompson, A.B. (1976a) Mineral reactions in pelitic rocks: I. Prediction of P-T-X (Fe-Mg) phase relations. Am. J. Sci. 276, 401-424.

Thompson, A.B. (1976b) Mineral reactions in pelitic rocks: II. Calculation of some P-T-X (Fe-Mg) phase relations. Am. J. Sci. 276, 425-454.

Thompson, A.B., Lyttle, P.T., and Thompson, J.B., Jr. (1977a) Mineral reactions and A-Na-K and A-F-M facies types in the Gassetts schist, Vermont. Am. J. Sci. 277, 1124-1151.

Thompson, A.B., Tracy, R.J., Lyttle, P.T., and Thompson, J.B., Jr. (1977b) Prograde reaction histories deduced from compositional zonation and mineral inclusions in garnet from the Gassetts schist, Vermont. Am. J. Sci. 277, 1152-1167.

Thompson, J.B., Jr. (1957) The graphical analysis of mineral assemblages in pelitic schists. Am. Mineral. 42, 842-858.

Thompson, J.B., Jr. (1979) The Tschermak substitution and reactions in pelitic schists. In V.A. Zharikov, V.I. Fonarev, and S.P. Korikovskii, Eds., Problems in Physicochemical Petrology. Academy of Sciences, Moscow, 146-159 (in Russian).

Thompson, J.B., Jr. (1981) An introduction to the mineralogy and petrology of the biopyriboles. In D.R. Veblen, Ed., Amphiboles and Other Hydrous Pyriboles-Mineralogy. Rev. Mineral. 9A, 141-188.

Thompson, J.B., Jr. (1982a) Composition space: an algebraic and geometric approach. In J.M. Ferry, Ed., Characterization of Metamorphism through Mineral Equilibria. Rev. Mineral. 10, 1-31.

Thompson, J.B., Jr. (1982b) Reaction space: an algebraic and geometric approach. In J.M. Ferry, Ed., Characterization of Metamorphism through Mineral Equilibria. Rev. Mineral. 10, 33-52.

Thompson, J.B., Jr., Laird, J., and Thompson, A.B. (1982) Reactions in amphibolite, greenschist and blueschist. J. Petrol. 23, 1-27.

Thompson, P.H. and LeClair, A.D. (1987) Chloritoid-hornblende assemblages in quartz-muscovite pelitic rocks of the Central Metasedimentary Belt, Grenville Province, Canada. J. Metamorphic Geol. 5, 415-436.

Tracy, R.J., Robinson, P., and Wolff, R.A. (1984) Metamorphosed ultramafic rocks in the Bronson Hill anticlinorium, central Massachusetts. Am. J. Sci. 284, 530-558.

van Bosse, J.Y. and Williams-Jones, A.E. (1988) Chemographic relationships of biotite and cordierite in the McGerrigle thermal aureole, Gaspé, Quebec. J. Metamorphic Geol. 6, 65-75.

Veblen, D.R. (1983) Microstructures and mixed layering in intergrown wonesite, chlorite, talc, biotite, and kaolinite. Am. Mineral. 68, 566-580.

Velde, B. (1965) Phengite micas: synthesis, stability and natural occurrence. Am. J. Sci. 263, 886-913.

Velde, B. and Rumble, D., III (1977) Alumina content of chlorite in muscovite-bearing assemblages. In Geophysical Laboratory Annual Report 1976-1977, 621-623.

Wang, G.-F., Banno, S., and Takeuchi, K. (1986) Reactions to define the biotite isograd in the Ryoke metamorphic belt, Kii Peninsula, Japan. Contrib. Mineral. Petrol. 93, 9-17.

White, S.H., Huggett, J.M., and Shaw, H.F. (1985) Electron-optical studies of phyllosilicate intergrowths in sedimentary and metamorphic rocks. Mineral. Mag. 49, 413-423.

Winchester, J.A. (1972) The petrology of Monian calc-silicate gneisses from Fannich Forest, and their significance as indicators of metamorphic grade. J. Petrol. 13, 405-424.

Yardley, B.W.D., Leake, B.E., and Farrow, C.M. (1980) The metamorphism of Fe-rich pelites from Connemara, Ireland. J. Petrol. 21, 365-399.

Yokoyama, K., Brothers, R.N., and Black, P.M. (1986) Regional eclogite facies in the high-pressure metamorphic belt of New Caledonia. In B.W. Evans and E.H. Brown, Eds., Blueschists and Eclogites. Geol. Soc. Am. Memoir 164, 407-423.

Zen, E-An (1966) Construction of pressure-temperature diagrams for multicomponent systems after the method of Schreinemakers-a geometric approach. U.S. Geol. Surv. Bull. 1225.

Zen, E-An (1974) Prehnite- and pumpellyite-bearing mineral assemblages, west side of the Appalachian metamorphic belt, Pennsylvania to Newfoundland. J. Petrol. 15, 197-242.

Zhong, W.J.S., Hughes, J.M., and Scotford, D.M. (1985) The response of chlorite to metasomatic alteration in Appalachian ultramafic rocks. Canadian Mineral. 23, 443-446.

Chapter 12

C. de la Calle & H. Suquet

VERMICULITE

INTRODUCTION

Vermiculites, 2:1 phyllosilicates, are generally composed of macroscopic particles, similar in appearance to micas, from which they originate by alteration. "True" vermiculites are in the great majority of cases trioctahedral minerals with layers similar to those in talc. Isomorphic substitutions create a deficit of positive charge, compensated by interlayer cations that can be easily exchanged and solvated. The swelling properties of vermiculites with different liquids (water, glycerol (G), ethylene glycol (EG)), are characteristic and can be used as the first test to identify them.

These minerals should be considered with reference to their genesis; mineralogy cannot be separated from geology. That is why in this chapter we make a brief reference to the geological origin and the principal known deposits of vermiculites. Their crystal chemistry and structure will be studied in detail because of their importance in the understanding and use of these minerals.

The structure of vermiculite has been the subject of numerous studies, justified by the great number of crystal structures that these minerals can present as a function of different variables. The configuration of the interlayer space, the bonds between the interlayer species and the oxygenated surfaces of the layers, as well as certain characteristics of the texture of the mineral can determine the stacking mode of the layers. We shall see that in accordance with variations in these parameters, the same vermiculite can present a great variety of structures, from a three-dimensionally ordered structure to one almost completely disordered.

The macroscopic nature of "true" vermiculites facilitates structural studies. Although it is not possible to consider vermiculite "crystals" as monocrystals, and because of the difficulty this implies in the use of the automated methods for determining their structure, the classical Weissenberg and precession techniques provide an adequate basis for three-dimensional determination that is much simpler than those involving microcrystalline powders.

In the X-ray diffraction diagrams of vermiculites there are many diffraction maxima that cannot be interpreted by the simple application of Bragg's law. In lamellar systems it is very common to observe disordered stacking of layers. The proper treatment of such disorder will be discussed. In spite of the limitations inherent in diffraction methods (structural analysis presupposes ideal periodicity in the crystal, which leads to an averaging of the atomic parameters over the total volume of the crystal), these continue to be the basic methods for the study of structures. Spectroscopic methods (Mössbauer, IR, EPR, NMR, etc.) are a source of local structural information independent of whether the distribution is periodic or not. Thus a combination of two methods, diffractometric and spectroscopic, is the most efficient and desirable for the study of these materials.

GEOLOGY OF VERMICULITE

Parentage

In general macroscopic vermiculites are of secondary origin and result from alteration of micas (phlogopites or biotites), chlorites, pyroxenes, or other similar minerals as a result of weathering, hydrothermal action, percolating ground water, or a combination of the three effects (Basset, 1963). There are not many laboratory studies on the alteration of chlorites, but see Makumbi and Herbillon (1972) and Ross and Kodama (1974). The latter suggest the existence of two stages in the chlorite-vermiculite transition: (1) oxidation of

456

Fe^{2+} and dehydroxylation of the brucite-like interlayer sheet, which could take place at high temperatures during metamorphic processes, and (2) acidic weathering giving rise to the chlorite-vermiculite transition. On the other hand, the phenomenon of vermiculitization in nature has been amply studied, e.g., Barshad (1948), Roy and Romo (1957), Foster (1963), Robert and Pedro (1966), Robert (1968), Robert and Barshad (1972), and Hoda and Hood (1972).

In large crystals of altered phlogopite, it is possible to observe a nucleus of un-altered mineral, surrounded by a transition zone, and true vermiculite at the exterior. Justo (1984) followed the phlogopite-vermiculite transition on single particles of vermiculite samples from Andalucia, Spain. He found that the alteration is accompanied by a progres-sive increase of the b parameter, the Fe^{3+}/Fe^{2+} ratio, the water content, and the Mg content. Also, Robert et al. (1987) and Justo et al. (1987) showed the presence in the Santa Olalla deposit, in addition to vermiculite crystals, of a fine fraction (< 2 mμ) of altered vermiculite that behaves like a smectite of charge x = 0.5, per half cell. This confirms the original observations of Gonzalez Garcia and Garcia Ramos (1962) on the same deposit. Gennaro and Franco (1975) have made similar observations on the Italian Monte Ernici deposit. This suggests a continuous process of alteration mica-vermiculite-smectite.

The alteration of micas implies the replacement of the interlayer K by a hydrated cation, generally Mg. This process is reversible. The fact that the crystal structure of the vermiculite contracts in the presence of K has been amply demonstrated (Gruner, 1939; Barshad, 1948, 1950, 1954a,b; Walker, 1949; de la Calle et al., 1976). The size of the K cation is such as to permit it to enter partially into the pseudo-hexagonal cavities on the surface of the vermiculite layers. Neighboring layers are thus linked by strong electrostatic bonds. The resulting structure is equivalent to that of a trioctahedral mica.

In view of the reversible character of the K-Mg exchange, the mica-vermiculite parentage can be demonstrated experimentally. Thus, de la Calle et al. (1976) showed that about 90% of natural vermiculites are of 1M parentage. Among these may be cited the vermiculites from Santa Olalla and Benahavis, Spain, Malawi, Nyasaland, Africa, Llano, Texas, and Prayssac, France. On the other hand, the vermiculite from Kenya, studied by Mathieson and Walker (1954), has a $2M_1$ parentage (de la Calle et al., 1976). Further, Levinson and Heinrich (1954) showed as the result of the X-ray examination of about 200 natural trioctahedral mica crystals that the great majority of them crystallize in the 1M structural form, and only a few in the $2M_1$ or 3T forms. These results are in accord with the supposition that most vermiculites are of parentage 1M.

Occurrences of macroscopic vermiculites

The problem of the origin of vermiculite is still subject to discussion; is it formed by the action of hydrothermal solutions or formed by the action of supergene solutions on biotite and phlogopite? The evidence in favor of the hydrothermal genesis of macroscopic vermiculite is principally: (a) association with high-temperature minerals and rocks and (b) its presence at depths over 200 ft. The evidence in favor of supergene origin is (a) biotite and phlogopite alter to vermiculite easily at room temperature, (b) the inhibiting effect of K even at low concentration in the formation of vermiculite, (c) the increase in the proportion of biotite or phlogopite with depth, (d) the presence of unaltered biotite in the interior of pyroxenite, and (e) the presence of vermiculite in proximity to pegmatites. Studies on the stability of vermiculite (Roy and Romo, 1957) in hydrothermal conditions show that in the region of 300°C and 700 atm pressure a portion of the octahedral Mg passes into the inter-layer space to form a brucite-like sheet. Komarneni and Roy (1981) studied the behavior of vermiculite in comparable conditions and arrived at the same conclusions. This indicates that vermiculite is unstable, even in weakly hydrothermal conditions, and so could not readily crystallize in such solutions.

The occurrence of macroscopic vermiculites can be divided into four categories, based on the host rock (Basset, 1963).

(1) Ultramafic and mafic. In all major commercial deposits: Transvaal, Palabora, West Chester, Tigerville, South Carolina, and Libby, Montana, the material is mixed layer vermiculite-biotite or vermiculite-phlogopite. In these deposits the vermiculite does not occur in veins, but is dispersed among the pyroxenite.

(2) Gneiss and schist. Basset (1963) described various types of vermiculite occurring in gneiss, and maintained that their characteristics are similar whatever their origin. Although associated with pegmatites, one does not find the zoning characteristic of those deposits where serpentine occurs.

(3) Carbonate rocks. In this case the vermiculites are practically the pure magnesian end member with very little iron, and occur in the external zones of magnesite, marble, and calcite bodies.The internal, unaltered part of these rocks contains phlogopite. The vermiculite may thus have formed under the influence of supergene solutions. The Llano occurrence may come in this category. The Llano vermiculite was formed by the surface weathering of phlogopite contained in a dolomitic and magnesitic marble (Clabaugh and Barnes, 1959).

(4) Granitic rocks. In this group the vermiculites are small in size and generally not pure, but are interstratified biotite-vermiculite. The origin proposed by most authors is supergene alteration. Justo et al. (1986) studied two vermiculites from southern Spain resulting from the alteration of granites (Ronquillo and Real de la Jara, Seville).

<u>Characteristics of some vermiculite deposits</u>

Vermiculite from Santa Olalla, Spain.. The origin and mineralogy of the Santa Olalla deposits and the whole region of Andalucia (Spain) have been extensively studied. We may cite among others Gonzalez Garcia and Garcia Ramos (1960), Velasco et al. (1981), Justo et al. (1983), Justo (1984) and Luque et al. (1985). This vermiculite occurs as lamellar flakes of variable size, from very small up to 12 cm diameter. The Santa Olalla vermiculite formed from a mica identified as a phlogopite resulting from the alteration of pyroxenites. These micas are found in fractures and fissures and are the products of the alteration of the magnesian skarns of La Garrenchosa (Santa Olalla, Huelva, Spain).

Three stages were involved in the evolution of the magnesian skarns of La Garrenchosa. First, a high-temperature stage caused development both of frontal exoskarns between magma and the dolomitic host rock and bimetasomatic ones between marble and interlayered Ca-rich samitic beds. Both skarns contain as main components an Al-rich (fassaitic) clinopyroxene + hercynite. The second stage consisted of the development of phlogopite either as replacement of the pyroxene skarn or as a filling of fractures. The last stage was characterized by the transformation of phlogopite into vermiculite. This stage has been assigned either to late hydrothermal activity or to supergenic post-skarn processes. Several data show that this second possibility is the right one (Velasco et al., 1981).

Vermiculite from Benahavis, Spain. The Benahavis (Malaga, Spain) vermiculite occurs in elongate veins. The sample consists of large packs of layers, not very thick, deep green in color, and up to 10 cm in diameter. The host rock is principally serpentine (Lopez Gonzalez and Barrales Rienda, 1972; and Justo, 1984).

Vermiculite from Malawi, Nyasaland. Morel (1955) studied the deposit of Kapirikamodzi, in the south of Malawi, in Africa. An occurrence is described in which hydrothermal agencies have altered biotite to vermiculite. The biotites are situated in gneisses and granulites of a Basement Complex. The host rocks of the Basement Complex are referred to the amphibolite and granulite facies of regional metamorphism. The field relations and the distribution and petrology of biotite bodies show that they are a series of metamorphosed ultrabasic intrusions. The vermiculite varies in size from flakes less than 1 mm in size to massive honey-yellow crystals up to 20 cm in diameter and 5 cm thick, associated with pegmatites. Occasional residual patches of black hydrobiotite indicate that the vermiculite was formed by alteration of biotite.

Vermiculite from Kenya. Vermiculite deposits in Kenya are described by Varley (1948). It appears that the best quality vermiculite in Kenya is associated with dunite-pegmatite contacts, but no large deposits of this type appear to have been discovered. Vermiculite originating from the alteration of biotite bands and slip zones in the crystalline schists is much more abundant, though usually of somewhat inferior quality.

Vermiculite from Prayssac, France. This deposit is near the village of Prayssac in the Aveyron. It is situated in or near a zone of contact gneiss-serpentine (Monchoux, 1961; Andre, 1972).

OPTICAL PROPERTIES

The optical properties of vermiculite can be measured more easily than in other clay minerals, but it must be realized that the type of interlayer cation and its degree of hydration lead to a great range of optical parameters. Some vermiculites have a considerable iron content and their refractive indices are high. In general natural vermiculites have optical parameters close to those of phlogopite. Large crystals of vermiculite often show a pseudo-hexagonal aspect resembling micas, but vermiculite crystals are more flexible and less brilliant. Also, the exfoliation planes of vermiculites are less directed along 001. An abnormal birefringence on the basal planes is noted.

The most useful characteristic property for differentiating vermiculites from micas is the optic angle 2V; it is larger in vermiculites. For the Llano vermiculite (Shirozu and Bailey, 1966) 2V varies between 5 and 15°, along a major plane of exfoliation. Velasco et al. (1981) give a 2V of 18° for Santa Olalla vermiculite. The refractive indices for Llano vermiculite are $\alpha = 1.520$, $\beta = \gamma = 1.5304$, and for Santa Olalla vermiculite $\alpha = 1.535$, $\beta = \gamma = 1.560$. The optical properties of Malawi vermiculite are very similar to those described by Gevers (1948) from Loolekop, Transvaal, for golden yellow vermiculite derived from biotite.

CHEMICAL COMPOSITION AND CRYSTAL CHEMISTRY

Macroscopic vermiculites are always trioctahedral, i.e., all the atomic sites of the octahedral sheet are in principle occupied. Mg is the principal octahedral cation, but it may be replaced by various divalent cations. Isomorphic substitution of trivalent Fe and Al gives rise to an excess of positive charge that can vary from 0.14 to 0.61 per formula unit (this high octahedral positive charge may differentiate vermiculites from saponites). The octahedral charge is neutralized by the negative charge arising from isomorphic substitutions in the tetrahedral sheet (Si/Al and sometimes Si/Fe). Since the tetrahedral negative charge is larger than the octahedral positive charge, the 2:1 layer has a net negative charge. The overall electrical neutrality of the layers is accomplished by the presence of cations in interlayer positions. The interlayer cations in natural vermiculites are mainly Mg, but Ca and Na are found in some samples. The original interlayer cations may easily be exchanged by K, Ca, Na, Rb, Cs, Li, NH_4. Certain cations are easier to replace than others, e.g., an NH_4 vermiculite may be prepared from natural vermiculite, and the NH_4 can then be exchanged for Na, Ca, but not K (Allison et al., 1953; Barshad, 1954a,b). In general, one may say that Na, Ca, Mg, and K are exchangeable between each other, but K, NH_4, Rb, and Cs are not.

The chemistry of vermiculite is closely linked to that of mica. The two factors on which the difference is based are (1) vermiculite has a lower layer charge and (2) the iron is oxidized by comparison with the original mica. Norrish (1973) showed that these differences are linked: the reduction of charge is related to the oxidation of iron. According to Foster (1963), however, there is not a 1:1 relationship between the charge reduction and the ferric iron content of vermiculites, which implies the existence of other changes of the composition, for example that linked to the redox reaction:

$$Fe^{2+} + OH \text{ (structural)} \rightarrow Fe^{3+} + O \text{ (structural)}$$

Chemical analyses

The great volume of data on chemical composition of vermiculites published in the literature does not show any relationship between the content of major elements (Mg, Al, Fe) and the geological location but that a relationship does exist with the content of minor elements. The chemical composition of 23 vermiculites from different geological locations was studied by Justo et al. (1986). The only net difference that can be established is between vermiculites localized in ultramafic zones relative to these in other locations. The former are enriched in Ti, Ni, Cr, and Zn and the latter are enriched in Mn.

Structural formula

The structural formula of a standard vermiculite may be written approximately as follows: $(Si,Al)_4(Mg,Al,Fe)_3O_{10}(OH)_2Mg_x(H_2O)_n$, $0.9 > x > 0.6$, x = layer charge per formula unit.

Several methods exist for converting chemical analyses into structural formulae (Caillere et al., 1982). Making no assumption about the structural formula of the mineral, one can make an experimental determination of the mass of the unit cell ($M = \rho V$, the volume V of the cell being calculated from the unit cell parameters as determined by X-rays, and the specific mass ρ determined pycnometrically). From the weight percentages of the different elements, one determines the exact content of each atomic type in the unit cell. This method has been used recently to establish the structural formulae of a saponite and a nontronite (Suquet et al., 1987).

Generally, the formula is calculated using a basis of 10 structural oxygens plus two OH or F-, and a total negative charge equal to 22 per formula unit. Using this principle, Köster (1977) gives a simple mathematical method for deriving the structural formulae from the chemical analyses. In the 2:1 layer silicates, the total cationic charge of 22 per formula unit must be distributed between the tetrahedral cations (15-16), the octahedral cations (5-6), and the interlayer cations (0-1). This distribution of the cationic charge in the structure can be represented by a triangular diagram (Köster, 1982) in which the corners represent the terminal species of the trioctahedral series phlogopite-taeniolite-talc or the dioctahedral series muscovite-celadonite-pyrophyllite. Each point of the triangular diagram represents the total cationic charge divided among the three structural positions of the cations. Thus the vermiculites would be characterized by having 6-7 charges in octahedral position and 15 in tetrahedral position.

The ternary diagram of Figure 1 represents the composition of the octahedral sheet for vermiculites of different origins (Llano, Santa Olalla, Kenya, Malawi, Benahavis, Prayssac). It may be seen that the compositions of all these samples lie in the phlogopite zones (Foster, 1963), which could indicate that all these vermiculites come from trioctahedral micas of phlogopite type.

By NMR spectroscopy, Lipsicas et al. (1984), Sanz and Serratosa (1984) and Thomson (1984) differentiated between tetrahedral and octahedral Al. For a sample of Llano vermiculite, Thomson (1984) confirms the Al(IV)/Al(VI) ratio determined by XRD for Norrish's sample 3 (1973), namely 12:1. The amount of Al(IV) in the Llano vermiculite has been determined by NMR by Herrero et al. (1985a), and their value is in agreement with that determined by chemical analysis, between 8.5 and 13.8% according to the origin of the sample.

The number of interlayer ions is variable and depends on the charge density of the layer. The d(001) layer distance of the pure n-alkylammonium derivatives enables one to determine the charge density in the silicate layers (Weiss, 1963). This density is generally expressed by the cation exchange capacity (CEC) and varies between 120 and 200 meq/100 g "air dry matter," or 140 to 240 meq/100 g calculated on the sample calcined at 1000°C

460

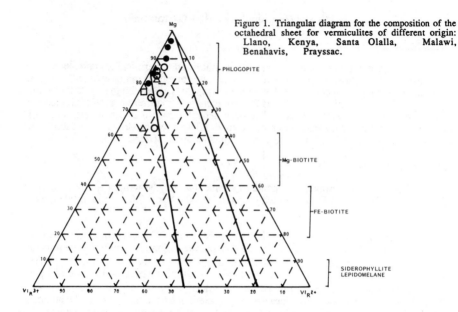

Figure 1. Triangular diagram for the composition of the octahedral sheet for vermiculites of different origin: Llano, Kenya, Santa Olalla, Malawi, Benahavis, Prayssac.

during 10-15 h. The importance of a reference state must be emphasized in defining the properties of a mineral (Tessier, 1984). Following other authors, we consider that reference to a mineral calcined at 1000°C is preferable to the "air dry" basis.

Distribution of Si-Al substitutions

The distribution of cations in the different structural sites and in particular the Si-Al distribution is an important crystal-chemical factor that determines the thermodynamic stability (Saxena, 1973) and the physicochemical behavior of these minerals (Méring and Pedro, 1969; Serratosa et al., 1984). An understanding of the Si-Al substitutions can also be of use in establishing the conditions of formation of the mineral and in elaborating the various polytypes.

It has been shown by ^{29}Si NMR (Herrero et al., 1985b) that in micas and vermiculites (Si/Al = 3) the Si-Al distribution depends principally on two electrostatic factors: (1) the Loewenstein rule (Loewenstein, 1954) that excludes the occurrence of two contiguous tetrahedra containing aluminum, and (2) local charge compensation. The latter implies a dispersion of the aluminums, in order to reduce the interaction between the charge deficits and to compensate for the charge of the interlayer cations. Further, the electrostatic interactions between neighboring layers play an important role in the Si-Al distribution (Herrero et al., 1986). In the final model proposed by those authors, in accord with their experimental data, the Al ions are distributed randomly among the tetrahedral sites of the pseudo-hexagonal surface rings.

X-ray diffraction studies on a Llano vermiculite sample (Bailey, 1986) give results that agree with those derived from NMR. In particular it has been shown clearly that the Si-Al substitutions are not more ordered in vermiculite than in the parent phlogopite from which it has arisen by alteration (Bailey, 1984). We will see in what follows regarding the crystalline order of vermiculites that the absence of order in the Si-Al substitutions can be considered as related directly to the frequency of appearance of semi-ordered layer sequences in layer silicates in general and in vermiculites in particular.

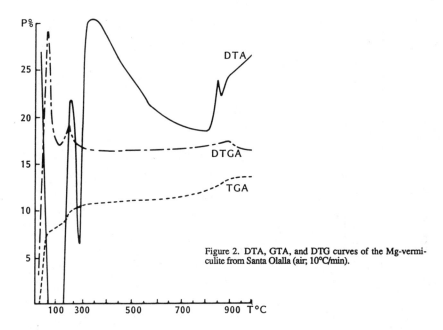

Figure 2. DTA, GTA, and DTG curves of the Mg-vermiculite from Santa Olalla (air; 10°C/min).

Thermal properties

Figure 2 gives, as an example, the DTA, TGA, and DTP (differential thermoponderal) curves of natural vermiculite from Santa Olalla. The first feature in the heat treatment is the loss of physically adsorbed (interparticulate) water. Two peaks at 140° and 240°C mark the endothermic transformation from the two-layer (d(001) = 14.3 Å) to the one-layer hydrate (d(001) = 11.6 Å) form, and from the one-layer hydrate to the anhydrous state (d(001) = 9.1 Å). In succession, one observes the oxidation of ferrous to ferric iron and then progressive dehydroxylation with the appearance of a non-swelling phase—a biopyribole—formed by an enstatite in epitaxial growth on the 010 plane of a non-swelling layer silicate of spacing 9.1 Å (Suquet et al., 1984). After heating at 550°C in air Mg vermiculite can recover the two-layer hydration state simply by air contact. Above 550°C the biopyribole appears and the vermiculite no longer swells. Enstatite must be formed from the hydroxyls linked to trivalent cations, since, according to Serratosa and Sanz (1985), the dehydroxylation temperature follows the order: $Fe^{2+} < Fe^{3+} < Al < Mg$.

The transformation of biopyribole to oxides (enstatite, cristobalite, forsterite, olivine, etc.) is shown in DTA by two endothermic peaks (800° and 900°C) separated by an exothermic peak (850°C). The intensity of these peaks varies according to the sample, its texture, and the nature of the interlayer cations (Walker and Cole, 1957; Andre, 1972). There are disagreements in the literature regarding the exact nature of the oxides formed and the order of their appearance.

SAMPLES FOR CRYSTAL STRUCTURE STUDIES

The samples used in our study of vermiculites are of two types: natural vermiculite and "artificial" vermiculite (see Table 1). The "artificial" vermiculites are derived from 1M phlogopites in which the K has been exchanged by hydratable cations. The exchange was carried out on crystals cut to dimensions of about 1 x 1 x 0.1 mm, at a temperature of 80°C, and with a concentration of 1 mg of the mineral per liter of a 2N solution of the appropriate cation. The exchange takes an average time of 21 d with nine to 10 changes of solution.

Table 1. Structural formula of vermiculites of different origins.

ORIGIN	STRUCTURAL FORMULA	charge density
Phlogopite A (Madagascar)	$K_{0.86}Na_{0.02}(Si_{2.68}Al_{1.32})(Mg_{2.59}Al_{0.15}Fe^{3+}_{0.10}Ti_{0.04})O_{10}(OH)_2$	0.88
Phlogopite B (Madagascar)	$K_{0.83}Na_{0.07}(Si_{2.77}Al_{1.17}Fe^{3+}_{0.06})(Mg_{2.19}Al_{0.48}Fe^{2+}_{0.08}Fe^{3+}_{0.06}Ti_{0.04})O_{10}(OH)_{1.85}F_{0.15}$	0.90
Vermiculite Santa Olalla	$Mg_{0.39}Ca_{0.02}(Si_{2.72}Al_{1.28})(Mg_{2.59}Al_{0.06}Fe^{3+}_{0.24}Fe^{2+}_{0.03}Ti_{0.08})O_{10}(OH)_2$	0.82
Vermiculite Praysacc (André, 1972)	$Mg_{0.25}K_{0.051}Na_{0.001}Ca_{0.0005}(Si_{2.82}Al_{1.17})(Mg_{2.36}Al_{0.20}Fe^{3+}_{0.43}Fe^{2+}_{0.004})O_{10}(OH)_2$	0.54
Vermiculite Benahavis (López González and Barrales Rienda, 1972)	$Mg_{0.24}Ca_{0.03}(Si_{2.81}Al_{1.10}Fe^{3+}_{0.09})(Mg_{2.46}Ti_{0.11}Fe^{3+}_{0.43})O_{10}(OH)_2$	0.53
Vermiculite Llano (Norrish, 1973)	$Ba_{0.45}(Si_{2.78}Al_{1.22})(Mg_{2.94}Ti_{0.02}Fe^{3+}_{0.01}Al_{0.2})O_{10}(OH)_2$	0.90
Vermiculite Malawi (Nyassaland) (Norrish, 1973)	$Ca_{0.30}K_{0.05}(Si_{2.89}Al_{1.04}Fe^{3+}_{0.12})(Mg_{2.53}Ti_{0.06}Mn_{0.01}Fe^{3+}_{0.41})O_{10}(OH)_2$	0.65

The natural vermiculite crystals were treated with a 1N solution of the chosen cation. An average of four treatments was sufficient, for a period of 48 h and at 80°C.

To obtain a particular hydrated phase, the samples were left in the RH selected between two and four weeks before X-ray examination. The diagrams were obtained from platelets introduced into capillaries, which were left at the equilibrium RH, then sealed off.

CRYSTAL STRUCTURE

For some years it has been known that in hydrated smectites, where the positive charge deficit is in the tetrahedral sheet, ordered and semi-ordered structures are generally found (Hofmann, 1956; Zvyagin, 1957; Glaeser et al., 1967; Suquet et al., 1975; Méring, 1975). The fact is that in this mineral type the excess negative charge is located on the oxygen atoms of the AlO_4 tetrahedra. These surface oxygens have the role of anchorage points for interlayer links using cation-dipole interactions between exchangeable cations and water molecules, and also of direct interactions between the surface oxygens of the layers and the interlayer water molecules (sometimes also the cations). The existence of such links between the interlayer space and particular surface oxygens causes a certain order in the stacking of layers.

The study of the crystalline order in vermiculite is easier than that in smectite because of its larger net layer charge and higher degree of crystallinity. It has thus been possible to identify more than ten modes of stacking of the layers by X-ray diffraction using single crystals (de la Calle et al., 1978, 1985).

Structural studies of vermiculite and the determination of factors that influence the layer stacking sequences have been investigated in order to understand the effects caused by water and various organic solvents (de la Calle, 1977; de la Calle et al., 1980; Slade and Stone, 1984; Slade et al., 1985). The behavior of soil materials has also been simulated in the laboratory in order to determine the mechanism involved. The behavior of soil vermiculites of different geographic origin can be compared if a reference state has been defined (Ben Rhaiem et al., 1986). This reference state corresponds to the condition in which the sample occurs in the soil, e.g., with Mg as interlayer cation and hydrated with two-water layers (d(001) = 14.3 Å), (de la Calle et al., 1988).

The study of the real crystal structure of vermiculites consists in an application of a complex of various physical methods to the same sample. However, in spite of their limitations, diffraction methods remain the basic ones for study of the structures. The results obtained by other methods should always be interpreted in the light of diffraction data (Drits, 1985).

X-ray diffraction studies

Since vermiculite crystals have one dimension much smaller than the others (namely the thickness), they give diffraction maxima that are elongate and not points. They also suffer from defects of crystallization (distortions, cracks, etc.), so that totally automated methods are not the most appropriate for their study.

An examination of Weissenberg diagrams (levels h0l, 0kl, 1kl) lead to the following conclusions:

(1)   If all the levels give only discrete reflections, one has a ordered phase. By an ordered phase, we understand one in which the position of any one layer in the crystal is given unequivocally by the translations defining the crystal lattice. Phlogopites show two types of ordered stacking; one of them (1M) is very common, the other ($2M_1$) much rarer. In both cases the pseudohexagonal cavities on the surface of adjacent layers face each other and enclose the interlayer K atoms. In vermiculites ordered stacking is rare. The cases that have been found (de la Calle et al., 1977; Slade and Stone, 1984; Slade et al., 1985) show that in order to obtain

464

ordered stacking the interlayer material must determine unequivocally the relative position of the layers. As with phlogopites, this type of stacking occurs when the pseudohexagonal cavities in adjacent layers are opposite each other. In this case the average structure can be described using classic crystallographic methods (Fourier transforms, least-squares refinements, electron density sections in various orientations).

(2)   If certain discrete reflections lying along rows parallel to Z^* (levels 0kl and 1kl) are replaced by diffuse streaks, the phase is partially-ordered. In this paper we will only be concerned with so called planar disorder due to layer translations in the plane of the layer. We will not consider the case of interstratification due to the random insertion of layers of different thickness (see Chapter 15). The structure of a vermiculite will be called semi-ordered when, in a pile of M layers, the passage from the nth layer to the (n+1)th in the Y direction can be made by means of two or more different translations (Fig. 3); the translations along X and Z are always unequivocal. For semi-ordered stacks, the reciprocal space cannot be described by a set of hkl reciprocal spots (h,k,l integer) but by reciprocal rods (h,k) with a continuous variation of the intensity along the rod depending on the nature of the layers and the way they are stacked (Méring, 1949; Guinier, 1964; Plançon, 1981).

The methods for studying such structures have been presented in recent publications (de la Calle et al., 1984, 1988). Essentially one proceeds as follows: (a) A classic approach by finding the one-dimensional electron density projection along Z and the map of electron density corresponding to the ordered XOZ projection. (b) An indirect approach that consists of comparing the observed intensities with theoretical intensities derived from a structural model for the more or less diffuse rods of the levels 0kl and 1kl (Plançon, 1981; de la Calle et al., 1984, 1988).

Appearance of the levels h0l, 0kl, and 1kl for a semi-ordered stucture. For a semi-ordered vermiculite, an examination of the Weissenberg X-ray films (Fig. 4) containing the intensity distribution along the h0, 0k, and 1k reciprocal rods shows they fall into two groups: those containing sharp diffraction spots and those containing more-or-less diffuse bands. Generally speaking, the rods of the h0l level are discrete. The ok and 1k rods may be classified as discrete, more-or-less diffuse, or diffuse.

The discrete or diffuse nature of various rods, for different kinds of stacking, can be shown schematically by a projection on the a^*b^* plane of the reciprocal lattice (Fig. 5). An examination of such a scheme allows us to group the discrete or almost discrete spots into a family of planes P(n) parallel to the (h0l) plane, which will be called P(0). For Mg-vermiculite (d(001) = 14.3 Å), the P(n) family of planes—discrete planes—is defined by the condition k = 3n (de la Calle et al., 1988). For Na-vermiculite (d(001) = 11.85 Å), for example, one may suppose such a family of P(n) planes to exist: the P(0) plane does have discrete spots, but the (hkl) planes for which $k = nk_1$ are only fictive Bragg planes because k_1 is not a whole number (e.g., $k_1 = 3.26$), (de la Calle et al., 1984).

We shall see in what follows that this simple calculation of the value of k_1 is important for the quantitative determination of the component along Y of the translation on passing from one layer to another in the stack of layers.

Quantitative analysis of diagrams of semi-ordered structures

Analysis of the h0l reflections corresponding to the ordered XOZ projection.

(i) One-dimensional projection of electron density along the Z axis. This projection is obtained from a one-dimensional Fourier transform of the F(00l) structure factors after correction of the intensities for the Lorentz and polarization factors and absorption effects. Phases for the 00l reflections are computed from the known configuration of the silicate part of the structure and assigned to the structure amplitudes. This assumes that the contribution to the total scattering from the interlayer material is comparatively small. Then the interlayer material is introduced. The positional parameters for the atoms of the silicate

465

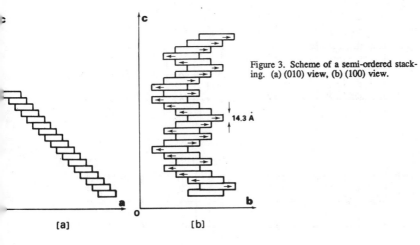

Figure 3. Scheme of a semi-ordered stacking. (a) (010) view, (b) (100) view.

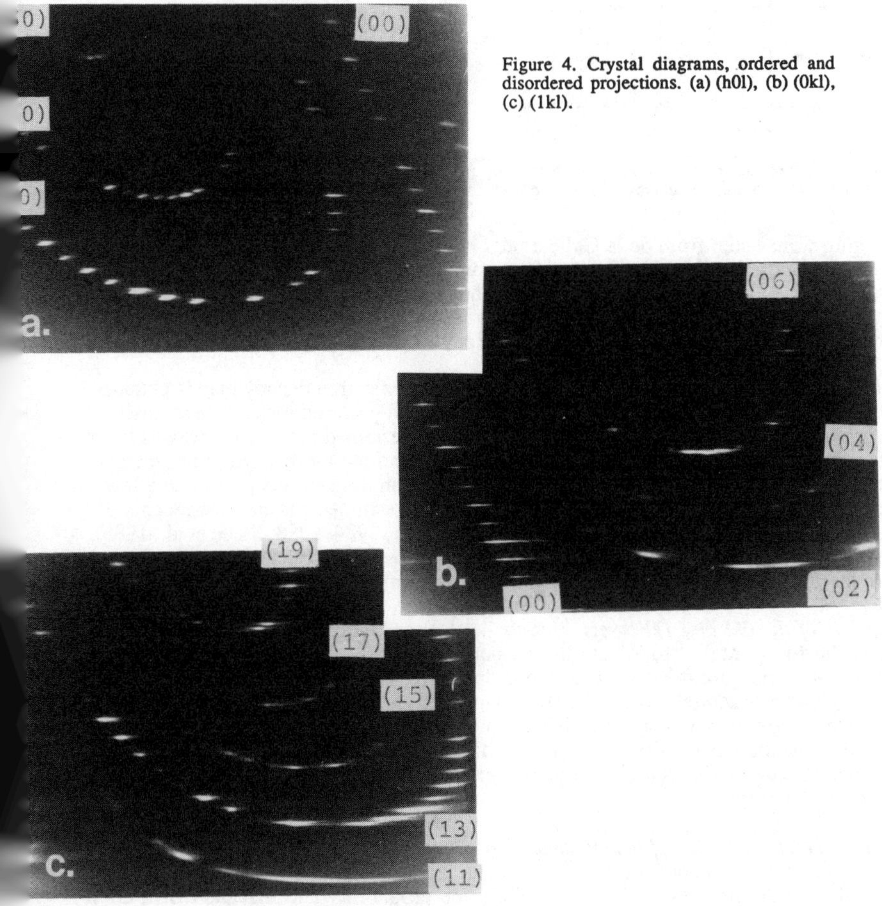

Figure 4. Crystal diagrams, ordered and disordered projections. (a) (h0l), (b) (0kl), (c) (1kl).

466

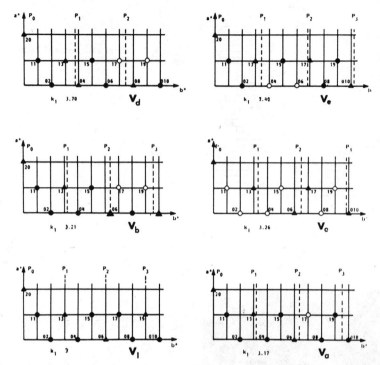

Figure 5. Schematic representation of the discrete or diffuse nature of different "rods" in different stacking sequences of layers. hkl discrete rows, hk diffuse "rods," more or less diffuse "rods."

structure were taken from de la Calle et al. (1977) and modified for the unit cell of each different material. The experimental electron density curve is then compared with the calculated one. In most cases this allows the interlayer cations to be positioned and also the z-height of the water or other interlayer molecules to be fixed (Le Renard and Mamy, 1971; Rausell Colom and Fornes, 1974; Slade and Raupach, 1982; de la Calle et al., 1988).

(ii) Electron density map for the h0l level. The electron density map is obtained by Fourier transform of all the F(h0l) structure factors, after corrections of the intensity (Lp, absorption). The phases for the h0l reflections are determined in the same way as for one-dimensional projection. The x, z, atomic positions and the thermal vibration factors are determined by minimizing the reliability factors. The analysis of this projection allows the z coordinates to be refined, and the x coordinates determined for the components of the interlayer region (Telleria et al., 1977; de la Calle et al., 1984, 1988; Slade et al., 1985). If there is sufficient contrast in the interlayer space it is possible to calculate a difference series in order to bring out its structure more exactly (Slade et al., 1987).

Analysis of the 0kl and 1kl levels. As we have already explained, if certain reflections along the rows parallel to Z* are drawn-out and diffuse, the method of approach will consist in comparing the observed intensities along the rod with theoretical intensities derived from a structural model. The first stage is to carry out a simple calculation to see if the positions of the maxima of the diffuse rods obtained theoretically are compatible with the experimental results; the second stage consists in applying the complete formulae for calculating the continuous intensity I(l) for the different rods, and comparing it with experimental results.

Research on the position of the maxima. We have considered two models of semi-ordered stacking that may account for the experimental reciprocal lattice. First of all we suppose that all translations parallel to Y have the same probability and that the distribution is arbitrary. This sort of stacking leads us to consider a family of planes P(n) parallel to the (h0l) planes, near which reflections from the planes show discrete spots, and of which the plane P(1) is a plane (hk1l) (cf. the paragraph above), k1 being determined experimentally (Fig. 5).

(i) Hypothesis of a model with two translations, arbitrarily distributed, and differing only in the sign of the component along the Y direction (+Ty and -Ty). Let us consider a pair of layers separated by the interlayer material. Their relative positions are not arbitrary. They will be such that the energy of this crystal element is a minimum. In this element, one passes from the lower to the upper layer by means of a translation +T (or -T) parallel to Y. It may be seen that the symmetry of the surface oxygens is such that the environment of the interlayer material is the same for the two translations +T and -T.

As has been shown by Méring (1949) the diffracted intensity due to a pile of layers, where the only interaction is between first neighbors, may be obtained by taking into consideration only the interference between two neighboring layers. The phase term describing the interference is

$$\psi = 2\pi(htx + kty + Zd(001)),$$

where Z corresponds to the coordinate in A perpendicular to the basal plane a^*b^*, and d(001) is the basal distance. We put $Tx = atx$ and $Ty = bty$. Taking the supposition of two translations +Ty and -Ty, one obtains, for any pair of adjacent layers, the following values for the phase difference:

$$\psi = 2\pi(htx + kty + Zd(001)),$$

$$\psi = 2\pi(htx - kty + Zd(001)).$$

In order that the reflections hkl with $k = k_1, 2k_1, ..., nk$ may appear discrete, the phase term ψ_1 and ψ_2 must necessarily differ by a multiple of 2π; hence:

$$\Delta\psi = -\psi_1 = \psi_2 m2\pi$$
$$ty = ty_1 = m/2k_1 = 4\pi k_1 ty,$$

and a second value of ty must be taken into account with

$$1 - 2ty = mk_1; \quad ty_2 = 1/2 (1 - m/k_1).$$

If the phase change is taken with reference to the level hk_1l, the whole number m must be unity.

The position of the maxima on the rods 0,k and 1,k is given by (Méring, 1949; Guinier, 1964):

$$Z = 1/d(001) (1 - kTy)$$

thus for $l' = (1 - kTy)$, with l an integer.

Thus in the case, for example, of two translations $Ty = (1/3)b$, having the same probability 1/2, for the indices k that are not multiples of three, the maxima appear for half-whole-number values of l' (this is the case of Mg vermiculite, d(001) = 14.3 Å).

(ii) Hypothesis of a model with three translations +Ty, - Ty, and 0. The possibility of such a model is linked to the ditrigonal symmetry of the surface of the layers. If we suppose these three translations are distributed arbitrarily, the reasoning above concerning the model with two translations may be repeated. Three values are obtained for the phase difference between neighboring layers:

$$\psi_1 = 2\pi(htx + kty + Zd(001))$$
$$\psi_2 = 2\pi(htx - kty) + Zd(001))$$
$$\psi_3 = 2\pi(htx + Zd(001)) \ .$$

In order that the hkl reflections with $k = k_1, 2k_1, ..., nk_1$ may appear discrete the phase terms ψ_1, ψ_2 and ψ_3 are only allowed to differ by a multiple of $2\pi m$ (m integral). For the case $k = k_1$, this involves

$$\psi_3 - \psi_2 = \psi_2 - \psi_1 = 2\pi k_1 ty = 2\pi m \ ,$$

where $ty = m/k_1$. Further, $\psi_3 - \psi_1 = 4\pi k_1 ty = 2\pi m$, which introduces no additional condition for the parameter ty. In this case the calculation of the position of the maxima is more complicated and one has to follow the complete development described by Méring (1949).

Calculation of the diffracted intensity I(l). The methods developed by various authors, (Hendricks and Teller, 1942; Méring, 1949; Kakinoki and Komura, 1952; Maire and Méring, 1970; Plançon, 1981), enable us to calculate the continuous intensity I(l) on various rods. The intensity diffracted along an h,k rod can be calculated from the matrix expression developed by Plançon (1981):

$$Ihk(s) = Trace(Re((F_hk)(w)(I) + 2\sum_1^n ((M - n)/M)(Q)^n))$$

Re   real part of...   M   the number of layers in the stacking
F_hk matrix of the structure factors   I   unit matrix
w   matrix containing the proportion of the different translations between the layers
Q   matrix describing the interference phenomena between adjacent layers in the stacking.

In the case of a vermiculite structure with two translations +T and -T and two different elements called "A layer" and "B layer," the matrix Q is given by:

$$(Q) = \begin{vmatrix} p(++) \exp(-2\pi is(+T)), & p(+-) \exp(-2\pi is(+T)) \\ p(-+) \exp(-2\pi is(-T)), & p(--) \exp(-2\pi is(-T)) \end{vmatrix} \ ,$$

where s is the scattering vector.

If w+ and w- are the relative proportions of the +T and -T translations, the relationship between proportion and probabilities, p, can be written:

$$(w+) + (w-) = 1; \quad p(++) + p(+-) = 1; \quad p(-+) + p(--) = 1; \quad (w+)p(+-) = (w-)p(-+);$$

Since there are four equations and six parameters, two of the parameters will be independent. If α is the probability of having a +T translation followed by another +T and if w+ is chosen as independent parameter, the six parameters are then given by the expressions:

$$(w-) = 1 - (w+); \quad p(+-) = 1 - \alpha; \quad p(-+) = (w+)/(1 - (w+))((1 - \alpha)); \quad p(--) = 1 - p(-+) \ .$$

So the calculated intensity essentially depends on two factors: one is the nature (F) and the number (w) of the two-dimensional elements that are stacked and the other is the relative position of these elements within this stacking (Q).

For vermiculites, the basic element that constitutes any stacking arrangement is the 2:1 silicate layer and the interlayer region. In the case of Ty = b/3 (Mg-vermiculite), for example, two elements s (terminology of Mathieson and Walker, 1954) called "A layer" and "B layer" are considered. The expression "layer" here indicates the structural unit consisting of a silicate layer and its interlayer space. A and B involve different locations of the water coordination polyhedron in relation to the silicate surface. The "A layer" requires a shift of +b/3 between adjacent layers to permit the establishment of hydrogen bonds between the water coordination polyhedron and the surface oxygens. In the "B layer," a - b/3 shift permits these hydrogen bonds to be established. Four probabilities, p(++),p(+),p(-+),p(--), are defined, e.g., p(+-) represents the probability for a +b/3 shift to be followed by a -b/3 shift or, in other words, the probability for an "A layer" to be followed by a "B layer." These four probabilities allow statistical characterization of the stacking model of basic elements ("A" or "B layer") in a layer sequence. We also define w+ and w- as the relative proportions corresponding to the +b/3 and -b/3 shifts, or in other words, as the proportions of "A" and "B layer," α is the fault probability in a two-layer polytype of adjacent layers being displaced alternately by ±b/3 in relation to the preceding one (Shirozu and Bailey, 1966). Assuming that w+ = w- = 0.5,the probability values for the shifts are: p(++) = p(--) = α, and p(+-) = p(-+) = 1 - α . A schematic presentation of the fault probability α of the silicate layer having an ordered or disordered layer stacking is given in Table 2.

Figure 6 represents the appearance of the 0,k and 1,k rods for an ordered structure (polytype with two layers, α = 0), for a disordered structure (α = 0.5), and with segregation of the two phases (α = 1). In making trial calculations, one can mainly vary the relative proportion of translations (+T and -T), and the values of the p parameters (p(++), p(+-), ...). The calculated I(l) profiles are compared with the observed intensities.

Analysis of abnormal diffraction of X-rays

It must be noted that the structures determined by XRD can only give a mean distribution of the compensating cations in the crystallographic sites corresponding to the equivalent positions defined by the space group. Since the number of cations is less than the number of sites, we have to find out whether their distribution is truly statistical, i.e., if the cations are distributed randomly in the sites attributed to them, or if there exists a short-range order in the distribution. The study of Bragg reflections is not sufficient for this, and one has to consider also the scattering between the Bragg sites (Guinier, 1964). If this scattering is continuous and diffuse (Laue scattering) the randomness of the distribution is confirmed, but if this scattering is concentrated in discrete spots, then it is possible by studying these "extra spots" to determine the nature of the order in the positioning of the compensating cations.

In order to bring out the existence of such superstructures, one uses an experimental procedure in which the crystal is stationary during the whole exposure, convergent monochromatic Laue technique, (Alcover et al., 1973; Rousseaux et al., 1984). Using this principle, Alcover et al. (1973) and Alcover and Gatineau (1980a,b) proposed structures that should approach closer to the real structure for various phases of hydrated vermiculites. Raupach et al. (1975), Slade and Stone (1983), Slade et al. (1978), and Slade and Raupach (1982) have made the same determinations for interlayer organic complexes.

Suquet and Pezerat (1987) state that all the "extra spots" between the Bragg spots can be related to the distribution of interlayer cations, which tend to situate themselves as far as possible from each other. The resultant superstructure can be detected experimentally when the charge density of the layer and the charge corresponding to the superstructure are compatible. We may note, however, that linear diffractions have been observed between certain Bragg reflections. Kodama (1975, 1977) proved that this type of diffraction is observed in smectites, vermiculites, and micas, and that it is due to fluctuations in the positions of the surface oxygen atoms around their equilibrium positions.

470

Table 2. Different tendencies of the silicate layers to have ordered or disordered stacking ($W_+ = W_- = 0.5$, $P_{++} - \alpha$).

	P_{++}	P_{+-}	P_{-+}	P_{--}
Ordered Stacking → 2 Layers Polytype	0	1	1	0
Tendency 2 Layers Polytype	0.25	0.75	0.75	0.25
Disordered Stacking → Random	0.50	0.50	0.50	0.50
Tendency Segregation	0.75	0.25	0.25	0.75
Ordered Stacking → Segregation	1	0	0	1

$\downarrow$

α

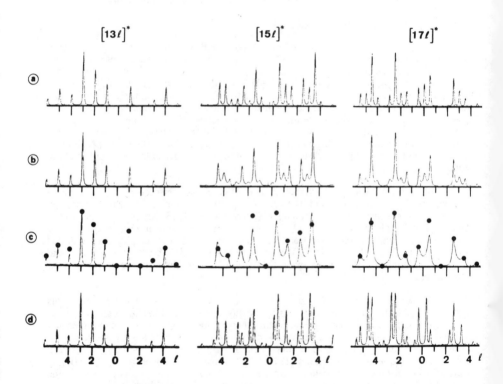

Figure 6. Calculated I(l) profiles for (a) an ordered Mg-vermiculite structure ($\alpha = 0$, polytype with two layers), (b) an ordered tendency ($\alpha = 0.25$), (c) a disordered structure ($\alpha = 0.5$), (d) a segregation of two phases ($\alpha = 1$). (1,3), (1,5), and (1,7) "rods."

Description of stacking of layers in ordered vermiculite structures

An ordered stacking of layers, giving rise to a three-dimensionally-ordered structure is only found in cases where the configuration of the interlayer material brings the ditrigonal cavities of adjacent layers opposite one another. Such is the case:

(1)   for the vermiculites hydrated with two water layers and with Na and Ca cations (r < 1.3 Å) for high relative humidity (de la Calle et al., 1977, 1978; Slade et al., 1985);

(2)   for anhydrous vermiculites saturated with large alkali cations (K, Rb, Cs, r >1.3 Å)

(3)   for the "zero layer" Li-vermiculite (Suquet et al., 1982). In this state (d(001) = 10.1 Å), one water molecule is trapped with Li in the ditrigonal cavities. The radius of the Li-H$_2$O group is > 1.3 Å.

(4)   for vermiculites intercalated with aniline (Slade and Stone, 1984; Slade et al., 1987), for vermiculite intercalated with a peptide (Pons, pers. comm.).

The powder diagrams of vermiculites with ordered stacking is schematically shown in Figure 7. On the same figure, for purposes of comparison, are shown the positions and intensities of reflections (series 021, 111 and 201, 131) of semi-ordered stackings. The ordered state may be distinguished, in particular, by the existence of 021, 111 reflections.

Ca vermiculite (d(001) = 14.92 Å). V3 stacking type. The nature of this structure, including the interlayer space, was studied by de la Calle et al. (1977) and their results were confirmed by Slade et al. (1985). The structure deduced from the X-ray data on this crystal is a simple one in which the pseudo-hexagonal surface cavities are opposite each other, on either side of the interlayer material. The space group is C2/m. The dimensions of the monoclinic cell are: a = 5.35 Å;   b = 9.26 Å;   c = 15.03 Å;   ß = 96.83°. The angle of rotation of the tetrahedra is about 7.5°.

This ordered structure corresponds to the chlorite structure Ia-2, following the nomenclature of Bailey and Brown (1962). The calcium atoms are situated in the center of the interlayer in two positions (Fig. 8 and 9); on the one hand (CaT) between the faces of the two tetrahedra; on the other hand (CaH) between the two pseudo-hexagonal surface cavities of the two adjacent half-layers. The two sites are approximately equally occupied. The water molecules, about 7-7.5 per Ca atom are distributed among crystallographically distinct positions, thus defining six species: A, B, C, D, E and F (Fig. 8). Five of them contribute to form water coordination polyhedra around the CaT and CaH atoms. The final species, B, probably serves as a link between the polyhedra. The species A and B form octahedra around the CaT, rather similar to those postulated for the polytype Ia-2. The species C, D, and E, on the other hand, form deformed cubes around the CaH (Fig. 9a,b), cubes with the corners EE incompletely occupied. Consequently, the Ca ion is slightly off center, being situated further away from the incompletely occupied corner. At the two ends of the "vertical" diagonal (practically parallel to c*) of the coordination pseudo-cube, the two water molecules of type C stand one above the other, practically at opposite extremes from the CaH. These molecules fit into the ditrigonal cavities of the facing layers, thus determining their "face to face" position and unequivocally determining the layer stacking.

The three-dimensional order thus appears, in this mineral, to be linked to the existence of a particular water polyhedron in the interlayer space. This ordered structure disappears at a RH < 40% (~7H$_2$O per Ca). This may be due to the reorganization of the interlayer space and a general rearrangement of the stacking. This structural change is probably linked with the fall of electric conductivity noted by Mamy and Gaultier (1973) for the same RH value.

Further, de la Calle et al (1978) postulate a distribution of the Ca(H$_2$O)n polyhedra in which they are situated as far apart as possible and form chains. The "choice" between various possible solutions probably depends on the charge and the distribution of the Si/Al substitutions in the layers. The diversity of possible solutions no doubt explains why these

472

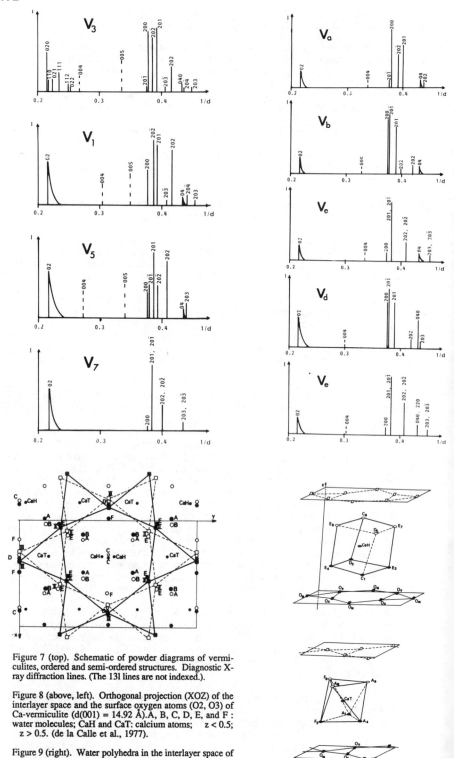

Figure 7 (top). Schematic of powder diagrams of vermiculites, ordered and semi-ordered structures. Diagnostic X-ray diffraction lines. (The 13l lines are not indexed.).

Figure 8 (above, left). Orthogonal projection (XOZ) of the interlayer space and the surface oxygen atoms (O2, O3) of Ca-vermiculite (d(001) = 14.92 Å).A, B, C, D, E, and F : water molecules; CaH and CaT: calcium atoms;   z < 0.5; z > 0.5. (de la Calle et al., 1977).

Figure 9 (right). Water polyhedra in the interlayer space of a two layer Ca-vermiculite.

authors were not able to find superstructures corresponding to domains in which the $Ca(H_2O)n$ polyhedra would be distributed in an ordered manner. Alcover and Gatineau (1980c) also failed to observe superstructure spots with this hydrated vermiculite phase, but they did see "rings" more or less broken up into spots. Using a mathematical model these authors give a semi-quantitative interpretation of the "rings." Their results taken as a whole indicate the presence of groups of cations + water molecules distributed randomly in the interlayer space, but with a minimum approach distance of 8-9 Å. These results are compatible with the mean structure proposed by de la Calle et al., (1977). The possible distances between groups and their relative abundance deduced from this mean structure lead to an important proportion of the distances between 8.2 and 9.2 Å.

Fornes et al. (1980) studied this phase by IR. Their study was concerned with the breaking up of the group of O-D bending frequencies centered around 2.500 cm^{-1}, shown by partly deuterated samples. The model of the decomposition into four bands proposed, corresponding to four different types of water, is the only one that can be put forward, taking into account the imprecision of the low frequency part of the spectrum and the limits imposed by the method of decomposition. However, the authors admit that it is not impossible that the real model would imply a larger number of bands, which is suggested by the structure given by X-rays.

A study of QNS carried out by Poinsignon et al. (1987) shows that the axis of rotation of the $Ca \cdot nH_2O$ complex (Fig. 8a) is along the line C1-C6 instead of along the Z axis of the layers. This study confirms the crystallographic conclusions of de la Calle et al. (1977), while bringing supplementary information on the mobility of the H atoms of the water molecules.

Na vermiculite (d(001) = 14.83 Å). V3 stacking type. This ordered structure corresponds to the polytype Ia-2. The transition mica Na vermiculite does not imply any sliding of the layers over each other. The pseudo-hexagonal surface cavities remain face to face, as in the original mica. The space group is C2/m. The dimensions of the monoclinic cell vary slightly according to the origin of the specimen. Thus, for "artificial" Na vermiculite (from Madagascar phlogopite): a = 5.35 Å; b = 9.28 Å; c = 14.93 Å; ß = 96.56° (de la Calle et al., 1978). For Na vermiculite from Llano: a = 5.26 Å; b = 9.23 Å; c = 14.97 Å; ß = 96.82° (Slade et al., 1985).

A preliminary study of the interlayer material was made by Bradley et al. (1963) using a one-dimensional Fourier projection, the water molecules being located at 5.9 Å from the origin (octahedral cations), and the Na at 7.4 Å: The complete structure of the interlayer material was described by Slade et al. (1985). Using Fourier projections these authors show the octahedral coordination of the water molecules around the Na cations. A comparison of Figures 10a and 10b shows the difference between the interlayer material in the two-layer hydrates of Na and Ca vermiculites. While the Ca is situated in two different positions with respect to the adjacent silicate surface, with two different types of coordination of water molecules, the Na occupies only one position with octahedral coordination.

A study by continuous wave (CW) wide line NMR and by pulse NMR of the H nucleus in a wide temperature range has been carried out by Hougardy et al. (1976). The results show that an octahedral distribution of water molecules around the Na cation fits the experimental data, the six rotation axes around which the water molecules are spinning rapidly are tilted by about 60° with respect to the c* axis. Water molecules within the hydration shell and/or water molecules "between" the hydration shells exchange protons with a frequency of the order of 10^{-5} sec^{-1} at room temperature.

Anhydrous K, Rb, Cs, and Ba vermiculites (d(001) = 10 Å). Anhydrous vermiculites that are saturated with cations of ionic radius > 1.3 Å also have an ordered structure. Their mode of stacking corresponds to that of the polytype 1M of the original phlogopite. With these cations the b parameter increases with the ionic radius (Suquet et al., 1981) and the stretching frequency of the OH bond diminishes as the ionic radius increases (Rausell-

474

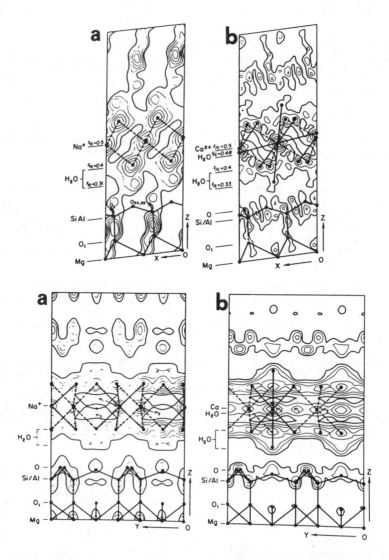

Figure 10 (above). $(F_o\text{-}F_c)$ projections onto (010) and (100) cell faces of the structures for (a) Llano Na-vermiculite and (b) for Llano Ca-vermiculite. (Slade et al., 1985).

Figure 11 (right). Honeycomb-shaped reflections around the Bragg spots on an electron diffraction pattern of Ba-vermiculite (Alcover and Gatineau, 1976).

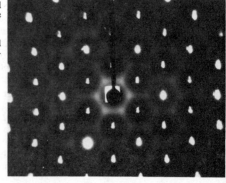

Colom and Serratosa, 1985). These results can be explained by the position of the interlayer cation. When these large cations are situated in the center of the ditrigonal cavities of adjacent layers lying face to face, they act on the surface oxygens and deform the layers. The b parameter, d(001), and the cation-OH distances increase.

Gatineau and Alcover (1976), in order to study the two-dimensional structure of the interlayer material, examined in vacuo an immobile crystal of a (previously dehydrated) Ba-vermiculite. The diagrams obtained show what are described as "honeycomb-shaped reflections" (i.e., hexagonal) around the Bragg spots (Fig. 11). The authors show that the distribution of Ba cations held in the two ditrigonal cavities is characterized by:

(1)   Given a ditrigonal cavity occupied by a Ba atom, the probability of finding another Ba atom in any of the six neighboring such cavities is practically zero.

(2)   There are three preferential directions, at 120° to each other, where the probability of finding a Ba is much greater than the chemical proportion of these ions.

The honeycomb diffractions characteristic of this distribution are very obvious with vermiculites saturated with barium, but hardly visible with Sr or Ca saturation.

Vermiculites with Li.H$_2$O (d(001) = 10.1 Å). The 1M structure is also found when the ionic radius of the interlayer cations is small enough for them to occupy a ditrigonal cavity with one molecule of coordinated water (Suquet et al., 1981, 1982). A recent study shows that the water content in a Li vermiculite, at 150°, is 0.4 molecules per cation; these molecules being located at 5.02 Å from the origin (octahedral cations) (Pozzuoli et al., to be published).

Vermiculite-aniline complex. The anilinium-intercalated Llano vermiculite, prepared by cation exchange of Na vermiculite with a 1% aniline hydrochloride solution, has a three-dimensionally ordered structure (Slade and Stone, 1984; Slade et al., 1987). The Weissenberg diagrams show only Bragg reflections in the entirety of reciprocal space; even those with k ≠ 3n are discrete. This implies that the relative position of adjacent layers is well defined by the characteristics of the interlayer material. There are no random translations of ±b/3 of the sort present in the starting material (crude vermiculite). The ditrigonal cavities of adjacent layers in the aniline-vermiculite compound are face to face. This position is determined by the aniline cations that form specific links with the surface oxygens of the cavities. "Extra spots" appear in the a*b* plane and point to an ordered distribution of organic cations in the interlayer space. This two-dimensional order reflects the existence of a centered 2a x 2b lattice, a and b being the dimensions of the ordinary vermiculite lattice.

Description of stacking layers in semi-ordered vermiculite structures

In semi-ordered vermiculite structures the position of one layer with respect to the preceding one is not unequivocal (as in the case where the ditrigonal cavities are face to face), but there are at least two well-defined possibilities that occur with arbitrary distribution in the crystal structure.

The X-ray diffraction results show that in these semi-ordered structures, the stacking shows translations from layer to layer of ±b/n, with the signs + and - randomly distributed. Figure 12 represents the surface oxygens on either side of the interlayer. In (a) the layer to layer translation is +b/3, in (b) -b/3. M represent the metallic atoms situated between two tetrahedral faces and surrounded with a water polyhedron. If there is no order in the Si-Al substitutions (Bailey, 1986), Figures 12(a) and 12(b) are symmetrical with respect to the trace plane called P. The situation for the hydrated cation M(H$_2$O)$_6$ will thus be energetically equivalent whether the translation is +b/3 or -b/3 (or, in general, +b/n or -b/n). There is thus no reason why there should be any order in the succession of these translations, especially since the links between the layer and the interlayer material are weak, and will not have any long-range effects. The sign of the translations from layer to

476

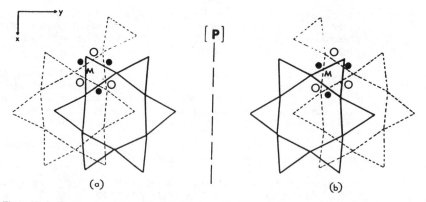

Figure 12. Surface oxygen on either side of the interlayer (a) +b/3 translation, (b) -b/3 translation. M: interlayer cation.

layer (n + 1) to (n + 2) will thus be independent of that from n to (n + 1). It will thus be understood that, except for special cases, the stacking of the layers in vermiculites will always be semi-ordered, since the translations +b/n and -b/n lead to equivalent situations. The special case will evidently be that with Ty = 0, in which the pseudo-hexagonal cavities are face to face (ordered stacking 1M, Ia-2).

The full description of a semi-ordered structure, as has been said , necessitates a complete study of the reciprocal lattice with one- and two-dimensional Fourier projections for the ordered structural projection, XOZ, and a study based on models for disordered projection, OYZ. Such a total study has so far been carried out only for two phases, the Mg vermiculite (d(001) = 14.3 Å), i.e., natural vermiculite, and the Na vermiculite (d(001) = 11.85 Å). For Sr vermiculite (d(001) = 11.85 Å) and Ba vermiculite (d(001) = 12.20 Å), the ordered XOZ projection has been studied by two-dimensional Fourier series, and qualitative studies have been made of the disordered OYZ projections. Li vermiculite (d(001) = 12.2 Å), Ca vermiculite (d(001) = 11.90 Å), Mg vermiculite (d(001) = 11.6 Å), Sr vermiculite (d(001) = 12.08 Å) has been studied in the direction perpendicular to the layer by one-dimensional Fourier series. Other phases such as Ca vermiculite (d(001) = 14.63 Å) and Mg vermiculite (d(001) = 13.8 Å) have only been studied qualitatively.

For certain phases complementary information to that from normal diffraction has been obtained by the interpretation of abnormal diffraction around the Bragg spots, and by spectroscopic studies, chiefly IR.

Hydrated vermiculite phases studied quantitatively along the three axes X, Y, Z.

Mg vermiculite (d(001) = 14.3 Å). V1 stacking type. Mg vermiculite has been the subject of several structural studies by XRD. Among the most important are Mathieson and Walker (1954) and Shirozu and Bailey (1966), who have studied both the stacking of layers and the organization of the interlayer material. They reported that the structure of the two water layer Mg vermiculite is three-dimensionally ordered. In fact, the observation of sharp k = 3n and of diffuse k ≠ 3n reflections on single crystal photographs indicates that the layer sequences are semi-random rather than completely ordered. De la Calle et al. (1976) reported that the observed intensities and diffuseness of the k ≠ 3n reflections indicate a large number of stacking faults to be present. These faults break up the regular alternation of +b/3 and -b/3 interlayer shifts.

The semi-ordered structure of the Santa Olalla Mg vermiculite has been determined (de la Calle et al., 1988). The Y axis ordered projection of the structure is shown in Figure 13. The electron-density contour maps provide accurate parameters for the atoms of the silicate layers and locate the water molecules at x = 0.142, z = 0.42 and probably x = 0.45, z = 0.35 (water in hexagonal hole) and the exchangeable cations at x = 0.5, z = 0.5. The final reliability factor for h0l reflections was 14% (70 reflections).

Calculations of the diffracted intensities along the 0,k and 1,k rods allows a single model to be derived (de la Calle et al., 1988). It is proved that the layer displacements are not regular. The disorder is created by the random distribution of two translations parallel to the Y-axis: +b/3 and -b/3. Figure 6 shows a comparison of calculated profiles of (1,3), (1,5) and (1,7) rods, (a) model with ordered translations +b/3 and -b/3 (two-layer polytype. Shirozu and Bailey's model), (c) model with random translations +b/3 and -b/3. The experimental data agree with the results of the calculations for the random model. There is agreement in both the positions and values of the observed intensities. Further, from the diffraction data, the position of the exchangeable cation can be determined. The interlayer Mg can occupy m1, m2, or m3 sites (terminology of Mathieson and Walker, 1954). Calculated values of intensities for a random model with Mg in m1 sites agree reasonably well with those obtained experimentally. The Mg atoms are situated between the faces of the tetrahedra, with the points directed downwards, of the two adjacent layers and are six-coordinated in the middle of an octahedron of water molecules.

Alcover et al. (1973) and Alcover and Gatineau (1980a) discovered traces of abnormal diffraction on their X-ray diffraction diagrams and showed that their origin lies in an ordered distribution of the interlayer cations. Based on the positions of these extra diffraction maxima, and in particular on the conditions for their extinction, the authors showed that the cations and the water molecules surrounding them tend to be distributed according to a two-dimensional superstructure having a centered cell of dimensions 3a x b (Fig. 14). The water molecules surround each cation in well-defined groups of composition $Mg_{12}H_2O$. The two-dimensional character of the observed order means that the cation distribution between two neighboring layers is not related to that in the next interlayer space. This superstructure exists only in limited regions of the interlayer space, and one passes from one domain to a neighboring one by step displacements. Further, the types of links that Alcover and Gatineau (1980a) propose between interlayer molecules and the oxygens of the layers confirm the existence of interlayer shifts as determined by XRD (Mathieson and Walker, 1954; Shirozu and Bailey, 1966; de la Calle et al., 1988).

In addition, Fornes et al. (1980) made an IR study similar to that made on the two-layer Ca vermiculite. This study shows good agreement between IR and XRD structural studies on the large proportion of strong hydrogen bonds. This is what is found in the chain structure (Fig. 14), where there are four strong hydrogen bonds (2.6 and 2.74 Å) for three weak or medium bonds (2.92 and 2.97 Å). The strong bonds give the interlayer structure a certain coherence and rigidity. As a limited assumption one may think of these interlayer water molecules as forming something similar to a structure of anionic links.

Na vermiculite (d(001) = 11.85 Å). Vc stacking type. The 11.85 Å monolayer phase of Na vermiculite was studied in detail by de la Calle et al. (1984). The conclusion is that there is a semi-ordered stacking mode in which one passes from one layer to the next by three translations, +T, -T and T0, distributed arbitrarily. +T and -T have the y component ±0.307, and T0 no shift. NMR studies (Sanz, pers. comm.) show a single type of environment for the protons of the water molecules in the interlayer space. The environment of the interlayer material is the same whatever the translation (+T, -T or T0) between the layers.

Bradley et al. (1963) and Le Renard and Mamy (1971) have calculated one-dimensional Fourier projections for this phase. They place the Na in the middle of the interlayer space, and the water molecules at a distance of 5.6 Å, from the octahedral cation, the number of water molecules per cation being two. The electron density projection on the ordered plane XOZ (Fig. 15) give the x values for Na and water and confirm the z values found from the one-dimensional series. The Na is located at x = 0.5, z = 0.5 and the water molecules at x = 0.46, z = 0.47 (de la Calle et al., 1984). Taking into account the x and z coordinates of Na and of the three possible relative positions of the superposed layers, the most probable situations are represented in Figure 16. For translation T0 the oxygens atoms of type O3 are coordinated to the Na, and the type O2 atoms are coordinated to the

478

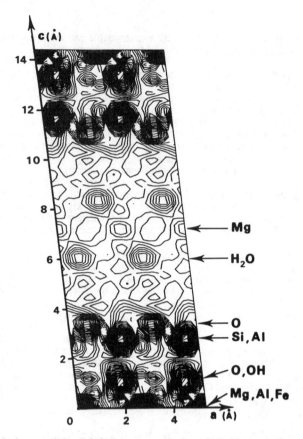

Figure 13. Projection onto (010) cell face of the structure for Santa Olalla Mg-vermiculite. (de la Calle et al., 1988).

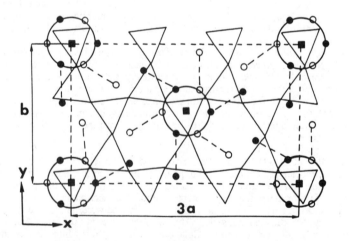

Figure 14. Interlayer space in a Mg-vermiculite. Only the tetrahedra of the lower layer are drawn; the upper layer was shifted by + b/3; Mg in the median plane (z = 0.5); H_2O (z > 0.5) (Alcover and Gatineau, 1980a).

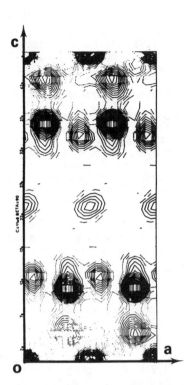

Figure 15. Projection onto (010) cell face of the structure of Santa Olalla Na-vermiculite (d(001) = 11.85 Å). (de la Calle et al., 1984).

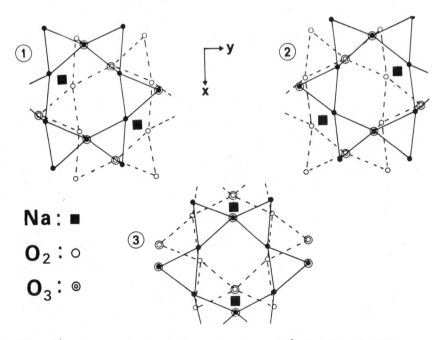

Na: ■
O_2: ○
O_3: ◎

Figure 16. Interlayer space of one layer Na-vermiculite (d(001) = 11.85 Å). (de la Calle et al., 1984).

Na for the translations +T and -T. The x and z coordinates of the water molecules are known fairly well and they are very close to those of the Na atoms, so it is possible to calculate their y position if we take the Na-O (H_2O) distance as being 2.40 Å.

Hydrated vermiculites studied quantitatively in two dimensions

$Ba.6H_2O$ vermiculite (d(001) = 12.2 Å). Vd stacking type. The type of stacking of this hydrated vermiculite phase is specific to barium, a divalent cation of large radius. From indexing the 20l series of XRD diagrams and from exact measurements of the values of d(00l) and d(20l) a layer to layer x shift of -0.78 Å is deduced. Also, good powder diagrams show the presence of discrete 041 and diffuse 061 reflections (cf. Fig. 7). The parameters of the planar lattice for a Ba saturated Santa Olalla vermiculite are a = 5.33 Å; c = 12.22 Å; ß = 93.67°. The b period, calculated from the discrete 040 reflection, is 9.23 Å (de la Calle et al., 1985).

Telleria et al. (1977), from h0l reflections obtained from Weissenberg diagrams, propose a plane (XOZ) lattice, leading to a layer-to-layer translation of -0.75 Å, which is a value close to that proposed by de la Calle et al. (1985).

Figure 5 showing the a*b* projection of the reciprocal lattice of this phase illustrates the aspect of the rods (discrete and diffuse) and the traces of the projections on the planes h0l, hk_1l, hk_2l, and hk_3l. It is seen that a value of k_1 of the order of 3.70 can explain the almost discrete nature of the spots observed for the rods with k = 4, 8, 3, 7, as well as the diffuse reflections for other values of k.

The x,z positions of the interlayer cations and water molecules were determined by means of a one-dimensional projection (Rausell-Colom et al.,1980) and a two-dimensional projection (Telleria et al., 1977). The water molecules are in the median plane, and the cations on either side are at a distance of 0.98 Å. Taking account of steric hindrances, the Ba is situated in one of the ditrigonal cavities at the coordinate x = 0.42 and the water at x = 0.45. Further, by IR spectroscopy Rausell-Colom et al. (1980) observed that Ba vermiculite is distinguished by a certain number of characteristic properties from other one-layer hydrates (Ca, Sr, Mg and Ni), in particular by the position of the compensatory cation, and by the appearance of a supplementary band at 3720 cm^{-1}. The Ba is immersed in a single ditrigonal cavity, and it is possible to place six molecules of water around each cation. The presence of a band at 3720 cm^{-1} is due to the hydrogen of a group being repelled by the cation towards the oxygen. The intensity of this supplementary band shows that each Ba atom only affects one OH group.

The existence of non-Bragg diffraction spots allowed Alcover and Gatineau (1980b) to give additional details on the organization of the interlayer space. The distribution of the small groupings ($Ba.6H_2O$) is periodic, with a 2a x b centered lattice as viewed in projection perpendicular to the layer. This organization is developed in identical domains oriented at 120° from each other. This study was carried out with a Ba vermiculite from Llano. The chemical analysis leads statistically to 0.9 Ba atoms per ab unit cell. The two-dimensional ordered arrangement described by Alcover and Gatineau leads to 1 atom of Ba per cell. The inconsistency between these values can be explained by the supposition that, outside the ordered zones, there are zones with a deficiency of Ba.

Besides this, the model proposed by Telleria et al. (1977) for the interlayer structure of this phase is based on the existence of a layer of water molecules at the points of a hexagonal network, the Ba being statistically distributed among the spaces in the center of the hexagons. This would imply for the interlayer material a centered a x b lattice, which is not compatible with the diffraction phenomena observed by Alcover and Gatineau (1980b).

$Sr·3.3 H_2O$ vermiculite (d(001) = 11.85 Å). Va stacking type. XRD diagrams suggest that this is a different type of semi-ordered structure (de la Calle et al., 1985). The parameters of the planar lattice for the Santa Olalla sample give the following: a = 5.34 Å, c =

481

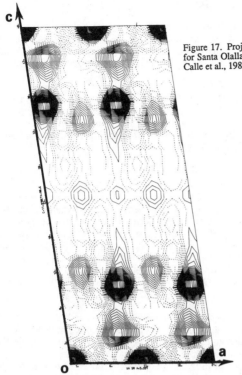

Figure 17. Projection onto (010) cell face of the structure for Santa Olalla Sr-vermiculite (d(001) = 11.85 Å). (de la Calle et al., 1985).

11.98 Å; ß = 98.57°. The b period, calculated from the discrete 060 reflection, is 9.24 Å. The component Tx, along X, of the interlayer translation is -a/3.

Study of the two-dimensional Fourier projection (Fig. 17) gives the x and z coordinates of the interlayer material and the number of water molecules per cation (3.3 H$_2$O/Sr). This value is identical with that obtained by Rausell-Colom et al. (1980) on Sr vermiculite from Llano using a one-dimensional Fourier projection. The water molecules are in the median plane, and the cations at about 0.3 Å from this plane. Sr is at x = 0.25, z = 0.48 and H$_2$O at x = 0.5, z = 0.5. Furthermore it is possible, in view of the ratio of the Sr and H$_2$O peaks, that some of the water is displaced to x = 0.25 and a z value of about 0.5.

In Figure 5 it can be seen that a value of k$_1$ of the order of 3.17 can account for the almost discrete character of the observed spots in the planes k = 3 and 6, and for the diffuse reflections for the other values of k.

Diagrams with a stationary crystal show large and diffuse spots associated with streaks between the Bragg spots. These diagrams have not proved usable for testing model structures of the organization of the interlayer material, because of poor definition (Alcover and Gatineau, 1980c). Furthermore, the IR absorption spectra cannot give any indications about the position of the interlayer cation. With Mg, Ca, or Sr as exchangeable cations in the one-layer structures, there is no modification of the stretching vibration of the structural hydroxyl at 3675 cm^{-1} (Rausell-Colom and Serratosa, 1985).

Hydrated and anhydrous vermiculite phases studied in the direction perpendicular to the layers, using one-dimensional Fourier projections

Le Renard and Mamy (1971) have used one-dimensional Fourier projections to study altered phlogopite with interlayer cations Ca, Na, and Ni. Also Rausell-Colom et al. (1980) proposed structures for weakly hydrated vermiculites (Mg, Ni, Ca, Sr, Ba) using the same methods and IR plus considerations based on steric hindrance for the interlayer and layer atoms.

Infrared absorption is of special use with anhydrous vermiculites for locating the cation with respect to the ditrigonal cavities. In trioctahedral minerals, including vermiculite, the OH groups have their O...H vectors oriented perpendicular to the layer. The study of the stretching vibrations of these hydroxyls would allow any interaction between the hydroxyl and the exchange cation to be detected.

The following results, among others, have been obtained:

Li·2.4H$_2$O vermiculite (d(001) = 12.2 Å), Vb stacking type. This phase appears with most vermiculites, in a wide range of RH (5-90%). The z height of the oxygens of the water molecules is determined as 5.75 Å and the number of water molecules per cell as 4.28 (Le Renard and Mamy, 1971). The water-cation-layer interactions were studied in this phase by IR spectroscopy (Suquet et al.,1982). Recently, Pozzuoli et al. (to be published) studied the modifications of this phase as a function of the temperature (between 25 and 70°C). These authors show that there are three different phases (12.17 Å, 12.02 Å and 11.89 Å) with an equivalent water content but with a different location in the interlayer space.

Sr.4H$_2$O vermiculite (d(001) = 12.08 Å). Ve stacking type. This phase is characterized by a Sr cation associated with four water molecules. The cations are in the median plane and the water molecules on the two sides and at about 0.5 Å from this plane. The Sr.4H$_2$O group is supposed to have the form of a very flattened tetrahedron. The cation is at the center and the four oxygens of the water molecules are at the vertices of this tetrahedron, and the two oxygens belonging to the neighboring layers complete the 6-fold coordination for Sr (Rausell-Colom et al., 1980).

Ca·4H$_2$O vermiculite (d(001) = 11.74 Å). Ve stacking type. The electron content of the interlayer space indicates 4 molecules of water per cation. Rausell-Colom et al. (1980) have observed three planes of atoms, one being the median plane, the two others at 0.3 Å on either side. The authors were not able to determine the respective positions of the planes of cations and of the water molecules.

Sr vermiculite (d(001) = 9.78 Å). Ca vermiculite (d(001) = 9.50 Å). In these anhydrous phases the cation is in the median plane and interacts with the hydroxyls of the neighboring layers.

Na vermiculite (d(001) = 9.6 Å). Li vermiculite (d(001) = 9.4 Å). The cations, as in the previous case, are in the middle of the interlayer space. Since their ionic radius is less than 1.3 Å, the layers slide over each other in order to ensure coordination with the surface oxygens. The displacement of the ditrigonal cavities postulated by Rausell-Colom and Serratosa (1985) is +a/3 for Na vermiculite and -a/3 for Li vermiculite.

Hydrated vermiculites studied qualitatively

Ca vermiculite (d(001) = 14.70 Å). V5 stacking type. The 14.70 Å Ca vermiculite phase has semi-ordered stacking. The state of hydration, although it corresponds to a double layer of water, is poorer in water that the ordered structure studied above (Ca vermiculite, d(001) = 14.92 Å). The parameters of the planar (XOZ) lattice are: a = 5.35 Å; c = 14.73 Å; ß = 93.53°. The b period is 9.26 Å.

Single crystal diagrams show, for the planes k ≠ 3n in place of discrete Bragg spots, a continuous diffusion containing maxima. The 0kl level, and especially the (0,2) and (0,4) rods, show that the $I(l)$ function is different from that of the other semi-ordered structures. Of course, the h0l plane shows only Bragg reflections and permits the study of the XOZ projection. Comparison of calculated and observed intensities leads to a model with ß = 93.5° and hence a Tx component along the X axis, for the translation between adjacent layers, of -a/6.

A qualitative study of the disordered OYZ projection confirms the following shifts for passing from one layer to the other (de la Calle et al., 1980):

$$-a/6, \quad -b/6; \quad -a/6, \quad +b/6; \quad +a/3, \quad 0 \quad (or -a/6; \quad +b/2)$$

These correspond to three layer shifts of equal amplitude, in three directions at 120° to each other, the succession of shifts being arbitrary.

Mg vermiculite (d(001) = 13.8 Å). V7 stacking type. This semi-ordered structure occurs with Mg as the interlayer cation and very low RH. The parameters of the XOZ plane are: a = 5.33 Å; c = 13.8 Å; ß = 90°. The b period is 9.26 Å. It is difficult to make observations on monocrystals of this phase because of the narrowness of its stability range as a function of RH. It forms a brief transition between the 14.3 Å phase and the single layer hydrate with d(001) = 11.6 Å.

Following Walker (1956), de la Calle et al. (1978) showed that the positions of the 201 and 131 reflections necessitates ß = 90°. This leaves only a single possible XOZ projection with the component Tx along the X axis for an interlayer shift equal to zero.

Since the passage from the 14.3 Å phase to the 13.8 Å one necessitates a layer shift of +a/3, it is probable that this shift cannot be completely achieved at the limit of the domain of stability of the 14.3 Å phase. An appreciable proportion of the pairs of layers will remain in the relative positions of the 14.3 Å phase, which is one important factor leading to disorder. Furthermore, being at the limit of the domain of stability of the one-layer hydrate, some of the pairs of layers will have the 11.6 Å spacing, and this is a second cause of disorder. The two types of disorder will diminish considerably the zones of coherence in the Z direction, leading to a broadening of the reflections in the l direction, and a lowered intensity.

A parallel study to that on the 14.3 Å phase is in progress in order to define the interlayer structure (de la Calle et al., to be published).

Vermiculite-organic interlayer complexes. The study of the structural organization of intercalated vermiculites is interesting. Lagaly (1987) wrote: "The adsorption of organic molecules by clay minerals is more strongly governed by structural aspects than previously assumed."

Among the studies published to this date on these interlayer complexes, only the one on the vermiculite-aniline complex gives clear results showing the existence of three-dimensional order in the stacking. The introduction of aniline between the layers causes a relative movement of the original vermiculite layers that brings the pseudo-hexagonal cavities face to face (Slade and Stone, 1984; Slade et al., 1985).

The ornithine-vermiculite complex (d(001) = 16.3 Å) was studied by Rausell-Colom and Fornes (1974) using one-dimensional series and by Slade et al. (1976) using two-dimensional series. It also has been characterized as presenting a three-dimensionally ordered structure corresponding to a one-layer polytype, but the results of the three-dimensional Fourier analysis are not entirely conclusive.

Investigations about the stacking have been done on other interlayer compounds with different organic molecules: 6-aminohexanoic acid (Slade et al., 1976); benzidine

(Slade and Raupach, 1982); cetylpyridinium bromide (Slade et al., 1978); piperidine (Iglesias and Steinfink, 1974); lysine (Raupach et al., 1975); benzylammonium (Pons et al., work in progress). These compounds show more or less disordered stacking, with shifts in the Y direction between adjacent layers.

The problem of the organization of the interlayer material in these intercalated vermiculites has also been studied by various authors. Raupach et al. (1975) investigated lysine-vermiculite complexes. Lysine and vermiculite form a stable, ordered interlayer complex that has a well-defined superlattice structure with respect to the ab vermiculite plane. The lysine molecules lie at a narrow angle to the silicate surfaces and establish a double layer network between them. Superlattice reflections provide evidence for a high degree of ordering of the organic molecules of cetylpyridinium in the interlayer region of vermiculites (Slade et al., 1978). For benzidine-vermiculite Slade and Raupach (1982) concluded that the patterns could be understood in terms of a primitive unit cell with the same a and b dimensions as the vermiculite cell.

Range of stability for various crystalline phase of vermiculite

The regions of stability for the different stacking modes of vermiculite have been defined in the literature (de la Calle et al., 1978, 1985) as a function of different factors : RH, nature of interlayer cations, charge of mineral. The values for the d(001) lattice dimension and the regions of stability for the different cations defined below were obtained with the Santa Olalla vermiculite (x = 0.82) using powder XRD. It was found that, for a given RH, crystals have a tendency to show a higher hydration state than powders, leading to a larger region of stability for the high hydration states (two layers). For a given interlayer cation and RH, low-charge vermiculites often show the same mode of stacking as high-charge vermiculites, but with a larger quantity of arbitrary stacking faults.

Table 3 indicates the cations with which the different stacking modes are obtained, their regions of stability, apparent basal spacing, the approximate number of water molecules per cation in the first coordination sphere, and the structural type: ordered or semi-ordered. In Table 4 is shown, for the different types of stacking identified, the values of the parameters of the planar lattice (a, c and ß), the b value for each layer, and Tx and Ty (component along X and Y, respectively, of the translation between two adjacent layers).

Unit-cell parameters were obtained by powder XRD (Guinier de Wolff camera, and monochromatized Co Kα radiation; Si as standard) and refined by least-squares. The value of b was obtained from the Y periodicity that is a function of the silicate layer, and is obtained from the 060 reflection (or 040 for the Vd stacking, for which the 060 reflection is not discrete).

Cation exchange and layer shifts

The above results show that, in vermiculites, the existence of an ordered or semi-ordered stacking of the layers is not only an "inherited" character, i.e., transmitted from the original mica structure. The nature of the stacking is related also to the internal reactivity of the surface of the mineral. The fact is that cation exchange reactions in vermiculite do not consist simply of an exchange of the interlayer cation, together with a variation of the basal spacing, but are accompanied by important changes in the relative positions of the layers.

Experiment shows that some structural transformations are easier than others. For example, the change from Ia-2, (V3), to Ia-4 + Ia-6 (random), (V1), seems to occur very easily (Na-vermiculite treated with a Mg salt). On the other hand, the inverse transformation seems less easy. There are probably differences of energy between the Mg (Ia-4 + Ia-6) structure and the Na or Ca (Ia-2), so that the conditions for exchange are not equivalent in the two directions Mg -- Na. The phenomenon of cationic selectivity shown by vermiculites has its origin no doubt in this difference of structural stability. Furthermore it is known that the energies involved in the phenomena of ionic diffusion in vermiculites are different for different inorganic cations (Walker, 1959; 1963). In addition, the kinetics of

Table 3. Stacking modes as function of cation, RH, basal spacing, number of water molecules per cation, and structural type.

Stacking mode	Cation	Stability interval for powders RH (%)	nH$_2$O/cation 1st coord. sphere	d(001) (A)	Ord-Dis
V1 (Ia4-Ia6)	Mg	7-100	6	14.30	S-O
V3 (Ia2)	Ca	45-100	6 et 8°	14.92	O
	Na	50-100	5.8^	14.83	O
V5	Ca	20-45	-	14.70	S-O
V7	Mg	2-7	<6	13.8	S-O
Va	Sr	0-2	3.3	11.85	S-O
	Na	43-47	-	12.21	S-O
Vb	Li	2-95	2.4'	12.20	S-O
Vc	Na	2-43	2$^{\infty}$	11.85	S-O
Vd	Ba	2-55	5.6`	12.20	S-O
Ve	Sr	2-18	4`	12.15	S-O
	Mg	0-2	3*	11.6	S-O
	Ca	0-13	4`	11.74	S-O
			3.5'	11.90	S-O

^   Slade et al. (1985)
'   Le Renard and Mamy (1971)   S-O: semi-ordered
x   de la Calle et al. (1977)   O: ordered
xx de la Calle et al. (1984)
*   Walker and Cole (1957)

Table 4. Stacking modes: parameters of the planar lattice, b value, and translations between two adjacent layers.

Planar stacking mode	d(001)	Lattice XOZ	Tx	Ty	b (Å)
V1 (Mg)	14.30	a = 5.348 Å c = 14.41 Å ß = 97°	-a/3	±b/3	9.260
V3 (Ca)	14.92	a = 5.351 Å c = 15.03 Å ß = 96.83°	-a/3	0	9.262
(Na)	14.83	a = 5.355 Å c = 14.93 Å ß = 96.56°	-a/3	0	9.280
V5 (Ca)	14.66	a = 5.346 c = 14.73 ß = 93.53°	-a/6	±b/6,+b/2	9.256
V7 (Mg)	13.8	a = 5.333 Å c = 13.8 Å ß = 90°	0	?	9.258
Va (Sr)	11.85	a = 5.337 Å c = 11.98 Å ß = 98.57°	-a/3	?	9.242
Vb (Li)	12.20	a = 5.345 Å c = 12.23 Å ß = 94.19°	-a/6	?	9.259
Vc (Na)	11.85	a = 5.334 Å c = 11.85 Å ß = 90°	0	±0.307,0	9.233
Vd (Ba)	12.2	a = 5.330 Å c = 12.22 Å ß = 93.67°	-0.78	?	9.227
Ve (Sr)	12.15	a = 5.336 Å c = 12.15 Å ß = 90°	0	?	9.245

486

exchange are probably not the same as the kinetics of layer shift. Exchange is a relatively rapid phenomenon, determined by the relative concentrations of the ions and their diffusion velocity, while the layer shifts are a structural rearrangement that may be impeded by all sorts of barriers dependent on the structural defects existing in the minerals.

Other factors must intervene to determine the ease of layer shifts, and thus the degree of structural perfection in the stacking of layers. Thus the lower the charge the more difficult the interlayer shifts are, and the difficulty of such shifts probably increases with the number of octahedral vacancies.

Factors that determine the crystal structure of a vermiculite

The attempt to demarcate those factors that influence the "choice" of a particular stacking mode has formed the subject of various works (de la Calle et al., 1975, 1979, 1980; Suquet and Pezerat, 1987). These factors can be divided into three major categories: those depending on the interlayer material, "hereditary" factors, and texture effects.

Factors derived from the genesis of the mineral ("inherited" factors and texture) cannot be interfered with and cannot be changed on passing from one type of stacking to another. On the contrary, factors that depend on the interlayer material can be altered at will. It is by varying these factors that one may bring out the influence of structural features derived from the genesis of the mineral.

Role of the interlayer material. The influence of the interlayer material on the disposition of the layers has already been mentioned several times in this chapter. Since this material forms a junction zone between the layers, it has a preponderant role in determining their mode of superposition.

The water content, which depends on the RH, determines the degree of swelling. The corresponding changes of state are accompanied by a structural transformation of the interlayer material, which carries with it a rearrangement of the relationship between layers.

The nature of the interlayer cations plays a major role in the "choice" of the stacking sequences. The degree of hydration, and the whole architecture of the interlayer material, and hence its manner of joining up with the neighboring layers, are dependent on the polarizing power of the cation (radius and charge).

Our results enable the following conclusions to be drawn:

(1)   An ordered structure occurs when the architecture of the interlayer material is such as to bring about the face to face positioning of the surface ditrigonal cavities of the two adjacent layers, so as to bring about coherent bonding with minimum energy. In this case, the position of the one layer with respect to the other is unequivocal. In general the large anhydrous cations, K and Ba, are the organizing elements of the structure. Even with low charge density they condition the layer stacking with the ditrigonal cavities face to face. So the radius of the interlayer cation, trapped between the two cavities, directly influences the a and b parameters of the planar lattice (Suquet et al., 1981).

(2)   A semi-ordered structure occurs when the cation-water polyhedron or the organic cation need another type of environment. There exist at least two possible (symmetric) relative positions of the layers, which occur with equal frequency. In this case the relative positions of next-neighbor layers are not unique.

(3)   In general, cations with high polarizing power—such as Mg—can form compact octahedral groups of water molecules with O-O distances sufficiently small to permit hydrogen bonding between the water molecules and between these molecules and the surface oxygens of the silicate layers (Fornes et al., 1980). Cations with lower polarizing power and larger radii such as, Na, Ca, Ba ions form less compact hydration groups that tend to lie opposite hexagonal holes of the sili-

cate layers. Consequently, they link successive layers together in somewhat the same way as the K ions in micas with hexagonal holes of adjacent layers facing each other. Thus cation exchanges like changes of hydration state, are often accompanied by interlayer shifts.

The number of the exchangeable cations (an inherited factor) also plays a role. The organizing influence of the interlayer materials depends, in part, on the number of "charge bearers." In general the lower the charge the greater the tendency to disorder.

Role of "inherited" factors

The faults in mica genesis, in particular the angular shifts from layer to layer, form an element of the "inherited" factors that reproduce themselves unmodified in the states that are hydrated or interlayered by organic cations.

The influence of the charge density on the hydration state of the layer silicates is known. Suquet and Pezerat (1987) showed, for a given interlayer cation and RH, that the state of hydration of a trioctahedral 2:1 phyllosilicate is higher as its layer charge is lower. The stacking of the layers will thus depend on the charge of the vermiculite. Also, when the number of interlayer cations diminishes as a function of the reduction of the layer charge one may note an increasing difficulty in producing the layer shift; cation exchange does not necessarily lead to the expected layer shift.

The localization of the negative charge on the surface oxygens also plays an important role because it has an influence on the "anchorage points" of the interlayer material. From this point of view, two factors are important. First of all, the possible existence of a charge deficit in the octahedral sheet (either by Mg-Li substitutions or by the existence of vacancies) will tend to diffuse the negative charge over all the surface oxygens, not only those of the AlO_4 tetrahedra. This factor will probably lead to less ordered stacking, which will be more difficult to study.

The more or less ordered distribution of the Si-Al substitutions may also play a role in the stacking mode. We may recall in this respect the work of Thomson (1984), Herrero et al. (1985b, 1986), and Bailey (1986), who have demonstrated complete disorder over long distances in the distribution of the Si-Al substitutions. It may thus be postulated, as has been suggested already (cf. paragraph, Semi-ordered structure), that the absence of any order in the Si-Al substitutions can be directly linked to the frequency of appearance of semi-ordered structures in vermiculite.

Role of texture. Insofar as texture is involved in the mode of superposition of the layers, it is certain that dry grinding has a tendency to create disorder that is difficult to reverse. In fact, a severe dry grinding breaks down the crystalline structure to yield an amorphous phase. Further, it has already been observed that there is a shift of the range of stability of different structural stackings depending on whether one uses a powder or a crystal. The crystal size, or more generally, the texture of layer silicates seems thus to play a role in the swelling behavior. One significant example of differences of swelling behavior as a function of crystal size has been published by de la Calle et al. (1979) in a study of Li vermiculites. This was done on single crystals and powders of different origins: Santa Olalla, Benahavis, and Prayssac, also on vermiculites prepared in the laboratory from phlogopites. The behavior of the powder samples of Li-vermiculite is as expected and varies as a function of the charge of the sample. On the contrary, the behavior of the single crystals of dimensions greater than 1 mm is very strange. The swelling differences observed with single crystals of different origin cannot be correlated with the difference of charge. For instance, the charge of the Santa Olalla and Llano vermiculite is practically the same and relatively large. Yet these sample do not behave in the same way. An examination in the SEM of surfaces of the crystals shows a marked difference of texture between the two vermiculites. The wrinkled and porous appearance of the surface and sections of Llano vermiculite leads us to suppose that this sample has an internal porosity that considerably increases the interlayer surface to such an extent that the interface/volume ratio may be comparable with that in powders.

The hypothesis of a direct influence of crystallite size on the hydration of the inter-layer space may be justified by continuous exchange with the water molecules in the ambient air. The texture factor can act by modifying the threshold of stability of the interlayer material with respect to the surrounding RH. Further, the totality of the defects that interrupt the continuity of the crystal should probably be considered in evaluating the diameter of coherence of the crystallites parallel to the layers. Among these defects we probably must count the boundaries between domains of growth interrupted by helicoidal disloca-tions (Baronnet, 1976). It is possible that the density of helicoidal dislocations has a direct influence on the state of swelling or the mode of layer stacking for a given "n-layer hydrate."

On the other hand, the macroscopic swelling behavior of the butylammonium com-plexes with respect to particle size (Garret and Walker, 1962) suggests that the very exten-sive planar surfaces obtainable only with macroscopic crystals are necessary for swelling to occur in this system. A gradual reduction in swelling ability develops in particles of about 100 μm maximum diameter and becomes more marked with further decrease in size.

SAXD and TEM studies on the powder samples of Mg saturated Santa Olalla ver-miculite, fully water saturated, show low values of interparticular porosity. The particles are very closely joined, and show no inter-particle lenticular pores. The inter-aggregate porosity, on the contrary, accounts for more than 50% of the volume of the hydrated sample (Touret et al., 1988).

In conclusion, these considerations regarding texture recall the historic investiga-tions of Méring et al. (1950, 1956) and Mathieu-Sicaud et al. (1951) on the importance of the association of particles in smectites. They show the importance that is attached to effects linked to the particle size and to external surface (Pedro, 1976; Tessier and Pedro, 1976; Pons, 1980). According to these authors, the mode of association of the particles conditions the external surface, the deformability, and finally the degree of hydration of the clay.

Jelly-like structure of vermiculite: Macroscopic swelling

The study of macroscopic swelling in layer silicates, the conditions of formation, and properties of gels (Pons, 1980; Pons et al., 1980), has been developed principally on montmorillonite and vermiculite. A comparative study of swelling in the two types of silicates shows the roles of particle size and charge density. Thus Norrish and Rausell-Colom (1963) have compared the swelling of a Na-montmorillonite and a Li-vermiculite (macroscopic swelling of Li-vermiculite had been noted previously by Walker and Milne, 1950). Qualitatively the minerals behave similarly when swollen in salt solutions. Quanti-tatively, however, their swelling is very different. (1) While the swelling of vermiculite appears to be reversible both with respect to electrolyte concentration and applied pressure, the swelling of montmorillonite shows a marked hysteresis. (2) In montmorillonite, swelling depends very strongly on pH or chemical treatment. The results of Norrish and Rausell-Colom (1963) confirm that double layer repulsive forces cause the swelling of Na-montmorillonite and Li-vermiculite and that the magnitude of Van der Waals' attraction at the observed interlayer distances is insufficient to balance this repulsion. Recently, Pons and Rausell-Colom have studied Li-vermiculite using SAXD and confirm the previous conclusions about the properties of the double layer (pers. comm.).

In montmorillonite edge to face bonds provide a resistance that limits the swelling of the clays. In vermiculite crystals, because the layers are considerably larger, the force arising from edge to face bonds appears to be small or absent. Swelling is therefore greater than in montmorillonite and is reversible.

Swelling of some vermiculite-organic complexes in water has been studied by Garret and Walker (1962) and Rausell-Colom and Salvador (1971b). Macroscopic vermi-culite crystal complexes with butylammonium ions absorb large quantities of water and swell in the direction perpendicular to the plane of the silicate layers. Small-angle X-ray examination confirmed that interlayer separations of several hundred A units are involved and indicated that individual interlayer distances do not deviate markedly from the mean value (Rausell-Colom, 1964).

Garret and Walker (1961), Rausell-Colom and Salvador (1971a), Rausell-Colom and Fornes (1974), and Saez-Auñon et al. (1983) observed that macroscopic vermiculite crystals complexed with various amino-acids (ornithine, lysine, etc.) also swell anisotropically to many times their original size. In the jelly-like swollen crystal, the silicate layers remain effectively parallel to one another and, under optimum swelling conditions, may be some hundreds of Angstroms apart.

Structures belonging to vermiculites with 2M$_1$ parentage

Hitherto we have considered only those types of disorder that can be produced by displacements of layers in those types of minerals where the stacking does not introduce any angular displacement of the layers (phlogopites 1M and their parentage). However, in mica genesis, the growth of the crystals quite frequently involves angular displacements between next neighbors; regular displacements, +120°, -120°, +120°... in the 2M$_1$ micas; random displacements of ±120° in several minerals, in particular the biotites.

In those vermiculites that arise from the 2M$_1$ phlogopites or biotites, one finds the same types of layer displacements as in those arising from the 1M phlogopites, but in addition, the stacking includes angular displacements of ±120°, as in the original mica. (The term rotation is often used, but it is preferable to use the term "angular displacement" in order to distinguish between a fictive displacement and a genuine rotary movement). We have never observed any transition between one parentage and the other (de la Calle et al., 1978).

In some rare natural Mg vermiculites, de la Calle et al. (1976) verified that the diffraction is modulated along the (0,2) rod following a function equivalent to that given by Mathieson and Walker (1954) for a Kenya Mg vermiculite. In this case, the XOZ projection is the same as that for natural vermiculites of parentage 1M. It is considered that the passage from one layer to another follows the sequence:

$$120° \pm b/3, \quad -120° \pm b/3, \quad +120° \pm b/3....$$

the + and - signs in front of b/3 being distributed randomly. Thus there is a periodic angular displacement of the layers of the original mica and superposed on that a lateral translation of constant value but with random direction. After treatment with a K salt and light heat treatment one obtains crystals that diffract similarly to the 0kl level of a 2M$_1$ phlogopite. This gives a direct proof of the parentage (de la Calle et al., 1976).

One can find three-dimensionally ordered structures corresponding to the theoretical p model of Mathieson and Walker (1954) after Na or Ca treatment of samples of Mg Kenya vermiculite (parentage 2M$_1$). The type of stacking conserves the regular alternation of +120° and -120° angular displacements and the pseudohexagonal surface cavities of adjacent layers are face to face. There are not studies on other phases of this parentage. It is certain that they can be deduced from the corresponding structures of parentage 1M by adding to the translations belonging to this type of structure the regular angular displacement of the 2M$_1$ type.

COMPARISON OF PROPERTIES OF VERMICULITES

AND HIGH CHARGE SMECTITES

A comparative study of the crystallochemical properties of vermiculites and high charge smectites, i.e., saponites, was recently carried out by Suquet and Pezerat (1987, 1988). According to these authors, there is no difference either in the structural stability of the layers or in the crystallochemical nature of the layers that would justify an abrupt division of trioctahedral minerals with tetrahedral substitutions at a layer charge density of 0.6 (Méring and Pedro,1969).

490

Table 5. Mean values of basal spacings [d(001), in Å] of water, glyceraol, and ethylene glycol solvation complexes of trioctahedral 2:1 phyllosilicates.

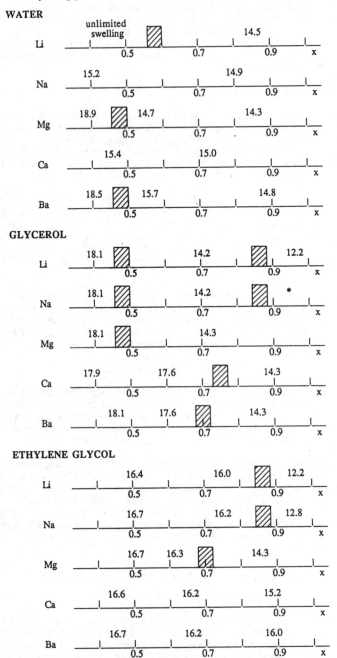

* indicates irregular layer sequence;   x = charge per formula unit;
bar indicates discontinuity in swelling behavior.

491

Table 6. Stacking modes as a function of charge density.

	x		10	20	30	40	50	60	70	80	90	100	water
	< 0.5	1M	DISORDER				*	V_3					$n > 3$
Li	0.5 << 0.7	1M	DISORDER					*	V_3				$n > 3$
	< 0.7	1M	V_b								V_3		V_3
	< 0.5	V_o	1M	V_a			*	V_3					V_3
Na	0.5 << 0.7	V_o	V_c			V_a		*	V_3				V_3
	< 0.7	V_o	V_c				V_a	*	V_3				V_a
	< 0.5		1M			*	V_a						V_a
K	0.5 << 0.7		1M										*
	< 0.7		1M										1M
	< 0.6	V_e	V_7	V_1				V_3					*
Mg	0.5 << 0.7	V_e	V_7	V_1									V_1
	< 0.7	V_e	V_7	V_1									V_1
	< 0.5	V_a	V_e	*	V_5		V_3						*
Ca	0.5 << 0.7	V_a	V_e	*	V_5		V_3						V_3
	< 0.7		V_e	*	V_5		V_3						V_3
	< 0.5	1M		V_d			V_3						*
Ba	0.5 << 0.7	1M		V_d			V_5						V_3
	< 0.7	1M		V_d				*	V_5				V_3

V_1, V_3, V_5, V_7 : two layer hydrates

V_a, V_b, V_c, V_d, V_e : one layer hydrate

1M, V_o : zero layer "hydrate"

* interestratified hydrates

1M, V_3 ordered structures

492

The swelling properties of saponites and vermiculites (Table 5) showed no single discontinuity, but rather several discontinuities at layer charges per formula unit of about 0.5, 0.6, 0.7, and 0.8. The discontinuity was dependent on the nature of the interlayer cation (Li, Na, Mg, Ca, or Ba) and the solvation liquid (water, ethylene glycol, or glycerol), but not on the type of mineral (saponite vs. vermiculite). These values represent limits for the stability domains of some interlayer structures as functions of the charge density, the specific nature of the balancing cation, and the nature of the solvation liquid, but not as a function of a crystallochemical difference of the layers of saponites and vermiculites.

Saponites may exhibit all layer stacking types previously described for vermiculites. Table 6 shows the stacking modes as a function of the charge density for saponites and vermiculites. The effect of the layer charge density on the transitions between the different states of hydration appears at the low and the high values of RH. Thus, the lower the charge density is, the easier the phyllosilicates dehydrate at low values of RH. The lower the number of charge bearers, i.e., of interlayer cations, is the more fragile the interlayer structure is (whatever its nature) so it can easily hydrate or dehydrate.

Inasmuch as saponite and low-charge vermiculites commonly exhibit the same layer stacking sequences for a given interlayer cation and a given RH, these phyllosilicates should not present any crystallochemical difference (e.g., difference in order-disorder of Si-Al substitutions) and should belong to a single family.

ACKNOWLEDGMENTS

The authors are indebted to Prof. S.W. Bailey and Prof. D.M.C. MacEwan for critical reviews of the manuscript and to Dr. C.H. Pons, University of Orleans, for helpful discussions.

REFERENCES

NOTE: Clays and Clay Minerals = C&CM.

Alcover, J.F. and Gatineau, L. (1980a) Structure de l'espace interlamellaire de la vermiculite Mg bicouche. Clay Minerals 15, 25-35.
---- and ---- (1980b) Structure de l'espace interlamellaire des vermiculites Ba monocouches. Clay Minerals 15, 193-203.
---- and ---- (1980c) Facteurs determinant la structure de la couche interlamellaire des vermiculites saturees par des cations divalents. Clay Minerals 15, 239-248.
----, ---- and Méring, J. (1973) Exchangeable cation distribution in nickel and magnesium vermiculite. C&CM 21, 131-136.
Allison, F.E., Roller, E.M. and Doetsch, J.M. (1953) Ammonium fixation and availability in vermiculite. Soil Science 75, 173-185.
Andre, L. (1972) Contribution a l'etude des mecanismes d'echange de cation dans les vermiculites trioctaedriques. Ph.D. thesis, Univ. Paul Sabatier, Toulouse, France.
Bailey, S.W. (1984) Review of cation ordering in micas. C&CM 35, 85-92.
---- (1986) Re-evaluation of ordering and local charge balance in Ia-chlorite. Can. Mineral. 24, 649-654.
---- and Brown, B.E. (1962) Chlorite polytypism: I Regular and semi-random one layer structures. Am. Mineral. 47, 819-850.
Baronnet, A. (1976) Polytypisme et polymorphisme dans les micas. Contribution a l'etude du role de la croissance cristalline. Ph.D. thesis, Univ. Aix-Marseille III, France.
Barshad, I. (1948) Vermiculite and its relation to biotite as revealed by base exchange reactions. X ray analyses differential thermal curves, and water content. Am. Mineral. 33, 655-678.
---- (1950) The effect of interlayer cations on the expansion of the mica type of crystal lattice. Am. Mineral. 35, 225-238.
---- (1954a) Cation exchange in micaceous minerals. I Replaceability of the interlayer cations of vermiculites with ammonium and potassium ions. Soil Science 77, 463-472
---- (1954b) Cation exchange in micaceous minerals.II Replaceability of ammonium and potassium from vermiculite, biotite and montmorillonite. Soil Science 78, 57-76.
Basset, W.A. (1963) The geology of vermiculite occurrences. C&CM 10, 61-69.

Ben Rhaiem, H., Tessier, D. and Pons C.H. (1986) Comportement hydrique et evolution structurale et texturale des montmorillonites au cours d'un cycle de dessication-humectation. I Cas des montmorillonites calciques. Clay Minerals 21, 9-29.

Bradley, W.F., Weiss, E.J. and Rowland, R.A. (1963) A glycol sodium vermiculite complex. C&CM 10, 117-122.

Caillere, S., Henin, S. and Rautureau, M. (1982) Mineralogie des Argiles. I. Structure et proprietes physico-chimiques. Masson, Paris.

Calle, C. de la. (1977) Structure des vermiculites. Facteurs conditionnant les mouvements des feuillets. Ph.D. thesis, Univ. Paris VI, France.

----, Glaeser, R. and Pezerat H. (1979) Effect of texture on vermiculite structure: lithium minerals. In Proc. Int'l Clay Conf., Oxford, 1978, M.M. Mortland and V.C. Farmer, ed., Elsevier, Amsterdam, 37-44.

----, Pezerat, H. and Gasperin, M. (1977) Problemes d'ordre - desordre dans les vermiculites. Structure du mineral calcique hydrate a deux couches. Journal de Physique, Colloque C.7 Suplement au nx12, 38, 128-133.

----, Suquet, H. and Pezerat, H. (1985) Vermiculites hydratees a une couche. Clay Minerals 20, 221-230.

----, ---- and ---- (1975) Glissement des feuillets accompagnant certaines echanges cationiques dans les monocristaux de vermiculites. Bull. Groupe franç. Argiles 27, 31-49.

----, ---- and Pons, C.H. (1988) Layer stacking order in a 14.30 A Mg vermiculite. C&CM, in press.

----, Dubernat, J., Suquet, H. and Pezerat, H. (1980) Ordre-desordre dans l'empilement des feuillets des phyllosilicates 2:1 hydrates. Bull. Minéral. 103, 419-428.

----, Suquet, H., Dubernat, J. and Pezerat, H.(1978) Mode d'empilement des feuillets dans les vermiculites hydratees a deux couches. Clay Minerals 13, 275-297.

----, Dubernat, J., Suquet, H., Pezerat, H., Gaultier, J. P. and Mamy, J. (1976) Crystal structure of two layer Mg-vermiculites and Na, Ca-vermiculites. In Proc. Int'l Clay Conf., Mexico, 1975, S.W. Bailey, ed., Applied Publishing Ltd., Illinois, 201-209.

----, Plan on, A., Pons, C.H., Dubernat, J., Suquet, H. and Pezerat, H. (1984) Mode d'empilement des feuillets dans la vermiculite sodique hydratee a une couche (phase a 11.85 A). Clay Minerals 19, 563-578.

Clabaugh, S.E. and Barnes, V.E. (1959) Vermiculite in central Texas. Texas Univ. Bur. Econ. Geol. Rept. Invest. 40.

Drits, V.A. (1985) Some aspects of the study of the real structure of clay minerals. In Proc. European Clay Groups, Prague, 1983, J. Konta, ed., Charles University, Prague, 33-42.

Fornes, V., Calle, C. de la , Suquet, H. and Pezerat, H. (1980) Etude de la couche interfoliaire des hydrates a deux couches des vermiculites calciques et magnesiennes. Clay Minerals 15, 399-411.

Foster, M.D. (1963) Interpretation of the composition of vermiculites and hydrobiotites. C&CM 10, 70-89.

Garret, W.G. and Walker, G.F. (1961) Complexes of vermiculite with amino-acids. Nature 191, 1389-1390.

---- and ----. (1962) Swelling of some vermiculites organic complexes in water. C&CM 9, 557-567.

Gatineau, L. and Alcover, J.F. (1976) Relations ordre-desordre dans les phyllosilicates. In Proc. Int'l Clay Conf., Mexico, 1975, S.W. Bailey, ed., Applied Publishing Ltd., Illinois, 131-137.

Gennaro, M. and Franco, E. (1975) Su alcuni casi di alterazione della biotite in vermiculite. Estratto dal Rediconto dell' Accad. Sci. Fisiche e Matematiche Soc. Nazionale di Sci., Lettere e Arti in Napoli ser. 4, 42, 1-18.

Gevers, T.W. (1948) Vermiculite at Loolekop, Palabora, North East Transvaal. Trans. Geol. Soc. S. Africa 133-173.

Glaeser, R., Mantin, I., and Méring, J. (1967) Observations sur la beidellite. Bull. Groupe franç. Argiles 19, 126-130.

Gonzalez Garcia, F. and Garcia Ramos, G. (1960) On the genesis and transformations of vermiculite. In Trans. 7th Int'l Congress Soil Science, Madison, 1960, ed., Int'l Soc. Soil Science 4, 482-491.

---- and ---- (1962) Procesos de genesis y degradacion de vermiculita: yacimiento de Santa Olalla. IV Transformacion de la vermiculita por erosion meteorica. Anales Edafafologia Agrobiol. 21, 183-204.

Gruner, J.W. (1939) Ammonium mica synthetised from vermiculite. Am. Mineral. 24, 428-435.

Guinier, A. (1964) Theorie et technique de la radiocristallographie. Ch. 13, p 490-663. Dunod, Paris.

Hendricks, S.B. and Teller, E. (1942). X-ray interference in partially ordered layer lattices. J. Chem. Phys. 10, 147-67.

Herrero, C.P., Sanz, J. and Serratosa, J.M. (1985a) Si, Al distribution in micas: analysis by high-resolution Si NMR spectroscopy. J. Phys. C, Solid State Phys. 18, 13-22.

----, ---- and ---- (1985b) Tetrahedral cation ordering in layer silicates by 29Si NMR spectroscopy. Solid State Commun. 53, 151-154.

----, ---- and ---- (1986) The electrostatic energy of micas as a fonction of Si, Al tetrahedral ordering. J. Phys. C: Solid State Phys. 19, 4169-4181.

Hoda, S.N. and Hood, W.C. (1972) Laboratory alteration of trioctahedral micas. C&CM 20, 343-358.

Hofmann, U. (1956) Intracristalline swelling, cation exchange of minerals of the montmorillonite group and of kaolinite. C&CM 456, 273.

494

Hougardy, J., Stone, W.E.E. and Fripiat, J.J. (1976) NMR study of absorbed water. 1. Molecular orientation and protonic motions in the two layer hydrate of a Na-vermiculite. J. Chem. Phys. 64, 3840-3851.

Iglesias, J.E. and Steinfink, H. (1974) A structural investigation of a vermiculite-piperidine complex. C&CM 22, 91-95.

Justo, A., Perez Rodriguez, J.L. and Maqueda, C. (1983) Estudio mineralogico de una vermiculita de Ojen (Malaga). Bol. Soc. Española Mineral. 7, 59-67.

---- (1984) Estudio fisico-quimico y mineralogico de vermiculitas de Andalucia y Badajoz. Ph.D. thesis, Univ. Sevilla, Spain.

----, Maqueda, C. and Perez Rodriguez, J. L. (1986) Estudio quimico de vermiculitas de Andalucia y Badajoz. Bol.Soc. Española Mineral. 9, 123-129.

----, Maqueda, C., Perez Rodriguez, J.L. and Lagaly G. (1987) An unusually expandable low-charge vermiculite. Clay Minerals 22, 319-328.

Kakinoki, J. and Komura, Y. (1952) Intensity of X-Ray diffraction by one-dimensionaly disordered crystal. J. Phys. Soc. Japan 7, 30-35.

Kodama, H. (1975) Diffuse scattering by X-ray and electrons in micas and mica like minerals. In Contributions to Clay Minerals, in honor of Prof. Toshio Sudo, 7-13.

---- (1977) An electron-diffraction study of a microcrystalline muscovite and its vermiculitized products. Mineral. Mag. 41, 461-468.

Komarneni, S. and Roy, R. (1981) Hydrothermal transformations in candite overpack materials and their effects on cesium and strontium sorption. Nuclear Techn. 54, 118-122.

Köster, H.M. (1977) Die Berechnung kristallchemischer Strukturformeln von 2:1-Schichtsilikaten unter Berucksichtigung der gemessenen Zwischenschichtladungen und Kationenumtauschkapazitaten, sowie die Darstellung der Ladungsverteilung in der Struktur mittels Dreieckskoordinaten. Clay Minerals 12, 45-54.

---- (1982) The crystal structure of 2:1 layer silicates. In Proc. Int'l Clay Conf., Bologna, Pavia, 1981, H. Van Olphen and F. Veniale, eds., Elsevier, Amsterdam, 41-72.

Lagaly, G. (1987) Clay-organic interactions: problems and recent results. In Proc. Int'l Clay Conf., Denver, 1985, L.G. Schultz, H. van Olphen and F.A. Mumpton, eds., Clay Minerals Soc., Bloomington, Indiana, 343-351.

Le Renard, J. and Mamy, J. (1971) Etude de la structure des phases hydratees des phlogopites alterees par des projections de Fourier monodimensionnelles. Bull. Groupe franç. Argiles 23, 119-127.

Levinson, A.A. and Heinrich, E.W. (1954) Studies in the mica group: single crystal data on phlogopites, biotites and manganophyllites. Am. Mineral. 39, 937-945.

Lipsicas, M., Raythatha, R.H., Pinnavaia, T.J., Johnson, I.D., Giese, R.F., Constanzo, P.M. and Robert, J.L. (1984) (1984) Silicon and aluminium site distribution in 2:1 layered silicate clays. Nature 5969, 604-607.

Loewenstein, W. (1954) The distribution of aluminium in the tetrahedra of silicates and aluminates. Am. Mineral. 39, 92-96.

Lopez Gonzalez, J.D. and Barrales Rienda, J.M. (1972) Caracterizacion y propiedades de una vermiculita de Benahavis (Malaga). Anales de Quimica 68, 247-262.

Luque, F.J., Rodas, M. and Doval, M. (1985) Mineralogia y genesis de los yacimientos de vermiculite de Ojen. Bol. Soc. Española Mineral. 8, 229-238.

Maire, J. and Méring, J. (1970) Graphitization of soft carbons. Chem. Phys. Carbon 6, 125-139.

Makumbi, L. and Herbillon, A.J. (1972) Vermiculitisation experimentale d'une chlorite. Bull. Groupe franç. Argiles 24, 153-164.

Mamy, J. and Gaultier, J.P. (1973) Application des mesures de conductivite eletrique a l'etude de l'eau fixee par des phyllites gonflantes. Bull. Groupe franç. Argiles 25, 43-52.

Mathieson, A.M. and Walker, G.F. (1954). Crystal structure of magnesium-vermiculite. Am. Mineral. 39, 231-255.

Mathieu-Sicaud, A., Méring, J. and Perrin-Bonnet, J. (1951) Etude au microscope eletronique de la montmorillonite et de l'hectorite saturee par differents cations. Bull. Soc. franç. Minéral. Cristallogr. 74, 439-460.

Méring, J. (1949) Interference des rayons X dans les systemes a interestratification desordonnee. Acta Crystallogr. 2, 371-377.

---- (1975) Smectites. In Encyclopedia of Soil Science, Springer-Verlag, New York, 4, 97-119.

---- and Pedro, G. (1969). Discussion a propos des criteres de classification des phyllosilicates 2:1. Bull. Groupe franç. Argiles 21, 1-30.

----, Mathieu-Sicaud, A. and Perrin-Bonnet, J. (1950) Observations au microscope eletronique de la montmorillonite saturee par differents cations. Trans. 4eme. Congres Int'l Scienes du Sol, V, III, 29 Amsterdan.

----, Oberlin, A. and Villiere, J. (1956) Etude par eletrodeposition de la morphologie des montmorillonites. Effet des cations Ca. Bull. Soc. franç. Mineral. Cristallogr. 74, 515-526.

Monchoux, P. (1961) Bulletin de la Societe d'Histoire Naturelle, 96, 300-360.

Morel, S.W. (1955) Biotitite in the basement complex of Southern Nyasaland. Geol. Mag. 92, 3, 241-255.

495

Norrish, K. (1973) Factors in the weathering of mica to vermiculite. In Proc. Int'l Clay Conf., Madrid, 1972, J.M. Serratosa, ed., Division de Ciencias, CSIC, Madrid, 417-432.
---- and Rausell-Colom, J.A. (1963) Low angle X-ray diffraction studies of the swelling of montmorillonite and vermiculite. C&CM 10, 123-149.
Pedro, G. (1976) Sols argileux et argiles. Elements en vue d'une introduction a leur etude. Sience du Sol Bull. de l'A.F.E.S. 2, 1-85.
Plançon, A. (1981) Diffration by layer structures containing differents kinds of layers and stacking faults. J. Applied Crystallogr. 14, 300-304.
Poinsignon, C., Estrade, H., Conard, J. and Dianoux, A.J. (1987) Water dynamics in the clay-water systems: a quasielastic neutron scattering study. In Proc. Int'l Clay Conf., Denver, 1985, L.G. Schultz, H. Van Olphen and F.A. Mumpton, ed., Clay Minerals Soc., Bloomington, Indiana, 284-291.
Pons, C.H. (1980) Mise en evidence des relations entre la texture et la structure dans les systemes eau-smectite aux petits angles du rayonement X synchrotron. Ph.D. thesis, Univ. Orleans, France.
----, Tchoubar, D. and Tchoubar, C. (1980) Organisation des molecules d'eau a la surface des feuillets dans un gel de montmorillonite. Bull. Minéral. 103, 452-456.
Raupach, M., Slade, P. G., Janik, L. and Radoslovich, E.W. (1975) A polarized infrared study and X-Ray study of lysine-vermiculite. C&CM 23, 181-186.
Rausell-Colom, J.A. (1964) Small angle X-ray diffraction study of the swelling of butylammonium vermiculite. Trans. Faraday Soc. 60, 190-201.
---- and Fornes, V. (1974) Monodimensional Fourier analysis of some vermiculites L-ornithine complexes. Am. Mineral. 59, 790-798.
---- and Salvador, P. (1971a) Complexes vermiculites-aminoacids. Clay Minerals 9, 139-149.
---- and ---- (1971b) Gelification de vermiculites dans de solutions d'acide aminobutyrique. Clay Minerals 9, 193-208.
---- and Serratosa, J.M. (1985) Perturbation of OH infrared frequencies by interlayer cations in homoionic vermiculites. Structural implications. Mineral. Petrogr. Acta 29, 409-423.
----, Fernandez M., Serratosa, J.M. Alcover, J.F. and Gatineau, L. (1980) Organisation de l'espace interlamellaire dans les vermiculites monocouches et anhydres. Clay Minerals 15, 37-57.
Robert, M. (1968) Etude experimentale sur les processus de vermiculitisation des micas trioctaedriques. Bilan de l'evolution et conditions de genese des vermiculites. Bull. Groupe franç. Argiles 20, 153-171.
---- and Barsahd, I.(1972) Transformation experimentale des micas en vermiculites ou smectites. Proprietes des smectites de transformation. Bull. Groupe franç. Argiles 24 137-151.
---- and Pedro, G. (1966) Transformation d'une phlogopite en vermiculite par extraction du potasium. Bull. Groupe franç. Argiles 17, 3-17.
----, Ranger, J., Malla, P.B., Tessier, D. and Lopez Rodriguez, J.L. (1987) Variation in microrganization and properties of Santa Olalla vermiculite with decreasing size. In Summaries-Proc. European Clay Groups, Sevilla, 1987, E. Galan, J.L. Lopez Rodriguez, and J. Cornejo, eds., Sociedad Española de Arcillas, 456-458.
Ross, G. and Kodama, H. (1974) Experimental transformation of a chlorite into vermiculite. C&CM 22, 205-211.
Rousseaux, F., Moret, R., Guerard, D., Lagrange, P. and Lelaurain, M. (1984) X-ray photographic study of the phase transitions in KC24 single crystals. J. Phys. Lett. 45, 1111-1118.
Roy, R. and Romo, L.A. (1957) Weathering studies. New data in vermiculite. J. Geol. 65, 603-610.
Saez-Auñon, J., Pons, C.H., Iglesias, J.E. and Rausell-Colom, J.A. (1983) Etude du gonflement des vermiculites-ornithines en solution saline par analyse de la diffusion des rayons X aux petits angles. Methode d'interpretation et recherche des parametres d'ordre. J. Applied Crystallogr. 16, 439-448.
Sanz, J. and Serratosa, J.M. (1984) Distribution of tetrahedrally and octahedrally coordinated Al in phyllosilicates by NMR spectroscopy. Clay Minerals 19, 113-115.
Saxena, S.K. (1973) Thermodynamics of rock-forming crystallines solutions. Springer, Heidelberg.
Serratosa, J.M., Rausell-Colom, J.A. and Sanz, J. (1984) Charge density and its arrangements and reactivity of adsorbed species. J. Molecular Catalysis 27, 225-234.
---- and Sanz, J. (1985) Dehydroxilation of micas and vermiculites. The effect of octahedral composition and interlayer saturating cations. Mineral. Petrogr. Acta 29-A, 399-408.
Shirozu, H. and Bailey, S.W. (1966) Crystal structure of a two-layer Mg-vermiculite. Am. Mineral. 51, 1124-1143.
Slade, P.G. and Raupach, M. (1982) Structural model for benzidine-vermiculite. C&CM 30, 297-305.
----, ---- and Emerson, W.W. (1978) The ordering of cetylpiridinium bromide on vermiculite. C&CM 30, 297-305.
---- and Stone, P.A. (1983) Structure of a vermiculite aniline intercalate. C&CM 31, 200-206.
---- and ---- (1984) Three-dimensional order and the structure of aniline-vermiculite. C&CM 32, 223-226.

----, ---- and Radoslovich, E.W. (1985) Interlayer structures of the two layer hydrates of Na- and Ca-vermiculites. Clays Clay Minerals 33, 51-61.
----, Telleria, M.I. and Radolovich, E.W. (1976) The structure of ornithine and 6-aminohexanoic acid-vermiculite. C&CM 24, 134-141.

496

----, Dean, C., Schultz, P.K. and Self, P.G. (1987) Crystal structure of a vermiculite-anilinium intercalate. C&CM 35, 177-188.

Suquet, H. and Pezerat, H. (1987) Parameters influencing layer stacking types in saponites and vermiculites. C&CM 35, 353-362.

---- and ---- (1988) Comments on the classification of trioctahedral 2:1 phyllosilicates. C&CM 36, 184-186.

----, Calle, C. de la and Pezerat, H. (1975) Swelling and structural organisation of saponite. C&CM 23, 1-9.

----, Malard, C. and Pezerat, H. (1987) Structure et proprietes des nontronites. Clay Minerals 22, 157-167.

----, Prost, R. and Pezerat, H. (1982) Etude par spectroscopie infrarouge et diffraction X des interactions eau-cation-feuillet dans les phases a 14.6A, 12.2A et 10.1A d'une saponite-Li de synthese. Clay Minerals 17, 225- 235.

----, Malard, C., Copin, E. and Pezerat, H. (1981) Variation du parametre b et de la distance basale d001 dans une serie de saponites a charge croissante. II. Etats "zero-couche". Clay Minerals 16, 181-193.

----, Malard, C., Quarton, M., Dubernat, J. and Pezerat, H. (1984) Etude du biopyribole forme par chauffage des vermiculites magnesiennes. Clay Minerals 19, 217-227.

Telleria, M.I., Slade, P.G. and Radolosvich, E.W. (1977) X-Ray study of the interlayer region of a barium-vermiculite. C&CM 25, 119-125.

Tessier, D. (1984) Etude experimentale de l'organisation des materiaux argileux. Ph.D. thesis, Univ. Paris VI, France.

---- and Pedro, G. (1976) Les modalites d'organisation des particules dans les materiaux argileux. Evolution des principaux argiles Ca au cours du phenomene de retrait. Science du Sol Bull. de l'AFES n°2, 85-100.

Thomson, J.G. (1984) 29Si and 27Al nuclear magnetic resonance spectroscopy of 2:1 clay minerals. Clay Minerals 19, 229-236.

Touret, O., Pons, C.H., Tessier, D. and Tardy, Y. (1988) Etude de la repartition de l'eau dans les argiles magnesiennes a saturation. Clay Minerals, in press.

Varley, E.R. (1948) Vermiculite deposits in Kenya. Imperial Inst. Bull. 46 348-352.

Velasco, F., Casquet, C., Ortega Huertas, M. and Rodriguez Gordillo, J. (1981) Indicio de vermiculita en el skarn magnesico (Aposkarn flogopitico) de la Garrenchosa (Santa Olalla, Huelva). Soc. Española Mineral. 2, 135-149

Walker, G.F. (1949) Water layers in vermiculites. Nature 163, 726- 727.

---- (1956) The mecanism of dehydration on Mg-vermiculite. C&CM 4, 101-115.

---- (1959) Diffusion of exchangeable cations in vermiculite. Nature 184, 1392-1394.

---- (1963) Ion exchange in clay minerals. Introductory speech. In Proc. Int'l Clay Conf., Stockholm, 2, 259-261.

---- and Cole, W.F. (1957) The vermiculite minerals. In R.C. Mackenzie ed., The differential thermal investigations of clays. Ch. VII., Mongr. 1, Mineral Society, London, 191-206.

---- and Milne, A. (1950) Hydration of vermiculites saturated with various cations. Trans. IVth Int'l Congres Soil 2, 62-67.

Weiss, A. (1963) Mica type layer silicates with alkylammonium ions. C&CM 10, 191-224.

Zvyagin, B.B. (1967) Electron diffraction analysis of clay minerals structures. Plenum Press, New York.

SMECTITES

INTRODUCTION

The term "smectite" is derived from the Greek word smectos, σμζγμα, meaning soap; and saponite was the first clay mineral that was described as a "smectite" by Cronstedt in 1788 (Ross and Hendricks, 1945). The intermediate state of matter between crystalline solids and amorphous liquids is defined as the mesomorphic state, and three types of mesomorphic phases, nematic, smectic, and cholestric, were distinguished by Friedel in 1922 (Gray, 1962). The "smectic" phase represents stacks of parallel layers consisting of rod-like molecules that are oriented parallel to each other with random azimuthal rotations. The term "turbostratic" (unordered layers) was suggested by Biscoe and Warren (1942) for solid "smectic" mesophases. "Smectic" solids consist of turbostratic groups comprising a number of layers with two-dimensional periodicity. These layers are "stacked together roughly parallel and equidistant but with a completely random azimuthal orientation, just like a deck of cards thrown down in a disorderly fashion upon a table." To mineralogists, smectites represent 2:1 type layer silicates with an expandable structure carrying a certain amount of excess negative layer charge. Some structural and textural similarities between smectic mesophases and smectite clay minerals are rather striking (in addition to their soap-like and viscous behavior) but this relationship has not yet been explored by clay mineralogists.

The first comprehensive review of smectites which laid a firm foundation for the development of modern smectite mineralogy was presented by Ross and Hendricks (1945). They also gave an in-depth historical perspective of the discovery of smectite species and clarified the smectite nomenclature from what had been a haze of confusing terms. For instance, saponite was described up to that time as smectis, fuller's earth, kessikel, steatite, piotin, lucuanite, thalite, etc. Smectites have been the subject of periodic reviews, the comprehensive ones in English were those by McEwan (1961), Grim (1968), Méring (1975), and Brindley (1980). These reviews provided a wealth of information on the structure, chemistry, morphology, thermal and hydration behavior of smectites. In addition, smectites in the soil environment were discussed by Borchardt (1977), and industrial uses of smectites were summarized by Odom (1984). This present review will, therefore, focus on recent (1980's) contributions to smectite mineralogy. First, the basic structural components of the smectites will be described.

BASIC STRUCTURAL ELEMENTS OF SMECTITES

The atomic structure of smectite is partially responsible for its unusual physical properties, and its various technological applications are commonly related to some structural features of smectite. For instance, the catalytic reactions of smectites are sometimes related to the immediate environment of certain interlayer cations. The rheological behavior of smectite can be controlled by the short-range interactions between interlamellar surfaces and water molecules. Basic components of smectite structure, namely the octahedral and tetrahedral sheets and interlayer configurations, must be understood first. They will be described in detail next.

Octahedral sheets

The octahedral sheet consists of two planes of spherical anions (O,OH) that are closely packed in AB or AC sequence. AB close packing generates a set of three octahedral sites (called set "I" by Bailey, 1984) and AC close packing of anions creates a set of alternate octahedral sites (set II in Bailey's notation). The octahedral sheet in 2:1 layer silicates has been traditionally considered to be the backbone of the structure and to strongly affect the other structural elements. This old concept may still have merit for understanding structural features of the smectites where the characteristics of the octahedral networks provide the major criteria for their identities. The classical view on the crystal growth mechanism of smectites (to be discussed later) is based on the idea that during the formation of smectites a metal hydroxyl precipitates first and serves as a template onto which silica monomers are epitaxially anchored. It is therefore rather important to understand the basic features of octahedral sheets. The main characteristics of these networks are related to the distribution (ordering) of metal ions and vacancies over the available sites and distortions in the geometry of the sheets.

Octahedral networks are called dioctahedral when only two out of three sites are occupied by mainly trivalent cations and trioctahedral when all three sites are populated by predominantly divalent cations. Smectites are accordingly divided into two subgroups as dioctahedral and trioctahedral smectites. Octahedral ordering involves the distribution of octahedral cations and vacancies over the the available sites. The octahedral site on the mirror plane in Figure 1 is designated M1 and this site is usually vacant in dioctahedral 2:1 layer silicates. The other two octahedral sites to the left and right of the mirror plane are designated M2. It is important to note that the hydroxyl configuration around the two types of octahedral sites (M1 and M2) is different. M1 sites have a trans-configuration (i.e., hydroxyls are across the site), whereas M2 sites display a cis-configuration (i.e., hydroxyls are

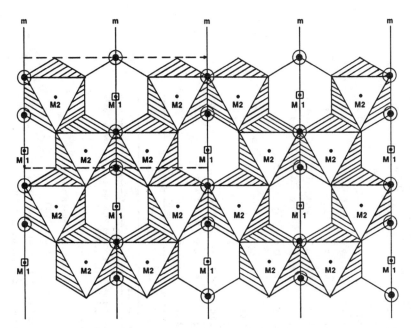

Figure 1. Distribution of _cis_ (M2) and _trans_ (M1) sites, and hydroxyl ions, ⬤ , in an ideal octahedral sheet. Mirror planes (m) and the unit-cell for the space group C2/m are also indicated.

adjacent on one side of the coordination polyhedron). Therefore, the M1 sites (or the octahedra around them) are often referred to as _trans_ sites (or _trans_ octahedra) and the M2 sites as _cis_ sites (or _cis_ octahedra).

Octahedral distortions have been discussed recently in detail by Toraya (1981), Bailey (1984), and Weiss et al. (1985). The octahedral distortions can be quantified by parameters such as the ratio of the unshared to shared edges e_u/e_s or by the octahedral (flattening) angle which is the angle between the normal to the octahedral sheet and a body diagonal of the octahedron. The latter can be easily calculated by the relationship: $\cos \psi = (t_o/2)/d_{M-O}$, where t_o is the thickness of the octahedron and d_{M-O} is the mean metal-anion distance. In order to indicate the magnitude and the range of octahedral distortions in smectites, a set of micas representing the compositional limits of common di- and trioctahedral smectites was selected. No brittle micas were included in the discussion because of apparent dissimilarities between them and smectites. Selected true micas and the related layer structures such as talc, brucite, pyrophyllite, and gibbsite are listed in Table 1. The octahedral distortions and other parameters of these minerals have been rather helpful in understanding the similar structural features of smectites since no crystal structure refinements have been carried out on smectites.

The octahedral distortions in the brucite structure are rather significant with an e_u/e_s ratio of 1.12 and an octahedral angle ψ of 59.5°, as compared with $e_u/e_s = 1.0$

Table 1. Characteristics of tri- and dioctahedral layer structures related to smectites

Mineral	Composition*	b_{obs} (Å)	$\bar{d}_{T-O}$ (Å)	$\bar{d}_{M-O}$ (Å)	Sheet Thickness (Å) Tetrahedral	Octahedral	Distortion Parameters Tetrahedral α_{obs}	α_{calc}	Octahedral ψ_{obs}	(e_u/e_s)	Misfit Index $\bar{d}_{M-O}/\bar{d}_{T-O}$
brucite	$Mg(OH)_2$	9.43	---	2.10	---	2.11	---	---	59.5*	1.12	---
talc	$Mg_3Si_4O_{10}(OH)_2$	9.17	1.621	2.07	2.18	2.17	3.6°	0°	58.4	1.09	1.28
phlogopite	#11	9.19	1.649	2.07	2.26	2.13	7.5	9.9	59.0	1.11	1.26
biotite	#32	9.25	1.656	2.13	2.25	2.19	5.3	9.1	59.1	---	1.29
annite	#12	9.32	1.659	2.11	2.25	2.21	1.5	6.7	58.4	1.09	1.27
phlogopite	#18	9.09	1.625	2.06	2.24	2.19	1.4	8.6	57.9	1.08	1.27
biotite	#TR	9.22	1.659	2.07	2.28	2.06	7.6	10.8	60.2	1.10	1.25
gibbsite	$Al(OH)_3$	8.68	---	1.91	---	2.04	---	---	57.5	1.14	---
pyrophyllite	$Al_2Si_4O_{10}(OH)_2$	8.97	1.618	1.91	2.15	2.08	10.2	11.5	57.0	1.18	1.18
paragonite	#20	8.90	1.653	1.91	2.24	2.08	16.2	17.9	57.0	1.15	1.16
phengite	#9	9.04	1.627	1.96	2.22	2.13	6	10.8	57.1	1.07	1.21
celadonite	#13	9.06	1.635	2.05	2.25	2.25	1.3	11.6	56.7	1.05	1.25
"undistorted" 2:1 layer silicate	---	---	---	---	---	---	---	0°	54.7°	1.00	1.33

References

Samples #9-32 are listed with the same numbers in Table 1 and 2 in Bailey (1984)

#TR: Takeda and Ross (1975)

pyrophyllite: Lee and Guggenheim (1981)

talc: Rayner and Brown (1973)

brucite: Zigan and Rothbauer (1967)

gibbsite: Saalfeld and Wedde (1974)

* Composition of minerals per $O_{10}(OH)_2$

#9 $(Al_{1.43}Fe^{3+}_{.05}Mg_{.50}Fe^{2+}_{.09}Ti_{.01})$ $(Si_{3.39}Al_{.61})$ $(K_{.87}Na_{.07}Ca_{.02}Ba_{.01})$

#11 Mg_3 $(Si_{2.95}Al_{1.05})$ $(K_{.77}Na_{.16}Ba_{.05})$

#12 $(Fe^{2+}_{2.22}Mg_{.12}Mn_{.05}Fe^{3+}_{.19}Al_{.09}Ti_{.22})$ $(Si_{2.81}Al_{1.19})$ $(K_{.88}Na_{.07}Ca_{.03})$

#13 $(Fe_{1.15}Fe^{3+}_{.36}Al_{.05}Mg_{.41}Ti_{.01})$ $(Si_{3.94}Al_{.06})$ $(K_{.83}Na_{.01}Ca_{.04})$

#18 $Mg_{2.56}$ Si_4 $K_{.88}$

#20 $(Al_{1.99}Fe^{3+}_{.03}Mg_{.01})$ $(Si_{2.94}Al_{1.06})$ $(Na_{.92}K_{.04}Ca_{.02})$

#32 $(Fe^{2+}_{1.39}Mg_{1.16}Al_{.12}Mn_{.01}Ti_{.32})$ $(Si_{2.79}Al_{1.21})$ $(K_{.99}Na_{.01})$

#TR $(Mg_{1.68}Fe^{3+}_{.71}Fe^{2+}_{.19}Mn_{.01}Ti_{.34}Al_{.19})$ $(Si_{2.95}Al_{1.14})$ $(K_{.78}Na_{.16}Ba_{.02})$

and ψ = 54.73° for an ideal octahedral network. These distortion parameters are slightly reduced in the talc structure where two silica tetrahedra sheets are attached to the brucite-like octahedral network. The mean octahedral metal-anion distance in brucite is 2.10 Å as compared to 2.07 Å in talc. The distortion parameters in the other trioctahedral micas (phlogopite, biotite, annite) are rather similar or a little smaller than in brucite. In fact, the thickness of the octahedral sheet shows more variation than the other parameters. Table 1 contains a rather small set of data to warrant any conclusions. Therefore, the ψ parameter and thickness of octahedral sheets is compared with the extensive data on the trioctahedral micas compiled by Bailey (1984). The ψ angle ranged from 57 to 61 degrees in 47 trioctahedral micas, angles $\geq$60° were mostly related to M1 octahedra whereas a weighted average of ψ angles for all the octahedral sites (two M2 and one M1) is given in Table 1. The variation in octahedral thickness of trioctahedral micas in Bailey's list were similar to those in Table 1 provided the lithium-rich micas are excluded. We may therefore state that the distortions in the octahedral networks of trioctahedral "true" micas are rather similar to those in brucite where no tetrahedral sheets or interlayer cations exist. Tetrahedral networks seem to adjust to their octahedral sheets primarily by rotations of tetrahedra (α) ranging from 1 to 8 degrees (Table 1). Additional adjustments seem to be accomplished by changes in the thickness of tetrahedral sheets.

The gibbsite structure, as determined by Saalfeld and Wedde (1974), shows rather severe distortions in the dioctahedral network of aluminum hydroxide. The octahedral angle (around the M2 sites) is 57.5° and the ratio of unshared to shared edges (in M2 octahedra) is about 1.14 with the unshared edges being stretched 14%. The mean Al-OH distance is 1.91 Å, and it is the same as the mean Al-anion distance in pyrophyllite and paragonite. The octahedral distortions in the key dioctahedral micas (paragonite, phengite, and celadonite) are rather similar to those in gibbsite (Table 1). The 22 dioctahedral micas in Bailey's compilation show an octahedral angle ψ ranging from 56 to 58 degrees for the M2 octahedra, as compared 57.5° in gibbsite. The only significant difference between the octahedral sheet of gibbsite and those of the other dioctahedral micas is related to the b-dimension which is about 0.2-0.3 Å shorter in gibbsite. This is caused by the shorter unshared edges around the vacancies (M1 sites) in gibbsite than in dioctahedral micas. Specifically, the mean unshared edge in the M1 vacant octahedron is 3.20 Å in gibbsite whereas it is 3.40 Å in paragonite (Lin and Bailey, 1984); the larger number of protons pulling towards the octahedral vacancies in gibbsite may play a role in this matter.

The above observations tentatively suggest that the octahedral distortions in 2:1 layer silicates may have been inherent and the role of tetrahedral sheets and interlayer cations may be rather limited. The congruency between tetrahedral and octahedral sheets seems to be mainly established by the adjustment of tetrahedral networks through tetrahedral rotations. In any case, Table 1 definitely indicates the realistic limits of octahedral distortions in smectites. Trioctahedral smectites are expected to have a ψ angle about $59\bar{+}1$ degrees and a range of e_u/e_s on the order of 1.10 ± 0.05. These estimates seem to be more reliable than the calculations assuming ideal models. Such models based on chemical composition and $\underline{b}$-dimensions often yield rather erroneous data. For instance, such calculations yield tetrahedral rotation angles (α) that are two to nine times larger than the observed ones in the selected 2:1 layer silicates in Table 1.

Smectites or most other 2:1 layer silicates seldom possess an ideal octahedral network with $e_u/e_s = 1.0$ and $\psi = 54.7^\circ$. A single octahedral coordination polyhedron may exist as an ideal octahedron but when they are polymerized into an ionically bonded sheet by sharing edges, the situation drastically changes. Due to strong repulsions between the octahedral cations the shared edges become shorter and the unshared ones longer. The distortions are, thus, generated as a direct result of the well-known "cation avoidance rule" with or without the presence of tetrahedral networks. The octahedral sheets are as much distorted in free hydroxides as in the silicate layers. The popular idea that octahedral sheets are distorted in order to acquire congruency with the tetrahedral networks does not seem to reflect reality. The observations from true micas indicate that the tetrahedral sheets undergo all possible adjustments (rotation, elongation, and tilting) to accomplish congruency with a smaller octahedral sheet, and they bend over (invert-modulate) to do the same with a larger octahedral sheet. The octahedral sheet seems to greatly predetermine as a template the distortional features of a 2:1 layer silicate.

Octahedral substitutions and related layer charges. Substitutions of divalent cations for trivalent ones in dioctahedral smectites create an excess negative charge whereas substitutions of a trivalent cation for a divalent one in trioctahedral smectite generates an excess positive charge in the octahedral network. These charges have far-reaching implications for the physical properties of smectites, such as swelling, hydration, and rheological behavior. We will examine the short-range aspects of such substitutions in a dioctahedral smectite with vacant trans sites and with octahedral compositions of ($Al_{1.5}Mg_{0.5}$). This composition is close to the upper limit of divalent cation substitutions in dioctahedral smectites. In addition to creating an excess layer charge, the substitution (R^{2+} for R^{3+}) can generate a ($2\underline{a}$, $\underline{b}$) superlattice, if short range ordering prevails as illustrated in Figure 2. The anions bonded to R^{2+} in each such octahedron carry an excess charge of $-1/6$ each. The charge density on each side of the octahedral sheet is $-1/2$ units per $\underline{a}$ x $\underline{b}$ area. The charged anion triads (O, OH, O)

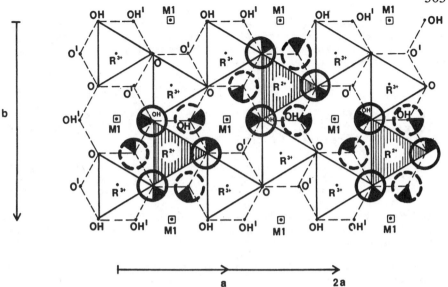

Figure 2. Distribution of octahedra with R^{2+} substitutions (hatched) and the anions with excess charges (-1/6) are marked with full heavy circles for the ones at the plane above metal sites and with heavy dashed circles for the ones at the plane below the metal sites.

are 9.0 Å apart in the Y direction and about 7.0 Å apart along the <110> directions. One half of the vacant octahedral sites (M1) are surrounded by four anions with excess charges (Fig. 2), and they become favorable sites for monovalent cations of suitable ionic radii up to 0.8 Å depending on the size of vacant sites. The remaining half of the trans-sites, M1, form a row along the X-direction and are surrounded by only two oxygens with excess charge. Thus, this distribution of R^{2+} subsitutions in M2 octahedra creates two types of trans-octahedral environments in a dioctahedral smectite. Two of the oxygens of an R^{2+}-octahedron are shared with tetrahedra on either side of the octahedral sheet. For those cases where the tetrahedron contains an R^{3+} substitution, the shared apical oxygen has an additional charge excess of -1/4. Thus, excess charge is expected to be highly localized, 7 to 9 Å apart within the octahedral sheets.

Tetrahedral sheets

Tetrahedral sheets of smectite are composed of six-fold hexagonal rings of silica tetrahedra similar to those in micas. They are assumed to have condensed on both sides of the octahedral sheets. The bridging oxygens connecting the tetrahedra are called basal oxygens (O_b). The main geometrical features of the tetrahedral sheets are tetrahedral rotations, described by the angle α, tetrahedral tilt (Δz), and tetrahedral sheet thickness (t_t) as explained by Bailey (1984). Tetrahedral rotation angle α directly shows the deviation of an existing tetrahedral network from an ideal hexagonal network, and it can be easily calculated by the observed $\underline{a}$-dimension ($\underline{a}_{obs}$) and calculated ideal $\underline{a}$-dimension of the tetrahedral network ($\underline{a}_{tetr}$). The latter is obtained by a set of simple relationships with $\underline{e}_b$ being the mean basal edge of tetrahedra:

$$\underline{a}_{tetr} = 2\underline{e}_b \quad \text{and} \quad \cos \alpha = \underline{a}_{obs}/2\underline{e}_b.$$

The basal edge, $\underline{e}_b$, for an ideal tetrahedron (with no elongation), and $\underline{d}_{T-O}$, the mean tetrahedral cation–oxygen distance, is related by the equation: $\underline{e}_b = \underline{d}_{T-O} \cdot \sqrt{8/3}$.

Commonly, these relationships are given in terms of the $\underline{b}$ unit-cell parameters since $\underline{b}_{tetr} = \underline{a}_{tetr}\sqrt{3} = 4\sqrt{2}d_{T-O}$ and for a tetrahedral network with the composition $(Si_{4-x}Al_x)$: $\underline{b}_{tetr} = 9.15 + 0.185x$, and $\cos \alpha = \underline{b}_{obs}/\underline{b}_{tetr}$. The tetrahedral rotation angle, calculated by the above equation, usually is not in good agreement with the observed angle unless $\underline{e}_b$ values are available from a structural analysis. In fact, the calculated values in Table 1 are in some instances as much as two to nine times larger than the observed rotations. Therefore it is preferable to estimate the range of tetrahedral rotations in smectites by examining the observed data in Table 1. The observed tetrahedral rotation angles in Table 1 are 3.6° for talc, 1.5° for annite, 7.5° for phlogopite and 7.6° for biotite. Trioctahedral smectites, within the compositional limits of talc-phologopite-annite-biotite, are expected to have tetrahedral rotation angles in the range of 1 to 8 degrees. Similar variations in the key dioctahedral 2:1 minerals are listed in Table 1, and they range from 1 to 16 degrees. The Al for Si tetrahedral substitutions in aluminous dioctahedral smectites are mostly limited below $(Si_{3.5}Al_{0.5})$. They are expected to have smaller α angles than paragonite with a tetrahedral composition $(Si_{3.0}Al_{1.0})$. A reasonable range for α in dioctahedral smectites may be estimated to be 1–10 degrees.

The ratio d_{M-O}/d_{T-O} can be considered as a measure of the dimensional discrepancy between the octahedral and tetrahedral sheets in 2:1 layer silicates. The $\underline{b}$-parameter of an undistorted octahedral network ($\underline{b}_o$) and the same parameter of an undistorted tetrahedral network ($\underline{b}_t$) are given by the following relationships:

$$\underline{b}_o = 3\sqrt{2}\, d_{M-O} \quad \text{and} \quad \underline{b}_t = 4\sqrt{2}\, d_{T-O}.$$

Therefore, the ratio d_{M-O}/d_{T-O} is equal to 4/3 for no dimensional discrepancy, and this ratio can be regarded as a "misfit index." The observed values of the misfit index in Table 1 vary in the range of 1.25–1.29 for trioctahedral 2:1 layer silicates, and in the range of 1.16–1.25 for the dioctahedral ones. The misfit index for these minerals is significantly different from 1.33 for the "perfect" 2:1 layer silicate with no discrepancy between the dimensions of the component sheets.

No ordering of tetrahedral cations have been reported for true dioctahedral micas with 1M or $2M_1$ stacking

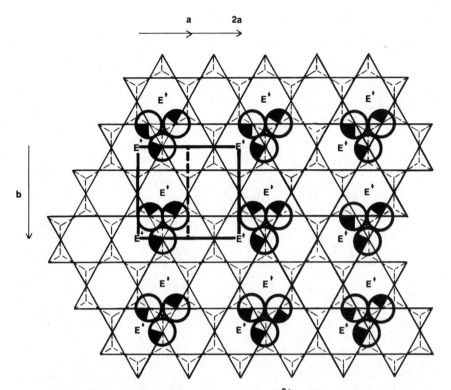

Figure 3. Distribution of tetrahedra with R^{3+} substitutions and basal oxygens with excess charges (-1/4). E^+ represents the favorable interlayer cation positions, and they form chains parallel to $\underline{b}$ in this short-range ordering mode.

sequences except for partial ordering in a few micas (Bailey, 1984). The tetrahedral R^{3+} substitution for Si^{4+} creates an excess charge on three basal oxygens and one apical oxygen, with an additional -1/4 charge for each. Such tetrahedra occur $4/x$ times, where x represents the tetrahedral subsitution of $(Si_{4-x} R_x^{3+})$. For a tetrahedral sheet composition of $x = 1/2$, one out of eight tetrahedra will have these oxygens with excess charge. This may result in a large number of possibilities for tetrahedral ordering. A simple example of a specific short-range ordering pattern, as illustrated in Figure 3, is considered in order to provide an insight into the problem. Figure 3 shows tetrahedral ordering generating a non-centered superlattice with $2\underline{a}$ x $\underline{b}$ dimensions. This ordering produces a charge distribution on the basal surfaces with excess charges (localized on triads of basal oxygens in Fig. 3) $\underline{b}$-distance apart along the $\underline{Y}$-direction and $2\underline{a}$ distance apart in the $\underline{X}$-direction. The three basal oxygens may have a total excess charge of -3/4 and their geometric center, which projects at the top of the tetrahedral cation, represents the highest charge density. Distribution of charge in smectite layers is expected to profoundly affect the configuration of the interlayer region, as discussed below.

Interlayer configurations

Most of the technological uses of smectite are related to reactions (e.g., catalytic, hydration, swelling) which take place in the interlayer. The interlayer region is more complex and can be more easily manipulated than the interlayers of micas. The significant crystal-chemical factors that affect the interlayer configuration are briefly discussed and restricted to only interlayer cations without hydration complexes. Interlayer complexes with organic and water molecules and "pillaring" with other complex ions are excluded. Comprehensive reviews of these latter topics are available (MacEwan and Wilson, 1980; Newman, 1987; Rausell-Colom and Serratosa, 1987; Thomas, 1984; Lagaly, 1984; and Fripiat et al., 1984). The number of interlayer cations is assumed to be stoichiometric with respect to the net negative layer charge. Although cations with higher valencies may substitute in the interlayer, the univalent and divalent cations are more common and they are considered here. For a total layer charge of 0.5 per formula unit, monovalent cations can occupy one half of the available interlayer sites, whereas the divalent cations can fill one fourth of these sites. The favorable sites for an interlayer cation without a hydration shell are shown in Figures 2 and 3 for octahedral and tetrahedral charge sites. The short-range ordering pattern (Fig. 3) for tetrahedral cations make two sites especially attractive for an interlayer cation. One of these is situated in the interlayer at the center of the oxygen triad with excess charges and coincides in projection with the tetrahedral R^{3+} cation. The other site is partially within the hexagonal cavities and probably closer to two basal and one apical oxygens of the tetrahedra with R^{3+} substitution. As indicated in Figure 3, the monovalent interlayer cations E^+ in hexagonal cavities can form chains parallel to the $\underline{Y}$-direction, and the alternate rows of hexagonal cavities are then left vacant. When these monovalent cations are hydrated they are expected to form similar chains of hydrated ions parallel to the $\underline{b}$-axis. The basal surface of a smectite with a tetrahedral charge as high as 0.5 does not have enough sites with sufficient charge for the development of a continuous plane of water molecules. This can, however, occur if other mechanisms, such as H-bond and dipole forces, provide an attraction between the basal oxygen surface and the water molecules.

In any case, monovalent cations of suitable size can penetrate into the hexagonal cavities, thereby bonding to only one smectite layer. Alternatively, they can move to a position above the hexagonal cavities by establishing bonds with two oxygens from the adjacent layer, and, thereby, provide cohesion between adjacent layers. This seems to be more favorable for larger monovalent cations such as K and Cs that are too large to penetrate into hexagonal cavities.

The most favorable positions for the divalent compensating cations are the center of the oxygen triads with excess charge. In fact, two similar oxygen triads with excess charge from adjacent layers can form [6]-fold coordination in various geometries (octahedron, trigonal prism, etc.) around the divalent interlayer cation. This arrangement is similar to the configuration of $Mg^{2+}(H_2O)_6$ complex in Mg-vermiculite structure (Shirozu and Bailey, 1966). Divalent cations, with or without their first hydration shell, are expected to be distributed at the corners of the (2a, b) superlattice between the adjacent layers. Divalent cations can also occur in the hexagonal cavities by coordinating with two charged oxygens from each adjacent layer. The latter arrangement is expected to be less stable as it can only provide a total -1 unit negative charges from four oxygens (each with an excess charge of -1/4). In any case, the first-coordination water molecules of divalent interlayer cations are expected to form islands between the smectite layers even for a tetrahedral charge of 0.5. A continuous water plane for interlayer cations indicates that water molecules attach themselves to silicate surfaces by other forces.

The above model discussed for the octahedral and tetrahedral charge distributions are to be considered as two of several possibilities.

Stacking of layers

Another characteristic of smectite structure is related to the stacking of elementary layers. Méring (1975) made the distinction between three stacking modes in smectites:
1. "Regular stacking" with zero degree rotations between the successive layers much like in a 1M dioctahedral mica; this stacking arrangement generates pseudohexagonal coordination cages. A perfect 3-dimensional periodicity is produced which generates hkl x-ray reflections as observed in Black Jack Mine beidellite or in some saponites. The regular stacking is rarely observed for other dioctahedral smectites.
2. "Semi-random stacking" of adjacent smectite layers with random n x 60° rotations while the large pseudohexagonal coordination cavities remain intact between the layers. From such a semi-random stacking only the hkl reflections with k = 3n are observed, whereas the reflections with k ≠ 3n are too diffuse to be observed.
3. "Turbostratic stacking" represents completely random rotations and translations between adjacent layers comparable to "a deck of cards thrown upon a table". This stacking mode does not generate hexagonal coordination cavities; interlayer positions with different stereochemistries and smaller holes can appear. Turbostratic stacking destroys the third dimensional periodicity for the stack of layers. Consequently, 001 reflections and broad hk diffraction bands only appear on diffraction patterns.

Mamy and Gaultier (1976) demonstrated that the turbostratic sequence can be modified to a "semi-random" stacking sequence by subjecting the K^+-saturated clay to 80-100 cycles of alternate wetting and drying. This pretreatment, which improves the stacking order, made it possible to pursue more detailed x-ray and electron diffraction analyses of smectites. Tsipursky and Drits (1984), however, noted that one-third of the K-saturated dioctahedral smectites treated did not respond. Larger cations of higher atomic number than K^+ were also effective for the same purpose. Besson et al. (1983a) noted that Cs saturation improved the stacking mode equal to 100 cycles of wetting and drying with K^+.

As noted by Méring (1975) dioctahedral smectites with the net layer charge resulting from the octahedra and with small monovalent interlayer cations, usually display turbostratic sequences. Monovalent cations smaller than K^+ are believed to be attached only to one layer whereas larger monovalent cations like K^+ seem to induce the elementary layers to a more regular stacking mode. Besson et al. (1983a) found that Cs saturation enhanced the crystallinity of smectite while simultaneously producing a large number of extremely small and thin clay particles (much below 0.2 microns). The above features of smectite interlayers are directly related to the rheological behavior of smectites, especially to the particle association and dissociation in suspensions containing different cations.

Significant contributions to the understanding of hydration and swelling behavior and stacking sequences of smectites and vermiculites have been recently made by Suquet et al. (1975, 1977, and 1987) as presented in the previous chapter in this monograph. For this purpose, synthetic saponites with increasing layer charges from 0.33 to 1.0 were hydrothermally prepared with the compositions

$$Na_x \ (Si_{4-x-q}Al_{x+q}) \ (Mg_{3-q}Al_q) \ O_{10} \ (OH)_2$$

with x and q varying in the ranges $0.33 \leq x \leq 1.0$ and q = 0 to 0.2 respectively. The synthetic smectites, and natural Kozakov saponite and vermiculites were systematically examined after saturating with eight different cations (Li, Na, K, Ca, Mg, Sr, Ba, and Ni) and subsequently solvation with water, ethylene glycol, and glycerol under relative humidities from 0 to 100%. Factors affecting hydration, swelling, and layer stacking in smectite structure under the above conditions were listed as:

1. layer charge density and location of the charge
2. nature of interlayer cation
3. nature of solvation molecules
4. di- or trioctahedral character of the mineral
5. mean crystallite size
6. relative humidity
7. the extent of (Si, Al) ordering in tetrahedral networks, and

8. ordering of interlayer cations over the available hexagonal cavities.

The interaction of the above factors in a rather complex and sometimes irrational manner leads to various water hydrates with 3, 2, 1, zero, or an indeterminate number of water layers in smectite interlayers. Similarly one or two layer complexes of ethylene glycol and glycerol can form between the smectite layers depending on the above conditions. Different stacking sequences with the various hydrates can develop between the layers. In fact, 11 ordered or semi-ordered layer stacking modes were found between tetrahedrally charged trioctahedral 2:1 layers. They are designated as V_I, V_{II}, V_V, V_{VII} for double-layer hydrates and V_a, V_b, V_c, V_c, V_e, and V_o for single layer hydrates. It is interesting to note that the most regular, triperiodic structure develops between two-layer hydrates of the smectites and not between one-layer hydrates. No regular stacking sequences were observed in octahedrally charged smectites probably because the electrostatic fields of octahedral charges are less effective in the interlayer region.

Classification

The structural and crystal chemical characteristics of smectite, discussed in this section, provide the following criteria to differentiate the species of smectites:
1. di- or tri-octahedral nature of the octahedral sheets, which divides smectites into dioctahedral and trioctahedral subgroups,
2. sources and sites of excess layer charges: the relative amounts of excess charges in tetrahedral (x_t) and octahedral (x_o) networks,
3. predominant octahedral cation, and
4. in some instances, proxy ions in octahedra like Mg in montmorillonite, and Li in hectorite.

Obviously, the chemical and structural features of the octahedral sheets determine the identity of the smectite species. The excess (negative) layer charges in smectites arise by the mechanisms:
1. tetrahedral substitutions of R^{3+} cations for Si^{4+},
2. octahedral substitutions of R^{1+} (Li^+ specifically) for R^{2+} in trioctahedral smectites, and R^{2+} for R^{3+} in dioctahedral smectites,
3. octahedral vacancies (defects), and
4. deprotonation of hydroxyls, i.e., loss of H^+ from the structure.

The total layer charges in smectites range from 0.2 to 0.6 units per $O_{10}(OH)_2$ with some exceptions in both ends. Smectites with high layer charges in the range of 0.5-0.7 show similar swelling and hydration behavior as vermiculites and they become undistinguishable in these properties then. In such instances, the layer charge boundary at 0.6 does not seem to have much diagnostic value, as discussed by Suquet et al. (1987). In the layer charge range 0.5-0.7; additional criteria such as crystallite size, paragenesis,

Table 2. Classification of natural and synthetic smectites

ratio between tetrahedral (x_t) and octahedral (x_o) charges	DIOCTAHEDRAL SMECTITES		TRIOCTAHEDRAL SMECTITES	
	predominant octahedral cation(s)	smectite species	predominant octahedral cation(s)	smectite species
$x_o/x_t > 1.0$ (octahedral charges predominant)	$Al(R^{2+})$*	montmorillonite	Mg Mg(Li) AlMgLi *	stevensite hectorite swinefordite
			single or mixed transition metals	transition metal "defect" trioc. smectites
$x_t/x_o > 1.0$ (tetrahedral charges predominant)	Al Fe^{3+} Cr^{3+} V^{3+}	beidellite nontronite volkonskoite vanadium smectite	Mg Fe^{2+} Zn Co Mn	saponite iron saponite sauconite cobalt smectite manganese smectite
			single or mixed transition metals	transition metal trioc. smectites

* octahedral substitutions

note: reasonably well-defined species (synthetic or natural) are given as separate entries.

and geological setting may then be called upon for the identification of the mineral.

On the basis of the three criteria mentioned above, a tentative classification of smectites is presented in Table 2. The main purpose of the table is to give an overview of the smectites whose structures and other characteristic features will be discussed in the following pages.

DIOCTAHEDRAL SMECTITES

Dioctahedral aluminous smectites are represented by the montmorillonite-beidellite series according to the structural formula

$$(Al_{2-y}Mg_y^{2+})(Si_{4-x}Al_x)O_{10}(OH)_2E_{x+y}^+ \cdot nH_2O \quad ,$$

where the amount of E^+ represents the interlayer cation, x and y the octahedral and tetrahedral substitutions, respectively. The smectites with y>x are called montmorillonite, and those with y<x beidellites. Appreciable amounts of trivalent Fe often occur in octahedra. Montmorillonites are commonly the main constituents of the rocks known as bentonites whereas beidellites are frequently found in soils as weathering products of detrital micas. Soil beidellites have been

recently surveyed by Wilson (1987) and are excluded from the present discussion.

Montmorillonite was first described by Damour and Salvetat in 1847 (Ross and Hendricks, 1945) as nodules in a brownish shale from Montmorillon in France. Larsen and Wherry in 1925 gave the name beidellite to a clay from the Beidell locality in Colorado (Ross and Hendricks, 1945). The validity of the name was questioned by various investigators, including Ross (1960) who suggested the term "aluminous montmorillonite" for beidellite. Weir and Greene-Kelly (1962), however, established the identity of beidellite through their studies of the sample from the Black Jack Mine locality.

The iron-analog of beidellite is called nontronite. The name was proposed by Berthier in 1827 (Ross and Hendricks, 1945) for the clay that was found with the manganese ore of Perigeux in the Nontron district near the village of Saint Pardoux in France. The other two lesser known species of dioctahedral smectite are volkonskite and vanadium smectite carrying Cr^{3+} and V^{3+} as the predominant octahedral cations, respectively; they both occur in rather special geological environments. The extent of the solid solutions between the three major dioctahedral smectites (montmorillonite-beideillite-nontronite) will be discussed after the structural and morphological features of the dioctahedral smectites are described.

Structural and Morphological Variations

Montmorillonite. The general configuration of the montmorillonite structure was established as a 2:1 dioctahedral layer silicate with an expandable lattice in the late 1930's and early 1940's. Extensive electron diffraction studies were carried out on the montmorillonite structure during the period of 1949-1978. These studies were discussed critically by Grim and Güven (1978) and will not be repeated here except for the conclusions about the different configurations of octahedral sheets in montmorillonite:

1. A pattern with vacant trans-sites (M1) as illustrated in Figure 1 was proposed by Zvyagin and Pinsker (1949). This configuration produces an elementary layer with a C2/m symmetry when the cis-sites are occupied by the same elements.

2. Another arrangement where the trans M1 site and one of the cis sites are occupied gives the layer C2 symmetry (Mering and Oberlin, 1971).

Drits et al. (1984) developed a general method in which the diffraction effects of stacking faults and the whole range (0-100%) of occupancies of M1 and M2 sites were considered. The geometrical relationship between the intralayer shift (Δs) and octahedral ordering was also included as a diagnostic criterion to determine the trans or cis character of the vacancies in the smectite layer.

Table 3. Atomic coordinates calculated by Tsipursky and
Drits (1984) for two models for montmorillonite.

Atoms	C2/m			C2		
	x/a	y/b	z/c	x/a	y/b	z/c
M1	0.0	0.0	0.0	0.0	0.0	0.0
M2	0.0	0.333	0.0	0.0	0.321	0.0
M3	0.0	0.667	0.0	0.0	0.654	0.0
T1	0.417	0.329	0.270	0.432	0.333	0.270
T2	0.417	0.671	0.270	0.432	0.662	0.270
K	0.5	0.0	0.5	0.5	0.0	0.5
O1	0.481	0.5	0.320	0.489	0.496	0.335
O2	0.172	0.728	0.335	0.173	0.725	0.335
O3	0.172	0.272	0.335	0.170	0.268	0.320
OH(O4)	0.419	0.0	0.105	0.334	-0.024	0.105
O5	0.348	0.691	0.110	0.417	0.656	0.109
O6	0.348	0.309	0.110	0.343	0.347	0.109

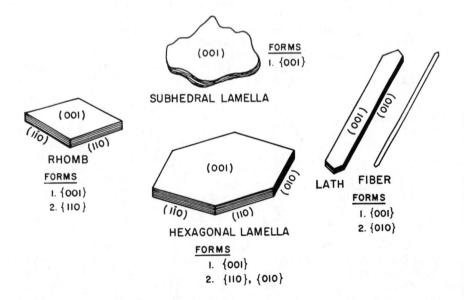

Figure 4. Perfect hexagonal lamella and the other habits of smectite
crystallites derived from it.

Intralayer shift (Δs) is caused by the stagger between tetrahedral sheets within a 2:1 layer, and it has magnitude of $\underline{a}/3$ for an octahedral sheet with all equal-size octahedra. Bailey (1975) noted that the intralayer shift is larger than $\underline{a}/3$ for dioctahedral mica layers with large (vacant) M1 octahedra and less than $\underline{a}/3$ for the layers with small M1 octahedra. The β-angle in a 1M mica is directly related to the intralayer shift by β = arccos ($-\underline{a}/3\underline{c}$) or arccos ($-\underline{b}/\underline{c}$ 3) provided the interlayer cations are not offset from the center of the ditrigonal silica rings. The latter can be assumed to be the case for dioctahedral smectites, and the relationship between the angle and the population of the octahedral sites will then have a diagnostic value as long as the stacking faults do not complicate this relationship.

Using the above method, Tsipursky and Drits (1984) examined a suite of 30 samples of dioctahedral smectites consisting of nontronites, montmorillonites, and a beidellite. These samples were pretreated according to Mamy and Gaultier's (1976) procedure to improve stacking order in them. The individual layers were found to possess monoclinic unit cells with β-angles ranging from 99° to 101°. Atomic coordinates were derived from the unit-cell parameters and the chemical composition for models with different octahedral ordering patterns. Table 3 gives the atomic coordinates for $\underline{C}2/\underline{m}$ and $\underline{C}2$ models for K-saturated Ascan smectite with the composition:

$$[(Al_{1.64}Fe^{3+}_{0.05}Fe^{2+}_{0.01}Mg_{0.31})(Si_{3.71}Al_{0.29})O_{10}(OH)_2]K^+_{0.58}.$$

The tetrahedral and octahedral excess negative charges are about equal in this smectite, which compositionally lies at the montmorillonite-beidellite boundary. Tsipursky and Drits (1984) found this sample to have $\underline{C}2/\underline{m}$ symmetry, with completely vacant trans octahedra. The other two Ascan smectites with almost end-member montmorillonite composition were assumed to have 75-100% occupancy of the trans octahedra. On the other hand, four montmorillonites with zero tetrahedral substitutions (i.e., truly end-member montmorillonites) were found to have completely vacant trans octahedral sites. One beidellite with appreciable trivalent iron in the octahedra was, however, found to have 50-75% occupancy of the trans site.

Recently, Cardile (1987) examined five montmorillonite samples with ^{57}Fe Mössbauer spectroscopy and assigned the octahedral Fe^{3+} to trans sites. The octahedral Fe^{3+} content of the montmorillonites was in the range of 0.29-0.36 per $O_{10}(OH)_2$.

In conclusion, montmorillonites may apparently have a wide range of occupancy (0-100%) in trans configurations; this seems to be even the case for the samples from the same deposit. However, well-crystalline dioctahedral smectites such as nontronites and beidellites, with a few exceptions, display strict octahedral ordering with vacant <u>trans</u> octahedral sites.

<u>Morphological Characteristics of Montmorillonite.</u>
Morphological variations of smectites can be derived from a hexagonal lamella (Fig. 4) which represents the ideal crystal form of a layer silicate. The hexagonal lamella exhibits {001}, {010}, and {110} forms with the predominance of the basal forms as expected from the structure of a layer silicate. The degeneration of a perfect hexagonal form into rhombs, laths, and fibers may be related to external factors during crystallization that impede the growth along certain directions, as discussed by Grim and Güven (1978). They examined morphological features of montmorillonites and beidellites in 80 bentonite samples from well-known deposits around the world. The main habits of smectite crystallites were found as follows.

1. Euhedral smectite lamellae in rhombic outline or elongated platelets with hexagonal angles (Figs. 5a and 5b).

2. Subhedral lamellae as thin platelets with irregular outlines. These platelets range in thickness from a single layer up to 200 Å. Individual subhedral lamellae as in Figure 5c are found in large quantities in the <0.2 micron fraction of almost every smectite sample; these minute platelets are often aggregated together through face-to-face associations.

3. Laths and ribbons of smectites occur mainly in beidellites, nontronites, and other smectites with predominantly tetrahedral charges. Ribbons are differentiated from laths as being wider and more flexible; such a typical ribbon is displayed in Figure 5d from Black Jack Mine beidellite.

Laths, ribbons, and rhomb-shaped lamellae are usually found in small amounts in smectites; the rest consist of aggregates with the following textures:

1. Lamellar aggregates that consist of thin but discernable smectite films or of minute subhedral platelets. These aggregates typically occur in Wyoming-type montmorillonites. The aggregate can be voluminous and porous due to curling and folding of thin films. It can also be compact when thin platelets are associated face-to-face. They give rise locally to lattice fringes in an incoherent matrix of curving smectite layers in high resolution transmission electron microscope images (Tessier and Pedro, 1987; Vali and Köster, 1986). Thus, there are two types of lamellar aggregates:

   a) foliated smectite aggregates as displayed in Figure 6a, and

   b) compact smectite aggregates as illustrated in Figure 6b.

2. Mossy aggregates that are distinguished by a felt-like (wooly) texture (Fig. 6c) where the smectite

515

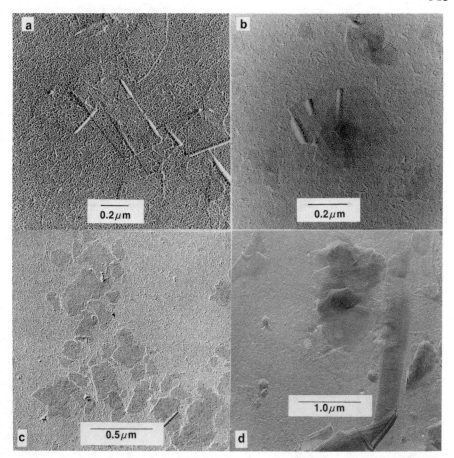

Figure 5. Different morphologies of individual crystallites in
montmorillonite-beidellite series: (a) equant and elongated platelets
(lamellae) with hexagonal angles, (b) a stack of three platelets with
rhombohedral outlines, and (c) subhedral platelets with irregular outlines,
and (d) a long ribbon of beidellite.

crystallites appear as acicular units due to the
curling of extremely thin films. The acicular
unit may also be thin and narrow laths consisting
of incoherent strips of layer silicate. Examples
of mossy aggregates are found in Cheto-type
montmorillonite along with lamellar aggregates.
3. Globular (granular) aggregates that display a
mosaic of disc-like platelets with a diameter in
the range of 150–250 Å (Figure 6d). They are
found in Otay-type montmorillonites.

Smectite particles in a sample are often
morphologically non-uniform; they display a variety of
habits. The individual crystallites are mostly concentrated
in the <0.2 microns fraction.

516

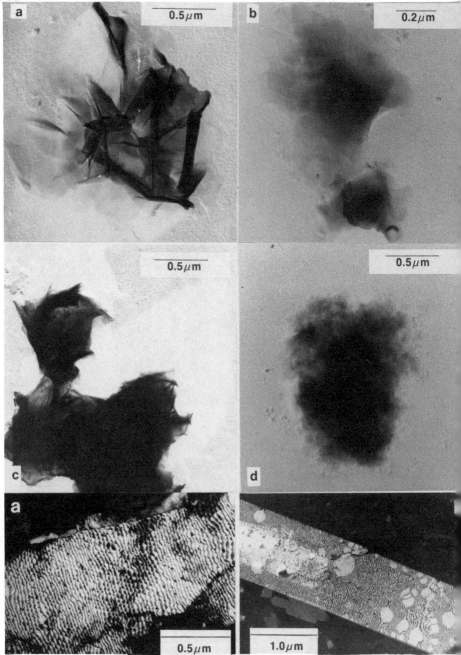

Figure 6 (upper four micrographs). Smectite aggregates with different textures: (a) foliated lamellar aggregate, (b) compact lamellar aggregate, (c) mossy aggregate with a felt-like texture, and (d) globular (granular) aggregate with mosaic of disc-like minute platelets.

Figure 7 (lowest two micrographs). (a) Dark-field image of BJM beidellite displaying dislocations. (b) Image of dislocations and dislocation-free domains in beidellite crystallite.

Beidellite. Weir and Greene-Kelly (1962) established the identity of beidellite as a distinct species with detailed chemical and x-ray diffraction analyses of a sample from the Black Jack Mine (BJM), Idaho. The BJM beidellite, U.S. National Museum R4762, is an unusual dioctahedral smectite with respect to its chemistry, morphology, and crystallinity. The chemical composition of the Na-saturated mineral is:

$$[(Al_{1.98}Mg_{0.01}Fe^{3+}_{0.02})(Si_{3.48}Al_{0.52})O_{9.98}(OH)_2](Na_{0.45}K_{0.01})$$

The interlayer cation was, however, Ca in the natural sample. The layer charge is derived solely from tetrahedral substitutions, as the octahedra are occupied almost exclusively by Al. BJM beidellite represents the end-member of the montmorillonite-beidellite series; such a pure end-member beidellite has not been reported elsewhere. The large crystallite size and better crystal forms of BJM beidellite makes it morphologically distinct from other dioctahedral smectites. BJM beidellite occurs as platelets, laths, and ribbons in the size range of 0.1 to 30 microns (Fig. 5d). Such large platelets (without any curled edges) and laths are commonly observed in dioctahedral micas but not in dioctahedral smectites. Weir and Greene-Kelly's (1962) x-ray diffraction data showed the presence of distinct hkl reflections indicating the existence of a truly regular (1M) stacking sequence. Detailed structural analysis of BJM beidellite with selected area electron diffraction (SAED) was reported by Güven and Pease (1975) and by Grim and Güven (1978). These studies clearly showed the total vacancy of the trans octahedra and C2/m layer symmetry in beidellite. The octahedral configuration was later confirmed with x-ray diffraction by Besson et al. (1983a).

Güven et al. (1979) reported rather unusual dark-field images of dislocations in beidellite, as displayed in Figures 7a and 7b. A set of parallel dislocation lines, 150-250 Å apart, form zig-zag patterns through the crystal. These defect lines are believed to represent boundaries of domains that are mostly triangular or rhombic in shape. In addition, oval shaped regions without any dislocations were distinctly visible in the dark-field images (Fig. 7b). These dislocations may be caused by structural or compositional mismatch between the domains.

Besson and Tchoubar (1972) and Besson et al. (1983a) reported a beidellite from Unterrupsroth (Germany) with the chemical composition

$$[(Al_{1.76}Mg_{0.27}Fe^{3+}_{0.01})(Si_{3.56}Al_{0.44})O_{10}(OH)_2Ca_{0.30}$$

to have a layer symmetry C2 with occupied trans octahedral sites. Nadeau et al. (1985) reported significant changes in tetrahedral charges in another sample of Unterrupsroth beidellite depending on the crystallite sizes. The ratios of tetrahedral to octahedral charges were found to be 4.35 for the >3.0 μm fraction, 1.79 for the 1.4–3.0 μm fraction, and 1.27 for the <1.4 μm fraction. The morphological features of these beidellites in different fractions of the same sample were drastically different. The crystallites in the >3.0 μm microns were predominantly ribbons or elongated platelets, whereas the particles in the finer fractions consisted of irregular masses of crystalline aggregates.

Another dioctahedral smectite, referred to as Ascan beidelite, and with a composition exactly at the montmorillonite–beidellite boundary

$$[(Al_{1.64}Mg_{0.31}Fe^{3+}_{0.05}Fe^{2+}_{0.01})(Si_{3.71}Al_{0.29})O_{10}(OH)_2]$$

$$(K_{0.27}Na_{0.21}Ca_{0.06})$$

was analyzed by Tsipursky and Drits (1984). K^+ was inserted into the interlayers by applying Mamy and Gaultier's pretreatment of wet–dry cycles, and oblique-texture electron diffraction was used. The Ascan beidellite was found to have vacant trans-octahedral sites. Hence, of the three structures discussed, two representing the lower and upper compositional limits of beidellite possessed a C2/m structure and cis-octahedral characters, whereas the beidellite with an intermediate composition (Unterupsroth) displayed trans-octahedral character, i.e., with trans sites occupied.

Nontronite. Nontronite occurs as a wide-spread authigenic clay in the recent submarine sediments and as hydrothermal precipitations from the hot brines discharged from spreading centers. Nontronite is a common weathering and alteration product of basalts and ultramafic rocks. It is also found in association with sedimentary iron ore deposits.

Nontronites have been studied in more detail than any other smectite; it therefore gives excellent insight into the structural complexities of smectites. The x-ray diffraction analysis by Eggleton (1977), which was based on only five observable basal reflections, clearly excluded the possibility of having inverted silica tetrahedra in the nontronite structure. Furthermore, Eggleton showed the presence of five hydration states in nontronite leading to 9.6, 12.2, 14.8, 15.2, and 17.4 Å layer thicknesses. A linear relationship was shown with a strong positive correlation between the tetrahedral charges and the number of water molecules in the unit cell.

The nontronite structure from the Garfield locality (Washington State) was later analyzed by Besson et al. (1983b) with electron diffraction, Mössbauer spectroscopy, and x-ray diffraction techniques. The main conclusions of this investigation are:

1. Octahedral ordering with vacant <u>trans</u> (M1) sites was found with $\underline{C2/m}$ symmetry and with coherent domains ranging 160-260 Å in diameter.

2. The <u>b</u>-parameter of Garfield nontronite was found to be 9.14 $\pm$ 0.02 Å, and the tetrahedral rotations were measured to be $6 \pm 1^\circ$.

3. Mössbauer spectra of the Garfield nontronite was interpreted in terms of three doublets; two belonging to octahedral Fe^{3+} in <u>cis</u> sites and the third one to the presence of tetrahedral Fe^{3+}. Accordingly the following structural formula was proposed for K-saturated Garfield nontronite:

$$(Fe^{3+}_{1.84}Al_{0.15}Mg_{0.02})(Si_{3.46}Al_{0.38}Fe^{3+}_{0.16})$$
$$O_{10}(OH)_2K_{0.57}$$

<u>Mössbauer spectroscopic analyses of nontronite.</u> ^{57}Fe Mössbauer spectra were initially interpreted as indicating octahedral Fe^{3+} occurring in <u>cis</u> and <u>trans</u> sites in nontronite (Goodman et al., 1976; Russell et al., 1979; Heller-Kallai and Rozenson, 1981). A better spectral fit was later obtained in terms of three doublets; two of them related to octahedral Fe^{3+} in <u>cis</u> sites and the third one to tetrahedral Fe^{3+} in the nontronite structure (Goodman et al., 1978; Besson et al., 1983b; Cardile and Johnston, 1985; Daynyak and Drits, 1987; and Murad, 1987). Although the amount of tetrahedral Fe^{3+} was found to be different by each, they were all in agreement about assigning Fe^{3+} to <u>cis</u> octahedral sites. Murad (1987) noted the presence of goethite as an impurity in three nontronite samples including the Garfield nontronite. He emphasized also that the Mössbauer spectra should be taken at $77^\circ K$ in order to correct for goethite contributions to the Mössbauer spectra of nontronite. By doing so he obtained a better fit and proposed the following structural formula for nontronite:

$$Na_{0.47}(Si_{3.60}Al_{0.30}Fe^{3+}_{0.10})(Al_{0.25}Fe^{3+}_{1.70}Mg_{0.05})O_{10}(OH)_2$$

Bonnin et al. (1985) objected to the occurrence of tetrahedral Fe^{3+} in nontronite after examining it with EXAFS, NMR, and optical spectroscopy, in addition to Mössbauer Spectroscopy. They concluded that the percentage of tetrahedral Fe^{3+} in Garfield nontronite must be less than 1% of the total iron and they proposed the structural formula:

$$Na_{0.48}(Si_{3.47}Al_{0.53})(Fe^{3+}_{1.98}Mg_{0.01}Ti_{0.01})O_{10}(OH)_2$$

Bonnin et al. (1985) confirmed the octahedral ordering of Fe^{3+} with vacant <u>trans</u> sites in nontronite structure.

<u>Morphological characteristics of nontronites.</u> Electron microscopic examination of four nontronite samples from Garfield (Washington), Manito (Washington), Hohen Hagen (Germany) and Flateyri (Iceland), indicate that laths and ribbons are the predominant morphologies in these samples,

520

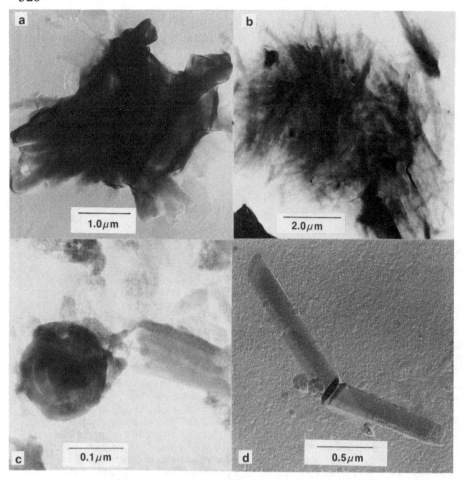

Figure 8. Morphological characteristics of nontronites: (a) aggregate of laths and ribbons in the sample from Manito, Washington, (b) aggregate of ill-defined laths and ribbons in the sample from Garfield, Washington, (c) a spheroidal (cabbage-like) aggregate in Flateyri nontronite, Iceland, and (d) typical well-developed ribbons in Manito nontronite, Washington.

except in the Flateyri nontronite. The laths and ribbons are usually intergrown into aggregates of about 5.0 microns in size.(Fig. 8a). Individual well-developed laths up to 10 microns in length and 1.0 micron in width occur rather frequently. Well-developed ribbons such as the one in Figure 8d are occasionally found; it is 1.0 micron long, 0.25 microns wide, and about 100 Å thick. Garfield nontronite consists mainly of heaps of ill-defined laths and ribbons. The laths which range in length from 2 to 5 microns are fuzzy at the edges and often split into fine fiber-like units consisting probably of strips of layer silicate. Minute and dense particles with prismatic to irregular outlines ranging in size from 0.1 to 0.3 microns are disseminated within the nontronite aggregate (Fig. 8b). The x-ray spectra obtained from dense particles show only

Fe- and O-lines, indicating the presence of an iron oxide, that is intimately mixed with nontronite.

Flateyri nontronite (Iceland) occurs mainly as spheroidal aggregates (cabbage-like) 0.1-0.2 microns in diameter (Figure 8c). Such forms are usually found in halloysites but not in smectites.

Synthesis of nontronite and other ferric smectites. Iron bearing smectites have been synthesized in the laboratory at low temperatures (Harder, 1976; Decarreau and Bonnin, 1986). They realized the complications caused by the oxidation of iron; Harder (1976) could only synthesize nontronite from a solution which was initially under reducing conditions. Decarreau and Bonnin (1986) and Decarreau et al. (1987a) synthesized ferric smectites by ageing the precipitates obtained from mixed solutions of sodium metasilicate and iron (II)-sulfate in the presence of H_2SO_4. The metal (Fe)-silica precipitate was separated from solution and dispersed into a 1% suspension in distilled water. The suspension was aged from 15 days under reducing conditions at 75°C. The precipitate remained amorphous for 15 days at 75°C under reducing conditions but upon oxidation it crystallized rapidly into a ferric smectite. The initial and final reaction products were characterized in detail with XRD, Mossbauer and infrared spectroscopy and its structural formula was determined to be $Si_4Fe^{3+}_{1.95}O_{10}(OH)_2Na_{0.05}$. The layer charge of ferric smectite is extremely low, and it is derived solely from octahedral defects (vacancies). The oxidation process which accelerated the crystallization of smectite was shown to take place in the solid state without dissolving the precipitate particles according to the reaction.

$$(Fe^{2+}_{3-x})\text{-smectite}+H_2O = (Fe^{3+}_{2-x})\text{-smectite}+Fe^{3+}OOH + 3H^+ + 3e^-,$$

where the parentheses represent the octahedral state of iron in smectite. The iron hydroxide (or oxide) was found to be cryptocrystalline and intergrown with the smectite; it was only detectable by Mössbauer spectroscopy. This ferric smectite was structurally closer to nontronite than to ferripyrophyllite (Chukhrov et al., 1979) although it carried very little layer charge. Decarreau et al. (1987) were also able to synthesize a similar ferric smectite under oxidizing conditions (without the initial reducing state) but only at temperatures of 100° and 150°C. The coherent domains in the a-b plane of this smectite was 64 Å at 100°C and 90 Å at 150° which are significantly smaller than a well-crystalline nontronite (e.g. 160-260 Å in Garfield nontronite). The b-dimension of the ferric smectite was determined to be 9.07 Å and its structural formula was given as: $Si_4Fe^{3+}_{1.83}O_{10}(OH)_2Ca_{0.26}$. Thus, the ferric smectite can be considered as a "defect" nontronite with the octahedral vacancies generating the layer charges.

Volkonskoite. The trivalent Cr^{3+} ion is similar in size (r = 0.62 Å) and charge to Fe^{3+}; it has been found as a proxy for Fe^{3+} in dioctahedral smectites such as nontronites (Besnus et al., 1975), and in montmorillonites and beidellites (Foord et al., 1987). Smectites with as little as 1% Cr_2O_3 were identified in the past as volkonskoite perhaps due to the distinctive emerald-blue color. The mineral name "volkonskoite" is reserved only for smectites carrying chromium as the dominant trivalent octahedral cation; others should be described using the appropriate adjectival modifiers, such as chromian beidellite, chromian montmorillonite etc. The Cr^{3+} bearing smectites occur as weathering or alteration products of Cr^{3+}-rich ultramafic or mafic rocks. An extensive list of world-wide occurrences and previous research was compiled by Foord et al. (1987).

The chromium-bearing clays are commonly intermixed with impurities, sometimes in colloidal allophane-like phases and often in amorphous precipitates of $CrO(OH)$, $FeO(OH)$, and others, thereby making their studies difficult. The recognition of these impurities and their proper removal is often a frustrating experience, and it has created confusion about the structural formulae of this mineral. In order to clarify the situation and to establish the structural formula for volkonskoite, Foord et al. (1987) reexamined two prominent volkonskoite samples from USSR, namely samples from Okhansk, Siberia (USNM #R4820) and from Mt. Efimiatsk, Ural, Siberia (USNM #16308). Their studies involved x-ray diffraction, Mössbauer spectroscopy, scanning electron microscopy, and detailed chemical analysis. Although Foord et al. (1987) did not detect any impurities with XRD and SEM for these samples, they were able to obtain a reasonable formula only after assuming an excess of Cr at 9% $CrO(OH)$ in sample R4820 and 3% $CrO(OH)$ in sample 16308. The following structural formulae were proposed for the neotype volkonskoites:

Okhansk sample, USNM #R4820:

$$[(Cr^{3+}_{1.07}Mg_{0.75}Fe^{3+}_{0.35})(Si_{3.59}Al_{0.43})O_{10}(OH)_2]$$
$$(Ca_{0.25}Mg_{0.05}K_{0.03}Fe^{2+}_{0.01}Mn_{0.01})$$

Mt. Efimiatsk sample, USNM #16308:

$$[(Cr_{1.18}Mg_{0.78}Fe^{3+}_{0.29}Ca_{0.02})(Si_{3.50}Al_{0.51})]$$
$$(Ca_{0.11}Mg_{0.11}Fe^{2+}_{0.03}K_{0.02})$$

Mackenzie (1984) located five similar chromian smectites from the Okhansk area in the Russian literature, and he calculated the following "mean" formula:

$$[(Cr^{3+}_{1.18}Mg_{0.75}Fe^{3+}_{0.27})(Si_{3.67}Al_{0.33})O_{10}(OH)_2]M^+_{0.50}$$

This "mean" formula is similar to those derived by Foord et al. (1987), and it seems to represent the composition of type locality volkonskoites.

An unusual chromian smectite was found by Khoury et al. (1984) in the Daba region of Jordan; the smectite was free of iron and contained more Mg than Cr:

$$[(Cr_{1.10}Mg_{1.26})(Si_{3.70}Al_{0.30})O_{10}(OH)_2]M^+_{0.53}$$

The smectite has an unusual excess of octahedral cations (with a total of 2.36) and Khoury et al. (1984) suggested an intermediate di-trioctahedral character.

Our examination of volkonskoite samples (USNM #R4820 and #16308) with analytical electron microscopy showed that the smectites in both samples are very similar in morphology. They occur as dense aggregates with a felt texture that consist of fine laths (Figs. 9a and 9b). The aggregates are about 5.0 microns in size and the thin laths range in width from 50 to 200 Å. These laths are often split into finer threads less than 50 Å in width; they probably consist of strips (about 5 unit cells wide) of 2:1 silicate with discontinuities between them. Bundles of such fine strips make up flexible ribbons up to 3.0 microns in length.

Distinct chlorite platelets and dense and subhedral silica particles are also found in the sample. It is, however, not reasonable to estimate even semiquantitatively the amount of these impurities with the electron microscope. A typical silica particle appears in Figure 9a, and it is about 1.0 microns in size. The x-ray spectra obtained from the particle gives only Si- and O-spectra.

Vanadium smectite. A dioctahedral smectite with V^{3+} as the predominant octahedral cation was reported by Guven and Hower (1979). The vanadium clay, from the Wasatch formation, occurs in small pockets with pitchblende in a uranium mine in Northern Converse County, Wyoming. The smectite has a basal spacing of 15.0 ± 0.1 Å in the air-dried state and carries Ca as the interlayer cation. The Na-saturated clay shows a strong 12.6 ± 0.1 Å basal reflection which expands to 16.7 ± 0.1 Å upon glycol saturation. The b-lattice parameter has been redetermined to be 9.08 ± 0.02 Å. The elemental composition of the Na-clay is listed in Table 4. Trivalent vanadium was oxidized to the quadrivalent state during the dissolution of the sample for chemical analysis. A reasonable structural formula was however obtained by assuming trivalent vanadium:

$$[(V^{3+}_{1.76}Mg^{2+}_{0.19}Fe^{3+}_{0.08})(Si_{3.28}Al_{0.46}V^{3+}_{0.26})O_{10}(OH)_2]$$

$$(Na_{0.68}Ca_{0.08})$$

The smectite is dioctahedral with predominantly tetrahedral charge. Typical morphological features of the vanadium smectite are displayed in Figures 9c and 9d. The smectite aggregates consist of curled films and "fanned"

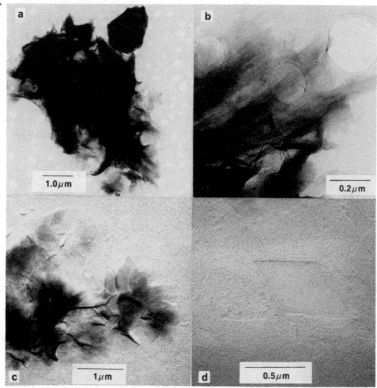

Figure 9. (a) Volkonskoite aggregates with felt-like (mossy) texture, (b) extremely thin and narrow (50 Å) laths in the previous volkonskoite aggregate, (c) aggregates of vanadium smectite consisting of thin foils and minute rhombs, and (d) a single euhedral platelet of vanadium smectite with rhombic outline.

Table 4. Chemical composition of the vanadium smectite (dried at 105°C) calculated for two oxidation states of vanadium.

| | Oxide composition weight % | |
	V^{3+}	V^{4+}
SiO_2	40.18	40.18
V_2O_3	30.75	34.03
Al_2O_3	4.76	4.76
MgO	1.54	1.54
Fe_2O_3	1.32	1.32
CaO	0.87	0.87
Na_2O	4.26	4.26
LOI	14.92	14.92
TOTAL	98.60	101.88

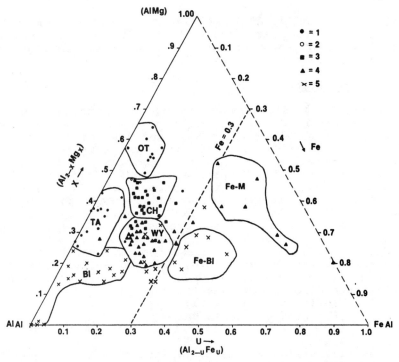

Figure 10. Ternary plot of the main octahedral cations in dioctahedral aluminous smectites. See Table 5 for symbols and designations.

Table 5. Designations and sources of samples in the ternary diagrams in Figures 10 and 11.

Graphical Symbol	Number	Smectite	Sources of Analyses
●	1	Tatatilla-type (TA) montmorillonite	Weaver and Pollard (1973), Grim and Kulbicki (1961), Schultz (1969)
○	2	Otay-type (OT) montmorillonite	Grim and Kulbicki (1961), Schultz (1969), Weaver and Pollard (1973)
■	3	Cheto-type (CH) montmorillonite	Grim and Kulbicki (1961), Schultz (1969), Weaver and Pollard (1973), Bystrom-Brusewitz (1976)
▲	4	Wyoming-type (WY) montmorillonites, and Fe-rich varieties (Fe-M)	Grim and Kulbicki (1961), Schultz (1969), Bystrom-Brusewitz (1976)
X	5	Beidellite (BI) and Fe-rich beidellites (Fe-BI)	Grim and Kulbicki (1961), Schultz (1969), Weaver and Pollard (1973), Chen et al. (1976), Bystrom-Brusewitz (1976), Hamilton (1971)
△	6	Ferric smectites in Figure 11 only	the smaller numbers (1-10) next the samples are the original numbers in Brigatti (1983)
☒	7	Nontronites (NT)	Eggleton (1977), Weaver and Pollard (1973)

stacks of rhombohedral lamellae. Perfectly rhomb-shaped
smectite platelets are found in large numbers, smectite
particles occur as subhedral thin platelets in the size
range below 0.5 microns with a few laths between them.

Chemical variations

Significant differences in chemical composition and
thermal behavior of dioctahedral aluminous smectites were
noted by Grim and Kulbicki (1961) and Schultz (1969). Grim
and Kulbicki (1961) pointed out the existence of two
distinct varieties of montmorillonites as Cheto- and
Wyoming-type. They attributed the different high-
temperature behavior of these smectites to a higher Mg-
content in the Cheto-type montmorillonite and to a possible
octahedral Mg-ordering. Schulz (1969) differentiated
aluminous smectites into seven groups according to (a) ionic
substitutions, (b) octahedral vs tetrahedral charges, and
(c) dehydroxylation temperatures; "normal" montmorillonites
dehydroxylate between $650-750^{\circ}C$, whereas Fe-rich "non-ideal"
ones dehydroxylate between $550-600^{\circ}C$. Schulz' subdivisions
of aluminous dioctahedral smectites were:
1. Wyoming-type montmorillonite
2. Otay-type montmorillonite
3. Chambers-type montmorillonite (same as the Cheto-
   type)
4. Tatatilla-type montmorillonite
5. "Non-ideal" (Fe-rich) montmorillonite
6. Beidellite
7. "Non-ideal" (Fe-rich) beidellite
The total octahedral Fe-contents of these "non-ideal"
montmorillonites and "non-ideal" beidellites were found to
be above 0.3; and the presence of excess structural iron in
the structure was suggested by the observed lower
dehydroxylation temperatures.

Statistical analysis of the chemical data. Chemical
data on aluminous dioctahedral micas (montmorillonite-
beidellite series) have been subjected to statistical
analyses by Weaver and Pollard (1973) on 101 selected
samples, Grim and Güven (1978) on 152 samples, Brigatti and
Poppi (1981) on 122 samples, and Alberti and Brigatti
(1985). There is good agreement among these studies
regarding the mean values of the structural cations, their
range of variation, and their correlation coefficients. The
mean values in Table 6, summarizing the results obtained by
Grim and Güven (1978) and those by Weaver and Pollard
(1973), indicate noteworthy characteristics of dioctahedral
aluminous smectites. When both mean values in Table 6 are
averaged, Mg appears to be the predominant substitution in
octahedra with a mean value of 0.36 whereas Fe^{3+} in
octahedra is on the average 0.18, about half of the
magnesium. The mean amount of tetrahedral Al is 0.18, again
half of the amount of octahedral Mg. The matrix of the
correlation coefficents among these compositional variables
is given in Table 7 for 152 chemical analyses selected by
Grim and Güven (1978). In this correlation matrix, the high

Table 6. Mean values and standard deviations for tetrahedral and
octahedral cations in dioctahedral aluminous smectites.

| | Grim and Guven (1978)* | | Weaver and Pollard (1973)** | |
	Mean	Standard deviation	Mean	Standard deviation
Tet. Si	3.80	0.17	3.84	0.14
Tet. Al	0.20	0.17	0.16	0.13
Oct. Al	1.47	0.16	1.49	0.16
Oct. Fe^{3+}	0.17	0.11	0.19	0.13
Oct. Fe^{2+}	0.01	0.02	0.01	0.02
Oct. Mg	0.36	0.16	0.35	0.15
Oct. Mn	0.0005	0.003	n.d.	n.d.
Oct. Li	0.0001	0.001	n.d.	n.d.

n.d.: not determined
  * : set of 152 samples
 ** : set of 101 samples

Table 7. The correlation matrix of tetrahedral and octahedral
cations in dioctahedral smectites.

	Si	Al-tet.	Al-oct.	Fe^{3+}	Fe^{2+}	Mg	Mn	Li
Si	1.00							
Al-tet.	-0.99	1.00						
Al-oct.	-0.18	0.18	1.00					
Fe^{3+}	-0.11	0.11	-0.31	1.00				
Fe^{2+}	0.10	-0.10	0.05	-0.05	1.00			
Mg	0.24	-0.24	-0.76	-0.37	-0.14	1.00		
Mn	-0.04	0.04	-0.09	-0.14	0.09	0.16	1.00	
Li	0.04	-0.04	0.11	-0.12	-0.04	-0.03	-0.02	100

528

negative correlation (-.99) between tetrahedral Al and Si is
trivial because of the constant-sum constraint (4.0) for
these tetrahedral ions. No other strong correlation is
indicated between tetrahedral Si and octahedral cations. In
fact, the only significant correlation in the matrix was
found between octahedral Mg and octahedral Al. R-mode
factor analysis was then applied to understand the
implications of the correlation matrix in Table 7 and to
reveal the underlying relationships in the variations of the
tetrahedral and octahedral cations. The R-mode factor
analysis yielded four independent factors that account for
96.0% of the variance in the chemistry of the dioctahedral
smectites (Grim and Güven, 1978). Tetrahedral Si and Al
bear heavily on Factor 1 which represents 37.0% of the total
variance. Similarly, octahedral Al and Mg heavily load on
Factor 2, and octahedral Fe^{3+} and Fe^{2+} on factors 3 and 4
respectively. All of these factors are orthogonal, i.e.
independent, indicating that tetrahedral cations, octahedral
Al + Mg, and two valence states of Fe are statistically not
correlated with each other. The actual amount of divalent
iron in dioctahedral aluminous smectites is rather small
(0.01 on the average) and total iron is often determined
without differentiation between the two valency states.
Octahedral Al, Mg, and total Fe can therefore be considered
as the principal variants in differentiating these
smectites. In fact, when the total of these three
octahedral cations are normalized to 2.0, then plotted on a
Al_2-$AlFe^{3+}$-$AlMg$ ternary diagram, significant relationships
appear (Fig. 10). The latter shows distinguishable clusters
for Tatatilla-, Otay-, Cheto-, and Wyoming-type
montmorillonites. The Cheto- and Wyoming-montmorillonite
fields have a higher density of samples, suggesting their
predominance among the natural dioctahedral aluminous
smectites. Most of Otay- and Tatatilla-montmorillonites
have a total iron below 0.1 per $O_{10}(OH)_2$; and Otay-type
smectites have higher Mg (in the range of 0.45-0.67) per
$O_{10}(OH)_2$. The Cheto- and Wyoming-montmorillonites display a
continuous series with increasing substitution of $(AlFe^{3+})$
for Mg^{2+}. When the relative magnitudes of tetrahedral and
octahedral charges are considered, a few of the
montmorillonites in the Wyoming field were found to be
actually beidellites in "senso lateral"; they are designated
as x in the ternary diagrams (Figs. 10 and 11). The total
Fe in the common aluminous dioctahedral smectites is below
0.3 as indicated by the dashed (Fe = 0.3) line in Figure 10.
A few samples contain a relatively higher Fe-content, and
are designated as Fe-rich montmorillonites (Fe-M) or Fe-rich
beidellites (Fe-BI). They plot to the right of the Fe = 0.3
line in Figure 10.

Brigatti and Poppi (1981) subjected 122 selected
smectite analyses to Q-mode factor analyses to examine the
similarities between the dioctahedral smectites. The Q-mode
factor analysis successfully revealed distinct clusters of
dioctahedral smectites that were statistically
differentiated based upon their chemistry. In fact, the Q-
mode factor analysis indicated 4 distinct groups, similar to

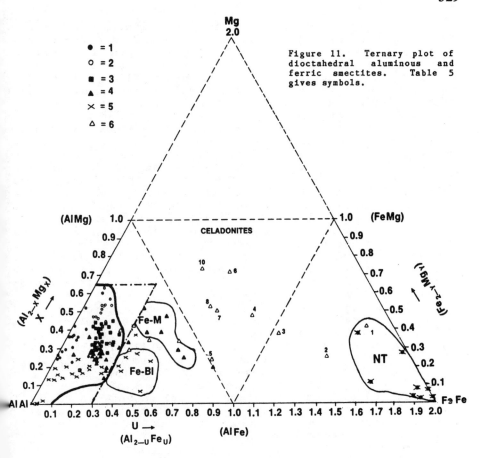

Figure 11. Ternary plot of dioctahedral aluminous and ferric smectites. Table 5 gives symbols.

those described above, namely Wyoming-, Cheto (Chambers)-, Tatatilla-, and Otay-types. Alberti and Brigatti (1985) re-examined the chemical variations in the above smectites with multivariate analysis of variance and discriminant analysis. They confirmed the subdivisions of dioctahedral aluminous smectites, and found Al, Fe, Mg, and interlayer Ca as the most important discriminant variables in them.

Recently some Fe-rich dioctahedral smectites with unusual chemical compositions were reported by Brigatti (1983). These smectites occur as weathering products of basaltic rocks in Northern Italy and they contain as much as 20% amorphous iron impurity. Structural formulae were calculated after correcting for these impurities. The smectites showed a strong correlation between their $\underline{b}$-parameters and Fe^{3+}-contents indicating that Fe^{3+} is not an impurity but is in the structure. These smectites were interpreted by Brigatti (1983) as being possibly the intermediate members between nontronites and non-ideal (Fe-rich) montmorillonites that were reported by Schultz (1969). In order to see this relationship we extended the ternary diagram in Figure 10 to its end-member limits $(Al_2-Fe_2^{3+}-Mg_2)$ in Figure 11. Brigatti's samples are designated with open triangles with their original numbers in Figure 11. A set of nontronites are also added to the same diagram, where they form a cluster marked NT around the Fe_2 corner. The ternary diagram $Al_2-Fe^{3+}-Mg$ in Figure 11 has four subfields represented by four smaller triangles. No dioctahedral smectites occur above the celadonite line; below it, three subfields can be distinguished. Aluminous smectites appear within the $Al_2-AlFe-AlMg$ triangle, whereas nontronites are confined to a limited section (NT) next to the Fe^{3+} corner of the $AlFe-Fe_2-FeMg$ triangle. The subfield $FeMg-AlMg-AlFe^{3+}$ contains the celadonites along the leucophyllite (AlMg) - celadonite $(Fe^{3+}Mg)$ line. The smectites with unusual chemistries (designated 3, 4, 5, 6, 7, 8, and 10 as in Brigatti's notation) with their high Fe^{3+} and Mg contents show similarities to celadonites. It raises the question whether these smectites are partially oxidized celadonites that lost some of the original layer charge during the postcrystallization-oxidation. In any case, the ternary diagram in Figure 11 shows the solid-solution behavior between dioctahedral smectites:

1. Along the Al_2-Fe_2 joint between nontronites and beidellites, Al and Fe^{3+} substitutions in octahedra are extremely limited to about 0.1 in both ends. However, along the Al_2-Mg_2 line, Mg^{2+} seems to substitute up to the compositional limit $(Al_{1.35}Mg_{0.65})$ despite the fact that Mg^{2+} (r = 0.72 Å) is larger than Fe^{3+} (r = 0.65 Å).

2. Coupled octahedral substitutions such as $(MgFe^{3+})$ for Al or (AlMg) for Fe^{3+} seem to be favored over single substitutions such as Al for Fe^{3+} in dioctahedral smectites.

Considering the radii of the ions (0.54 Å for Al, 0.65 Å for Fe^{3+}, 0.72 Å for Mg, and possibly 0.80 Å for vacancies, these substitutions may create large strains in the octahedral sheet since all four of these octahedra with different sizes must share edges.

TRIOCTAHEDRAL SMECTITES

Trioctahedral smectites can be divided into three subdivisions considering the predominant octahedral cation(s) and the paragenetic relationships:

1. Trioctahedral magnesian smectites with predominantly octahedral charges derived from monovalent Li^+ substitutions or defects (vacant octahedra). They are represented by the hectorite-stevensite series in nature as products of magnesian-rich environments.

2. Trioctahedral smectites of the transition metals (Ni, Co, Zn, Cu, Mn, Fe ...) that are found around ore bodies and in laterites and saprolites of ultramafic rocks. They are easily synthesized in the laboratory with single metal or mixed metal

octahedral sheets. These smectites may display
unusual catalytic properties and have the
potential for new technological applications. The
smectites of transition metals may also play a
major role in environmental waste control. These
transition metal smectites may, therefore, deserve
consideration as a separate group. The smectites
in this group are the ones that contain a
transition metal as the predominant octahedral
cation.

3. Ferromagnesian trioctahedral smectites (saponite/
iron saponite series) with Mg and Fe^{2+} as the
predominant octahedral cations. They are widespread in nature, especially as the main
authigenic clay minerals produced by the
alteration of oceanic and continental basalts and
other basic volcanic material.

Syntheses of trioctahedral smectites especially those
at low temperatures have yielded rather fundamental data for
understanding the formation as well as the crystal-chemical
features of these minerals. The results of these
experimental studies will be briefly summarized in order to
provide a background for discussing the trioctahedral
smectites.

Smectite synthesis at low temperature

Smectites have been synthesized at temperatures ranging
from $3°$ to $850°C$ using various starting materials such as
gels, oxides, and other compounds; fine powders of minerals
and rocks were also subjected to hydrothermal alterations.
The hydrothermal experiments were very significant in
determining the stability and compositional limits of
smectites and in producing highly crystalline materials
within a short time. In contrast, the low temperature
synthesis takes a longer time and often gives poorly
crystalline products. However, it yields smectites at
various stages of crystallization from nucleation to final
crystalline form. These observations prove to be helpful in
understanding the crystal chemical features of smectites.
Furthermore, the experimental conditions ranging in
temperature (from $3°$ to $150°C$) and starting materials (gels,
solutions) simulate more closely the sedimentary
environments where some of these minerals form. The two
most significant series of experimental studies where a
large number of smectites with various octahedral cations
were synthesized are described next.

(a) Smectite synthesis through ageing of freshly
precipitated metal hydroxides. This has been a traditional
approach at low temperature (Caillère and Hénin, 1962).
Harder (1972, 1976, 1977, and 1978) clarified the main
factors controlling this process and demonstrated
crystallization of smectites with numerous octahedral
cations of transition metals, as well as with Al and Mg.

Harder's investigations therefore, deserve more description. Metal hydroxides were slowly precipitated from dilute silica solutions (10-60 ppm SiO_2) by gradually changing pH and Eh conditions. The slowly flocculating gel of metal hydroxide was found to adsorb and/or to entrap silica in the pores of the gel. The initial gel was assumed to be amorphous but, when aged in contact with the solution, it crystallized into a smectite within a few days to several months at room temperature or even at $3^{\circ}C$. The critical factors for the crystallization of smectites were:

1. The initial precipitate must have a similar silica/metal hydroxide ratio to that of the smectite.

2. The metal hydroxide precipitate should develop a brucite- (or gibbsite-) like two-dimensional octahedral network. The octahedral sheet was assumed to act as a template for the condensation of silica monomers. This idea was the central theme of Caillère and Hénin's (1962) approach to the clay synthesis whereas Harder's clarified the next condition.

3. The most critical factor in the crystallization of smectite was the silica concentration in the solution. The silica concentration should be kept within the range of 10-100 ppm SiO_2 so that the silicic acid remains as monomers in solution. At higher concentrations silica can polymerize and inhibit the crystallization of the gel to smectite.

Gels with Mg^{2+}, Ni^{2+}, Co^{2+}, and Zn^{2+} were found to be rather favorable for smectite formation whereas the gels with Mn^{2+}, Cu^{2+}, Cr^{3+}, Fe^{2+}, and Fe^{3+} were not favorable. This difficulty was, however, overcome by preparing mixed-gels with small additions of other metals like Mg^{2+} or Ni^{2+} to the gels of "difficult" metals. For instance, no pure copper smectite was crystallized, but copper smectite with octahedral substitutions of Mg was synthesized. Similarly, aluminous dioctahedral smectites could only be formed with at least 6% MgO in the initial gel. Harder (1977, 1986) was also able to synthesize a ferroan trioctahedral smectite with pure Fe^{2+} octahedra under reducing and alkaline conditions. The synthesis products in Harder's experiments were identified mainly with x-ray diffraction and chemical analyses. They were usually in small amounts and mixed with amorphous or other phases.

(b) **Direct precipitation of smectite nuclei at low temperature**. This method was developed by Decarreau (1980, 1981, 1985) and it showed a better control over the homogeneity of the reaction products. The syntheses were easily reproducible and generated large quantities of homogeneous smectites. Moreover, Decarreau and his co-investigators characterized the synthetic smectites in more detail using rather sophisticated methods such as TEM, SAD, NMR, Mössbauer and IR spectroscopy in addition to chemical and XRD analyses. The results have been extremely valuable for understanding crystal chemical features, chemical variations, and crystal growth mechanism of smectites as discussed below.

This series of studies demonstrated that large amounts of homogeneous (single-phase) smectites can be synthesized at low temperatures by stochiometrically mixing the appropriate solutions of sodium metasilicate (Na_2SiO_3) and metal salts (e.g. $MgCl_2$) in an equally appropriate acidic solution, e.g., HCl. The silica-metal coprecipitate was found after separation and drying not to be amorphous but, indeed, to have a poorly developed crystalline structure, i.e., the smectite nuclei. The precipitate gave a distinct 001 basal reflection of a smectite, and it was composed of smectite nuclei with dimensions of 20 Å in the a-b plane and with a thickness of two to four layers. Ageing the precipitate in an aqueous suspension at 25 to 98°C improved the crystallinity and give rise to 02, 11 and 13, 20 diffraction bands in addition to a stronger 001 reflection. The crystal growth mechanism was different than in Harder's method, and it proceeded with the lateral development of coherent domains without any further increase in thickness, i.e., with no growth along the Z-direction. The enlargement of coherent domains was achieved by the lateral coalescence (edge-to-edge condensation) of the initial smectite nuclei. Any dissolution/reprecipitation and new nucleation were ruled out in the growth process. Within three weeks, the coherent domains were found to reach 100 Å in size (in the a-b plane) at 75°C and 200 Å at 90°C.

Decarreau (1981) showed that three types of smectites can be precipitated by adding a sodium aluminate to Na-silicate and mixed metal salts solutions:
1. Trioctahedral smectites with only tetrahedral substitutions (saponite-type).
2. Trioctahedral smectites with no tetrahedral substitutions but with octahedral vacancies (stevensite-type).
3. Trioctahedral smectites with both octahedral and tetrahedral substitutions.

Complete solid solutions were found between Mg, Fe, Co, Ni, and Zn and restricted substitution between Mg and Cu. The b-dimensions of the smectites were linearly changed depending on the amount of the substitutions up to 50%, e.g., Fe/Mg + Fe = 0.5. Above that level the b-dimension remained unchanged. No correlation was found between the crystal field stabilization energies of the transition metals and the octahedral substitutions in trioctahedral smectites. Furthermore, tetrahedral (AlSi) substitutions in synthetic saponite were independent from the octahedral substitutions.

The partitioning of the transition metal ions (Ni^{2+}, Co^{2+}, Zn^{2+}, Fe^{2+}, and Cu^{2+}) between a synthetic magnesian smectite precipitate and water was determined by Decarreau (1985). The partition coefficient (D_{Ni-Mg}) for an equilibrium between a magnesian smectite and a transition metal like Ni can be expressed as the ratio of their concentrations (Ni/Mg) in smectite and water:

$$D_{Ni-Mg} = \frac{(Ni/Mg)_{smectite}}{(Ni/Mg)_{water}} \; .$$

The partition coefficients were about the same at 25° and $75^{\circ}C$ and they were near 10^{3} for Ni, Co, and Zn, indicating the high selectivity of stevensite for these ions over Mg. The D_{Cu-Mg} was found near 10^{4} but the partition coefficient was lower for Fe with D_{Fe-Mg} = 300 and Mn with D_{Mn-Mg} = 30. All transition metals were strongly stabilized in both the stevensite and saponite structures, and they can readily occur as octahedral substitutions in these smectites. The transition metals Cu, Ni, Co, Zn, and Fe are, therefore, expected to form their own trioctahedral smectites.

In the presence of Ni, Co, and Zn, the crystal growth rate was significantly faster than in magnesian smectites, and the products were better crystallized than the magnesian members. The size of the nuclei in the a-b plane in the initial silico-metal precipitates were about 45 Å for Zn, 32 Å for Ni, 30 Å for Co as compared to 15-20 Å for Mg. The crystal growth process for Zn, Ni and Co took place at a higher rate (2.5 Å , 1.3 Å, 0.7 Å per day, respectively) than for Mg with a 0.5 Å lateral growth per day.

Difficulties were encountered with Fe^{2+}, Cu^{2+}, and Mn^{2+} and, indeed, the syntheses of the smectites with only these metals in the octahedra were not successful. In the case of Fe^{2+}, smectite nuclei 30 Å in size were found in the initial precipitates under reducing conditions, but no further growth on these nuclei were observed at 75° after 15 days.

The nuclei formed from the initial Si-Cu coprecipitates did not have a smectite structure but rather that of chrysocholla; therefore, no pure copper smectite has been synthesized. It was, however, possible to synthesize mixed (Cu-Mg) smectites only up to a range of $Cu/(Cu + Mg)$ = 0.5. This was not, however, successful for Fe^{2+} and Mn^{2+}. In the case of pure manganese, the synthesis required reducing conditions, but no (crystalline) nuclei were precipitated and no crystallization was observed during the ageing process. Manganese was found to oxidize rapidly and the precipitate remained amorphous.

Trioctahedral micas of the transition metals were synthesized by Hazen and Wones (1972 and 1978). They showed a similar trend with Ni being the most stable octahedral cation and Mn being the least favorable. They interpreted the compositional limits of the transition metal micas in terms of the ionic radii of the metals and the related dimensional misfit between octahedral and tetrahedral sheets in micas. The limiting size of the octahedral cation was calculated to be about 0.76 Å, which would cause a zero degree rotation in tetrahedral sheets with the (Si_3Al) composition. This would limit the ionic radii of the octahedral cation to 0.73 Å for a pure silica tetrahedral sheet as in the above smectite composition. Smectites with pure silica tetrahedra were however, synthesized at low temperatures with cations like Zn (0.74 Å), Co (0.75 Å), and even with Fe^{2+} (0.78 Å). The smectite structures seem to show a little more tolerance to the size of the metals. Their ability to polymerize in the form of octahedral (brucite-like) sheets that can absorb silica monomers seems to be more critical.

A nomenclature issue should be addressed before the individual members of transition metal smectites with the saponite-type structure versus those with the stevensite-type structure are described. We need to reach a consensus as to whether separate designations for them are to be adopted. For example, the zincian smectite may qualify simply as a (normal) sauconite for the saponite-type zinc smectite. The term "defect" sauconite could be used when octahedral defects are the main source of layer charges, i.e. stevensite-type. Experimentalists seem to group all the trioctahedral smectites whose charges are derived from octahedral vacancies as "stevensites" and to qualify them specifically with their predominant octahedral cation such as zincian stevensite, nickeloan stevensite etc.

Several other syntheses were carried out for a specific smectite and they will be mentioned while discussing individual smectite minerals. The major natural and/or synthetic trioctahedral smectites will be described in the next section.

Trioctahedral magnesian smectites with Li-substitutions

and/or vacancies in octahedra

Pure magnesian smectites with small or no tetrahedral substitutions but with octahedral vacancies (stevensite), and those with limited Li-octahedral substitutions (hectorite) are closely related. Both Li^+ substitutions and vacancies can occur together in the octahedral sheets. A comparison of the magnitude of charges created by Li^+-substitutions with those related to vacancies may permit distinction to be made between hectorite and stevensite. Swinefordite, carrying Li, Al, and Mg as the predominant octahedral cations, is a unique species. This mineral does not have any similarity with the hectorite-stevensite type smectites, it is included in this section mainly for convenience of discussion.

Hectorite. Hectorite is a rather unusual smectite in that it carries significant substitutions of Li^+ for Mg^{2+} and F^- for $(OH)^-$ in the octahedral sheets. Hectorite requires rather special chemical conditions for its formation; therefore, it is relatively scarce. The Hector deposit in California is believed to have formed by the alteration of pyroclastics by hot spring solutions enriched in Mg^{2+}, Li^+, and F^- (Ames et al., 1958). There are numerous and accurate chemical analyses available for the sample from the Hector locality. The structural formula derived by Ames et al. (1958) is:

$$(Mg_{2.65}Li_{0.33}Al_{0.02})Si_4O_{10}(OH)_{1.35}F_{0.65}(Na_{0.28}K_{0.01}Ca_{0.01}).$$

Hectorite occurs as extremely thin, but well defined laths, ranging from 0.1 to 2 microns in length and from 500 Å to 0.2 microns in width (Fig. 12a). The thickness of these laths is difficult to determine because of metal coating along the edges as illustrated in Figure 12a; it is estimated to be a few layers. These thin laths easily break down into minute fragments and they also tend to aggregate by edge-to-edge or end-to-end associations with other laths. Such agglomerations of hectorite laths were demonstrated by McAtee and Lamkin (1979) and they form radial arrangements or bundles of parallel laths. Oriented clay films, prepared for XRD analysis of Na-saturated samples, give rather sharp basal reflections with a 12.4 Å spacing. The stacks of hectorite layers in the coherently scattering domains were found to be 10–12 layers thick, indicating that the thin laths develop coherent face-to-face associations on the slides. Similar observations were made by Vali and Köster (1986) in the high resolution electron microscopic images of the hectorite, showing coherent stacks of 10–15 layers. The thin laths represent the typical morphology of hectorites in the Hector sample. In addition, another morphology that occurs in subordinate amounts appears as an intimate mixture of thin films and laths (Fig. 12b). Numerous saponite impurities were also observed in the hectorite clay. Saponites occur as large foliated aggregates of thin films and flakes in the size range of 5 microns (Fig. 12c). The x-ray spectral analysis of these flakes with the KEVEX quantum detector gives the following elemental ratios after converting the spectral intensities to atomic ratios under the thin film approximation: Mg/Si = 0.67 and Al/Si = 0.07. The laths, however, have the atomic ratio: Mg/Si = 0.66, and no aluminum.

Hectorite is a very valuable clay because of its high viscosity and other rheological properties. In response to industrial demand, it has been synthesized in large quantities and several methods of synthesis have been developed. The conditions of syntheses and the properties of the products in some cases are worth noting. A commercial product of synthetic hectorite is known by the trade name laponite, and it has the composition (Neumann and Sampson, 1970):

$$(Mg_{2.55}Li_{0.30}\square_{0.15})Si_4O_{9.70}(OH)_{2.30}Na_{0.30}.$$

The laponite has both vacancies and Li^+-substitutions in the octahedral sheet; apparently the layer charge is reduced by Si-OH silanol groups. Typical laponite particles occur as porous aggregates with a wooly texture consisting of twisted

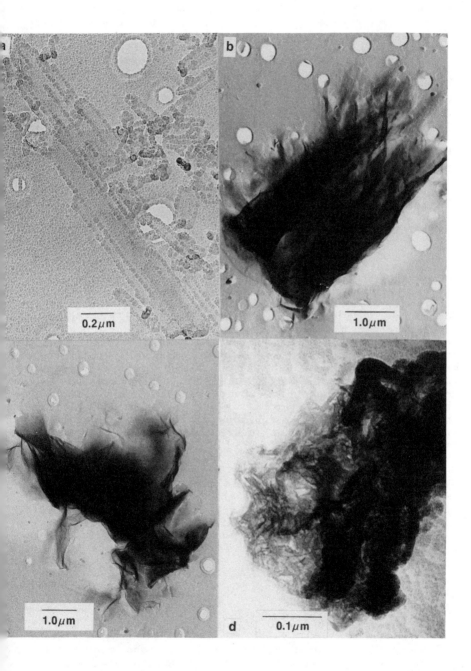

Figure 12. Typical natural and synthetic hectorite particles: (a) common thin laths, (b) aggregate of laths and thin smectite films, (c) foliated aggregate of a saponite impurity, and (d) synthetic hectorite (laponite) aggregate with a wooly texture.

538

and curled thin layers (Fig. 12d). The aggregates are about
0.5 microns in size and contain dense oval or donut shaped
islands.

A hectorite containing 1.84% Li_2O and 1.53% F was
synthesized by Granquist and Pollack (1960) by mixing fresh
silica and $Mg(OH)_2$ gels into a 10% aqueous suspension. The
suspension was then hydrothermally treated in the presence
of NaOH and LiOH (or LiF) at $300^{\circ}C/1200$ psi for 7 days. The
presence of LiF was found to greatly accelerate the
crystallization of hectorite. Baird et al. (1971, 1973)
synthesized hectorites using Granquist and Pollack's
procedure by refluxing the freshly prepared gels of silica
and magnesia in the presence of LiF at $300^{\circ}C/12,000$ psi.
They examined the morphological and structural evolution of
the hectorite with electron microscopy. $Mg(OH)_2$ particles
in the initial gel mixture after freeze-drying at $0^{\circ}C$ were
already crystalline disc-like platelets about 0.2 microns in
size; the silica particles, however, were amorphous. The
brucite particles progressively changed with increasing time
and temperature during the ageing process. After a short
refluxing time (1-1.5 hours), donut-shaped domains about 0.1
micron in diameter were developed on brucite particles,
apparently by the condensation of silica. After seven
hours, smectite particles appeared as flakey aggregates in
association with still-visible, oval-shaped domains. After
an hour, the SAD patterns showed the characteristic spots of
smectite, which were superimposed on the brucite rings.
With increasing time of reflux, the brucite SAD patterns
were completely replaced by the sharp continuous rings of
smectites. These observations suggest that the
crystallization of hectorite takes place by gradual fixing
of silica monomers on a previously precipitated crystalline
brucite sheet and that brucite gradually converts to
smectite.

Fluorhectorites (i.e., those without structural
hydroxyls) were synthesized in the solid state at $800^{\circ}C$
within 24 hours or in melts at 850° within 2 hours using
reagent grade chemicals such as pure silica, MgO, MgF_2, LiF,
and Na_2CO_3 (Barrer and Jones, 1970). These hectorites were
able to rehydrate after syntheses at $800^{\circ}C$ to $850^{\circ}C$ and were
found to have the compositions for the solid state product:

$[(Mg_{2.65}Li_{0.35})Si_4O_{10}F_2]Li_{0.35}$ with CEC = 90meq/100g,
and for the melt product:

$[(Mg_{2.42}Li_{0.58})Si_4O_{10}F_2]Na_{0.58}$ with CEC = 150meq/100g.

Decarreau (1980) synthesized hectorite by ageing the
silica-magnesia coprecipitate in an aqueous suspension in
the presence of LiF (or LiCl) at $75^{\circ}C$ within 2-3 weeks. The
hectorite crystallites had a platy morphology; no laths or
fibers were found. The minute platelets were composed of
coherent domains of about 100 Å in size (lateral), and NMR
spectra indicated that Li was indeed in the octahedral
coordination.

Stevensite. The first reported stevensite (in 1877) was a pseudomorph after pectolite – $HNaCa_2(SiO_3)_3$ – found in the Watchung Basalt near Springfield, New Jersey. Stevensite was later redefined as a member of the smectite group by Faust and Murata (1953). X-ray diffraction of data from this type-locality sample and those from other samples (Shimoda, 1971; Imai et al., 1970) show a distinct superlattice reflection at 24–25 Å in addition to the characteristic set of smectite reflections. The clay mineral in these samples can be visualized as a mixture of domains with stevensite structure containing octahedral vacancies and kerolite domains with no such vacancies. Superposition of these domains in adjacent layers may give rise to diffraction effects of mixed-layering. Random kerolite/stevensite mixed-layers were found abundantly as lake deposits in the Amargosa Desert of southern Nevada by Eberl et al. (1982). The structural formula for a mixed-layer sample containing 80% stevensite and 20% kerolite is given below and shows the presence of both Li^+ substitutions and vacancies in the stevensite layers:

$$[(Mg_{2.56}Al_{0.13}Fe^{2+}_{0.06}Li_{0.15})(Si_{2.92}Al_{0.08})O_{10}(OH)_2]E^+_{0.26}$$

The clay had a CEC of 67 meq/100 g, and 0.1 octahedral vacancies. The average composition of five representative stevensites was calculated by Köster (1982), which gives 0.12 vacancies and 0.03 Li in octahedral sheets:

$$(Mg_{2.77}Al_{0.03}Fe^{3+}_{0.02}Ni_{0.03}Li_{0.03})(Si_{3.99}Al_{0.01})O_{10}(OH)_2E^+_{0.23}$$

The stevensite was synthesized early at low temperatures but the synthetic mineral shows a lower layer charge, often below 0.1 per formula unit. Perruchot (1976) circumvented this problem by incorporating small amounts of alkaline earths like Ca and Sr into the initial silico-metallic precipitate in Decarreau's procedure. During the ageing process, Ca^{2+} was expelled from the smectite structure in the initial precipitate, thereby creating vacancies in the structure. The reaction product then showed a layer charge in the magnitude of 0.3. Stevensite was synthesized hydrothermally by Sakamoto et al. (1982) by altering pectolite, wollastonite, and bustamite in basic magnesium carbonate solutions at temperatures of 100–450°C.

Stevensite was found in various ancient and modern saline lakes (Bradley and Fahey, 1962; Bradley and Eugster, 1969; Tardy et al., 1974; Dyni, 1976; Tettenhorst and Moore, 1978; Badaut and Risacher, 1983; and Darragi and Tardy, 1987). Stevensite was also found as a common alteration product of sepiolite in nature (Randall, 1956). Sepiolite was easily converted to stevensite in the laboratory (Güven and Carney, 1979); the conversion starts at 150°C and takes place topotactically at low temperatures but by dissolution/reprecipitation at a higher temperature. Typical stevensites formed by such reactions are displayed in Figures 13a and 13b. Stevensite formed below 200°C

540

displays a texture with faint striations (Fig. 13a) reflecting the bundles of parallel fibers of the original sepiolite (Fig. 13c). This texture suggests a topotactic reaction involving an $\underline{n}$-glide [$(1/2)\underline{a}$ + $(1/2)\underline{c}$] between the alternating silicate ribbons in the sepiolite structure, as schematically shown in Figure 14. Minute and mostly oval shaped platelets are disseminated over the topotactic stevensite. These platelets range from 250 Å to 0.1 micron. More of these platelets were found at higher temperature; Figure 13c shows the abundance of these equant to hexagonal platelets that developed during the hydrothermal treatment at 314°C. These platelets represent a dissolution and reprecipitation mechanism for the formation of stevensite at higher temperatures.

Swinefordite (pronounced 'Swainfedait'). A lithium-rich clay with a jelly like appearance occurs as a pseudomorph after spodumune in a pegmatite of the Carolina tin-spodumene belt. The clay was studied by Tien et al. (1975) with x-ray diffraction, DTA, IR spectroscopy, and chemical analysis; it was named swinefordite in honor of Ada Swineford, a distinguished clay mineralogist. Six states of hydration were recorded from the air-dried clay slide ranging from fully hydrated clay to the relatively stable 13.0 Å phase below 60% RH with the appearance of the intermediate hydrated phases with 40, 20, 16, and 14.5 Å. Upon glycerol saturation a rational sequence of 10 basal reflections was observed starting with a 18.0 Å reflection. The DTA curve showed three endothermic reactions at 125°, 200°, and 625° indicating the loss of absorbed water in two stages, a phenomenon attributed by Tien et al. (1975) to be a possible segregation of divalent (Ca, Mg) and monovalent (Li) cations in the interlayers. IR absorption spectra of the clay were similar to those of the other smectites. The CEC values of two samples from the Li-rich clay were found to be 102 and 93 meq/100g respectively. However, the total of interlayer cations extracted by NH_4Cl-leaching were much higher than the CEC. The extract contained large quantities of Li except when the sample was dried at 150°C prior to cation extraction. These observations led the investigators to assume that about one third of Li in the analysis belonged to the interlayer and they derived the following structural formula for swinefordite:

$$(Si_{3.83}Al_{0.17})(Li_{0.88}Al^{3+}_{0.94}Fe^{3+}_{0.08}Fe^{2+}_{0.05}Mg_{0.66})$$

$$O_{10}(OH)_{1.68}F_{0.32}(Ca_{0.12}Mg_{0.02}Na_{0.06}Li_{0.36}K_{0.02})$$

The total octahedral occupancy of the smectite according to the formula was 2.61. Swinefordite was assumed, therefore, to have an intermediate di-trioctahedral character. However, Köster (1982) readjusted the formula by taking into account the CEC measurement of the clay and obtained:

$$(Si_{3.87}Al_{0.13})(Li_{1.06}Al_{0.99}Fe^{3+}_{0.10}Mg_{0.70})O_{10}(OH)_2E^{+}_{0.40}$$

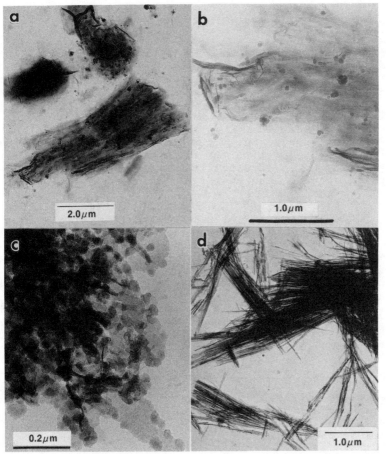

Figure 13. (a) Stevensite aggregate displaying a fibrous texture inherited from the sepiolite precursor, (b) stevensite that was converted topotactically from sepiolite, (c) minute stevensite platelets precipitated after the dissolution of sepiolite, and (d) a bundle of original sepiolite fibers before their conversion to stevensite.

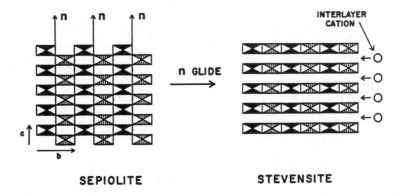

Figure 14. Topotactic transformation of sepiolite to stevensite through n-glide displacements, as projected along the fiber direction (a).

The new structural formula for swinefordite with a total octahedral occupancy of 2.85 is similar to that of trioctahedral smectites. Our electron microscopic examination of the sample showed that the smectite occurs in two different morphologies: (a) Extremely thin films that are unusually large up to 15-20 microns (Fig. 15a). The central sections of these thin films without wrinkles give an SAD pattern with sharp spots (Fig. 15b), indicating a regular stacking sequence. (b) Stacks of thin lamellae and curled films, which are on the order of 10 microns in size, indicate a highly crystalline smectite. Our x-ray diffraction pattern from the air-dried clay slide, prepared from less than 0.5 micron fraction of the clay, gave a basal spacing of 14.1 Å at 55% relative humidity. This basal spacing expanded to 16.7 Å upon complexing with ethylene glycol. In addition, a distinct superlattice reflection appeared at 27.4 Å for the air-dried clay; the latter expanded to 34.5 Å upon glycolation. These superlattice reflections suggest a mixed-layering of two interlamellar hydrates; one with two water layers (15 Å) and the other with one water layer (12 Å). The two water layer hydrates may be related to the interlayer cations of Ca, Mg, and the one water layer hydrate may be formed with Na and Li. Thus, a segregation of interlayer cations seems to take place in the clay mineral as suggested by Tien et al. (1975).

Trioctahedral smectite of transition metals

Smectites with Ni, Co, Zn, Mn, or Fe as the predominant octahedral cations were found either in nature and/or were synthesized in the laboratory. They will, subsequently, be described in more detail.

Sauconite. A clay from a zinc mine in Saucon Valley (Pennsylvania) with the composition

$$[(Zn_{1.95}Mg_{0.12}Al_{0.17}Fe^{3+}_{0.58})(Al_{0.61}Si_{3.34})O_{10}(OH)_2]E^+_{0.33}$$

was named sauconite by Roepper and reported by Genth in 1875 (in Ross, 1946). A set of rather pure zinc smectites from various zinc mines including the Saucon Valley sample were studied by Ross (1946). These investigations showed that the clays were trioctahedral smectites with the octahedral zinc content ranging from 1.48 to 2.89 and with the total octahedral occupancy varying from 2.70 to 3.06. The layer charges in these smectites were derived from tetrahedral substitution, and calcium was found as the predominant interlayer cation. One of the formulae with complete octahedral occupancy is given below for the smectite from Zinc Village, Boone County, Arkansas:

$$[(Zn_{2.64}Mg_{0.11}Al_{0.12}Fe^{3+}_{0.13})(Si_{3.27}Al_{0.73})O_{10}(OH)_2]$$

$$(Ca_{0.20}Na_{0.04}K_{0.04})$$

"Normal" sauconite and sauconite with octahedral vacancies were synthesized by Decarreau (1981) as described above. A

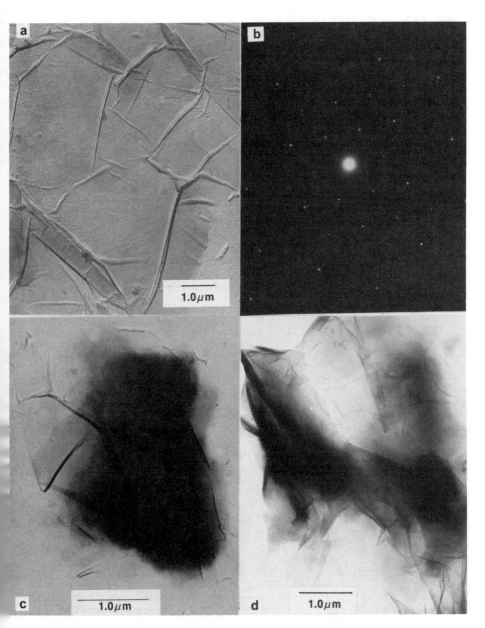

Figure 15. (a) Typical swinefordite films with polygonal wrinkles (b) SAED pattern obtained from the flate section of the smectite film, (c) typical saponite flake in Ballarat sample, and (d) foliated thin films of saponite in Ballarat sample.

544

zinc-bearing smectite was recently found as fillings in the
karstic cavities in the Qin Khamonda area of Central Tunisia
(Rivière et al., 1985). Zincian clay is admixed with a
nearly amorphous zinc hydroxide. The x-ray spectra obtained
from the individual smectite particles showed only strong
lines of Si, Zn, and Al. The smectite was, therefore,
assumed to be a sauconite; however, no structural formula
was calculated.

Nickel Smectite. Calculations by Manceau and Calas
(1985) indicate that the crystal field stabilization energy
(CFSE) of Ni in stevensite is about 1.0 kcal/mole lower then
CFSE of Ni in pimelite (a nickel talc). It is, however, yet
to be proven whether this difference in CFSE may be a factor
in the predominant formation of pimelite over Ni-smectite.

So far no smectite with Ni as the predominant
octahedral cation has been reported although pure nickel
smectites have been synthesized at low temperature. A
smectite with apparently Ni-dominated octahedra was found in
a laterite as a weathering product of Jacuba pyroxenes in
ultramafics of Niquelandia, Goias State, Brazil (Decarreau
et al., 1987b). The usual diagnostic x-ray diffraction data
indicated the presence of a smectite in the sample. The
smectite appeared in the form of typical flakes as small as
3000 Å in diameter and its electron microprobe analysis
gives the composition:

$$[(Al^{3+}_{0.17}Fe_{0.50}Mg_{0.48}Ni_{1.47}Cr_{0.02})(Si_{3.92}Al_{0.08})$$
$$O_{10}(OH)_2]E^+_{0.07}$$

Decarreau et al. (1987b) carried out extensive studies
on this smectite using XRD, EMPA, TEM/EDS, and spectroscopic
analyses with IR, Mössbauer, EXAFS, and diffuse reflectance.
The results revealed the true nature of the smectite
particle which appeared homogeneous at the 0.3 micron scale.
It was actually composed of a mixture of domains with
nontronite-like structures containing (Al, Fe) clusters and
those with pimelite-like domains consisting of (Ni, Mg)
clusters. These domains were found to be about 30-40 Å in
lateral dimensions. Although the flakes appeared like a
smectite with a diameter of 3000 Å and a thickness of five
layers, they actually consisted of a mosaic of domains with
different structures, with Ni primarily residing in
pimelite-like domains.

Cobalt Smectite. Hydrothermal synthesis of a cobalt
smectite at 250°C was recently accomplished by Bruce et al.
(1986). The authors also supplied a detailed
characterization of the smectite. The synthesis was
completed by mixing $Co(NO_3)_2 \cdot 6H_2O$ with a sodium silicate
solution which was freshly prepared by adding colloidal
silica to a solution of NaOH. The resultant mixture was
autoclaved under H_2 atmosphere at 250°C for 16 hours. A Na-
cobalt smectite was found in the precipitate with a basal
spacing of 12.4 Å. The basal spacing expanded to 16.9 Å

upon saturation with ethylene glycol and to 17.3 Å with glycerol. The chemical analysis of the precipitate after washing and dialyzing yielded a structural formula $Na_{0.06}Co_{3.07}Si_{3.95}O_{10}(OH)_2$ which indicated a very low layer charge for a smectite. The investigators offered several possibilities for the excess of Co and the deficiency of Si. One of them favors the presence of Co^{2+} in the tetrahedra; the effective ionic ratio of high spin Co^{2+} in tetrahedral coordination is 0.574 Å and is similar to that of Mg^{2+} in tetrahedral coordination (0.57 Å). The substitution of Mg^{2+} for Si^{4+} in tetrahedra has been previously reported in 2:1 layer silicates. The absorption spectra of this smectite strongly indicated that small amounts of Co^{2+} were located in tetrahedral environments. Another explanation for the excess Co^{2+} in the formula was related to the possibility of hydroxy-cobalt species (Co-hydroxide or oxyhydroxide) being adsorbed at the edges of the clay particles. The cobalt smectite developed as discs or hexagonal platelets up to 0.5 microns in lateral dimensions. The morphological features and SAD patterns of the particles exhibited an extremely good crystallinity and the combined x-ray and electron diffraction data gave the lattice parameters for the dehydrated cobalt smectite: $\underline{a}$ = 5.37, $\underline{b}$ = 9.34, $\underline{c}$ = 9.34 Å; β = 97°.

Manganese smectite. A Mn^{2+}-rich trioctahedral smectite was discovered at the initial weathering stage of a tephroite (Mn-olivine) under the tropical conditions of the Ivory Coast (Nahon et al., 1982a). Tephroite and spessartine (a Mn-Al garnet) occur as the predominant components of the metamorphic country rock of Ziemougoula area, northwestern Ivory Coast. Smectite has developed at the initial stage of the weathering and partially replaced the tephroite crystals. The electron microprobe analysis of the smectite flakes gave the average composition for five particles:

$$(Mn^{2+}_{2.38}Mg_{0.30}Fe^{3+}_{0.004}Cr^{3+}_{0.007}Al_{0.20})(Si_{3.92}Al_{0.08})$$

$$O_{10}(OH)_2Ca_{0.05}$$

The smectite was found to be poorly crystalline and it only remained stable at the contact in the microenvironment of the parent mineral (tephroite). When the environment was conducive to oxidation of Mn^{2+}, smectite was rapidly replaced by manganese oxyhydroxides. Thus, the smectite had a rather ephemeral existence, which was limited to a microenvironment of the precursor olivine grain at the early stage of weathering.

In the same area and under the same weathering conditions, forsterites (Mg-olivine) were grain-by-grain replaced by a saponite, indicating the composition of the smectite was directly related to the chemistry of the microenvironment determined by the individual grains.

Smectites with appreciable Mn and other transition metals in octahedra have been found at the early stages of lateritic weathering of ultramafic rocks. Examples of such smectites were described by Calas et al. (1984), Nahon et al. (1982a and 1982b), and Paquet et al. (1982) among others.

Trioctahedral ferromagnesian smectites

Trioctahedral ferromagnesian smectites can be defined as trioctahedral smectites with Mg and Fe as predominant cations, and they can be represented by a complete Mg/Fe solid solution series from the end-member saponite to its pure iron analog which will be tentatively called iron saponite:

$$Mg_3(Si_{4-x}Al_x)O_{10}(OH)_2E_x^+ \ -- \ Fe_3^{2+}(Si_{4-x}Al_x)O_{10}(OH)_2E_x^+$$

where the layer charge x is assumed to be 0.6 or less. The magnesian end-member is well-known as saponite but the iron-end member is not well understood and has been a subject of speculation and confusion.

Trioctahedral ferromagnesian smectites along with nontronites were found to be the predominant authigenic clay minerals in Recent submarine sediments and the underlying crustal basalts. Comprehensive studies under the Deep Sea Drilling Project (DSDP) were conducted on the smectites from cores of submarine sediments and crustal basalts. Only a few representative investigations will be cited here, namely those by Cole and Shaw (1983), Badaut et al. (1985), Scheidegger and Stakes (1977), Lawrence et al. (1979), Seyfried et al. (1978), Hein and Sholl (1978), and McMurtry et al. (1983). A few experimental studies on DSDP basalt cores were also conducted on the alteration of basalt with sea water (Seyfried and Bischoff, 1979) and with various hydrothermal solutions (Kurnosov et al., 1982). These investigations revealed the physico-chemical conditions of smectite formation in the above environments:

1. The temperature range of smectite crystallization varied from 3° at the bottom of the ocean up to 500^oC during the deuteric alteration of crustal basalts.
2. Interactions of basalts and other volcanic materials with juvenile water and seawater were responsible for the smectite formation.
3. Smectites were also directly precipitated from hydrothermal solutions around the oceanic spreading centers.
4. The low temperature reactions between iron oxyhydroxides, biogenic silica and the magnesia from sea water in the oceanic depressions also led to the crystallization of ferromagnesian smectites generally under reducing and alkaline conditions.

Moreover, authigenic ferromagnesian smectites have been
found as the main alteration and weathering products of
continental basalts and ultramafic rocks. (Nahon et al.,
1982b; Nahon and Colin, 1982; Kodama et al., 1988, among
others.) Chang (1986) reported the diagenetic formation of
ferromagnesian smectites in Cretaceous sediments derived
from mafic source rocks in Cassipore Basin offshore Brazil.

Pioneering work on iron-rich smectites was carried out
by Sudo (1954, 1978) on the alteration products of basalts
and pyroclastics in the Miocene Green Tuff Formation of
Japan. Among these smectites, Sudo reported an iron-
saponite with a Mg:Fe ratio of almost 1:1 in octahedral
sheets. Sudo (1954) defined the term "lembergite" for this
composition of trioctahedral smectite. Kodama (personal
communication) pointed out that another mineral, a synthetic
Na-nepheline hydrate, was also named by Lagorio in 1895 as
lembergite (Hey, 1955). This nomenclature problem should be
clarified. An iron saponite with a deep bluish-green color
was found at Oya (Kohyama et al., 1973) replacing rock
fragments in glassy rhyolitic tuffs from the Green Tuff
Formation. The Mg/Fe ratio in this smectite was about 1:2,
more than in Sudo's lembergite. A term "griffithite" was
introduced by Larsen and Steiger in 1928 (mentioned in Ross,
1960) for a saponite with Mg/Fe_{total} ratio of about 2:1 from
Griffith Pass, California.

Ferromagnesian smectites appear as members of a
complete solid solution series in various geological
environments. In view of this fact, a proper definition of
this series of smectites is needed, even if a pure iron end-
member must be hypothesized for the present. It is
desirable to have a name for this end-member. In the
meantime smectites with octahedral $Mg>Fe_{total}$ may be called
saponite and the others with $Fe_{total}>Mg$ iron saponite.

Saponite. Various natural saponites close to the
magnesian end-member composition were reported by Cahoon
(1954), Kitahara (1957), MacKenzie (1957), Midgley and Gross
(1956), Schmidt and Heysteck (1953), Weiss et al. (1955),
Wilson et al. (1968), and Whelan and Lepp (1961). The
average chemical composition of these saponites give a
structural formula (Köster, 1982):

$$(Mg_{2.92}Fe^{3+}_{0.02}Al_{0.03})(Si_{3.53}Al_{0.47})O_{10}(OH)_2E^+_{0.48}.$$

The formula indicates a total octahedral occupancy of 2.97
and a mean layer charge of 0.48, which is almost entirely
derived from tetrahedral substitutions.

A clay from Kozakov, Czechoslovakia, has been used as a
prototype saponite in various studies by Suquet et al.
(1975, 1977, 1987). The saponite apparently forms as an
alteration product in vesicles of a basalt at the Kozakov
locality. The chemical composition of the saponite is
represented by:

$$(Mg_{2.50}Fe^{2+}_{0.26}Fe^{3+}_{0.24})(Si_{3.30}Al_{0.68}Fe^{3+}_{0.02})O_{10}(OH)_2$$

$$(Ca_{0.22}K_{0.01}Na_{0.01})$$

The octahedral iron in this saponite is significant with a Mg/Fe_{total} ratio of about 5:1. The smectite carries excess tetrahedral charges of 0.46, and its cation exchange capacity was determined to be 119 meq/100 g. The crystal lattice was C-centered monoclinic with unit-cell dimensions of $\underline{a}$ = 5.333, $\underline{b}$ = 9.233, $\underline{c}$ = 15.42 A, β = 96.66° (for a Na-saturated sample). The smectite occurred as laths up to 15 microns long and in equant platelets ranging in size from 1-5 microns. Three-, two-, and one-layer homogeneous hydrates of the saponite were found depending on the interlayer cation and the relative humidity. Stacking disorder in the hydrated layers varied depending on the number of water layers and on the interlayer cation.

Some unusual saponites with macroscopic 'asbestiform' morphology consisting of fibers as long as 10 mm were reported by Cowking et al. (1983) in Scotland, by Quakernaat (1970) in Pfalz, West Germany, and by Curtis (1976) in Derbyshire, England. These saponites with a distinct fibrous texture, which are sometimes referred to by the varietal name bowlingite, occur as hydrothermal alteration products in veins and vesicles of basaltic rocks. The saponite fibers range in length from 0.1 to 10 mm but these macroscopic fibers actually consist of a bundle of distinct laths with $\underline{a}$-axis elongation. The Mg/Fe_{total} ratios in the octahedral composition of the fibrous saponites mostly range from 3:1 to 10:1.

Another saponite with a different origin and in a different geological setting was described by Post (1984) near Ballarat, California. The saponite was formed by the alteration of a dolomitic limestone by hydrothermal solutions. The structural formula as derived from the x-ray fluorescence analysis indicates a saponite composition with a very small iron content:

$$(Mg_{2.61}Fe^{3+}_{0.06}Al_{0.15})(Si_{3.77}Al_{0.23})O_{10}(OH)_2$$

$$(Ca_{0.10}Na_{0.16}K_{0.02})$$

The sample, available from the Source Clay Repository of the Clay Minerals Society at the University of Missouri, Columbia, was examined with x-ray diffraction and electron microscopy. The bulk sample contains about 80% saponite with small amounts of quartz, diopside, feldspars, tremolite, and carbonates. The Na-saturated saponite has a 15.2 Å basal spacing in the air-dried state, which expands to 16.8 Å upon glycolation. Two morphological types of saponite particles occur; one as flakes with curled edges as large as 4.0 microns (Fig. 15c) and the other as typical crumpled smectite foils (Fig. 15d).

Iron saponite. Ferrous smectite will obviously be rather unstable under atmospheric conditions. It may oxidize immediately after removal from its native reducing environment. Such a ferrous smectite was synthesized by Harder only under reducing and highly alkaline conditions. A natural smectite with a composition close to the end-member ferrous smectite was recently reported by Badaut et al. (1985). Although the identification is still to be considered 'tentative', the properties and the behavior of this smectite are noteworthy at least to show the technical difficulties involved with its characterization. This ferrous trioctahedral smectite was found in Recent sediments of Atlantis II Deep in the Red Sea. The smectite occurs in a dark brown "soupy" mud containing about 80-90% interstitial brine at the top of the sediment column. These sediments, including the smectite, are assumed to have directly precipitated from the hot brines that are currently discharging in Atlantis II Deep (Bischoff, 1972). The clay was found to be very unstable when removed from the mud, and it rapidly changed its color from black to red upon air-drying. In order to minimize the oxidation of ferrous smectites, special pretreatments of the clay were selected, and "transmission" x-ray diffraction was used. The wet mud, after washing off the salts, was packed between Mylar sheets and analyzed. The $\underline{b}$-dimension of the smectite, as determined in the bulk sample, was 9.34 Å. The basal spacing of the smectite was 13.5 Å, which expanded to 16.6 Å upon glycolation.

A small portion of the original mud was dispersed in distilled water and transfered onto the grids for electron microscopic examination. The smectite particles were found to have two different morphologies: a) crumpled foils about 1-2 microns in size, and b) thin elongated platelets with 0.1 x 0.05 microns dimensions. SAED pattern of the crumpled smectite foils consisted of uniform hk rings and gave a $\underline{b}$-dimension of 9.33 Å. The SAED patterns of the thin platelets displayed the superposition of two sets of spots; one giving a $\underline{b}$-dimension of 9.32 Å (trioctahedral smectite), and the other set providing a $\underline{b}$-dimension of 9.09 Å (dioctahedral smectite). Thus, the thin platelets were composed of domains with either trioctahedral or dioctahedral character.

The x-ray spectra of the smectite particles indicated only Fe and Si as the major elements. From the presence of Fe as the predominant octahedral cation and the $\underline{b}$-dimension of 9.33 Å the investigators conclude that the smectite foils were the ferrous trioctahedral smectite with the approximate structural formula:

$$(Si_{4-x}Al_x) \; Fe_3^{2+}O_{10} \; (OH)_2 \; E_x^+.$$

The exact composition of the smectite is yet to be determined, but the observations discussed above are rather convincing for the presence of a trioctahedral iron smectite close to the iron end-member. The appearance of a second (dioctahedral iron) smectite phase within the thin platelets was related to the oxidation of Fe^{2+}, which seemed to start along the edges of particles and progress to the center of the platelets. The oxidation took place more extensively during the preparation of samples for Mossbauer spectroscopy. The later analysis showed that all iron in smectite was in octahedral sheets. Significant amounts of free iron oxides were also detected when the samples were left in the air for about one week. The above observations show that:

1. The ferrous smectite is extremely unstable under ambient conditions, and its structure rapidly changes due to oxidation of Fe^{2+}. The oxidation reaction can simply be written as:

$$3 \ Fe^{2+}_{clay} + 0.5 \ O_2 \ ---> \ 2 \ Fe^{3+}_{clay} + 0.5 \ Fe_2O_3.$$

The iron oxide (or oxyhydroxide) was found to be closely associated with the smectite particles, and sometimes migrated to the edges of the particles. This free iron oxide was only detectable with Mössbauer spectroscopy.

2. The instability of ferrous smectite when removed from its native environment creates major difficulties for its identification with ordinary methods and sample preparation techniques. Upon oxidation, the original ferrous smectite gradually transforms to an intimate mixture of a ferric smectite (nontronite-like) and iron oxides (or oxyhydroxides) without the destruction of the original lattice. The oxidation product may sometimes appear as an intermediate di- and trioctahedral smectite that often manifests itself with an excess of Fe atoms for a dioctahedral smectite. This is a frequent conclusion in the literature, and inferences are made for the existence of smectites with intermediate character between dioctahedral nontronites and ferrous trioctahedral smectites.

However, a few examples of correctly characterizing the nature of iron-rich smectites will be mentioned. Kodama et al. (1988) found a ferromagnesian smectite in the saprolite of a ferrogabbro at Mont Megantic, Quebec. The structural formula of the smectite was derived from the chemical analysis of the sample after removal of amorphous iron oxides by the dithionite-citrate-bicarbonate method, and considering the Mössbauer spectra, as:

$$(Mg_{1.39}Fe^{3+}_{0.85}Al_{0.17}Mn_{0.03})(Si_{3.49}Al_{0.51})O_{10}(OH)_2Na_{0.61}$$

The above structural formula with a total octahedral occupancy of 2.44 strongly suggests the presence of a smectite of intermediate di- and trioctahedral character. However, Kodama et al. (1988) considered the possibility of the oxidation of an originally ferroan saponite, and revealed the original trioctahedral nature of the smectite. In fact, the original Fe^{2+} content of the structure can be calculated by multiplying the present Fe^{3+} content by 1.5. This will lead to an amount of $Fe_{1.28}$ in the above formula and increase the total octahedral occupancy to 2.87.

Another example of ferromagnesian smectites with apparently di- and trioctahedral character was discussed by Daynyak et al. (1980). A smectite formed as the alteration product of a continental basalt from Tunguska trap rocks and another smectite occurring in an oceanic basalt core were examined and compared. Both smectites were determined to have b-dimensions of 9.22 Å, typical of a trioctahedral smectite. When the chemical analyses were converted to the structural formulae under the assumption of a perfect anionic framework with $O_{10}(OH)_2$, the following results were obtained:

Continental smectite

$$(Al_{1.07}Fe^{3+}_{0.22}Fe^{2+}_{1.26}Mg_{0.07})(Si_{3.21}Al_{0.47}Fe^{3+}_{0.32})O_{10}(OH)_2E^+_x \quad .$$

Oceanic smectite

$$(Fe^{3+}_{0.55}Fe^{2+}_{0.46}Mg_{1.79})(Si_{3.40}Al_{0.59}Fe^{3+}_{0.01})O_{10}(OH)_2E^+_x \quad .$$

The formulae indicate an intermediate di- and trioctahedral character for the smectites with total octahedral occupancies of 2.62 and 2.80, respectively. However, the Mössbauer spectra of the smectites showed significantly high quadruple splitting ($\Delta \approx 1$ mms^{-1}) as compared to 0.4-0.6 mms^{-1} for octahedral Fe^{3+} in the other smectites. This anomalously high quadruple splitting was similar to that of Fe^{3+} in dehydroxylated trioctahedral smectites. The replacement of OH by O in the octahedral coordination of Fe^{3+} and the ensuing distortions were found to be responsible for the high Δ in Mössbauer spectra of these samples. This hydroxyl deficit was confirmed by the combined gas chromatographic-thermogravimetric analysis of the smectites. The hydroxyl content was found to be 1.1% for the continental smectite and 2.4% for the oceanic smectite, considerably lower than the expected values of 4.3 and 4.2% under the assumption of $O_{10}(OH)_2$, respectively. When the anionic compositions were adjusted to the observed hydroxyl content and part of the Fe^{3+} was recalculated as Fe^{2+} the following structural formulae were obtained for the continental smectite:

$$(Al_{1.26}Fe^{3+}_{0.38}Fe^{2+}_{1.29}Mg_{0.07})(Si_{3.42}Al_{0.50}Fe^{3+}_{0.08})$$
$$O_{11.36}(OH)_{0.64}E^+_x$$

and for oceanic smectite:

$$(Al_{0.14}Fe^{3+}_{0.59}Fe^{2+}_{0.47}Mg_{1.80})(Si_{3.52}Al_{0.48})O_{10.78}(OH)_{1.22}E^+_x \quad .$$

Both smectites were perfectly trioctahedral with a total octahedral occupancy of 3.00.

552

ACKNOWLEDGMENTS

Special appreciations are due to Hideomi Kodama for his review of the section on trioctahedral smectites and to Steve Guggenheim for reviewing an early version of the part on dioctahedral smectites. I also thank B. L. Allen for his correction of English in the text and his overall review of the manuscript. My graduate students Siva Sivalingham, Charlie Landis, and Darrell Brownlow have been of great assistance during the preparation of the manuscript. This work has been supported by the Texas Advanced Technology Research Program.

I am especially grateful to the following gentlemen for generously sharing with me their invaluable smectite samples for my electron microscopic examination: Joseph W. Stucki for Garfield nontronite, Eugene E. Foord for volkonskoites, Pei-lin Tien for swinefordite, and M.I. Knudson, Jr. for laponite.

REFERENCES

Alberti, A. and Brigatti, M.F. (1985) Crystal chemical differences in Al-rich smectites as shown by multivariate analyses of variance and discriminant analysis. Clays and Clay Minerals 33, 546–558.

Ames, L.L., Jr., Sand, L.B. and Goldrich, S.S. (1958) A contribution on the Hector, California bentonite deposit. Econ. Geol. 53, 22–37.

Badaut, D. and Risacher, F. (1983) Authigenic smectite on diatom frustules in Bolivian saline lakes. Geochem. Cosmochim. Acta 47, 363–375.

_____, Besson, G., Decarreau, A. and Rautureau, R. (1985) Occurrence of a ferrous, trioctahedral smectite in recent sediments of Atlantis II Deep, Red Sea. Clay Minerals 20, 389–404.

Bailey, S.W. (1975) Cation ordering and pseudosymmetry in layer silicates. Am. Mineral. 60, 175–182.

_____ (1984) Crystal chemistry of the true micas. Ch. 2 in Reviews in Mineralogy v. 13 Micas S.W. Bailey ed. Mineral. Soc. Am. 13–60.

Baird, T., Cairns-Smith, A.G. and MacKenzie, D.W. (1973) An electron microscopic study of magnesium smectite synthesis. Clay Minerals 18, 17–26.

_____, _____, _____ and Snell, D. (1971) Electron microscope studies of synthetic hectorite. Clay Minerals 9, 250–252.

Barrer, R.M. and Jones, D.L. (1970) Synthesis and properties of fluorhectorites. J. Chem. Soc. A. 1531–1537.

Besnus, Y., Fusil, G., Janot, C., Pinta, M. and Siefferman, G. (1975) Characteristics of some weathering products of chromitic ultrabasiac rocks in Bahia State, Brazil: nontronites, chlorites, and chromiferous talc. Proc. Inter. Clay Conf. Mexico, 27–34.

Besson, G. and Tchoubar, C. (1972) Determination du group de symmetry du feuillet elementaire de la beidellite. C.R. Acad. Sci. Paris 275, 633–636.

_____, Bookin, A.S., Dainyak, L.G., Rautureau M., Tsipursky, S.I., Tchoubar, C. and Drits, V.A. (1983b) Use of diffraction and Mössbauer methods for the structural and crystallochemical characterization of nontronites. J. Appl. Cryst. 16, 374–383.

_____, Glaeser, R. and Tchoubar, C. (1983a) Le cesium, revelateur de structure des smectites. Clay Minerals 18, 11–19.

Bischoff, J.L. (1972) A ferroan nontronite from the Red Sea geothermal system. Clays and Clay Minerals 20, 217–223.

Biscoe, T. and Warren, B.E. (1942) An x-ray study of carbon black. J. Appl. Physics, 364–371.

Bonnin, D., Calas, G., Suquet, H., and Pezerat, H. (1985) Sites occupancy of Fe^{3+} in Garfield nontronite. A spectroscopic study. Phys. Chem. Minerals 12, 55–64.

Borchardt, G.A. (1977) Montmorillonite and other smectite minerals in: Minerals in Soil Environments, J.B. Dixon and S.B. Weed eds. Soil Science Soc., Madison, Wisconsin, 293–330.

Bradley, W.H. and Fahey, J.J. (1962) Occurrence of stevensite in the Green River Formation of Wyoming. Am. Mineral. 47, 996–998.

_____ and Eugster, H.P. (1969) Geochemistry and paleolimnology of the trona deposits and associated authigenic minerals of the Green River Formation of Wyoming. U.S. Geol. Surv. Prof. Paper 496B, 71p.

Brigatti, M.F. (1983) Relationships between compositon and structure in Fe-rich smectites. Clay Minerals 18, 177–186.

_____ and Poppi, L. (1981) A mathematical model to distinguish the members of the dioctahedral smectite series. Clay Minerals 16, 81–89.

Brindley, G. (1980) Order-disorder in layer silicates. Ch. 3 in: Crystal Structures of Clay Minerals and their X-ray Identification, G.W. Brindley and G. Brown eds. Mono. 5 Mineral. Soc., London, 1–124.

Bruce, L.A., Sanders, J.V. and Turney, T.W. (1986) Hydrothermal synthesis and characterization of cobalt clays. Clays and Clay Minerals 34, 25–36.

Bystrom-Brusewitz, A.M. (1976) Studies on the Li test to distinguish between beidellite and montmorillonite. Proc. Int'l Clay Conf. Mexico City, 1975, S.W. Bailey, ed. Applied Publishing Ltd. Wilmette, Illinois, 419–428.

Cahoon, H.P. (1954) Saponite near Milford, Utah. Am. Mineral. 39, 222–230.

Caillère, S. and Henin, S. (1962) Vues d'ensemble sur le probleme de la synthese des mineraux argileux a basse temperature. Collogue No. 105 du C.N.R.S. 32–41.

Calas, G., Manceau, A., Novikoff, A. and Boukili, H. (1984) Comportement du chrome dans les mineraux d'alteration du gisement de Campo Formoso (Bahia, Brazil) Bull. Mineral. 107, 755–766.

Cardile, C.M. (1987) Structural studies of montmorillonites by ^{57}Fe Mössbauer Spectroscopy. Clay Minerals 22, 387–394.

_____ and Johnston, J.H. (1985) Structural studies of nontronites with different iron contents by Fe Mössbauer spectroscopy. Clays and Clay Minerals 33, 295–300.

Chang, H.K., Mackenzie, F.T. and Schoonmaker, J. (1986) Comparisons between the diagenesis of dioctahedral and trioctahedral smectite, Brazilian Offshore Basins. Clays and Clay Minerals 34, 407–423.

Chen, P.Y., Wan, H.M. and Brindley, G.W. (1976) Beidellite clay from Chang-Yuan Taiwan: geology and mineralogy. Clay Minerals 11, 221–233.

Chukhrov, F.V., Zvyagin, B.B., Drits, V.A., Gorshkov, A.I., Ermilova, L.P., Golio, E.A., and Rudnitskaya, E.S. (1979) The ferric analog of pyrophyllite and related phases. Proc. Int'l. Clay Conference, Oxford, 55–64.

554

Cole, T.G. and Shaw, H.F. (1983) The nature and origin of authigenic smectites in some recent marine sediments. Clay Minerals 18, 239–252.

Cowking, A., Wilson, M.J., Tait, J.M. and Robertson, R.H.S. (1983) Structure and swelling of fibrous and granular saponitic clay from Orrock Quarry, Fife, Scotland. Clay Minerals 18, 49–64.

Curtis, C.D. (1976) Unmixed Ca^{2+}/Mg^{2+} saponite at Calton Hill, Derbyshire. Clay Minerals 11, 85–89.

Darragi, F. and Tardy, Y. (1987) Authigenic trioctahedral smectites controlling pH, alkalinity, silica and magnesium concentrations in alkaline lakes. Chem. Geol. 63, 59–72.

Daynyak, L.G. and Drits, V.A. (1987) Interpretation of Mossbauer spectra of nontronite, celadonite, and glauconite. Clays and Clay Minerals 35, 363–372.

_____, _____, Kudryavtsev, I., Simanovich, I.M. and Slonimskaya, M.V. (1980) Crystallochemical specificity of trioctahedral smectite containing ferric iron, the secondary alteration products of oceanic and continental basalts. Doklady Akademii Nauk SSSR, 259, 1458–1462.

Decarreau, A. (1980) Cristallogenese experimentale des smectites magnesiennes: hectorite, stevensite. Bull. Mineral. 103, 579–590.

_____ (1981) Cristallogenese a basse temperature de smectites trioctahedrique par vieillisement de coprecipites silico metallique de formule$(Si_{4-x})M_3O_{10} \cdot nH_2O$. C.R. Acad. Sci. Paris 292, 61–64.

_____ (1985) Partitioning of divalent transition elements between octahedral sheets of trioctahedral smectites and water. Geochim. Cosmochim. Acta 49, 1537–1544.

_____ and Bonnin, D. (1986) Synthesis and crystallogenesis at low temperture of Fe(III)-smectites by evolution of coprecipitated gels: experiments in partially reducing conditions. Clay Mineral. 21, 861–877.

_____, _____, Badaut-Trauth, D., Couty, R. and Kaiser, P. (1987a) Synthesis and crystallogenesis of ferric smectite by evolution of Si-Fe coprecipates in oxidizing conditions. Clay Minerals 22, 207–223.

_____, Colin, F., Herbillon, A., Manceau, A., Nahon, D., Paquet, H., Trauth-Badaud, D. and Trescases, J.J. (1987b) Domain segregation in Ni-Fe-Mg-smectites. Clays and Clay Minerals 35, 1–10.

Drits, V.A., Plancon, A., Sakhorov, B.A., Besson, G., Tsipursky, S.I. and Tchoubar, C. (1984) Diffraction effects calculated for structural models of K-saturated montmorillonite containing different types of defects. Clay Minerals 19, 541–561.

Dyni, J.R. (1976) Trioctahedral smectite in the Green River Formation, Duchesne County, Utah. U.S. Geol. Surv. Prof. Paper 967, 14 p.

Eberl, D.D., Jones, B.F. and Khoury, H.N. (1982) Mixed-layer kerolite/stevensite from the Amargosa Desert, Nevada. Clays and Clay Minerals. 30, 321–326.

Eggleton, R.A. (1977) Nontronite: chemistry and x-ray diffraction. Clay Minerals 12, 181–194.

Faust, G.T. and Murata, K.J. (1953) Stevensite, redefined as a member of the montmorillonite group. Am. Mineral. 38, 973–987.

Foord, E.F., Starkey, H.C., Taggart Jr., J.E. and Shawe, D.R. (1987) Reassessment of the volkonskoite-chromian smectite nomenclature problem. Clays and Clay Minerals 35, 139–149.

Fripiat, J., Letellier, M. and Levitz, P. (1984) Interaction of water with clay surfaces. Phil Trans. Royal Soc. London A 311, 287–299.

555

Goodman, B.A. Russell, J.D. and Fraser, A.R. and Woodhams, F.W. D. (1976) A Mossbauer and IR Spectroscopic study of the structure of nontronite. Clays and Clay Minerals 24, 53-59.

Granquist, W.T. and Pollack, S.S. (1960) A study of the synthesis of hectorite. Clays and Clay Minerals 8, 150-169.

Gray, G.W. (1962) Molecular Structure and the Properties of Liquid Crystals, Ch. 2, Academic Press, London, 17-54.

Grim, R.E. (1968) "Clay Mineralogy", 2nd Ed., McGraw-Hill Book Co., New York, 596 P.

_____ and Guven, N. (1978) Bentonites, Geology, Mineralogy, Properties and Uses. Elsevier Scientific Publ. Co. Ch. 4, 161-214.

_____ and Kulbicki, G. (1961) Montmorillonite: high temperature reactions and classification. Am. Mineral. 46, 1329-1369.

Güven, N. and Carney, L.L. (1979) The transformation of sepiolite to stevensite and the effect of added chloride and hydroxides. Clays and Clay Minerals 27, 253-260.

_____ and Hower, W.F. (1979) A vanadium smectite. Clay Minerals 14, 241-245.

_____ and Pease, R.W. (1975) Selected area electron diffraction studies on beidellite. Clay Minerals 10, 427-436.

_____, _____, and Murr, L.E. (1977) Fine structure in the selected area diffraction patterns of beidellite, and its dark-field images. Clay Minerals 12, 67-74.

Hamilton, J.D. (1971) Beidellitic montmorillonite from Swansea, New South Wales, Clay Minerals 9, 107-123.

Harder, H. (1972) The role of magnesium in the formation of smectite minerals. Chem. Geol. 10, 31-39.

_____ (1976) Nontronite synthesis at low temperatures. Chemical Geology 18, 169-180.

_____ (1977) Clay mineral formation under lateritic weathering conditions. Clay Minerals 12, 281-288.

_____ (1978) Synthesis of iron layer silicates under natural conditions. Clays and Clay Minerals 26, 65-72.

_____ (1986) Synthesis of iron-rich clays in environments with little or no oxygen in Clay Minerals and the Origin of Life. eds. A.G. Cairns-Smith and H. Hartman, Cambridge University Press, 91-96.

Hazen, R.M. and Wones, D.R. (1972) The effect of cation substitutions on the physical properties of trioctahedral micas. Am. Mineral. 57, 103-129.

_____ and _____ (1978) Predicted and observed limits of trioctahedral micas. Am. Mineral. 63, 885-892.

Hein, J.R. and Scholl, D.W. (1978) Diagenesis and distribution of late Cenozoic volcanic sediment in the southern Bering Sea. Geol. Soc. Am. Bull. 89, 197-210.

Heller-Kallai, L. and Rozenson, I. (1981) The use of Mossbauer spectroscopy of iron in clay mineralogy. Phys. Chem. Minerals 7, 223-238.

Hey, M.H. (1955) An Index of Mineral Species and Varieties. 2nd edition. British Museum, p. 140.

Imai, N., Otsuka, R., Nakumara, T. and Tsunashima, A. (1970) Stevensite from the Akatani Mine, Niigata Prefecture, Northeastern Japan. Clay Science 4, 11-29.

Khoury, H.N., MacKenzie, R.C. Russell, J.D. and Tait, J.M. (1984) An iron-free volkonskoite. Clay Minerals 19, 43-57.

Kitahara, Jun-Ichi (1957) Saponite from the Wakamatsu Mine, Tari district, Tottori Prefecture. J. Japanese Assoc. Mineral. Petr. Econ. Geol. 41, 148-151.

556

Kodama, H., De Kimpe, C.R. and Dejou, J. (1988) Ferrian saponite in a gabbro saprolite at Mount Megantic, Quebec. Clays and Clay Minerals 36, 102-110.

Kohyama, N., Shimoda, S. and Sudo, T. (1973) Iron-rich saponite (ferrous and ferric forms). Clays and Clay Minerals 21, 229-237.

Köster, H.M. (1982) The crystal structure of 2:1 layer silicates. Int'l. Clay Conf. 1981 eds. H.V. Olphen and F. Veniale, Elsevier Publ. Co., pp 41-71.

Kurnosov, V.B., Kholodkevich, I.V., Kokoria, L.P., Kotov, N.V. and Chudaev, O.V. (1982) The origin of clay minerals in the oceanic crust revealed by natural and experimental data. Int'l Clay Conf. 1981, eds. H.V. Olphen and F. Veniale, Elsevier Publ. Co. 547-556.

Lagaly, G. (1984) Clay-organic interactions. Phil. Trans. Royal Soc. London A 311, 315-332.

Lawrence, J.R., Drever, J.I., Anderson, T.F. and Brueckner, H.K. (1979) Importance of alteration of volcanic material in the sediments of deep sea drilling site 323: Chemistry, $^{18}O/^{16}O$, $^{87}Sr/^{86}Sr$. Geochem. Cosmochim. Acta 43, 573-588.

Lee, J.H. and Guggenheim, S. (1981) Single crystal x-ray refinement of pyrophyllite-1Tc. Am. Mineral. 66, 350-357.

Lin, C-yi and Bailey, S.W. (1984) The crystal structure of paragonite - $2M_1$. Am. Mineral. 69, 122-127.

Mackenzie, R.C. (1957) Saponite from Allt Ribbein, Fiskavaig Bay, Skye. Mineral. Mag. 31, 672-680.

_____ (1984) Constitution of and relationships among volkonskoites. Clay Minerals 19, 669-671.

Mamy, J. and Gaultier, J.P. (1976) Evolution de l'ordre crystalline dans la montmorillonite en relation avec la diminution d'exchangebilite' du potassium. Proc. Int'l Clay Conference 1975 Mexico City, S.W. Bailey, ed., Applied Publishing Ltd. Wilmette, Illinois. 149-155.

Manceau, A. and Calas, G. (1985) Hetrogeneous distribution of nickel in hydrous silicates from New Caledonia Ore deposits. Am. Mineral. 70, 549-558.

McAtee, J.L. and Lamkin, G. (1979) A modified freeze-drying procedure for the electron microscopic examination of hectorite. Clays and Clay Minerals 27, 293-296.

McEwan, D.M.C. (1961) Montmorillonite minerals. Ch. 4 in the X-ray Identification and Crystal Structures of Clay Minerals. G. Brown, ed., Mineral. Soc., London, 143-207.

_____ and Wilson, M.J. (1980) Interlayer and intercalation complexes of clay minerals. Ch. 3 in Crystal Structures of Clay Minerals and Their X-Ray Indentification. G.W. Brindley and G. Brown, eds. Mineral. Soc. Monogr. 5. London, 197-248.

McMurtry, G.M. and Yeh, H.W. (1981) Hydrothermal clay mineral formation of East Pacific Rise and Bauer Basin sediments. Chem. Geol. 32, 189-205.

Mering, J. (1975) Smectites in Soil Components 2, Inorganic components, J.E. Gieseking, ed., Springer Verlag, New York 98-120.

_____ and Oberlin, A. (1971) The smectites, in The Electron-Optical Investigations of Clays, J.A. Gard, ed., Mineral. Soc. Monogr. 3, 193-229.

Midgley, H.G. and Gross, K.A. (1956) Thermal reactions of smectities. Clay Minerals Bull. 3, 79-90.

Murad, E. (1987) Mössbauer spectra of nontronites: structural implications and characterization of associated iron oxides. Zeits. Pflanzenernaehr. Bodenk. 150, 279-285.

557

Nadeau, P.H., Farmer, V.C., McHardy, W.J. and Bain, D.C. (1985) Compositional variations in the Unterrupsroth beidellite. Am. Mineral. 70, 1004–1010.

Nahon, D.B. and Colin, F. (1982) Chemical weathering of orthopyroxenes under lateritic conditions. Am. J. Sci. 282, 1232–1243.

_____, Colin, F. and Tardy, Y. (1982a) Formation and distribution of Mg, Fe, Mn smectites in the first stages of the lateritic weathering of fosterite and tephroite. Clay Minerals 17, 339–348.

_____, Paquet, H. and Delvigne, L. (1982b) Lateritic weathering of ultramafic rocks and the concentration of nickel in the western Ivory Coast. Econ. Geol. 77, 1159–1175.

Neumann, B.S. and Sansom, K.G. (1970) The formation of stable sols from laponite, a synthetic hectorite–like clay. Clay Minerals 8, 389–404.

Newman, A.C.D. (1987) The interaction of water with clay mineral surfaces. Ch. 5 in: Chemistry of clays and Clay Minerals, A.C.D. Newman ed. Mono. 6 Mineral. Soc., John–Wiley, New York, 237–274.

Odom, I.E. (1984) Smectite clay minerals: Properties and uses. Phil. Trans. R. Soc. London, series A, 311, 391–409.

Paquet, H., Duplay, J. and Nahon, D. (1982) Variations in the composition of phyllosilicate monoparticles in a weathering profile developed on ultrabasic rocks. Int'l Clay Conf. 1981 H.V. Olphen and F. Veniale, eds. Elsevier Publ. Co., 595–603.

Perdikatsis, B. and Burzlaff, H. (1981) Struktur verfeinerung am Talk $Mg_3[(OH)_2Si_4O_{10}]$. Zeits. Kristallogr. 156, 177–186.

Perruchot, A. (1976) Contribution a l'etude des echanges d'ions dans les gels silicates. Comportement des elements alcalino-terreux et de quelques elements de transition, Mn^{2+}, Fe^{2+}, Co^{2+}, Ni^{2+}, Cu^{2+}, Zn^{2+}. Bull. Soc. Fr. Mineral. Cristallogr. 99, 234–242.

Post, J.L. (1984) Saponite from near Ballarat, California. Clays and Clay Minerals 32, 147–153.

Quakernaat, J. (1970) A new occurrence of a macrocrystalline form of saponite. Clay Minerals 8, 491–493.

Randall, B.A.O. (1956) Stevensite from the Whin Sill in the region of the North Tyne. Mineral. Mag. 32, 218–229.

Rausell–Colom, J.A. and Serratosa, J.M. (1987) Reactions of clays with organic substances. Ch. 8 in: Chemistry of Clays and Clay Minerals, A.C.D. Newman, ed. Mono. 6 Mineral. Soc., John–Wiley, New York, 371–422.

Rayner, J.H. and Brown, G. (1973) The crystal structure of talc. Clays and Clay Minerals 21, 103–114.

Rivière, M., Rautureau, M., Besson, G., Steinberg, M. and Amouri, M. (1985) Complementarite des rayons X et de la microscopic electronique pour la determination des diverses phases d'une argile zincifere. Clay Minerals 20, 53–67.

Ross, C.S. (1946) Sauconite – a clay mineral of the montmorillonite group. Am. Mineral. 31, 411–424.

_____ (1960) Review of the relationships in the montmorillonite group. Am. Mineral. 31, 411–424.

_____ and Hendricks, S.B. (1945) Minerals of the montmorillonite group; their origin and relation to soils and clays. U.S. Geol. Survey Prof. Paper 205–B, 23–79.

Russell, J.D., Goodman, B.A. and Fraser, A.R. (1979) Infrared and Mossbauer studies of reduces nontronites. Clays and Clay Minerals 27, 63–71.

Saalfeld, H. and Wedde, M. (1974) Refinement of the crystal structure of gibbsite $Al(OH)_3$. Zeits. Kristallogr. 139, 129–135.

558

Sakamato, T., Koshimizu, M., Shimoda, S. and Otsuka, R. (1982) Hydrothermal transformation of some minerals into stevensite. Int'l. Clay Conf. 1981, H.V. Olphen and F. Veniale, eds. Elsevier Publ. Co., 537-546.

Scheidegger, K.F. and Stakes, D.S. (1977) Mineralogy, chemistry, and crystallization sequence of clay minerals in altered tholeiitic basalts from the Peru Trench. Earth Planet. Sci. Letters 36, 413-423.

Schmidt, E.R., and Heysteck, H. (1953) a saponite from Krugersdorp district, Transvaal. Mineral. Mag. 30, 201-210.

Schultz, L.G. (1969) Lithium and potassium absorption, dehydroxylation temperature and structural water content of aluminous smectites. Clays and Clay Minerals 19, 137-150.

Seyfried, W.E., Jr. and Bischoff, J.L. (1979) Low Temperature basalt alteration by seawater: an experimental study at 70° and 150°. Geochim. Cosmochim. Acta 43, 1937-1947.

_____, Shanks, W.C., III and White, D. E., (1978) Clay mineral formation in DSDP Leg 37 basalt. Earth Planet. Sci. Letters 41, 265-276.

Shimoda, S. (1971) Mineralogical studies of a species of stevensite from the Obori Mine, Yamagata Prefecture, Japan. Clay Minerals 9, 185-192.

Shirozu, H. and Bailey, S.W. (1966) Crystal structure of a two-layer Mg-vermiculite. Am. Mineral. 51, 1124-1143.

Sudo, T. (1954) Iron-rich saponite from the Tertiary iron sand beds of Japan (re-examination of "lembergite") J. Geol. Soc. Japan, 60, 18-27.

_____ (1978) An outline of clays and clay minerals in Japan. in Clays and Clay Minerals of Japan, T. Sudo and S. Shimoda, eds. Developments in Sedimentology 26, Elsevier Sci. Publ. Co., New York 1-104.

Suquet, H., De La Calle, C. and Pezerat, H. (1975) Swelling and structural organization of saponite. Clays and Clay Minerals 23, 1-9.

_____, Iiyama, J.T., Kodama, H. and Pezerat, H. (1977) Synthesis and swelling properties of saponites with increasing layer charge. Clays and Clay Minerals 25, 231-242.

_____ and Pezerat, H. (1987) Parameters influencing layer stacking types in saponite and vermiculite: a review. Clays and Clay Minerals 35, 353-362.

Takeda, H. and Ross, M. (1975) Mica polytypism: dissimilarities in the crystal structures of co-existing 1M and $2M_1$ biotite. Am. Mineral. 60, 1030-1040.

Tardy, Y., Cheverry, C. and Fritz, B. (1974) Neoformation d'une argile magnesienne dans les depressions interdunaires du Lac Tchad - Application aux domaines de stabilite des phyllosilicates alumineux, magnesiens et ferriferes. C. R. Acad. Sci. Paris 278, 1999-2002.

Tessier, D. and Pedro, G. (1987) Mineralogical characterization of 2:1 clays in soils: Importance of the clay structure. Proceed. Intl. Clay Conference 1985, L.G. Shultz, van Olphen, H. and Mumpton, F.A., eds. Clay Minerals Soc., Bloomington, Indiana, 78-84.

Tettenhorst, R. and Moore, C.E. (1978) Stevensite oolites from the Green River Formation of Central Utah. J. Sed. Pet. 48, 587-594.

Thomas, J.M. (1984) New way of characterizing layered silicates and their intercalates. Phil. Trans. Royal Soc. London. A311, 271-285.

559

Tien, Pei-Lin, Leavens, P.B. and Nelen, J.A. (1975) Swinefordite, a dioctahedral-trioctahedral Li-rich member of teh smectite group form Kings Mountain, North Carolina. Am. Mineral. 60, 540–547.
Toroya, H. (1981) Distortions and octahedral sheets in 1M micas and the relation to their stability. Zeits. Kristallgr. 157, 173–190.
Tsipursky, S.I. and Drits, V.A. (1984) The distribution of octahedral cations in the 2:1 layers of dioctahedral smectites studied by oblique-texture electron diffraction. Clay Minerals 19, 177–193.
Vali, H. and Köster, H.M. (1986) Expanding behavior, structural disorder, regular and random interstratification of 2:1 layer silicates studied by high resolution images of transmission electron microscopy. Clay Minerals 21, 827–859.
Weaver, C.E. and Pollard, L.D. (1973) The Chemistry of Clay Minerals. Elsevier, Amsterdam. 213.
Weir, A.H. and Greene-Kelly, R. (1962) Beidellite. Am. Mineral. 47, 137–146.
Weiss, A., Koch, G. and Hofmann, U. (1955) Zur Kenntnis von saponit. Ber. Deutsch. Keram. Ges. 32, 12–17.
Weiss, Z., Rieder, M., Chmielova, M. and Krajicek, J. (1985) Geometry of the octahedral coordination in micas: a review of refined structures. Am. Mineral. 70, 747–757.
Whelan, L.A. and Lepp, H. (1961) An occurrence of saponite near Silver Bay, Minnesota. Am. Mineral. 46, 430–433.
Wilson, M.J. (1987) soil smectites and related interstratified minerals: recent developments. Proc. Int'l. Clay Conference, Denver, 1985, L.G. Schultz, H. van Olphen, and F.A. Mumpton, eds. Clay Minerals Soc., Bloomington, Indiana, 167–183.
_____, Bain, D.C. and Mitchell, W.A. (1968) Saponite from the Dalvadian meta-limestone of Northeast Scotland. Clay Minerals 7, 343–349.
Zigan, F. and Rothbauer, R. (1967) Neutronbeugungs-messungen am Brucit. N. Jahrb. Mineral. Monats., 137–143.
Zvyagin, B.B. and Pinsker, Z.G. (1949) Electron diffraction study of the montmorillonite structure. Dokl. Acad. Nauka SSR 68, 30–35.

VECTOR REPRESENTATION OF PHYLLOSILICATE COMPOSITIONS

INTRODUCTION

Geologists like graphical representations, and I am a geologist. The representation of rock and mineral compositions is conventionally done barycentrically, in terms of lines, triangles, or tetrahedra, with elements, oxides, or mineral endmembers at the corners, and the compositions of interest plotted internally. One thing that has always bothered me about this procedure ("triangle worship") is the counting of (even the naming of) components. In mineralogy or petrology, a one-component system is a point, a two-component system is a line, a three-component system is a plane, and so on. In vector mathematics, on the other hand, a single vector of one component has both a magnitude and direction, and thus defines a line (actually, a family of lines unless a starting point is specified), two vectors or two components define a plane, and so on. We can remove this discrepancy in nomenclature and counting by treating mineralogic and petrologic components as vectors (a procedure mentioned by J.B. Thompson, Jr. in 1979 and explained more fully in 1981 and 1982; cf. Smith, 1959; Bragg et al., 1965, p. 22-23).

The first step to understanding this procedure is to understand "exchange operators", components with negative quantities of certain elements that perform the operation of exchange (which is why I named them that: Burt, 1974, 1979; they have also been called "exchange components"). An example familiar to most petrologists is $FeMg_{-1}$, which substitutes an Fe for an Mg; I got started on this road with F_2O_{-1} (Burt, 1972). For solid solutions these components are "actual" in the terminology of J. Willard Gibbs (see Brady, 1975), in that they can be added or subtracted without destroying the homogeneity of a phase. Although others consider them "imaginary" or "negative" (something like the square root of -1), because they are physically unrealizable, they are useful algebraically and, as we shall see, geometrically.

In particular, ionic substitutions in minerals, as represented by exchange operators such as $FeMg_{-1}$, have both a magnitude and direction (the direction opposite to $FeMg_{-1}$ is $MgFe_{-1}$), so why not treat them as vectors? Such exchange vectors can be combined, yielding coupled substitution vectors. We can then generate an entire chemical composition space by the operation of these exchange vectors on a single point (termed the "additive component" by Thompson, 1982), in much the same way that a crystal lattice is generated by unit cell vectors operating on a single point. For complex systems we may want to use more than one additive component (although as shown by the first example below, one will suffice), but a good strategy is to keep the number of such components to a minimum. For a single mineral group, such as the chlorites, or a series of closely related groups, such as the phyllosilicates, a single additive component should generally suffice. In what follows, all of the common phyllosilicate compositions are generated via exchange vectors from a single starting point, the composition of brucite. This can be done on ordinary graph paper (no triangles needed).

Compare this approach with the conventional approaches of Stoessel (1984) and Walsche (1986), for example, who each require six (somewhat different) endmembers to describe compositional variations in a single mineral, chlorite. Note that, according to Bragg et al. (1965, p. 22), such a casting into endmembers (part of what I shall call the "tyranny of the triangle") is "based on a principle which is fundamentally wrong" because "it implies the existence of a chemical 'molecule'"; molecules "do not exist in ... minerals, which are continuous". Furthermore "in cases of complex substitution, the variations of replacement are so extensive that it is almost impossible to frame a list of endmembers which covers all the permutations. On the other hand, when the structure is taken into account the apparently bewildering varieties of composition are seen to fall into an extremely simple scheme" because (p. 23) "instead of a list of 'endmembers' it is simpler to indicate the nature and extent of the substitution ..." Inspired by these quotations, shall we now continue?

In order to simplify further what follows, let us largely omit from discussion simple or homovalent ionic substitutions, such as $FeMg_{-1}$, $MnMg_{-1}$, $NiMg_{-1}$, $ZnMg_{-1}$, $Fe^{3+}Al_{-1}$, $Cr^{3+}Al_{-1}$, BAl_{-1}, $F(OH)_{-1}$, and so on. We shall "condense down" these vectors onto the system $MgO-Al_2O_3-SiO_2-H_2O$, later adding Li_2O, K_2O (Na_2O), CaO (BaO), TiO_2, FeO, and Fe_2O_3. Adding axes for simple exchange vectors is a relatively simple procedure, demonstrated at the end with regard to iron. Let us first consider the complex coupled or heterovalent substitutions that complicate the existence of mineralogists and petrologists.

With one exception, all of the vectors derived below are oxygen-conservative (that is, involve cation exchange alone, if H^+ is included as a cation). An implication is that nearly all of the composition diagrams are expressed in gram-oxygen units (or combining minor F with OH, gram-anion units), according to the terminology of Brady and Stout (1980). This feature allows us to show only cation contents of formulas on many diagrams (the anion contents and some cation contents, if not indicated, remain constant). Furthermore, we can use multiples of normal phyllosilicate formula units, in order to avoid fractional subscripts (insofar as possible, at least).

VECTOR REPRESENTATIONS

As an introduction to the method, Figure 1 gives three alternative representations of mineral and other interesting compositions in the system $MgO-Al_2O_3-SiO_2$ (cf. Newton, 1987). Figure 1A is the conventional triangular representation, in terms of barycentric coordinates. The position of each phase or composition is weighted according to the proportions of each oxide in the mineral. Thus pyrope garnet, $Mg_3Al_2Si_3O_{12}$, abbreviated Prp, has 3 moles (gram formula units) of MgO, 1 of Al_2O_3, and 3 of SiO_2, and is correspondingly plotted at the center of gravity of a triangle with weights of 3, 1, and 3 on each of the corners, as shown. You can also plot compositions outside the triangle, with negative quantities of the elements, if you consider the analog of upward forces (e.g., the center of gravity of systems that include helium balloons). Two such useful compositions in this system are the exchange operators $MgSiAl_{-2}$ (the "Tschermak substitution:" cf. Tschermak, 1890; Thompson, 1979; Miyashiro and Shido, 1985) and $[]Al_2Mg_{-3}$ (the dioctahedral-trioctahedral substitution, where [] is an octahedral vacancy). These respectively consist, in terms of oxide components, of MgO + SiO_2 $-Al_2O_3$, and of Al_2O_3 - 3 MgO. The latter point plots negatively (beyond infinity to the right; cf. Korzhinskii, 1959; Thompson, 1982) Note the large number of collinearities or compositions that can be related to each other by $MgSiAl_{-2}$. These include the unstable composition $MgSi_2O_5$, obtained by performing the Tschermak substitution on Al_2SiO_5 (this composition turns up again in connection with the formation of phyllosilicates). Other than the basal oxides, only forsterite can be related to pyrope by Al_2Mg_{-3}. The unstable oxide composition $Mg_3Al_2O_6$ is shown because it is half dioctahedral, half trioctahedral (a phenomenon that occurs in a number of phyllosilicates, including, e.g., sudoite, a di,trioctahedral chlorite-group mineral, as detailed below).

Given the points $MgSiAl_{-2}$ and Al_2Mg_{-3} on Figure 1A, we might be tempted to use them as corners of a barycentric triangle (more "triangle worship"). A possible result is Figure 1B, for which the composition corner MgO is multiplied by 3 to make corundum plot at $X(Al_2Mg_{-3})$ = 0.5 (halfway across the base). The physically accessible part of the triangle is unchanged from Figure 1A and is the portion marked by periclase, corundum, and quartz at the corners. Note again the several lines

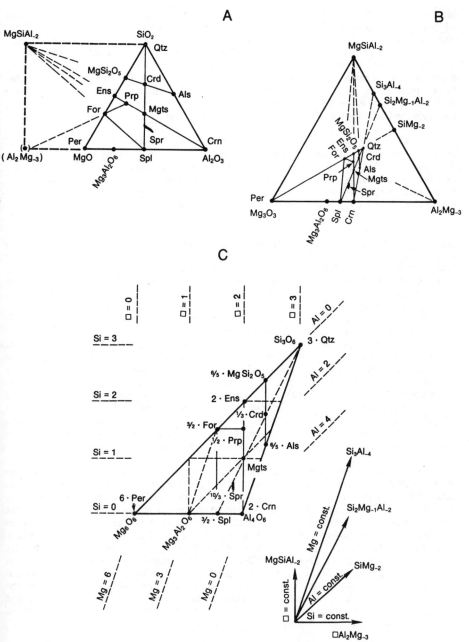

Figure 1. Alternate depictions of phase compositions in the system MgO-Al₂O₃-SiO₂. (A) Conventional barycentric oxide triangle, showing the compositions of the exchange operators MgSiAl₋₂ and Al₂Mg₋₃ (which plots negatively). (B) Barycentric triangle with 3MgO, Al₂Mg₋₃, and MgSiAl₋₂ at the corners. (C) Vector representation, in terms of the exchange vectors MgSiAl₋₂ and []Al₂Mg₋₃ as basis vectors, and 6MgO as the additive component, so that mineral formulas are normalized to 6 oxygens. Inset to lower right shows the vector scale and the orientation of the derived vectors SiMg₋₂ (line of constant Al), Si₃Al₋₄ (line of constant Mg), and Si₂Mg₋₁Al₋₂ (discussed in text).
Abbreviations: Per, periclase; Crn, corundum; Qtz, quartz; Als, Aluminosilicate (unspecified); For, forsterite; Ens, enstatite; Crd, cordierite; Prp, pyrope; Spr, sapphirine; Mgts, magnesiotschermaks component, MgAl₂SiO₆.

intersecting at the Tschermak substitution, $MgSiAl_{-2}$. By adding the composition Al_2Mg_{-3} to $MgSiAl_{-2}$, we get a composition halfway between them marked $SiMg_{-2}$ (the Al's cancel). Similarly, by adding Al_2Mg_{-3} to three times $MgSiAl_{-2}$, we get Si_3Al_{-4} (the Mg's cancel). The first is collinear with the periclase-quartz line on the figure, the second with the corundum-quartz line. A third exchange operator, $Si_2Mg_{-1}Al_{-2}$, equal to Al_2Mg_{-3} plus twice $MgSiAl_{-2}$, marks a point collinear with the spinel-quartz line. Can you figure out the points collinear with the line joining periclase, pyrope, cordierite or joining periclase, Mg-tschermak's pyroxene, aluminosilicate on Figure 1B? It is not easy on this type of representation.

Drawing such figures marked the limits of my understanding of the graphical uses of exchange operators from 1972 until 1981, when (independently of J.B. Thompson, Jr.,'s 1981 paper on amphiboles) I realized that it was simpler to use exchange operators as vectors, rather than points (although I had long been using their chemical potentials as vectors). I realized this while wrestling with the compositions of the lithium micas (cf. Cerny and Burt, 1984; also Burt, 1988b).

To illustrate this insight, Figure 1C gives a third representation in terms of $MgSiAl_{-2}$ and $[]Al_2Mg_{-3}$ as orthogonal exchange vectors of unit length. (There is no inherent reason why these two basis vectors need to be made orthogonal and of the same length, although doing so facilitates the plotting of points or analyses on ordinary graph paper or via standard computer plotting routines.) The additive component from which all of the other compositions on the composition diagram are generated, is Mg_6O_6, or 6 times the formula of periclase, and the formula of every other phase is recalculated on the basis of 6 oxygens (any other number could have been chosen; 6 is convenient for this example). Thus pyrope, mentioned above, is $Mg_{1.5}AlSi_{1.5}O_6$, or half of its normal formula with 12 oxygens, and the Al-silicates are $Al_{2.4}Si_{1.2}O_6$ (or 6/5 times Al_2SiO_5). The position of pyrope, for example, is generated from periclase by going 2 unit vectors horizontally and 1.5 vertically (because $Mg_{1.5}AlSi_{1.5}O_6 = Mg_6O_6 + 2Al_2Mg_{-3} + 1.5MgSiAl_{-2}$), that of quartz, Si_3O_6, by going 3 units horizontally and 3 vertically (because $Si_3O_6 = Mg_6O_6 + 3Al_2Mg_{-3} + 3MgSiAl_{-2}$), and that of Al-silicate by going 2.4 units horizontally and 1.2 vertically (because $Al_{2.2}Si_{1.2}O_6 = Mg_6O_6 + 2.4Al_2Mg_{-3} + 1.2MgSiAl_{-2}$). The other points on the composition diagram are plotted similarly. The resulting triangle of accessible compositions is somewhat deformed, but still recognizable.

Plotting points on such composition diagrams is greatly aided by contouring lines of equal cation contents, as shown. Lines of equal Si are horizontal, and lines of equal octahedral vacancies (not significant for this diagram, but important when we apply the same two vectors to generate phyllosilicate compositions) are vertical. Two of the other vectors shown to the lower right, Si_3Al_{-4} and $SiMg_{-2}$, can be derived as described above by taking linear combinations of the two basis vectors so as to cancel, respectively, Mg and Al (operations shown graphically on the vector insert of Fig. 1C). The slopes of the vectors Si_3Al_{-4} and $SiMg_{-2}$ give, respectively, the slopes of lines of equal Mg (+3) and of equal Al (+1) on the composition diagram. The third vector, $Si_2Mg_{-1}Al_{-2}$, of slope +2, is, as mentioned above, the vector that defines the compositional collinearity among spinel, magnesiotschermaks component, cordierite, and quartz. This vector turns up in several places below (first in the discussion of chlorites) as the direction of constant octahedral Al.

In similar vector diagrams that follow, the vectors are labelled by the element(s) alone, with the suffix " = const." being understood. These vectors likewise define the slopes of lines of constant content of the element(s) in question (element contours) on the accompanying composition diagram.

Application of the vector Si_3Al_{-4} directly converts corundum, Al_4O_6, into quartz, Si_3O_6, generating aluminosilicate along the way. Similarly, $SiMg_{-2}$ applied three times converts periclase, Mg_6O_6, into quartz, generating forsterite and enstatite along the way. Incidentally, the point SiO_2 on Figure 1A, although it exists physically, can also be thought of as the acidic exchange operator $(SiO_4)O_{-2}$, that converts periclase to forsterite, as $(SiO_3)_2(SiO_4)_{-1}$, that converts forsterite to enstatite, or as $(SiO_5)O_{-3}$, that converts corundum to aluminosilicate. The point H_2O on barycentric diagrams discussed below is similarly the acidic anion exchange operator $(OH)_2O_{-1}$, which converts oxides into hydroxides. All simple exchange operators are intrinsically acidic or basic in the electronic or Lewis (1938) sense (Burt, 1974; 1979), a subject best left to other papers.

The physically accessible part of Figure 1C could be described as that subject to $0<Mg<6$, $0<Al<4$, and $0<Si<3$. Similar inequalities, generally referring to confusing subscripts x, y, and z in mineral formulas, such as, for Figure 1, $(Mg_{6-x}Al_{4-y}Si_{3-z})O_6$, are commonly used to describe accessible formula ranges for the stability of a given crystal structure. The vector type of representation is most useful for this type of discussion, because the exchange vectors directly and graphically reflect the substitutions in crystals (cf. the quotation above from Bragg, 1965, p. 23). For minerals as diverse as those on Figure 1, it offers no real advantage, although some interesting relations become apparent.

SIMPLE PHYLLOSILICATES

Two triangular diagrams similar to Figure 1A are given in Figures 2A and 2B for MgO-SiO_2-H_2O and Al_2O_3-SiO_2-H_2O, respectively (see Thompson, 1978, p. 247 for more detail on the former). Not shown by Thompson is the exchange operator common to both, $Si_2O_5(OH)_{-2}$ (or $Si_2O_3H_{-1}$), equal to $2SiO_2$ $-H_2O$ [to see this, recall that $H_2O = (OH)_2O_{-1}$], that plots at an $X(SiO_2)$ of $+2$ along the H_2O-SiO_2 line, as shown. This fundamental operator represents the addition of a tetrahedral $(Si_2O_5)^{2-}$ sheet to a sheet of hydroxides. Using it once converts brucite or gibbsite into a 1:1 or T-O serpentine mineral (trioctahedral) or kaolinite (dioctahedral), with a single tetrahedral sheet alternating with a hydroxide sheet. Note that these 7 Å phases are compositionally similar to the 2:1 or T-O-T-O or 14 Å chlorites which consist of a 2:1 or T-O-T tetrahedral layer separated by an extra hydroxide sheet. Consequently serpentine and kaolinite have been called "septechlorites" (a misleading term in view of their distinct structures and properties). Applying the operator again converts a 1:1 or T-O silicate into a 2:1 or T-O-T tetrahedral layer silicate, such as trioctahedral talc or dioctahedral pyrophyllite. Applying it a third time yields the unstable anhydrous compositions $MgSi_2O_5$ and $Al_{0.67}Si_2O_5$, as shown on Figures 2A and 2B, respectively (recall that $MgSi_2O_5$ also appeared on Fig. 1). The direction along which these conversions occur is labelled the "sheet addition line" on Figures 2A and 2B.

Algebraically, we can write the sequence as

$Mg_6(OH)_{12}$ + $2Si_2O_5(OH)_{-2}$ --> $Mg_6Si_4O_{10}(OH)_8$ + $2Si_2O_5(OH)_{-2}$ -->
Brucite   Serpentine

$Mg_6Si_8O_{20}(OH)_4$.
Talc

Starting with $Al_4(OH)_{12}$ (gibbsite) to get kaolinite, then pyrophyllite, is analogous. The fact that the sheet addition operator is equal to $2SiO_2$ - H_2O implies that sheet addition is favored by increasing silica activity, acidity (because we are subtracting hydroxyls), and temperature (because we are dehydrating). Trioctahedral 1:1 phyllosilicates (serpentine-group minerals) seem to be incompatible with quartz at any temperature - they react with it to form the 2:1 silicate talc (e.g., Velde, 1985).

Another interesting exchange operator on Figure 2A is H_2Mg_{-1}, which lies at infinite distance to the right. This successively converts enstatite to anthophyllite to talc, as follows:

566

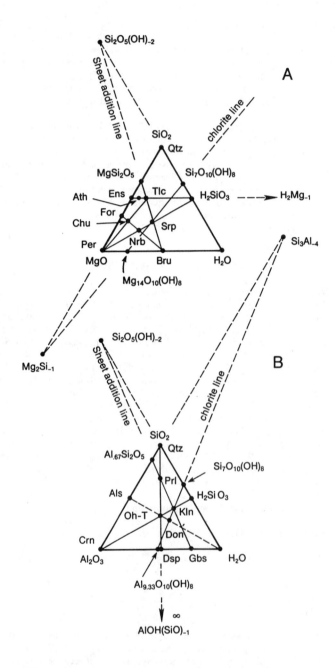

Figure 2. Composition triangles showing the exchange operator $Si_2O_5(OH)_{-2}$ that adds one, then two tetrahedral sheets to hydroxide sheets via changes that occur along the sheet addition line. See text for a discussion of the chlorite line on each triangle. (A) System $MgO-SiO_2-H_2O$. Abbreviations not on Figure 1: Ath, anthophyllite; Tlc, talc; Srp, serpentine; Bru, brucite; Chu, clinohumite; Nrb, norbergite (intermediate phases clinohumite and chondrodite omitted for simplicity). (B) System $Al_2O_3-SiO_2-H_2O$. Abbreviations not on Figure 1: Prl, pyrophyllite; Kln, kaolinite; Dsp, diaspore; Gbs, gibbsite; OH-T, hydroxyl-topaz (unstable); Don', donbassite of special composition discussed in text.

$Mg_8Si_8O_{24}$ + H_2Mg_{-1} --> $Mg_7Si_8O_{22}(OH)_2$ + H_2Mg_{-1} -->
Enstatite   Anthophyllite

$Mg_6Si_8O_{20}(OH)_4$.
Talc

This compositional degeneracy implies a simple dependence of structure type on $\log[a(H^+)^2/a(Mg^{2+})]$ in coexisting hydrothermal fluids (although I am unaware of any experimental work that would verify this prediction). In terms of ions, it could be written $[](OH)_2(MgO_2)_{-1}$ (cf. Bragg et al., 1965, p. 267).

A final compositional degeneracy on Figure 2A involves the humite-group minerals, discussed also by Thompson (1978), whose compositions lie along a line between forsterite and brucite, in the order clinohumite, humite, chondrodite, and norbergite (only the two extreme members are shown in Fig. 2A). Their compositions can therefore be derived from that of forsterite in terms of the exchange operator $(OH)_4(SiO_4)_{-1}$, which could also be written H_4Si_{-1}, or from brucite in terms of its inverse, $SiO_4(OH)_{-4}$. This exchange operator would plot at an $X(SiO_2)$ coordinate of -1 along the SiO_2-H_2O line (not shown; more or less on top of the label "B"). Incidentally, the same operator describes the formation of hydrogarnet from silicate garnet; the hydrogrossular endmember has a composition of $Ca_3Al_2(OH)_{12}$. (As a vector, it also describes lines of constant Al on Fig. 11 below for pyrophyllite solid solutions.)

Another operator that affects pyrophyllite solid solutions also would appear on Figure 2B, at infinite distance downwards, along the vertical line that joins diaspore to quartz. Starting with quartz, $AlOH(SiO)_{-1}$ successively forms pyrophyllite, OH-topaz (an unstable composition, mentioned here because we are condensing down the $F(OH)_{-1}$ vector), and diaspore, as follows:

Si_3O_6 + $AlOH(SiO)_{-1}$ --> $AlSi_2O_5(OH)$ + $AlOH(SiO)_{-1}$ -->
3Quartz   1/2Pyrophyllite

$Al_2SiO_4(OH)_2$ + $AlOH(SiO)_{-1}$ --> $Al_3O_3(OH)_3$
(OH)-topaz   3Diaspore.

The same operator and related derived operators can cause minor variations in the hydroxyl content of clay minerals, as discussed below.

The exchange operator $Si_2O_5(OH)_{-2}$ that appears as a point on Figures 2A and 2B can also be treated as a vector, as shown on Figure 3, for which the "additive component" is $Mg_6(OH)_{12}$ or 6 brucites. The vertical vector is $2Si_2O_5(OH)_{-2}$, which successively makes 1:1 or T-O silicates, then 2:1 or T-O-T silicates, then anhydrous (and unstable) Si_2O_5 compositions, as mentioned above. The horizontal unit is the "dioctahedral converter" vector $[]Al_2Mg_{-3}$, which converts brucite into gibbsite, serpentine into kaolinite, and talc into pyrophyllite, as shown. Lines of equal octahedral vacancies, Al, and Mg are vertical, and lines of equal Si and (OH) are horizontal. For want of a better term, I shall refer to this as the "sheet addition plane".

The $2Si_2O_5(OH)_{-2}$ vector of the "sheet addition plane", incidentally, is the only one we shall use that is not oxygen conservative. From the 12 oxygens in the first (brucite-gibbsite) level, it yields 18 in the second (serpentine-kaolinite) level, 24 in the third (talc-pyrophyllite) level, and 30 in the metastable OH-free level. I have nevertheless used it because it yields conventional phyllosilicate formulas and a visually-pleasing diagram. The fact that it is not oxygen-conservative does not matter for most of what follows, inasmuch as we shall be looking mainly at sections (planes and volumes) of constant O-content. For cases in which O-conservation might be important, the phyllosilicate formulas can be recast to 36 (or 72) oxygens, and the vectors derived on Figure 4 (an alternate depiction of the "sheet addition plane") used. The multipliers needed to convert the formulas in Figure 3 to those of Figure 4 are indicated to the left of Figure 4. The O-conserving "sheet addition operator" for 18 brucites is the somewhat awkward $Si_4O_{10}Mg_{-3}(OH)_{-10}$, or $Si_4Mg_{-3}H_{-10}$, the vector of constant Al. Vectors of constant Mg and OH are indicated on Figure 4.

Fibrous palygorskite (formerly also called attapulgite) and sepiolite, both made up of chain-like ribbons of 2:1 sheets of tetrahedra, cannot be derived simply by using either this operator or the exchange operator $Si_2O_5(OH)_{-2}$ on brucite. Neglecting interlayer cations and zeolitic water, palygorskite can be considered roughly as $5Mg(OH)_2 + 8SiO_2$; sepiolite as $8Mg(OH)_2 + 2Si_2O_5(OH)_{-2} + 8SiO_2$. Clearly, both are more OH-rich than continuously-layered phyllosilicates. In other words, their compositions (not shown), would lie to the right of the "sheet addition line" on Figure 2A.

Note that Figures 3 and 4 do not break out the serpentine "polymorphs". In particular, antigorite has a variable composition very slightly above the point labelled "serpentine" (for chrysotile and lizardite), along the line towards talc (cf. Mellini et al., 1987) on both figures.

1:1 PHYLLOSILICATES AND ISOCHEMICAL 2:1 CHLORITES

Figure 2A also shows the point Mg_2Si_{-1}, to the lower left, and Figure 2B the point Si_3Al_{-4}, to the upper right. Lines drawn between these points and the compositions of serpentine and kaolinite are labelled "chlorite line" on each figure. The chlorite line on Figure 2A intersects the edge of the triangle at the compositions $Mg_{14}O_{10}(OH)_8$ and $Si_7O_{10}(OH)_8$ (which are respectively obtained from the serpentine composition, $Mg_6Si_4O_{10}(OH)_8$, by first adding 4 times, then subtracting 8 times, the operator Mg_2Si_{-1} such that Si, then Mg, is eliminated). The compositions $Al_{9.33}O_{10}(OH)_8$ (or $Al_{14}O_{15}(OH)_{12}$) and $Si_7O_{10}(OH)_8$ on Figure 2B are similarly obtained.

The two lines together define a chlorite plane, as shown on Figure 5 (actually, it is a chlorite-serpentine-kaolinite plane). This intersects the sheet addition plane (of Figs. 3 or 4) along the line between serpentine and kaolinite, as shown.

This plane is better seen in a vector representation (Fig. 6) that uses the same two vectors as Figure 1C, namely $MgSiAl_{-2}$ and $[]Al_2Mg_{-3}$. (Note that the shape of this plane is the same as that of Fig. 1C.) The interesting part of this plane, along which stable or potentially stable chlorite compositions occur, is shown in more detail in Figure 7. The 14 Å chlorite structure, as opposed to the 7 Å structure of serpentine and kaolinite, is stable on both sides of the clinochlore-donbassite line (on the trioctahedral side, between Si contents of about 2.34 and 3.45, according to Foster, 1962, as cited in Bailey, 1988, this volume), whereas amesite again has a serpentine structure (cf. reviews by Bailey, 1980, Newman and Brown, 1987; Bailey, 1988, this volume). At any composition, chlorite formation is favored by increasing temperature. Sudoite, a di,trioctahedral chlorite (analogous to the composition $Mg_3Al_2O_6$ on Fig. 1C) is on the clinochlore-donbassite line at its intersection with the vertical line [] = 1. The condensed composition of brindleyite, a cation-deficient Ni-Al serpentine (Maksimovic and Bish, 1978) from bauxites, would plot about half-way between clinochlore and sudoite, and again illustrates the polymorphism between aluminous serpentines and chlorites.

Donbassite has a range of compositions along the slanting line of Mg = 0 (tetrahedral Al lies between 0.6 and 1.3, corresponding to Si between 2.7 and 3.4, according to Bailey, 1988). The special composition labelled Don' on Figures 2B and 7

569

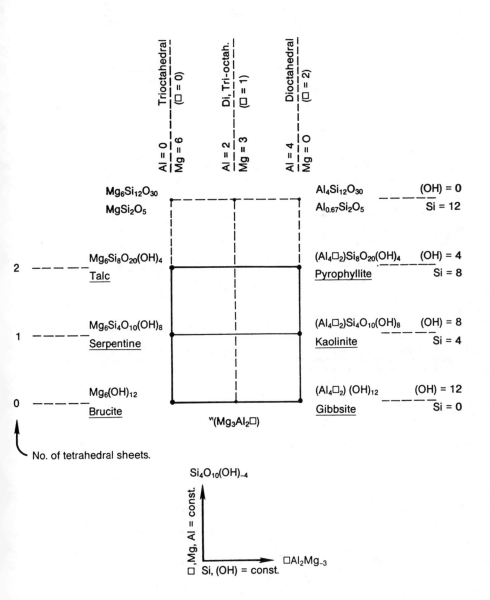

Figure 3. Vector representation ("sheet addition plane") showing how the compositions of the simple phyllosilicates can all be generated from the brucite composition, $6Mg(OH)_2$, by operation of the orthogonal exchange vectors $2Si_2O_5(OH)_{-2}$ and $[]Al_2Mg_{-3}$. The first vector is not oxygen-conservative.

570

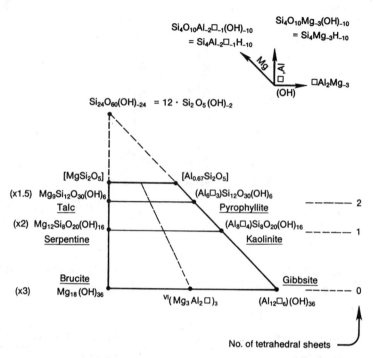

Figure 4. Alternate, oxygen-conservative representation of the sheet addition plane, showing how simple phyllosilicate compositions (modified to have 36 O plus OH) can be generated from the brucite composition $18Mg(OH)_2$ using the vectors $Si_4O_{10}Mg_{-3}(OH)_{-10}$ (or $Si_4Mg_{-3}H_{-10}$) and $[]Al_2Mg_{-3}$.

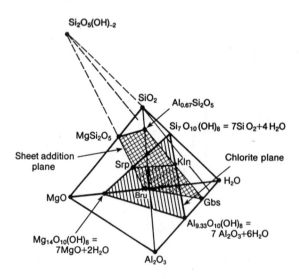

Figure 5. Composition tetrahedron $MgO-Al_2O_3-SiO_2-H_2O$ showing how the sheet addition plane (brucite-gibbsite-$Si_2O_5(OH)_{-2}$) intersects the chlorite plane along the serpentine-kaolinite line. See text for discussion.

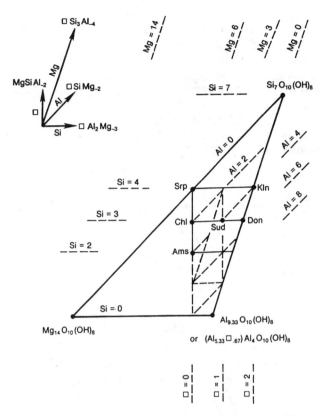

Figure 6. Vector representation of the chlorite (-serpentine-kaolinite) plane showing how typical chlorite and related compositions can be generated from a single composition, such as that of serpentine, by operation of the orthogonal exchange vectors MgSiAl_2 and []Al_2Mg_-3. Abbreviations: Srp, serpentine; Kln, kaolinite; Chl, clinochlore; Sud, sudoite; Don, donbassite; Ams, amesite. See Figure 7 for more detail.

is that subject to the added restriction that Al:Si = 2:1. A donbassite very close to this special composition formed as a hydration product of pegmatitic andalusite from Szabo Bluff, Antarctica (Ahn et al., 1988); the original Al:Si ratio of andalusite was preserved during the alteration.

Figure 7 also shows the expanded forms of the various exchange operators, with coordination changes. Thus []SiMg_-2 actually involves loss of a tetrahedral Al and gain of an octahedral Al; the two cancel in writing the operator. This exchange therefore does not occur in Al-free serpentine (along the line Al = 0 on Fig. 5). The exchange []Si_3Al_-4, on the other hand, does actually occur along the line Mg = 0 on Figures 5 and 6; this fact explains the variable composition of donbassite noted above.

TALC AND PYROPHYLLITE: 2:1 PHYLLOSILICATES

Given the success with the 1:1 layer silicates and chlorites, it is tempting to try an analogous treatment for the 2:1 or T-O-T silicates talc and pyrophyllite. By drawing a line from the point Mg_2Si_{-1} through the composition of talc on Figure 2A, and doing the same from Si_3Al_{-4} to pyrophyllite on Figure 2B, we could respectively generate a "talc line" and a "pyrophyllite line" (not shown), which would combine to define a "talc-pyrophyllite plane" on Figure 5 (also not shown). Its corners would have

572

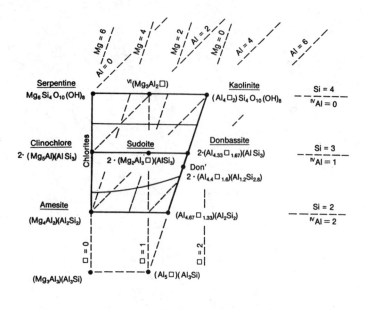

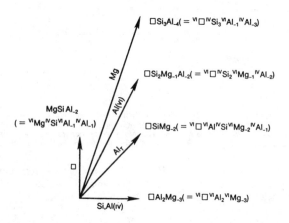

Figure 7. Detail of the chlorite plane, showing accessible compositions in terms of the orthogonal exchange vectors $MgSiAl_{-2}$ and $[]Al_2Mg_{-3}$. Don' = donbassite of Al:Si ratio 2:1 formed by alteration of andalusite (Ahn et al., 1988).

the compositions $Mg_{22}O_{20}(OH)_4$, $Al_{14.67}O_{20}(OH)_4$, and $Si_{11}O_{20}(OH)_4$, and it would intersect the sheet addition plane along the line between talc and pyrophyllite. The corresponding vector diagram, analogous to Figure 6, is Figure 8, which is shown in more detail in Figure 9. Nature is ignorant of our cleverness in drawing such diagrams, however, and observed talc and especially pyrophyllite compositions only vary slightly off their respective points (see summary in Newman and Brown, 1987). Nevertheless, both substitutions do occur to a limited extent, especially in talc (e.g., Fawcett, 1963); this justifies my drawing the diagram, which I shall mention again below.

On the talc side, Al-substitution is probably small because too much of it generates interlayer charge and leads to the formation of Mg-smectites or even Mg-vermiculite (e.g., Velde, 1973, p. 306 and 1985, p. 147). As discussed below,

573

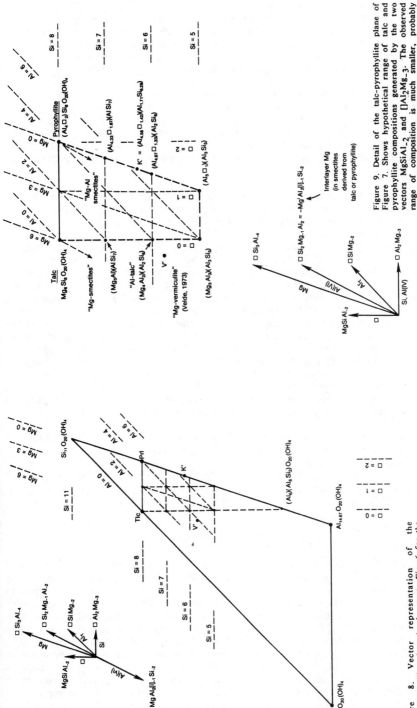

Figure 9. Detail of the talc-pyrophyllite plane of Figure 7. Shows hypothetical range of talc and pyrophyllite compositions generated by the two vectors MgSiAl$_{-2}$ and [ ]Al$_2$Mg$_{-3}$. The observed range of composition is much smaller, probably due to the formation of Mg-Al smectites or vermiculite. Point K' has an Al:Si ratio of 1:1 and is equivalent to dehydrated kaolinite; point V' is the composition of the vermiculite-like phyllosilicate synthesized by Velde (1973).

Figure 8. Vector representation of the talc-pyrophyllite plane (analogous to Fig. 6 for the chlorite plane). Abbreviations: Tlc, talc; Prl, pyrophyllite. Point K' has an Al:Si ratio of 1:1; point V' corresponds to the vermiculite-like phyllosilicate synthesized by Velde (1973).

interlayer charge as Mg^{2+} would be added along the vector $MgAl_2[i]_{-1}Si_{-2}$, where [i] is an interlayer vacancy; this is the opposite of a vector already on the diagram (as indicated on Fig. 8). On the pyrophyllite side, too much Si-deficiency generates either dioctahedral Mg-Al smectites (also synthesized by Velde, 1973) or a 1:1 composition (labeled K' on Fig. 8) equivalent to dehydrated kaolinite (i.e., kaolinite instead of pyrophyllite would grow at low temperatures; andalusite at high temperatures).

Addition of too much interlayer charge in the form of Mg or Al can cause the formation of (Mg,Al)-hydroxide layers between those of talc or pyrophyllite; we then are talking about chlorite. Figure 10 shows how the composition of trioctahedral chlorite (clinochlore) lies between that of talc and Tschermak-substituted brucite (an impossible composition by itself). It also shows the composition of the mixed-layer phyllosilicate kulkeite (Schreyer et al., 1982), an ordered alternation of clinochlore and talc.

Analyses of phyllosilicates commonly yield OH-contents that vary somewhat from the theoretical values. Rosenberg and Cliff (1980), for example, synthesized a series of pyrophyllites in which they detected the coupled substitution $HAlSi_{-4}$, or $AlOH(SiO)_{-1}$, mentioned above in connection with Figure 2B. This, combined with the substitution $[]Si_3Al_{-4}$ (actually, I used its inverse $Al_4[]_{-1}Si_{-3}$) yields a plane of potential pyrophyllite compositions (Fig. 11). Lines of constant octahedral vacancies are vertical; of constant OH-content horizontal. Lines of constant Al have a slope of -4 and of constant Si, -3. Either substitution yields Si:Al ratios of less than 2 (such ratios range down to 1.75 in natural specimens and to 1.722 in synthetics, according to data cited in Newman and Brown, 1987). Line K-K'-K" marks Si:Al ratios of 1:1, equivalent to dehydrated kaolinite, as mentioned above. Starting with pyrophyllite, the vector $AlOH(SiO)_{-1}$ generates the composition of (OH)-topaz above and that of quartz below, as shown. Given the independent vector $[]Si_3Al_{-4}$, this is actually a somewhat deformed depiction of the triangle or plane of Figure 2B, but expressed as $Al_{16}O_{24}$-$Si_{12}O_{24}$-$H_{48}O_{24}$. In other words, the vector $AlOH(SiO)_{-1}$ and its derivatives (discussed below) are redundant (i.e., linearly dependent) with regard to the vectors already used. They are nevertheless valid (if minor) substitutions in phyllosilicates, and their vector representation is likewise a valid alternative depiction.

Minor $AlOH(SiO)_{-1}$ substitution could occur in other Al-rich phyllosilicates; if Mg is present, the exchange $MgOH(AlO)_{-1}$, derived by adding the Tschermak exchange to $AlOH(SiO)_{-1}$, could also occur. In Fe-bearing phases, this exchange has an interesting analog, $Fe^{2+}OH(Fe^{3+}O)_{-1}$, or $HFe^{2+}Fe^{3+}_{-1}$, which can be further shortened to H^0 (atomic H), a most peculiar, but nevertheless valid, exchange operator (cf. Thompson, 1982) that is discussed in a separate section below.

OCTAHEDRAL LITHIUM

Lithium theoretically can replace octahedral Mg in any of the minerals discussed above via the coupled substitution $LiAlMg_{-2}$, which does not affect the number of vacancies. Minor (OH) variation would then be possible via $LiOH(MgO)_{-1}$, (equal to $MgOH(AlO)_{-1}$ + $LiAlMg_{-2}$). Although the substitution $LiAlMg_{-2}$ or its ferrous analog is extensive in the true micas (cf. Cerny and Burt, 1984) and undoubtedly occurs to a minor extent in all the Mg-silicates discussed so far, only in the chlorite group has a separate lithium mineral been defined, cookeite. This is derived from, of all phases, sudoite (making it also di,trioctahedral), as shown in Figure 12, a vector representation involving $LiAlMg_{-2}$ and $[]Al_2Mg_{-3}$, with the composition clinochlore the additive component. Carrying the vertical operation through to completion leads to a trioctahedral Li-chlorite endmember of theoretical composition $(Li_{2.5}Al_{3.5})(Si_3Al)O_{10}(OH)_8$ (a chlorite analog of trilithionite).

575

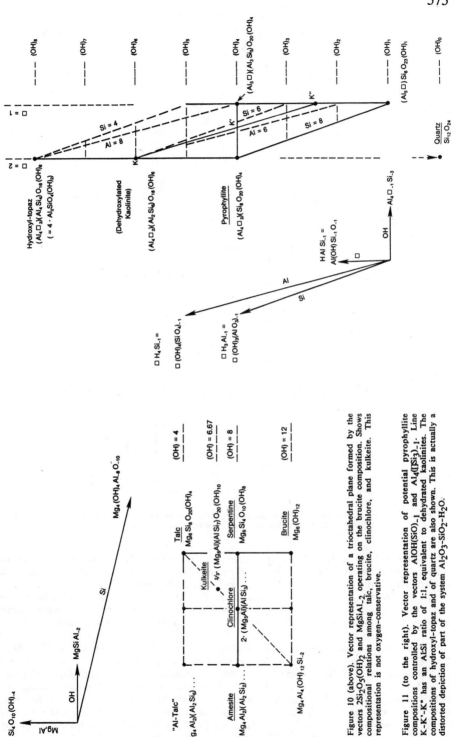

Figure 10 (above). Vector representation of a trioctahedral plane formed by the vectors $2Si_2O_5(OH)_2$ and $MgSiAl_{-2}$ operating on the brucite composition. Shows compositional relations among talc, brucite, clinochlore, and kulkeite. This representation is not oxygen-conservative.

Figure 11 (to the right). Vector representation of potential pyrophyllite compositions controlled by the vectors $AlOH(SiO)_{-1}$ and $Al_4([]Si_2)_{-1}$. Line K-K'-K" has an AlSi ratio of 1:1, equivalent to dehydrated kaolinites. The compositions of hydroxyl-topaz and of quartz are also shown. This is actually a distorted depiction of part of the system Al_2O_3-SiO_2-H_2O.

576

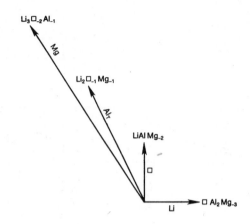

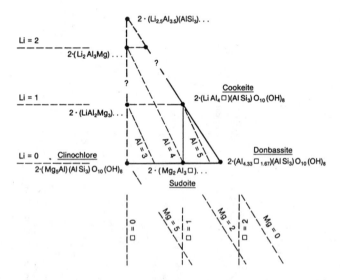

Figure 12. Vector representation of a lithium-chlorite plane showing how cookeite can be generated from clinochlore by a combination of the orthogonal vectors $LiAlMg_{-2}$ and $[]Al_2Mg_{-3}$.

Such a Li-rich clinochlore is presumably less stable than some combination of micas, holmquistite, and Li-aluminosilicates in the silica-saturated, alkali-rich environment of the average Li-rich pegmatite. It may well also be unstable thermodynamically (on its own composition); if it is stable and occurs in nature, this occurrence would be likely in a high-lithium, low-silica, low-alkali environment, such as might be provided where a Li-pegmatite cuts a serpentinite or meta-bauxite.

I am tempted to draw diagrams analogous to this for the other Mg-phyllosilicates discussed. This would lead to the generation of unknown Li-Al phyllosilicate compositions, such as for serpentine and talc, of compositions $(Li_3Al_3)Si_4O_{10}(OH)_8$ and $(Li_3Al_3)Si_8O_{20}(OH)_4$, respectively (the latter would be a K-free analog of polylithionite). Given the lack of natural or synthetic examples reported to date, however, I shall leave this as an exercise for the reader (or would-be experimenter). A complete treatment would be to make various planar vector diagrams involving $LiAlMg_{-2}$ in combination with $Si_2O_5(OH)_{-2}$, $[]Al_2Mg_{-3}$, and $MgSiAl_{-2}$, then to combine these into 3-D depictions (such as I did for the Li-micas in 1984; cf. Cerny and Burt, 1984).

INTERLAYER CHARGE (POTASSIUM AND SODIUM)

Speaking of micas, let us make a few. The justification (this volume, after all, is supposed to exclude micas) is the existence of the smectites (mentioned above), vermiculites, and mixed-layer 2:1 clay minerals. These have a variable interlayer charge intermediate between the 0.0 of talc and pyrophyllite and the 1.0 of true micas (2.0 with the "double formulas" I am using). We can generate them in our model by adding the large interlayer cation K^+ (or the smaller Na^+, inasmuch as we are still "condensed"), via the substitution $KAl[i]_{-1}Si_{-1}$, where [i] denotes an interlayer vacancy (with values between 0.0 in true micas and 2.0 in "doubled" talc and pyrophyllite). The symbol [i] is distinct from the [] that I am using to denote an octahedral vacancy (with values also between 0.0 in trioctahedral phyllosilicates and 2.0 in dioctahedral phyllosilicates). Substitutions of the type $KAlSi_{-1}$ are also important in the amphiboles (cf. Thompson, 1981) and in framework silicates (e.g., relating quartz to albite to nepheline).

Allowing this substitution vector and the dioctahedral substitution vector to operate on the talc composition yields Figure 13, nearly as fundamental as the "sheet addition plane" of Figure 3 (cf. the back plane of Giggenbach's [1985] Fig. 1, p. 233). Let's call it the "interlayer charge addition plane". Along the true-mica top, intermediate di,trioctahedral micas in solid solution with phlogopite were synthesized at high pressure by Green (1981), as shown, but are unknown in nature. Note that the smectites on this plane really should stick up along a vertical axis of $+H_2O$ (addition of variable interlayer water of hydration), owing to their nature as expanding clays; this variable H_2O content is conventionally neglected.

The Tschermak substitution complicates the true micas even more than it does serpentine and kaolinite (in the chlorite plane of Figs. 6 and 7). (Recall that talc and pyrophyllite, shown on Figs. 8 and 9, have little tolerance for this substitution.) This effect is shown on Figure 14, the true mica plane (after my Fig. 5 in Cerny and Burt, 1984, with the operation of $MgFe_{-1}$). Neither the composition at the lower left corner nor the one above it are now referred to as "eastonite"; the Na-mica preiswerkite is a recognized species. I have referred to the Mg-Al celadonite at the upper right corner as "leucophyllite", following Guidotti (1984). The composition at the lower right corner was apparently synthesized by Francke et al. (1981); the Mg-analog of montdorite at the upper left corner was synthesized by Seifert and Schreyer (1971), who also synthesized Al-free mica solid solutions with compositions approaching 0.5 tetrahedral Mg, half way to the composition $^{VI}(Mg_6)^{IV}(MgSi_7)$, as indicated. A natural mica with considerable tetrahedral Mg is reported by Skodyreva et al. (1985); possible tetrahedral Mg in other phyllosilicates has not been considered.

We may consider the true mica composition plane of Figure 14 as a roof, with the smectite group minerals sheltering in an irregular tapering polyhedron under it (the city hall in Tempe, Arizona, is shaped something like this--it is a strange-looking upside-down pyramid). The accessible polyhedron narrows as interlayer charge decreases, reaching nearly zero width at "ground level", the talc-pyrophyllite line (which conceptually is the plane of Fig. 9). We can see this happening for the dioctahedral 2:1 silicates in Figure 15, equivalent to looking down the right side of Figure 14. Note how the single point available at the pyrophyllite composition widens to allow the Tschermak substitution as interlayer charge increases to 2 at the mica-leucophyllite level. A somewhat similar triangle was drawn by Yoder and Eugster (1955, p. 257) and has been reproduced in modified forms many times since to explain the compositions of dioctahedral smectite and mixed-layer minerals.

578

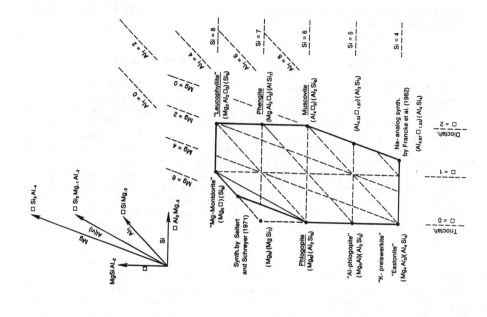

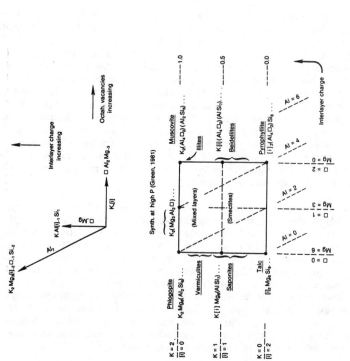

Figure 13 (above). Vector representation showing the generation of true mica compositions by way of smectite-group minerals, vermiculites, and mixed-layer minerals from talc and pyrophyllite via the operation of the vector $KAl[i]_{-1}Si_{-1}$, which adds interlayer charge. For all formulas, $...O_{20}(OH)_4$.

Figure 14 (to the right). Vector representation of the true mica plane in terms of $[]Al_2Mg_{-3}$ and $MgSiAl_{-2}$ (modified from Cerny and Burt, 1984). See text for discussion. For all formulas, $K_2...O_{20}(OH)_4$.

579

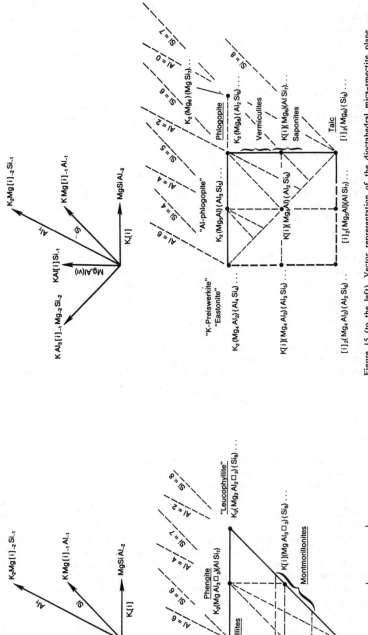

Figure 15 (to the left). Vector representation of the dioctahedral mica-smectite plane, showing increased operation of the Tschermak substitution with increasing interlayer charge. For all formulas, ...O₂₀(OH)₄.

Figure 16 (to the right). Vector representation of the trioctahedral mica-smectite-vermiculite plane, showing reverse operation of the Tschermak substitution with increasing interlayer charge. The Na-analog of this plane contains wonesite. For all formulas, ...O₂₀(OH)₄.

We can also consider the trioctahedral clays. Figure 16 is a diagram analogous to Figure 15, but looking down the left side of Figure 14. Here the limited (but finite) range of Tschermak substitution available to talc widens backwards (in reverse) as interlayer charge is added towards phlogopite. The Na-analog of this plane contains wonesite, and was drawn as a triangle by Veblen (1983); conceptually it is a quadrilateral, as shown in dashed lines.

Another plane (Fig. 17), which I haven't seen drawn previously, is the tetrasilicic plane (of Si = 8), in which interlayer charge must be added by the vector $KMg[i]_{-1}Al_{-1}$, inasmuch as tetrahedral Al is excluded. Figures 13 and 14 show that $KMg[i]_{-1}Al_{-1}$ is the vector sum of $KAl[i]_{-1}Si_{-1}$ plus $MgSiAl_{-2}$. Lithium-bearing clays such as hectorite can similarly add interlayer charge via substitutions such as $KLi[i]_{-1}Mg_{-1}$ (equivalent to $KMg[i]_{-1}Al_{-1}$ plus $LiAlMg_{-2}$).

A whole family of possible smectite planes exists, for varying levels of interlayer charge. One such plane, drawn at the half-way point (K = 1) is shown on Figure 18. Note how the Tschermak substitution can only act over half of its range in micas, especially in the upwards direction; for more typical smectite compositions (K < 1) its range would be restricted even further.

I could show (and have drawn) other planes, such as the Mg-free plane with pyrophyllite, muscovite, and compositions yielded by the substitution $[]Si_3Al_{-4}$, or a plane with talc, phlogopite, and Mg-montdorite at the corners of a triangle, but let us instead jump to Figure 19, a three-dimensional representation of the "upside-down pyramid" which is cut in half by the planes of Figures 13 (vertically) and 18 (horizontally) and contains Figures 14 through 17 as faces. Figure 14, the true mica plane, is the top, Figure 15, the dioctahedral plane, is the right front face, Figure 16, the trioctahedral plane, is the left rear face, and Figure 17, the tetrasilicic plane, is slanting under the rear. For simplicity I have drawn the talc-pyrophyllite join as a merely a line; actually, as mentioned above, the Tschermak substitution would probably be sufficient to allow a model in this form to balance precariously on its basal plane (Fig. 9).

DIVALENT INTERLAYER CATIONS (CALCIUM AND BARIUM)

Smectite group minerals commonly have divalent interlayer cations, which will be modeled as Ca^{2+} (except that it is rarer, the larger Ba^{2+} would serve as well). Starting from the talc-pyrophyllite line, divalent interlayer charge could be added via the vector $CaAl_2[i]_{-1}Si_{-2}$ (analogous to that relating anorthite to quartz). This would generate a rectangle or square analogous to Figure 13. In the case of Mg, which also occurs as interlayer charge in smectites (see above), the vector $MgAl_2Si_{-2}$, or its inverse, is already part of our system (see, e.g., Fig. 8), so that we could instead add Ca via $CaMg_{-1}$, but to do so would ignore the different sizes and sites of the two ions in phyllosilicates. Many others (e.g., Thompson, 1981) would introduce Ca via the plagioclase substitution $CaAlNa_{-1}Si_{-1}$ (or its Ba and K-analog in our condensed system), which also relates paragonite to margarite compositions. I shall, however, be a little unconventional and use $Ca[i]Na_{-2}$ and $Ba[i]K_{-2}$ as basis vectors, because these substitutions preserve interlayer charge. The results are shown as severely deformed triangles on Figures 20A (left side, for trioctahedral micas) and 20B (right side, for dioctahedral micas). The other Ca-vectors mentioned above are seen to be linear combinations (of $NaAl[i]_{-1}Si_{-1}$ and $Ca[i]K_{-2}$ on Fig. 20B). The main advantage of this representation is that it demonstrates graphically the doubled interlayer charge of the brittle micas (which in turn explains their brittleness).

The above arguments give me an excuse to draw the brittle mica plane, Figure 21, which is closely analogous to the true mica plane, Figure 14 (and would lie directly above it on a three-dimensional diagram involving the vertical vector

581

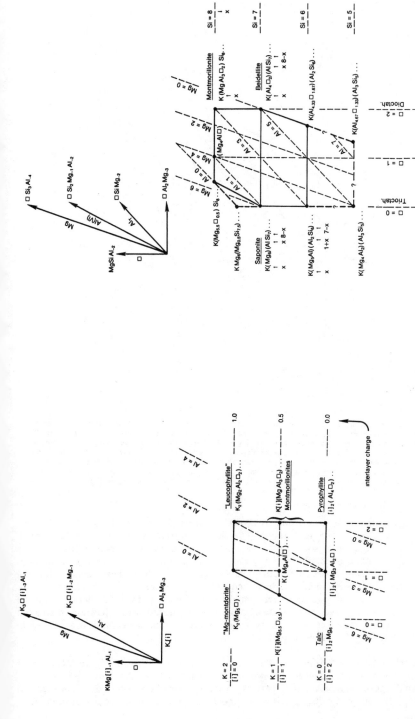

Figure 17. Vector representation of the tetrasilicic (Si=8) mica-smectite plane, with interlayer charge added by the vector $KMg[i]_{-1}Al_{-1}$. For all formulas, $...O_{20}(OH)_4$.

Figure 18. Vector representation of a smectite plane at the half-level of interlayer charge (K = 1 or midway between the talc-pyrophyllite line and the true mica plane). Most smectites have less interlayer charge, which would further restrict the operation of the Tschermak vector (especially in the up direction). For all formulas, $...O_{20}(OH)_4$.

582

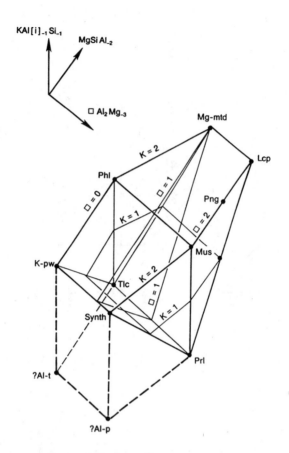

Figure 19. Three-dimensional vector representation of "smectite space" as an inverted irregular pyramid. Interlayer charge is added vertically starting with the talc-pyrophyllite line (conceptually a plane: Fig. 9) and stopping with the true mica plane (Fig. 14). Figure 13 bisects the volume vertically, and Figure 18 horizontally; Figures 15-17 are side faces. Abbreviations: Tlc, talc; Prl, pyrophyllite; Mus, muscovite; Phl, phlogopite, Mg-mtd, Mg-montdorite; Lcp, leucophyllite, png, phengite; Synth, synthetic mica $K_2(Al_{4.67}[]_{1.33})(Al_2Si_2)$; K-pw, K-preiswerkite; Al-t, hypothetical Al-rich talc of composition $(Mg_4Al_2)(Al_2Si_6)$; Al-p, hypothetical Al-rich pyrophyllite of composition $(Al_{4.67}[]_{1.33})(Al_2Si_6)$.

$CaAlK_{-1}Si_{-1}$). This is a condensed diagram, so that I have included the Ba-mica kinoshitalite. Its Ca-analogue is unstable (Olesch and Seifert, 1976), but the Tschermak substitution yields clintonite solid solutions, whose natural range is more restricted than the compositions experimentally produced by Olesch and Seifert (1976). Numerous experimental studies of margarite (cited in Guggenheim, 1984) have dealt with the Al-endmember and seem not to have considered the effect of the Tschermak substitution (phengite analogues). There might also be montdorite analogues and an extended composition range of unknown extent, as indicated with question marks on the figure. The "X" indicates the approximate composition of a synthetic Ca-Mg-Al-Si fluormica reported by Lantukh et al. (1984).

Although I have not drawn it, the similarity of Figure 21 to Figure 14 suggests that the "smectite space" beneath it would be analogous to the "inverted pyramid" depicted in Figure 19, with the vertical vector being $CaAl_2[i]_{-1}Si_{-2}$ instead of $KAl[i]_{-1}Si_{-1}$. In other words, possible compositions would pinch out severely with

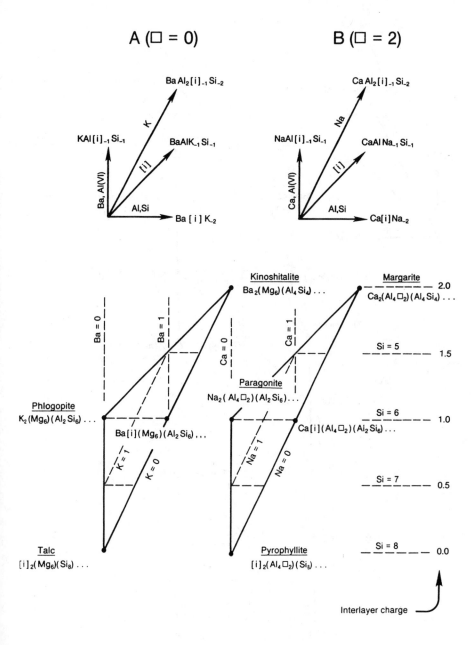

Figure 20. Vector representations of how divalent cations double the interlayer charge. (A) Trioctahedral plane ([] = 0) using vector Ba[i]K_{-2}, showing formation of phlogopite and kinoshitalite from talc. (B) Dioctahedral plane ([] = 2) using vector Ca[i]Na_{-2}, showing formation of paragonite and margarite from pyrophyllite.

584

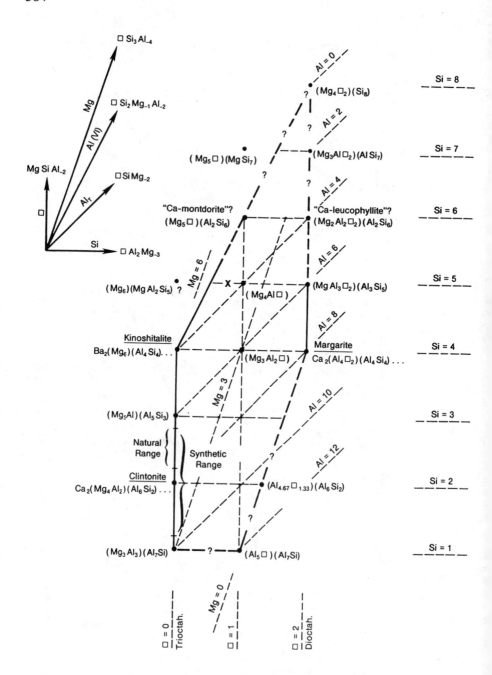

Figure 21. Vector representation of the condensed brittle mica plane in terms of []Al₂Mg₋₃ and MgSiAl₋₂. "X" = approximate composition of synthetic Ca-fluormica reported by Lantukh et al. (1984). Natural range of clintonite micas according to Guggenheim (1984); synthetic range according to experiments of Olesch and Seifert (1976). For all formulas, (Ca,Ba)₂...O₂₀(OH)₄.

decreasing interlayer charge, until reaching the talc-pyrophyllite line (or very restricted plane).

Small trivalent cations (Al^{3+}) could be introduced into interlayers via substitutions such as $^IAl^{IV}Al_3[i]_{-1}Si_{-3}$ (an exchange operator already present on, e.g., Fig. 8), but these, along with Mg^{2+}, are likely to get hydrated, and then we are talking about the structure of chlorite (or vermiculite), as mentioned above. Similarly, we can add small interlayer Li^+, via substitutions such as $LiAl[i]_{-1}Si_{-1}$ (= $LiAlMg_{-2}$ - $MgSiAl_{-2}$ - $[]Al_2Mg_{-3}$) or $LiMg[i]_{-1}Al_{-1}$ (= $LiAlMg_{-2}$ - $[]Al_2Mg_{-3}$). Such a smectite is the di,trioctahedral swinefordite (Tien et al., 1975), which I suspect could similarly turn into the di,trioctahedral chlorite cookeite.

OCTAHEDRAL TITANIUM

Although titanium is usually thought of in connection with biotite analyses (e.g., Robert, 1976; Dymek, 1983; Labotka, 1983; Hewitt and Abrecht, 1986) there is no reason why it cannot substitute in other phyllosilicates, and in fact, it does. For example, Arima et al. (1985) report a Ti-rich berthierine (a ferroan serpentine-group mineral). Inasmuch as Ti^{4+} is assumed to be mainly octahedral, the best vector to introduce it would appear to be one that affects only octahedral sites without introducing vacancies (I did the same for Li via $LiAlMg_{-2}$), namely, $TiMgAl_{-2}$. This is called the "spinel" substitution by Dymek (1983), although it also relates, e.g., geikelite to corundum. Its ferrous-ferric analog, $TiFe^{2+}Fe^{3+}_{-2}$, relates ilmenite to hematite, and can be shortened to $TiFe_{-1}$ (cf. Thompson, 1981). As pointed out by, e.g., Dymek (1983), other Ti-substitutions, such as $TiAl_2Mg_{-1}Si_{-2}$ and $[]TiMg_{-2}$, are linear combinations of this and those already used, as can be seen in Figures 22 and 23. One not shown, Hewitt and Abrecht's (1986) $TiO_2Mg_{-1}(OH)_{-2}$ (cf. Evans and Trommsdorff, 1983) is a linear combination of $TiMgAl_{-2}$ minus 2 times $MgOH(AlO)_{-1}$, which itself is a linear combination of $MgSiAl_{-2}$ plus the $AlOH(SiO)_{-1}$ introduced in connection with Figure 11.

Figure 22 shows the analogous Ti substitution ranges in trioctahedral serpentine-chlorite (Fig. 22A) and phlogopite (Fig. 22B). The greater the octahedral Al introduced via the Tschermak substitution, the more Ti can replace it, so that Si-deficient trioctahedral phyllosilicates have room for more Ti. If we push the substitution too far, we get tetrahedral Ti, as shown. Robert (1976) was unable to synthesize phlogopite of composition $^{VI}(Mg_6)^{IV}(TiAl_2Si_5)$ outside of the triangle in Figure 22B.

Figure 22A shows that in the case of chlorites, it might be undesirable to say that Ti enters via the vector $TiAl_2Mg_{-1}Si_{-2}$, inasmuch as this substitution would yield chlorite compositions coming from the serpentine structural field and then returning to the serpentine field (recall for Fig. 6 that 7 A serpentine structures occur both above and below 14 A chlorite structures). In the case of biotite solid solutions (Fig. 22B), the substitution could be described either way (or in any other way); Hewitt and Abrecht (1986) conclude that ternary (and more complicated) solid solutions such as biotite cannot be described in terms of a unique set of binary exchanges (or ionic substitutions). This is hardly a surprising conclusion! (That is, any two linearly independent vectors define a plane, any three a space, and so on.)

Figure 23 shows a second planar section through Ti-mica space; it shows that dioctahedral silicates such as muscovite, having more octahedral Al, should have more room for Ti-substitution. The same is theoretically true of kaolinite and pyrophyllite, but they seem to be even more ignorant of this substitutional opportunity than muscovite is. (Incidentally, Dymek, 1983, refers to the dioctahedral composition at the top of the triangle as "Ti-phlogopite", perhaps because Forbes and Flower, 1974, named their synthetic intermediate "titanophlogopite".) Figure 24 emphasizes this point for both

586

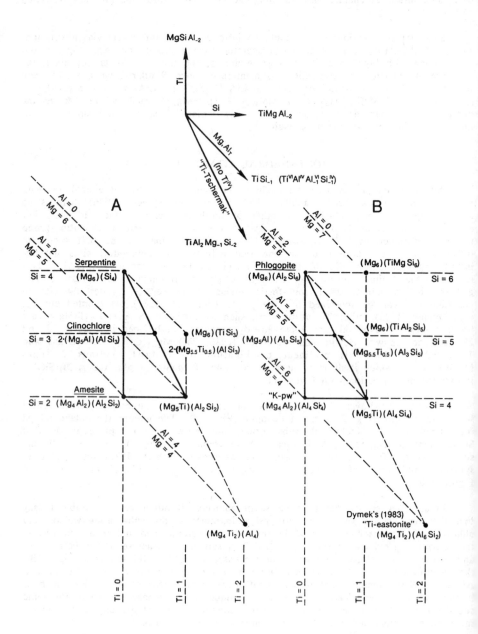

Figure 22. Vector diagrams in terms of MgSiAl$_{-2}$ and TiMgAl$_{-2}$ that show the substitution of Ti in trioctahedral phyllosilicates. (A) In serpentine and chlorite solid solutions. For all formulas, ...O$_{10}$(OH)$_8$. (B) In phlogopite solid solutions. For all formulas, K$_2$...O$_{20}$(OH)$_4$.

587

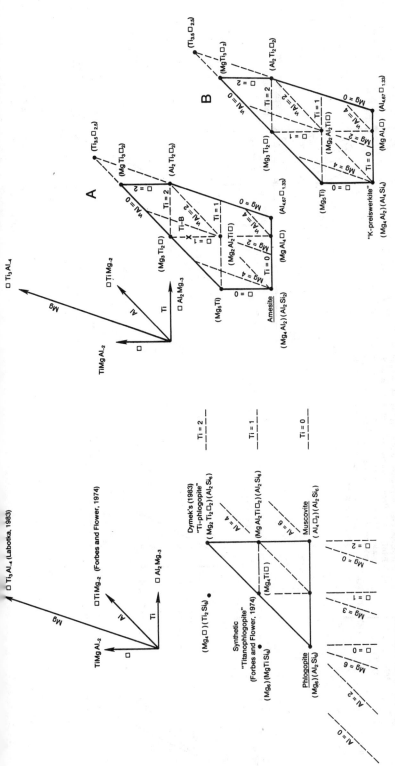

Figure 23. Vector diagram in terms of $[]Al_2Mg_{-3}$ and $TiMgAl_{-2}$ that shows the theoretically greater extent of Ti-substitution in Al-rich dioctahedral phyllosilicates such as muscovite as compared with Al-poor trioctahedral phyllosilicates such as phlogopite. For all formulas, $K_2...O_{20}(OH)_4$.

Figure 24. Vector diagram in terms of $[]Al_2Mg_{-3}$ and $TiMgAl_{-2}$ that shows that a wide range of Ti substitution (up to 3 Ti's per formula unit) is theoretically possible for Al-rich phyllosilicates. Such extensive Ti-substitution has not been found in nature. (A) Compositions derived from amesite. "Ti-B" is the approximate composition of the titanian berthierine of Arima et al. (1985). For all formulas, $...(Al_2Si_2)O_{10}(OH)_8$. (B) Compositions derived from K-preiswerkite. For all formulas, $K_2...(Al_4Si_4)O_{20}(OH)_4$.

588

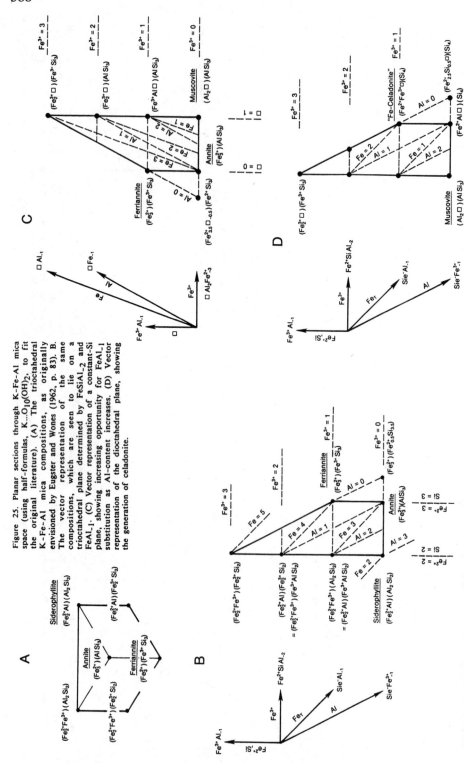

Figure 25. Planar sections through K-Fe-Al mica space (using half-formulas, $K...O_{10}(OH)_2$, to fit the original literature). (A) The trioctahedral K-Fe-Al mica compositions, as originally envisioned by Eugster and Wones (1962, p. 83). B. The vector representation of the same compositions, which are seen to lie on a trioctahedral plane determined by $FeSiAl_{-2}$ and $FeAl_{-1}$. (C) Vector representation of a constant-Si plane, showing increasing opportunity for $FeAl_{-1}$ substitution as Al-content increases. (D) Vector representation of the dioctahedral plane, showing the generation of celadonite.

amesite (serpentine) and preiswerkite (mica) by deriving compositions with 3 Ti per dioctahedral formula; again, natural muscovites seem indifferent to this "Ti-freedom" (that is, analyses by, e.g., Kwak, 1968, and Guidotti et al., 1977, consistently show more Ti in biotite than in coexisting muscovite). Point "Ti-B" on Figure 24A is the approximate composition of the di,trioctahedral serpentine-group mineral described by Arima et al. (1985). Figure 24B is a third plane through condensed Ti-mica space, which space I shall not depict because I am not supposed to be talking about micas anyway. (I used micas only because they are well-characterized for Ti.)

FERROUS AND FERRIC IRON, PLUS HYDROGEN

At first glance one might reject a discussion of Fe^{2+} substitution for Mg and of Fe^{3+} for Al (via the vectors $FeMg_{-1}$ and $FeAl_{-1}$) as being too straightforward for discussion (although the analytical discrimination of Fe^{2+} from Fe^{3+} in phyllosilicates is hardly straightforward, and a 900-page book "Iron in Clays and Clay Minerals" recently appeared: Stucki et al., 1988). This conclusion may be valid, if the substitutions occur individually. However, phyllosilicates commonly contain both ferrous and ferric iron, and then certain interesting properties (degeneracies) involving oxidation and reduction become apparent and worthy of a theoretical treatment. I apologize for starting off with the micas again, but these have been the most studied, and they exhibit phenomena that occur in other phyllosilicates.

The classic study is that of Eugster and Wones (1962) on annite; they originally (p. 83) envisioned the compositions of the trioctahedral K-Fe-Al biotites as lying on the corners of a trigonal prism, as shown in Figure 25A. The vector representation (Fig. 25B) shows that all of these compositions lie instead on a single plane. If you are still a "triangle worshipper", note that this plane could be made into a triangle with a corner at $KFe_3(Fe_{0.5}Si_{3.5})O_{10}(OH)_2$, an improbable but physically-attainable mica composition with tetrahedral ferrous iron (its Mg-analog appears on Fig. 14). Note also that the electron (e^-) appears for the first time in vectors (exchange operators) of constant Fe(total) and Al. (It is what remains after you subtract ferric from ferrous iron.) Figures 25C and 25D give two other sections through K-Fe-Al mica space; they mainly show that the more Al, the greater the opportunity for its replacement by Fe^{3+} (a somewhat similar observation was made for Ti above, and Ti and Fe^{3+} commonly accompany each other in phyllosilicates; the degenerate vector $TiFe_{-1}$ is mentioned in the section on Ti above). On Figure 25C note also the degenerate vector $[]Fe_{-1}$, via which Fe is replaced by a vacancy. The same substitution relates maghemite to magnetite.

Diehard triangle worshippers should note that these sections can be made triangular, the first (Fig. 25C) by using a composition not only with improbable Fe^{2+}(IV), but also with impossible negative [] (i.e., an overstuffed structure) and the second (Fig. 25D) by using a composition with octahedral Si. These compositions are physically attainable, but it is highly unlikely that they would correspond to a mica.

The same (physical attainability) cannot be said for Wones and Eugster's (1965, p. 1232) "oxybiotite", which is a composition containing negative hydrogen, and which allowed them to represent biotite compositions on a triangle (Fig. 26A, a good example of "triangle worship"). Of course, there is nothing wrong with physically unattainable compositions containing negative quantities of the elements (e.g., exchange operators) and I have made many triangular representations with them myself (e.g., Fig. 1B). The point is that for crystal-chemical purposes a vector representation is generally easier and more convenient.

A vector treatment of the same range of biotite compositions is given in Figure 26B. The vertical vector is minus H^0 (atomic hydrogen), which is, as mentioned above, the shorthand form of the vector $Fe^{3+}O(Fe^{2+}OH)_{-1}$, or $Fe^{3+}(HFe^{2+})_{-1}$. Unlike its

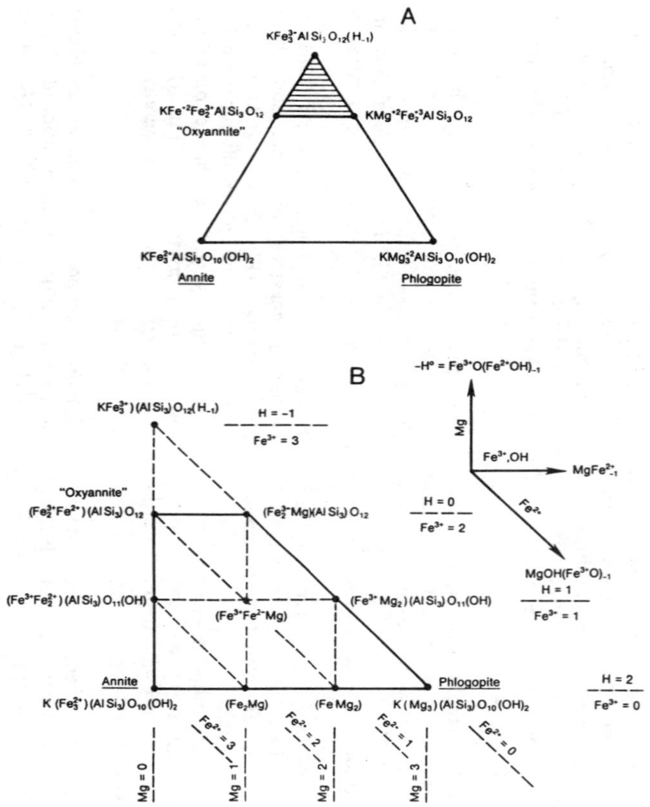

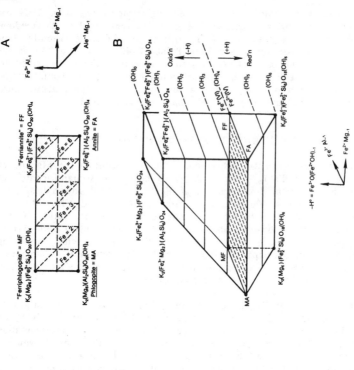

Figure 26. Two representations of K-Mg-Fe biotites, including "oxyannite" (using half formulas, $K_{...}O_{10}(OH)_2$, to fit the original literature). (A) Triangular representation of Wones and Eugster (1965). (B) Vector representation in terms of $MgFe_{-1}$ and $-H^0$, or $Fe^{3+}O(Fe^{2+}OH)_{-1}$.

Figure 27. Diagrams for potential H-absorption or desorption by Mg-Al-Fe biotites. (A) Reciprocal ternary plane. (B) Three dimensional vector diagram with vertical axis $-H^0$.

analogue $AlO(MgOH)_{-1}$ and related substitutions such as the $SiO(AlOH)_{-1}$ depicted in Figure 11, this really is an important substitution in minerals. Why? Because rather than interchanging Mg and Al, or Al and Si, you have only to move electrons and protons through the structure. The implication is that ferrous iron-rich hydroxylated phases can easily lose (desorb) hydrogen, oxidizing ferrous to ferric iron at the same time, and ferric iron-rich phases could, under sufficient pressure, absorb hydrogen, reducing ferric to ferrous iron at the same time. (Such H-absorption is presumably less likely if the ferric iron to be reduced is in a tetrahedral site.) This phenomenon, along with thermal dehydroxylation, is reviewed by Brindley and LeMaitre (1987; cf. Scott and Amonette, 1988); I have suggested that it may be of interest if iron-rich clays occur on Mars (Burt, 1988a).

Rebbert and Hewitt (1986) suggest that anion vacancies might also occur in ferroan biotites, via a reaction such as $Fe^{2+} + 1/4O_2 = Fe^{3+} + 1/2O^{2-}$. As a vector this would be $Fe^{2+}_2[A]Fe^{3+}_{-2}O^{2-}_{-1}$, or $- O^0$, where [A] is an anion vacancy and O^0 is atomic oxygen. Such a substitution might affect biotites during growth, but it presumably could have little influence on a biotite, once crystallized (because oxide ions are much more difficult to move around than protons). Anion vacancies are also structurally unfavorable.

Micas, being relatively deficient in hydroxyls, are likely to run out of OH-groups before they run out of ferrous iron during hydrogen evolution; this is why Wones and Eugster's (1965) "oxybiotite" is physically unattainable on Figure 26A. Vector diagrams can depict graphically the theoretical capability of different clays to absorb or desorb hydrogen (act as "hydrogen sponges"). Figure 27A gives the planar reciprocal diagram provided by the simultaneous substitution of Fe^{2+} for Mg and of Fe^{3+} for Al, starting with phlogopite (all such diagrams are reciprocal rectangles and they are therefore the boring part of this depiction); Figure 27B is Figure 26B extended to the third dimension of $FeAl_{-1}$. The vector $AlO(MgOH)_{-1}$ lies in the space of Figure 27B; it is equal to $Fe^{3+}O(Fe^{2+}OH)_{-1} - Fe^{3+}Al_{-1} - Fe^{2+}Mg_{-1}$. Although annite, having no ferric iron, has no ability to absorb hydrogen, ferriannite and ferriphlogopite, with equal amounts of ferric iron, should have similar H-absorption capacities, as shown by the wedge below the ruled surface. However, all of their ferric iron is tetrahedral, so that the actual H-absorption capacity of non-Tschermak substituted biotites (i.e., those derived from phlogopite) is likely to be small. This barrier to H-absorption (no tetrahedral Fe^{2+}) is indicated by the ruled plane on the figure.

Figures 28A and 28B are analogous representations for "leucophyllite" and celadonite; there is less room for substitution and thus less possibility of H gain or loss. The fact that the Fe^{3+} is octahedral means that there should less of a barrier to H-gain than for biotite.

Figures 29A and 29B give analogous representations for 2:1 chlorite-group minerals, derived from clinochlore. The ferri-chlorites should be able to absorb some H, inasmuch as half their ferric iron is octahedral. The ruled plane indicates the division between "easy" and "hard" H-absorption. On oxidation (H-desorption) they should run out of ferrous iron before they run out of hydroxyls.

The compositions derived from the more Al-rich amesite (1:1 type structure) are depicted in Figure 30A and their H-absorption and desorption capacities are depicted in Figure 30B. Note that cronstedtite or ferriamesite should have a relatively large H-absorption capacity, again, half easy and half hard (separated by the ruled surface). This statement is only theoretically-based; I am not aware of anyone experimentally working with high pressures of H_2 to try to form such "extra" hydroxyls via H-absorption. Most studies (Brindley and LeMaitre, 1987) have instead concentrated on H-loss in a vacuum and its replacement at low pressures of H_2.

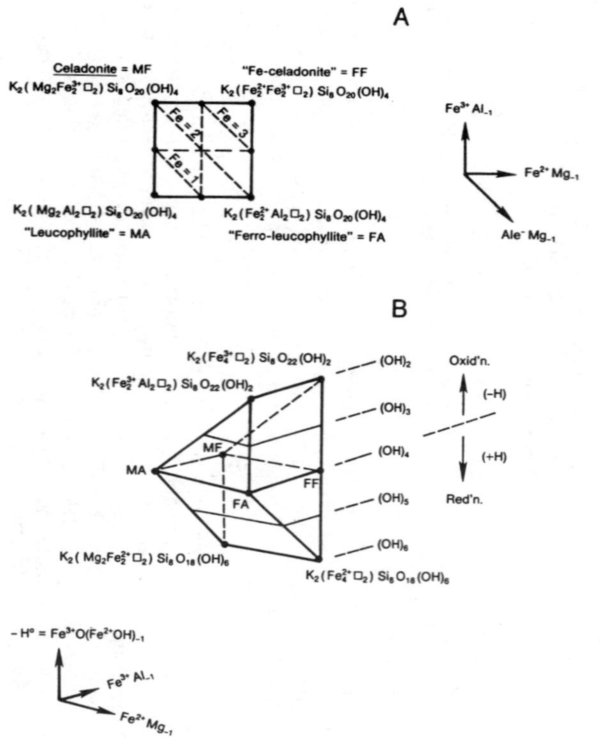

592

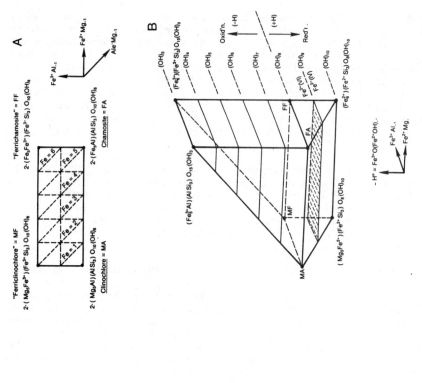

Figure 28. Diagrams for potential H-absorption or desorption by Mg–Al–Fe celadonites. (A) Reciprocal ternary plane. (B) Three dimensional vector diagram with vertical axis –H°.

Figure 29. Diagrams for potential H-absorption and desorption by Mg–Al–Fe chlorites related to clinochlore. (A) Reciprocal ternary plane. (B) Three dimensional vector diagram with vertical axis –H°.

593

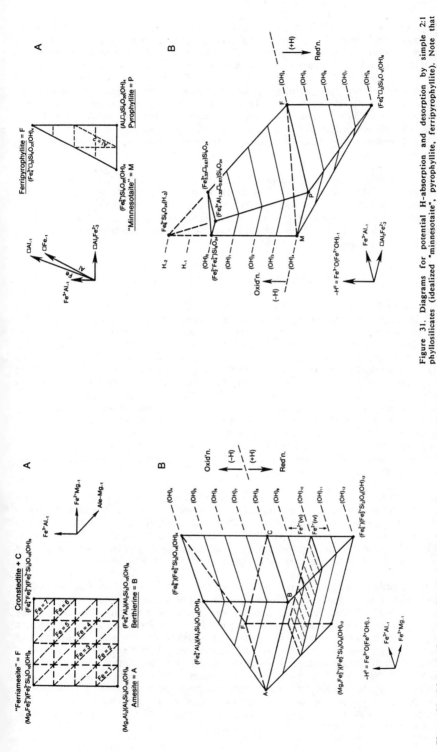

Figure 31. Diagrams for potential H-absorption and desorption by simple 2:1 phyllosilicates (idealized "minnesotaite", pyrophyllite, ferripyrophyllite). Note that natural minnesotaite is a modulated 2:1 structure (Guggenheim and Eggleton, 1986). (A) Planar vector diagram with [JAl_3Fe_{-3} and $FeAl_{-1}$ acting on "minnesotaite". (B) Three dimensional vector diagram with vertical axis -H^0.

Figure 30. Diagrams for potential H-absorption and desorption by Mg-Al-Fe phyllosilicates related to amesite. (A) Reciprocal ternary plane. (B) Three dimensional vector diagram with vertical axis -H^0.

Figures 31A and 31B depict the case of the simple 2:1 phyllosilicates "minnesotaite" (idealized), pyrophyllite, and ferripyrophyllite, which, to show my lack of prejudice, do lie on a triangle (Fig. 31A). "Minnesotaite" (like annite) can only lose H, until it runs out of hydroxyls, ferrimuscovite can only gain it (all of its Fe^{3+} is octahedral), and pyrophyllite can do neither. The polygonal shapes on Figure 31B are only significant as regards the H-exchange capacities of mechanical mixtures of the phases (as in a soil); their mutual solid solubility appears to be small. Note that, structurally, minnesotaite is not really the Fe-analog of talc; instead, because of the too-large size of octahedral Fe^{2+}, it is a modulated 2:1 structure and thus is hardly "simple" (Guggenheim and Eggleton, 1986). I am here using its idealized (simplified) composition.

The last two diagrams, Figures 32 and 33, are analogous representations for dioctahedral (Fig. 32) and trioctahedral (Fig. 33) smectites derived from montmorillonite and saponite respectively. The dioctahedral composition is, as usual, Fe^{2+}-poor and could mainly gain hydrogen; the trioctahedral composition, on the other hand, is Fe^{2+}-rich and could mainly lose it. This "ferrosaponite" phase is sometimes referred to as "lembergite" (e.g., Stucki, 1988, p. 627).

SUMMARY AND CONCLUSIONS

It works! The phyllosilicate composition space in the system $MgO-Al_2O_3-SiO_2-H_2O$ can be generated starting from brucite and using as vectors the three exchange operators $Si_2O_5(OH)_{-2}$ (the tetrahedral sheet addition operator; an O-conserving alternative is $Si_4Mg_{-3}H_{-10}$), $[]Al_2Mg_{-3}$ (the dioctahedral conversion operator) and $MgSiAl_{-2}$ (the Tschermak substitution operator). Later we added $LiAlMg_{-2}$ (octahedral Li-addition operator), $KAl[i]_{-1}Si_{-1}$ (the interlayer charge addition operator, leading through smectite space to the plane of the true micas), $Ca[i]K_{-2}$ (divalent charge substitution operator) or $CaAl_2[i]_{-1}Si_{-2}$ (divalent interlayer charge addition operator), either leading through smectite space to the plane of brittle micas, and $TiMgAl_{-2}$ (octahedral Ti-addition operator). (The portentous-sounding names in parentheses are themselves insignificant.) Redundant (linearly dependent) substitutions involving $AlOH(SiO)_{-1}$ or $MgOH(AlO)_{-1}$ or $LiOH(MgO)_{-1}$ or $TiO_2(MgOH_2)_{-1}$ (a single one of these is sufficient to vary the hydroxyl content) could be added, although their importance in Fe-free systems is probably minor.

At the end we added Fe, via the simple substitutions $FeMg_{-1}$ and $FeAl_{-1}$, but the result was not as simple as might be expected, owing to complex behavior involving the migration of electrons and protons in hydroxylated minerals containing both ferrous and ferric iron. In particular, the coupled substitution $Fe^{3+}O(Fe^{2+}OH)_{-1}$, or minus H^0, makes it easy instead of hard for ferruginous phyllosilicates to gain or lose hydrogen, thereby varying their hydroxyl content and ferrous to ferric ratio.

The resulting composition space has neither gained nor lost dimensions, as compared with a more conventional representation, although in the case of Figure 25A the vector method allowed the deletion of an extraneous (nonexistent) dimension. Many easy-to-depict composition planes and unnamed endmember compositions became evident; the composition planes are potentially useful for plotting natural phase compositions (e.g., Fig. 7 for chlorites and serpentines), because they are so readily contoured for individual elements. In addition, the vector inserts that accompany each composition diagram yield insights into the mutually-dependent nature of coupled substitutions in phyllosilicates. Of course, just because a given substitution is vectorially possible, does not mean that crystal-chemical factors allow it to occur. This is best illustrated by the lack of titanium in dioctahedral phases such as muscovite, discussed in connection with Figure 23.

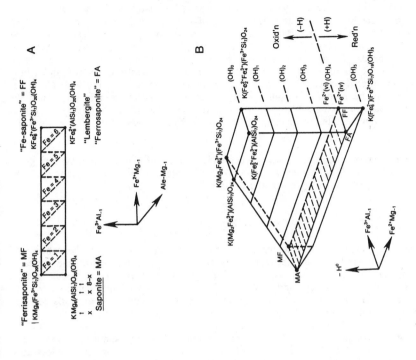

Figure 32. Diagrams for potential H-absorption and desorption by Mg–Al–Fe montmorillonites. (A) Reciprocal ternary plane. (B) Three dimensional vector diagram with vertical axis –H⁰.

Figure 33. Diagrams for potential H-absorption and desorption by Mg–Al–Fe saponites. (A) Reciprocal ternary plane. (B) Three dimensional vector diagram with vertical axis –H⁰.

596

Obviously, the choice of vectors given for each diagram is not unique; the figures demonstrate how numerous other vectors of constant Al, constant octahedral Al, constant Mg, etc. can be derived as linear combinations of those used. These or other vectors (derived from scratch) could be used as basis vectors to satisfy particular presentational needs. I emphasize that in coupled substitution space, or even on a plane, there is no unique set of exchange vectors or substitutions for describing complex compositional variations. I also emphasize that the vectors derived above for phyllosilicates can be used for a wide variety of other rock-forming minerals. Thus the phyllosilicate space is extended to anhydrous phases by the vector $O(OH)_{-2}$ or minus H_2O, starting from brucite, or to amphiboles and pyroxenes via MgH_{-2}, starting from talc.

An important implication of using vectors is that all of the vector diagrams given above could be plotted on ordinary graph paper. In other words, "the tyranny of the triangle" is over (and not only for phyllosilicates).

ACKNOWLEDGMENTS

I remain grateful to J.B. Thompson, Jr., for first suggesting to me the concept of exchange operators (in his 1982 paper he cites W.L. Bragg for originating the concept) and for inspiring me to continue experimenting with them. I am also grateful to S.W. Bailey, P.H. Ribbe, R.J. Tracy, and C.V. Guidotti for reviews of this paper at various stages in its preparation, and to Pam Thompson and especially Deborah Barron for preparing the complex figures.

This work was performed while the author was a Visiting Scientist at the Lunar and Planetary Institute, which is operated by the Universities Space Research Association under Contract No. NASW-4066 with the National Aeronautics and Space Administration. This paper is Lunar and Planetary Institute Contribution No. 668.

REFERENCES

Ahn, J.H., Burt, D.M. and Buseck, P.R. (1988) Alteration of andalusite to sheet silicates in a pegmatite. Am. Mineral. 73, 559-567.

Arima, M., Fleet, M.E., and Barnett, R.L. (1985) Titanian berthierine: a Ti-rich serpentine-group mineral from the Picton ultramafic dike, Ontario. Can. Mineral. 23, 213-220.

Bailey, S.W. (1980) Structures of layer silicates. In G.W. Brindley and G. Brown, Eds., Crystal Structures of Clay Minerals and Their X-ray Identification, Mineral. Soc. (Gt. Br.) Monogr. 5, p. 1-123.

Bailey, S.W. (1988) Chlorites: Structures and crystal chemistry. This volume, Chap. 10.

Brady, J.B. (1975) Chemical components and diffusion. Am. J. Sci. 275, 1073-1088.

Brady, J.B. and Stout, J.H. (1980) Normalizations of thermodynamic properties and some implications for graphical and analytical problems in petrology. Am. J. Sci. 280, 173-189.

Bragg, L., Claringbull, G.F. and Taylor, W.H. (1965) Crystal Structures of Minerals. Cornell Univ. Press, Ithaca, New York.

Brindley, G.W. and LeMaitre, J. (1987) Thermal, oxidation, and reduction reactions of clay minerals. In A.C.D. Newman, Ed., Chemistry of Clays and Clay Minerals, Mineral. Soc. Monogr. 6, p. 318-370. Wiley-Interscience, New York.

Burt, D.M. (1972) The influence of fluorine on the facies of Ca-Fe-Si skarns: Carnegie Inst. Wash. Year Book 71, 443-450.

Burt, D.M. (1974) Concepts of acidity and basicity in petrology--The exchange operator approach (extended abs.). Geol. Soc. Am., Abstr. Programs 6, 674-676.

Burt, D.M. (1979) Exchange operators, acids, and bases. In V.A. Zharikov, W.I. Fonarev, and S.P. Korikovskii, Eds., Problems in Physico-chemical Petrology, v. 2,

p. 3-15. "Nauka" Press, Moscow (in Russian).

Burt, D.M. (1988a) Iron-rich clay minerals on Mars: Potential sources or sinks for hydrogen and indicators of hydrogen loss over time (extended abs.). Lunar Planet. Sci. 19, 148-149.

Burt, D.M. (1988b) Vector representation of lithium and other mica compositions. In L.L. Perchuk, Ed., Advances in Physical Geochemistry (D.S. Korzhinskii Memorial Vol.), in press. New York, Springer-Verlag.

Cerny, P. and Burt, D.M. (1984) Paragenesis, crystallochemical characteristics, and geochemical evolution of micas in granitic pegmatites. Rev. Mineral. 13, 257-297.

Dymek, R.F. (1983) Titanium, aluminum, and interlayer cation substitutions in biotite from high-grade gneisses, West Greenland. Am. Mineral. 68, 880-899.

Eugster, H.P. and Wones, D.R. (1962) Stability relations of the ferruginous biotite, annite. J. Petrol. 3, 82-125.

Evans, B.W. and Trommsdorff, V. (1983) Fluorine hydroxyl titanian clinohumite in Alpine recrystallized garnet peridotite: compositional controls and petrologic significance. Am. J. Sci. 283-A (Orville Vol.), 355-369.

Fawcett, J.J. (1963) The alumina content of talc. Carnegie Inst. Wash. Year Book 62, 139-140.

Forbes, W.C. and Flower, M.F.J. (1974) Phase relations of titano-phlogopite, $K_2Mg_4TiAl_2Si_6O_{20}(OH)_4$: a refractory phase in the upper mantle? Earth Planet. Sci. Lett. 22, 60-66.

Foster, M.D. (1962) Interpretation of the composition and a classification of the chlorites. U.S. Geol. Survey Prof. Paper 414-A, 1-33.

Francke, W., Jelinski, B., and Zarei, M. (1982) Hydrothermal synthesis of an ephesite-like sodium mica. N. Jahrb. Mineral. Monatsh., 337-340.

Giggenbach, W.F. (1985) Construction of thermodynamic stability diagrams involving dioctahedral potassium clay minerals. Chem. Geol. 49, 231-242.

Green, T.H. (1981) Synthetic high-pressure micas compositionally intermediate between the dioctahedral and trioctahedral mica series. Contrib. Mineral. Petrol. 78, 452-458.

Guggenheim, S. (1984) The brittle micas. Rev. Mineral. 13, 61-104.

Guggenheim, S. and Eggleton, R.A. (1986) Structural modulations in iron-rich and magnesium-rich minnesotaite. Can. Mineral. 24, 479-497.

Guidotti, C.V., Cheney, J.T. and Guggenheim, S. (1977) Distribution of titanium between coexisting muscovite and biotite in pelitic schists from northwestern Maine. Am. Mineral. 62, 438-448.

Guidotti, C.V. (1984) Micas in metamorphic rocks. Rev. Mineral. 13, 357-467.

Hewitt, D.A. and Abrecht, J. (1986) Limitations on the interpretation of biotite substitutions from chemical analyses of natural samples. Am. Mineral. 71, 1126-1128.

Korzhinskii, D.S. (1959) Physicochemical Basis of the Analysis of the Paragenesis of Minerals. Consultant's Bureau, New York (translated from 1957 Russian edition).

Kwak, T.A.P. (1968) Ti in biotite and muscovite as an indicator of metamorphic grade in almandine amphibolite facies rocks from Sudbury, Ontario. Geochim. Cosmochim. Acta 32, 1222-1229.

Labotka, T.C. (1983) Analysis of the compositional variations in biotite in pelitic hornfelses from northeastern Minnesota. Am. Mineral. 68, 900-914.

Lantukh, V.I., Pavlikov, V.N. and Lugovskaya, E.S. (1984) Study of a system formed by calcium fluorphlogopite and potassium tetrasilicic mica. In V.V. Skorokhod, Ed., Strukt. Svoistva Poroshk. Mater. Osn. Tugoplavkikh. Met. Soedin., p. 71-74. Akad. Nauk Ukr. SSR, Inst Probl. Materialoved, Kiev, USSR (in Russian, not seen, abstr. in Chem. Abstr. 105-140716z).

Lewis, G.N. (1938) Acids and bases. J. Franklin Inst., 226, 293-313.

Maksimovic, Z. and Bish, D.L. (1978) Brindleyite, a nickel-rich aluminous serpentine mineral analogous to berthierine. Am. Mineral. 63, 484-489.

Mellini, M., Trommsdorff, V. and Compagnoni, R. (1987) Antigorite polysomatism: behaviour during progressive metamorphism. Contrib. Mineral. Petrol. 97, 147-155.

Miyashiro, A. and Shido, F. (1985) Tschermak substitution in low- and middle-grade

598

pelitic schists. J. Petrol. 26, 49-487.

Newman, A.C.D. and Brown, G. (1987) The chemical constitution of clays. In A.C.D. Newman, Ed., Chemistry of Clays and Clay Minerals, Mineral. Soc. Monogr. 6, p. 1-128. Wiley-Interscience, New York.

Newton, R.C. (1987) Thermodynamic analysis of phase equilibria in simple mineral systems. Rev. Mineral. 17, 1-33.

Olesch, M. and Seifert, F. (1976) Stability and phase relations of trioctahedral calcium brittle micas (clintonite group). J. Petrol. 17, 291-314.

Rebbert, C.R. and Hewitt, D.H. (1986) Biotite oxidation in hydrothermal systems: An experimental study. Internat. Mineral. Assoc., 4th General Mtg., Stanford, CA, Abstr. Program, 207.

Robert, J.-L. (1976) Titanium solubility in synthetic phlogopite solid solutions. Chem. Geol. 17, 213-227.

Rosenberg, P.E. and Cliff, G. (1980) The formation of pyrophyllite solid solutions. Am. Mineral. 65, 1217-1219.

Schreyer, W., Medenbach, O., Abraham, K., Gebert, W. and Muller, W.F. (1982) Kulkeite, a new metamorphic phyllosilicate mineral: ordered 1:1 chlorite/talc mixed layer. Contrib. Mineral. Petrol. 80, 103-109.

Scott, A.D. and Amonette, J. (1988) Role of iron in mica weathering. In J.W. Stucki, B.A. Goodman and U. Schwertmann, Eds., Iron in Soils and Clay Minerals, p. 537-623. D. Reidel Publ. Co., Dordrecht.

Seifert, F. and Schreyer, W. (1971) Synthesis and stability of micas in the system $K_2O-MgO-SiO_2-H_2O$ and their relations to phlogopite. Contrib. Mineral. Petrol. 30, 196-215.

Smith, J.V. (1959) Graphical representation of amphibole compositions. Am. Mineral. 44, 437-440.

Skodyreva, M.V., Vlasova, E.V., Zhukhlistov, A.P. and Bagdasarov, Yu. A. (1985) First occurrence of natural mica ([tetra]ferriphlogopite) with magnesium in tetrahedral coordination. Dokl. Akad. Nauk SSSR, 285, 208-211 (in Russian, not seen, abstr. in Chem. Abstr. 104-189860k).

Stoessel, R.K. (1984) Regular solution site-mixing model for chlorites. Clays and Clay Minerals 32, 205-212.

Stucki, J.W. (1988) Structural iron in smectites. In J.W. Stucki, B.A. Goodman and U. Schwertmann, Eds., Iron in Soils and Clay Minerals, p. 625-675. D. Reidel Publ. Co., Dordrecht.

Stucki, J.W., Goodman, B.A. and Schwertmann, U., Eds. (1988) Iron in Soils and Clay Minerals. (NATO ASI Series, Ser. C, Vol. 217). D. Reidel Publ. Co., Dordrecht, 900 p.

Thompson, J.B., Jr. (1978) Biopyriboles and polysomatic series. Am. Mineral. 63, 239-249.

Thompson, J.B., Jr. (1979) The Tschermak substitution and reactions in pelitic schists. In V.A. Zharikov, W.I. Fonarev, and S.P. Korikovskii, Eds., Problems in Physico-chemical Petrology, v. 1, p. 146-159. "Nauka" Press, Moscow (in Russian).

Thompson, J.B., Jr. (1981) An introduction to the mineralogy and petrology of the biopyriboles. Rev. Mineral. 9A, 141-188.

Thompson, J.B., Jr. (1982) Composition space: An algebraic and geometric approach. Rev. Mineral. 10, 1-31.

Tien, P.-L., Leavens, P.B. and Nelen, J.A. (1975) Swinefordite, a dioctahedral-trioctahedral Li-rich member of the smectite group from Kings Mountain, North Carolina. Am. Mineral. 60, 540-547.

Tschermak, C. (1890) Die Chloritgruppe I. Teil. Sitzungsber. Akad. Wiss. Wien (Math.-Naturwiss. Kl.) 99, 174-264.

Veblen, D.R. (1983) Exsolution and crystal chemistry of the sodium mica wonesite. Am. Mineral. 68, 554-565.

Velde, B. (1973) Phase equilibria studies in the system $MgO-Al_2O_3-SiO_2-H_2O$. Chlorites and associated minerals. Mineral. Mag. 39, 297-312.

Velde, B. (1985) Clay Minerals. A Physico-Chemical Explanation of Their Occurrence (Developments in Sedimentology, 40). Elsevier, Amsterdam, 421 p.

Walsche, J.L. (1986) A six-component chlorite solid solution model and the conditions of chlorite formation in hydrothermal and geothermal systems. Econ. Geol. 81, 681-703.

Wones, D.R. and Eugster, H.P. (1965) Stability of biotite: experiment, theory, and applications. Am. Mineral. 50, 1228-1272.

Yoder, H.S. and Eugster, H.P. (1955) Synthetic and natural muscovites. Geochim. Cosmochim. Acta 8, 225-280.

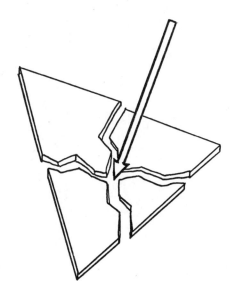

MIXED LAYER CHLORITE MINERALS

INTRODUCTION

Chlorite forms two-component interstratified minerals with all of the other common phyllosilicate layer types. If the operational definition of a chlorite layer is accepted as a 14 Å, 2:1 phyllosilicate that cannot be expanded or collapsed by "normal" laboratory procedures, then some type of chlorite can form between room temperature and the chlorite zone of regional metamorphism.

Composition and ordering possibilities for a given mixed-layered chlorite species are more restricted than those displayed by some other minerals, such as illite/smectite. There are few (none?) chlorite mixed-layered suites that provide examples of a complete range in composition from one end member to the other. To date, almost all reported instances of mixed-layered chlorite minerals in rocks can be classified into two types, (1) nearest-neighbor ordered types (R = 1) containing very nearly equal proportions of the two components, and (2) randomly interstratified (R = 0) minerals whose compositions lie near one end of the range or the other. Soil chlorites are much more complex, and interstratifications with more than two components have been reported.

The interesting aspect of chlorite mixed-layering lies in the richness of the mixed-layered varieties. Chlorite interstratifications have been reported with swelling chlorite, serpentine, kaolinite, talc, vermiculite, smectite, and mica (see Tables 2 and 3). Either di- or trioctahedral chlorite may be present in some of these mixed-layered minerals. In this chapter, the name chlorite implies tri-tri-chlorite unless otherwise indicated.

MIXED-LAYERED CLAY MINERALS

Statistical description of ordering

Most interstratified minerals contain two components, and the first step in naming the mineral requires an identification of the two layer types. The next most important qualifiers describe the type of order (or lack thereof) in the stacking sequence of the two components along Z, and the proportions of the two in the mixed-layered mineral. The term Reichweite (Jadgodzinski, 1949) has achieved wide acceptance, and is abbreviated, for example, R = 1 or R1. The number denotes the "reach-back" or the correlation distance over which the occurrence of a layer type affects the probability of occurrence of a given layer. The designation R = 0 means that nearest neighbors have no effect, so the probability of occurrence of a given type is simply equal to its molar proportion in the mixed-layered mineral. This is termed random interstratification.

R = 1 means that only adjacent layer types affect the probability of occurrence of a given layer. This describes nearest-neighbor ordering and gives rise, for equal proportions of two components, to an alternating structure of the type ABABAB--etc., where the two types are designated as A and B. R1 structures also are possible when A and B are present in unequal proportions. For example, if the proportion of A (P_A) = 0.6, then P_B = 0.4. In such a structure, all B layers are preceded and followed by A types, but because this arrangement does not utilize all of the A types, some B layers are separated by more than one A. Consequently, such a structure has some disorder, whereas the 50/50 composition (R = 1) does not. The designation R = 1, P_A = 0.6 can be thought of as a random interstratification of an AB superstructure with excess A layers. The proportion of excess A layers is zero at the 50/50 composition (no disorder) and increases as P_A increases (increasing disorder). For such structures, the single unequivocal conclusion is that the layer pair BB is not present.

Partial disorder can occur even for 50/50 compositions, and for the other R1 structures. Such a condition can conveniently be described by, for example, R = 0.5. For such structures, the tendency for A and B to occur in pairs is imperfect, but is greater than for the random case, and the pair BB exists, though it's frequency is less than it would be in a randomly interstratified structure.

Longer range ordering, that is, R > 1, has been identified for mixed-layered illite/smectite (Reynolds and Hower, 1970) and glauconite/smectite (Thompson and Hower, 1977) but has not yet been reported for mixed-layered chlorite minerals. A discussion of R > 1 structures is given by Reynolds (1980).

The recent work of Nadeau and coworkers (e.g. Nadeau et al.,1984) has, in the opinion of some, cast doubt on the validity of the Reichweite scheme for the description of ordering in mixed-layered clay minerals. The present writer does not share these concerns. It does not matter how we define the unit cells or unit layers that are the fundamental particles, or are aggregated into fundamental particles. As the X-ray beam sees it, a unit cell is a fundamental particle, as is an aggregate of unit cells or an aggregate of fundamental particles, supposing of course that the alignments of these are sufficiently accurate to constitute X-ray coherent domains. We can still describe the structures in terms of the stacking sequences of two kinds of unit layers. Application of Nadeau's ideas to statistical assemblages of unit layers produce some stacking sequences whose terminations differ from those predicted by conventional Markov statistical descriptions (Reynolds, 1980), and these effects are observed in the experimental diffraction patterns (Tellier and Reynolds, 1987). But the end-effects do not invalidate the Reichweite nomenclature. Using illite/smectite as an example, $P_I = 0.6$, R = 1 describes a structure in which the layer pair smectite-smectite does not occur, that is, no 10-Å fundamental particles are present. All smectite layers are separated from each other by one or more illite layers, in other words, the fundamental illite particles are 20, 30, 40 .... Å thick. The Reichweite and fundamental particle approaches are simply two different ways of describing the same structure.

Names of mixed-layered chlorite minerals

Randomly interstratified (R = 0) mixed-layered chlorite minerals are designated simply by the names and proportions of layer types. A convention, not yet formalized, places first the name of the component with the smallest value for d(001) after treatments that cause the maximum expansion of any expandable layer type. The decimal fraction of the first component in the interstratification is added in parenthesis. A typical example is chlorite (0.6)/smectite. The designations talc/chlorite, chlorite/smectite, chlorite/ vermiculite, mica/chlorite, and serpentine/chlorite follow this rule.

Perfectly ordered structures (R = 1) that contain equal proportions of the two components are given mineral names. The logic here is that for a perfect alteration of two layer types, ABABAB---etc., a new cell can be defined as the unit AB whose c-axis dimension is equal to the sums of these dimensions for the layer types A and B. A crystal composed of the units AB is not a mixed-layered mineral, and its fixed stoichiometry and structure qualify it as a distinct mineral species.

Identification of random and ordered interstratification

A simple chlorite mineral produces 00l reflections at diffraction angles that correspond exactly to d(001)/1, d(001)/2, d(001)/3--etc. Such an X-ray diffraction pattern is termed *rational*. In addition, the breadths of the reflections, corrected for $K\alpha 1$-$K\alpha 2$ broadening, are proportional to $1/\cos\theta$. If another layer type with a different d(001) is randomly interstratified with chlorite, the rationality of the 00l pattern is lost and diffraction line breadths may follow no monotonic relation with respect to θ. The best way to think about such diffraction patterns is to use the concepts of Méring (1949). He pointed out that diffraction maxima lie between the positions of reflections from each of the two components involved. The positions of the mixed-layered peaks depend on the proportions of the two, and the peaks are broad if the two end-member reflection angles are widely separated. In the general case, the diffraction pattern is different from that of a simple mechanical mixture of the two components.

Figure 1 shows a calculated[1] X-ray diffraction pattern for randomly interstratified chlorite(0.5)/ E.G. smectite. The abbreviation E.G. is used here to signify ethylene glycol solvation. The solid vertical lines mark the positions of the smectite 00l series, and the dashed lines depict the positions of the chlorite maxima. The continuous profile is the mixed-layered clay diffraction pattern. Note that the reflections have different breadths, and that these are proportional to the separations between the smectite and chlorite 00l diffraction positions. The very broad peak at d = 4.97 Å is a composite of the smectite 003 and chlorite 003, *and* the chlorite 003 and the smectite 004 peaks (termed the 003/004 to correspond to the sequence of words in the name chlorite-smectite). The mixed-layered maxima lie, as they should, between the positions of reflections for pure smectite and chlorite. This hypothetical mixed-layered mineral contains equal proportions of the two components, and according to Méring's principles, reflections should be equidistant from the diffraction angles of the adjacent pure components. The peaks at 15.8, 7.82, 3.46, and 2.83 Å follow this principle. The reflection at d = 4.97 Å is really a composite of three component reflections, so the rule does not apply to it. But it is premature to conclude that composition can be accurately determined by linear interpolations in 2θ-composition space, because plots of d versus proportions of components (the so-called migration curves) are typically S-shaped. This means that Mering's principle of peak position behaves best at the middle of the compositional range, which is the case here. The best method for determining proportions of components, at the present time, is to compare experimentally observed peak positions with those from calculated diffraction patterns.

A test of rationality for the diffraction pattern is provided by a calculation of the CV or coefficient of variability (Bailey, 1982), and the results are shown by Table 1. The AIPEA Nomenclature Committee has defined a rational diffraction pattern on the basis of a maximum CV of 0.75% (Bailey, 1982). As expected, the large CV (6.5%) for this example places it in the irrational category, indicating significant disorder in the stacking sequence of chlorite and smectite layers.

Figure 2 shows a calculated diffraction pattern of chlorite(0.5)/E.G. smectite, R = 1. The peak positions correspond to those predicted by the Bragg Law, $l \lambda = 2d \sin\theta$, where l is the order of diffraction, and d = d(001) for a superstructure whose d(001) is equal to the sum of d(001) for chlorite (14.2 Å) and E.G. smectite (16.9 Å), or 31.1 Å. The first-order reflection from this superstructure is often called a long-spacing, and its presence is strong evidence for an ordered structure of some kind in the sample.

The rationality of the pattern is imperfect, given the inaccuracies in measuring d on the scale of Figure 2, but the CV is less than 0.5%; such a pattern identifies R = 1 ordering, and a 50/50 proportion of components. These parameters qualify such a mineral for a distinct name, in this case, corrensite, provided that ten orders are used for the calculation, and that odd and even orders have closely similar diffraction line breadths (Bailey, 1982).

---

[1] Calculated X-ray diffraction patterns are used throughout this chapter. They were generated by the computer program NEWMOD (Reynolds, 1985). Relevant input parameters are summarized in the Appendix. For all patterns, Cu$K\alpha$ is assumed.

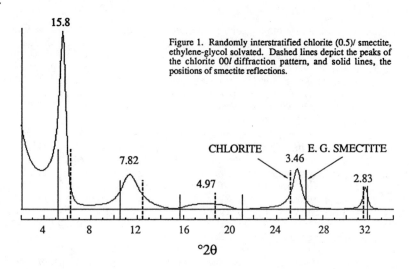

Figure 1. Randomly interstratified chlorite (0.5)/ smectite, ethylene-glycol solvated. Dashed lines depict the peaks of the chlorite 00*l* diffraction pattern, and solid lines, the positions of smectite reflections.

Table 1. Coefficient of variability (CV) for randomly interstratified chlorite/E.G. smectite, 50% chlorite.

Apparent Order	d (Å)	Order x d	(Mean- Order x d)^2
1	15.8	15.80	0.54
2	7.82	15.64	0.33
3	4.97	15.91	0.71
4	3.46	13.84	1.51
5	2.83	14.15	0.84
		Mean 15.07	Sum 3.93

Standard Deviation = sqr(Sum/(N-1)) = sqr(3.93/4) = 0.98

CV = 100 x 0.98/15.07 = 6.5%

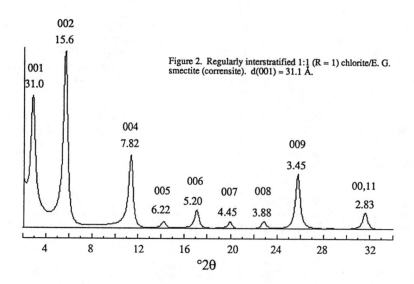

Figure 2. Regularly interstratified 1:1 (R = 1) chlorite/E. G. smectite (corrensite). d(001) = 31.1 Å.

Table 2. Mixed-layered chlorite minerals, 1:1 ordered interstratifications (R = 1).

Name	Description	References
Kulkeite	Talc/chlorite	Schreyer et al. (1982)
Corrensite High-charge	Chlorite/vermiculite trioctahedral	Johnson (1964) Gradusov (1969)
Corrensite Low-charge	Chlorite/Smectite trioctahedral	Schlenker (1971)
Tosudite	Chlorite/smectite dioctahedral	Brown et al. (1974)
	Li-chlorite/smectite dioctahedral	Nishiyama et al.(1975)
	Illite/chlorite	Lee and Peacor (1985)
	[Biotite(0.44)/chlorite] $P_{C.B} = 0.8$	Eroshchev-Shak (1970)
	Biotite/chlorite	Maresch et al. (1983)

Table 3. Randomly interstratified chlorite minerals. Chlorite* denotes hydroxy-Al interlayers

Type	Analytical methods	References
Kaolinite/chlorite	Single crystal XRD[1]	Brindley and Gillery, 1954
Serpentine/chlorite	HRTEM[2] and AEM[3]	Ahn and Peacor, 1985
Biotite/chlorite	HRTEM	Veblen and Ferry, 1983
Wonesite/chlorite	HRTEM	Veblen, 1983
Muscovite(illite)/chlorite	HRTEM and AEM	Lee et al., 1984
Chlorite/smectite	Powder XRD; microprobe	Bettison and Schiffman, 1988
Chlorite*/vermiculite	Powder XRD	Tamura, 1955 Sawhney, 1960

[1]X-ray diffraction
[2]Analytical electron microscopy
[3]High resolution transmission electron microscopy

MIXED-LAYERED CHLORITE MINERALS

A perfectly ordered, alternating structure containing two layer types stacked in the sequence ABABAB----etc. is qualified for a mineral name. Table 2 shows reported instances of such minerals in which chlorite is one component, in addition to other ordered 1:1 structures that have not been documented sufficiently at present to be acceptable as named species.

Randomly interstratified chlorite minerals are less frequently reported than the ordered varieties. They are also, in general, more poorly defined, because they tend to occur in mixtures with discrete chlorite which produces severe diffraction line interference, and compromises the interpretation of chemical analyses. The X-ray identification of such species is unreliable if it is based on only one or two reflections at low diffraction angles, and many reports in the literature are based on such poor data. Studies based on HRTEM lattice-fringe imaging are becoming more frequent, and promise definitive conclusions on the validity of some of these occurrences, though the lattice fringe-images are sometimes difficult to reconcile with X-ray diffraction results. The problem is exacerbated by the fact that few of these studies show or discuss X-ray diffraction patterns that correspond to the materials photographed. (There are notable exceptions, for example Schreyer et al., 1982). One gets the impression that X-ray diffractionists and electron microscopists are looking at different ends of the same reality, and that like the case of the elephant and the blind men, the different views are difficult to correlate. The writer has observed that the mixed-layered clay minerals that are best known and most often studied by X-ray diffractionists all contain the expandable layer types smectite or vermiculite (e.g., chlorite/smectite, illite/smectite, mica/vermiculite, kaolinite/smectite). The mixed-layered phyllosilicates that receive the most attention by the TEM specialists are the non-expandable types (e.g., mica/chlorite, serpentine/chlorite). This situation may arise because the expandable types produce often spectacularly different X-ray diffraction patterns when treated with suitable solvating agents, whereas the expandable layers collapse in the vacuum chamber of the electron microscope and may be difficult to differentiate from non-expandable layer types. It may be that the randomly interstratified chlorite minerals have compositions that tend to be near one end member or the other. Diagnosis of such non-expandable minerals is difficult by X-ray methods because irrational patterns are not obvious, and the interpretation must be based on line breadths which require higher quality diffraction patterns than are routinely obtainable. Small proportions of different layer types, however, are easily observed with the electron microscope. Table 3 lists some of the reported occurrences of randomly interstratified chlorite minerals.

THE STRUCTURES OF MIXED-LAYERED CHLORITE MINERALS

Assignment of atoms to the unit cell

Vermiculite and smectite expand when saturated with water and polar organic solvents. The charge on the 2:1 layer controls the expansion, and it is the tetrahedral charge that is crucial for chlorite interstratified with other 2:1 layer types. Figure 3 shows a schematic depiction of chlorite/smectite. For this species, a tetrahedral charge of -1 is associated with an octahedral hydroxide interlayer, and a charge of -0.3 gives rise to an interlayer structure that consists of water and hydrated cations. At the outset, we see a difficulty in assigning the atoms that belong respectively to chlorite and smectite. This difficulty arises from the intuitively sound location of the chlorite (or smectite) layer about the center of symmetry (on projection to Z) in the 2:1 octahedral sheet. This location of the origin of the structure produces a 2:1 layer with half of a hydroxide interlayer on one end, and half of a water layer on the other, and this is neither chlorite nor smectite. The problem is solved by centering the structures on the interlayer which is another center of symmetry

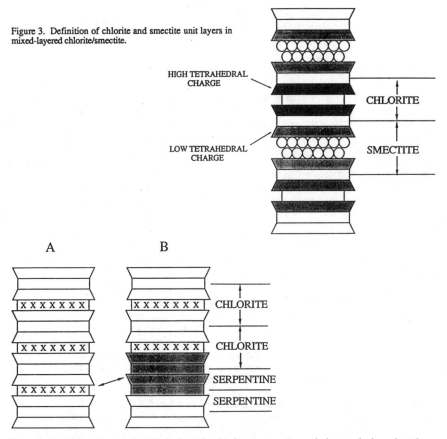

Figure 3. Definition of chlorite and smectite unit layers in mixed-layered chlorite/smectite.

HIGH TETRAHEDRAL CHARGE

LOW TETRAHEDRAL CHARGE

CHLORITE

SMECTITE

A   B

CHLORITE

CHLORITE

SERPENTINE

SERPENTINE

Figure 4. Derivation of serpentine/chlorite from the chlorite structure. Arrow designates the inversion of a tetrahedral sheet to form two serpentine layers.

on projection to Z. Then the chlorite and smectite layers share the cations in the 2:1 silicate octahedral sheet. This arrangement defines a 2:1 silicate layer that lacks a center of symmetry on projection to Z because of the asymmetrical distribution of tetrahedral Al.

The structural assignment shown by Figure 3 is convenient for chlorite interstratifications with all of the other 2:1 layer types, and it is equally useful for 7-Å components. Figure 4 shows a normal chlorite structure (A) and a mixed-layered chlorite/serpentine stacking sequence (B) that has been derived by the inversion (arrow) of one 2:1 tetrahedral sheet. Figure 4B shows the familiar 1:1 serpentine I-beam symbols for the serpentine layers, but the atomic assignments for the asymmetrical 1:1 layers are different and can be centered on a tetrahedral sheet that separates *two* octahedral hydroxide sheets. The arrangement seems awkward because it is impossible to assign all tetrahedral sheets to either chlorite or serpentine. For the purposes of X-ray diffraction studies, we can think of chlorite, serpentine, and interstratified chlorite/serpentine structures as stacks of hydroxide sheets separated from each other by 7 Å. The spaces between these sheets are filled with single tetrahedral sheets that have one of the two possible orientations shown by Figure 4B. Of course the three kinds of octahedral sheets (2:1 octahedral, interlayer, and 1:1 octahedral) have different compositions. But the differences are due to the ratios of O/OH, and the single electron from the hydrogen atom has almost no effect on X-ray scattering amplitudes. Brindley and Gillery (1954) first suggested that 7-Å layers can be generated by the inversion of tetrahedral sheets in chlorite.

IDENTIFICATION AND CHARACTERIZATION OF
MIXED-LAYERED CHLORITES

Sample preparation

X-ray powder diffraction remains the most powerful and widely used method for the identification and characterization of mixed-layered phyllosilicates. HRTEM and SEM studies and chemical analyses are useful and can provide different and important kinds of information, but, with the exception of the SEM, they cannot at present be applied routinely to the analysis of large numbers of samples. Most occurrences of mixed-layered chlorite minerals are fine-grained (silt to clay size) and the chloritic material is invariably mixed with other minerals. Samples are simply too impure for valid bulk chemical analysis, and the microprobe often has insufficient resolution to deal with individual minerals in clay-sized aggregates. Analytical electron microscopy (AEM) and HRTEM investigations are vital as research tools, but they are too time-consuming for routine analysis. The method of choice for the average analyst will be X-ray powder diffraction, consequently, the discussion here is concentrated on this subject.

Definitive identifications of mixed-layered chlorite species depend on good diffraction patterns of the 00l series of reflections. The few reflections at low diffraction angles are often insufficiently diagnostic; the analyst should have 4 or 5 reflections to work with. The higher angle peaks are weak for many species, so good preferred orientation and thick specimens (> 20 mg/cm^2) are required. Modern high-intensity X-ray tubes provide excellent counting statistics which can be improved by computer controlled step-scanning procedures that use long count times. The common peak interferences from discrete chlorite, kaolinite, and illite can be dealt with by peak-stripping or deconvolution procedures, algorithms for which are included as parts of the standard packages for several commercial powder diffraction systems. The interferences can be almost eliminated by magnetic separations (Senkayi et al., 1981; Tellier et al., in press) because the high Fe contents of mixed-layered chlorites make them suitable for purification by this method. Attention to these details will provide good analyses for mixed-layered chlorites in rocks, but they are less effective with soil clays. The latter, unfortunately, are both more complicated and generally possess poor crystallinity which produces broad and ill-defined high angle intensities that often cannot be resolved from the background. Poor crystal morphology and admixed organic matter and Fe oxy-hydroxides minimize good preferred orientation, causing further deterioration of already poor diffraction patterns. Organic matter and Fe oxides can be removed chemically (Jackson, 1969), but such procedures are risky. The expansion behavior of mixed-layered chlorites in water and organic solvents depends on layer charge, and chlorites contain large amounts of ferrous Fe which can be oxidized by the hydrogen peroxide or sodium hypochlorite used to remove organic matter. As Tamura (e.g., 1955) has shown, some chlorite types in soils have hydroxy-Al layers which can be removed by the citrate-dithionite method (Jackson, 1969) for iron oxide removal. For these reasons, mixed-layered chlorites in soils are likely to remain more poorly understood than they are in rocks. Table 4 lists the layer types that are likely to occur as interstratifications in chlorite, and the procedures for identifying them.

The analysis of mixed-layered chlorite species requires multiple diffraction analysis of a series of sample aliquots that have received different chemical or physical treatments. The first analysis will almost always be of an air-dried sample, or at least one that has come to equilibrium with the moisture in laboratory air. This serves as a baseline pattern against which the results of the other treatments are compared. Common, though not universal circumstances, are the natural occurrence of Ca-smectite which forms a two-water-layer complex with d(001) ≈ 15 Å, or water Mg-vermiculite for which d(001) = 14.3 Å.

Cation saturation is accomplished by exchange with 1M KCl and 0.1M MgCl$_2$. A solution that contains about ten times the exchange capacity of the mineral sample is added, and the sample is mixed and centrifuged, after which the supernatant liquid is discarded. This process is repeated four times--three is probably sufficient for X-ray work, but not for accurate measurements of exchange capacity. After exchange, the sample is washed with water by means of centrifugation. Four or five cycles is usually sufficient. The last few washings should be made in a 50/50 water-ethanol solution to minimize hydrolysis which can cause the substitution of H^+ for the exchange cations (Jackson, 1969). Testing the supernatant liquid with AgCl is a convenient method for establishing that the exchange salt has been removed.

Heat treatments are for 1 hr, and no special precautions are necessary except for the choice of substrate. Byström-Brusewitz (1976) has shown that Na can be absorbed from glass at high temperatures, so it seems prudent to use fused silica slides for these preparations.

Table 4. Chemical and physical treatments used to identify non-chlorite components in mixed-layered chlorite minerals.

Treatment	Layer-type	d(001) (Å)	Reference
Ethylene Glycol	Smectite	16.9	Bradley (1945)
Glycerol	Smectite	17.7	MacEwan (1944)
Ethylene Glycol	Vermiculite	14.3-16.3	Brindley (1966)
Mg-Sat.-Glycerol	Vermiculite	14.3	Walker (1957, 1958)
Mg-Sat.-Glycerol	Smectite	17.7	Walker (1957, 1958)
K-Sat.	Vermiculite	10.4	MacEwan and Wilson (1980)
K-Sat.	Smectite	11.0-12.5	MacEwan and Wilson (1980)
K-Sat., 350°C	Smectite	10	Martin-Vivaldi et al. (1963)
K-Sat., 350°C	Vermiculite	10.4	Barshad (1950)

Expandable layers in chlorite, and some pure trioctahedral smectites solvate slowly with glycerol and ethylene glycol and the final products may not be reproducible. (Novich and Martin, 1983). Clear-cut results require drastic methods for applying these reagents. The standard vapor methods which suspend the sample in ethylene glycol vapor at 60°C or glycerol vapor at 100°C for 12 or more hours may be insufficient to achieve full expansion for some minerals. Novich and Martin recommend the method suggested by Kunze (1955) in which the organic reagent is added directly to the aqueous clay suspension. They add the reagent in amounts that are roughly equal, by weight, to the total amount of clay mineral. The sample is prepared by normal methods and allowed to air-dry. Samples solvated with ethylene glycol must be stored in an atmosphere of the vapor, but glycerol-solvated preparations are stable indefinitely in laboratory air.

A mineral sample should be treated by the various methods described above (Table 4) and analyzed by X-ray diffraction methods. The use of Méring's principles will lead to a rationale for the sample that fits the diffraction data to the kinds of layer spacings produced by the various treatments. Of course, if the sample is unaffected by all of them, then any non-chlorite layers must be mica, talc or serpentine (or kaolinite), and the diffraction patterns should be explainable in these terms.

Swelling chlorite is a layer type not included in Table 4, yet it has been widely reported interstratified with smectite and with chlorite (e.g. Lippmann, 1954, 1956; Stephen and MacEwan, 1950; Martin-Vivaldi and MacEwan, 1960; MacEwan and Wilson, 1980). Discrete swelling-chlorite has d(001) ≈ 14 Å, with rational higher order reflections. These spacings are stable to at least 430°C. Glycerol-solvation yields a rational 00l diffraction pattern with d(001) = 17.8 Å (Steven and MacEwan, 1950). It is difficult to reconcile these diffraction results with any alternative structure, and the correctness of the interpretation of MacEwan and his co-workers seems unassailable. Nevertheless, the structure and chemical composition of swelling-chlorite remain poorly defined, and the layer-type is not acceptable for inclusion within the term corrensite, as has been done in some of the older literature (Bailey, 1982). Martin-Vivaldi and MacEwan (1960) suggest that swelling-chlorite has an incomplete hydroxide interlayer, perhaps consisting of islands of brucite-like material. Remaining clear interlayer space can be penetrated by glycol or glycerol to give smectite-like expansion. A mixed-layered chlorite/swelling-chlorite should behave like chlorite/smectite, with ethylene glycol solvation, yet produce a chlorite-like diffraction pattern upon K-saturation and heating.

Swelling-chlorite is grouped with other 14-Å structures that are sensitive to aggressive chemical treatments. These are called the labile chlorites (Martin-Vivaldi and MacEwan, 1960). Slaughter and Milne (1960) produced 14-Å structures that are stable to heating at moderate temperatures (200°C) by precipitating Al, Mg, and Ni hydroxides in the interlayer space of smectite. Natural examples of hydroxy-Al minerals are provided by the work of Tamura (1955), and Ildefonse et al.(1986) have identified hydroxy-Cu vermiculite that formed by the weathering of biotite associated with a stratiform copper deposit in Brazil. Natural hydroxy-Al material behaves like a chlorite with respect to the various treatments listed in Table 4, but the interlayer Al can be removed by treating with dithionite-citrate or sodium citrate (Sawhney, 1960), after which it behaves like smectite or vermiculite, whichever is the case.

Once a qualitative identification has been made and the ordering type assigned, the next step in mineral characterization is to estimate the proportions of the layer types. This is done by comparisons of the spacings of selected peaks with those tabulated for the same peaks from diffraction patterns that have been calculated for specified compositions.

Ordered interstratified minerals require demonstration of a CV < 0.75% if a specific mineral name needs to be assigned. A suitably small CV is often difficult to achieve because of errors in d(001), even though the rest of the pattern seems to eminently qualify the sample for name status (G. W. Brindley, pers. comm.). There are several reasons why d(001) is very difficult to measure accurately. First of all, the diffraction angle is very small, typically near 3°2θ (for CuKα radiation). Very small errors in 2θ translate into large errors in d. For example, an error of 0.05°2θ, gives an error in d of 1.7% at 3°2θ. Assuming that the diffractometer is properly aligned, the largest source of error is the sample displacement error, which is the displacement of the sample surface from the axis of rotation of the goniometer. The relevant relation is Δ2θ = 2Scosθ/R, where Δ2θ is expressed in radians, S is the sample displacement in cm, and R is the goniometer radius in cm (Parrish and Wilson, 1959). For a goniometer radius of 20 cm, this means that a sample displacement of 0.08 mm causes an error of 1.7% in d at 3°2θ. For any surface that is not perfectly flat, the effective surface has only an average location which is difficult to define. Consequently, only very smooth and planar samples, perfectly positioned in the diffractometer, can be expected to give d(001) values that are accurate enough to pass the stringent requirements for the CV. Sample position problems can be eliminated by incorporating a standard into the mixed-layered mineral aggregate. Pyrophyllite or talc seem obvious choices. They will produce interferences on the higher order reflections, but that is inconsequential because it need be used only for the calibration of d(001). The pyrophyllite or talc should first be calibrated by checking it's 00l spacings with respect to those of a suitable standard material such as powdered quartz , with which it is mixed.

Elimination of the sample-position error helps but does not eliminate more fundamental sources of error in measuring d at very low diffraction angles. If the peaks are broad, the Lorentz-polarization factor has the effect of shifting the peak center to lower diffraction angles. The 001 peak should be recorded as a digital file, and the intensity at each value of 2θ divided by $(1 + \cos^2 2\theta)/\sin\theta$. This procedure removes the effect of the single crystal Lorentz-polarization factor, and d(001) is measured after such a correction. The correction is a conservative one, for the random powder form of the Lorentz factor (for a continuous profile) is proportional to $1/\sin^2\theta$, and correcting by means of that shifts the diffraction maximum to even higher values of 2θ.

Lorentz factor displacement is unimportant if the 001 reflection is sharp, but then imperfections in the diffractometer optical system become important. Flat-specimen and vertical divergence distortions (Klug and Alexander, 1974) are more extreme at low diffraction angles, and both of these shift the peak center toward lower diffraction angles. The effects of both can be minimized by using a fine divergence slit (0.3° or less) and incorporating two soller slits in the optical arrangement, even if your diffractometer has only one as standard equipment.

00*l* X-RAY DIFFRACTION PATTERNS FOR MIXED-LAYERED CHLORITES

Ordered (R1) structures

Figure 5 shows calculated diffraction patterns for high and low charge corrensites (trioctahedral) and for tosudite (dioctahedral). The low-charge corrensite patterns (C and D), and the tosudite (B) represent the ethylene-glycol solvated state for which d(001) of the smectite component = 16.9 Å. The high charge variety (A) assumes the two-water layer Mg-vermiculite structure, with d(001) for vermiculite = 14.3 Å. The reflections produce rational diffraction patterns based on d(001) = 31.1 Å for low charge corrensite and tosudite, and d(001) = 28.5 Å for the high charge variety. All intensities have been normalized to a constant value for the 002 peaks. The tosudite pattern was calculated on the basis of no Fe and the others assume one Fe per three octahedral positions for both the 2:1 silicate layer and the interlayer hydroxide sheet (cf. Brigatti and Poppi, 1984a).

Figure 5D depicts chlorite(0.55)/E.G. smectite, R = 1. The structure contains some disorder as evidenced by the broadening and weakening of the 001 peak, in comparison to that of the 1:1 interstratification portrayed by Figure 5C. A CV of 0.61% was computed for the pattern shown by Figure 5D, demonstrating that a 5% departure from 1:1 component proportions is insufficient to remove this structure from the category of corrensite. Measured d(001) = 31.4 Å, instead of the nominal value of 31.1 Å, attesting to the effect of Lorentz-polarization induced shift of this relatively broad reflection.

Tosudite (Fig. 5B) produces a pattern very similar to corrensite, except that the relative intensities differ. The intensity ratio of the 004/006 reflections is a useful index of total octahedral scattering, and if I(006) > I(004), the investigator should suspect a low-Fe dioctahedral species. Note, however, that a low-Fe corrensite and a high-Fe tosudite (if such exists) could have similar octahedral scattering, and in such a case, produce identical 006/004 intensity ratios. The identification of tosudite requires verification, and such can be provided by a dioctahedral 060 spacing or, better yet, a chemical analysis.

The high-charge corrensite of Figure 5A is easily distinguished from the others by peak position and a relatively strong 003 reflection. Ethylene glycol-solvated low-charge corrensite has an undetectable 003 reflection, so a cursory examination of the diffractogram gives useful information in this regard. But mica must be absent, because the mica 001 produces interference at this diffraction angle. By chance, Mg-saturated and glycerol-

612

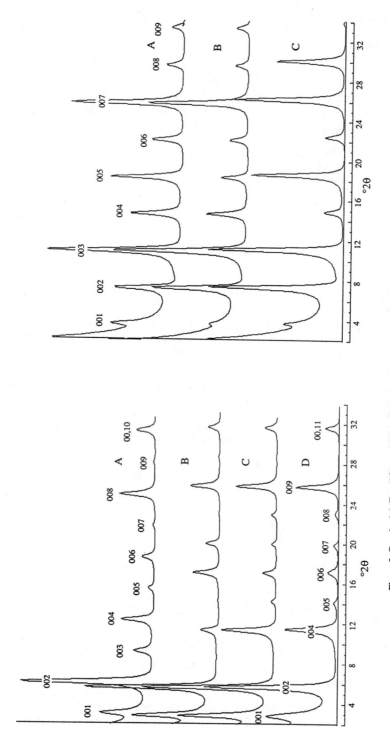

Figure 5. Regular 1:1 (R = 1) interstratified chlorite minerals. Trace A simulates air-dried chlorite/Mg-vermiculite (high-charge corrensite) with d(001) = 28.5 Å. Trace B is ethylene glycol-solvated tosudite (di-dichlorite/dismectite), d(001) = 31.1 Å. Trace C depicts ethylene glycol-solvated low charge corrensite (d(001) = 31.1 Å) and trace D shows the effects of increasing the proportion of chlorite layers to 0.55.

Figure 6. Regular 1:1 (R=1) interstratified chlorite minerals. Pattern A represents muscovite/tri-trichlorite (d(001)=24.2 Å), B is biotite/chlorite, and C is kulkeite (talc/chlorite) with d(001) = 23.7 Å.

solvated high-charge corrensite produces a diffraction pattern that is very similar to that of the air-dried preparation (unlike low-charge corrensite). Uncertainties in the structure of the interlayer Mg-glycerol complex preclude showing calculated patterns here.

Potassium saturation and heating at 350°C collapses the expandable layers in corrensite to values near 10 Å, and then the diffraction patterns resemble those shown by Figure 6A and B. This is a useful and necessary test to verify the identification of corrensite, and to eliminate the possibility of swelling chlorite layers, for the latter collapse only to 14 Å with such treatment.

Figure 6 shows calculated diffraction patterns for kulkeite (talc/chlorite) and 1:1 regular (R = 1) interstratifications of muscovite (illite)/chlorite and biotite/chlorite. For these calculations, one Fe per three octahedral sites is assumed for biotite/chlorite, the kulkeite contains no Fe, and the muscovite/chlorite is based on alternating dioctahedral (no Fe) and trioctahedral (1 Fe per three sites) unit layers (Lee and Peacor, 1985). The trioctahedral hydroxide interlayers contain one Fe per three octahedral sites.

These minerals should be easy to recognize. Their diffraction characteristics are unaffected by moderate heat treatment, K-saturation, and ethylene glycol or glycerol solvation. There are many relatively strong reflections based on a superstructure spacing of approximately 24 Å. The characteristic kulkeite spacing is best established by the positions of the relatively strong 007 and 008 reflections, although the 007 is reliable only in the absence of quartz and mica because of interferences. Note that the long-spacing (the 001), which is so useful for the identification of ordered 1:1 interstratifications, is relatively weak for these minerals compared to the intensities of long-spacing reflections from chlorite/smectite and chlorite/vermiculite. It would probably be undetectable for talc/chlorite and biotite/chlorite in significant admixture with other minerals.

Figure 7 shows calculated diffraction patterns for two three-component mineral structures. Figure 7A depicts a normal 1:1 chlorite/E.G. smectite (corrensite), shown here for purposes of comparison. Figure 7B is an example of chlorite (0.3)/Mg-vermiculite(0.2)/E.G. smectite. Vermiculite is randomly interstratified with chlorite, and both chlorite and vermiculite are ordered (R = 1) with respect to smectite. This mineral is difficult to distinguish from corrensite (Fig. 7A); only the very strong 002 reflection of the three-component mineral differentiates the two, and this criterion would be difficult to apply with confidence. K-saturation and heat treatment, on the other hand, provide definitive results. Figure 8 shows a comparison between the simulated heated material (Fig. 8A), and a heat-treated corrensite (Fig. 8B). For each of these, d(001) = 10 Å was assumed for both smectite and vermiculite layers. Absence of a normal corrensite structure is evidenced by the irrational spacings and the non-systematic line broadening (Fig. 8A).

Figure 7C demonstrates a different possible (though as yet unreported) structure, consisting of serpentine(0.1)/ chlorite (0.4)/E.G. smectite. The serpentine layers are randomly interstratified with chlorite, and both are ordered (R = 1) with respect to the smectite component. Except for a slight shift of the 001 reflection, the serpentine-bearing structure would be very difficult to differentiate from the other two (Fig. 7A and B), and here, heat treatment would not produce diagnostic results. But line broadening is distinctive. Table 5 shows these data for the pattern of Figure 7C. Note that very small amounts of serpentine (10%), cause an easily measured line-broadening pattern. The *absolute values* for β depend on the crystallite thickness or the defect-free distance, but the *pattern* of broadening is distinctive for this hypothetical mineral species.

Random (R = 0) structures

The next series of figures show calculated diffraction patterns of randomly interstratified chlorite minerals. All but talc/chlorite have been identified by either HRTEM

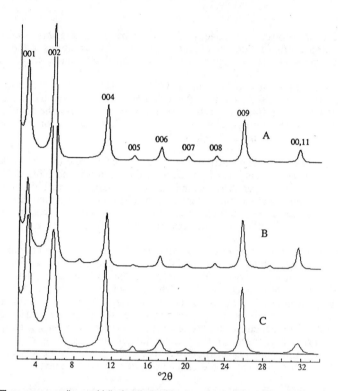

Figure 7. Three-component "corrensitic" minerals (R with respect to smectite = 1). Trace A is a simulated normal corrensite (chlorite/E. G. smectite). Pattern B shows the results for a three-component mineral containing 0.3 chlorite randomly interstratified with 0.2 vermiculite and 0.5 E. G. smectite that is ordered (R = 1) with respect to the chlorite-vermiculite components. Vermiculite layers are assumed to have the same spacing and scattering amplitudes as for the air-dried two-water layer Mg model. Trace C represents 0.4 chlorite randomly interstratified with 0.1 serpentine, both of which are ordered (R = 1) with respect to 0.5 E. G. smectite.

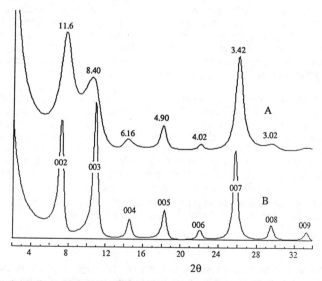

Figure 8. Simulated effects of dehydration (350 °C) of the chlorite/vermiculite/smectite shown by Fig. 7B (pattern A), compared to the dehydration of normal corrensite (pattern B).

Table 5. Line breadths (β) for the 00*l* series of the three-component mineral chlorite (0.4)/serpentine (0.1)/ E.G. smectite, R = 1.

Reflection	β
001	0.49
002	0.71
004	0.46
005	0.47
006	0.71
007	0.69
008	0.49
009	0.44
00,11	0.79

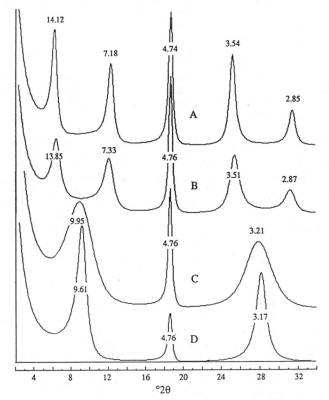

Figure 9. Randomly interstratified talc/chlorite, 0.1 talc (A), 0.25 talc (B) 0.75 talc (C) and 0.9 talc (D).

or X-ray diffraction studies, but to the writer's knowledge, there are no published diffraction patterns for biotite/chlorite, muscovite/chlorite, serpentine/chlorite, or di-chlorite/di-smectite. The patterns are shown here to hasten the documentation for such minerals and thus add to our knowledge of their occurrences. Examination of the figures shows that some identifications require the correct interpretation of subtle characteristics of the diffraction patterns whose critical details may be obscured by admixture with other minerals, or by poor-quality diffraction patterns. For other mixed-layered systems, the diffraction patterns are surprisingly sensitive to small amounts (a few percent) of a second interstratified layer type.

Randomly interstratified talc/chlorite compositions are treated by Figure 9. The irrational patterns indicate randomness in the structures, and Méring's rules should provide easy identification of 9.5 and 14.2 Å layer thicknesses. The most striking feature of these patterns is the non-uniform breadths of the reflections on each. Indeed, Figure 9C is strange-looking. It has extremely broad peaks for the talc/chlorite 001/002 and 003/004 reflections, yet the 002/003 peak is very sharp. It is easy to explain the sharp 002/003 reflections on all of these patterns. The talc 002 reflection has a spacing of 9.5/2 = 4.75 Å, and the chlorite 003 spacing is 14.2/3 = 4.73 Å. Thus both components produce reflections at very nearly the same diffraction angle, and the resulting peak maintains almost constant breadth and d-spacing regardless of the proportions of talc and chlorite. Careful measurements of line-breadths for similar minerals should allow the investigator to detect very small amounts of talc in chlorite, or, very small amounts of chlorite in talc (a few percent), based on diffraction patterns that casual inspection would classify simply as talc or chlorite. A simple test for predicting the positions of such sharp reflections is to divide the largest d(001) by the smallest, and then multiply this quotient by the integers 2, 3, 4, etc. If the result is nearly integral, then the multiplier gives the l index for a sharp reflection from the layer type with the smallest d(001). For example, 14.2/9.5 = 1.495, and 1.495 x 2, x3, x4, x5, x6, x7, and x8 give, respectively, 2.990, 4.49, 5.98, 7.475, 8.97, 10.465, and 11.96. The values 2.99, 5.98, 8.97, and 11.96 are nearly integral, and these correspond to the talc reflections 002, 004, 006, and 008. Randomly interstratified talc/chlorite produces sharp reflections for these diffraction orders, and, for all talc reflections for which l is even. The broadest peaks, and those that migrate the most with respect to component proportion, are those for which the remainder of the numerical division described above is closest to 0.5.

Calculated diffraction patterns for randomly interstratified biotite/chlorite are shown by Figure 10. The spacings are irrational and composition can be accurately estimated by using the positions of any of several reflections. Our rule for peak sharpness does not work very well here because, for the first 10 orders of biotite reflections, the quantity (14.2/10) x l is nearly integral only for the biotite 007 reflection, which is not shown on Figure 10. The biotite 006 should be broad and the 007 reflection sharp if chlorite layers are present. Calculations for biotite(0.9)/chlorite show that the peak breadth ratio for the 006 to the 007 amounts to about 1.7, suggesting that very small amounts of interlayered chlorite can be detected by this method. The low angle region, however, also shows different line breadths, and small amounts of chlorite in biotite (or biotite in chlorite) can be detected by means of the line breadths of the 001/002 with respect to the 002/003 (see Fig. 10).

Figure 11 shows calculated patterns for randomly interstratified serpentine/chlorite. The results would be very similar for kaolinite/chlorite. In this case, the ratio of d(001) values is very nearly equal to 2, so peaks corresponding to all serpentine reflections are sharp, and those corresponding to odd-order chlorite positions are broad. Chlorite odd-order line broadening is a sensitive indicator of interstratified 7-Å layers (see Dean, 1983), but unfortunately appreciable (10%, see Figure 11A) chlorite interstratification can occur without diffraction evidence because of the weakness of the chlorite odd-order reflections for all compositions save the most Mg-rich types.

The randomly interstratified chlorite minerals described above are identified and quantified on the basis of peak position and peak breadth; they are unaffected by K-saturation and/or solvation with glycerol or ethylene glycol, and this behavior provides useful diagnostic information. The randomly interstratified expandable chlorite minerals can be interpreted by similar methods, but, in addition, appropriate sample treatments produce different structural varieties whose diffraction data provide redundant information that can confirm diagnosis of the type and composition of a given mineral species.

Figure 12 shows calculated diffraction patterns of randomly interstratified chlorite/smectite. Figures 12D and 12B correspond to tri-tri-chlorite/tri-smectite and Figures

617

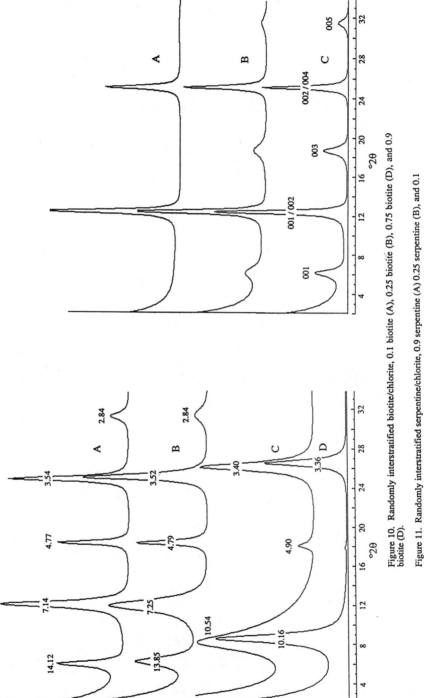

Figure 10. Randomly interstratified biotite/chlorite, 0.1 biotie (A), 0.25 biotite (B), 0.75 biotite (D), and 0.9 biotite (D).

Figure 11. Randomly interstratified serpentine/chlorite, 0.9 serpentine (A) 0.25 serpentine (B), and 0.1 serpentine (C).

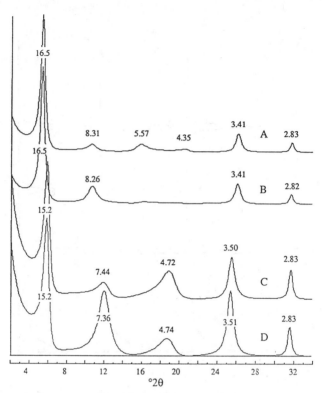

Figure 12. Randomly interstratified di and trioctahedral chlorite and ethylene glycol-solvated smectite. Trace A simulates di-dichlorite(0.25)/dismectite, B is tri-trichlorite(0.25)/trismectite, C is di-dichlorite(0.75)/dismectite, and D represents tri- trichlorite(0.75)/trismectite.

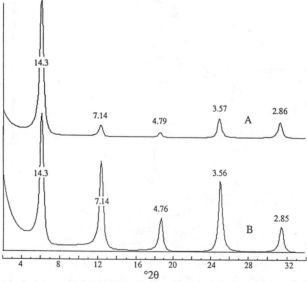

Figure 13. Randomly interstratified chlorite/vermiculite. The two-water layer Mg vermiculite structure is assumed. Trace A represents chlorite(0.25)/vermiculite and B is chlorite(0.75)/vermiculite.

12A and 12C represent di-di-chlorite/di-smectite. Identification should be easy because of the markedly irrational diffraction patterns and the large differences in peak breadth within each pattern. The intensity ratio of the reflections near 12 and 19°2θ is related to octahedral scattering and can be used to differentiate di- and trioctahedral varieties. The most useful reflection for estimating composition in this system is the 004/005 peak near 25°2θ which displays typical migration characteristics with respect to composition.

If the smectite component contains Ca or Mg as the exchangeable cation, the smectite layers in the air-dried preparation will have a spacing similar to but not identical to air-dried vermiculite. Then, the material depicted by Figure 12D will resemble that shown by Figure 13B, and 12B will be close to the pattern shown by Figure 13A. The dioctahedral chlorite/smectite minerals follow the same relations, but for them, the 003/003 reflections near 18°2θ are much more intense.

Saturating chlorite/smectite or chlorite/vermiculite (Fig. 13) with K and heating to 350°C for one hour causes the collapse of the expandable layers to values near 10 Å. After this treatment, the diffraction patterns of the minerals shown by Figures 12 and 13 should resemble those of randomly interstratified mica/chlorite (see Fig. 10). Dioctahedral varieties are again distinguishable from trioctahedral types by a relatively strong peak near 18°2θ.

Table 6 lists the spacings and positions of reflections that are useful for estimating the composition of randomly interstratified chlorite/E.G. smectite. The quantity Δ2θ represents the difference in diffraction angle of the two reflections listed in Table 6. Srodon (1980) has demonstrated the power of this approach. It tends to minimize errors due to goniometer zero misalignment and sample positioning, and it compensates for departures from the nominal value of 16.9 Å for the ethylene glycol-solvated smectite layers. In addition, the sensitivity for determining composition is enhanced because the two reflections move in opposite directions with respect to changes in the proportions of chlorite and smectite in the mixed-layered mineral.

Table 6. The positions (CuK_α) of useful reflections for estimating percent chlorite in chlorite/E. G. smectite, R = 0.

| % Chlorite | 002/002 | | 004/005 | | °Δ2θ |
	d(Å)	°2θ	d(Å)	°2θ	
10	8.39	10.54	3.39	26.29	15.75
20	8.29	10.67	3.40	26.18	15.51
30	8.15	10.86	3.42	26.05	15.19
40	7.98	11.09	3.43	25.95	14.86
50	7.80	11.34	3.45	25.80	14.46
60	7.59	11.66	3.47	25.64	13.98
70	7.40	11.96	3.50	25.48	13.52
80	7.28	12.16	3.52	25.32	13.16
90	7.18	12.33	3.53	25.20	12.88

Diffraction patterns for randomly interstratified chlorite/vermiculite are shown by Figure 13. These are based on a Mg trioctahedral vermiculite component that has been air-dried to produce the two-water-layer structure. The value of d(001) for this layer type is 14.3 Å, which is very close to the chlorite value of 14.2 Å, so that such minerals cannot be identified confidently without K-saturation and heat treatment. The pattern of Figure 13A superficially resembles pure vermiculite, and Figure 13B a chlorite whose heavy metals are concentrated in the 2:1 silicate layer.

GEOLOGICAL OCCURRENCES OF NON-EXPANDABLE MIXED-LAYERED CHLORITE MINERALS

Non-expandable mixed-layered chlorite minerals include talc/chlorite (Schreyer et al., 1982), biotite/chlorite (Eroshchev-Shak, 1970; Veblen, 1983; Veblen and Ferry, 1983; Maresch et al., 1983; Baños and et al., 1983), illite/chlorite (Lee et al., 1984; Lee and Peacor, 1985), wonesite (Na-trioctahedral mica)/ chlorite (Veblen, 1983) and serpentine/ chlorite (Dean, 1983; Ahn and Peacor, 1985). Veblen (1983) also reports, based on HRTEM lattice-fringe images, three-component random interstratifications of normal chlorite, talc, and abnormal chlorite that contains two hydroxide interlayers. Occurrences seem to be separable into four distinct categories, namely, formation by weathering of a mica precursor, formation during prograde metamorphism or diagenesis of pelitic rocks, retrograde metamorphism of igneous rocks, and hydrothermal activity.

Kulkeite, or ordered 1:1 talc/chlorite (Schreyer et al., 1982), has the most unusual geologic occurrence of the species listed above. The mineral, which contains a Na-Al talc component, crystallized during the progressive metamorphism of a dolomite rock that is associated with a Triassic evaporite sequence. The maximum temperature is estimated at 400°C. The report suggests that the structure is possible only with the Na-Al talc layer type, which in this case is explained by the likely presence of fluids from the evaporites that contained high activities of dissolved Na. The authors report that the rocks contain discrete chlorite and discrete Na-Al talc in mutual contact, and conclude that kulkeite might be a metastable intermediary in a reaction series. Kulkeite is a very low-Fe mineral. An example has the formula (Schreyer et al., 1982),

$$(Na_{0.38}K_{0.01}Ca_{0.01})(Mg_{8.02}Al_{0.99})(Si_{6.57}Al_{1.43}O_{20})(OH)_{10}.$$

The formation of chlorite layers in biotite has been ascribed to the precipitation of hydroxide-like interlayers in interlayer spaces that were opened by bending and kinking of mica during deformation (Baños, et al., 1983). Such a reaction mechanism leads to a volume increase because the chlorite hydroxide sheet is thicker than the K plane in mica. In addition, K is lost and Mg, Fe, Al, and H must be added. Veblen and Ferry (1983) show HRTEM lattice-fringe images that demonstrate an alternative reaction mechanism. Post-tectonic retrograde metamorphism of biotite in a quartz monzonite yielded randomly interstratified biotite/chlorite, but in this case the reaction occurred by the dissolution of both tetrahedral sheets from some of the 2:1 silicate skeletons, yielding randomly interstratified biotite/chlorite. The original octahedral sheets from the 2:1 silicate layers have apparently remained to form the interlayer hydroxide sheets of the newly formed chlorite units. Such a reaction diminishes the volume and leads to a loss of K Si, and Mg and the addition of H. As Veblen and Ferry point out, these two reaction mechanisms have very different chemical consequences for the rock and for associated fluids.

Neglecting Fe, and using ideal formulae, the two reactions are as follows:

$$KMg_3Si_3AlO_{10}(OH)_2 + 2Mg^{2+} + Al^{3+} + 6H_2O \Rightarrow (Mg_5Al)Si_3AlO_{10}(OH)_8 + K^+ + 6H^+ ,$$

and $$KMg_3Si_3AlO_{10}(OH)_2 + 4H^+ \Rightarrow Mg_2Al(OH)_6^+ + 3SiO_2 + Mg^{2+} + K^+ .$$

The reaction stoichiometry demonstrates the important role of pH in conditioning the stability of biotite and chlorite for these reaction mechanisms (Veblen and Ferry, 1983).

Eroshchev-Shak (1970) described the formation of an ordered 1:1 biotite/chlorite as a two-step process. Weathering of biotite in a biotite gneiss led to the formation of regularly interstratified biotite/vermiculite, and later hydrothermal alteration converted the vermiculite component to chlorite. Maresch et al. (1983) reported ordered 1:1 biotite/chlorite from a metabasalt. In this instance, chlorite is apparently the primary phase, and its alteration has produced the interlayered biotite component.

Peacor and coworkers (Lee and Peacor, 1983; Lee et al., 1984; Ahn and Peacor, 1985; Lee and Peacor, 1985; Lee et al.,1984; Lee et al., 1985) have studied the prograde metamorphic transformations of phyllosilicates in pelitic rocks, using the techniques of HRTEM, AEM, and electron diffraction. The picture that emerges is one in which crystallites of mica/smectite, dioctahedral mica, chlorite, and lesser amounts of kaolin or serpentine layers show a range of organization that extends from randomly interstratified unit layers to crystallites of each of these phases that show intercrystallite registry at the sub-unit cell level. Within zones of interstratified mica/chlorite, regular 1:1 ordering has been observed (Lee and Peacor, 1985). The general thesis is that interstratified types are metastable, and that reaction progress causes the development of single-phase crystals during the metamorphic transformation of shale to slate (Lee et al., 1984), or the high grade burial diagenesis of, for example, Gulf Coast shales (Ahn and Peacor, 1985). For impermeable argillaceous rocks, within which the pore fluids are rock-dominated, Lee et al., (1985, p. 540) state:

"These changes collectively express a trend from a highly imperfect, disordered phase with significant random solid solution of several components, to a collection of well-ordered, homogeneous, translation-periodic and defect-free phases. Such changes represent a steady, consistent trend toward more highly ordered phases which consequently have lower relative potential energies."

Chemical analyses, based on AEM, suggest that the 7-Å mineral is berthierine. The micas involved in the interstratifications are dioctahedral (illite, phengite or muscovite), and the chlorites are ripidolites. Interstratification of dioctahedral mica and trioctahedral chlorite is difficult to reconcile with the crystal structure model shown by Figure 3, particularly for the ordered 1:1 structure, although it is easy enough to visualize a regular alternation of di- and trioctahedral 2:1 silicate layers if the assignment of interlayer material to each is ignored.

The appearance of many of the HRTEM lattice-fringe images suggests that X-ray diffractionists may need to change some of the rules for the interpretation of X-ray diffraction patterns. The picture described above is one of large zones of one mineral or the other, interspersed with domains in which the interlayering becomes more intimate until the frequency of layer type transition is best characterized as an interstratified mineral. The basal X-ray diffraction pattern for such a material would consist of the mixed-layered reflections and the reflections from one or both of the pure components. The traditional interpretation would label such a material as a physical mixture of discrete and mixed-layered minerals. The term "mixture" could lead one to believe that these minerals can be separated by physical means such as particle-sizing or magnetic separation, when in fact, such techniques would prove fruitless. More importantly, for sedimentary rocks, the term physical mixture could mislead an investigator into rationalizations that involve provenance, when the better view might be one of a "mixture" of stable and metastable phases.

A concrete example involves mixed-layered chlorite/smectite. X-ray diffraction studies of corrensite from sedimentary rocks invariably disclose the presence of discrete chlorite with its annoying peak interferences. Perhaps if we "saw" the crystallites, we would observe coherent domains of chlorite that contain random patches of corrensitic interstratification. This would throw a different light on the interpretation, and we would stop looking for a source of the "detrital" chlorite present in the samples.

Ahn and Peacor (1985) describe random interstratifications of high-Fe ripidolite and berthierine in Gulf Coast shales. Their observations were made on samples that covered the estimated temperature range of 70° to 165°C, and no significant changes were noted in the relative abundance of berthierine layers in that interval. Studies of a presumably higher-temperature Ordovician mudstone and slate disclosed fewer occurrences of 7-Å layer types (Lee et al., 1984). They argue that the berthierine is a metastable precursor to chlorite, and that 7-Å layers become rare as diagenesis proceeds and are transformed altogether at the temperatures of low grade metamorphism.

Brindley and Gillery (1954) described a mixed-layered kaolinite/chlorite and showed, by Fourier synthesis of the 00l diffraction pattern, the necessary inversion of some tetrahedral sheets (Fig. 4). In the background discussion following the oral presentation of that work, W. F. Bradley concurred with the interpretation and remarked that such structures are common in chlorites from sedimentary rocks. That was 34 years ago, yet there is only a single report of a 7/14 Å interstratification (Dean, 1983) based on X-ray diffraction data. Such structures produce only subtle distortions of the normal chlorite diffraction pattern, namely, odd-order 00l line-broadening or "streaking" (Fig. 11 and Ahn and Peacor, 1985, Fig. 6), and it is possible that Bradley was correct; perhaps they are common and we have been missing them.

GEOLOGICAL OCCURRENCES OF EXPANDABLE MIXED-LAYERED

CHLORITE MINERALS

By far the largest number of documented occurrences of mixed-layered chlorite minerals are chlorite/smectite and chlorite/vermiculite, and of these, the dominant types are the ordered 1:1 trioctahedral varieties, or corrensites. The original definition of corrensite was proposed by Lippmann (1954, 1956), who identified and described the distribution of the mineral in the Triassic Röt sandstone of Germany. Since that work, corrensite has been described by many others, and today, it is second only to illite/smectite (of the mixed-layered phyllosilicates) in the attention it has attracted, and in the diversity of its occurrences. Indeed, the geological associations of corrensite are so varied that the only discriminatory environments seem to be magmatic and metamorphic grades that exceed the chlorite zone of regional metamorphism (cf. Velde, 1977). The dioctahedral chlorite/smectite, tosudite, is much more restricted; its occurrence is almost limited to the hydrothermal alteration of intermediate to acidic igneous rocks (Sudo et al., 1954; Sudo and Kodama, 1957; Shimoda, 1969; Nishiyama et al., 1975; Ichikawa and Shimoda, 1976; Creach et al., 1986) Merceron et al., 1988). However, Morrison and Parry (1986) found tosudite in a Permian sandstone, and concluded that it formed at moderate temperatures (~100°C) by reaction of the sandstone with percolating acid pore waters.

Corrensite has been identified in the contact metamorphic zones of shales (Blatter et al., 1973; April, 1980; Vergo and April, 1982), in old (Paleozoic) carbonate sequences (Stephen and MacEwan, 1951; Bradley and Weaver, 1956; Earley et al., 1956; Peterson, 1961; Suchecki et al., 1977), in clastic and volcanoclastic sedimentary sequences (Schultz, 1963; Shover, 1964; Whitney and Northrop, 1986), and in Lake Superior iron ores (Bailey and Tyler, 1960). It has formed by the hydrothermal alteration of intermediate to basic (ophiolitic) volcanic rocks (Alietti, 1958; Furbish, 1975; Brigatti and Poppi, 1984b; Evarts and Schiffman, 1983; Bettison and Schiffman, 1988;) and dolomite (DeKimpe et al., 1987), by burial diagenesis of (trioctahedral) clay mineral sediments (Hoffman and Hower, 1979; Chang et al., 1986), and by weathering (Johnson, 1964; Post and Janke, 1974). Disordered chlorite/smectite (R = 1 with P_C > 0.5, or R = 0) is found in the same kinds of rocks as corrensite.

The most consistent occurrence of chlorite/smectite is with saline deposits, both terrestrial and marine. Indeed, a good working hypothesis is to assume that corrensite is present in one of the associated facies of any young evaporite sequence. It may be that diagenesis converts corrensite to chlorite (Bodine and Madsen, 1985), and that may account for the scarcity of corrensite in old (e.g. Silurian) evaporite sequences.

April has described the type and distribution of corrensite from the Jurassic sedimentary rocks of the Connecticut Valley. The high-charge type is found in shales that underlie basalt flows (April, 1980; April, 1981a) and in shales from the contact metamorphic aureoles of dike and sills (Vergo and April, 1982; see also Blatter et al., 1973). The origin is ascribed to moderately elevated temperatures and the local circulation of Mg-rich fluids derived from the basalt and from the decomposition of dolomite in the sediment.

Low-charge corrensite occurs in lacustrine facies of the Jurassic sedimentary rocks (April, 1981a, b). It is restricted to gray mudstones that overlie black shales rich in saponite. The mudrocks were deposited in non-marine, alkaline saline lakes whose salinities changed cyclically. April argues that local high pH values in the lakes, in conjunction with Mg-rich fluids from the underlying shales, raised the activity ratio Mg^{2+}/H^+ sufficiently to precipitate some brucite-like interlayers within a precursor trioctahedral smectite. Such a transformation has been demonstrated in the laboratory (Slaughter and Milne, 1960). They slowly titrated a suspension of montmorillonite in 1-2N $MgCl_2$ with LiOH and precipitated brucite in the smectite interlayer spaces. The formation of chlorite from smectite, regardless of the mechanism, requires the precipitation of a brucite-like phase, and that reaction can be written as

$$Mg^{2+} + 2H_2O \Rightarrow Mg(OH)_2 + 2H^+ \; .$$

The stability of the hydroxide layer in chlorite is related to the activity quotient $Mg^{2+}/(H^+)^2$. The important role of Mg^{2+}/H^+ in corrensite formation had been previously pointed out by Grim et al.(1960). Bodine (1985), in his discussion of corrensite formation in marine evaporites, points out that progressive evaporation causes the Mg^{2+} activity to increase faster than the H^+ activity, so the ratio moves progressively toward the stability field of corrensite or chlorite.

Bodine (1983, 1985) argued persuasively that corrensite in marine evaporites is the result of Mg metasomatism of precursor minerals such as authigenic saponite or detrital montmorillonite. The evidence is summarized as follows:

1. The phyllosilicate mineralogy changes cyclically with changes in the evaporite facies. That is, the silicates reflect the ambient water chemistry.

2. Talc and talc/saponite are present in some facies, yet are rare in normal marine or terrestrial sediments.

3. The phyllosilicate assemblages are very similar in evaporite sequence that are widely separated in space and time (cf. Fournier, 1961; Grim et al., 1960).

These considerations would seem to rule out detrital input as a controlling factor in the occurrence of corrensite (cf. Kopp and Fallis, 1974) though Lippmann (1956) has otherwise interpreted his data on corrensite in the Röt sandstone.

Bodine (1985) and Bodine and Madsen (1985) suggest that the compositions are set by equilibration with ambient fluids, and that the final mineralogical character of the phyllosilicate assemblage is achieved by mild burial diagenesis. More profound diagenesis results in the transformation of corrensite to chlorite The evidence for these views, respectively, is the absence of corrensite in modern marine saline assemblages, and its absence in the Silurian evaporites of the eastern U. S. which, based on the illite/smectite reaction extent, have been exposed to diagenetic conditions comparable to those at the deepest levels studied in Gulf Coast Tertiary sediments. However, smectite (probably saponite) and corrensite have recently (Hluchy and Reynolds, 1987) been reported in the Salina Group (Silurian) of New York State. These isolated occurrences have yet to be reconciled with an otherwise well-documented and well-reasoned model.

Corrensite in clastic and carbonate marine sedimentary rocks is likely the result of high local activities of Mg and high pH. In clastic rocks it is associated with the sand and/or silt fraction which may contain labile, high-Mg tuffaceous material (Schultz, 1963). The requisite conditions for carbonate rocks could be attained by the reflux of saline, Mg-rich aqueous fluids common to sabkha environments. In this connection, it is interesting and perhaps significant that corrensite is commonly reported in dolomites or dolomitic limestones.

Chang et al.(1986) document the formation of corrensite by the burial diagenesis of trioctahedral clay minerals in Cretaceous shales and sandstones. In the wells studied, associated dioctahedral smectites showed the classical sequence of transformation to illite/smectite. Corrensite formed from trioctahedral smectite, via a randomly interstratified chlorite/smectite reaction series. The appearance of corrensite in shales occurred at slightly shallower depths than the onset of ordering in illite/smectite. They estimate that corrensite in sandstones formed at 60°C, whereas the first appearance in shales suggests formation at 70°C. Both temperatures are in good agreement with the value of 80°C suggested by Hoffman and Hower (1979) for the first appearance of corrensite in sandstones. Chang et al. (1986) suggest that the different temperatures of formation in shales and sandstones might be accounted for by differences in the chemical compositions of the pore fluids. Rock-dominated fluid compositions are likely in shales, but the fluids in the more permeable sandstones were probably exogenous and could have had chemical compositions that accelerated the reaction rate, or lowered the temperature of corrensite stability.

Corrensite may form directly, that is, the transformation may not not take place via an intermediate randomly interstratified series. Examination of published reports on randomly interstratified chlorite/smectite reveals that they document few (but see Chang et al.,1986, and Bettison and Schiffman, 1980) instances of chlorite/smectite that contain more than 50% smectite layers. The data of Chang et al. (1986) show that in the diagenetic formation of chlorite/smectite the composition moves very rapidly from smectite to corrensite, and shows only small further increases in chlorite content with increasing depth. Schultz (1963) mentions specifically in his study of the Triassic sedimentary rocks of the Colorado plateau, that chlorite/smectites range in composition from 50/50 (corrensite) to pure chlorite with all intermediate compositions represented. Yet he notes the absence of examples that contain > 50% smectite layers. These findings suggest that corrensite is a single phase, in a thermodynamic sense, with a discrete stability field in PT-chemical space (cf.Velde, 1977). It is not a mixed-layered phyllosilicate any more than chlorite is mixed-layered brucite/talc. If this reasoning is correct, then chlorite/smectite with > 50% chlorite layers is a metastable mixture of corrensite and chlorite in the sense of Lee et al.(1985) as described above. The hypothesis is testable, because it means that randomly interstratified chlorite/smectite does not exist. The problem, though, is that it is very difficult to detect ordering by X-ray diffraction studies (unlike illite/smectite) for compositions much different from 50/50, though it is possible with good diffraction patterns free from interferences (a rare case). The intensity of the first order superstructure reflection (the long-spacing) is markedly reduced by even a slight amount of disorder caused by compositions a bit greater or less that 50%, or by values for the ordering parameter, $P_{C.S.}$, that deviate slightly from unity. Consequently, identifications of ordered types are difficult, and at the present time we simply do not know how common the occurrences are of randomly interstratified chlorite/smectite. The excellent diffraction pattern shown by Schultz (1963) for chlorite (0.7)/smectite shows evidence of ordering, and indeed is noted as such by Schultz. H. Roberson (pers. comm.), has concluded that diffraction patterns of his specimens of highly chloritic randomly interstratified chlorite/smectite from ophiolites can be explained by the alternative model of small amounts of corrensite (too small to observe the odd-order superstructure reflections) "mixed" with chlorite. The nature of "randomly interstratified" chlorite/smectite needs further elucidation.

Attempts to correlate geologic environment with the occurrence of low charge (chlorite/smectite) and high charge (chlorite/vermiculite) corrensite are fraught with difficulties, because many of the older published works used laboratory procedures that are insufficiently definitive for such a subdivision. Saturation with Mg and glycerol solvation is required, although corrensites that expand to 31.1 Å with glycol are likely (but not definitely) chlorite/smectites. X-ray diffraction studies of air-dried preparations give no information on this subdivision.

Corrensites and chlorite/expandable minerals form by the oxidative weathering of chlorite in low pH soil environments. Johnson (1964) reported corrensite in a soil profile developed from a greenstone in Pennsylvania. The corrensite formed from chlorite in the parent material. Similarly, Post and Janke (1974) have documented the formation of corrensite from chlorite during the weathering of a slate from northern California. The process is visualized as proposed by Jackson (1963), who suggested that acidic weathering of chlorite causes the removal of alternate brucite-like interlayers from the chlorite structure. Ross and Kodama (1976) and Senkayi et al. (1981) have demonstrated experimentally that the oxidation of structural ferrous iron is an important factor in the transformation of chlorite to chlorite/smectite and chlorite/vermiculite. They used bromine water near 100 °C as an oxidant and succeeded in converting chlorite to the mixed-layered chlorite minerals. The oxidation of Fe^{2+} lowered the silicate layer charge to that of vermiculite, and the low pH produced by the dissociation of hypobromous acid caused dissolution of the hydroxide interlayer. Ross and Kodama (1976) found that this experimental treatment of a diabantite (high Fe) produced vermiculite, but brunsvigite (intermediate Fe) was converted to corrensite. Senkayi et al. (1981) suggest that the distribution of Fe between the silicate layer and the hydroxide interlayer may be a key factor in controlling the final product. Oxidation of Fe in the silicate layer is slow, and if most of the Fe is contained in the silicate layer, chlorite is transformed into corrensite. Alternatively, an Fe-rich hydroxide layer is quickly oxidized and removed by the acidic solution to produce vermiculite or smectite.

The formation of 14-Å hydroxy interlayers in smectite and vermiculite can produce interstratified chlorite minerals. The most common example is the widespread occurrence in acidic soils of hydroxy-Al interlayers that form the so-called chlorite intergrades. The identification of such layer types is made by the removal of the hydroxy-Al interlayer, which restores expandability to the previously stable 14-Å component. Hydroxy-Al interlayers are removed by treatment with sodium-dithionite-citrate (Tamura, 1955) or sodium citrate (Sawhney, 1960). A very large literature in this field precludes its inclusion here. The interested reader is referred to a review paper by Barnhisel (1982).

APPENDIX

The 00l X-ray diffraction patterns contained in this chapter were calculated by the computer program NEWMOD (Reynolds, 1985). The three-component calculations (Figs. 7 and 8) were produced by a modification of NEWMOD prepared for this purpose. Instrumental conditions are based on a 20 cm goniometer radius, one 2° soller slit, a divergence slit with 1° divergence, and a sample length of 4 cm. A 12° standard deviation of the orientation function was used to compute the Lorentz-polarization factor (Reynolds, 1986). Crystallite size is based on the defect-broadening theory of Ergun (1970). All calculations are based on a mean defect-free distance of 70 Å, and a maximum crystallite thickness of approximately five times the defect-free distance. Unless otherwise indicated, all trioctahedral sheets contain one Fe per three sites, and for micas, one K is assumed per Si_4O_{10}.

626

REFERENCES

Ahn, J. H. and Peacor, D. R. (1985) Transmission electron microscopic study of diagenetic chlorite in Gulf Coast argillaceous sediments. Clays & Clay Minerals 33, 228-236.

Alietti, A. (1958) Some interstratified clay minerals of the Taro Valley. Clay Minerals Bull. 19, 207-211.

April, R. H. (1980) Regularly interstratified chlorite/vermiculite in contact metamorphosed red beds, Newark Group, Connecticut Valley. Clays & Clay Minerals 28, 1-11.

_____ (1981a) Clay petrology of the upper Triassic/lower Jurassic terrestrial strata of the Newark Supergroup, Connecticut Valley, U. S. A.. Sed. Geol. 29, 283-307.

_____ (1981b) Trioctahedral smectite and interstratified chlorite/smectite in Jurassic strata of the Connecticut Valley. Clays & Clay Minerals 29, 31-39.

Bailey, S. W. (1982) Nomenclature for regular interstratifications. Amer. Mineral. 67, 394-398.

_____ and Tyler, S. A. (1960) Clay minerals associated with the Lake Superior iron ores. Econ. Geol. 55, 150-175.

Baños, J. O., Amouric, M., De Fouquet, C. and Baronnet, A. (1983) Interlayering and slip in biotite as seen by HRTEM. Amer. Mineral. 68, 754-758.

Barnhisel, R. I. (1982) Chlorites and hydroxy interlayered vermiculite and smectite. Ch. 10. in: Minerals in Soil Environments, J. B. Dixon and S. B. Weed, eds. Soil Sci. Soc. Amer., Inc., Madison, Wisconsin.

Barshad, I. (1950) The effect of interlayer cations on the expansion of mica-type crystal lattice. Amer. Mineral. 35, 225-238.

Bettison, L. A. and Schiffman, P. (1980) Compositional and structural variations of phyllosilicates from the Point Sal ophiolite, California. Amer. Mineral. 73, 62-76.

Blatter, C. L., Roberson, H. E. and Thompson, G. R. (1973) Regularly interstratified chlorite-dioctahedral smectite in dike-intruded shales, Montana. Clays & Clay Minerals 21, 207-212.

Bodine, M. W. (1983) Trioctahedral clay Minerals assemblages in Paleozoic marine evaporite rocks. in: Sixth International Symposium on Salt, vol. 1, B. C. Schreiber and H. L. Harner, eds., The Salt Institute, Alexandria, Virginia, pp. 267-284.

_____ (1985) Clay mineralogy of insoluble residues in marine evaporites. In: Mineralogy - Applications to the Minerals Industry, D. M. Hausen and O. C. Kopp, eds. Amer. Inst. Mining, Metallurgical Petroleum Engineers, New York.

_____ and Madsen, B. M. (1985) Mixed-layered chlorite/smectites from a Pennsylvanian evaporite cycle, Grand County, Utah. Proc. Int'l Clay Conf., Denver, 85-93.

Bradley, W. F. (1945) Diagnostic criteria for clay minerals. Amer. Mineral. 30, 704-713.

_____ and Weaver, C. E. (1956) A regularly interstratified chlorite-vermiculite clay mineral. Amer. Mineral. 41, 497-504

Brigatti, M. F. and Poppi, L. (1984a) Crystal chemistry of corrensite: a review. Clays & Clay Minerals 32, 391-399.

_____ and Poppi, L. (1984b) 'Corrensite-like minerals' in the Taro and Ceno Valleys, Italy. Clay Minerals 19, 59-66.

Brindley, G. W. (1966) Ethylene glycol and glycerol complexes of smectites and vermiculites. Clay Minerals 6, 237-260.

_____ and Gillery, F. H. (1953) A mixed-layer kaolin-chlorite structure. Clays & Clay Minerals 2, 349-353.

Brown, G., Bourguignon, P. and Thorez, J. (1974) A lithium-bearing aluminian regular mixed layer montmorillonite-chlorite from Huy, Belgium. Clay Minerals 10, 135-144.

Byström-Brusewitz, A. M. (1976) Studies on the Li test to distinguish between beidellite and montmorillonite. Proc. Int'l Clay Conf., Mexico City, Mexico, 419-428.

Chang, H. K., Mackenzie, F. T. and Schoonmaker, J. (1986) Comparisons between the diagenesis of dioctahedral and trioctahedral smectite, Brazilian offshore basins. Clays & Clay Minerals 34, 407-423.

Creach, M., Meunier, A. and Beaufort, D. (1986) Tosudite crystallization in the kaolinized granitic cupola of Montebras, Creuse, France. Clay Minerals 21, 225-230.

Dean, R. S (1983) Authigenic trioctahedral clay minerals coating Clearwater Formation sand grains at Cold Lake, Alberta, Canada (abstract). 20th Ann'l Clay Minerals Conf., Buffalo, New York, p. 79.

DeKimpe, C. R., Miles, N., Kodama, H. and Dejou, J. (1987) Alteration of phlogopite to corrensite at Sharbot Lake, Ontario. Clays & Clay Minerals 35, 150-158.

Earley, J. W., Brindley, G. W., McVeagh, W. J. and Vanden Heuvel, R. C. (1956) A regularly interstratified montmorillonite-chlorite. Amer. Mineral. 41, 258-267.

Ergun, S. (1970) X-ray scattering by very defective lattices. Phys. Rev. B, 131, 3371-3380.

627

Eroshchev-Shak, V. A. (1970) Mixed-layered biotite-chlorite formed in the course of local epigenesis in the weathering crust of a biotite gneiss. Sedimentology 15, 115-121.

Evarts, R. C. and Schiffman, P. (1983) Submarine hydrothermal metamorphism of the Del Puerto ophiolite, California. Amer. J. Sci. 283, 289-340.

Fornier, R. O. (1961) Regular interlayered chlorite-vermiculite in evaporite of the Salado Formation, New Mexico. U. S. Geol. Surv. Prof. Paper 424-D, 323-327.

Furbish, W. J. (1975) Corrensite of deuteric origin. Amer. Mineral. 60, 928-930.

Gradusov, B. P. (1969) An ordered trioctahedral mixed-layer form with chlorite and vermiculite packets. Doklady Akad. Nauk, SSSR 186, 140-142 (English translation).

Grim, R. E., Droste, J. B. and Bradley, W. F. (1960) A mixed-layer clay mineral associated with an evaporite. Clays & Clay Minerals 8, 228-236.

Hluchy, M. M., and Reynolds, R. C., Jr. (1987) Chlorites and mixed-layered chlorite-smectite in Salina Group evaporites of New York State (abstr.). 24th Ann'l Clay Minerals Conf., Socorro, New Mexico, p. 71.

Hoffman, J. and Hower, John (1979) Clay Minerals assemblages as low grade metamorphic geothermometers: application to the thrust faulted disturbed belt of Montana, USA. In: Aspects of Diagenesis. P. A. Scholle and P. R. Schluger, eds., SEPM Spec. Publ. 26, 55-79.

Ildefonse, P., Manceau, A., Prost, D. and Groke, M. C. T. (1986) Hydroxy-Cu-vermiculite formed by weathering of Fe-biotites at Salobo, Carajas, Brazil. Clays & Clay Minerals 34, 338-345.

Ichikawa, A. and Shimoda, S. (1976) Tosudite from the Hokuno Mine, Hukono, Gifu Prefecture, Japan. Clays & Clay Minerals 24, 142-148.

Jackson, M. L. (1963) Interlayering of expansible layer silicates in soils by chemical weathering. Clays & Clay Minerals 11, 29-46.

_____ (1969) Soil Chemical Analysis-Advanced Course. 2nd Ed., Published by the author, Madison, Wisconsin, 53705, U.S.A., 895 pp.

Jadgozinski, H. (1949) Eindimensionale Fehlordnung in Kristallen und ihr Einfluss auf die Röntgeninterferenzen. I. Berechnung des Fehlordnungsgrades aus der Röntgenintensitäten. Acta Crystallogr. 2, 201-207.

Johnson, L. J. (1964) Occurrence of regularly interstratified chlorite-vermiculite as a weathering product of chlorite in a soil. Amer. Mineral. 49, 556-572.

Kopp, O. C. and Fallis, S. M. (1974) Corrensite in the Wellington Formation, Lyons, Kansas. Amer. Mineral. 59, 623-624.

Klug, H. P. and Alexander, L. E. (1974) X-Ray Diffraction Procedures, 2nd ed.. John Wiley and Sons, New York, 966 p.

Kunze, G. W. (1955) Anomalies in the ethylene glycol solvation technique used in X-ray diffraction. Clays & Clay Minerals 3, 88-93.

Lee, J. H. and Peacor, D. R. (1983) Interlayer transitions in phyllosilicates of the Martinsburg shale. Nature, 608-609.

_____ and Peacor, D. R. (1985) Ordered 1.1 interstratification of illite and chlorite. A transmission and analytical electron microscopy study. Clays & Clay Minerals 33, 463-467.

_____, Ahn, J. H. and Peacor, D. R. (1985) Textures in layered silicates. progressive changes through diagenesis and low-temperature metamorphism. J. Sed. Petrol. 55, 532-540.

_____, Peacor, D. R., Lewis, D. D. and Wintsch, R. P. (1984) Chlorite-illite/muscovite interlayered and interstratified crystals. a TEM/STEM study. Contrib. Mineral. Petrol. 88, 372-385.

Lippmann, F. (1954) Über einen Keuperton von Kaiserweiher bei Maulbron. Heidelb. Beitr. Mineral. Petrogr. 4, 130-144.

_____ (1956) Clay minerals from the Röt member of the Triassic near Göttingen, Germany. J. Sed. Petrol. 26, 125-139.

MacEwan, D. M. C. (1944) Identification of the montmorillonite group of minerals by X-rays. Nature 154, 577-578.

_____ and Wilson, M. J. (1980) Interlayer and intercalation complexes of clay minerals. Ch. 3. In: Crystal Structures of Clay Minerals and their X-Ray Identification, G. W. Brindley and G. Brown, eds. Mineralogical Society, London.

Maresch, W. V., Massone, H. J., and Czank, M. (1983) An ordered 1.1 chlorite/biotite mixed-layered mineral as an alteration product of chlorite. Fortschr. Mineral. 61, 139-141.

Martin-Vivaldi, J. L. M. and MacEwan, D. M. C. (1957) Triassic chlorites from the Jura and Catalan Coastal Range. Clay Minerals Bull. 3, 177-183.

_____ and MacEwan, D. M. C. (1960) Corrensite and swelling chlorite. Clay Minerals Bull. 4, 173-181.

_____, _____ and Rodriguez, G. M. (1963) Effects of thermal treatment on the c axial dimension of montmorillonite as a function of the exchange cation. Proc. Int'l. Clay Conf., Stockholm, 45-51.

628

Merceron, T., Inoue, A., Bouchet, A. and Meunier, A. (1988) Lithium-bearing donbassite and tosudite from Eschassieres, Massif Central, France. Clays & Clay Minerals 36, 39-46.

Méring, J (1949) L'Intérference des Rayons X dans les systems à stratification désordonnée. Acta Crystallogr. 2, 371-377.

Morrison, S. J. and Parry, W. T. (1986) Dioctahedral corrensite from Permian red beds, Lisbon Valley, Utah. Clays & Clay Minerals 34, 613-624.

Nadeau, P. H., Tait, J. M., McHardy, W. J. and Wilson, M. J. (1984) Interstratified X-ray diffraction characteristics of physical mixtures of elementary clay particles. Clay Minerals 19, 67-76.

Nishiyama, T., Shimoda, S., Shimosaka, K. and Kanaoka, S. (1975) Lithium-bearing tosudite. Clays & Clay Minerals 23, 337-342.

Novich, B. E. and Martin, R. T. (1983) Solvation methods for expandable layers. Clays & Clay Minerals 31, 235-238.

Parrish, W. and Wilson, A. J. C. (1959) Precision measurements of lattice parameters of polycrystalline specimens, In: International Tables for X-Ray Crystallography, vol. 2. K. Lonsdale, ed. Kynoch Press, Birmingham, pp. 216-234.

Peterson, M. N. A. (1961) Expandable chloritic clay minerals from upper Mississippian carbonate rocks of the Cumberland Plateau in Tennessee. Amer. Mineral. 46, 1245-1269.

Post, J. L. and Janke, N. C. (1974) Properties of 'swelling' chlorite in some Mesozoic formations of California. Clays & Clay Minerals 22, 67-76.

Reynolds, R. C., Jr. (1980) Interstratified clay Minerals. Ch. 3. In: Crystal Structures of Clay Minerals and their X-Ray Identification, G. W. Brindley and G. Brown, eds. Mineralogical Society, London.

_____ (1985) NEWMOD© a Computer Program for the Calculation of One-Dimensional Diffraction Patterns of Mixed-Layered Clays. R. C. Reynolds, 8 Brook Rd., Hanover, New Hampshire 03755, U.S.A.

_____ (1986) The Lorentz-polarization factor and preferred orientation in oriented clay aggregates. Clays & Clay Minerals 34, 359-367.

_____ and Hower, John (1970) The nature of interlayering in mixed-layer illite-montmorillonite. Clays & Clay Minerals 18, 25-36.

Ross, G. J. and Kodama, H. (1976) Experimental alteration of a chlorite into a regularly interstratified chlorite-vermiculite by chemical oxidation. Clays & Clay Minerals 24, 183-190.

Sawhney, B. L. (1960) Weathering and aluminum interlayers in a soil catena. Hollis-Charlton-Sutton-Leicester. Soil Sci. Soc. Amer. Proc. 28, 221-226.

Schlenker, B. (1971) Petrographische Untersuchungen am Gipskeuper und Lettenkeuper von Stuttgart. Oberrheinische Geologische Abhandlungen 20, 69-102.

Schreyer, W., Medenbach, O., Abraham, K., Gebert, W. and Muller, W. F. (1982) Kulkeite, a new metamorphic phyllosilicate Minerals. ordered 1.1 chlorite/talc mixed-layer. Contrib. Mineral. Petrol. 80, 103-109.

Schultz, L. G. (1963) Clay minerals in the Triassic rocks of the Colorado Plateau. U. S. Geol. Surv. Bull. 1147-C, 1-71.

Senkayi, A. L., Dixon, J. B. and Hossner, L. R. (1981) Transformation of chlorite to smectite through regularly interstratified intermediates. Soil Sci. Soc. Amer. J. 45, 650-656.

Shimoda, S. (1969) New data for tosudite. Clays & Clay Minerals 17, 179-183.

Shover, E. F. (1964) Clay-mineral environmental relationships in Cisco (U. Penn.) clays and shales, north central Texas. Clays & Clay Minerals 12, 431-443.

Slaughter, M. and Milne, I. H. (1960) The formation of chlorite-like structures from montmorillonite. Clays & Clay Minerals 8, 114-124.

Srodon, J. (1980) Precise identification of illite/smectite interstratifications by X-ray powder diffraction. Clays & Clay Minerals 28, 401-411.

Stephen, I. and MacEwan, D. M. C. (1950) Some chloritic clay minerals of unusual type. Clay Minerals Bull. 1, 157-162.

Suchecki, R. K., Perry, E. A. and Hubert, J. F. (1977) Clay petrology of Cambro-Ordovician continental margin, Cow Head klippe, western Newfoundland. Clays & Clay Minerals 25, 163-170.

Sudo, T., Takahashi, H. and Matsui, H. (1954) Long spacing of 30 Å from a fire clay. Nature 173, 161.

_____ and Kodama, H. (1957) An aluminian mixed-layer mineral of montmorillonite-chlorite. Z. Kristallogr. 109, 379-387.

Tamura, T. (1955) Weathering of mixed-layered clays in soils. Clays & Clay Minerals 4, 413-422.

Tellier, K. E. and Reynolds, R. C., Jr. (1987) Calculation of one-dimensional X-ray diffraction profiles of interstratified illite/smectite as "fundamental particle" aggregates (abstract). 24th Ann'l Clay Minerals Conf., Socorro, New Mexico, p. 127.

_____, Hluchy, M. M., Walker, J. R. and Reynolds, R. C., Jr. (In press) Application of high gradient magnetic separation (HGMS) to structural and compositional studies of clay mineral mixtures. J. Sed. Petrol.

Thompson, G. R. and Hower, John (1975) The mineralogy of glauconite. Clays & Clay Minerals 23, 289-300.

Veblen, D. R. (1983) Microstructures and mixed layering in intergrown wonesite, chlorite, talc, biotite and kaolinite. Amer. Mineral. 68, 566-580.

_____ and Ferry, J. M. (1983) A TEM study of the biotite-chlorite reaction and comparison with petrologic observations. Amer. Mineral. 68, 1160-1168.

Velde, B. (1977) Clays and Clay Minerals in Natural and Synthetic Systems. Developments in Sedimentology 21, Elsevier Sci. Pub. Co., New York, 218p.

Vergo, N. and April, R. H. (1982) Interstratified clay minerals in contact aureoles, West Rock, Connecticut. Clays & Clay Minerals 30, 237-240.

Walker, G. F. (1957) On the differentiation of vermiculites and smectites in clays. Clay Minerals Bull. 3, 154-163.

_____ (1958) Reactions of expanding lattice minerals with glycerol and ethylene glycol. Clay Minerals Bull. 3, 302-313.

Whitney, G. and Northrop, H. R. (1986) Vanadium-chlorite from a sandstone-hosted vanadium-uranium deposit, Henry Basin, Utah. Clays & Clay Minerals 34, 488-495.

SEPIOLITE AND PALYGORSKITE

INTRODUCTION

The name palygorskite was first used by von Ssaftschenkov (or Savchenkov) in 1862, from palygorsk (Martin et al., 1970; Hay, 1975), where it was found for the first time. According to Ovcharenko and Kukovsky (1984) "when mountain-leather deposits were prospected in the Palygorsk Division Mine near Popovka River, Perm Province (USSR), this unusual mineral was assumed to represent a variety of asbestos." Nevertheless, the mineral appears to have been known since Theophrastus' time, ca. 314 B.C. (Robertson, 1963). Later, Fersman (1913) applied the name palygorskite to a family of fibrous hydrous silicate minerals forming an isomorphous series between two end-members: an aluminum form, called paramontmorillonite, and a magnesium form, called pilolite.

The name attapulgite was first given by Lapparent (1935) to a clay mineral found in fuller's earth from Attapulgus, Georgia, and Mormoiron, France. This name has enjoyed wider recognition than palygorskite because "attapulgite clays" from the U.S.A. have been widely commercialized. However, the name palygorskite has priority (Bailey et al., 1971). Two other terms used for palygorskite have been pilolite (Heddle, 1879; Friedel, 1907; Fersman, 1913) and lassalite (Friedel, 1901). Both have been abandoned in favor of palygorskite. Among palygorskite varieties can be cited Mn-palygorskite and Mn-ferropalygorskite (Semenov, 1969). Two new minerals, yofortierite (Perrault et al., 1975) and tuperssuatsiaite (Karup-Moller and Petersen, 1984) have been accepted as belonging to the palygorskite group. The former contains a high percentage of Mn and a definite quantity of Zn substituting for Al; it can be considered a manganese analogue of palygorskite, similar to that described by Semenov (1969). The latter can be considered a Na-Fe palygorskite.

The name sepiolite was first applied by Glocker in 1847, and is derived from the Greek for "cuttlefish," the bone of which is as light and porous as the mineral. Before Glocker, Cronsted (1758) had described the "Keffekill Tartarorum," which was probably sepiolite. Werner, in 1788, used the name Meerschaum for a magnesium mineral, perhaps sepiolite. Later, Kirwan (1794) used the names Myrsen and Meerschaum for sepiolite; Hauy (1801) named sepiolite as Ecume de Mer, a magnesium and siliceous carbonate. Brochant (1802) described low-density and white magnesium silicates with the names Meerschaum, Ecume de Mer, and Talcum Plasticum. Also, Brongniart (1807) used the term "Plastic Magnesite" for a probable sepiolite. Fersman (1908, 1913) distinguished two sepiolite varieties, one fibrous, α-sepiolite or parasepiolite, and another of laminar morphology, ß-sepiolite. Efremov (1939) thought he found further varieties, γ-δ and ε-sepiolites which were distinguished by their $MgO:SiO_2:H_2O$ ratios.

More recent studies have demonstrated that lamellar sepiolites, such as those studied by Davis et al. (1950), are not sepiolite (Martin Vivaldi and Robertson, 1971) and that the laminar variety of palygorskite of Attapulgus, described by Marshall et al. (1942), is better considered an illite (Davis et al., 1950). No other lamellar palygorskite or sepiolite has been described. In contrast, in our research on these minerals carried out over several years, lamellar morphologies have never been observed; thus we assume that sepiolite and palygorskite occur only with a fibrous habit.

Several sepiolite varieties have been described: Fe-sepiolite, corresponding to the obsolete names of gunnbjarnite (Boggild, 1951), xylotile (Hermann, 1845), and ferrisepiolite (Strunz, 1957); nickeliferous sepiolite (Caillere, 1936), now officially named falcondoite (Springer, 1976); Na-sepiolite described by Fahey et al. (1960) and named loughlinite; Al-sepiolite (Rogers et al., 1956; Firman, 1966); Mn-sepiolite and Mn-ferrisepiolite (Semenov, 1969), etc.

Minerals of the palygorskite-sepiolite group have also been called cardboard, paper, mountain leather, mountain wood, wool, etc. Most of these occurrences have been identified as palygorskite and some as sepiolite, but many others were chrysotiles (Brauner and Preisinger, 1956).

Palygorskite and sepiolite clays are rare in nature, but they have been used by man for centuries because of their many diverse and useful properties. The pre-Columbian Indians manufactured the highly prized organic pigment known as "Maya Blue" with palygorskite of Yucatan Peninsula (Mexico) and indigo. The mineral was used also in the past and present for ceramic ware, and was termed Sac lu'um (white earth) by the indigenous potters. Sacalum (a Spanish corruption of sac lu'um), a village in the northern peninsula of Yucatan, has served as a source of this clay for over 800 years (Van Olphen, 1966; Arnold, 1971; Isphording, 1984). In southern Spain a sepiolite marl from Lebrija, Sevilla province, has been used for hundreds of years for the purification of wine (sherry wine), and thus was called "Tierra del Vino" (Galan and Ferrero, 1982).

Sepiolite has been used for making curved tobacco pipes for centuries (Turkey, Hungary, Germany). The Vallecas (Spain) sepiolite has been mined since the late 1600s. The softer and more compact variety of the ore was also used first in the manufacture of pipes and cigarette filters, and later as a building material (Villanova, 1875). Between 1735 and 1808, ceramic paste for the famous "porcelains" of the Buen Retiro, Madrid, was prepared by mixing Vallecas sepiolite with clay from Capodimonte, near Naples, Italy (Prado, 1864).

At present, the two most important deposits of these minerals are in the Meigs-Attapulgus-Quincy district, Georgia and Florida (U.S.A.) for palygorskite, and in Vallecas-Vicalvaro, Madrid (Spain) for sepiolite. The former has been commercialized at least since 1895 (Patterson and Buie, 1974) and the Vallecas sepiolite since 1945 (Lacazette, 1947). There are not many important deposits of these minerals, but recently other prospects have been reported; some of them could be of a great interest, i.e. Cherkassk palygorskite deposit (to the south of Kiev, USSR) (Ovcharenko and Kukovsky, 1984); Hyderabad palygorskite deposit (Andhra Pradesh district, India) (Siddiqui, 1984), and Liling sepiolite deposit (Hunan province, China) (Zhang et al., 1985), among others.

STRUCTURES

Sepiolite and palygorskite are included in the phyllosilicates because, according to the definition of this mineral group (Brindley and Pedro, 1972), they contain a continuous two-dimensional tetrahedral sheet of composition T_2O_5 (T = Si, Al, Be, ...) but they differ from the other layer silicates in lacking continuous octahedral sheets. The structures of these minerals can be regarded as containing ribbons of 2:1 phyllosilicate structure, one ribbon being linked to the next by inversion of SiO_4 tetrahedra along a set of Si-O-Si bonds. Ribbons extend parallel to the X-axis, and they have an average width along Y of three linked pyroxene-like single chains in sepiolite and two linked chains in palygorskite. The linked ribbons thus form a 2:1 layer continuous along X but of limited lateral extent along Y.

The tetrahedral sheet is continuous across ribbons but with apices pointing in different directions in adjacent ribbons, whereas the octahedral sheet is discontinuous. Consequently, with this framework rectangular channels also run parallel to the X-axis between opposing 2:1 ribbons.

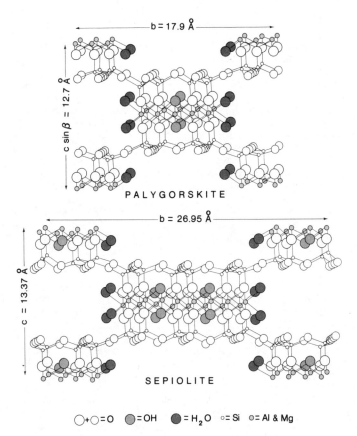

O+O=O O=OH $\bullet$=H$_2$O $\circ$=Si $\circ$=Al & Mg

Figure 1. Schematic structure of palygorskite (after Bradley, 1940) and sepiolite (after Brauner and Preisinger, 1956).

TABLE 1

SOME CRYSTALLOGRAPHIC DATA FOR SEPIOLITE AND PALYGORSKITE

	a	b	c or c sinβ	β	space group
Sepiolite					
Nagy & Bradley (1955)	5.30	27.0	13.4	?	A 2/m
Brauner & Preisinger (1956)	5.28	26.80	13.40	90º	Pnan
Brindley (1959)	5.25	26.96	13.50	90º	——
Zvyagin et al. (1963)	5.24	27.2	13.4	90º	Pnan
Bailey (1980) (average)	5.28	26.95	13.37	90º	Pnan
Galán (unpublished, Vallecas sepiolite)	5.23	26.77	13.43	90º	Pnan
Palygorskite					
Bradley (1940)	5.2	18.0	12.9	?	A 2/m
Zvyagin et al (1963)	5.22	18.06	12.75	95.83º	P 2/a
Christ et al (1969) (Sapillo)	5.24	17.87	12.72	90º	P n
" " " "	5.24	17.83	12.78	95.78º	P 2/a
Bailey (1980) (average)	5.20	17.90	12.70	90º,96º,107º	

As the octahedral sheet is discontinuous at each inversion of the tetrahedra, oxygen atoms of the octahedra at the edge of the ribbons are coordinated to cations on the ribbon side only, and coordination and charge balance are completed along the channel by H, H_2O (bound), and a small number of exchangeable cations. Also, a variable amount of zeolitic water is contained in the channels.

First attempts to determine the sepiolite structure were carried out by Migeon (1936), Longchambon and Migeon (1936), Longchambon (1937), and Caillere (1951), who established that sepiolite was a well-defined mineral from a structural point of view, that both sepiolites described by Fersman (1913) were identical, and that the structurally important element in this mineral was the amphibole chain oriented with its long direction parallel to the X-axis. Also, they pointed out that sepiolite contains "zeolitic water" (water lost at low temperature, <300°C), as well as hydroxyl water lost at higher temperatures.

The first structural pattern for sepiolite was proposed by Nagy and Bradley (1955). The interpretation from an X-ray fiber photograph (using Okl reflections) was not very conclusive. The atomic parameters parallel to the fiber length were not given, so the structure could be either orthorhombic or monoclinic. They thought the C2/m (A2/m) space group was most appropriate (Table 1).

Later, Brauner and Preisinger (1956) and Preisinger (1959) proposed another model for sepiolite with space group Pnan (Fig. 1, Table 1). The fundamental difference between both models lies in whether the tetrahedral inversion at the edge of the ribbons occurs along the middle of the zig-zag Si-O-Si chains (Nagy and Bradley) or along their edges (Brauner and Preisinger). Consequently, in the Brauner and Preisinger (1956) model, adjacent inverted ribbons are joined by a single basal oxygen instead of two, and there are eight octahedral sites in a ribbon instead of nine as in the Nagy-Bradley model, four OH instead of six, and eight zeolitic water molecules instead of six.

A theoretical monoclinic structure of space group A2/a is also possible if the distribution of filled octahedral cation positions (I or II) is the same in each 2:1 ribbon, instead of a regular alternation between the two sets in adjacent ribbons. But to date, only evidence for the orthorhombic symmetry has been obtained. Electron diffraction patterns from single fibers by Brindley (1959), Zvyagin (1967), and Gard and Follet (1968) have confirmed that the extinctions are in accord with the space group Pnan. The Brauner-Preisinger model for sepiolite has also been confirmed and refined by Rautureau et al. (1972), Rautureau and Tchoubar (1974), Rautureau (1974), and Yucel et al. (1981).

The unit cell parameters determined for the sepiolite are (Table 1) $a = 5.28$ Å; $b = 26.95$ Å; $c = 13.37$ Å; $ß = 90°$. Channels in the structure are 3.7 Å x 10.6 Å in dimension and run parallel to the fiber length.

Regarding the palygorskite structure, Bradley (1940) proposed a model (Fig. 2) with a probable A2/m space group (Table 1). The main difference from the sepiolite model is the shorter b dimension because only two linked pyroxene-like single chains are in the ribbon. Later, Drits and Sokolova (1971) confirmed the Bradley model and measured a b-angle of 107°. So, the side linkage by two oxygens appears excluded for both sepiolite and palygorskite.

Preisinger (1963) reported an orthorhombic model for palygorskite similar to the orthorhombic sepiolite of Brauner and Preisinger (1956) except for ribbon width. Christ et al. (1969) studied five palygorskite samples by X-ray diffraction and found three orthorhombic (Pn) and two different monoclinic cells. Although there are not sufficient data to define exactly the difference between them, it is clear that at least two symmetries are possible for palygorskite, one orthorhombic and another monoclinic (Table 1). Monoclinic structures have an n-glide plane parallel to (100). One of the monoclinic symmetries is close to that determined by Zvyagin et al. (1963) (P2/a, $ß = 95.83°$). The other, with the

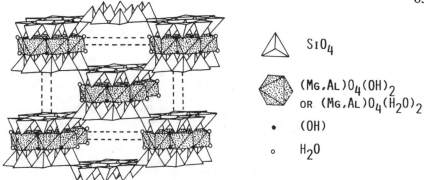

Figure 2. Crystal structure of palygorskite, *c* axis vertical (Zoltai and Stout, 1985).

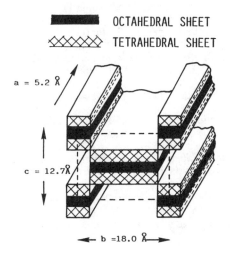

Figure 3. Gard-Follet's (1968) model for palygorskite.

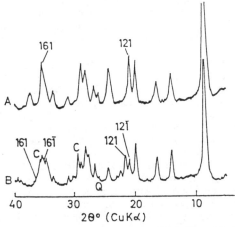

Figure 4. X-ray diffractometer pattern of A, the orthorhombic, and B, the monoclinic variety of palygorskite (c = calcite, Q = quartz; after Christ et al., 1969).

Z-axis as the monoclinic axis and $\gamma = 92.23°$ for the monoclinic angle, has been given no structural interpretation up to the present time.

Another structural model was proposed by Gard and Follet (1968) on the basis of P2/a symmetry (Fig. 3). In this pattern, they proposed an alternation of 2:1 ribbons containing three zig-zag tetrahedral single chains (like sepiolite ribbons) with those containing only one chain. This model is closer to a normal 2:1 phyllosilicate, as only 1/4 of the tetrahedra are inverted. Half of the channels in the a-direction have the same cross-section as those in sepiolite, but the others are much narrower. An orthorhombic symmetry is possible also from the alternation of I and II octahedral cation sets in each adjacent 2:1 ribbon. This model of Gard and Follet could explain the intergrowths of different ribbon widths proposed by Martin Vivaldi and Linares (1962) to explain observed intermediate X-ray spacings.

According to the studies referenced here, at least two orthorhombic and two monoclinic palygorskite structures must be considered (Table 1):

(a)   orthorhombic: Pnmb (Preisinger, 1963), Pn (Christ et al., 1969)

(b)   monoclinic: A2/m (Bradley, 1940; Drits and Sokolova, 1971), P2/a (Zvyagin et al., 1963; Christ et al., 1969; Gard and Follet, 1968)

Parameters (average): $a = 5.2$ Å; $b = 17.9$ Å; $c \sin \beta = 12.7$ Å; $\beta = 90°, 96°$ or $107°$.

Channels in the structure are 3.7 Å x 6.4 Å in dimension and they run parallel to the fiber length.

X-ray powder diffraction patterns of palygorskite and sepiolite are shown in Figures 4 and 5. Table 2 presents the basic X-ray diffraction data.

Because of the fibrous morphology and partly because of the chain-like structure, the name "hormites" has been suggested for the sepiolite-palygorskite group (Robertson, 1962) derived from the Greek ορμos, a chain, but this name has not been accepted by the IMA.

More recently, Zoltai (1981) has described palygorskite and sepiolite structures as biopyriboles (bio = biotite, pyr = pyroxenes, iboles = amphiboles) built of tri-di-octahedral modules (Fig. 6), where tri-module = $M_3A_2Si_4O_{10}$, and di-module = $M_2A_2Si_2O_{10}$. M is the octahedral cation and A is the anion not bonded to Si within the module; it can be oxygen when bonded to Si and is (OH) when bonded to more than one M cation. One half of each A anion is an H_2O molecule when the anion is bonded to only one M cation.

The width of these modules is one tetrahedral chain, and their height (t) is four times the height of an ideal polyhedral layer. Combinations of these modules can form complete crystal structures with a vertical displacement between the modules = nt; n = 0, 1/2, 3/4. If n = 0, the major layered silicates are produced. A sequence of n = 0 and 3/4 between modules produces palygorskite, and the sequence 0, 0, and 3/4 gives the sepiolite structure (Fig. 7). Symbol 0 is relative to the orientation of tetrahedral chains, indicating that the faces of adjacent tetrahedra point in opposite directions ("0" chains).

Although the complex description by Zoltai (1981) could be crystallographically attractive for some investigators, we think that palygorskite and sepiolite should be considered phyllosilicates (as noted above) with special characteristics, and not biopyriboles, because they have structure, physicochemical properties, and genetic environments closer to the phyllosilicates (especially to the clay minerals) than to the other biopyriboles.

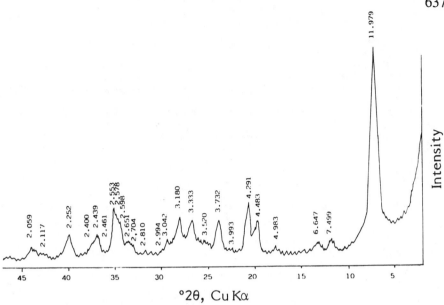

Figure 5. X-ray diffractometer pattern of Vallecas sepiolite.

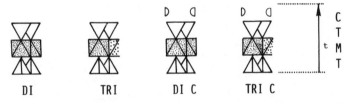

DI   TRI   DI C   TRI C

Figure 6. The four basic layer modules of the biopyriboles. The tetrahedra are completely rotated so that the anions are in close-packed sheets. Di = dioctahedral, Tri = trioctahedral, C = interlayer cation. The second octahedra in Tri and Tri C modules are dotted. Intermediate layer module, not shown, is 1/2 Tri, intermediate between Di and Tri (after Zoltai, 1981).

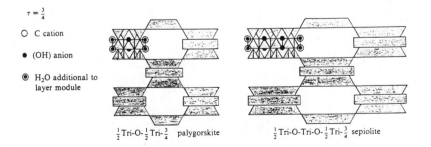

Figure 7. Palygorskite and sepiolite structures (after Zoltai, 1981).

638

Table 2. X-ray powder data for sepiolite and palygorskite.

SEPIOLITE FROM VALLECAS, SPAIN
(1) Brindley, 1959   (2) Galán, unpublished

PALYGORSKITE FROM ATTAPULGUS, Ga. (Bradley, 1940) (1) and
FROM SAPILLO, N.M. (Christ at al. 1969) (2)

SEPIOLITE

hkl	(1) d (obs)	(1) I	(2) d (obs)	(2) I
011	12.1	100	12.11	99
031	7.5	7	7.49	3
002 } 040 }	6.7	4B	6.64	3
051	5.04	3B	4.99	2
060	4.49	25	4.49	9
131	4.29	35	4.30	14
033	4.02	7	3.98	1
062	3.73	25	3.74	10
142	3.50	5	3.51	2
080	3.339	45	3.339	13
133	3.181	15B	3.182	9
162	----	---	3.044	3
073	2.950	5	2.938	1
180	----	---	2.815	1
124	----	---	2.740	1
0,10,0 } 015 }	2.66	8NR	2.662	4
182	2.59		2.586	
035				
220		45NR		10NR
211				
173				
191	2.56		2.562	
231	----	----	2.474	1
202 } 240 }	2.43		2.437	
0,11,1 }		20NR		9NR
222	2.395		2.395	
164 }				
260 } 213 } 1,10,2 }	2.256	20	2.254	10
062,075,233	----	--	2.204	1
046 } 0,12,2 } 0,10,4 }	2.117	4B	2.116	1
280	2.060	10	2.060	4
	1.873	4	1.873	1
	1.716	7	1.72	1
	1.691	10	1.691	1
	1.598	4	1.598	3
	1.578	7	-----	-
	1.540	8	1.55	2
	1.517	15	1.52	1
	1.465	3	-----	-
	1.406	4	1.411	2

PALYGORSKITE

(1) hkl	(1) d(obs)	(1) I (obs)	(2) hkl	(2) d(obs)	(2) I
011	10.50	ss	011	10.44	100
002	6.44	ms	002	6.36	13
031	5.42	m	031	5.39	9
022	----	--	040	4.466	20
040	4.49	s	121	4.262	22
013	4.18	w	013 131	4.129 sh	2
042	3.69	m	122	3.679	15
033 } 051 }	3.50	w	051	3.440	2
			132	3.348	7
004	3.23	ss	113 004	3.129 sh	12
024	3.03	ww	123	3.096	16
053	----	--	133	2.889	4
044	2.61	s	114	sh	—
015	2.55	w	152	2.679	8
035	2.38	w	124	sh	—
080	----	--	210	sh	—
006	2.15	m	044 } 160 }	2.589	10
055	----	--			
084	----	--	201	2.567	12
093	1.82	ww	211 } 161 }	2.539	20
066	----	--			
008	1.62	ww			
086	1.56	w			
0,12,0	1.50	m			

B = broad
NR = Not resolved
sh = peak present, but as shoulder on adjacent, more intense peak

Note: For parameters, see Table I

CHEMISTRY

From an historical point of view, it may be interesting to note that Wiegleb (before 1784) was the first to attempt the chemical analysis of a meerschaum pipe from Turkey, but Klaproth (1794) made a better analysis of a sepiolite from Eskisehir. Since that time, chemical analyses have been presented in most of the papers regarding these minerals.

Based on the model of Drits and Sokolova (1971) for palygorskite, the approximate structural formula is

$$(Mg_{5-y-z} R_y^{3+} \square_z) (Si_{8-x} R_x^{3+}) O_{20} (OH)_2 (OH_2)_4 R^{2+}_{(x-y+2z)/2} \cdot (H_2O)_4$$

Tetrahedral occupancy for Al ranges from 0.01 to 0.69 for eight positions (Weaver and Pollard, 1973) or from 0.12 to 0.66 (Newman and Brown, 1987). The sum of octahedral cations lies between 3.76 and 4.64 (Newman and Brown, 1987) with a mean value of 4.00, indicating that the mineral can be considered dioctahedral. Octahedral cations (Mg, Al, Fe^{2+}, Fe^{3+}) are ordered, with the vacant sites at the center of the ribbons, according to the IR study carried out by Serna et al. (1977). Four H_2O molecules are present in the channels (zeolitic water) and four others are bound to octahedral cations. On heating (see thermal behavior) these latter water molecules are lost in two stages; when the first two water molecules are lost, the structure collapses by alternate rotation of ribbons (folding; Van Scoyoc et al., 1979).

The structural formula for sepiolite based on the model of Brauner and Preisinger (1956) is:

$$(Mg_{8-y-z} R_y^{3+} \square_z) (Si_{12-x} R_x^{3+}) O_{30} (OH)_4 (OH_2)_4$$

$$R^{2+}_{(x-y+2z)/2} \cdot (H_2O)_8$$

Sepiolite may contain more zeolitic water than palygorskite but proportionately less water coordinated to the octahedral cations at the edges of the ribbons. This coordinated water is also lost in two stages on heating, and the structure folds when half of the coordinated water has been lost (Serna et al., 1975).

The theoretical SiO_2/MgO ratio of sepiolite is 2.23, with $SiO_2 = 55.6\%$; $MgO = 24.99\%$; or on an anhydous basis, $SiO_2 = 61.7\%$; $MgO = 27.6\%$. But usually SiO_2 falls in the range of $53.9\pm1.9\%$, and MgO between 21-25%.

Tetrahedral occupancy ranges from 0.05 Al per 12 positions to 0.53 Fe^{3+} and .24 Al^{3+} per 12 sites. The total number of octahedral cations ranges from 7.01 to 8.01 (Newman and Brown, 1987). These cations are predominantly Mg, with minor Mn^{2+}, Fe^{2+}, Fe^{3+}, and Al. If Ni > Mg, the species is known as falcondoite (Springer, 1976); if it is an iron-rich or an aluminum-rich variety, the sepiolite is named Fe-sepiolite or Al-sepiolite, respectively. Loughlinite is Na-sepiolite, where 2 Na substitute for 2 Mg, with 2 Na in the channels.

Table 3 presents some typical chemical analyses of palygorskites and sepiolites. Cation exchange capacity for these minerals is low. Values can range from 5-40 meq/100 g, but high values probably are related to smectite impurities.

Regarding the loughlinite first described by Fahey and Axelrod (1948), it is very difficult to understand a sepiolite structure containing Na in Mg sites which can be removed by water leaching, leaving a normal sepiolite (Table 3). Echle (1978) indicates that in loughlinite the marginal Mg of the sepiolite structure is replaced by 2 Na, one of them placed with water in the channels of the structure. The X-ray diffraction pattern given by

640

TABLE 3

SOME CHEMICAL ANALYSES OF PALYGORSKITES AND SEPIOLITES

	1	2	3	4	5	6	7	8
SiO_2	55.03	53.75	51.50	52.85	52.50	53.70	52.43	50.80
Al_2O_3	10.24	10.23	10.03	1.03	0.60	1.15	7.05	0.66
Fe_2O_3	3.53	1.83	2.36	0.04	2.99	0.64	2.24	1.85
FeO		0.26	0.52	0.01	0.70	0.02	2.40	1.51
MgO	10.49	9.39	12.28	23.74	21.31	23.31	15.08	16.18
CaO		2.29	1.81	0.51	0.47	0.03		0.12
Na_2O			0.12			0.67		8.16
K_2O	0.47	0.02	0.13			0.61		
H_2O^+	10.13	12.04	14.43	9.04	21.27	9.83	10.48	7.12
H_2O^-	9.73	10.16	7.36	12.67		9.76	9.45	13.68
Total	99.62	99.97	100.54	99.89	99.84	99.72	100.01 [1]	100.08 [2]

Formulas for palygorskites

1) $(Si_{7.80}Al_{0.20})$ $(Al_{1.51}Fe^{3+}_{0.38}Mg_{2.22})_{4.11}$ $K^+_{0.09}$ On the basis of

2) $(Si_{7.82}Al_{0.18})$ $(Al_{1.57}Fe^{3+}_{0.20}$ $Fe^{2+}_{0.03}$ $Mg_{2.04})_{3.84}$ $Ca^{2+}_{0.36}$ O_{20} $(OH)_4$ $(OH_2)_2$

3) $(Si_{7.64}$ $Al_{0.36})$ $(Al_{1.44}$ $Fe^{3+}_{0.26}$ $Mg_{2.46})_{4.14}$ $Ca^{2+}_{0.09}$ $Na^+_{0.02}$ $K^+_{0.01}$ (3)

Formulas for sepiolites

4) $(Si_{11.78}$ $Al_{0.22})$ $(Al_{0.06}$ $Fe^{3+}_{0.01}$ $Mg_{7.89})_{7.96}$ $Ca^{2+}_{0.12}$ On the basis of

5) $(Si_{11.8}$ $Al_{0.16}$ $Fe^{3+}_{0.04})$ $(Fe^{3+}_{0.47}$ $Fe^{2+}_{0.13}$ $Mg_{7.14})_{7.74}$ $Ca^{2+}_{0.11}$ O_{30} $(OH)_4$ $(OH_2)_2$

7) $(Si_{11.54}$ $Al_{0.46})$ $(Al_{1.37}$ $Fe^{3+}_{0.37}$ $Fe^{2+}_{0.44}$ $Mg_{4.95})_{7.13}$ $(NH_4)^+_{0.45}$

(1) Included in total 0.88% $(NH_4)_2 = 0.58\%$ NH_3; 0.45 monovalent cations as NH_4^+

(2) Included in total 0.38% dolomite, 0.21% magnesite

(3) After elimination of impurities from the analyses (see original paper)

1.- Palygorskite, Attapulgus, Georgia , USA (Bradley, 1940)

2.- Palygorskite, Kuzuu District, Tochigi Pref, Japan (Imai et al. 1969)

3.- Palygorskite, Torrejón, Cáceres , Spain (Galán et al. 1975)

4.- Sepiolite, Kuzuu District, Tochigi Pref. Japan (Imai et al. 1969)

5.- Sepiolite, Ampandrandara, Madagascar (Caillere, 1951)

6.- Sepiolite, Amboseli, Kenya (Stoessell and Hay, 1978)

7.- Aluminium sepiolite, Tintinara, South Australia (Rogers et al. 1956)

8.- Loughlinite, Sweetwater Country, Wyoming, USA (Fahey et al., 1960). Leached 58 day Na_2O is 0.06 and formula

is: $(Si_{11.96}$ $Al_{0.04})$ $(Al_{0.17}$ $Fe^{3+}_{0.46}$ $Fe^{2+}_{0.26}$ $Mg_{6.78})$ $_{7.67}$ $Na^+_{0.03}$

TABLE 4

THEORETICAL VALUES FOR TOTAL CONTENT OF OCTAHEDRAL CATIONS PER
FORMULA UNIT ON THE BASIS OF O_{10} (Martín Vivaldi and Fenoll, 1970)

Mineral	Octahedral cations
Talc	3.0
Sepiolite	2.7
Palygorskite	2.5
Pyrophyllite	2.0

Fahey et al. (1960) for loughlinite is very similar to that of sepiolite. Echle claims that the two are distinctive, but relies primarily on the difference between 12.9 and 12.2 Å, respectively, for the principal reflection. Other properties, i.e., specific gravity, DTA response, electron microscope morphology, etc., are also very similar to sepiolite. For us the distinction of loughlinite as a different mineral from sepiolite is still to be conclusively demonstrated.

Chemical analyses of Vallecas sepiolite crystals by energy-dispersive analysis (Galan and Perez Rodriquez, unpublished) have demonstrated that most of the crystals contain no aluminum. Analyses show only Si, Mg, and sometimes Fe, but impurities in the samples contain aluminum (illite, smectites, feldspars). Cation exchange capacity for this sepiolite was 9.5 meq/100 g.

When compared with other phyllosilicates on the same atomic basis, sepiolite and palygorskite appear to be intermediate between dioctahedral and trioctahedral, with palygorskite nearer the former and sepiolite nearer the latter (Martin Vivaldi and Cano, 1956; Mackenzie, 1966; Martin Vivaldi and Fenoll, 1970) (Table 4).

Martin Vivaldi and Cano (1956) suggested that change in structure from lamellar to fibrous takes place according to the number of octahedral vacancies. If the number of vacancies is little or very great, the habit is lamellar, and in the intermediate cases, the habit is fibrous. A verification and explanation of this proposed theory have not yet been given.

Paquet et al. (1987) have reported analyses by energy dispersive scanning-transmission electron microscopy of 145 individual particles from palygroskite-smectite and sepiolite-smectite assemblages. The results show great variability and significant overlap in the octahedral composition fields of the fibrous clays and the smectites. The sepiolite field is definitely in the trioctahedral smectite domain, but the palygorskite field is both in the dioctahedral domain and extends into the region separating dioctahedral and trioctahedral smectites. Nevertheless, a distinct octahedral compositional gap between palygorskite and sepiolite, such as originally documented by Martin Vivaldi and Cano (1956), can be recognized. The gap extends from about 0.3 in ratio of trivalent octahedral cation to magnesium, the same as found by Martin Vivaldi and Cano (1956), to about 0.6, which is about half the value of the earlier study. The new data, based on individual particles rather than bulk clay fraction, clearly indicate that palygorskites can contain more octahedral substitution than originally thought. Actually, analysis #7 (Table 3) falls very close to the upper limit of the compositional gap for sepiolites. This is probably due to the high iron, particularly ferrous, content.

PHYSICAL, PHYSICO-CHEMICAL AND THERMAL PROPERTIES

Physical properties

Sepiolite and palygorskite are characteristically of a fibrous *habit* when seen under the electron microscope (Figs. 8a,b; 9). *Sizes of fibers* are quite variable, ranging from ~100 Å to a 3-4 μm in length, ~100 to ~300 Å wide, and ~50 to ~100 Å thick. An exception is the Ampandrandara (Madagascar) sepiolite whose crystals can be millimeters in length.

For the sepiolite of Vallecas, Robertson (in Martin Vivaldi and Robertson 1971) found as typical dimensions 8,000 x 250 x 40 Å, from which he deduced that these laths, broken only in length by milling, are 1500 unit cells in length, eight or nine of them in width, and only three in thickness. For palygorskite, Bates (1958) found a typical fiber dimension of 10,000 x 150 x 75 Å. It seems that sepiolite crystals have a width/thickness ratio greater than palygorskite crystals. The latter show a practically cylindrical habit. Generally, sepiolite fibers have dimensions greater than palygorskite.

642

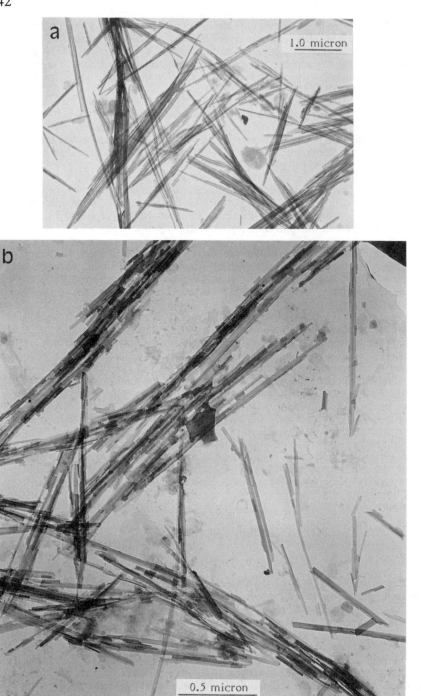

Figure 8. Transmission electron micrographs of palygorskite of (a) Caceres and (b) Vallecas, Spain.

643

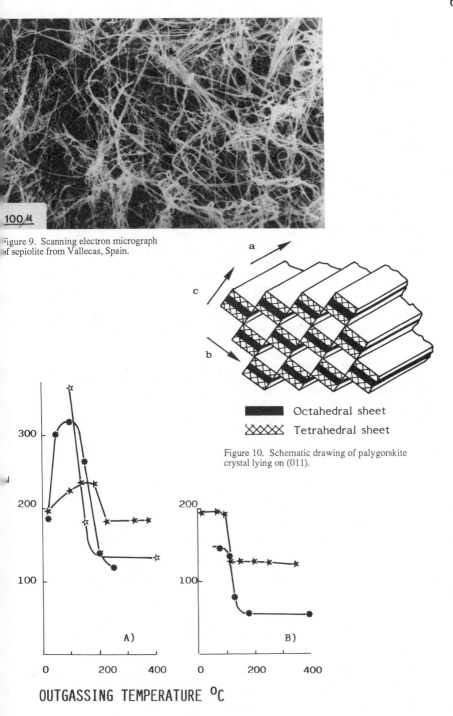

Figure 9. Scanning electron micrograph of sepiolite from Vallecas, Spain.

■■■ Octahedral sheet

XXXX Tetrahedral sheet

Figure 10. Schematic drawing of palygorskite crystal lying on (011).

A)

B)

OUTGASSING TEMPERATURE °C

Figure 11. Effect of outgassing temperature on surface area (BET-N2) of sepiolite (A) and palygorskite (B), according to various authors (in Serratosa, 1979).

644

Specific gravity varies between 2.29-2.36 for palygorskite and 2-2.2 for sepiolite (Henin and Caillere, 1975; Galan, 1979). The minerals can be white, yellowish-white, greenish, grey, or pink in color. Sepiolite occurs with compact (meerschaum), earthy, or fibrous fabric. Palygorskite can also occur in massive and earthy form, such as with mountain leather, cardboard, etc. They are refractory minerals: sepiolite has a *melting point* close to 1400°C; *hardness* is about 2.2.5; good *cleavage* (011) is usually apparent (Fig. 10).

Sorptive properties

The structure of the chain-structure minerals possesses three kinds of active sorption centers: (a) Oxygen ions on the tetrahedral sheet of the ribbons; (b) water molecules coordinated to magnesium ions at the edges of structural ribbons (two H_2O for each Mg^{2+}); and (c) SiOH groups along the fiber axis (Serratosa, 1979). The silanol groups are formed because broken Si-O-Si bonds at external surfaces compensate their residual charge by accepting a proton or a hydroxyl and becoming Si-OH groups. These groups can interact with molecules adsorbed on external surfaces (Ahlrichs et al., 1975). They occur at intervals of approximately 5 Å along the fiber axis, and their abundance can be related to the dimension of the fibers and crystal imperfections. Relative abundances of Si-OH groups have been determined by IR (Ahlrichs et al., 1975; Serna et al., 1975). In palygorskite they are theoretically less abundant than in sepiolite, due to the smaller external surface area of the palygorskite. Nevertheless, Serna and Van Scoyoc (1979) have determined more Si-OH groups in palygorskite than in sepiolite, probably because of more crystal imperfections in sepiolite than in palygorskite. These last results agree with those obtained by Hermosin and Cornejo (1986). They obtained a value of 2.2 Si-OH/100 $Å^2$ for sepiolite and 4.5 Si-OH/100 $Å^2$ for palygorskite.

Palygorskite and sepiolite can be considered polydisperse structures. Palygorskite has a 300 Å effective pore size, 40-80 Å intermediate pores (mesopores), and about 20 Å micropores. The specific surfaces of the inner channels and pores make up 50% of the total surface (224 m^2/g by BET-method) (Ovcharenko and Kukovsky, 1984). The sepiolite of Vallecas (Spain) has a mean micropore diameter of 15 Å, the predominant radii of the mesopores are about 15 and 45 Å, and the surface area is 284 m^2/g (BET-method for samples outgassed at room temperature) of which 139 m^2/g is the external surface area and 145 m^2/g is the surface area of the micropores (Galan, 1987).

In these minerals, external surfaces consist predominantly of (011) crystal faces (Fig. 10), although crystal habit (001) external face area is also abundant (Serna et al., 1977). Theoretical average values for external and internal surfaces have been estimated as 400 m^2/g external and 500 m^2/g internal for sepiolite (Vallecas), and 300 m^2/g external and 600 m^2/g internal for palygorskite (Attapulgus; Serna and Van Scoyoc, 1979).

The surface area available in palygorskite and sepiolite depends mainly on the nature of the molecules (size, shape, polarity) used as sorbate to penetrate the intra-crystalline channels. So, the surface area calculated using hexane is 330 m^2/g (Robertson, 1957); 60 m^2/g using cetylpyridinium bromide (Greeland and Quirk, 1964); 275 m^2/g using pyridine (Ruiz-Hitzky et al., 1983); 276 m^2/g by BET-method (Ruiz-Hiztky and Fripiat, 1976), or 470 m^2/g for sepiolite using ethylene glycol (Fenoll and Martin Vivaldi, 1968).

The sorptive abilities of these minerals also depend greatly on the outgassing pretreatment. Sorption of N_2 is enhanced by removal of water from channels by outgassing in vacuo at 70-80°C (Dandy, 1968; Fernandez Alvarez, 1978) (Fig. 11). Nevertheless, it seems probable that N_2 cannot penetrate far into the channels structure and is predominantly sorbed on external surfaces. The maximum surface specific to N_2, found after drying, is about 380 m^2/g sepiolite (Dandy, 1968) and 150 m^2/g for palygorskite (Fernandez

Alvarez, 1978), both only fractions of the total theoretical surface. The sepiolite surface available to NH_3 under the same conditions is 610 m^2/g (Dandy, 1971).

Molecules of organic sorbates mainly interact with the edges of the ribbons, either in the channels or on the external surfaces. Compounds of a small molecular size and high polarity can penetrate into the structural channels, whereas large, non-polar molecules are generally confined to external surfaces. In the case of non-polar compounds, interactions are mainly of Van der Waals type, although some interaction with the Si-OH groups probably adds to the adsorption forces. Sorption of polar organic compounds (i.e., alcohol) is produced just after zeolitic water has been outgassed and molecules interact with the coordinated water through H-bonding, as demonstrated by IR studies (Serna and Van Scoyoc, 1979). After repeat adsorption-desorption cycles, a portion of the coordinated water is replaced by the alcohol, and molecules are now coordinated directly to exposed Mg^{2+}, which plays a role similar to that of interlayer, exchangeable cations in smectites. Only short chain alcohol (methanol and ethanol) can penetrate into the internal channels according to these cited authors, and adsorption can occur by replacement of the coordinated water (ion - dipole interaction) and by hydrogen bonding to the water.

In general, polar molecules, chiefly water and ammonium, and to a lesser extent, methyl and ethyl alcohols, can enter the channels of palygorskite and sepiolite, while nitrogen, oxygen, and other non-polar gasses and organic compounds cannot. Without steric hindrance, intracrystalline penetration appears to be strongly related to the magnitude of the dipole moment of the molecule and is apparently related to the vapor pressure (Serna and Van Scoyoc, 1979). The sorption capacity diminishes when minerals are heated because structural changes cause decreasing surface area mainly by destruction of micropores (see next heading).

Sepiolite and palygorskite minerals are not expandable by intercalation of organic compounds into their channels. Nevertheless, some "expandability" has been described after ethylene glycol solvation. Expanding palygorskite (Watts, 1976; Jeffers and Reynolds, 1987) and sepiolite (Fleischer, 1972; Guven and Carney, 1979) have been described. This expansion has also been observed after dimethylsulfoxide solvation (Galan, unpublished).

Glycollation of palygorskite produces a shift in the reflection containing the c parameter. Diagnostic peaks at 10.4, 6.34 and 3.18 Å shift to 10.7, 6.6 and 3.3 Å, respectively (Jeffers and Reynolds, 1987). Lesser shifts (about 0.1-0.2 Å) have been observed by Watts (1976) and Galan (unpublished). Similar behavior has been observed for sepiolite, both upon glycolation with ethylene glycol and with dimethylsulfoxide (Galan, unpublished). In this case, all the samples tested (Vallecas and Yunclillos, Spain; Las Vegas, U.S.A.) expanded. The d-values for X-ray diffraction peaks for solvated samples heated to 100-150°C returned to those found for the air-dried samples, indicating that no irreversible structural changes had occurred.

Expansion of palygorskite and sepiolite with polar molecules suggests a mechanism by which the chain structure may be weakened or ruptured, allowing rearrangement into a regular sheet phyllosilicate. Such a hypothesis opens new possibilities of explanation for certain transformations (palygorskite-montmorillonite, sepiolite-stevensite) after the interaction of fibrous clay minerals with organic constituents of high polarity.

Surface area modifications

Chain-structure clay surface area can be modified by heating (Fernandez Alvarez, 1970; 1978). The area for sepiolite is maximum at 150°C, when about 10% of the hydroscopic zeolite water has been lost (~280 m^2/g BET-method) and then decreases between 200 and 400°C, reaching 140 m^2/g at 300°C. The increase in surface area with the elimination of hydroscopic zeolitic water parallels the increase in micropores (most of them are in the structural channels). As the mineral is heated above 300°C, the surface area continues to decrease.

The first structural changes in these minerals occur as about half of the bound water is driven off, and alternate ribbons rotate positively and negatively to close the channels, forming what are known as folded structures (Fig. 12) (Nagata et al., 1974; Serna et al., 1975; Van Scoyoc et al., 1979). This change begins at about 300°C *in vacuo*. Further heating to 600°C causes the loss of the second part of the bound water, and also the dehydroxylation of palygorskite (Mifsud et al., 1978). Folding is associated with a decrease in surface accessible to N_2 (Fernandez Alvarez, 1978) because of a sintering of channels or grooves on the superficial surface in which it is believed some N_2 sorption occurs. For example, specific surface for pores are 156 m^2/g and 15 m^2/g at 200°C and 300°C, respectively, for sepiolite; thus, BET-surface area decrease is largely due to destruction of micropores.

The surface area of sepiolite can be increased by acid treatment with 5% HCL (Fernandez Alvarez, 1972), which is attributed to changes in the surface texture, decreasing the number of pores with a radius below 10 Å, and increasing those with a radius between 10 and 50 Å. Treatment with HNO_3 at different concentrations and heating at 200°C-300°C increases the number of pores, and thus surface areas may increase up to as much as 449 m^2/g (Jimenez Lopez et al., 1978; Lopez Gonzalez et al., 1981). In these treatments, surface acid centers are also increased (Bonilla et al., 1981).

Acid treatment practically destroys sepiolite and palygorskite structures because these minerals are particularly susceptible to hydrogen ion attack. Treatments to increase surface area by acids are rather treatments to "amorphize" these minerals. In this way, Gonzalez et al. (1984) and Cornejo and Hermosin (1986) recently have obtained a porous amorphous product of fibrous morphology with a surface area of 500 m^2/g by treating a sepiolite with 6N HC1 until all magnesium was extracted.

On the other hand, sepiolite whose Mg was located at the edges of the octahedral sheets and had been exchanged by monovalent or divalent cations, shows a decreasing surface area (Mifsud et al., 1987). After an exchange of Mg by about 150 MEQ/100G of other cations, the original area of sepiolite (BET-method) changed from 295 square meter/G to 205, 196, 170, 108, 165, 155 square meter/G for Ca-, Ba-, Li-, Na-, K-, and Cs-sepiolite, respectively.

Rheological properties

Palygorskite and sepiolite are two of the most important gel-forming clays. They give stable suspensions of high viscosity at relatively low concentrations compared to other clays. During dispersion, bundles of their needle-shaped crystals separate to form a random structure that entraps liquid and increases viscosity of the system. These minerals can thicken fresh and salt water and other solvents of high or medium polarity. The suspensions have a generally non-Newtonian behavior that depends on many factors, such as concentration of clay, shear-stress, pH, and electrolyte composition.

Sepiolite and palygorskite may also form suspensions in non-polar solvents, but it is necessary to previously modify the hydrophilic surface of the minerals with surface active agents. Suspensions of these clays show a rapid increase in viscosity as the shear rate increases. The same occurs when shearing time is increased at a constant shear rate (Table 5) (Alvarez, 1984). The Vallecas sepiolite produces a stable suspension of high viscosity (1,000-4,000 cps at 5 rpm in a Brookfield viscometer) at relatively low concentrations (2-10%) in water or other liquids of high or medium polarity (Alvarez, 1984). Palygorskite suspensions show analogous behavior (Haden and Schwint, 1967).

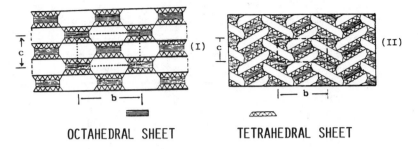

Figure 12. General scheme for an unfolded (I) folded (II) fibrous clay mineral.

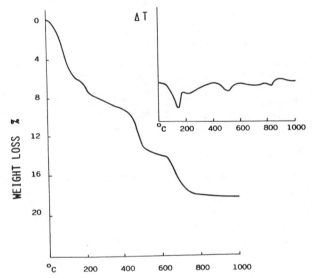

Figure 13. DTA and TG curves of palygorskite of Caceres, Spain.

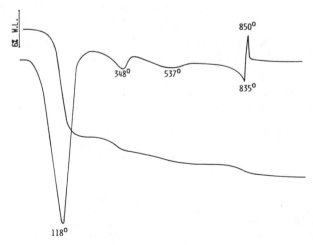

Figure 14. DTA and TG curves of sepiolite of Vallecas, Spain.

648

Table 5. Development of viscosity of sepiolite suspensions.

Shear Rate (rpm)	Time of shearing (minutes)	Brookfield viscosity at 5 rpm (cps)
500	5	400
	10	840
	20	1,480
1,000	5	2,280
	10	3,400
	20	3,800
2,000	5	4,000
	10	4,200
	20	4,400

The optimum pH to get the best suspension characteristics for these clays in water is 8-8.5. At pH higher than 9, they can flocculate. Extrusion of clays increases the viscosity of water dispersions by as much as 50%. However, the effect is less in water containing sodium chloride. In general, the viscosity of aqueous suspensions of palygorskite is influenced only slightly by sodium chloride, other electrolytes, and conventional clay flocculating agents (Haden and Schwint, 1967).

Thermal behavior

Both palygorskite and sepiolite have similar thermal behavior (Caillere and Henin, 1957; Hayashi et al., 1969; Imai et al., 1969; Martin Vivaldi and Fenoll, 1970; Nagata et al., 1974; Serna et al., 1975; Mifsud et al., 1978). DTA curves (Figs. 13 and 14) can be divided into three parts: (1) the low-temperature region; (2) the central region; and (3) the high-temperature region. Structural and morphology changes occurring on heating have been followed also by IR electron microscopy (Serna et al., 1974; Rautureau and Mifsud, 1977).

In the low temperature region (<300°C) the curves show a large endothermic reaction at about 150°C in palygorskite and at a somewhat lower temperature (~120°C) in sepiolite. This lower peak corresponds to the loss of adsorbed and zeolitic water. In palygorskite a second peak usually appears at about 280-300°C, smaller than the other, and probably due to zeolitic water loss. Weight losses during this interval are variable, mainly depending on the experimental conditions. In palygorskite, these losses are close to 9% and 3%, respectively, and in sepiolite the loss is about 10% or more. In this temperature range, there are no important structural changes.

In the central region (300-600°C) two endothermic effects appear at about 350°C and 500-550°C in sepiolite. The first is attributed to loss of the first two molecules of combined water. Theoretically, the weight lost due to coordination water should be 6.25%. Experimental results gave an average of 6.20% (Serna et al., 1975; Perez Rodriquez and Galan, unpublished), 3.30% and 2.90%, respectively, for the two endotherms. While the peaks at 350°C and 550°C are very different in shape, the amount of water lost at the two temperatures is only slightly different.

According to Serna et al. (1975), at 330-350°C the crystal folds when approximately half of the water of hydration, which is coordinated at the edge magnesium atoms inside of the channels, is removed (Fig. 15). All the remaining water is in collapsed channels from which it is removed only at higher temperatures. The final dehydration produces

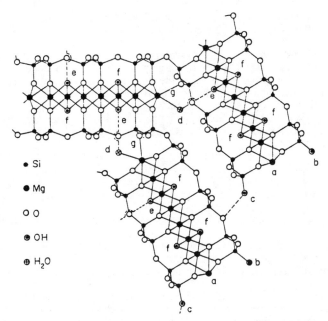

- ● Si
- ● Mg
- ○ O
- ◉ OH
- ⊕ H₂O

Figure 15. End view of the right-hand edge of a model sepiolite crystal which folded after losing approximately half of its water of coordination (a) dehydrated Mg; (b) unperturbed OH of Si; (c) OH of Si perturbed by bond to surface oxygen; (d) remaining internal water; (e) perturbated octahedral OH; (f) unperturbed octahedral OH; (g) Mg to oxygen bond.

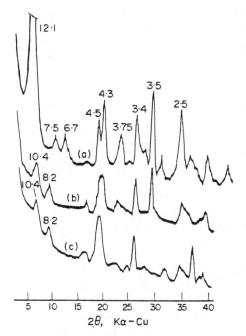

Figure 16. X-ray powder diffraction pattern of sepiolite under vacuum at (a) 25°C, 4 h; (b) 200°C, 4 h; (c) 530°C, 6 h (after Serna et al., 1975).

no important structural change (sepiolite anhydride). Under vacuum, more than two of the four coordinated waters for half-cell are gone by 200°C.

In palygorskite the DTA curve usually appears as one peak in the central region at about 450-500°C. A detailed study of palygorskite by DTA (Mifsud et al., 1978) has shown that there is no agreement between the water percentages calculated experimentally and those derived from the Bradley model. The coordination water is partially lost through the three typical stages of the curve, i.e., the elimination starts when the zeolitic water is lost and ends when dehydroxylation begins at about 400°C under pressure. Thus, it is impossible to assign exactly the water percentage calculated from the TG curve to the different types of water.

In the region of high temperature (>600°C), it is common to observe at about 800°C an endothermic peak immediately followed by an exothermic maximum. The endothermic effect represents dehydroxylation of the structure together with a change in entropy due to structure collapse. Dehydroxylation is confirmed by the weight loss shown on the TG curve. The exothermic peak appears to be due to the formation of clinoenstatite.

Usually weight loss in sepiolite at 800°C is slightly higher than the theoretical amount predicted by the model of Brauner and Preisinger (1956) -- about 2.7%. This may be because part of the SiOH of the edges have also been destroyed (Serna et al., 1975).

All the sepiolites yield a very sharp exothermic peak just above 800°C, but this peak is variable in size on palygorskite curves, which could be due to the variation in Mg:Al ratio. When this is smaller than one (Al-rich members), the temperature range of stability is very small (Kulbicki, 1959), whereas in Mg-rich members, the temperature range of stability of clinoestatite is very wide.

The total water content of these minerals is considerably below that required for a monolayer containing 0.28 mg H_2O per m^2 of surface covering, for example, 900 m^2/g in sepiolite, equivalent to about 26% water. For sepiolite, the total water is usually less than 20%, including 2.4-3% H_2O from dehydroxylation. For palygorskite the total water is 22%, including two at 6% from dehydroxylation (Weaver and Pollard, 1973). Therefore, Mifsud et al. (1978) concluded that there was usually a deficit of zeolitic water in palygorskites. It seems clear the packing of water on the surface of palygorskite and sepiolite is very much less dense than that in the "normal monolayer" formed for other layer silicates. This fact should be taken into account in calculating apparent specific surfaces from water sorption studies.

On heating, structural changes lead to decreases in the intensity of principal XRD peaks. So in sepiolite (Fig. 16), the reflections at 12, 4.5, 3.8, and 3.4 Å decrease after heating (at 250°C for 1 h) while new reflections appear at 10.4 and 8.2 Å. Further heating to 450°C increases these new reflections, which are persistent up to 700°C (Hayashi et al., 1969; Fernandez Alvarez, 1970; Nagata et al., 1974). In palygorskite the intensity of the reflections at 10.5, 4.5 and 3.23 Å decrease on heating and new peaks at 9.2 and 4.7 Å also appear. On heating at 325°C, these changes are more marked. Heating at 600°C completely eliminates the 10.5 Å reflection and the 9.2 Å peak also decreases in intensity (Hayashi et al., 1969) and shifts to 8.7 Å. At 700°C, palygorskite is practically amorphous.

The decreasing of first reflections occurs because structural disorder produced by heating is more prominent along the principal cleavage face (011) and less along the (040) plane (Lokanatha and Bhattacherjee, 1984). In palygorskite, the crystallite size slightly increases in the 200-300°C range with the expulsion of the channel water molecules, but crystallite sizes decrease markedly at 600°C when the anhydride stage is reached. Parameter a along the fiber axis increases, while the size decreases, until the structure collapses.

Regarding rehydration after heating at different temperatures, it is to be noted that palygorskite and sepiolite can rehydrate after the structure is folded; but it is very difficult if the anhydride sepiolite or palygorskite is reached. The new bonds originating in the anhydride structure resist rehydration. Nevertheless, if sepiolite was heated at 60°C and 100% r.h., rehydration occurred slowly over several weeks and at 100°C and 100% r.h., it was completed in 21 h (Fig. 17) (Serna et al., 1975).

OCCURRENCES AND ORIGINS

Locations and time distribution

In addition to the pioneering overview of chain-structure clay occurrence and significance given by Millot (1964) and the wide variety of contributions presented by Singer and Galan (1984), the most comprehensive reviews of sepiolite-palygorskite distribution and genetic implications have been given by Weaver and Beck (1977), Singer (1979), and Velde (1985). This section relies heavily on these references.

Both palygorskite and sepiolite have been identified in almost as wide a variety of environments as the more commonly recognized clay minerals, but seldom in as great an abundance. In fact, the chain-structure clays appear to be relatively rare. This situation may be in part due to failure to identify these minerals in very fine-grained mixtures, particularly because of the close correspondence of the principal X-ray diffraction maximum of palygorskite with that of illite, or that of sepiolite with alkali smectite. Another constraint is in the instability of these minerals under conditions of significant leaching, such as in weathering (Paquet and Millot, 1972). As noted by Velde (1985), the chain-structure clays have received more attention in recent work because of intensive studies of deep-sea deposits and an increased interest in soil mineralogy in semi-arid to arid areas. It also seems that the French have had a head start in recognizing these clays because of significant occurrences in their particular areas of interest, such as the Paris basin or North Africa.

Palygorskite has been almost the only chain-structure phase identified in truly deep-sea sediments. Indeed Velde (1985) asserts that Fleischer (1972) has reported the only authigenic occurrence of sepiolite in deep ocean floor sediments, but Couture (1977) cites the identification of sepiolite in volcanic ash in the northwest Pacific. Bowles et al. (1971) most effectively described authigenic marine palygorskite and sepiolite associated with active ridge and fracture zones in the Atlantic. Palygorskite predominates as the chain-structure species found in detrital sediments of shallow coastal marine basins and associated continental slope or margin deposits (Singer, 1979; Velde, 1985). Palygorskite and smaller amounts of sepiolite are apparently characteristic of the peri-marine, coastal lagoon environment adjacent to landmasses undergoing intensive chemical weathering, such as the Miocene of the southeastern U.S. (Weaver and Beck, 1977), or the earlier Tertiary of northwestern France (Esteoule-Choux, 1984) and southern Australia (Callen, 1977). Both phases can also be found in marine carbonates, phosphatic sediments, and salt deposits.

The chain-structure clays occur in continental deposits in three different primary modes but only in arid to semi-arid environments in any significant amounts. They have received most attention in calcareous soils, particularly paleosols and calcretes where these minerals take on considerable importance. Both phases can also be found in the alteration of igneous rocks and in lacustrine closed basin deposits.

It is of interest to note, as does Velde (1985), that the more common clay minerals can be completely absent from sepiolite-palygorskite assemblages. In many situations the chain-structure phases appear to be stably associated with smectites, but they also occur with kaolinite, serpentine, alkali zeolites, carbonates, sulfates, or other salts, and particularly with silica or chert. Perhaps the most controversial aspect in the occurrence of these

652

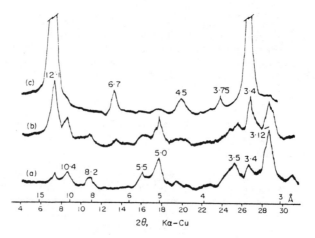

Figure 17. X-ray powder diffraction patterns of sepiolite anhydride as it alters to sepiolite upon rehydration (a) for 1/4 h in laboratory air, (b) 7 h, and (c) 21 h in 100% r.h. at 100°C.

Table 6. Hydrolysis reactions for clay minerals in the system magnesium-silica-water, and for the conversion of smectite to palygorskite. Stability constant values and references used in constructing the phase diagram (fig.18) are given in conjunction with the appropriate reaction.

Number of Equation	Reaction

1. Sepiolite
$Mg_2Si_3O_{7.5}(OH) \cdot 3H_2O + 4H^+ = 2Mg^{2+} + 3SiO_2(aq.) + 5.5H_2O$

   log K(crystalline) = 15.76; Stoessell, 1988.

   log K(amorphous) = 18.78; Wollast, et al., 1968.

2. kerolite
$Mg_3Si_4O_{10}(OH)_2 \cdot H_2O + 6H^+ = 3Mg^{2+} + 4SiO_2(aq.) + 5H_2O$

   log K = 25.79; Stoessell, 1988.

3. talc
$Mg_3Si_4O_{10}(OH)_2 + 6H^+ = 3Mg^{2+} + 4SiO_2(aq.) + 4H_2O$

   log K = 23.02; re-calculated from Bricker, et al., 1977.

4. chrysotile
$Mg_3Si_2O_5(OH)_4 + 6H^+ = 3Mg^{2+} + 2SiO_2(aq.) + 5H_2O$

   log K = 32.17; Hostetler and Christ, 1968.

5. palygorskite
$5X^+_{.48} Mg_{.92} Fe^{3+}_{.18} Al_{.93} Si_{3.90} O_{10}(OH) + 7.8H^+ + 12.2H_2O =$

   montmorillonite
   $3X^+_{.35} Mg_{.25} Fe_{.30} Al_{3.10} Si_{3.90} O_{10}(OH)_2 + 3.8Mg^{2+} + 7.8SiO_2(aq.) + 1.5X^+$

   log K = 22.43; re-calculated from Weaver and Beck, 1977.

unusual minerals is the primary question addressed by Singer (1979), that is, whether these phases owe their origin to transformation of precursor clay minerals or to precipitation directly from solution. It now seems inescapable that these clays commonly owe their formation to a dissolution-precipitation mechanism which incorporates components (primarily sesquioxide) of preexisting detrital material. It then also becomes apparent that these phases can contain a significant quantity of geochemical and hydrologic information about their formation environment.

The distribution of chain-structure clay deposits with time and space has been most extensively considered by Callen (1984). The major intervals of abundance are in the Late Cretaceous and Tertiary, irrespective of whether they occur as terrestrial or marine deposits. The average percentage of chain-structure clays in the total worldwide sediment accumulation since the Cretaceous appears to be relatively constant, regardless of the increase in total mass of sediments in the most recent epochs. The lack of these minerals in older sediments does suggest metastability on a geologic time scale, despite Velde's (1985) claim that there is no reason to believe they are unstable in all rocks of Paleozoic age. Based in part on paleogeographic reconstructions, Callen (1984) has shown a latitudinal concentration of palygorskite-sepiolite deposits falling between 30 and 40 degrees north or south. Obviously, this suggests a favorable sedimentary temperature effect on the processes leading to the formation of the chain-structure clays.

Synthesis and stability

The laboratory precipitation of sepiolite at room temperature and atmospheric pressure has been reported by Siffert and Wey (1962), Wollast et al. (1968), Couture (1977), and La Iglesia (1978). Wollast et al. (1968) obtained a poorly crystallized, but apparently single-phase precipitate, by additions of sodium metasilicate and NaOH to seawater. Though lacking the diagnostic 12 Å X-ray diffraction peak, the material was identified by chemical analysis and the infrared spectrum. Subsequently, Christ et al. (1973) reported equilibrium solubility data for a well-crystallized natural sepiolite at 51, 70, and 90°C from which a stability constant at 25°C could be estimated and compared with that derived from the solution data of Wollast et al. (1968). The reaction and appropriate equilibrium constants are given in Table 6. Christ et al. (1973) found a greater solubility for a poorly crystallized sepiolite at 51°C. indicating that poor crystallinity was the probable reason for the higher solubility obtained in the Wollast et al. (1968) study at 25°C. In the experiments of Siffert and Wey (1962), silica to magnesium ratios were fixed and pH varied. Sepiolite began to precipitate at the same pH of 8.5 as in the Wollast et al. (1968) experiments. At pH greater than 9, smectite and talc formed. Unfortunately, final concentrations of Mg and Si were not determined by Siffert and Wey (1962), so stability constants could not be derived. In the experiments of La Iglesia (1978), homogeneous precipitation of sepiolite sufficiently crystalline for reasonable X-ray diffraction and electron microscope characterization was achieved, but the formation of multi-phase precipitates prevented solubility determination. Calculation and three-dimensional plotting of the stability relations to be expected from the estimates of Christ et al. (1973) and Wollast et al. (1968) bracketed the trends in analytical data and crystallinity obtained.

Recently, Stoessell (1988) presented the results of sepiolite and kerolite (hydrous talc) dissolution experiments with five different solutions at 25°C for time periods approaching ten years. The results of sepiolite are in good agreement with those predicted by Christ et al. (1973) and confirm the excessive solubility of the Wollast et al. (1968) precipitate. No such previous experimental studies have been done with kerolite. The solubility difference between the two phases is relatively minor. Sepiolite is favored in solutions at equilibrium or supersaturated with respect to quartz, but it is metastable with respect to kerolite at lower silica concentrations. According to the data of Hostetler and Christ (1968) at such low silica contents, both phases are metastable with respect to serpentine (chrysotile).

From phase relations and field-based solute activity estimates, Jones (1986) had suggested that for typical natural waters kerolite rather than sepiolite would precipitate with increasing pH and solute Mg and decreasing aqueous silica content. Further, kerolite-stevensite interstratifications, and finally stevensitic smectite, were predicted to result from increasing sodium concentration in solution. Put another way, it is clear that higher concentrations of other constituents lower the activity of water in solution and will favor progressively lower solid hydration states, as in the sequence sepiolite-kerolite-stevensite. Taken all together, these considerations indicate regular phyllosilicates are preferred over chain-structure clays by highly alkaline or saline solutions. Finally, both Jones (1986) and Stoessell (1988) clearly point out that all the aforementioned phases are metastable with respect to talc at all silica concentrations in excess of 1 microgram per liter, emphasizing a distinct problem with reaction kinetics in the system.

With the addition of the aluminum in palygorskite, the situation becomes more complicated. Singer and Norrish (1974) derived a stability constant for the aqueous dissolution of an Australian pedogenic palygorskite, and Singer (1977) further examined Australian palygorskite dissolution rates and relative stability in weak acid, and relative stability as a function of aluminum content. The first study provided a reference value for further stability calculations by La Iglesia (1977) and Weaver and Beck (1977) employing the estimation scheme of Tardy and Garrels (1974) to obtain free energy of formation data for palygorskites of slightly different chemical composition. La Iglesia (1977) constructed a three-dimensional stability diagram within which to examine the evolution of solution composition with the homogeneous precipitation of palygorskite. Weaver and Beck (1977) used the free energies of formation derived for their Georgia Miocene clays to obtain an equilibrium constant for the transformation of montmorillonite to palygorskite, with conservation of alumina and iron. This relationship, when plotted on a magnesium-silica stability diagram (log activity Mg^{2+}pH vs. log activity aqueous silica for 25°C; Fig. 18) results in the intersection of the lower stability limits of sepiolite with those of kerolite and serpentine as well, at the same point (within analytical error). This point (log aMg^{2+}pH = 14.3; log $aSiO_2$ = -4.25) could well be taken as a locus for the composition of seawater free of siliceous organisms, and makes a strong argument for solute silica as the ultimate control of magnesium in the oceans.

Within the magnesia-silica-water activity diagram offered by Jones (1986), there was a secondary attempt to estimate stability fields for representative Al- and Fe-bearing smectites, chlorite, and kaolinite under conditions of sesquioxide conservancy and without consideration of palygorskite. Although other such phases were not specifically evaluated for Figure 18, aluminous smectite (plus solution) would occupy almost the entire lower part of the diagram. From this it is apparent how the addition of solute magnesium and silica to previously existing detrital smectite with increasing pH can be viewed effectively as transformation, as argued by Weaver and Beck (1977) and Jones (1986), even though the actual mechanism involves dissolution-precipitation. Indeed, incongruent dissolution and neoformation is required by the negligible sesquioxide solubility at optimal pH for palygorskite stability. In contrast, the energetics of solid-state alteration from the 2:1 smectite sheet to the chain-structure clay configuration is no doubt prohibitive, and higher temperature experiments readily accomplish the reverse (as discussed by Singer, 1979). Thus the only intermediate state is the mixed layer interstratification of kerolite-smectite 2:1 structures proposed by Eberl et al. (1982) and Jones and Weir (1983), probably formed through a kind of templating mechanism. Singer (1979) notes the lack of intermediate states in the alteration of smectite to palygorskite. Velde (1985) discusses very rapid dissolution and precipitation as a process passing through an aqueous stage, but having a very short residence time for the sparingly soluble species. He suggests that the control of the activity of the insoluble species is at the interface of the precursor mineral, which to us appears to be the same as the residual colloid of an incongruent dissolution reaction.

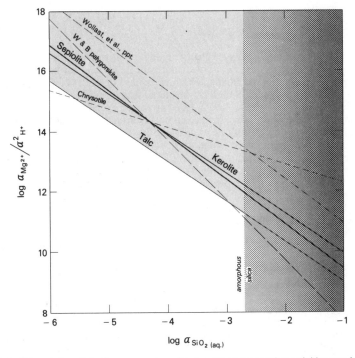

Figure 18. Stability relations for hydrous magnesium silicates with respect to solute activities, as calculated from references cited in the text. Shaded regions correspond to supersaturation with solid phases in a sesquioxide-free system, either Mg-silicate (light shade) or amorphous silica (dark shade). Field boundaries for sepiolite, kerolite, talc and chrysotile correspond to equations 1 through 4, respectively, in Table 6. The boundary for palygorskite corresponds to its formation from montmorillonite with conservation of Al and Fe in the solid, as given in equation 5, Table 7. The representative chemical compositions for these minerals are taken from Weaver and Beck (1977).

Environments

Marine. As pointed out by Singer (1979), views on the origin of chain-structure clays found in marine sediments deposited under conditions of normal seawater salinity and temperature have ranged between the extremes of (1) asserting, with extensive distributional detail, the detrital nature of all such deposits (Weaver and Beck, 1977) and (2) proposing that they all are diagenetic (Couture, 1977). The delicate appearance of these fibrous minerals originally argued against significant transport, but much evidence, particularly interspersal with other detrital clay phases such as illite and chlorite and the proximity of terrestrial sources, has indicated the likelihood of detrital origins for many oceanic occurrences. Palygorskite occurring in large quantities in the Late Cretaceous and the Early Tertiary of the continental African periphery provides the source for much material in the eastern Atlantic and Mediterranean. Similarly, the abundance of palygorskite in the northwest Indian Ocean can be related to sources in Arabia.

Although chain-structure clay is certainly widespread and most probably detrital in the oceans adjacent to landmasses in warm water areas, abyssal sediments in the central Pacific also contain significant deposits that Couture (1977, 1978) believes are authigenic in origin, particularly palygorskite, clinoptilolite, and chert in the Cretaceous and Paleocene, and smectite, philipsite and chert in the later Tertiary. As summarized by Singer (1979), Couture's (1977) arguments are based on compatibility of sediment interstitial fluid chemistry with palygorskite stability, association of palygorskite with high silica minerals such as chert, opal and clinoptilolite, and mobilization of silica. He proposes that possible precursors for diagenetic palygorskite formation could be volcanogenic material, or reaction of smectite or philipsite with biogenic silica.

Beck and Weaver (1978) argue effectively that local, sporadic or variable time distribution of oceanic palygorskite is not consistent with its general formation from normal seawater. However, all authors agree that a high silica environment and somewhat elevated temperatures will favor the precipitation of the chain-structure silicates from seawater. In this connection, Church and Velde (1979) have provided exchange ion and isotopic evidence that deep sea palygorskite has formed in part from detrital precursors at somewhat elevated temperatures. Perhaps both heat and slightly increased interstitial fluxes of aluminum or even magnesium might be added to the sediment-seawater interface by alteration of oceanic basalt below the sediments. Velde (1985) notes the low temperature alteration of basalt involves expulsion of the elements required to form both the chain-structure clays and zeolites. Formation becomes especially favorable with the addition of silica from dissolution of biogenic material. Direct diagenetic transformation of volcanic glass into palygorskite has been proposed by Gieskes et al. (1975) for basal sediments in the southwest Indian Ocean, and noted by Couture (1977) in the Pacific.

Weaver and Beck (1977) have presented the most extensive evidence for the formation of chain-structure clays in the shallow, lagoonal, "peri-marine" environment. Their highly detailed and well-documented studies of the palygorskite and sepiolite deposits of the Georgia-Florida Miocene indicated that these minerals were derived from the reaction of smectite with excess silica derived from siliceous organisms. As noted by Singer (1979), their contention was based on transmission and scanning electron microscope pictures, where palygorskite fibers appear to be forming from smectite. Aluminum and iron are conserved in the solid, whereas silica and magnesium are derived from the solution, as demonstrated in equation 5, Table 6. This can be considered an incongruent dissolution-precipitation process. The reaction can be combined with the alteration of dolomite or Mg-calcite to give pure calcite coexistent with palygorskite as follows:

$$Ca_{.25}(Mg_{.5}Al_{1.5})Si_4O_{10}(OH)_2 + 1.25CaMg(CO_3)_2 + 2SiO_2 =$$
montmorillonite

$$2Mg_{.87}Al_{.75}Si_3O_{7.5}(OH) + 1.5CaCO_3 + CO_2$$
palygorskite   calcite

This result is favored by the lower pCO_2 to be expected in the environment of a desiccating shallow lagoon of brackish water adjacent to a land mass undergoing intensive weathering. Under these general conditions, one can consider that the continent ultimately provides the silica, reactive sesquioxide colloids and pure H_2O, whereas the sea provides the magnesium and alkalinity. This scenario appears to fit the Tertiary deposits of many basins in France, as well as north Africa and south Australia.

Continental - soils, calcretes and alluvium. Singer (1978) has stated that soils, and especially paleosols, are the type of sediment in which palygorskite neoformation has been documented most convincingly. At the same time, however, Paquet and Millot (1972) have indicated that the chain-structure clays are unstable and that they weather to smectite when the mean annual rainfall exceeds 300 mm. Detailed study of this degradation in soil profiles over a wide area of west Texas was presented by Bigham et al. (1980). Thus the most extensive palygorskite-sepiolite occurrences in soils are reported from semi-arid to arid environments, often representing earlier, drier conditions.

Singer (1984) indicates that significant amounts of pedogenic palygorskite or sepiolite are usually associated with one of three specific situations in the continental environment. The first situation is modern soils that are now or have in the past been affected by rising ground water. These materials are commonly found on floodplains or low terraces, typically involving fine-textured calcareous alluvial or aeolian sediments. Calcite is ubiquitous, the pH is 7.5-8.0, and salinity is significant. Gypsum crystals are often present and can act as substrate for the chain-structure clay. The second situation involves soil strata that include sharp and distinct textural transitions, such as many paleosols. Accumulation occurs on the coarse-grained side of the size transition. The palygorskite will be found with secondary calcite, even in materials containing Mg-calcite or dolomite, recalling the smectite-dolomite, palygorskite-calcite antithesis of Weaver and Beck (1977) in the southeastern U.S. Miocene peri-marine deposits. The third situation is in the formation of calcretes, crusts and caliches.

The best documented pedogenic deposits of chain-structure clays are found in association with well-developed caliche or calcrete paleosols, which are often very extensive. The classic description of such material is that of Vanden Heuvel (1966) for a site near Las Cruces, New Mexico, where both chain-structure species were identified, palygorskite predominating. These studies were extended elsewhere in central and eastern New Mexico by Bachman and Machette (1977), who believed the abundance of sepiolite increased with age within calcretized geomorphic surfaces. Velde (1985) devotes considerable discussion to this occurrence as representative of the close association of chain-structure clay with carbonate formed by leaching and recementation through capillarity and evaporation. This is the process termed calcareous epigenesis by Millot et al. (1977). Chain-structure clay is characteristic of the carbonate soil zone (palygorskite can compose 100% of the clay-size fraction), whereas smectite dominates the clays below the calcareous layers. In southern Australia, Singer and Norrish (1974) found no precursor minerals, such as smectite or volcanic glass, in the soil material of paleosols with palygorskite now developed over the soil peds, thus indicating chain-structure clay precipitation directly from the soil solution. They also analyzed soil-water extracts giving a composition compatible with the stability requirements of the mineral. Others have noted the inverse relation between chain-structure clay abundance and that of smectite, suggesting the former developed at the expense of the latter (Paquet, 1970; Yaalon and Wieder, 1976; Watts, 1976, 1980). Singer (1984) points out that the frequent association of these two different clay mineral species does not necessarily imply a solid-phase transformation, and indeed no morphological evidence has been produced to substantiate it. As mentioned earlier, however, the relative insolubility of the aluminum and small amounts of iron incorporated in the palygorskite under the conditions of calcareous soils or crust formation argues very strongly for some kind of incongruent dissolution-precipitation process. Watts (1980) has indicated that the source of the magnesium for chain-structure clay formation in the Kalahari calcretes is in the recrystallization of high Mg-calcite to low Mg-calcite, with silica replaced by calcite precipitation. Millot et al. (1977) suggest that soil pH can get high enough to dissolve aluminosilicates congruently at the time of calcite epigenesis.

Martin de Vidales et al. (1986) propose the pedogenetic formation of palygorskite in paleosols developed on terraces of the Tajo River in Spain by the dissolution of montmorillonite concommitantly with development of calcic horizons. They suggest the process is aided by temporally wet conditions and recrystallization of higher Mg carbonate. While pointing out the difficulties in recognizing transitional morphologies, they note the presence of suspicious glomerules associated with cotton-wool-type remains of montmorillonite and abundant fibrous laths of palygorskite.

Bachman and Machette (1977) have underscored the widespread association of the chain-structure clays with calcic soils, pedogenic calcretes, and other surficial carbonates of the semi-arid southwestern United States. They showed that the youngest palygorskite occurs in upper Pleistocene soils and it becomes the predominant clay in middle Pleistocene and older soils of New Mexico. Sepiolite was noted only where palygorskite was dominant and smectite relatively depleted. In contrast, Hay and Wiggins (1980) and Jones (1983) found that sepiolite can be virtually the sole silicate component of calcrete in southwestern Nevada and vicinity, and is not restricted to deposits only where palygorskite is dominant, as reported by Bachman and Machette (1977) for other parts of the southwestern United States. Jones (1983) indicated that abundant solute magnesium and silica derived from dolomitic bedrock of Paleozoic age and siliceous volcanics of Tertiary age favored the formation of sepiolite over palygorskite. In surficial carbonate deposits developed on the siliceous volcanics of southwest Nevada, chain-structure clays are missing or poorly developed, suggesting that formation of intergrades with pre-existing smectite is interfering with sepiolite or palygorskite formation.

The most recent and detailed investigation of the mineralogy of the extensive indurated calcic deposits of the Llano Estacado of west Texas has been by McGrath (1984). He demonstrated that the dominant chain-structure clay was palygorskite and that the amount generally increased with the increase of calcite induration, as did the subordinate opal and sepiolite content. Further, McGrath noted that the greater the pedodiagenetic development, the less feldspar survived in the parent material. Actually, the highest concentration of palygorskite was in some of the vertical joints in the indurated deposits. Samples with dolomite were considered abnormal, and the youngest calcretes had no chain-structure clay in them at all. Sepiolite dominance, in the one area where it occurred, was referenced to the studies of Hay and Wiggins (1980), who found, as did Jones (1983), that this feature can be most readily attributed to the paucity of solute or colloidal alumina derived from suitable parent material. McGrath (1984) concluded that the principal mechanism of chain-structure clay formation was by direct precipitation, and specifically rejected the topotactic transformation ideas of Parry and Reeves (1968) or Bachman and Machette (1977). He supported the two principal controls on the formation of palygorskite and sepiolite identified by Jones (1983): (1) direct precipitation in surficial deposits through the same evaporative concentration of vadose solutions which promotes induration of calcic soils, and (2) that with the parent material of most calcic horizons, palygorskite is favored initially, and sepiolite only forms from solution on absence or complete immobilization of solute and colloidal aluminum.

Velde (1985) took the mineralogical detail of Vanden Heuvel's soil profile as a basis for examining phase relations for the clays identified therein. He maintained that, though the soil zone as a whole is obviously an open system, single horizons produce assemblages typical of closed systems where local equilibrium is attained. However, to us it is more straightforward to suggest that the chain-structure clays are formed at the expense of the detrital smectite, kaolinite, and perhaps even mica, as controlled by local hydrology. In fact, the profile shown by Velde (1985) illustrates the complete disappearance of kaolin in the zone of maximum sepiolite abundance. This indicates that a reduction in permeability by carbonate precipitation gradually halts the dissolution of detritus and colloidal transport, immobilizing the alumina necessary for palygorskite formation. With further evaporative solution concentration, this leaves only solute magnesium and silica to form sepiolite. Such a scenario for the literal "closing" of the system suggests that the clay mineral sequence in pedogenic calcrete can be at least semi-quantitatively related to the hydrologic properties of the sediment, and the kinetics of carbonate precipitation relative to silicate diagenesis.

Bachman and Machette (1977) considered sepiolite in soils to be a late-stage product of pedogenesis and noted its occurrence in the palygorskite-dominated medial portion of the profile in soils of middle Pleistocene to Pliocene age. They also noted, as have many of the aforementioned authors, that the younger soil horizons overlying well-developed pedogenic calcretes contain no recognizable palygorskite or sepiolite. Thus it appears that there is a distinct kinetic problem in the formation of the chain-structure species. In fact, Bachman and Machette (1977) suggested that sepiolite less than 200,000 years old does not occur in the calcretes of New Mexico. Paquet (1983) has asserted that palygorskite in calcretes does not remain stable and is ultimately replaced by calcite. In the uppermost leached horizons or under more humid conditions, palygorskite tends to transform into smectites or to disappear. In contrast, Fassi et al. (1988) have demonstrated the formation of palygorskite in a silicate residue resulting from organic decay-induced carbonate dissolution at the base of lacustrine limestone.

Continental - lacustrine. In some of his early studies of the Tertiary basins of north and west Africa, Millot (1964) worked out the classic clay mineral chemical sequences in closed-basin deposits. The most aluminous and iron-rich phases are at the periphery; they are succeeded basinward by more siliceous and magnesian clays; sepiolite was found at the center. Similarly, Garrels and Mackenzie (1967) presented a calculative model for the development of alkaline, saline (so called "soda") lakes by evaporative concentration of drainage from crystalline rock terranes. They proposed the direct precipitation of sepiolite from lake waters as the principal control on the concentration of silica and magnesium in solution. Subsequent studies of the geochemical evolution of waters in modern continental closed basins have further explored the reaction of solutes with silica and silicates (Jones and VanDenburgh, 1966; Eugster and Jones, 1979; Al-Droubi et al., 1980). Investigation of recent lacustrine sediments at Lake Chad in central Africa (Carmouze et al., 1977; Gac et al., 1977; Pedro et al., 1978), Lake Abert, Oregon (Jones and Weir, 1983) and Great Salt Lake, Utah (Spencer, 1982; Jones and Spencer, 1985) have confirmed the sequence of more magnesian and siliceous clays basinward, but failed to identify sepiolite. Also, no chain-structure clay has been recognized in any of the lacustrine sediments in modern saline lakes and playas of the western United States (Droste, 1961; Guven and Kerr, 1966; Jones et al., 1972).

In contrast, sepiolite and palygorskite have been found by McLean et al. (1972) to be common, sometimes the major, clay minerals in calcareous lacustrine deposits on the southern High Plains in west Texas and eastern New Mexico. Dolomite is usually associated with sepiolite and calcite with palygorskite. Volcanic ash was given as the source of the silica. Volcanic ash was also given as the source of silica for the sepiolite at the margins of the Pleistocene Lake Tecopa deposits in southeastern California by Blackmon and Starkey (1984), but it is possible in this case that both the Si and Mg, and perhaps some of the fibers themselves, were derived from trioctahedral smectite detritus carried into the lake by the ancestral Amargosa River from the Amargosa playa sepiolite-smectite deposits several miles upstream. These deposits have been described in most detail by Hay et al. (1986).

Some of the best-developed accumulations of sepiolite and palygorskite are in the intermontane continental Tertiary strata of Spain. Probably the most volumetrically and economically significant of these are from the lacustrine environment, particularly the Miocene of the Madrid (Tajo) basin. Galan and Castillo (1984) group the Spanish deposits based on geological setting and mineralogy into four fundamental types. The first is the Tajo: sepiolite is most prominent in distal alluvial fan deposits, but it is also present in perennial lacustrine sediments subordinate to trioctahedral smectite. Palygorskite is associated only with older arkosic deposits near the borders of the basin. Other precipitate minerals associated with the chain-structure clays include carbonate, chert, and a trace of zeolite. The second type of deposit occurs in the small tectonic Torrejon basin, where palygorskite apparently formed from slaty chlorite by dissolution-precipitation. The third type of occurrence is represented by palygorskite as a diagenetic cement in sandstone and conglomerates of the southern margin of the Duero basin at Segovia or Salamanca, and at the south of the Tajo at San Martin de Pusa, Toledo. The fourth type of occurrence is the Lebrija deposit near Cadiz which represents the brackish lacustrine (or peri-marine) environment in transition from marine to continental during the Pliocene. Sepiolite occurs with carbonate in the lower part of the deposit, whereas higher in the section palygorskite is present in distinct units with more detrital input. This sequence is similar in gross aspect to that described by Fontes et al. (1967) for the Paris basin.

Study of the areal and vertical distribution of clay minerals in the Tajo basin sediments is helpful in unraveling the principal controls and implications of the chain-structure clay occurrences. The sediment facies distribution follows the classic concentric scheme of the Millot (1964) model and occurs in successive belts from the border of the basin toward the center (Galan and Castillo, 1984; Doval et al., 1986). In contrast with the Millot model, most sepiolite does not occur at the center of the basin, but in rather marginal paleogeographical positions related to the old alluvial fan system (Calvo et al., 1986). Major facies distribution is as follows:

— Arkosic facies (fringing alluvial fan areas): the alluvial clay is dominated by dioctahedral smectite, with minor illite and kaolinite; sepiolite occurs both in calcrete within the alluvium and in deposits representing marginal ponds associated with fan toes, which contain only minor aluminous smectite.

— Transitional facies (mud flat-marginal lake complex): trioctahedral Mg-smectite dominates in this facies, made up of interbedded green shales and carbonates, with some sepiolite, and frequently, pink clays composed of interstratified kerolite-smectites.

— Central facies (saline lacustrine environment): evaporite deposits (mainly anhydrite, gypsum, and magnesite) typically contain illite, trioctahedral smectite and minor kaolinite.

The clay mineral associations noted above are consistent with a closed basin geomorphic and hydrologic setting where crystalline rocks of bordering ranges were the primary source for detritus and weathering-derived solute silica. At the same time, fluctuating lacustrine brackish to saline waters were the principal contributors of reactive magnesium, probably derived from the recycling of Mesozoic marine evaporite strata.

It should be apparent from the considerations presented above that marginal or transitional marine basins might very well possess all the essential features given for the Tajo basin saline lake system. Thus, the French and Morrocan basins and clay mineral suites investigated by Trauth (1977) seem to share many of the essential features of the closed lacustrine Tajo system. Indeed, it should be quite difficult to separate the effects of a modified marine evaporite solute matrix from a normal marine milieu. Similarly, the sedimentological and geomorphic characteristics described for the Miocene Tajo basin can very well fit the modern Great Salt Lake area.

Paquet et al. (1987) have documented a considerable overlap in the compositions of individual particles in chain-structure clay-smectite assemblages. What controls the type of clay minerals obtained in a given deposit? The modern smectitic saline lake sediments cited earlier are all from closed basins maintained by surface water inflow providing a source of colloidal detritus. The clay formulas given by Carmouze (1976) were obtained by subtracting the Chari River smectite composition from the analysis of the Lake Chad bottom sediments. Likewise, Jones and Spencer (1985) indicate that the Great Salt Lake clay chemistry can be achieved simply by the addition of magnesium silicate to average inflow river clay composition. Velde (1985) interpreted Trauth's (1977) smectite trace element data as indicating a dominant detrital component in the stevensite precipitate phase. He suggested this resulted from a dissolution-precipitation mechanism fixing magnesium and releasing alkalis. The same effect can be achieved by simple addition of kerolite interstratification to detrital smectite, as proposed by Jones and Weir (1983) to explain the clay compositional evolution at saline Lake Abert, Oregon.

In contrast, sepiolitic clays at the margins of a pan in the Kalahari Desert, southwest Africa (Kautz and Porada, 1976), in the Amargosa Desert, Nevada (Eberl et al., 1982), and in the Amboseli basin, Kenya (Stoessell and Hay, 1978; Hay and Stoessell, 1984) were formed under conditions of sustained ground-water inflow, which also provided a minimum of waterborne detritus. The final product at Amargosa and Amboseli was an almost pure trioctahedral 2:1 layer phase (kerolite), in addition to sepiolite. Although good textural evidence exists for the coalescent nucleation of sepiolite (Khoury, 1979), trioctahedral 2:1 phyllosilicates seems to require pre-existing sheet structure substrate, possibly because the simple phyllosilicate structure does not as readily accommodate water of hydration (see Guven and Carney, 1979). Equally important, of course, is the presence of excess alkali.

It is not known what concentration of sodium will destabilize sepiolite, but field examples, such as in the Amargosa playa, would suggest a minimum solute Na^+ concentration of one millimolar.

Hay et al. (1986) have indicated that the Mg clays of the Amargosa Desert were deposited in playas and associated marshland, with limestone and dolomite seepage mounds and caliche-breccia masses formed along zones of ground-water seepage providing the bulk of the inflow. Sepiolite was formed in waters of relatively low salinity, whereas Mg smectite was a product of higher salinity. This finding was supported by lower oxygen-18 contents in the sepiolite as compared to the Mg smectites. Sepiolite and Mg smectite disseminated in the carbonate deposits fall outside the oxygen isotopic limits of the purer clays, probably because of diagenetic recrystallization. However, these results suggest a great deal of hydrologic information resides in the isotopic compositions of such materials.

Association with igneous rocks. A number of occurrences of the association of palygorskite and sepiolite with veins or similar alteration of igneous rocks have been cited by Singer (1979) or Velde (1985), and not uncommonly attributed by the original authors to hydrothermal activity. However, Velde (1985) refers only to low temperature and Singer (1979) notes that it is often not possible to specify whether the chain-structure clay formed as a replacement of a pre-existing mineral or by precipitation from a solution obtained by alternation of the rock. We would add that it is often equally impossible to specify the temperature of formation, but would assert that no good case has been presented for such formation under true hydrothermal (as opposed to epigenetic) conditions. Indeed, Guven and Carney (1979) have demonstrated the transformtion of sepiolite to stevensite experimentally under hydrothermal conditions, and Yang and Xu (1987) have documented the transformation of Permian sepiolite to talc and stevensite at sub-hydrothermal temperatures determined by vitrinite reflectance of associated coal.

Summary

Velde (1985) has offered a low temperature (80°C) phase diagram for the hydrous system silicon-magnesium-aluminum (plus ferric iron) based on higher temperature mineral synthesis and analyses of natural phases by Trauth (1977). His diagram shows distinct fields for dioctahedral smectite, palygorskite and sepiolite covering a continuous compositional range up to two-thirds trivalent octahedral occupancy, and saponite, with tie lines suggesting stable coexistence of only one smectite phase with a chain-structure species. On similar trilinear diagrams, Figure 19 shows the distribution of chemical compositions for samples from key references discussed herein, as well as analyses of palygorskite, sepiolite and saponite compiled by Weaver and Pollard (1973). In contrast to the diagram of Velde, there is a substantial overlap of compositions between chain-structure clays and siliceous smectites with average octahedral occupancies significantly exceeding two (stevensites). This is not only suggestive of interstratification in the smectites, but of the importance of factors other than simple major element composition in determining the phase expected in a given natural environment.

The results of the 145 analyses presented by Paquet et al. (1987) illustrate important correlations with the genetic environment. Examination of the octahedral composition of chain-structure and smectite pairs from paleocalcrete, basin margin, and evaporative central basin locations in southeastern France reveals that the most magnesian palygorskite is found in calcrete, whereas the most aluminous stevensite is from the evaporatite association. There was a small but definite compositional gap found between the palygorskite and sepiolite population which is at a somewhat higher ratio of magnesium to octahedral trivalent ion than previously suggested. Over the entire range of octahedral composition, the average chain-structure clays have a very slightly lower value of this ratio than the smectites, whereas sepiolite was the purest magnesium silicate analyzed. These results strongly support the roles of salinity and hydrologic environment in determining the nature of the

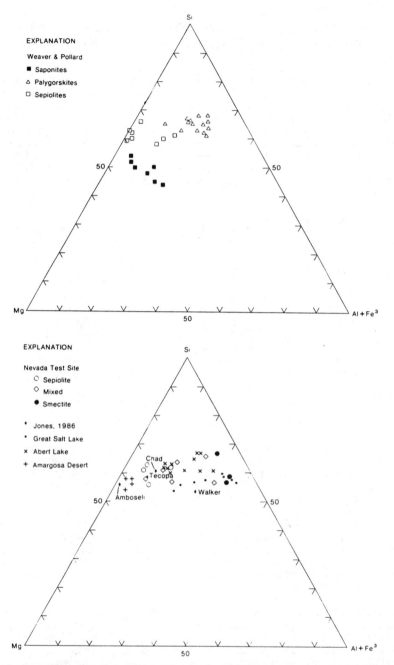

Figure 19. Chemical composition of natural sepiolites, palygorskites and magnesium smectites, as taken from (a) Weaver and Pollard (1973), or (b) other sources discussed in the text; named points are from Jones (1986, Table 1); Green River formation smectites cover the compositional range from that of Lake Tecopa to the central point of the L. Abert smectites (Jones and Weir, 1983). Amargosa Basin and Nevada Test Site points are taken from Eberl et al. (1982) and Jones (1985), respectively. Great Salt Lake smectite points are unpublished analyses referred to by Jones and Spencer (1985). Except for the siliceous palygorskite (impure?) from Salinelles, the range of compositions shown here completely overlap those given by Trauth (1977), or in Velde (1985). Similarly, the palygorskite-sepiolite compositions given in Table 3 (analyses 1-7) fall completely within the range presented in the two diagrams.

magnesian clay phase found in sediments. Thus palygorskite predominates in pedogenic calcretes and aquifer cements where the interstitial solution is recharged by dilute waters which never accumulate as much solute even on evaporative concentration as do basin sinks. Sepiolite in calcrete apparently forms only when reactive (probably largely colloidal) alumina is unavailable; it appears in association with closed basin solutions and sediments only when solute alkali is too low to form smectite and pH levels are not extremely alkaline. Its formation may be inhibited by the pre-existence of simple phyllosilicate or other sheet structures, as with waters with high sediment content or sediments with insufficient pore size. However, its water-accommodating structure is more suitable to direct precipitation from solution.

Weaver and Beck (1977) insisted that palygorskite formation in the marginal marine environment required less than normal salinity, as well as somewhat elevated levels of solute silica. In lacustrine closed basins or marginal continental basins transitional to marine, the occurrence of palygorskite indeed appears to indicate brackish water, and the presence of sepiolite reflects both salinity increases and the lack of abundant surficial sediment inflow.

A summary of environmental characteristics and associations of magnesian clays is given in Table 7.

APPLICATIONS

Palygorskite and sepiolite have a wide variety of industrial applications. Colloidal grades of these minerals are well-known thickening, gelling, stabilizing, and thyxotropic agents in products as diverse as paints or rubber, and drilling muds. Sorptive grades find use as decolorizing agents, animal waste absorbents, pesticide carriers, catalysts, refining aids, etc.

In comparison with other clays used in the chemical processing industries, palygorskite and sepiolite have some particularly desirable sorptive, colloidal-rheological, and catalytic properties. Slips or slurries of these minerals are viscous and thyxotropic, but unlike bentonite, they are not flocculated by electrolytes. Properties and applications of palygorskite have been reviewed by Grim (1962), Haden (1963), Ovcharenko (1964), Haden and Schwint (1967), and those of sepiolite, by Robertson (1957), Grim (1962), Alvarez (1984), Galan (1987), among others.

The physiocochemical properties of these minerals are the basis for their technological applications. Such properties are closely linked to mineral fabric, surface area, porosity, crystal morphology, structure and composition. These properties can vary with thermal, acid and mechanical treatments. Such treatments can lead to loss of different types of water, to alteration of the surface characteristics and porosity, to partial destruction of structure, etc. In these processes, some of the most useful mineral properties can be enhanced.

Sorptive applications

Palygorskite and sepiolite in their natural form are highly adsorbent clays and can retain up to 200% and 250%, respectively, of their own weight in water. The optimum sorption of water and other small polar molecules occurs when the minerals are heated between 200°C and 400°C (Fig. 18). The increase in sorptivity may arise from the liberation of zeolitic water from the channels, although these sites are not available equally to all molecules.

The order of sorptivity of different compounds on palygorskite is as follows: water > alcohols > acids > aldehydes > ketones > n-olefins > neutral esters > aromatics > cycloparaffins > paraffin (Haden and Schwint, 1967). Straight-chain hydrocarbons are

Table 7.

Environmental Characteristics and Association of Magnesian Clays

Association & Extent		Minerals		
Chemistry		Palygorskite	Sepiolite	Trioctahedral Smectite
pH and alkalinity	moderate, pH <8.5	*	+	—
	intermediate pH=8-9.5	+	*	+
	high, pH >9.5	—	—	*
major constituent ratios	high Mg+Si/Al	+	*	—
	high Mg+Fe/Si	—	—	*
sediment-water P_{CO_2}	high	—	—	*
	low	*	*	—
alkali salinity	high	—	—	*
	intermediate	+	*	+
	moderate	*	+	—
Environment				
pedogenic calcrete, or alluvial	siliciclastic or arkosic matrix	*	+	—
	carbonate or mafic matrix	+	*	—
closed basin lacustrine	groundwater input dominant	—	*	+
	surface runoff dominant	?	—	*
	hypersaline	—	—	*
marine	lagoon or tidal	*	+	—
	deep sea	+	—	+

Symbols: * favored
          + less favored
          — unfavored

TABLE 8

PROPERTIES OF ABSORBENT SEPIOLITE GRANULES, VALLECAS, SPAIN

	6/30 *	30/60 **
Bulk density (g /l)	560	615
Water absorption (%) (Ford test)	94	120
Oil absorption (%) (Ford test)	58	95
Shell index (Kg/cm^2) ***	3.9	—
Moisture (%)	12 ± 3	12 ± 3

  * 6-30 mesh

 ** 30-60 mesh

*** Mechanical strength index

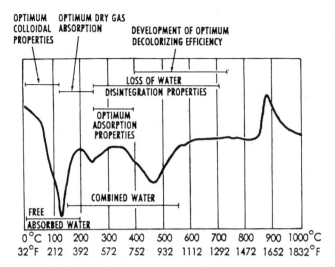

Figure 20. Development of optimal properties by heating of palygorskite as related to the specific thermal behavior (DTA curve) (after Haden and Schwint, 1967).

more readily sorbed than those with branched chains (Ovcharenko, 1964). The selectivity disappears when palygorskite is heated above 100°C (Barrer et al., 1954).

In general, the selectivity of fibrous clay minerals for sorption of organic compounds can be used for separation of mixed components. This selectivity in adsorption plays an important role in the decoloration of oils, and in separation processes, such as the refining of crude petroleum oils by selective adsorption of light fractions, or in cigarette filters, where the minerals adsorb selectively nitriles, acetone, and other dangerous polar gaseous hydrocarbons.

Sepiolite and palygorskite are effective decolorizing agents (Gonzalez Garcia and Peiro, 1963; Huertas, 1969; Huertas et al., 1979). They physically retain colored particles during filtration or percolation, and adsorb colored compounds and convert them catalytically into colorless substances. Clays activated above 200°C, and especially between 400 and 700°C, show an increased decolorizing activity (Fig. 20) which is mainly due to the increase in oxidation power of the surface and changes in porosity. Also these clays are substantially "activated" by extrusion under high pressure at low moisture content prior to heat treatment. Extrusion breaks up and tears apart the bundles of needles of these minerals and increases the pore volume and surface area.

Because of their sorptive properties, sepiolite and palygorskite can also be used as pesticide carriers, filter aids for sugar refining, water treatment and other industrial processes, cleaning products to remove water, grease, oil, dirt, dust, and odors from floors, factories, farms, garages, butcher shops, etc. Also, they are widely used as sorptive litter and bedding for animals (Table 8).

In pharmaceutical applications, these minerals can be used as excipients on which active products can be retained and released at suitable rates (Hermosin et al., 1981; Forteza, 1987). They can also be used as adsorbents for toxins, bacteria, and liquid in the gastrointestinal tract (Martindale, 1982).

Sepiolite has been investigated for animal nutrition applications because of its sorptive, free-flowing anticaking, and atoxic properties and its chemical inertness. The principal opportunities in this field appear to be in formulations for growth promotion, delivery of nutritional supplements, carrying, and feed binding.

666

Organo-mineral derivatives

This term includes compounds in which true covalent bonds are formed between the mineral substrate and the molecules of the organic reactant. Sepiolite and palygorskite contain large amounts of external Si-OH groups and react directly with organic reagents either in vapor phase or in solutions of inert organic solvents. In either case, the mineral framework is essentially preserved.

Grafting reactions with sepiolite have been achieved through Si-O-Si bonds, when the mineral reacts with organo-chlorosilanes (Ruiz-Hitzky and Fripiat, 1976), through Si-O-C bonds, by additions of alkyl or phenyl-isocyanates (Fernandez-Hernandez and Ruiz-Hiztky, 1979), by reaction with epoxides (Casal Piga and Ruiz-Hitzky, 1977) or by reaction with diazomethane (Hermosin and Cornejo, 1986).

Catalytic application

Because of their high surface area, mechanical strength, and thermal stability, sepiolite and palygorskite granules are increasingly used as catalyst carriers. The catalyst can be impregnated on the surface by treatments with salt of another metal and can also bring about substitution for part of the structural magnesium cations. Potential uses can be increased if Mg^{2+} ions are exchanged for catalytically important species such as Ni^{2+}, Pt^{n+}, Pd_m^{n+}, Zn^{2+}, Co^{2+}, Cu^{2+}, Ba^{2+}; La (different Japanese patents, Japan Kokai, 1978; Corma et al., 1985). Different catalytic processes, including dehydration, hydrogenation and dehydrogenation, desulfuration, denitrogenation, the production of butadiene from ethanol, and of hydrocarbons from methanol, and various hydrocracking processes have been achieved. Palygorskite is also used in carbonless copy paper.

The catalytic activity of clay minerals is primarily a function of their surface activity. Silanol groups present on the surface of sepiolite and palygorskite have a certain degree of acidity and can act as catalysts or reaction sites. Minerals are used in natural and in thermally or acid-modified forms. Heating at temperatures of 200-300°C lowers the surface area but increases the strength of the granules, which is an important property in fluid bed hydrocracking.

Rheological applications

The principal applications of palygorskite-sepiolite colloidal grades generally involve thickening, gelling, stabilizing, or other modifications of rheological properties. These minerals are used in paints, especially where highly thyxotropic properties are advantageous, in adhesives, sealants, fertilizer suspensions, and cosmetics (i.e., milks, masks), and as fluid carriers for pregerminated seeds.

They are also used in drilling muds. Their advantage over other clays, such as bentonite, is that palygorskite or sepiolite-based muds are less sensitive to salt, i.e., the desired rheological properties remain relatively constant despite high electrolyte concentration over a wide pH range below pH 8. At pH > 9, peptization causes a sharp decrease in viscosity, and the rheological behavior becomes Newtonian. Sepiolite has a "mud yield" of >150 bbl/ton in saturated salt water (Alvarez, 1984) and palygorskite of 100-125 bbl/ton (Haden and Schwint, 1967). Additives, such as magnesium oxide (alone or with lime) or barium oxide, increase the mud yield and the inertness to soluble salts.

Palygorskite and sepiolite are efficient thickeners for aqueous and organic liquids. They are particularly suited for high temperature uses because of their insolubility. They also have considerable binding power, which finds applications in oil-bonded foundry sands, molecular sieves, etc. After surface modification with organic compounds, they can be used as a reinforcing filler in rubber and foam.

New applications

Sepiolite and palygorskite are currently used in more than 100 different applications. Most of the applications are similar to those of the other clays more traditionally employed. New commercial products are being investigated, especially those that make use of organophillic materials in paints, plastics, greases, rubbers, plastisols, and cosmetics. The most important new areas are those involving catalysis, agriculture, and environmental protection.

According to Alvarez et al. (1985), modified sepiolite could be used to form a stable complex with type-I collagen, thereby allowing monomeric collagen to be isolated for the production of biomaterials capable of substituting wholly or in part for organic-supported tissues. Sepiolite may also find application as a noncarcinogenic substitute for asbestos (e.g., in asbestos-free cement products; Castell et al., 1987).

The effect of sepiolite and palygorskite on lysis of erythrocytes has been demonstrated by Oscarson et al. (1986). These authors have found that the activity of various silicates, such as lysing agents, is as follows: smectite > silica > palygorskite) sepiolite > chrysotile > kaolinite. Different chemical compositions of these minerals did not appreciably alter their hemolytic activity. Structural folding of fibrous mineral reduced lysis suggesting that edge surfaces and silanol groups are important in this process; on the other hand instead, the fibrous morphology did not appear to be relevant. Also, they found that when mineralogical surfaces became saturated with cellular components (usually after 1 h), minerals lost their lytic activity. Thus, it seems the lysis of erythrocytes by these minerals is mainly related to their surface physicochemical properties rather than to their particle shape.

In summary, there are numerous applications for possibilities of these industrial minerals, especially in the chemical industry in their natural form and as customized derivatives, and new applications are continually being discovered.

ACKNOWLEDGMENTS

We wish to thank Howard May and S.W. Bailey for review of this effort, and D.D. Eberl, J.P. Calvo, J.M. Serratosa, C. Serna, and especially H. Paquet for very helpful references. We are greatly indebted to C. Lee and C. Moss for invaluable aid with manuscript preparation and illustrations, respectively.

REFERENCES
Note: Clays and Clay Minerals = CCM

Ahlrichs, J.L., Serna, C. and Serratosa, J.M. (1975) Structural hydroxyls in sepiolites. CCM, 23, 119-124.
Al Droubi, A., Fritz, B., Gac, J.Y., and Tardy, Y. (1980) Generalized residual alkalinity concept: Application to prediction of the chemical evolution of natural waters by evaporation. Am. J. Sci., 280, 560-572.
Alvarez, A. (1984) Sepiolite: Properties and uses. In: A. Singer and E. Galan, eds. Palygorskite-Sepiolite. Occurrences, Genesis and Uses. Developments in Sedimentology 37, Elsevier, Amsterdam, 253-287.
_____, Castillo, A., Perez Castell, R., Santaren, J. and Sastre, J.L. (1985) New trends in the use of sepiolite. Proc. Abstrs, Int'l Clay Conf., Denver, 1985, p. 7.
Arnold, D. (1971) Ethnomineralogy of Ticul Yucatan potters. Am. Antiquity, 36, 20-40.
Bachman, G.O. and Machette, M.N. (1977) Calcic soils and calcretes in the southwestern United States. U.S. Geological Survey Open-File Rept. 77-794, 163 p.
Bailey, S.W. (1980) Structure of layer silicates. In: G.W. Brindley and G. Brown, eds., Crystal Structures of Clay Minerals and their X-ray Identification. Mineral. Soc., London 1-123
_____, Brindley, G.W., Johns, W.D., Martin, R.T. and Ross, M. (1971) Clay Minerals Soc. Report of Nomenclature Committee 1969-1970. CCM, 19, 132-133.

668

Barrer, R.M., MacKenzie, N. and MacLeod, D.M. (1954) Sorption by attapulgite: Part II. Selectivity shown by attapulgite, sepiolite and montmorillonite for n-paraffins. J. Phys. Chem., 58, 568-572.

Bates, T.F. (1958) Selected electron micrographs of clays, 45, 46. Circular no. 51, Mineral Industries Experiment Station, State Univ. Pennsylvania.

Beck, K.C. and Weaver, C.E. (1978) Miocene of the S.E. United States, a model for chemical sedimentation in a peri-marine environment-A reply. Sediment. Geol., 21, 154-157.

Bigham, J.M., Jaynes, W.T. and Allen, B.L. (1980) Pedogenic alteration of sepiolite and palygorskite on the Texas High Plains. Soil Sci. Soc. Am. J., 44, 159-167.

Bøggild, O.B. (1951) Gunnbjarnite, a new mineral from east Greenland. Medd. Grønland, 142, 3-11.

Bonilla, J.L., Lopez Gonzalez, J.D., Ramirez Saenz, A., Rodriguez Reinoso, F. and Valenzuela Calahorro, C. (1981) Activation of a sepiolite with dilute solutions of HNO_3 and subsequent heat treatments. II. Determination of surface acid centres. Clay Minerals, 16, 173-180.

Bowles, F.A., Angino, E.A., Hosterman, J.W. and Galle, O.K. (1971) Precipitation of deep sea palygorskite and sepiolite. Earth Planet. Sci. Lett., 11, 324-332.

Bradley, W.F. (1940) The structural scheme of attapulgite. Am. Mineral., 25, 405-410.

Brauner, K. and Preisinger, A. (1956) Struktur und Enstehung des Sepioliths. Tschermaks Miner. Petrog. Mitt., 6, 120-140.

Brindley, G.W. (1959) X-ray and electron diffraction data for sepiolite. Am. Mineral., 44, 495-500.

_____ and Pedro, G. (1972) Report of the AIPEA Nomenclature Committee. AIPEA Newsletter, 4, 3-4.

Brochant, A.S.M. (1802) Traite de Mineralogie. Villier, Paris, 1, 451.

Brongniart, A. (1807) Traite Elementaire de Mineralogie. Crapalet. Paris, 1, 490.

Caillere, S. (1936) Nickeliferous sepiolite. Bull. Soc. franc Min., 59, 163-326.

_____ (1951) Sepioite. In: G.W. Brindley, X-ray Identification and Structures of Clay Minerals. Mineral. Soc., London, 224-233.

_____ and Henin, S. (1957) The sepiolite and palygorskite minerals. In: R.C. Mackenzie, ed., The Differential Thermal Investigation of Clays. Mineral. Soc., London, 231-247.

Callen, R.A. (1977) Late Cainozoic environments of part of northeastern South Australia. J. Geol. Soc. Aust., 24, 151-169.

_____ (1984) Clays of the palygorskite-sepiolite group: Depositional environment, age and distribution. In: A. Singer and E. Galan, eds., Palygorskite-Sepiolite. Occurrences, Genesis and Uses. Developments in Sedimentology 37, Elsevier, Amsterdam, 1-37.

Calvo, J.P., Alonso, A.M. and Garcia Del Cura, M.A. (1986) Depositional sedimentary controls on sepiolite occurrence in Paracuellos de Jarama, Madrid Basin. Geogaceta, 1, 25-28.

Carmouze, J.P. (1976) La regulation hydrogeochemique du lac Tchad: Travaux et Documents de L'Office de la Recherche Scientifique et Technique Outre-Mer, 58, 418.

_____, Pedro, G. and Berrier, J. (1977) Sur la nature des smectites de neoformtion du lac Tchad et leur distribution spatiele en fonction des conditions hydrogeochemiques. C. R. Acad. Sci., Paris, ser. D., 284, 615-618.

Casal Piga, B. and Ruiz Hitzky, E. (1977) Reaction of epoxides on mineral surfaces. Organic derivatives of sepiolite. Proc. 3rd Europ. Clay Conf., Oslo, 35-37.

Castell, R.P., Santaren, J. and Alvarez, A. (1987) Improvements in FRC processability by using wet micronized sepiolite. Proc. EUROCLAY '87, Sociedad Espanola de Arcillas, Sevilla, 163-166.

Christ, C.L., Hathaway, J.C., Hostetler, P.B. and Shepard, A.O. (1969) Palygorskite: New X-ray data. Am. Mineral., 54, 198-205.

_____, Hostetler, P.B. and Siebert, R.M. (1973) Studies in the system $MgO-SiO_2-CO_2H_2O$ (III): The activity product constant of sepiolite. Am. J. Sci. 273, 65-83

Church, T.M. and Velde, B. (1979) Geochemistry and origin of a deep-sea Pacific palygorskite deposit. Chem. Geol., 25, 31-39.

Corma, A., Perez Pariente, J. and Soria, J. (1985) Physico-chemical characterization of Cu^{2+} exchanged sepiolite. CCM, 20, 467-475.

Cornejo, J. and Hermosin, M.C. (1986) Efecto de la temperatura en la acidez superficial del producto obtenido por tratamiento acido de sepiolita. Bol. Soc. Esp. Mineral., 9, 135-138.

Couture, R.A. (1977) Composition and origin of palygorskite-rich and montmorillonite-rich zeolite-containing sediments from the Pacific Ocean. Chem. Geol. 19, 113-130.

_____ (1978) Miocene of the S.E. United States--A model for chemical sedimentation in a peri-marine environment-Comments. Sediment. Geol., 21, 149-153.

Cronsted, A. (1758) In: J.D. Dana. A System of Mineralogy, 6th ed., 1892. John Wiley and Sons, 696.

Dandy, A.J. (1968) Sorption of vapours by sepiolite. J. Phys. Chem., 72, 334-339.

_____ (1971) Zeolitic water content and adsorption capacity for ammonia of microporous sepiolite. J. Chem. Soc. (A) 2383-2387.

_____, Brell, J.M., La Iglesia, A. and Robertson, R.H.S. (1975) The Caceres palygorskite deposit, Spain. In: S.W. Bailey, ed., 1975 Int'l Clay Conf., Mexico 1975. Applied Pub., Wilmette, Illinois, 91-94.

Gard, J.A. and Follet, E.A. (1968) A structural scheme for palygorskite. Clay Minerals, 7, 367-369.

Garrels, R.M. and MacKenzie, F.T. (1967) Origin of the chemical composition of some springs and lakes. In: W. Stumm, ed., Equilibrium Concepts in Natural Water Systems. Am. Chem. Soc. Advances in Chemistry, 67, 222-242.

Gieskes, J.M., Kastner, M. and Warner, T.B. (1975) Evidence for extensive diagenesis, Madagaskar Basin, Deep Sea Drilling Site 245. Geochim. Cosmochim. Acta, 39, 1385-1393.

Glocker, E.F. (1847) Synopsis, Halle, p. 190. In: J.D. Dana, A System of Mineralogy, 5th ed., 1868, Trubner and Co., London, p. 456.

Gonzalez Garcia, F. and Peiro, A. (1963) Aplicacion de las arcillas sedimentarias de Lebrija a la decoloracion de aceites minerales. Anal. Univ. Hispalense, 23, 39-49.

Gonzalez, L. Ibarra, L.M., Rodriguez, A., Moya, J.J. and Valle, F.J. (1984) Fibrous silica gel obtained from sepiolite by HCl attack. Clay Minerals, 19, 93-98.

Greeland, D.J. and Quirk, J.P. (1964) Surface area determination on soils. J. Soil Sci., 15, 178-191.

Grim, R.E. (1962) Applied Clay Mineralogy. McGraw-Hill, New York, 422 p.

Güven, N. and Carney, L.L. (1979) The hydrothermal transformation of sepiolite to stevensite and the effect of added chlorides and hydroxides. CCM, 27, 253-260.

_____ and Kerr, P.F. (1966) Selected Great Basin playa clays. Am. Mineral., 51, 1056-1067.

Haden, W.L. (1963) Attapulgite: Properties and uses. In: E. Ingerson, ed., Proc. 10th Nat. Conf. Clays Clay Miner., Pergamon Press, New York, p. 284.

_____ and Schwint, I.A. (1967) Attapulgite: Its properties and applications. Ind. Engine. Chem., 59, 58-69.

Hauy, R.J. (1801) Traite de Mineralogie. Paris vol. 4, p. 443, New edition in 1968. Culture et Civilisation. Bruxelles.

Hay, M.H. (1975) Chemical Index of Minerals. 2nd ed. British Museum (Nat. Hist.) London, 728 p.

Hay, R.L. and Stoessell, R.K. (1984) Sepiolite in the Amboseli Basin of Kenya: A new interpretation. In: A. Singer and E. Galan, eds., Palygorskite-Sepiolite. Occurrences, Genesis and Uses. Developments in Sedimentology 37, 125-137.

_____ and Wiggins, B. (1980) Pellets, ooids, sepiolite and silica in three calcretes of the southwestern United States. Sedimentology, 27, 559-576.

_____. Pexton, R.E., Teague, T.T., and Kyser, T.K. (1986) Spring-related carbonate rocks, Mg clays, and associated minerals in Pliocene deposits of the Amargosa Desert, Nevada and California. Geol. Soc. Am. Bull., 97, 1488-1503.

Hayashi, H., Otsuka, R. and Imai, N. (1969) Infrared study of sepiolite and palygorskite on heating. Am. Mineral., 53, 1613-1624.

Heddle, M.F. (1879) Pilolite. Mineral. Mag. 2, 206-219.

Hennin, S. and Caillere, S. (1975) Fibrous minerals. In: J.E. Gieseking, ed., Soil Components. Springer Verlag, Berlin, vol. 2, 335-349.

Hermann, R. (1845) Untersuchungen russischer mineralien, J prakt. Chem., 34, 177-181.

Hermosin, M.C. and Cornejo, J. (1986) Methylation of sepiolite and palygorskite with diazomethane. CCM, 34, 591-596.

_____, Cornejo, J., White, J.L., and Hem, S.L. (1981) Sepiolite, a potential excipient for drugs subject to oxidative degradation. J. Pharm. Sci., 70, 189-192.

Hostetler, P.B. and Christ, C.L. (1968) Studies in the system $MgO-SiO_2-CO_2-H_2O$ (I): The activity-product constant of chrysotile. Geochim. et Cosmochim. Acta, 32, 485-497.

Huertas, F. (1969) Minerales fibrosos de la arcilla. Su genetica en cuencas sedimentarias espanolas y sus aplicaciones tecnologicas. Ph.D. thesis, University of Madrid, 284 p.

_____, Pozzuoli, A. and Linares, J. (1979) Studio del potere decolorante della sepiolite con un semplice apparato di percolazione. Rendiconto Acad. Sci. Fisiche e Matem. Soc. Naz. Sci. Lett e Arti, Napoli. Serie IV, 46, 619-632.

Imai, N., Otsuka, R., Kashide, H. and Hayashi, R. (1969) Dehydration of palygorskite and sepiolite from Kuzuu district, Tochigi Pref., Central Japan. In: L. Heller, ed., Proc. Int'l Clay Conf., Tokyo, 1, 99-108.

Isphording, W.C. (1984) The clays of Yucatan, Mexico. A contrast in genesis. In: A. Singer and E. Galan, eds., Palygorskite-Sepiolite. Occurrences, Genesis and Uses. Developments in Sedimentology 37, Elsevier, Amsterdam, 59-73.

Japan Kokai (1978) 07592, 30996, 34691.

Jeffers, J.D. and Reynolds, R.C., Jr. (1987) Expandable palygorskite from Cretaceous-Tertiary boundary, Mangyshlak Peninsula, USSR. CCM, 35, 473-476.

670

Davis, D.W., Rochow, T.G., Rowe, F.G., Fuller, M.L., Kerr, P.F. and Hamilton, P.K (1950) Electron micrographs of reference clay minerals. Prelim. Rep. n° 6 of Reference Clay Minerals, A.P.I. Research Project, n° 49, Columbia Univ., New York.

Doval, M., Calvo, J.P., Brell, J.M. and Jones, B.F. (1986) Clay mineralogy of the Madrid basin: Comparison with other lacustrine closed basins. In: R. Rodriguez-Clemente and Y. Tardy, eds., Geochemistry of the Earth Surface and Processes of Mineral Formation (absts.), 188-189.

Drits, V.A. and Sokolova, G.V. (1971) Structure of palygorskite. Soviet Phys. Crystallogr., 16, 183-185.

Droste, J.B. (1961) Clay mineral composition of sediments in some desert lakes in Nevada, California, and Oregon. Sci., 113, 1928.

Eberl, D.E., Jones, B.F. and Khoury, H.N. (1982) Mixed-layer kerolite/stevensite from the Amargosa Desert, Nevada: CCM, 30, 321-326.

Echle, W. (1978) The transformation sepiolite-loughlinite: Experiments and field observations. N. Jb. Miner. Abh., 133, 303-321.

Efremov, N.E. (1939) Some magnesium silicates and their alumino-analogues. Dokl. Akad. Nauk. USSR, 24, 287-289 (Abstract in Am. Mineral., 25, 313, 1940).

Esteoule-Choux (1984) Palygorskite in the Tertiary deposits of the American Mossif. In: A. Singer and E. Galan, eds., Palygorskite-Sepiolite. Occurrences, Genesis and Uses. Developments in Sedimentology 37, Elsevier, Amsterdam, 75-86.

Eugster, H.P. and Jones, B.F. (1979) Behavior of major solutes during closed-basin brine evolution. Am. J. of Sci., 279, 609-631.

Fahey, J.J. and Axerod, J.M. (1948) Loughlinite, a new hydrous magnesium silicate. Am. Mineral., 33, 195.

_____, Ross, M. and Axelrod, D.J. (1960) Loughlinite, a new hydrous sodium magnesium silicate. Am. Mineral., 45, 270-281.

Fassi, D., Ildefonse, P. and Borquier, G. (1988) Alteration sous couverture calcaire avec genese de palygorskite, a l'origine d'une evolution du modele. C.R. Acad. Sci. Paris 306, 1277-1281.

Fenoll, P. and Martin Vivaldi, J.L. (1968) Contribucion al estudio de la sepiolita. IV. Superficie especifica de los cristales. Anales de Quimica 64, 77-82.

Fernandez Alvarez, T. (1970) Superficie especifica y estructura de poro de la sepiolita calentada a diferentes temperaturas. In: J.M. Serratosa, ed., Proc. Reunion Hispano Belga de Minerales de la Arcilla, Madrid, CSIC, 202-209.

_____ (1972) Activacion de la sepiolita con acido clorhidrico. Bol. Soc. Esp. Ceram. Vidr., 11, 365-374.

_____ (1978) Efecto de la deshidratacion sobre las propiedades adsorbentes de la palygorskita y sepiolita. Clay Minerals, 13, 375-386.

Fernandez Hernandez, M.N. and Ruiz-Hitzky, E. (1979) Interaccion de isocianatos con sepiolita. CCM, 14, 295-305.

Fersman, A.E. (1908) Ueber die Palygorskitgruppe. Bull. Acad. Sci., St. Petersbourg, 2, 255-274.

_____ (1913) Research on magnesium silicates. Zap. imp. Akad. Nauk 32, 321-430.

Firman, R.J. (1966) Sepiolite from the Malvern Hill. Mercian Geologist, 1, 247-253.

Fleischer, P. (1972) Sepiolite associated with Miocene diatomite, Santa Cruz basin, California. Am. Mineral., 57, 903-913.

Fontes, J.C., Fritz, P., Gauthier, J. and Kulbicki, G. (1967) Mineraux argileux, elements traces et compositions isotopiques ($^{18}O/^{16}O$ et $^{13}C/^{12}C$) dans les formations gypsiferes de l'Eocene Superieur et dans l'Oligocene de Cormeilles-en-Parisis. Bull. Cent. Rech. Pau - SNPA, 1, 315-366.

Forteza, M. (1987) Estudio de la interaccion de dexanetasona con sepiolita, palygorskita y montmorillonita. Ph.D. thesis, University of Seville, 227 p.

Friedel, G. (1901) Termietite et Lassalite. Bull. Soc. franc. Min., 24, 12-17.

_____ (1907) Sur un noveau gisement de pilolite. Bull. Soc. franc. Min., 30, 80-83.

Gac, J.Y., Al Droubi, A., Fritz, B. and Tardy, Y. (1977) Geochemical behavior of silica and magnesium during the evaporation of waters in Chad. Chem. Geol., 19, 215-228.

Galan, E. (1979) The fibrous clay minerals in Spain. In: J. Konta, ed., 8th Conf. Clay Mineral. Petrol. Teplice, 1979, Univ. Carolinae, Prague, 239-249.

_____ (1984) Sepiolite-palygorskite in Spanish Tertiary basins: Genetical patterns in continental environments. In: A. Singer and E. Galan, eds., Palygorskite-Sepiolite. Occurrences, Genesis and Uses. Developments in Sedimentology 37, Elsevier, Amsterdam, 87-124.

_____ (1987) Industrial applications of sepiolite from Vallecas-Vicalvaro, Spain: A review. In: L.G. Schultz, H. van Olphen and F.A. Mumpton, eds., Proc. Int'l Clay Conf., Denver, 1985. The Clay Mineral. Soc., Bloomington, Indiana, 400-404.

_____ and Ferrero, F. (1982) Palygorskite-Sepiolite clays of Lebrija, Southern Spain. CCM, 30, 191-199.

Jimenez Lopez, A., Lopez Gonzalez, J.D., Ramirez Saenz, A., Rodriguez Reinoso, F., Valenzuela Calahorro, C. and Zurita Herrero, L. (1978) Evolution of surface area in a sepiolite as a function of acid and heat treatments. Clay Minerals, 13, 375-386.

Jones, B.F. (1983) Occurrence of clay minerals in surficial deposits of southwestern Nevada. Sci. Geol. Mem., 72, 81-92.

Jones, B.F. (1986) Clay mineral diagenesis in lacustrine sediments. U.S. Geological Survey Bull. 1578, 291-300.

_____ and Spencer, R.J. (1985) Clay minerals of the Great Salt Lake basin. Proc. Int'l Clay Conf., Denver, Absts., 6.

_____ and Vandenburgh, A.S. (1966) Geochemical influences on the chemical character of closed lakes. In: Symposium of Garda, Hydrology of Lakes and Reservoirs. Int'l Assoc. Sci. Hydro. Publ., 70, 435-446.

_____, Vandenburgh, A.S., Deike, R.G., Rettig, S.L. and Truesdell, A.H. (1972) Mineralogy of some sediments from alkaline saline lakes. Proc. Int'l Clay Conf. Absts., Madrid, 80-81.

_____ and Weir, A.H. (1983) Clay minerals of Lake Abert, an alkaline, saline lake. CCM, 31, 161-172.

Karup-Moller, S. and Petersen, O.V. (1984) Tuperssuatsiaite, a new mineral species from the Ilimaussaq intrusion in South Greenland. N. Jb. Mineral Mh., 501-502 (In. Am. Mineral., 70, 1985, p. 1332).

Kautz, K. and Porada, H. (1976) Sepiolite formation in a pan of the Kalahari. N. Jb. Mineral Mh. 12, 545-559.

Khoury, H.N. (1979) Mineralogy and chemistry of some unusual clay deposits in the Amargosa Desert, southern Nevada. University of Illinois, Urbana, Ph.D. thesis, 171.

Kirwan, R. (1794) Elements of Mineralogy. 2nd ed., Elmsly, London, 144-145.

Klaproth, M.H. (1794) Beobachtungen und Entdeckungen an der Naturskunde. Berlin, 5, 149.

Kulbicki, G. (1959) High temperature phases in sepiolite, attapulgite and saponite. Am. Mineral., 44, 752-764.

Lacazette, F. (1947) Estadistica minera y metalurgica de Espana. Consejo de Mineria. Ministerio de Industria, Madrid.

LaIglesia, A. (1977) Precipitation por disolucion homogenea de dilicatos de aluminio y magnesio a temperature ambiente. Sintesis de la palygorskita. Estudios geol., 33, 535-544.

_____ (1978) Sintesis de la Sepiolita a temperatura ambiente por precipitacion homogenea. Bol. Geologico y Minero, T. LXXXXIX-III, 258-265.

Lapparent, J. de (1935) Sur un constituant essentiel des terres a foulon. C. R. Acad. Sci., Paris, 201, 481-482.

Lokanatha, S. and Bhattacherjee, S. (1984) Structure defects in palygorskite. Clay Minerals, 19, 253-255.

Longchambon, H. and Migeon, G. (1936) Sepiolites. C. R. Acad. Sci., Paris, 203, 431-433.

_____ (1937) Thermal properties of sepiolites. Bull. Soc. franc. Miner., 60, 232-276.

Lopez Gonzalez, J.D., Ramirez, A., Rodriguez, F., Valenzuela, C. and Zurita, L. (1981) Activacion de una sepiolita con disoluciones diluidas de HNO3 y posteriores tratamientos termicos. I. Estudio de la superficie especifica. Clay Minerals, 16, 103-113.

Mackenzie, R.C. (1966) Clay mineralogy. Earth-Sci. Rev., 2, 49-91.

Marshal, C.E., Humbert, R.P., Shaw, B.T. and Caldwell, O.G. (1942) Studies of clay particles with the electron microscope, II. Soil. Sci., 54, 149-158.

Martin de Vidales, J.L., Jimenez Ballestra, R. and Guerra, A. (1987) Pedogenic significance of palygorskite in paleosoils developed on terraces of the river Tajo (Spain). In: R. Rodriguez Clemente and Y. Tardy, eds., Geochemistry and Mineral Formation in the Earth Surface. C.S.I.C., Madrid, 535-548.

Martin Vivaldi, J.L. and Cano, J. (1956) Sepiolite. II. Consideration on the mineralogical formula. CCM, 4, 173-176.

_____ and Fenoll, P. (1970) Palygorskites and sepiolites (Hormites). In: R.C. Mackenzie, ed. Differential Thermal Analysis. Academic Press, London, I, 553-573.

_____ and Linares, J. (1962) A random intergrowth of sepiolite and attapulgite. CCM, 9, 592-602.

_____ and Robertson, R.H.S. (1971) Palygorskite and sepiolite (The Hormites). In: J.A. Gard, ed., Electron-optical Investigation of Clays. Mineral. Soc., London, 255-276.

Martindale, W. (1982) The Extra Pharmacopoeia. 28th ed., Pharm. Soc., Great Britain, Pharmaceutical Press, London, 2025 pp.

McGrath, D.A. (1984) Morphological and mineralogical characteristics of indurated caliches of the Llano Estacado. M.S. thesis, Texas Tech. University, 123 p.

McLean, S.A., Allen, B.L. and Craig, J.R. (1972) The occurrence of sepiolite and attapulgite on the southern High Plains. CCM, 20, 143-149.

Migeon, G. (1936) Sepiolites. Bull. Soc. franc. Mineral. 59, 6-134.

Mifsud, A., Garcia, I., and Corma, A. (1987) Thermal stability and textural properties of exchanged sepiolites. Proc. Euroclay '87. Sociedad Espanola de Arcilla, Sevilla, 392-394.

672

Mifsud, A., Rautureau, M. and Fornes, V. (1978) Etude de l'eau dans la palygorskite a l'aide des analyses thermiques. CCM, 13, 367-374.

Millot, G (1964) Geologie des Argiles. Masson and Cie, Paris, 510.

_____, Nahon, D., Paquet, H., Ruellan, A. and Tardy, Y. (1977) L'epigenie calcaire des roches silicatees dans les encroutements carbonates en pays subaride, Antiatlas, Maroc. Sci. Geol. Bull., 30, 3, 129-152.

Nagata, H., Shimoda, S. and Sudo, T. (1974) On dehydration of bound water of sepiolite. CCM, 22, 285-293.

Nagy, B. and Bradley, W.F. (1955) The structural scheme of sepiolite. Am. Mineral., 40, 885-892.

Newman, A.C. and Brown, G. (1987) The chemical constitution of clays. In: A.C.D. Newman, ed., Chemistry of Clays and Clay Minerals. Mineral. Soc., London, 1-128.

Oscarson, D.W., Van Scoyoc, G.E. and Ahlrichs, J.L. (1986) Lysis of erythrocytes by silicate minerals. CCM, 34, 74-86.

Ovcharenko, F.D. (ed.) (1964) The Colloid Chemistry of Palygorskite. Daniel Davey and Co., New York, 101 p.

_____ and Kukovsky, G. (1984) Palygorskite and sepiolite deposits in the USSR. In: A Singer and E. Galan, eds., Palygorskite-Sepiolite Occurrences, Genesis and Uses. Developments in Sedimentology 37, Elsevier, Amsterdam, 233-241.

Paquet, H. (1970) Evolution geochimique des mineraux argileux dans les alterations et les sols des climats mediterraneens tropicaux a saisons contrastees. Bull. Serv. Carte Geol. Als. Lorraine, 30, 1-212.

_____ (1983) Stability, instability, and significance of attapulgite in the calcretes of Meditteranean and tropical areas with marked dry season. Sci. Geol., Mem. 72, p. 131-140.

_____ and Millot, G. (1972) Geochemical evolution of clay minerals in the weathered products of soils of mediterranean climate. Proc. Int'l Clay Conf., Madrid, 199-206.

_____, Duplay, J., Valleron-Blanc, M.M. and Millot, G. (1987) Octahedral compositions of individual particles in smectite-palygorskite and smectite-sepiolite assemblages. Proc. Int'l Clay Conf., Denver, 1985, 73-77.

Parry, W.T. and Reeves, C.C., Jr. (1968) Sepiolite from Pluvial Mound Lake, Lynn and Terry Counties, Texas. Am. Mineral. 53, 984-993.

Patterson, S.H. and Buie, B.F. (1974) Field conference on kaolin and fuller's earth. November 14-15, 1974, Guide book 14. Georgia Geological Survey for the Soc. of Economic Geologists, Atlanta, 53 p.

Pedro, G., Carmouze, J.P. and Velde, B. (1978) Peloidal nontronite formation in recent sediments of Lake Chad. Chem. Geol., 23, 139-149.

Perez Rodriguez, J.L., Carretero, M.I. and Maqueda, C. (1987) Behaviour of sepiolite, vermiculite and bentonite as supports in anaerobic digesters. Proc. EUROCLAY'87, Seville. Sociedad Espanola de Arcillas, Sevilla, 432-434.

Perrault, G., Harvey, J. and Pertsowsky, R. (1975) La yofortierite, un noveau silicate de manganese de St. Hilaire. P.Q. Can. Mineral., 13, 68-74.

Prado, F. (1864) Descripcion fisiografica y geologica de la provincia de Madrid. Junta General de Estadistica. Imprenta Nacional. Madrid, p. 148.

Preisinger, A. (1959) X-ray study of the structure of sepiolite. CCM, 6, 61-67.

_____ (1963) Sepiolite and related compounds: Its stability and application. CCM, 10, 365-371.

Rautureau, M. (1974) Analyse structurale de la sepiolite par microdiffraction electronique. Theses Universite d'Orleans, 89 p.

_____, Tchoubar, C. and Mering, J. (1972) Analyse structurale de la sepiolite par microdiffraction electronique. C.R. Acad. Sci. Paris, 274 C, 269-271.

_____ and Tchoubar, C. (1974) Precisions concernant l'analyse structurale de la sepiolite par microdiffraction electronique. C.R. Acad. Sci. Paris, 278B, 25-28.

Rautureau, M. and Mifsud, A. (1977) Etude par microscope electronique des differents etats d'hydration de la sepiolite. Clay Minerals, 12, 309-318.

Robertson, R.H.S. (1957) Sepiolite: A versatile raw matrial. Chem. Ind., 1492-95.

_____ (1962) The acceptability of mineral group names. Clay Minerals Bull., 5, 41-43.

_____ (1963) Perlite and palygorskite in Theophrastus. Classical Rev. New Ser. 13, 132.

Roger, L.E., Quirk, J. and Norrish, K. (1956) Occurrence of an aluminum-sepiolite in a soil having unusual water relationships. J. Soil Sci., 7, 177-184.

Ruiz-Hitzky, E. and Fripiat, J.J. (1976) Organomineral derivatives obtained by reacting organochlorosilanes with the surface of silicates in organic solvents. CCM, 25, 25-30.

_____, Casal, B. and Serratosa, J.M. (1983) Sorption of pyridine on sepiolite and palygorskite. Abstracts EUROCLAY'83, Czechoslovak Group for Clay Mineralogy and Petrology. Prague, p. 125.

Semenov, E.I. (1969) Mineralogy of the Ilimaussaq Alkaline Massif, Southern Greenland. Inst. Mineral. Geokhim. Krystallokhim. Redk Elementov, Izdat. "Nauka" 1969, 164 p. (in Russian).

Serna, C. and Van Socyoc, G.E. (1979) Infrared study of sepiolite and palygorskite surfaces. In: M.M. Mortland and V.C. Farmer, eds., Proc. Int'l Clay Conf. 1978, Elsevier, Amsterdam, 197-206.

_____, Ahlrichs, J.L. and Serratosa, J.M. (1975) Folding in sepiolite crystals. CCM, 23, 452-457.

_____, Rautureau, M., Prost, R., Tchoubar, C. and Serratosa, J.M. (1974) Etude de ia sepiolite a l'aide des donnes de la microscopie electronique de l'analyse thermoponderale et de la spectroscopie infrarouge. Bull. Groups franc. Argiles, 26, 153-163.

_____, Van Scoyoc, G.E. and Ahlrichs, J.L. (1977) Hydroxyl groups and water in palygorskite. Am. Mineral., 62, 784-792.

Serratosa, J.M. (1979) Surface properties of fibrous clay minerals (palygorskite and sepiolite). In: M.M. Mortland and V.C. Farmer, eds., Proc. Int'l Clay Conf. 1978, Elsevier, Amsterdam, 99-109.

Siddiqui, M.H.K. (1984) Occurrence of palygorskite in the Deccan Trap Formation in India. In: A. Singer and E. Galan, eds., Palygorskite-Sepiolite. Occurrences, Genesis and Uses. Developments in Sedimentology 37, Elsevier, Amsterdam, 243-250.

Siffert, B. and Wey, R. (1962) Synthese d'une sepiolite a temperature ordinaire. C. R. Acad. Sci., Paris, 254, 1460-1463.

Singer, A. (1977) Dissolution of two Australian palygorskites in dilute acid. CCM, 25, 126-130.

_____ (1979) Palygorskite in sediments: Detrital, diagnetic or neoformed. A critical review. Geol. Rund., 68, 996-1008.

_____ (1984) Pedogenic palygorskite in the arid environment. In: A. Singer and E. Galan, eds., Palygorskite-Sepiolite. Occurrences, Genesis and Uses. Developments in Sedimentology 37, Elsevier, New York, 169-177.

_____ and Galan, E., eds. (1984) Palygorskite - Sepiolite. Occurrences Genesis, and Uses. Developments in Sedimentology 37, Elsevier, New York, 352.

_____ and Norrish, K. (1974) Pedogenic palygorskite occurrences in Australia. Am. Mineral., 59, 508-517.

Spencer, R.J. (1977) Silicate and carbonate sediment-water relationships in Walker Lake, Nevada. Univ. of Nevada, Reno, M.S. thesis, 98 p.

_____ (1982) The geochemical evolution of Great Salt Lake. Johns Hopkins Univ., Baltimore, Maryland, Ph.D. thesis, 308 p.

Springer, G. (1976) Falcondoite, a nickel analogue of sepiolite. Can. Mineral., 14, 407-409.

Ssaftschenkov, T.V. (1862) Definition of palygorskite. Zap. imp. Mineral. Obshch 102-104 (Verh. K. Ges. gesammt. Miner., St. Petersbourg).

Stoessell, R.K. (1988) .25 C and 1 atm. dissolution experiments of sepiolite and kerolite. Geochim. et Cosmochim. Acta, 52, 365-374.

_____ and Hay, R.L. (1978) The geochemical origin of sepiolite and kerolite at Amboseli, Kenya. Contrib. Mineral. Petrol., 65, 255-267.

Strunz, H. (1957) Mineralogische Tabellen, 3rd ed. Akademische Verlags geselleschaft. Leipzig.

Tardy, Y. and Garrels, R.M. (1974) A method of estimating the Gibbs energies of formation of layer silicates. Geochim. et Cosmochim. Acta, 38, 1101-1106.

Trauth, N. (1977) Argiles evaporitiques dans la sedimentation Carbonatee Continental et epicontinentale tertiaire. Sci. Geol. Mem., 49, 195.

Vanden Heuvel, R.C. (1966) The occurrence of sepiolite and attapulgite in the calcareous zone of a soil near Las Cruces, New Mexico. CCM, Proc. 13th Nat. Conf., 193-208.

Van Olphen, H. (1966) Maya Blue: Clay-organic pigment. Science, 154, 645-646.

Van Scoyoc, G.E., Serna, C. and Ahlrichs, J.L. (1979) Structural changes in palygorskite during dehydration and dehydroxylation. Am. Mineral., 64, 216-223.

Velde, B. (1985) Clay minerals: A physico-chemical explanation of their occurrence. Developments in Sedimentology, 40, Elsevier, New York, 427.

Villanova, J. (1875) Anal. Soc. Esp. Hist. Nat., IV Acta 46 (in Calderon, S. 1910. Los minerales de Espana. Junta para la ampliacion de estudios e investigaciones cientificas, Madrid).

Watts, N.L. (1976) Paleopedogenic palygorskite from the basal Permo-Triassic of northwest Scotland. Am. Mineral., 61, 299-302.

_____ (1980) Quaternary pedogenic calcretes from the Kalahari (South Africa): Mineralogy, genesis, and diagenesis. Sedimentology, 27, 661-686.

Weaver, C.E. and Beck, K.C. (1977) Miocene of the S.E. United States: A model for chemical sedimentation in a peri-marine environment. Sediment. Geol., 17, 1-234.

Weaver, C.E. and Pollard, L. (1973) The Chemistry of Clay Minerals. Elsevier, Amsterdam, 213 p.

Werner, A.G. (1788) Definition of the Species. Bergm. J., p. 377. In: J.D. Dana, A System of Mineralogy, 5th ed., 1868. Trubner and Co., London, p. 456.

Wiegel B. (before 1784) Neveste Entdeckungen in der Chemie, part 5, 3. In: Martin Vivaldi and Robertson, 1971.

674

Wollast, R., MacKenzie, F.T. and Bricker, O.P. (1968) Experimental precipitation and genesis of sepiolite at earth-surface conditions. Am. Mineral., 53, 1645-1662.

Yang Zhenqiang and Xu Junwen (1987) Diagenetic transformation of early Permian sepiolite and its relationship with coal metamorphism--an example in Pingle Depression and its vicinity, south China. Geochemistry 6, 65-75.

Yaalon, D.H. and Wieder, M. (1976) Pedogenic palygorskite in some arid brown (Calciorthid) soils of Israel. Clay Minerals, 11, 73-80.

Yucel, A.M., Rauterau, M., Tchoubar, D. and Tchoubar, C. (1981) Calculation of the x-ray powder reflection profiles of very small needle-like crystals. II. Quantitative results on Eskisehir sepiolite fibers. J. Appl. Clays, 14, 431-454.

Zhang, R., Qiu, C., Peng, Ch, Dai, G. and Yang, Z. (1985) The characteristics of magnesium-rich clay in Liling area, Hunan province and a discussion on its genesis. Bull. Yichang Inst. Geol. Mineral. Resources, CAGS, 9, 1-13.

Zoltai, T. (1981) Amphibole asbestos mineralogy. In: D.R. Veblen, ed., Amphiboles and Other Hydrous Pyriboles. Reviews in Mineralogy, vol. 9a. Mineral. Soc. Am., 237-278.

_____ and Stout, J.H. (1985) Mineralogy. Concepts and Principles. Burgess Pub. Co., Minneapolis, Minnesota, 505 p.

Zvyagin, B.B. (1967) Electron Diffraction Analysis of Clay Mineral Structures. Plenum Press, New York, 122 p.

_____, Mishchenko, K.S. and Shitov, V.A. (1963) Electron diffraction data on the structures of sepiolite and palygorskite. Soviet Phys. Crystallogr., 8, 148-153.

Chapter 17

S. Guggenheim & R.A. Eggleton

CRYSTAL CHEMISTRY, CLASSIFICATION, AND IDENTIFICATION OF MODULATED LAYER SILICATES

INTRODUCTION

Modulated layer silicates are those minerals in which there is a periodic perturbation to the basic layer silicate structure. These basic structures, either involving 1:1 or 2:1 layer configurations, are outlined extensively in other chapters and are not reviewed in detail here. All such structures have strong layer aspects that dominate their physical characteristics, such as, most notably, perfect (001) cleavage. In modulated layer silicates, this feature usually is evident also, especially for those cases where the octahedral sheet remains essentially intact. Layer-like qualities, however, diminish as the severity of the modulations increase. These modulations are usually complex and have many variations. Fortunately, however, they follow a geometric theme and it is this theme that is developed here. We have elected to outline the principles before presenting the individual mineral cases. Although this approach leads to some awkwardness by citing examples that appear later in the chapter, it serves to unify the discussion, thereby allowing overall structural diversity to be systematized on the basis of chemistry and geometry. Portions of the text are updated from Guggenheim and Eggleton (1987).

GEOMETRIC CONCEPTS

The *octahedral sheet* of a modulated layer silicate is invariably a brucite-like (trioctahedral) sheet. The two anion planes are dominated by O or OH, but presumably may include small amounts of F or Cl. Adjacent *tetrahedral sheets* involve Si (and minor Al) tetrahedra ideally linked into six-member rings by sharing three tetrahedral corners to form a *basal plane* of oxygens. The fourth oxygen of each tetrahedron (*apical oxygen*) is directed away from the basal plane and may point toward the octahedral sheet or away from it. In the former case, the apical oxygen belongs also to the anion plane of the octahedral sheet, thereby being in common with both the sheet of octahedra and the sheet of tetrahedra. In the latter case where the apical oxygen points away from the octahedral sheet, it may link to another tetrahedron. Details regarding the nature of this linkage usually differ for the various modulated layer silicates.

Where an ideal tetrahedral sheet forms a common junction with the octahedral sheet, each octahedral cation is coordinated to two apical oxygens (i.e., also linking to tetrahedra) and to a third anion, usually OH but possibly F or Cl. This third anion is not bonded to a tetrahedral cation, although it is coplanar with the apical oxygens. It is coordinated only to the octahedral cation. In *modulated 1:1 layer structures*, the octahedral cations ideally complete their six fold coordination with OH groups only, which form the other anion plane of the brucite-like sheet. In contrast, in *modulated 2:1 layer silicates*, the octahedral cations ideally complete their six-fold coordination with a second tetrahedral sheet inverted to oppose the first, so that a similar configuration of apical oxygens and OH groups surround the octahedral cation. It is emphasized that these are ideal configurations as the tetrahedral sheets have also tetrahedra with inverted apices that point away from the octahedra. In such regions, octahedral coordination is completed with an OH or F anion.

In regions where the apical oxygen is part of both the octahedral sheet and the tetrahedral sheet, the basal oxygen spacing is closely coupled with the anion to anion spacing of the common junction between these two sheets. The reason for this close association in spacing is that the common junction and the basal plane are strongly bonded through the silicon. Therefore, in order to maintain octahedral coordination in the brucite-like sheet, it is a requirement that the lateral dimensions of the octahedral sheet be equal to the lateral dimensions of the attached tetrahedral sheet. In addition to overall charge balance, this relationship represents the major constraint to variations in chemistry in both tetrahedral and octahedral sheets.

In modulated layer silicates, where the lateral dimensions of the octahedral sheet are considerably larger than the lateral dimensions of the ideal tetrahedral sheet, apical oxygen inversions and other structural perturbations allow the basal oxygen spacing to "reset" in phase with the anion to anion spacing of the common junction. In "normal" layer silicates, where the lateral dimensions of the octahedral sheet are smaller or nearly equal to the lateral dimensions of the ideal tetrahedral sheet, there are less severe structural adjustments that allow tetrahedral and octahedral sheets to be congruent. These adjustments are discussed briefly in the next section because some of the polyhedral distortions may be relevant in understanding similar possible distortions in the modulated structures.

Structures with continuous octahedral sheets

Reducing relatively large tetrahedral sheets. Tetrahedral rotation, the in-plane rotation of adjacent tetrahedra in opposite senses around the silicate ring (see, for example, Newnham and Brindley, 1956; Radoslovich, 1961; Guggenheim, 1984, and many others), is the most important distortion to reduce the lateral dimensions of a larger ideal tetrahedral sheet by as much as about 12% to conform to the dimensions of an octahedral sheet. This adjustment lowers the symmetry of the silicate ring from an ideally hexagonal configuration to ditrigonal, but it does not affect the size or shape of the individual tetrahedra.

Many layer silicates, however, exhibit octahedral and tetrahedral distortions that affect lateral dimensions by compensating with a vertical adjustment. Octahedra may expand laterally by thinning (see Donnay et al., 1964) and tetrahedra may expand vertically by shortening slightly in lateral directions (Radoslovich and Norrish, 1962; Eggleton and Bailey, 1967). Guggenheim and Eggleton (1987) noted, however, that these distortions may occur for reasons other than to maintain octahedral and tetrahedral sheet congruency only (see also Hazen and Wones, 1972; McCauley et al., 1973; Lee and Guggenheim, 1981; Toraya, 1981; Lin and Guggenheim, 1983). Nonetheless, if sheet congruency can be approached by these distortions for cases where the ideal tetrahedral sheet is large relative to the octahedral sheet, then similar distortions may be acting in modulated layer silicates to obtain the reverse effect. In the latter, octahedra may shorten and tetrahedra may expand laterally to achieve approximate congruency. Such distortions are generally not observed in modulated layer silicates, possibly because they are small

OBTAINING CONGRUENCY BETWEEN TETRAHEDRAL AND OCTAHEDRAL SHEETS

[where ideal tetrahedral sheet (TS) < ideal octahedral sheet (OS)]

PURELY CHEMICAL MECHANISMS

a) Substitution of larger cation (e.g., Al) for Si in TS

b) Substitution of smaller cation (e.g., Al) in OS

STRUCTURAL MECHANISMS

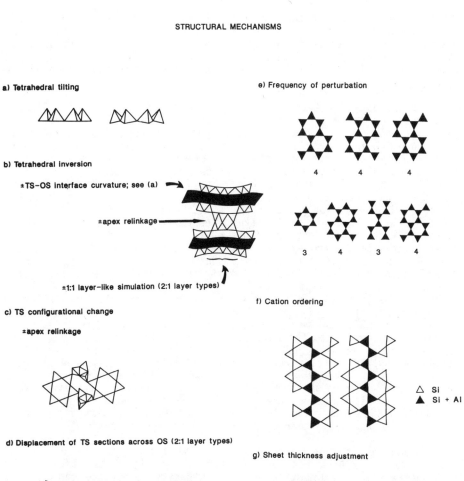

a) Tetrahedral tilting

b) Tetrahedral inversion

±TS-OS interface curvature; see (a)

±apex relinkage

±1:1 layer-like simulation (2:1 layer types)

c) TS configurational change

±apex relinkage

d) Displacement of TS sections across OS (2:1 layer types)

e) Frequency of perturbation

4   4   4

3   4   3   4

f) Cation ordering

△ Si
▲ Si + Al

g) Sheet thickness adjustment

Figure 1. Summary of chemical and structural mechanisms to obtain tetrahedral and octahedral sheet congruency for relatively small (ideal) tetrahedral sheets.

effects and consequently high quality X-ray data would be required to detect them. Modulated layer silicates are usually poorly crystallized. Furthermore, we note that these effects, if they exist, cannot alleviate major differences in ideal sheet sizes.

Enlarging relatively small tetrahedral sheets. There is no simple mechanism (Fig. 1), such as tetrahedral rotation, to minimize significant misfit between the component sheets. Instead, a severe re-adjustment is required and commonly results in a modulated tetrahedral "sheet", with tetrahedral inversions, tetrahedral omissions, tetrahedral additions, or combinations of the above. These configurational changes (details for each mineral are given below) are an obvious consequence of narrowly defining the discussion to structures with continuous octahedral sheets. It is shown below that the severity of the modulation is roughly correlated to the size of the octahedral sheet, although it is often difficult to define differences in "severity". Other structural adjustments also limit the usefulness of this correlation in predicting the nature of the tetrahedral perturbation.

Expansion of a tetrahedral sheet in a modulated layer silicate, compared to a "normal" layer silicate with six-member tetrahedral rings only, is obtained by a change in tetrahedral linkage. This variation in linkage often involves ("extra") tetrahedra between sheet sections of six-member rings or parts of six-member rings. Alternatively, if ring configurations are considered, variations may be thought of as "insertions" of n-member rings, where n = 4, 5, etc. In either case, the ratio of the number of tetrahedral to octahedral ions increases from the normal (minimum) value of 2:3 for trioctahedral layer silicates. The $T^{IV}:M^{VI}$ ratio is based on 1:1 type structures and, in order to make a meaningful comparison, it is necessary to take half the T^{IV} count of a 12 Å layer silicate and *all* the interlayer tetrahedra plus the tetrahedra on one side only of the octahedra sheet for 10 Å structures. Thus, examples include $T^{IV}:M^{VI}$ ratios of 2.25 for pyrosmalite (with four-member rings), 2.25 for stilpnomelane, 2.44 for minnesotaite, and 2.50 for ganophyllite (each of the latter with five-member rings). The difference in the compositions of these sheets results from differences in the frequency at which the modulations occur. (The interested reader may find it helpful to refer to the related figures of the examples given above; see individual mineral sections.) Widely spaced tetrahedral inversions (modulations) leave relatively large areas of "normal" tetrahedral sheet structure. These may be hexagonal/trigonal in shape and are referred to as *islands*, or they may be infinite in one dimension and are termed *strips*. In stilpnomelane, the islands extend over seven six-fold rings; in ganophyllite the strips are three six-fold rings wide and, hence, ganophyllite has the higher tetrahedral to octahedral ratio.

Expansion of the tetrahedral sheet is related to the octahedral cation radius. As the octahedra increase in size, the width of the tetrahedral islands or strips decreases because coherence can only be maintained over a short distance if the octahedra are much bigger than the tetrahedra. Concomitant with a reduction in island or strip width is an increase in the number of inversions needed to allow coordination, and therefore an increase in the number of tetrahedra relative to the number of octahedra. Figure 2 shows a plot of mean octahedral cation radius vs. the number of tetrahedra (Si+Al) per three octahedral cations. Because there are a variety of ways to accommodate misfit (Al for Si substitution, different n-ring arrangements, etc.) the relationship between average

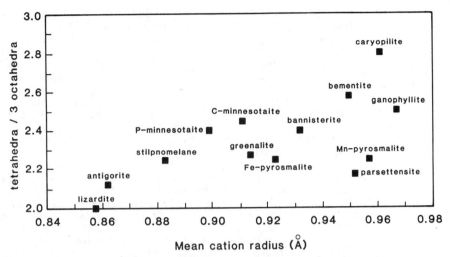

Figure 2. Plot relating mean octahedral cation radius to the number of tetrahedra per three octahedral cations. Comparisons are made to the 1:1 type structures as described in the text.

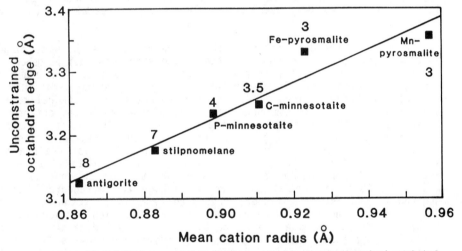

Figure 3. Graph relating mean octahedral cation radius to the unconstrained octahedral edge. Values with data points are the number of tetrahedra coordinating to the octahedral sheet between tetrahedral inversions, with fractional values representing averages.

octahedral cation size and the number of tetrahedra per three tetrahedra is not simple, although there is a general trend.

Guggenheim and Eggleton (1987) noted other adjustments to alleviate the incongruency in size between the ideal tetrahedral and octahedral sheets. In conjunction with tetrahedral inversions and relinkage across ideally vacant interlayer regions, limited curving of the tetrahedral-octahedral sheet interface produces a wave-like structure. Such curvature, (e.g., see antigorite or minnesotaite descriptions below) allows one side of the octahedral sheet to reduce in size to achieve registry with the tetrahedral sheet. The effect is achieved also by the out-of-plane tilting of the tetrahedra, which increases the lateral separation of the apical oxygens, allowing them to coordinate to the octahedral cations. Consequently, the opposing side of the octahedral sheet has an anion plane so expanded that it cannot coordinate to a tetrahedral sheet with six-fold rings. This side, therefore, involves OH only and gives some 1:1 character to any modulated layer silicate with wave-like corrugations of the octahedral sheet. It is for this reason also that tetrahedral sheet sections are displaced or staggered across the octahedral sheet so that rifts between strips (see minnesotaite, ganophyllite) or islands (see stilpnomelane) are offset. The wave-like nature of these structures allows for a continuously varying structural adjustment depending on the size of the radius of curvature, although such curvature is somewhat limited in variation (see below).

Modulated layer silicates with tetrahedral strips have distorted octahedra. Octahedra are relatively shortened along the strip axis because they are constrained to fit with the anion to anion spacing of the apical oxygens of the tetrahedra. In contrast, perpendicular to the strip, the misfit is periodically relieved by the modulation (tetrahedral inversion, etc.) and the octahedra are relatively expanded. Thus, unlike normal layer silicates, there is a significant departure from orthohexagonal geometry ($b \neq a\sqrt{3}$). A graph of mean cation radius vs. the unconstrained octahedral edge (= $b/3$ in most cases) is given in Figure 3, together with the number of tetrahedra spanning the islands or strips. This latter value decreases as the average octahedral cation size increases.

Because of the differences in the octahedral dimensions parallel and perpendicular to the strip axis, octahedral distortions must be generally related to the periodicity of the tetrahedral inversions. For example, minnesotaite has two closely related structural forms with similar tetrahedral configurations but slightly different mean octahedral cation sizes (see Fig. 3). To compensate for the differences in the octahedral cation size, the magnitude of the periodicity between tetrahedral inversions diminishes as octahedral size increases. However, although compensation appears to occur in only one direction, misfit along the strip axis must be reduced also by this mechanism. Octahedra have a greater ability to change size and shape if the oxygen configuration around the site is not fixed through linkage to tetrahedra. Thus, Guggenheim and Eggleton (1986a) suggested that local octahedral site distortions in the region near the tetrahedral inversion may account for the relief of strain parallel to the strip axis. As the magnitude of the periodicity between the tetrahedral inversions decreases in these minerals, a greater relief of strain may be accommodated along a strip axis. Presumably, a similar mechanism may be used also to explain the relief of misfit in antigorite.

In ganophyllite, Al preferentially enters the central continuous chain of the tetrahedral strip axis (see below). This substitution increases the chain dimension so that it may span the octahedra along the chain direction. Therefore, octahedra in ganophyllite can coordinate to a wider tetrahedral strip than its octahedral cation radius would suggest and it is not plotted in Figure 3.

Avoiding sheet-like tetrahedral linkages. In the previous section, various mechanisms are illustrated to describe ways in which primarily silicon-containing tetrahedra may link to a large octahedral sheet, one essentially containing Fe^{2+} and Mn. Structures are known, however, which

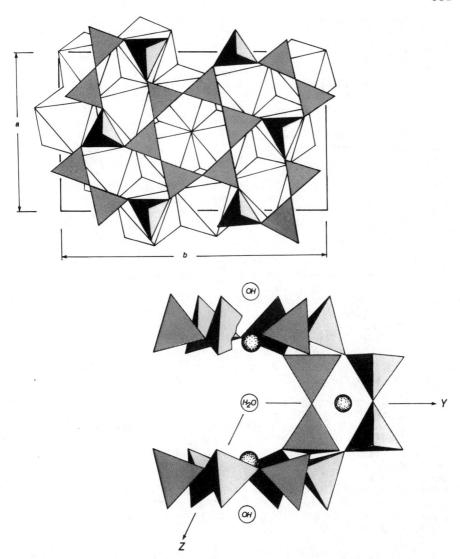

Figure 4. The fedorite structure illustrated (modified from Sokoleva et al., 1983). Fedorite, a member of the reyerite group, shows the limited extent of tetrahedral linkages to a primarily calcium octahedral plane, as shown on the left. Note the tetrahedral out-of-plane tilting in the perspective view on the right. Stippled spheres are partly occupied (Na + K) sites.

contain even larger cations, such as Ca, and the resulting tetrahedral perturbations occur at sufficiently short intervals that the pretense of calling the tetrahedral sections "sheet-like" is, at best, unconventional. These structures (the reyerite and tobermorite groups) have been noted to be "layer-like" because Ca occupies central planes (Fig. 4). Taylor and Howison (1956) first noted the layer-like qualities of tobermorite. Megaw and Kelsey (1956) and Hamid (1981) showed that dreierketten (3 unit repeat) chains of tetrahedra run parallel to Y. Hamid (1981) found that the calcium ions are coordinated to seven oxygens or (OH) groups rather than six as in an ideal octahedral sheet. Tobermorite represents a series of minerals with some varieties being "anomalous", with tetrahedra linking across the "interlayer" region (El-Hemaly et al., 1977; Mitsuda and Taylor, 1978; Wieker et al., 1982), and others with few or no interlayer linkages ("normal"). The reyerite group (including reyerite: Merlino, 1988; truscottite: Lachowski et al., 1979; gyrolite: Eberhard and Rahman, 1982; fedorite: Sokoleva et al., 1983; minehillite: Dunn et al., 1984; see also Taylor, 1964; Gard et al., 1975; 1981) has complex tetrahedral arrangements between calcium central planes.

A major feature of the geometric theme developed here is that the tetrahedral "sheet" is continuous in two dimensions, but the linkage between the octahedral and tetrahedral sheets is interrupted periodically in their common plane. Otherwise, a strained interface would develop and the resulting structure would be unstable. The large calcium present in a central plane in the tobermorite and reyerite group minerals prevents the formation of an extensive tetrahedral sheet. Therefore, common junctions between the tetrahedral and calcium components are very limited in extent.

The three unit repeat dreierketten chain in tobermorite readily achieves congruence with the calcium sheet along its one axis. The reyerite group minerals have not been characterized sufficiently to determine if the principles developed above apply in detail to the entire group. Certainly, however, some of the geometric devices explored above do apply (see Fig. 4). Because these minerals are not traditionally considered layer silicates, as their basic units are neither 1:1 or 2:1 layer derivations, they are not considered further.

Structures with discontinuous octahedral sheets

The minerals of the palygorskite/sepiolite group (Fig. 5) contain non-continuous octahedral sheets, better described as "ribbons" that extend along the 5.3 Å axis (defined as Z in the literature, but analogous to X if compared to an ideal layer silicate). Classification as a layer silicate originates from the continuous planar nature of the tetrahedral basal oxygens. The tetrahedral sheet is modulated from the ideal by the periodic reversal of tetrahedral apices along Y, thereby forming a 2:1 layer ribbon with the octahedra. Channels between ribbons may contain exchangeable cations and variable amounts of zeolitic-like water. More strongly bound water occurs at the edges of the octahedral units and these are depicted in Figure 5 as OH_2 groups.

Variations within the palygorskite/sepiolite group are based on ribbon widths. Palygorskite-type minerals have five octahedral cation sites per formula unit and sepiolite has eight, producing structural formulas of $R^{2+}_{(x-y-2z)/2}(H_2O)_4(Mg_{5-y-z}R^{3+}_y \Box_z)(Si_{8-x}R^{3+}_x)O_{20}(OH)_2(OH_2)_4$ and $R^{2+}_{(x-y+2z)/2}(H_2O)_8(Mg_{8-y-z}R^{3+}_y \Box_z)(Si_{12-x}R^{3+}_x)O_{30}(OH)_4(OH_2)_4$, respectively. Figure 5a illustrates the Brauner-Preisinger model (Brauner and Preisinger, 1956; Preisinger, 1959; Caillère and Hénin, 1961; Rautureau and Tchoubar, 1974; Rautureau, 1974) for sepiolite, which is a modification of an earlier model by Nagy and Bradley (1955). Palygorskite shows several different modifications to its basic structure, which was outlined by Bradley (1940) and later confirmed by Drits and Aleksandrova (1966) and Drits and Sokolova (1971). The Bradley model for palygorskite is given in Figure 5b. Several different unit cell geometries exist for palygorskite and may involve the way adjacent ribbons are oriented and linked and/or different ribbon widths (see Martin-Vivaldi and Linares-

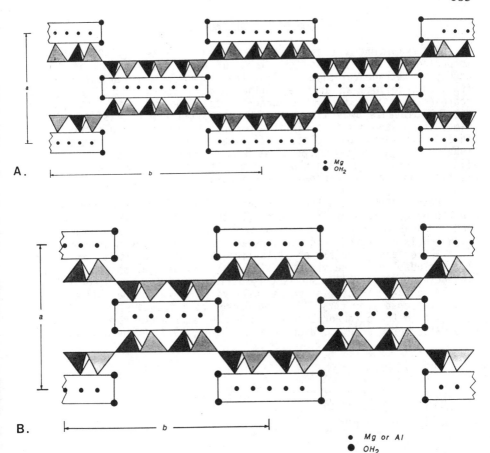

A.

Mg
OH₂

B.

Mg or Al
OH₂

Figure 5. A stylized model projection of the sepiolite (a) and palygorskite (b) structures down Z.

Gonzalez, 1962; Preisinger, 1963; Zvyagin et al., 1963; Christ et al., 1968; Gard and Follett, 1968). Bailey (1980) summarized these results further, and additional details of the palygorskite/sepiolite group are given in Chapter 16.

Lateral dimensional misfit between the sheets of tetrahedra and the ribbons of octahedra does not explain the tetrahedral inversion relationships observed in the palygorskite/sepiolite minerals. The geometric theme developed here requires a continuous octahedral sheet so that the oxygen to oxygen spacings of the common junction cannot be "reset" by the displacement of octahedra along Y. In the palygorskite/sepiolite minerals, the displacement occurs often and in phase with adjacent units, thereby further minimizing the development of an extensive octahedral-tetrahedral interface. Also, sufficient Al is sometimes present that lateral dimensional misfit is not required to account for these structures.

684

COMPOSITIONAL FACTORS

In examining 1:1 type layer silicates, Bates (1959) defined a morphological index (M) on the basis of the average radius of cations occupying tetrahedral sites ($\bar{r}_t$) vs the average octahedral cation radius ($\bar{r}_0$). He defined an arbitrary line in a $\bar{r}_t$ vs $\bar{r}_0$ plot (Fig. 6) that represents a "best fit" of component sheets and calculated the perpendicular distance away from this line as a measure of misfit. At the time, it was thought that misfit and morphology could be directly related in 1:1 structures, with platy serpentines having the best fit and coiled, tubular or cylindrical serpentines having poor fit. Although Bates did find a general relationship between misfit and morphology, he found that misfit alone is not responsible for serpentine morphology.

Bates (1959) noted the importance of two factors in determining the degree of curvature of 1:1 layers: (1) the misfit between the tetrahedral and octahedral sheets and (2) the strength of the interlayer bonds. Figure 6 is separated into three areas depending on the nature of misfit. Area II contains minerals of plate-like character (no misfit) whereas Area III includes coil-like structures (a small tetrahedral sheet relative to the octahedral sheet), such as chrysotile. Area I, according to Bates, should contain structures such as halloysite in which the coil is in the opposite sense from chrysotile and where the tetrahedral sheet is large relative to the octahedral sheet. Clearly, such a plot does not work well and later workers (e.g., Wicks and Whittaker, 1975; Guggenheim et al.,

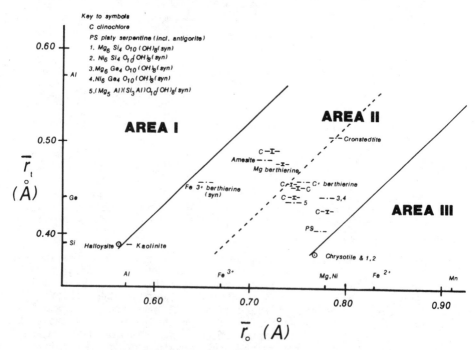

Figure 6. Kaolin/serpentine and chlorite minerals plotted on a $\bar{r}_t$ vs $\bar{r}_0$ diagram (after Bates, 1959). The nomenclature has been updated. The radii used are from Goldschmidt; see Bates (1959). Greenalite and caryopilite (not shown) plot well within Area III.

1982) confirmed that apparent compositional overlap between morphologically distinct structures is possible and that structural modulations can
account for apparently plate-like serpentines. Bates concluded that
misfit is probably not the major cause of variations in morphology in
1:1 layer silicates. Whereas morphology may be an expression of misfit,
more recent work emphasizes the relationship of misfit on structure
instead of morphology. Regardless of this distinction, however, misfit
is probably not the *sole* cause of the *nature* of structural modulations
in 1:1 layer silicates.

Bates (1959) argued that the strength of the interlayer bonds must
play an important role in how the 1:1 layers respond to misfit. He showed
that hydrogen content is roughly related to morphology also, which
suggests that the number of interlayer hydrogen bonds can constrain the
flexibility of an individual 1:1 layer. Bates (1959), Gillery (1959), and
later workers (e.g., Steadman and Nuttall, 1963; Chernosky, 1975; Mellini,
1982) recognized that electrostatic interactions exist between 1:1 layers
where, because of cation substitutions, the octahedral sheet acquires a
positive charge and the tetrahedral sheet a negative charge, although the
resultant 1:1 layer charge is neutral. Therefore, the nature of structural modulations in 1:1 layer silicates is interdependent not only on
the misfit between the tetrahedral and octahedral sheets, but also on
layer-to-layer interactions involving hydrogen bonding and on other
electrostatic interactions.

In contrast to 1:1 layer silicates, Guggenheim and Eggleton (1987)
suggested that the r_t vs r_o plot may be used to delineate geometrical fields
of misfit between the sheets of tetrahedra and octahedra for modulated 2:1
structures (Fig. 7). The modulated 2:1 layer silicates lack strong
layer-to-layer interactions, either through hydrogen bonding or by
electrostatic interactions caused by cation substitutions. Usually one
or more tetrahedral combinations are involved in linkages across the
interlayer. Tetrahedral sheets in a modulated 2:1 layer silicate face
tetrahedral sheets across the interlayer, thereby preventing octahedral
sheet to tetrahedral sheet interactions across the interlayer. In
addition, except for the tetrahedral connectors, layer-to-layer interactions are reduced because interlayer distances are quite large. Furthermore, because tetrahedra remain rigid owing to strong covalent bonding,
any adjustment by curving at the tetrahedral-octahedral sheet interface
is reflected by out-of-plane adjustments of basal oxygens also. Such
curvature enlarges the layer-to-layer distances further, thereby reducing
the likelihood of electrostatic interactions. In at least one case
(e.g., ganophyllite), the large interlayer separation allows H_2O and
cations in zeolite-like sites to exist adjacent to Al-substituted
tetrahedra to satisfy excess negative electrostatic charges on the basal
oxygens.

Figure 7 shows that neither theoretical considerations (line a) based
on simple geometrical considerations nor experimental studies (line b) can
be used to separate definitively regions of normal layer silicate
geometrical stability from that of modulated 2:1 layer silicates. Note
that there is overlap of both modulated structures (stilpnomelane) and
normal 2:1 layer silicates (talc). Guggenheim and Eggleton (1987)
suggested that such regions should not be well-defined because of
difficulties in assessing environmental factors (see below). In addition,
plotted positions are dependent on the compositions of the phases,
including the oxidation state of the constituent cations, and on the

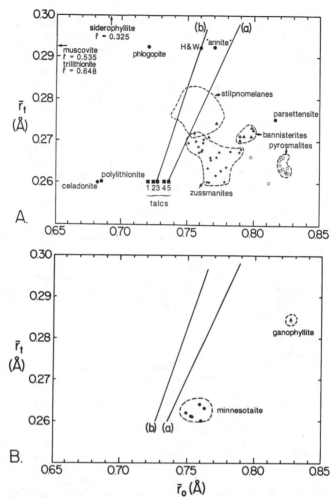

Figure 7. The modulated 2:1 layer silicates plotted on a $\bar{r}_t$ vs $\bar{r}_o$ diagram (after Guggenheim and Eggleton, 1987) with island structures in (a) and strip structures in (b). The radii are from Shannon (1976). Pyrosmalite, a modulated 1:1 layer silicate, is plotted also, as this structure has no hydrogen bonds between the layers. See original source for details.

proper allocation of elements into the tetrahedral and octahedral sites. Based on the nature of a continuous octahedral sheet and if data do not exist to allocate cation site occupancy, it is recommended that aluminum be assigned to the tetrahedral sites rather than octahedral sites, especially if octahedral sheets are primarily Fe^{2+} and/or Mn^{2+} rich (see Guggenheim and Eggleton, 1987, for additional details).

ENVIRONMENTAL FACTORS

The $\bar{r}_t$ vs $\bar{r}_o$ plot for modulated 2:1 layer silicates requires precise values for ionic radii at the temperature and pressure of mineral formation, if the plot is to represent the true upper limits of tetrahedral-octahedral sheet misfit. However, not only are these values difficult to obtain, it is often difficult to ascertain the environmental conditions of mineral formation. Therefore, the regions of delineating geometric stability are only approximations as the ionic radii used for these plots are based on ambient conditions.

High temperature studies of fluorophlogopite (Takeda and Morosin, 1975) and muscovite (Guggenheim et al., 1987) and high pressure studies of phlogopite and chlorite (Hazen and Finger, 1978) have demonstrated strong anisotropic temperature and pressure responses which relate to differences in bonding within and between 2:1 layers. For example, interlayer distances are characterized by weaker bonds in these layer silicates. Therefore, interlayer volume reduction is high when compressive forces are applied along Z^*. These results imply that hydrostatically applied stress to the system is not necessarily transmitted through the structure isotropically, but is generally controlled by and transmitted along bonds. In contrast, the inverse relationship generally holds for increasing temperature, with interlayer volume increasing greatly.

Unlike normal layer silicates, modulated layer silicates generally have tetrahedral connectors across interlayer space and, therefore, these structures should be significantly stiffer along Z^*. Thus, the interlayer region cannot act as an efficient pressure "buffer". Instead, the tetrahedral connectors would be expected to transmit pressure more efficiently along Z^*. Because the tetrahedra are very rigid, an octahedral site in a modulated structure with interlayer tetrahedral connectors would be expected to have a different compressibility than the analogous site in a normal layer silicate. Assuming that compressive forces are transmitted to the octahedral sites efficiently and are not greatly attenuated by T-O-T bending (see the suggestion by Vaughan in Hazen and Finger, 1982, p. 155-6), a high pressure, low temperature environment should effectively promote a reduced tetrahedral-octahedral sheet misfit. These conditions apparently prevailed during the formation of zussmanite (Agrell et al., 1965), which is found in the Franciscan Formation of California and is believed to be of blueschist-facies origins. Of course, although tetrahedral-octahedral sheet misfit may be a geometrically limiting feature of these structures, thermal stability is not based necessarily on this feature alone.

POLYSOMATIC AND HOMOLOGOUS SERIES

The systematics of the modulated layer silicate group may be conveniently described using the concepts of polysomatic and homologous series. For modulated 1:1 layer silicates, Mellini et al. (1985) described carlosturanite (see below) as composed of two chemically distinct modules (Thompson, 1978) and noted the possibility of structural intermediates (polysomes) by varying the ratios of the modules. In addition, antigorite (Spinnler, 1985; Livi and Veblen, 1987) may be thought of as polysomatic in nature, with serpentine-like slabs and reversal modules. In contrast, modulated 2:1 layer silicates have been described (Guggenheim and Eggleton, 1987) in terms of a homologous series, which is defined here by analogy to a biological series: a group of species (minerals) with closely corresponding structures derived from common genetic (e.g., component sheet chemistry) origins. The origin of the modulations is closely related to misfit between the tetrahedral and octahedral sheets.

These two concepts are not mutually exclusive. The carlosturanite and antigorite series, for example, are homologous series, although the apparent chemical compositions of the component sheets are not the only cause of the structural variations. It is more instructive, however, to emphasize the differences between the two concepts rather than the similarities.

In contrast to the polysomatic series in which compositional variations are limited, the homologous series allows the systematics of the group to be rationalized, even for structures with diverse chemistries. Although this approach is less successful for 1:1 modulated layer silicates, the genetic relationships regarding chemistry do generally hold: octahedral sheets with large cations such as Fe^{2+}, Mn^{2+} (e.g., greenalite, caryopilite) generally require greater tetrahedral sheet structural adjustments than octahedral sheets with smaller cations (e.g., antigorite, carlosturanite). The inability to determine structural constraints more precisely within a compositionally restrictive series in the serpentines (e.g., antigorite, carlosturanite) is related to the lack of precise understanding of what causes the modulations.

The early analysis by Bates (1959) of the importance of hydrogen bonding between layers, the later work of Wicks and Whitaker (1975) showing major element compositional overlap between lizardite, chrysotile, and parachrysotile, and the importance of water fugacity in understanding the phase relations of kaolinite (Yeskis et al., 1985) suggest that hydrogen bonding (in particular, the number and orientation of H bonding between layers) is a major parameter in understanding the systematics of the 1:1 modulated layer silicates, especially the Mg-rich varieties. It follows, therefore, that hydrogen (and water) fugacity differences should be considered as an important variable in the crystallization of such materials and in the development of polysomes. It should be noted that this suggestion is somewhat speculative and requires further study.

It is interesting to note further that such a suggestion is consistent with the polysomatic series as applied to the hydrous biopyriboles (e.g., Veblen and Buseck, 1981). In these minerals, Veblen and Buseck (1981) suggested the importance of hydration alteration reactions in the transformation of chain zippers of different widths (polysomes). Similarly, there is increasing evidence that a wide variety of intergrowth microstructures appear to occur in Mg-rich 1:1 modulated layer silicates in analogy to the hydrous biopyriboles. Such intimate intermixing would be common in a polysomatic series.

MODULATED 1:1 LAYER SILICATE STRUCTURES

Modulated 1:1 layer silicates are presented roughly in order of increasing octahedral cation size. Rolled forms such as chrysotile, although they follow similar principles as outlined above, are not periodic in structure and therefore, are not included in this chapter.

Antigorite

Antigorite is used here to designate a series of Mg serpentines in which there is a variable superstructure along the X direction owing to the reversal of apical oxygens and a relinkage of the tetrahedral sheet to the adjacent octahedral sheet (Fig. 8). The structure of antigorite

A.

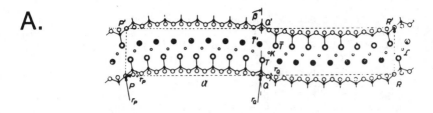

From Kunze (1961)

B.

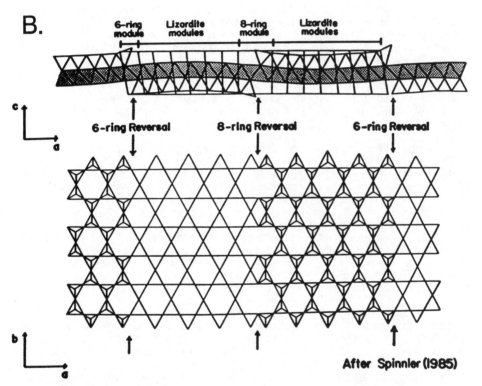

Figure 8. The antigorite alternating wave structure illustrated in projection down Y (from Kunze, 1961) in (A), and as a polysomatic model and in projection down Z in (B). Part (B) is from Livi and Veblen (1987), after Spinnler (1985).

(Zussman, 1954; Kunze, 1956; 1958; 1961) involves an alternating wave form and, for the sample depicted in Figure 8, each half wave has a slightly different curvature and different width. Note that there are 17 tetrahedra in a full wave with half wave widths of nine and eight tetrahedra, respectively. Because tetrahedral edges parallel a and are equal to 2.55 Å (= $a/2$) in length, the supercell edge, A', may be related approximately to m, the number of tetrahedra in a full wave by the equation $A' = m (a/2)$, or 43.3 Å for the 17 tetrahedral repeat shown in Figure 8. At certain inversion points, the six-fold tetrahedral ring configuration changes to four and eight-fold rings with apical oxygens reversed in opposite senses in adjacent half waves. Thus, an "extra" tetrahedron is inserted into the sequence relative to an ideal tetrahedral sheet. The wave-like nature of the tetrahedral component illustrates the mechanism of obtaining congruency between two sheets with apparent misfit. The out-of-plane tilt of the tetrahedra enlarges apical oxygen to apical oxygen distances, thereby allowing a curvature at the interface between the octahedral and tetrahedral sheets. The octahedral sheet accommodates such tilting by becoming wave-like also. The integrated system works because the radius of curvature differs slightly for the octahedral cations as compared with the tetrahedral cations.

Although the octahedral cation spacing is regular in Figure 8 with eight octahedral sites depicted per half wave, note that OH and apical oxygen topology differs at the two inversion lines (PP' or RR' vs QQ'). At QQ', one third of the octahedra have a composition of $MgO_4(OH)_2$ rather than the normal $MgO_2(OH)_4$. Kunze (1961) suggested that QQ' type inversions, resulting in four and eight fold rings, occur at all inversion points when adjacent half waves have even numbers of tetrahedra. As inversion lines of four and eight-fold rings increase in number, the sample becomes less hydrous. Antigorite clearly has a complex interlayer bonding system involving interlayer connections at inversion sites and, possibly some sort of hydrogen interlayer bonding network between connections.

Kunze (1958) related the chemical composition of antigorite to m by the equation $m[Mg_{3(1-1/m)}SiO_2O_5(OH)_{1+3(1-2/m)}]$. Zussman et al. (1957), Chapman and Zussman (1959), Kunze (1961) and Shitov and Zvyagin (1966) have found that m values are usually between 10 to 20 (approximately 25 to 50 Å repeats along X), resulting in chemical compositions of $Mg_{2.70}Si_2O_5(OH)_{3.40}$ to $Mg_{2.85}Si_2O_5(OH)_{3.70}$, respectively. Mellini et al. (1987) found that the a axis dimension and, hence, chemistry varies as a function of metamorphic grade and crystallinity, with $m = 17$ believed to be the most stable configuration at about $435^{\circ}C$ for the Franscia rocks (Alps) in their study. Samples with small m values have been shown (Mellini and Zussman, 1986) to be carlosturanite (see below) and not, as previously thought, picrolite, a fibrous variety of antigorite.

Although the model structure of antigorite as shown in Figure 8 is well established, Uehara and Shirozu (1985) suggested a variation of the model for antigorites with an odd number of octahedra and an even number of tetrahedra in the A' repeat. For these structures, the model as presented by Kunze (1961) predicts that some Mg cations would have ten-fold coordination groups. The model of Uehara and Shirozu for these structures involves a $b/2$ displacement along Y relative to the Kunze model, thereby making inversion points at the tetrahedral-octahedral sheet junction more like those of Figure 8, but with more reasonable six-fold coordination about the Mg. Such a model predicts a true periodicity of two waves (2 x A'), which is observed in some samples. By analogy to Kunze (1961), a subcell composition can be derived by using the ratio of $M = (m-1)/2$: $Mg_{6}Si_{4(1+M/2)}O_{10(1+M/2)}(OH)_{8-2/M}$. Further details of antigorite may be found in Chapter 5.

Livi and Veblen (1987) described an antigorite-like phase in which the modulations are non-periodic. In a similar way to that presented by Mellini et al. (1985) for carlosturanite (see below), they illustrated modulation variability in antigorites as a polysomatic series. The antigorite-like phase is composed of varying numbers of lizardite modules separated by inversions involving modules of either four and eight-fold rings (QQ' type inversions) or six-fold rings. Spinnler (1985) noted that the two inversion modules are not electrostatically neutral and are oppositely charged, and should occur in pairs, as is observed for this phase. Unlike normal antigorite, this phase shows modulations in a single grain along any of the three pseudo-hexagonal a axes as compared to an ideal serpentine structure. Paired offsets (half waves) are approximately 18 Å to 65 Å and correspond to 32 Å to 130 Å repeats along X for antigorite.

Carlosturanite

Carlosturanite was described by Compagnoni et al. (1985) as a low-grade metamorphic rock-forming mineral from the Monviso ophiolite, Sampeyre, Italy. It has been found also at Taberg, Sweden (Mellini and Zussman, 1986), where it had been described previously (Whittaker and Zussman, 1956) as picrolite, a variety of antigorite. Several unreported localities from the Western Alps (Ferraris, as cited in Mellini and Zussman, 1986) have been found for carlosturanite also, suggesting that the occurrence of the mineral is probably widespread.

Carlosturanite (Compagnoni et al., 1985) has a general structural formula, of $M_{21}[T_{12}O_{28}(OH)_4]OH_{30}.H_2O$ with M predominantly Mg with lesser amounts of Fe, Mn, Ti, Cr and T = Si, Al. Unit cell parameters are $a = 36.70$ Å, $b = 9.41$ Å, $c = 7.291$ Å, $\beta = 101.1°$ and it crystallizes in space group Cm, as do several serpentines. Carlosturanite differs from serpentines chemically by higher magnesium and water contents and by a lower silicon content.

Carlosturanite was described by Mellini et al. (1985) in terms of a polysomatic series. The model structure consists of a continuous planar octahedral sheet and a tetrahedral "sheet" modified to strips, six tetrahedra in width, running parallel to b (Fig. 9). The structure contrasts with antigorite by (a) the discontinuous nature of the tetrahedral arrangement and the lack of tetrahedral reversals across the interlayer region, (b) protons at tetrahedral corners located at strip edges, (c) H_2O molecules between tetrahedral strips, and (d) the planar octahedral sheet instead of a wave-like curvature. The model structure differs from other inophite (*ino*silicates = chains, *ophite* = serpentines) polysomes of this series by variations in strip width. The six-tetrahedra-wide strip of carlosturanite may be described as five connected (100) serpentine slabs (designated as "S" in Fig. 9) separated by interstrip regions (designated as "X"). Hence, carlosturanite may be designated as an S_5X polysome. Mellini et al. noted from (TEM) lattice fringes that (100) faulting is common to produce S_4X, S_6X and S_7 polysomes, although these periodicities do not extend for more than one or two unit cells.

The periodicities of the chain widths are roughly analogous to the periodicity of the wave-like form of antigorite, thereby producing similar $hk0$ diffraction patterns. Mellini and Zussman (1986) noted, however, that the patterns are readily distinguishable by comparing the sinusoidal distribution of intensities along X^* in the $hk0$ net. For antigorite, the continuous nature of the tetrahedral sheet with a strong hexagonal character plus the strong hexagonal character of the octahedral sheet imparts a similar symmetry to the $hk0$ net with intensity maxima clustering near normal serpentine nodes. In contrast, because of the discrete

692

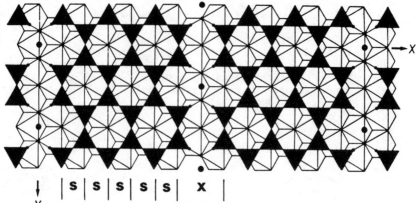

Figure 9. Polyhedral model of carlosturanite projected down Z with S and X modules illustrated (modified from Mellini et al., 1985).

tetrahedral strips, the hexagonal character of the $hk0$ net of carlosturanite is diminished owing to the less continuous modulation of the structure relative to antigorite. They noted also that samples with a axis repeats of less than 33.7 Å are likely to be carlosturanite rather than antigorite.

Baumite

Frondel and Ito (1975) described baumite as a serpentine with a chemical formula of $(Mg_{5.4}Mn_{2.2}Fe_{1.7}Zn_{1.0}Al_{0.7})(Si_{7.1}Al_{0.9})O_{20}(OH)_{16}$ from Franklin, New Jersey. Guggenheim and Bailey (1988) re-examined the original bulk material by transmission electron microscopy and powder X-ray diffraction techniques and found that it consists of a mixture of a serpentine phase similar to greenalite and caryopilite (see below) intergrown with chlorite and other phases. At the present time, it is unclear if the term baumite is to be re-defined or discredited (pending IMA action) and, therefore, it is included here as a distinct species based on the old definition. Although an accurate determination of baumite chemistry was not possible, Guggenheim and Bailey found that it contains significant amounts of Zn, but argued that baumite contains considerably less Al than the rock bulk chemistry suggests.

Baumite differs from greenalite and caryopilite by both its Zn-rich chemistry and the dimensions of its island-like domains. Baumite has island-like regions (approximately 30 Å) considerably larger than either greenalite (21.3 - 23.3 Å) or caryopilite (16.7 - 17.2 Å). These values represent a measure of the lateral extent to which the tetrahedral and octahedral sheets can maintain congruence before strain must be relieved by a structural perturbation.

Greenalite

Greenalite was recognized as a serpentine mineral by Gruner (1936) from X-ray powder data. Steadman and Youell (1958) determined the cell dimensions as relating to a one-layer orthogonal cell (=1T) of one sample and confirmed its 1:1 layer structure. Electron microprobe analysis by Floran and Papike (1975) indicated a pure Si tetrahedral sheet and a Fe^{2+}-rich octahedral sheet with lesser amounts of Mg (3.8% MgO) and very small amounts of Mn and Al. Guggenheim et al. (1982) examined a large number of samples and found that greenalite (and caryopilite) consistently show an excess of Si and a deficiency of total octahedral cations relative to an ideal serpentine composition.

Guggenheim et al. (1982) found that greenalite is always composed of coherent fixed structural intergrowths of a group-C polytype (see Bailey, 1969) with smaller volumes of a group-A polytype, with the latter rotated by $180°$ or inverted without rotation with respect to the former. Group-C structures ($1T$, $2T$, $3R$ polytypes) are characterized by the same occupancy of octahedral sites in every layer and by interlayer shifts of zero or $b/3$, whereas group-A structures ($1M$, $2M_1$, $3T$) have also the same occupancy of octahedral sites in each layer, but with interlayer shifts of $a/3$ along one, two, or three of the pseudohexagonal X axes. The volume of the group-A structure present is (non-linearly) dependent on the Mn content (see caryopilite) and is small even for very iron-rich samples. Data are not available to determine if it is present at the end-member composition.

Electron diffraction, high resolution transmission electron microscopy, and optical imaging techniques were used to determine the structure. Greenalite is composed of island-like regions of six-member rings with out-of-plane tilting that may be roughly described in three-dimensions as resembling a stack of saucers. The larger octahedral sheet is convex upward and caps the island-like tetrahedral six-member rings. Inverted four-member and three-member rings exist at island boundaries, and they give rise to the non-stoichiometric chemistry relative to true serpentines. Figure 10 represents an idealized structure. Guggenheim et al. suggested that entire islands may be inverted and presented data showing that the islands are not always equidimensional or four rings in diameter.

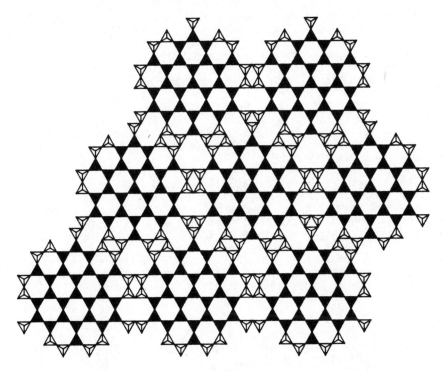

Figure 10. Idealized tetrahedral sheet in greenalite (from Guggenheim et al., 1982).

Details of these structures (greenalite, caryopilite, and baumite), especially with regard to the island boundary connectors, should be considered tentative until more data become available.

"*Tosalite*" (Yoshimura, 1967) was described as an intermediate member of the greenalite-caryopilite series. Guggenheim et al. (1982) examined one sample (Matsuo mine) of tosalite and found it to be structurally identical to greenalite. Therefore, if all tosalites are similar, they are classified as manganoan greenalites. In contrast, samples containing substantial amounts of Fe and Al have been shown to be *berthierine*, sometimes referred to as chamosite (or 7 Å chamosite) in the older literature. Presumably, because of large amounts of tetrahedral aluminum, berthierine is a non-modulated, platy 1:1 layer serpentine and is not considered further here.

Both greenalite and caryopilite (see below) have structures that are consistent with crystal growth from a low energy environment, such as that found in colloidal or gel-like media subjected to low grade metamorphic conditions. Island-like domains in the structure may represent (random) nucleation sites which extend laterally as a normal 1:1 layer until tetrahedral and octahedral sheet misfit requires a perturbation. This interpretation is consistent with (1) the variable size and shape and lateral disposition of the domains and the poorly defined interdomain region and (2) textural evidence for both greenalite (Jolliffe, 1935) and caryopilite (Huebner, 1986).

Caryopilite

Kato (1963) noted that two distinct Mn-rich minerals were being called bementite, and designated the friedelite-type species as bementite (see below) and the chamosite-type (7 Å chamosite is now called berthierine) as caryopilite. He noted that the diffraction pattern of caryopilite has satellite reflections indicating a complex structure. Shirozu and Hirowatari (1955) determined that caryopilite is serpentine-like and has a $1M$ structure. Later, based on X-ray powder patterns and electron microprobe data only, Peacor and Essene (1980) suggested that caryopilite is structurally similar to friedelite. Guggenheim et al. (1982), however, showed the close resemblance of the electron diffraction patterns of caryopilite to greenalite (see above) and reconciled differences on the basis of a model.

The structure of caryopilite has island-like regions of three silicate rings in diameter instead of four as in greenalite. This variation in structure is consistent with a homologous relationship between the iron-rich (greenalite) and manganese-rich (caryopilite) end of the compositional series, owing to the differences in lateral sizes of Fe-rich and Mn-rich octahedral sheets. In contrast to greenalite, caryopilite is composed of a dominant group-A polytype in addition to a small amount of group-C polytype. Like greenalite, the polytypes are coherent structural intergrowths and not just simple mixtures.

Pyrosmalite group

The pyrosmalite-type minerals may be described chemically by using the simplified formula (Frondel and Bauer, 1953) of $M^{2+}_{16}T_{12}O_{30}(OH,Cl)_{20}$, where M^{2+} = Mn^{2+}, Fe^{2+}, Mg, Zn and T = Si, Fe^{3+}, Al and possibly Ti^{4+}. *Pyrosmalite* is a general term for the Mn, Fe series, *manganpyrosmalite* refers to compositions with Mn > Fe, and *ferropyrosmalite* applies to Fe > Mn (Vaughan, 1987). Pyrosmalite crystallizes in space group *P3m1* (Frondel and Bauer, 1953). *Friedelite* is compositionally similar to manganpyrosmalite, but it has a rhombohedral lattice (Bauer and Berman, 1928) with c = 21.43 Å or c = $3xc_0$ of pyrosmalite (c_0 = 7.15 Å). *Mcgillite* (Donnay et al., 1980) has recently been described with c = 85.657 Å (c = $12xc_0$) in space group *R3m*. *Schallerite* (c = 14.311 Å = $2c_0$) has been found to be both chemically (Bauer and Berman, 1928) and structurally (Frondel and Bauer, 1953; McConnell, 1954) related to the pyrosmalite group minerals. Recently, however, Dunn et al. (1981) compiled additional chemical data. They suggested that arsenic is present in stoichiometric quantities, indicating a unique crystallographic arsenic site. The data are not substantial and additional work on schallerite is necessary to determine its relationship to the pyrosmalite group. *Nelenite*, formerly known as "ferroschallerite", is As-rich and high in iron. Although "ferroschallerite" was discredited by Frondel and Bauer (1953) as a variety of friedelite, Dunn and Peacor (1984) have proposed the new name for this mineral because they believe that the arsenic content is essential and non-variable. Nelenite is isostructural with friedelite. Peacor et al. (1986) have reported several arsenites with the same subcell as the pyrosmalite group minerals. *Manganarsite* is the arsenite analogue of manganpyrosmalite. In addition, two other arsenites appear to be related to schallerite and friedelite. Peacor et al. suggested that these arsenites are based on chains or rings instead of a pyrosmalite tetrahedral sheet. It is possible also that interlayering of the silicate and the arsenite phases may contribute to the chemical complexity of schallerite and nelenite.

The crystal structure of manganpyrosmalite has been determined by Takéuchi et al. (1969), partially refined by Kashaev (1968) using pyrosmalite, and more completely refined by Kato and Takéuchi (1983) with crystals of manganpyrosmalite and iron-rich manganpyrosmalite. Although this structure was defined based on a rhombohedral cell, a monoclinic cell was later found to be the true cell. This cell is similar to the unit layer in the mcgillite and friedelite structures (Ozawa et al., 1983; Iijima, 1982a). Details of the polytypic relation between the pyrosmalite minerals have been given by Kashaev and Drits (1970), Iijima (1982b) and Takéuchi et al. (1983).

The manganpyrosmalite structure (see Fig. 11) is a modified 1:1 layer structure with half of the tetrahedra coordinating to the octahedral sheet normally associated with a true serpentine. The remaining tetrahedra are inverted and link to adjacent octahedral sheets across what is normally the interlayer region. The tetrahedral sheet consists of laterally linked six-fold tetrahedral rings arranged in a three-fold array. Alternate tetrahedral rings are inverted. The same array may be visualized with 4-, 6-, and 12-fold rings of tetrahedra with half of the tetrahedra in the 4- and 12-fold rings inverted. Because there are an equal number of tetrahedra coordinating to the two adjacent octahedral sheets, it is possible to visualize the structure as a modified 2:1 layer structure. However, unlike 2:1 layer structures, such a scheme produces a 2:1 layer stacking sequence with no interlayer region. This result and the strong 7 Å subcell repeat along the stacking direction indicate a greater affinity to the 1:1 layer structures.

There are several important differences between the pyrosmalite structure and other modulated 1:1 types structures. Antigorite, which has

696

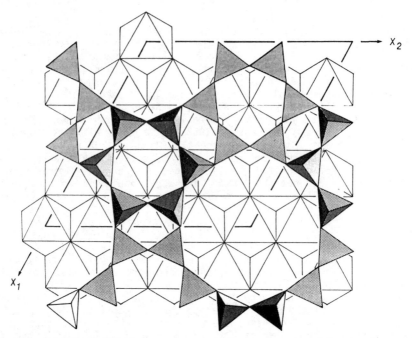

Figure 11. Polyhedral model of the manganpyrosmalite structure down Z (based on Kato and Takéuchi, 1983).

similar tetrahedral inversions across the "interlayer" region but in a different configuration, has an apparent wave-like character to both the tetrahedral and octahedral sheets. In contrast, the wave-like nature of manganpyrosmalite as described by Kato and Takéuchi (1983) appears confined to the tetrahedral sheet alone and without significant curvature at the interface between the octahedra and the tetrahedra. In other words, the octahedral cation plane is more ideally planar than in antigorite and more closely resembles the octahedral configuration in carlosturanite (Mellini and Zussman, 1986). This suggests that for the composition of manganpyrosmalite (Kato and Takéuchi, 1983), the tetrahedral configuration resolves the misfit between the two component sheets adequately.

In contrast to the platy serpentines, the pyrosmalite group minerals do not have hydrogen bonding between layers; adjacent layers are interconnected through inverted tetrahedra only. Therefore, it is of interest to compare the pyrosmalite group minerals to the 2:1 modulated structures on a $\hat{r}_t$ vs $\hat{r}_o$ plot (see Fig. 7), as these structures do not have interlayer hydrogen bonding either.

Bementite

Caryopilite (see above) and bementite were identified by Kato (1963) as separate species based on electron diffraction patterns. Caryopilite was found to be similar to serpentine ("chamosite-like") with a hexagonal-based unit cell, whereas bementite was found to have a rectangular-based cell with a = 14.5 Å, b = 17.5 Å, c = 7.28 Å x 4 = 28 Å, and space group $P222_1$ or $P222$. Chemistry has not been reliably established but was assumed by Kato (1963) to be $Mn_5Si_4O_{10}(OH)_6$. Kato (1963) and Kato and Takéuchi (1980) suggested a structural model in space group $P222_1$, with a chemistry of $Mn_7Si_6O_{15}(OH)_8$, Z = 16 by analogy to pyrosmalite.

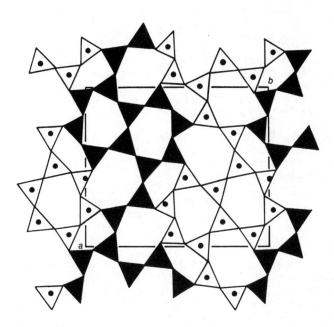

Figure 12. The terahedral sheet of bementite projected on the (001) plane (Eggleton and Guggenheim, 1988).

The Kato model, which was not supported with the requisite calculations to determine its compatibility with the data, involves adjacent octahedral sheets rotated relative to each other by 24° in the (001) plane. This relationship prevents the (3.3 Å) hexagonal subcell commonly observed in the other modulated layer silicate structures from forming. For a pyrosmalite-type tetrahedral sheet as proposed by Kato (1963), six tetrahedral edges span the 14.5 Å dimension. On the basis that O--O distances in Si tetrahedra are about 2.65 Å, the 14.5 Å dimension is too short and corresponds to O--O distances of 2.42 Å. Such apparent distances are possible if tetrahedra are rotated in-plane, but tetrahedral rotation, which reduces the lateral dimensions of a tetrahedral sheet, would be unlikely based on the large Mn-rich octahedral sheet.

Eggleton and Guggenheim (1988) presented data showing that bementite is monoclinic with a = 14.83 Å, b = 17.58 Å, c = 14.70 Å (2 x c_0), β = 95.5°, and space group $P2_1/c$. Like the model of Kato (1963), they suggested a structure with alternate octahedral octahedral sheets rotated by 24°, but with a unit cell containing a tetrahedral sheet with pairs of six-fold rings connected by five-fold and seven-fold rings (see Fig. 12). Tetrahedra within each pair of six-fold rings coordinate to one octahedral sheet, with adjacent pairs inverted and linking to the octahedral sheet either above or below. This model has a chemistry of $Mn_7Si_6O_{15}(OH)_8$ and is consistent with the observed X-ray data, electron diffraction patterns, and electron images. Additional data (Eggleton and Guggenheim, unpublished), however, showed variations in unit cell dimensions in some grains, thereby indicating structural complexities still not understood. Further work to explore these variations and possible differences in chemistry is needed.

MODULATED 2:1 LAYER SILICATE STRUCTURES

The modulated 2:1 layer silicate structures may be separated into two groups, (1) those having strips composed of tetrahedral rings on both sides of the continuous octahedral sheet or (2) those having regions of tetrahedral rings that are analogous to islands. Zussmanite and stilpnomelane belong to the latter whereas minnesotaite and ganophyllite comprise the former. Bannisterite does not clearly belong to either group but, for convenience, has been included with zussmanite and stilpnomelane.

Although the modulated 1:1 layer silicate structures may be separated into subgroups also, there is less virtue in this approach for these structures. In contrast to the modulated 2:1 layer silicates, extensive hydrogen bonding between layers in most 1:1 structures prevents a strictly geometric interpretation in explaining modulation diversity. It is this difference, of course, that leads in part to the conceptual differences between a polysomatic reaction series and a homologous series as given above.

Guggenheim and Eggleton (1987) noted that variations between strip and island arrangements of the tetrahedra must have important consequences in how misfit between the two sheets is relieved. The change in configuration of the tetrahedral sheet in strip structures relieves the strain along one axis. Whereas this mechanism is suitable for relieving misfit in one direction, cation ordering or small strip widths are required to help reduce strain in other directions. In contrast, island structures relieve misfit along both X and Y simultaneously by surrounding the (island) region developing strain by a tetrahedral configurational change.

Island structures

Zussmanite. Zussmanite was described briefly by Agrell et al. (1965) in which they proposed a chemical formula of: $(K_{0.92}Na_{0.07})(Fe^{2+}_{10.85}$ $Mg_{1.53}Mn_{0.46}Al_{0.34}Fe^{3+}_{0.11}Ti_{0.01})_{\Sigma = 13}(Si_{16.6}Al_{1.4})O_{42.2}(OH)_{13.8}$ (ideally: $KFe_{13}(Si_{17}Al)O_{42}(OH)_{14})$. Lopes-Vieira and Zussman (1969) determined and refined the structure of zussmanite in space group $R\bar{3}$, but noted also that many reflections were streaked parallel to Z^*. Jefferson (1976) showed that these reflections were caused by two types of stacking disorder, one involving the rhombohedral structure and the other relating to a severely strained structure with triclinic symmetry. There is an apparent compositional variation between the two structures with the triclinic form depleted in K and slightly enriched in Mn at the expense of Fe. Polytypes may form with both structures, and Jefferson (1976) has noted the occurrence of numerous one and two-layer varieties. Muir Wood (1980), in a study on the same material, did not find the inverse potassium to manganese variation in all cases as was reported by Jefferson (1976). In addition, he noted the occurrence of zussmanite-like (designated as "Zu2") material with cell parameters approximately 8% smaller and with a chemistry of approximately $KAlMn_{3-5}Fe^{2+}_{10-8}Si_{17}$ $O_{42}(OH)_{14}$ (cf. zussmanite).

The zussmanite structure (Lopes-Vieira and Zussman, 1969) contains T-O-T layers defined by six-fold rings of tetrahedra opposing a continuous octahedral sheet. Three-fold tetrahedral rings laterally connect the six-fold rings and connect adjacent T-O-T layers through a tetrahedral edge across what is the interlayer region in a normal layer silicate (see Fig.

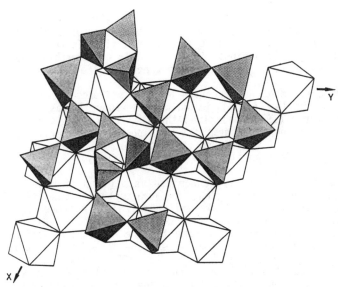

Figure 13. Zussmanite polyhedral model with K cations omitted. These cations reside in twelve-fold sites between opposing six-fold rings, one of which is illustrated (from Guggenheim and Eggleton, 1987 as modified from Lopes-Vieira and Zussman, 1969). Compare this figure to that of fedorite, figure 4.

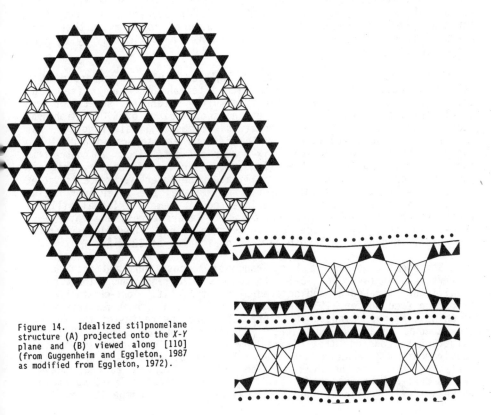

Figure 14. Idealized stilpnomelane structure (A) projected onto the X-Y plane and (B) viewed along [110] (from Guggenheim and Eggleton, 1987 as modified from Eggleton, 1972).

13). Alkali ions reside in the hole formed by two opposing six-fold rings in the interlayer region but the number of such sites is restricted by the presence of the three-fold rings; there are only one fourth as many alkali positions in zussmanite as compared with a normal mica.

Stilpnomelane. Modern structural work on stilpnomelane was initiated by Eggleton and Bailey (1965) in which the subcell was defined and described. Eggleton (1972) described the full cell and later (Eggleton and Chappell, 1978) compared chemistry with physical properties and cell data. Crawford et al. (1977) found evidence for two 2-layer and many long range periodicities from an electron microscope study.

Stilpnomelane is generally recognized as a group of minerals, with the bulk of component iron ranging from primarily ferrous (*ferrostilp-nomelane*) to primarily ferric (*ferristilpnomelane*). Although Mg-dominant stilpnomelanes had been known for some time, Dunn et al. (1984) designated this species as *lennilenapeite*. A name for the Mn-dominant stilpnomelane phase has not been designated. A manganese analogue known as *parsetten-site* (see below) was recently (Guggenheim, 1986; Ozawa et al., 1986) shown to be structurally different from stilpnomelane. The extent of manganese solid solution in the stilpnomelane structure is not presently known.

The end-member iron varieties have structural formulae dependent on the protonization of the hydroxyl groups: $K_5Fe^{2+}_{48}(Si_{63}Al_9)O_{168}(OH)_{48} \cdot 12H_2O$ for ferrostilpnomelane and $K_5Fe^{3+}_{48}(Si_{63}Al_9)O_{216} \cdot 36H_2O$ for ferri-stilpnomelane. Eggleton and Chappell (1978) suggested a simplified formula based on one eighth of the structural formula so that, for example, ferrostilpnomelane can be represented approximately by $K_{0.6}Fe_6(Si_8Al)(O,OH)_{27} \cdot 2H_2O$.

The structure of stilpnomelane (Eggleton, 1972) is composed of a continuous octahedral sheet and modulated tetrahedral sheets (see Fig. 14). The tetrahedral sheets articulate to the octahedral sheet, but form nearly trigonal islands of six-member hexagonal rings of SiO_4 tetrahedra. Each island consists of seven complete six-member silicate rings and has a diameter of seven tetrahedra. The islands are separated and linked laterally by an inverted single six-member silicate ring. In contrast to the hexagonal rings within the islands, these rings have a more trigonal configuration. In a normal layer silicate, the interlayer space is vacant or contains the alkali cation. In stilpnomelane, the interlayer region contains the alkali cation and the inverted six-member trigonal silicate rings, which link the islands and adjacent layers across the interlayer space. The interlayer space has two trigonal rings joined along Z through the apical oxygens. This results in a d_{001} value of 12.2 Å.

Stilpnomelanes have a large range of compositions and span across a $\bar{r}_t$ vs $\bar{r}_o$ plot (see Fig. 15a) into regions with (right) and without (left) significant tetrahedral and octahedral sheet misfit. Post-crystallization oxidation (see Hutton, 1938 and later workers) can account for this spread of points, as may be noted in Figure 15b where the radius of the ferrous ion is used in the calculations. Guggenheim and Eggleton (1987) were reluctant to conclude that misfit between the two component sheets is a requirement to produce a modulated 2:1 layer structure, as one Al-rich analysis falls to the left portion of the plot. Clearly, however, tetrahedral-octahedral sheet misfit is a geometrically *limiting factor*. Whereas (Fe^{2+}, Fe^{3+})-stilpnomelanes may have a large compositional range,

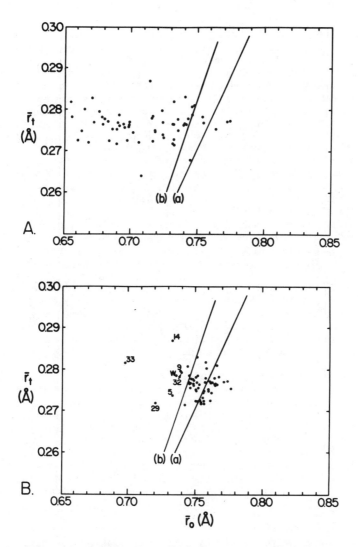

Figure 15. Average tetrahedral cation radii ($\bar{r}_t$) vs average octahedral cation radii ($\bar{r}_o$) for stilpnomelane. Analyzed chemistry is given in (A) and all iron assumed FeO in (B). Numbers and letters in (B) are based on reported analyses from Eggleton and Chappell (1978). See Guggenheim and Eggleton (1987) for details. The radii are from Shannon (1976).

there is a geometrical limit at which the misfit between the two component sheets prevents the formation of the structure (at higher average octahedral cation sizes).

Parsettensite. Parsettensite was first described by Jakob (1923) as a manganese-rich stilpnomelane. Dunn et al. (1984) confirmed that the chemistry of parsettensite is roughly analogous to stilpnomelane. However, closer comparison (Guggenheim and Eggleton, 1987) of the chemical composition of the two phases shows that parsettensite has a tetrahedral cation deficiency and an octahedral cation excess that is not compatible with the stilpnomelane model structure.

Guggenheim (1986) and Ozawa et al. (1986) presented data showing that parsettensite is structurally distinct from stilpnomelane. The diffraction data of Ozawa et al. showed that the $d(001)$ values of each were comparable at about 12.5 Å, but the lateral cell dimensions differ. Guggenheim (1986) and Guggenheim and Eggleton (1987) noted the plotted position of parsettensite in a r_t vs r_o diagram and suggested that parsettensite should have island-like areas composed of six-member hexagonal rings similar to stilpnomelane but smaller in lateral extent, in accord with the greater misfit expected for a manganese-rich octahedral sheet. Further details of the structure have not been given.

Bannisterite. Smith (1948) and Smith and Frondel (1968) recognized that two different minerals had been confused for ganophyllite. Smith and Frondel (1968) presented X-ray, chemical, and additional optical data for bannisterite that established it as a species in its own right. Threadgold (1979) presented a structural formula based on electron microprobe analysis and a refinement of the structure. However, full details have not yet been presented. Dunn et al. (1981) chemically analyzed material from Franklin, N.J., and Plimer (1977) reported on material from Broken Hill. Chemical formulae appear to range from: $Ca_{0.45}(K_{0.32}Na_{0.19})(Fe_{5.1}Mn_{4.8}Mg_{0.1})_{\Sigma=10}(Si_{14.4}Al_{1.6})_{\Sigma=16}O_{38}(OH)_8 \cdot 2(H_2O)$ for the Broken Hill material (Threadgold, pers. comm.) to $Ca_{0.43}(K_{0.41}Na_{0.055})(Fe_{1.46}Mn_{6.105}Mg_{1.425}Zn_{1.045}Fe^{3+}_{0.13})_{\Sigma=10.165}(Si_{14.27}Al_{1.53}Fe^{3+}_{0.20})_{\Sigma=16}O_{38}(OH)_8 \cdot 6.1(H_2O)$ for the Franklin material (Dunn et al., 1981) with a structural formula of: $K_{0.5}Ca_{0.5}M_{10}T_{16}O_{38}(OH)_8(H_2O)_{2-6}$ where M and T are the octahedral and tetrahedral cations, respectively.

Bannisterite (Threadgold, 1979) is a two layer structure with a quarter of the tetrahedra inverted toward the interlayer space. These tetrahedra share their apical oxygens with similarly inverted tetrahedra for adjacent layers. Six-fold rings are composed of tetrahedra which coordinate directly to the octahedral sheet. Five-fold rings, very distorted six-fold rings, and seven-fold rings are formed also, but each of these rings contains two inverted tetrahedra (Threadgold, pers. comm.). In this way, the close proximity of highly charged tetrahedral cations is eliminated in the five-fold and distorted six-fold rings. Similar five-fold and distorted six-fold rings are found in ganophyllite.

Strip structures

Minnesotaite. Gruner (1944) suggested that minnesotaite was the ferrous iron analogue of talc based on X-ray powder photographs of impure material. However, although the $d(001)$ value of approximately 9.6 Å suggests a talc structure, Guggenheim and Bailey (1982) concluded from single crystal photographs of a twinned crystal that the structure is modulated. Further X-ray analysis and a chemical and transmission electron microscopy (TEM) study by Guggenheim and Eggleton (1986a) supported this conclusion and outlined the details of the structure.

Guggenheim and Eggleton (1986a) showed that minnesotaite crystallizes in two structural forms, which may be designated P and C in accord with the Bravais lattice of the two respective unit cells. Crystallization of the structural forms apparently is related to chemistry, and the resulting misfit between the tetrahedral and octahedral sheets. A structural formula for the C-centered cell is $(Fe,Mg)_{27}Si_{26}O_{86}(OH)_{26}$. Of the seven specimens studied, only the most iron-rich sample was of the C-centered type and, furthermore, it was the only sample that crystallized suffi-

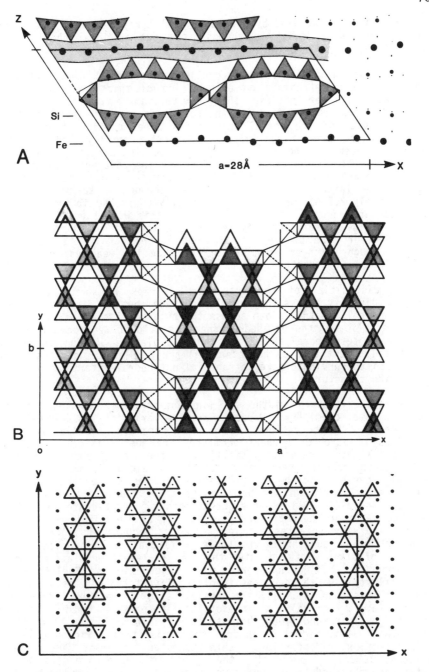

Figure 16. (A) Perspective view of the idealized *P*-cell structure of minnesotaite. The interlayer tetrahedra share corners to form a chain parallel to *Y*, which is more clearly seen in (B) in the (001) plane drawing. (C) Diagrammatic illustration of the *C*-centered cell of minnesotaite. Note the alternation of four- and three-tetrahedra-wide strips along *X*. The strips are joined by a corner-sharing chain along *Y* similar to that shown in (B) (from Guggenheim and Eggleton, 1986a).

ciently well to be studied by single crystal X-ray methods. The P-cell
structure varied somewhat in composition, but an ideal chemistry is
given by $(Fe,Mg)_{30}Si_{40}O_{96}(OH)_{28}$.

Like talc, minnesotaite has a continuous octahedral sheet. There are
adjacent silicon tetrahedra on either side of this sheet to form an
approximate 2:1 layer (see Fig. 16). However, in contrast to talc, strips
of linked hexagonal rings of tetrahedra are formed only parallel to Y.
A discontinuity of the normal tetrahedral sheet configuration results
along X because some of the tetrahedra invert partially (note Fig. 16a
and 16b). These tetrahedra form a chain extending parallel to Y in the
interlayer region. The chain serves to link the adjacent strips within
the basal plane and also links adjacent strips across the interlayer
space. The latter linkage is achieved through a tetrahedral edge
(approximately 2.7 Å), which is dimensionally equivalent to the interlayer
separation in a normal talc structure. In this way, the (001) d value
of minnesotaite closely approximates the predicted value of an iron-rich
talc. However, the bonding of adjacent layers across the interlayer
region is much stronger in minnesotaite than the weak van der Waals'
bonding in a talc structure.

The disposition of the tetrahedral strips is such that strips
superpose directly across the interlayer space, but are displaced parallel
to X by one-half a strip width across the octahedral sheet. In a normal
2:1 layer silicate, the octahedral sheet is held in tension between the
two opposing tetrahedral sheets and thus, is planar. In minnesotaite, the
strip displacement across the octahedral sheet relaxes such tension and
allows limited curving of the tetrahedral-octahedral interface to produce
a wave-like structure.

Tetrahedral strip widths in minnesotaite are dependent on the
component sheet chemistry. Hence, there are variations in unit cell size
and symmetry. As the cations in the tetrahedra are essentially all Si,
variations in octahedral sheet chemistry are most critical. When the
octahedral sheet is relatively small, as in the magnesium-enriched sheets,
a modulation occurs every four tetrahedra along X; ten tetrahedra
(4+1+4+1) span nine octahedra; a P-cell is produced. For the C-centered
cell, strip widths alternate between three and four tetrahedra so that
nine tetrahedra (4+1+3+1) span eight octahedra. The C-centered cell has
been found only in the most ferrous iron-rich sample. These relationships
suggest that minnesotaites with even greater amounts of magnesium than
those studied by Guggenheim and Eggleton (1986a) could produce regular
sequences of strip widths greater than the (4+1+4+1) modulations previously reported. It is noteworthy that Guggenheim and Eggleton found
occasional strip-width sequences involving five tetrahedra. Strip-width
sequences less than (4+1+3+1) are possible also if the average cation size
is large (e.g., Fe + Mn). A minnesotaite-like phase enriched in manganese
was noted by Muir Wood (1980) but structural information is lacking. In
minnesotaite, tetrahedral-octahedral sheet misfit is relieved along the
strip direction by distortions within the octahedral sheet (Guggenheim and
Eggleton, 1986a). Apparently, the structure can accommodate greater
misfit when strip edges are numerous because of narrow strip widths (3 and
4 tetrahedra wide). The octahedra would have a greater ability to distort
if the oxygen normally shared between both the tetrahedral and octahedral
sheets is not constrained in its position. Tetrahedral inversions require
that those oxygens are replaced by OH groups to complete octahedral
coordination about the iron ions. Thus, the OH groups have greater

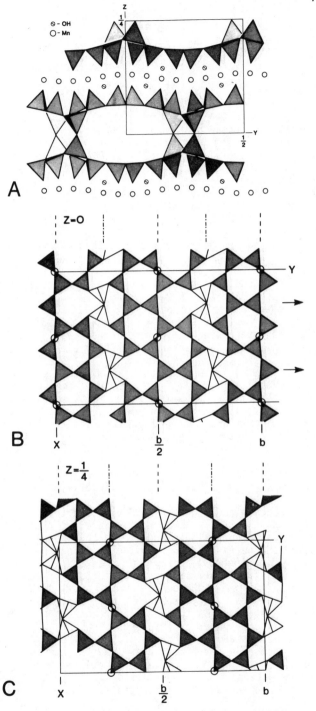

Figure 17. (A) Ganophyllite structure projected down *X*; (B) and (C) Configuration of the tetrahedral sheet for z = 0 and z = 1/4, respectively, for the monoclinic form. Open circles in (B) and (C) indicate symmetry centers for *A2/a* (from Eggleton and Guggenheim, 1986).

positional freedom. Presumably, such an explanation is consistent also with how tetrahedral apex inversions relieve strain in some modulated 1:1 layer silicates, such as in antigorite. Petrologic data for minnesotaite and additional information on iron-rich talc are given in Chapter 8.

Ganophyllite. The first careful optical study of ganophyllite was made by Smith (1948). He found evidence that the term "ganophyllite" was being applied to two separate layer silicates. Smith and Frondel (1968) differentiated these phases on the basis of X-ray data and designated them as ganophyllite and bannisterite (see above). The approximate chemical formula (Eggleton and Guggenheim, 1986) is: $(K,Na,Ca)_6^{+7.56}(Mn,Fe,Mg)_{24}$ $(Si_{32.8}Al_{7.8})O_{96}(OH)_{16}.21H_2O$ and $Z = 8$. Peacor et al. (1984) gave the name *eggletonite* to samples with Na > K. For the limited number of samples studied, Dunn et al. (1983) and Ounchanum and Morad (1987) found that there is little variation in the octahedral occupancy, although the latter authors found a slightly greater range of octahedral aluminum.

Smith and Frondel (1968) showed that ganophyllite crystallizes in the monoclinic space group $A2/a$ with cell parameters $a = 16.60(5)$ Å, $b = 27.04(8)$ Å, $c = 50.34(15)$ Å, $\beta = 94.2(2)^0$. There is a prominent monoclinic subcell with space group $I2/a$ and cell parameters $a_0 = 5.53$ Å $= a/3$, $b_0 = 13.52$ Å $= b/2$, $c_0 = 25.17$ Å $= c/2$, and $\beta = 94.2^0$. Jefferson (1978) noted a triclinic polytype $[a = 16.54(5)$ Å, $b = 54.30(9)$ Å, $c = 28.52(8)$ Å, $\alpha = 127.5(3)^0$, $\beta = 94.1(2)^0$, $\gamma = 95.8(2)^0]$ in space group $(P1$ or $P\bar{1})$ and interpreted variations in polytype as resulting from structural columns which can be stacked along {011} planes.

Kato (1980) determined the ganophyllite subcell structure by using single crystal X-ray data. He described the subcell (note resemblances in Fig. 17) as having a continuous octahedral sheet with tetrahedral strips on both sides of the octahedra. The strips are three tetrahedral chains wide and are extended parallel to X. The strips are staggered across the octahedral sheet so that the rifts between strips are offset. The rift pattern allows a marked sinusoidal curvature of the octahedral sheet and of attached tetrahedra along Y. According to the Kato model, large alkali cations connect adjacent layers by occupying sites where rifts oppose each other across the interlayer.

Guggenheim and Eggleton (1986b) found that the large cations can be readily exchanged. These results are in contrast to the predicted behavior based on the Kato model. In that model, there is a high residual negative charge associated with the coordinating anions of the interlayer alkali site, thereby making cation exchange unlikely. Eggleton and Guggenheim (1986) re-examined the structure of ganophyllite using the Kato data set, several hundred additional weak reflections, and electron diffraction data. They proposed a model (see Fig. 17) whose significant difference is based on the interlayer connectors. The triple chain strips are connected by pairs of inverted tetrahedra which serve to connect both adjacent layers and neighboring strips. The alkali cations reside in the tunnels formed in the "interlayer" region between the inverted tetrahedra. These sites are numerous and poorly defined and may be compared to the exchange cation sites in zeolites. Overall, the structure may be classified as a modulated mica structure.

Eggleton and Guggenheim (1986) suggested that Al preferentially substitutes in the tetrahedra of the (central) continuous tetrahedral chain of the triple chain in ganophyllite. This substitution sufficiently increases the chain dimension so that it may span the octahedra along X. Evidence that this substitution occurs comes from (a) tetrahedral cation size considerations and (b) the position of the interlayer cation, which is located predominantly near the tetrahedral rings at the center of the strip. These tetrahedra would have undersaturated basal oxygens. Tetrahedral chains lateral to the central Al-bearing tetrahedral chain are lengthened by the addition of the inverted tetrahedra (see Fig. 17). Ganophyllite represents the first modulated layer silicate refined sufficiently to reveal that cation ordering effects may play an important role in establishing the stability of a modulated structure.

Gonyerite. Frondel (1955) reported an aluminum-poor, iron- and manganese-rich chlorite from Langban, Sweden. By analogy to true chlorites, he determined the formula of gonyerite as $(Mn_{3.25}Mg_{1.95}Fe^{3+}_{0.64}Zn_{0.04}Pb_{0.02}Ca_{0.01})_{\Sigma} = 5.91(Si_{3.75}Fe^{3+}_{0.17}Al_{0.08})_{\Sigma} = 4[O_{10.20}(OH)_{7.80}]_{\Sigma} = 18.00$ by assuming all iron to be ferric. He also noted that gonyerite is a two-layer structure with an orthorhombic-shaped cell with $a = 5.47$ Å, $b = 9.46$ Å and $c = 28.8$ Å.

Guggenheim and Eggleton (1987) used the plotted position of gonyerite in an $\bar{r}_t$ vs $\bar{r}_o$ diagram as an indication that the misfit between the tetrahedral and octahedral sheets is too severe to allow congruence without structural modulations. Electron diffraction patterns were unlike those from normal chlorite and 00*l* intensities indicated that the two-layer cell was composed of two structurally different types of layers. A one-dimensional electron density projection from single-crystal X-ray data and HRTEM data (Eggleton and Guggenheim, unpublished) confirmed that the two-layer repeat is related to structural modulations and not simple stacking variations. Both data sets indicated that tetrahedra were inverted away from the octahedral sheet of the 2:1 layer to coordinate to what is normally the brucite-like interlayer in an ideal chlorite. The HRTEM data indicated also that 14 Å chlorite units are commonly interleaved with gonyerite, thereby making both bulk chemical analyses and microprobe data difficult to interpret.

IDENTIFICATION

Modulated layer silicates have strong (00*l*) diffraction maxima and measurement of these *d*-spacings is sufficient to classify the mineral as a 7 Å, 10 Å, 12 Å, or 14 Å structure type, by analogy with other layer silicates. No modulated layer silicate expands or contracts significantly on heating, hydration, or intercalation using routine clay-research techniques. Evidence for a modulated structure is often indicated in random powder X-ray diffraction (XRD) patterns by the presence of weak lines with large (>4 Å) spacings. XRD powder patterns for modulated layer silicates are given in the Appendix to this chapter.

Single-crystal diffraction patterns show increased *a* and *b* cell dimensions, resulting from perturbations to the tetrahedral sheets. Unfortunately, single crystal X-ray studies are often impossible, because these structures commonly occur in fine-grained mixes with other minerals. Electron diffraction patterns, however, allow both a rapid and a definitive test for the recognition of a modulated structure. Figure 18 shows electron diffraction patterns from oriented (001) cleavage fragments of various modulated layer silicates; we call these "basal plane"

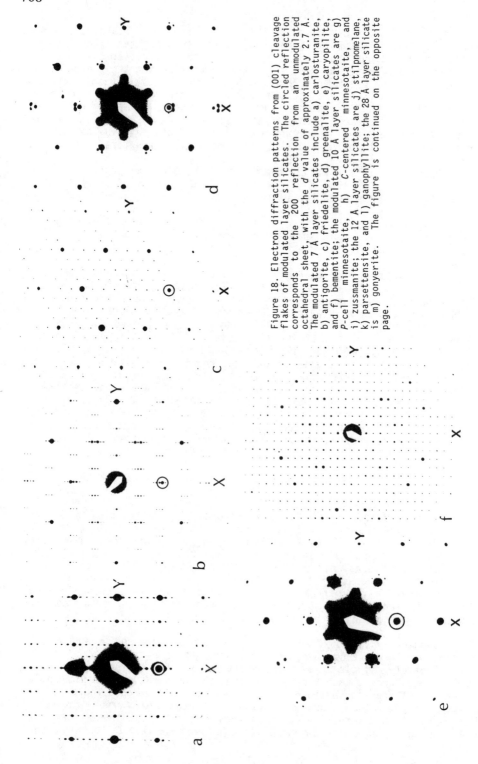

Figure 18. Electron diffraction patterns from (001) cleavage flakes of modulated layer silicates. The circled reflection corresponds to the 200 reflection from an unmodulated octahedral sheet, with the d value of approximately 2.7 Å. The modulated 7 Å layer silicates include a) carlosturanite, b) antigorite, c) friedelite, d) greenalite, e) caryopilite, and f) bementite; the modulated 10 Å layer silicates are g) P-cell minnesotaite, h) C-centered minnesotaite, and i) zussmanite; the 12 Å layer silicates are j) stilpnomelane, k) parsettensite, and l) ganophyllite; the 28 Å layer silicate is m) gonyerite. The figure is continued on the opposite page.

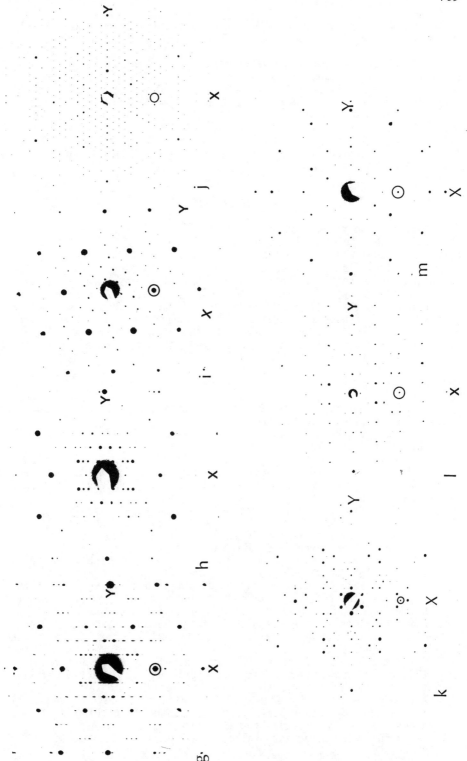

patterns, as many are not *hk*0 patterns because Z is not necessarily perpendicular to (001).

Further evidence of a modulated nature may be obtained from chemical analysis. A trioctahedral layer silicates having greater than two tetrahedral ions (Si + Al) per tetrahedral sheet relative to three octahedral cations is a common indication of a modulated layer silicate. Of the known and characterized modulated layer silicates, only minnesotaite would fail to be recognized by this feature. The average size of the octahedral cations compared to that of the tetrahedral cations is another indication of a modulated structure as described earlier and demonstrated by Figure 7.

CLASSIFICATION

Previous sections have emphasized structural characteristics, chemistry, the principles establishing systematics, and diffraction effects of modulated layer silicates. This section uses these principles and characteristics to develop a classification that serves to group the structures and potentially to be predictive. This classification concerns those structures with continuous octahedral sheets and consequently, emphasizes modulations and configurations of the tetrahedral sheet-like network. Such an approach parallels traditional schemes (e.g., the Bragg-Náray-Szabó classification; Liebau, 1985); we emphasize, however, that the development of this classification is based on the dimensions of the octahedral sheet.

General

The following represents the general outline used to establish the classification (see Tables 1 through 4):
-- Modulated layer silicates are modulated at the tetrahedral sheet; the octahedral sheet is continuous.
-- Modulations involve linked tetrahedral sheets with periodic inversions of the tetrahedra. Tetrahedra are referred to as direct (D) if they coordinate to the octahedral sheet or reversed (R) if they do not.
--In 7 Å structures, there are only D-tetrahedral. In 10 Å structures, a single plane of R-tetrahedra connect to D-tetrahedra across the interlayer and, in 12 Å structures, double planes of R-tetrahedra coordinate to D-tetrahedra.
--The terms 1:1 and 2:1 are retained and should be thought of as applying to limited parts of the infinite layer. A modulated 1:1 layer silicate has tetrahedra coordinating to only one side of the octahedral sheet for most or all of the octahedral sheet's extent. A modulated 2:1 layer silicate has tetrahedra coordinating to opposite sides of the octahedral sheet for at least some extent. Diffraction data (d_{001}) may be used to establish the nature of the basic layer.
--The modulated 2:1 layer silicates have interlayer R-tetrahedra that do not coordinate to any octahedra but are inverted relative to the D-tetrahedra coordinating to the octahedral sheet. The R-tetrahedra may be in a single plane, which do not affect the basal spacing (10 Å, minnesotaite), or they may be doubled in the Z direction, which increases the spacing along Z to 12 Å.
--The following definitions also express important features:

Table 1. Classification of modulated layer silicates with type **H2** layers.

TYPE H2:
　　Hexagonal or trigonal tetrahedral sheets. Two levels of D-tetrahedra.
　　All tetrahedra coordinate to octahedra.
　　Islands linked by 4-, 6-, and 12-rings.

tetrahedra
octahedra

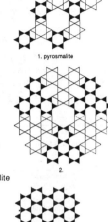

1. pyrosmalite

2.

	Island size	M_pT_q	$\underline{a}$	$\underline{d}_{001}$	Type	Island connectors	Example
1.	one 6-ring	$M_{16}T_{12}$	13.4	7	1:1 H2	4-12	pyrosmalite
2.	four 6-rings	$M_{49}T_{36}$	22.5	7	1:1 H2	4-8-12	none
3.	ten 6-rings	$M_{100}T_{72}$	54.6	7	1:1 H2	4-8-12	none
4.	infinite	M_3T_2	5.2	7	1:1 H2		lizardite

4. lizardite

Table 2. Classification of modulated layer silicates with type **H3** layers.

TYPE H3:
　　3 levels of tetrahedra in the modulated sheet.
　　Unequal numbers of tetrahedra (D) and (R). (D) coordinate to octahedra,
　　(R) coordinate to (D) of the next 2:1 layer.

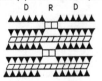

tetrahedra
octahedra

1. zussmanite

	Island size	M_pT_q	$\underline{a}$	$\underline{d}_{001}$	Type	Island connectors	Example
1.	one 6-ring	$M_{13}T_{18}$	13	9	2:1 H3	3-8	Zussmanite

712

Table 3. Classification of modulated layer silicates with type **H4** layers.

TYPE H4:

4 levels of tetrahedra in the modulated sheet.
Unequal numbers of tetrahedra (D) and (R). (D) coordinate to octahedra, (R) coordinate to inverted (R) of the next 2:1 layer.

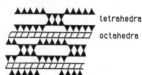

2.

3. stilpnomelane

	Island size	M_pT_q	$\underline{a}$	$\underline{d}_{001}$	Type	Island connectors	Example
1.	4 tetrahedra 4 (D) 6(R)		10.4	12	2:1 H4	5-9	?
2.	3 incomplete 6-rings 10 (D) 6(R)		13.5	12	2:1 H4	5-7	?
3.	7 6-rings 24 (D) 12(R)	$M_{48}T_{72}$	20.8	12.2	2:1 H4	5-8	Stilpnomelane

Table 4. Classification of modulated layer silicates with type **S** layers.

TYPE S:

a. Sheets composed of (D) tetrahedral strips

Strip width	M_pT_q	$\underline{a}$	$\underline{b}$	$\underline{d}_{001}$	Type	Strip connectors	Example
4 chains	$M_{24}T_{17}$	43.3	9.2	7.2	1:1 S2	4-8	antigorite

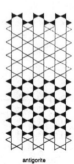

antigorite

b. (D) and (R) strips of unequal width.

Strip width	M_pT_q	$\underline{a}$	$\underline{b}$	$\underline{d}_{001}$	Type	Strip connectors	Example
2 chains	$M_{27}T_{36}$	28	9.4	9.6	2:1 S3	5-7	P-minnesotaite
3 chains	$M_{24}T_{40}$	16.6	27.0	12.5	2:1 S4	5-7	ganophyllite

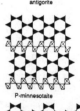

P-minnesotaite

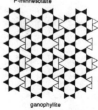

ganophyllite

--An *n-ring* is a ring of *n* tetrahedra linked at their corners. An hexagonal ring is a 6-ring. Also common are 4-, 5-, 8-, 9-, and 12-rings.
--An *island* is a group of one or more linked hexagonal rings.
--An *n-strip* is a pyribole-like chain of tetrahedra, *n* single chains wide. An amphibole double chain is therefore a 2-strip.
--*Lateral connectors* are located at the island and strip edges. Lateral connectors may be identical to the island or strip configuration but inverted (modulated 1:1 layer silicates) or inverted narrow strips, rings, or ring portions (modulated 2:1 layer silicates).

Discussion

The nature of the junction involving the lateral connector is determined by the extension necessary to achieve congruency between tetrahedral and octahedral sheets. In modulated 1:1 layer silicates restrictions are imposed by adjacent octahedral sheets and specifically by the requirement that all tetrahedra coordinate to octahedra. In contrast, tetrahedra involved as lateral connectors coordinate only to other tetrahedra in modulated 2:1 layer silicates. In modulated 1:1 layer silicates, the connecting tetrahedra link in 4-rings (2D,2R), 6-rings, or 8-rings (4D,4R). In modulated 2:1 layer silicates, the connecting tetrahedra are commonly linked as 5-rings (3D,2R) or 7-rings (5D,2R). These connecting tetrahedra may be designated, for example, in notation form as (4-8) for 4-ring and 8-ring combinations.

There are two general types of modulations known thus far: two-dimensional hexagonal or trigonal (designated in the tables as H) and one-dimensional, formed by the lateral linkage of n-strips (designated S). Three different thicknesses of combinations of tetrahedral sheets are known: two tetrahedra thick (D D, inverted relative to each other) as in antigorite (2), three as in minnesotaite and zussmanite (3), or four as in stilpnomelane and ganophyllite (4). Thus, antigorite is classified a 1:1 S2 modulated layer silicate, zussmanite is a 2:1 H3 modulated layer silicate, and stilpnomelane is a 2:1 H4 structure. Because of the tentative nature of our knowledge of the greenalite-caryopilite, bementite, and bannisterite structures, they are not included in Tables 1-4, although further divisions could easily accommodate either these structures or additional ones. We recognize that future work may show intermediate forms between strip and island categories; certainly, bannisterite does not clearly belong to either group. Therefore, this scheme is provisional and should be considered open-ended.

Many of the structural series described above involve a continuously varying structural adjustment to maintain congruence between an octahedral sheet and portions of attached tetrahedral sheets. These series, whether called polysomatic or homologous, are structural and differ significantly from chemical analogies, such as solid solutions, which do not involve major structural variations. Structural differences, however, are traditionally used to distinguish minerals and to designate new ones. These series can potentially lead to a proliferation of new mineral names, if a conservative approach is not taken. Just as defining intervals in a chemical solid solution series is undesirable, a new mineral name for each structural punctuation for these series should be avoided.

714

ACKNOWLEDGMENTS

We thank Dr. J. Zussman, Manchester University, for critically reviewing the manuscript and Dr. S. W. Bailey for editorial comments. D. R. Veblen and K. J. Livi kindly provided portions of Figure 8. Support was provided by the Petroleum Research Fund of the American Chemical Society under grant 17263-AC2 and the National Science Foundation under grant EAR8704681. We thank also Dr. B. G. Hyde, Research School of Chemistry, Australian National University, and the Research Resources Center, University of Illinois at Chicago, for use of the electron microscopes.

REFERENCES

Agrell, S.O., Bown, M.G. and McKie, D. (1965) Deerite, howieite and zussmanite, three new minerals from the Franciscan of the Laytonville District, Mendocino County, California. Am. Mineral. 50, 278 (abstract).

Bailey, S.W. (1969) Polytypism of trioctahedral 1:1 layer silicates. Clays & Clay Minerals 17, 355-371.

_____ (1980) Structures of layer silicates. Ch. 1 in Crystal Structures of Clay Minerals and their X-ray Identification, G.W. Brindley and G. Brown, eds., Mineral. Soc. Great Britain Monograph 5, London.

Bates, T.F. (1959) Morphology and crystal chemistry of 1:1 layer lattice silicates. Am. Mineral. 44, 78-114.

Bauer, L.H. and Berman, H. (1928) Friedelite, schallerite, and related minerals. Am. Mineral. 13, 341-348.

Blake, R.L. (1965) Iron phyllosilicates of the Cuyuna district in Minnesota. Am. Mineral. 60, 148-169.

Bradley, W.F. (1940) The structural scheme of attapulgite. Am. Mineral. 25, 405-410.

Brauner, K. and Preisinger, A. (1956) Struktur und Entstehung des Sepioliths. Tschermaks Mineral. Petrogr. Mitt. 6, 120-140.

Caillère, S. and Hénin, S. (1961) Sepiolite in The X-ray Identification and Crystal structures of Clay Minerals., G. Brown, ed., 2nd edition, Mineral. Soc. Great Britain, London.

Chapman, J.A. and Zussman, J. (1959) Further electron optical observations on crystals of antigorite. Acta Cryst. 12, 550-552.

Chernosky, J.V. (1975) Aggregate refractive indices and unit cell parameters of synthetic serpentine in the system $MgO-Al_2O_3-SiO_2-H_2O$. Am. Mineral. 60, 200-208.

Christ, C.L., Hathaway, J.C., Hostetler, P.B. and Shepard, A.O. (1969) Palygorskite: new X-ray data. Am. Mineral. 54, 198-205.

Compagnoni, R., Ferraris, G. and Mellini, M. (1985) Carlosturanite, a new asbestiform rock-forming silicate from Val Varaita, Italy. Am. Mineral. 70, 767-772.

Crawford, E.S., Jefferson, D.A. and Thomas, J.M. (1977) Electron-microscope and diffraction studies of polytypism in stilpnomelane. Acta Cryst. A33, 548-553.

Donnay, G., Morimoto, N., Takeda, H. and Donnay, J.D.H. (1964) Trioctahedral one-layer micas. I. Crystal structure of a synthetic iron mica. Acta Cryst. 17, 1369-1381.

_____, Betournay, M. and Hamill, G. (1980) McGillite, a new manganese hydro-oxychloro-silicate. Can. Mineral. 18, 31-36.

Drits, V.A. and Aleksandrova, V.A. (1966) The crystallochemical nature of palygorskite. Zap. vses. miner. Obshch. 95, 551-560.

_____ and Sokolova, G.V. (1971) Structure of palygorskite. Soviet Phys. Crystallogr. 16, 183-185.

Dunn, P.J., Leavens, P.B., Norberg, J.A. and Ramik, R.A. (1981) Bannisterite: new chemical data and empirical formulae. Am. Mineral. 66, 1063-1067.

_____ and Peacor, D.R. (1984) Nelenite, a manganese arsenosilicate of the friedelite group, polymorphous with schallerite, from Franklin, New Jersey. Mineral. Mag. 48, 271-275.

_____, _____, Leavens, P.B. and Wicks, F.J. (1984) Minehillite, a new layer silicate from Franklin, New Jersey, related to reyerite and truscottite. Am. Mineral. 69, 1150-1155.

_____, _____, Nelen, J. and Norberg, J.A. (1981) Crystal- chemical data for schallerite, caryopilite and friedelite from Franklin and Sterling Hill, New Jersey. Am. Mineral. 66, 1054-1062.

_____, _____, _____ and Ramik, R.A. (1983) Ganophyllite from Franklin, New Jersey; Pajsberg, Sweden; and Wales: new chemical data. Mineral. Mag. 47, 563-566.

_____, _____ and Simmons, W.B. (1984) Lennilenapeite, the Mg-analogue of stilpnomelane, and chemical data on other stilpnomelane species from Franklin, New Jersey. Can. Mineral. 22, 259-263.

Eberhard, S. and Rahman, S.H. (1982) Prinzip der Kristallstruktur von Gyrolith $Ca_{13}[Si_8O_{20}]_3$ $(OH)_2.{}^{-}22H_2O$. Z. fur Krist. 159, 34-36.

Eggleton, R.A. (1972) The crystal structure of stilpnomelane. Part II. The full cell. Mineral. Mag. 38, 693-711.

_____ and Bailey, S.W. (1965) The crystal structure of stilpnomelane Part I. The subcell. Clays & Clay Minerals 13, 49-63.

_____ and _____ (1967) Structural aspects of dioctahedral chlorite. Am. Mineral. 52, 673-698.

_____ and Chappell, B.W. (1978) The crystal structure of stilpnomelane. Part III. Chemistry and physical properties. Mineral. Mag. 42, 361-368.

_____ and Guggenheim, S. (1986) A re-examination of the structure of ganophyllite. Mineral. Mag. 50, 307-15.

_____ and _____ (1988) A new model for the structure of bementite, a modulated 1:1 layer silicate. Geol. Soc. Am. Abst. (in review).

El-Hemaly, S.A.S., Mitsuda, T. and Taylor, H.F.W. (1977) Synthesis of normal and anomalous tobermorites. Cement Conc. Res. 7, 429-438.

Floran, R.J. and Papike, J.J. (1975) Petrology of the low-grade rocks of the Gunflint Iron-Formation, Ontario-Minnesota. Geol. Soc. Amer. Bull. 86, 1169-1190.

Frondel, C. (1955) Two chlorites: gonyerite and melanolite. Am. Mineral. 40, 1090-1094.

_____ and Bauer, L.H. (1953) Manganpyrosmalite and its polymorphic relation to friedelite and schallerite. Am. Mineral. 38, 755-760.

Gard, J.A. and Follett, E.A.C. (1968) A structural scheme for palygorskite. Clay Minerals 7, 367-370.

_____, Luke, K. and Taylor, H.F.W. (1981) Crystal structure of K-phase, $Ca_7Si_{16}O_{40}H_2$. Kristallog. 26, 1218-1223.

_____, Mitsuda, T. and Taylor, H.F.W. (1975) Some observations on Assarsson's Z-phase and its structural relations to gyrolite, truscottite, and reyerite. Mineral. Mag. 40, 325-333.

Gillery, F.H. (1959) The X-ray study of synthetic Mg-Al serpentines and chlorites. Am. Mineral. 44, 143-152.

Gruner, J.W. (1936) The structure and chemical composition of greenalite. Am. Mineral. 21, 449-455.

_____ (1944) The composition and structure of minnesotaite, a common iron silicate in iron formations. Am. Mineral. 29, 363-372.

Guggenheim, S. (1984) The brittle micas. Ch. 3 in Micas, Bailey, S.W. ed., Rev. in Mineral. 13, 61-104.

_____ (1986) Modulated 2:1 layer silicates. Int'l. Mineral. Assoc. 14th Gen. Meet., Stanford Univ., 116-117 (abstract).

_____ and Bailey, S.W. (1982) The superlattice of minnesotaite. Can. Mineral. 20, 579-584.

_____ and _____ (1988) Baumite: A modulated serpentine related to the greenalite-caryopilite series. Am. Mineral. (in press).

_____, _____, Eggleton, R.A. and Wilkes, P. (1982) Structural aspect of greenalite and related minerals. Can. Mineral. 20, 1-18.

_____, Chang, Y-H. and Koster van Groos, A.F. (1987) Muscovite dehydroxylation: High temperature studies. Am. Mineral. 72, 537-550.

_____ and Eggleton, R.A. (1986a) Structural modulations in Mg-rich and Fe-rich minnesotaite. Can. Mineral. 479-497.

_____ and _____ (1986b) Cation exchange in ganophyllite. Mineral. Mag. 50, 517-520.

716

_____ and _____ (1987) Modulated 2:1 layer silicates: Review, systematics, and predictions. Am. Mineral. 72, 724-738.

Hamid, S.A. (1981) The crystal structure of the 11 Å natural tobermorite $Ca_{2.25}[Si_3O_{7.5}-(OH)_{1.5}]$ $1H_2O$. Z. Krist. 154, 189-198.

Hazen, R.M. and Finger, L.W. (1978) The crystal structures and compressibilities of layer minerals at high pressure. II. Phlogopite and chlorite. Am. Mineral. 63, 293-296.

_____ and _____ (1982) Comparative Crystal Chemistry. John Wiley and Sons, New York. 231 p.

_____ and Wones, D.R. (1972) The effect of cation substitution on the physical properties of trioctahedral micas. Am. Mineral. 57, 103-129.

Huebner, J.S., Flohr, M.J.K. and Matzko, J.J. (1986) Origin and metamorphism of manganese sediments in melange of the Franciscan complex, California. Int'l. Mineral. Assoc. 14th Gen. Meet., Stanford Univ., 131 (Abstract).

Hutton, C.O. (1938) The stilpnomelane group of minerals. Mineral. Mag. 25, 172-206.

Iijima, S. (1982a) High-resolution electron microscopy of mcgillite I. One-layer monoclinic structure. Acta Cryst. A38, 685-694.

_____ (1982b) High-resolution electron microscopy of mcgillite. II. Polytypism and disorder. Acta Cryst. A38, 695-702.

Jakob, J. (1923) Vier Mangansilikat aus dem Val d'Err (Kt Graubunden). Schwiezer. Mineral. Petrog. Mitt. 3, 227-237.

Jefferson, D.A. (1976) Stacking disorder and polytypism in zussmanite. Am. Mineral. 61, 470-483.

_____ (1978) The crystal structure of ganophyllite, a complex manganese aluminosilicate. I. Polytypism and structural variation. Acta Cryst. A34, 491-497.

Jolliffe, F. (1935) A study of greenalite. Am. Mineral. 20, 405-426.

Kashaev, A.A. (1968) The crystal structure of pyrosmalite. Soviet Physics-Cryst. 12, 923-924.

_____ and Drits, V.A. (1970) The polytype properties of pyrosmalite minerals. Soviet Physics-Cryst. 15, 40-43.

Kato, T. (1963) New data on the so-called bementite. Mineral J. (Japan) 6, 93-103.

_____ (1980) The crystal structure of ganophyllite; monoclinic subcell. Mineral. J. (Japan), 10, 1-13.

_____ and Takéuchi, Y. (1980) Crystal structures and submicroscopic textures of layered manganese silicates. Mineral. J. (Japan) 14, 165-178.

_____ and _____ (1983) The pyrosmalite group of minerals I. Structure refinement of manganpyrosmalite. Can. Mineral. 21, 1-6.

Kunze, G. (1956) Die gewellte Struktur des Antigorits. I. Z. Krist. 108, 82-107.

_____ (1958) Die gewellte Struktur des Antigorits. II. Z. Krist. Kristallgeom. 110, 282-320.

_____ (1961) Antigorit. Strukturtheoretische Grundlagen und ihre praktische Bedeutung fur weitere Serpentin Forschung. Fortschr. Mineral. 39, 206-324.

Lachowski, E.E., Murray, L.W. and Taylor, H.F.W. (1979) Truscottite: composition and ionic substitutions. Mineral. Mag. 43, 333-336.

Lee, J.H. and Guggenheim, S. (1981) Single crystal X-ray refinement of pyrophyllite-1Tc. Am. Mineral. 66, 350-357.

Liebau, F. (1985) Structural chemistry of silicates. Springer-Verlag, Berlin.

Lin, J.-C. and Guggenheim, S. (1983) The crystal structure of Li,Be-rich brittle mica: a dioctahedral-trioctahedral intermediate. Am. Mineral. 68, 130-142.

Livi, K.J.T. and Veblen, D.R. (1987) "Eastonite" from Easton, Pennsylvania: A mixture of phlogopite and a new form of serpentine. Am. Mineral. 72, 113-125.

Lopes-Vieira, A. and Zussman, J. (1969) Further detail on the crystal structure of zussmanite. Mineral. Mag. 37, 49-60.

Martin-Vivaldi, J.L. and Linares-Gonzalez, J. (1962) A random intergrowth of sepiolite and attapulgite. Clays & Clay Minerals 9, 592-602.

McCauley, J.W., Newnham, R.E. and Gibbs, G.V. (1973) Crystal structure analysis of synthetic fluorophlogopite. Am. Mineral. 58, 249-254.

McConnell, D. (1954) Crystal chemistry of schallerite. Am. Mineral. 39, 929-936.

Megaw, H.D. and Kelsey, C.H. (1956) Crystal structure of tobermorite. Nature 177, 390-391.

Mellini, M. (1982) The crystal structure of lizardite 1$\underline{T}$: hydrogen bonds and polytypism. Am. Mineral. 67, 587-598.

_____, Ferraris, G. and Compagnoni, R. (1985) Carlosturanite: HRTEM evidence of a polysomatic series including serpentine. Am. Mineral. 70, 773-781.

_____, Trommsdorff, V. and Compagnoni, R. (1987) Antigorite polysomatism: behavior during progressive metamorphism. Contrib. Mineral. Petrol. 97, 147-155.

_____ and Zussman, J. (1986) Carlosturanite (not "picrolite") from Taberg, Sweden. Mineral. Mag. 50, 675-679.

Merlino, S. (1988) The structure of reyerite, $(Na,K)_2Ca_{14}Si_{22}Al_2O_{58}(OH)_8.6H_2O$. Mineral. Mag. 52, 247-256.

Mitsuda, T. and Taylor, H.F.W. (1978) Normal and anomalous tobermorites. Mineral. Mag. 42, 229-235.

Muir Wood, R. (1980) The iron-rich blueschist-facies minerals: 3. Zussmanite and related minerals. Mineral. Mag. 48, 605-614.

Nagy, B. and Bradley, W.F. (1955) The structural scheme of sepiolite. Am. Mineral. 40, 885-892.

Newnham, R.E. (1961) A refinement of the dickite structure and some remarks on polytypism of the kaolin minerals. Mineral. Mag. 32, 683-704.

_____ and Brindley, G.W. (1956) The crystal structure of dickite. Acta Cryst. 9, 759-764.

Ounchanum, P. and Morad, S. (1987) Paragenesis of akatoreite and ganophyllite in the manganiferous rocks of the Haste field, Norberg ore district, central Sweden. N. Jb. Mineral. Abh. 157, 225-244.

Ozawa, T., Takahata, T. and Buseck, P. (1986) A hydrous manganese phyllosilicate with 12 A basal spacing. Int'l. Mineral. Assoc. 14th Gen. Meet. Stanford Univ., 194 (abstract).

_____, Takéuchi, Y., Takahata, T., Donnay, G. and Donnay, J.D.H. (1983) The pyrosmalite group minerals II. The layer structure of mcgillite and friedelite. Can. Mineral. 21, 7-17.

Peacor, D.R., Dunn, P.J. and Simmons, W.B. (1984) Eggletonite, the Na- analogue of ganophyllite. Mineral. Mag. 48, 93-96.

_____, _____, _____ and Wicks, F.J. (1986) Arsenites related to layer silicates: Manganarsite, the analogue of manganpyrosmalite, and unnamed analogues of friedelite and schallerite from Langban, Sweden. Am. Mineral. 71, 1517-1521.

_____ and Essene, E.J. (1980) Caryopilite - a member of the friedelite rather than the serpentine group. Am. Mineral. 65, 335-339.

Plimer, I.R. (1977) Bannisterite from Broken Hill, Australia. N. Jahr. Mineral. Monat. 504-508.

Preisinger, A. (1959) X-ray study of the structure of sepiolite. Clays & Clay Minerals 6, 61-67.

_____ (1963) Sepiolite and related compounds: its stability and application. Clays & Clay Minerals 10, 365-371.

Radoslovich, E.W. (1961) Surface symmetry and cell dimensions of layer-lattice silicates. Nature, 191, 67-68.

_____ and Norrish, K. (1962) The cell dimensions and symmetry of layer lattice silicates. I. Some structural considerations. Am. Mineral. 47, 599-616.

Rautureau, M. (1974) Analyse structurale de la sépiolite par microdiffraction electronique relations avec les propriétés physico-chimiques. PhD Thesis, Université D'Orléans.

_____ and Tchoubar, C. (1974) Précisions concernant l'analyse structurale de la sépiolite par microdiffraction electronique. C.r. hebd. Seanc. Acad. Sci., Paris 278B, 25-28.

Shannon, R.D. (1976) Revised effective ionic radii and systematic studies of interatomic distances in halides and chalcogenides. Acta Cryst. A32, 751-767.

Shirozu, H. and Hirowatari, F. (1955) Bementite from the Tokuzawa mine, Fukushima prefecture. J. Jap. Assoc. Mineral., Petr., Econ. Geol. 39, 241-248.

Shitov, V.A. and Zvyagin, B.B. (1966) Electron microdiffraction study of serpentine minerals. Soviet Physics-Cryst. 10, 711-716.

Smith, W.C. (1948) Ganophyllite from the Benallt mine, Rhiw, Caernarvonshire. Mineral. Mag. 28, 343-352.

Smith, M.L. and Frondel, C. (1968) The related layered minerals ganophyllite, bannisterite, and stilpnomelane. Mineral. Mag. 36, 893-913.

718

Sokoleva, G.V., Kashaev, A.A., Drits, V.A. and Llyukhin, V.V. (1983) The crystal structure of fedorite. Kristallog. 28, 170-172.

Spinnler, G.E. (1985) HRTEM study of antigorite, pyroxene-serpentine reactions, and chlorite. Ph.D. thesis, Arizona State Univ., Tempe.

Steadman, R. and Nuttall, P.M. (1963) Polymorphism in cronstedtite. Acta Cryst. 16, 1-8.

_____ and Youell, R.F. (1958) Mineralogy and crystal structure of greenalite. Nature 181, 45.

Stillwell, F. and McAndrew, J. (1957) Pyrosmalite in the Broken Hill lode, New South Wales. Mineral. Mag. 31, 371-380.

Takeda, H. and Morosin, B. (1975) Comparison of observed and predicted structural parameters of mica at high temperature. Acta Cryst. B31, 2444-2452.

Takéuchi, Y., Kawada, I., Irimaziri, S. and Sadanga, R. (1969) The crystal structure and polytypism of manganpyrosmalite. Mineral. J. (Japan) 5, 450-467.

_____, Ozawa, T. and Takahata, T. (1983) The pyrosmalite group of minerals III. Derivation of polytypes. Can. Mineral. 21, 19-27.

Taylor, H.F.W. (1964) The calcium silicate hydrates. in The Chemistry of Cements, H.F.W. Taylor, ed., 168-232, vol. 1, Academic Press, New York.

_____ and Howison, J.W. (1956) Relationships between calcium silicates and clay minerals. Clay Minerals. Bull. 3, 98-111.

Thompson, J.B., Jr. (1978) Biopyriboles and polysomatic series. Am. Mineral. 63, 239-249.

Threadgold, I. (1979) Ferroan bannisterite - a new type of layer silicate structure. Seminar on Broken Hill. Mineral. Soc. New South Wales & Mineral. Soc. Victoria (abstract).

Toraya, H. (1981) Distortions of octahedra and octahedral sheets in 1M micas and the relation to their stability. Z. Krist. 157, 173-190.

Uehara, S. and Shirozu, H. (1985) Variations in chemical composition and structural properties of antigorites. Mineral. J. (Japan) 12, 299-318.

Vaughan, J.P. (1986) The iron end-member of the pyrosmalite series from the Pegmont lead-zinc deposit, Queensland. Mineral. Mag. 50, 527-531.

_____ (1987) Ferropyrosmalite and nomenclature in the pyrosmalite series. Mineral. Mag. 51, 174.

Veblen, D.R. and Buseck, P.R. (1981) Hydrous pyriboles and sheet silicates in pyroxenes and uralites: intergrowth microstructures and reaction mechanisms. Am. Mineral. 66, 1107-1134.

Whittaker, E.J.W. and Zussman, J. (1956) The characterization of serpentine minerals by X-ray diffraction. Mineral. Mag. 31, 107-126.

Wicks, F.J. and Whittaker, E.J.W. (1975) A reappraisal of the structures of the serpentine minerals. Can. Mineral. 13, 227-243.

Wieker, W., Grimmer, A.-R., Winkler, A., Magi, M., Tarmak, M. and Lippmaa, E. (1982) Solid-state high-resolution ^{29}Si NMR spectroscopy of synthetic 14 Å and 11 Å and 9 Å tobermorites. Cement Conc. Res. 12, 333-339.

Yeskis, D., Koster van Groos, A.F. and Guggenheim, S. (1985) The dehydroxylation of kaolinite. Am. Mineral. 70, 159-164.

Yoshimura, T. (1967) Manganese mineralization, minerals, and ores (Part 1, supplement to Manganese Ore Deposits of Japan). Kyushu Univ. Sci. Rep., Fac. Sci. 9, Ser. D.

Zussman, J. (1954) Investigation of the crystal structure of antigorite. Mineral. Mag. 40, 498-512.

_____, J., Brindley, G.W. and Cromer, J.J. (1957) Electron diffraction studies of serpentine minerals. Am. Mineral. 42, 133-153.

Zvyagin, B.B., Mishchenko, K.S. and Shitov, V.A. (1963) Electron diffraction data on the structures of sepiolite and palygorskite. Soviet. Phys.-Cryst. 8, 148-153.

APPENDIX. POWDER X-RAY DIFFRACTION DATA FOR MODULATED LAYER SILICATES

Antigorite

hkl	I(calc)	d(obs) Å	I(obs)	d(calc) Å
001	100	7.30	vs	7.26
20T	2+	6.95	w	6.93
30T	26	6.52	w	6.56
401	8	5.80	w	5.96
710	3	5.11	vvw	5.15
810	6	4.67	mw	4.68
020	4	4.64	mw	4.63
910	2	4.27	w	4.28
81T	6	3.95	mw	3.95
102,10Z	300	3.63	s+	3.63
302,202	24	3.52	vvw	3.50
14,0T	2	2.88	mw	2.88
15,01	2	2.66	vvw	2.66
930	4	2.60	vw	2.60
17,00,16,0T	4	2.57	mw	2.55
16,01	70	2.53	vs	2.52
93T	9	2.46	m	2.46
003,18,00,	38	2.42	mw	2.42
17,01,303, 10,3T	9	2.39	w	2.39
+037	6	2.35	mw	2.35
15,02	5	2.24	mw	2.24
16,0Z	7	2.21	mw	2.20
83Z	22	2.169	ms	2.150
16,02	20	2.153	m	2.150
92Z	4	2.127	m	2.126
17,0Z	4	2.113	w	2.113
11,3Z	5	2.035	vvw	2.033
15,03	3	1.879	w	1.879
004,10T,833	12	1.832	m+	1.836
93T	23	1.813	E	1.815
93T	1+	1.782	E	1.782
10,3T	5	1.755	m-	1.756
17,03	10	1.738	mw	1.737
21,31	4	1.680	vw	1.680
22,31	3	1.638	vw	1.636
14,0T	2	1.587	w	1.583
2+,30	12	1.563	m+	1.561
060	9	1.541	E	1.541
24,3T	9	1.534	E	1.531
15,04,22,3Z, 16,0T	13	1.523	w	1.522
061	8	1.509	mw	1.508
17,0T,93T	10	1.497	m-	1.495
934	7	1.480	w	1.479
18,0T	6	1.468	w	1.467
10,3+	6	1.462	-	1.451
205	6	1.452	E	1.451
205	9	1.448	mw	1.446

a = 43.5   c = 7.28
b = 9.26   β = 91.4°
Composite Debye-Scherrer data from 5
samples; Whittaker and Zussman (1956)

Bannisterite

hkl	I	d(obs) Å	d(calc) Å
002	100	12.33	12.31
111	1	11.45	11.39
21T	2	8.78	8.77
211	2	8.43	8.42
202	1+	7.95	7.94
113	3	7.11	7.11
113	1	6.82	6.83
31T	2+	6.61	6.61
311	2	6.38	6.38
004	1	6.16	6.15
213	5	5.93	5.93
20T	5	5.56	5.56
40Z	5	5.196	5.196
402	1	4.911	4.912
015	10	4.709	4.715
22T	5	4.593	4.596
413	10	4.280	4.279
32T	1	4.211	4.211
006	15.b	4.1	4.103
040	8	3.793	3.795
333	2	3.691	3.690
600	6	3.571	3.573
340	20	3.436	3.436
117	10	3.401	3.403
044	1	3.360	3.360
117	3	3.237	3.237
051	12	3.212	3.216
217	2	3.129	3.139
+35T	3	3.077	3.079
008		3.021	3.024
20T		2.968	

16 lines I<2 omitted below line

hkl	I	d(obs) Å	d(calc) Å
	8	2.769	
68T	15	2.638	
	11	2.606	
	10	2.410	
	9	2.384	
	4	2.246	
	5	2.223	
	5	2.162	
	4	1.930	
		1.708	
	3	1.633	
	8	1.614	
	3	1.599	
		1.561	

+ 19 lines to 1.26

b: broad peak

a = 22.20 (7)   β = 94.3° (1)
b = 16.32 (5)   A2/a
c = 24.70 (8)
Franklin Furnace, N.J., U.S.A.; Fe/Mn.
114.6 mm camera; Smith and Frondel (1968)

Bementite

hkl	I	d(obs) Å	d(calc) Å
020	2	8.75	8.79
002	200	7.315	7.315
012	10	6.76	6.753
102	5	6.37	6.318
220	10	5.64	5.622
022			5.652
130	10	5.43	5.444
310	20	4.74	4.739
032	5	4.57	4.572
141	10	4.04	4.08
14T			4.02
330	20	3.77	3.766
042			3.768
004	60	3.658	3.658
014	100	3.583	3.581
024	30	3.375	3.377
500 ?	40	3.025	
160	20	2.955	2.944
44T	5	2.871	2.873
16T	20	2.822	2.814
520			2.829
521	20	2.601	2.799
44T			2.704
16Z	70	2.703	2.704
522	50	2.535	2.538
44+			2.527
16+			2.496
006	20	2.498	2.438
523	30	2.330	2.337
44+	40		2.321
16+			
008	20	2.279	2.260
	5		2.280
3,10,0	40	1.829	1.829
68T	30	1.655	1.655
680		1.640	1.640
			1.639
681	25	1.618	1.617
			1.620
92T	20	1.610	1.607
92+	5	1.584	1.584
3.10.3			1.583
	5	1.441	
	5	1.397	
	5	1.141	

a = 14.833
b = 17.572   β = 95.52°
c = 14.699
FeKα, 114.6 mm Debye-Scherrer camera,
plus diffractometer data (CuKα) for d_{00l};
data from authors

Carlosturanite

hkl	I	d(obs) Å	d(calc) Å
200	25	16.02	16.01
400	9	9.01	9.00
001	100	7.17	7.16
201			7.1+
201	10	6.28	6.25
40T			6.21
11T	5	5.67	5.76
510			5.72
111			5.50
601	15	5.15	5.11
401			5.11
601	5	4.22	4.22
801			4.19
020	10	4.71*	4.71
202	20	3.637	3.645
202	45	3.595	3.601
10.00			3.578
002	10	3.513	3.518
801			3.498
T0.01	55	3.397	3.387
202	15	3.096	3.106
802	10	2.988	3.001
12.00			2.994
10.01			2.979
T2.01	5	2.849	2.844
331			2.843
602			2.824
T0.02			2.824
730	5	2.818	2.670
802	5	2.67+*	2.584
T2.02	15	2.586	2.571
531	20	2.562	2.553
703	10	2.539	2.424
703		2.425	2.426
203	10	2.308	2.308
803	10	2.293	2.298
14.01	35	2.280	2.285
T6.01			2.276
12.02	5	2.101*	2.109
15.11			2.101
17.10			2.067
T7.03	10	2.065*	2.065
623			2.065
932	10	1.9373	1.9452
T7.03			1.9280
14.02	15	1.9223	1.9212
333			1.9211
15.30	5	1.9030	1.9096
20+	5	1.8170	1.8147
60+			1.8123
18.21	15	1.7098*	1.7117
10.23			1.7081
14.03			1.7081
060	15	1.6030	1.6024
T2.05	20	1.5679**	1.5583
T2.03			1.4005
27.03			1.3968
T4.05	5	1.3995	1.3671
T7.34			1.3675
16.04	5	1.3671	1.2654
463			1.2654
T2.04	5	1.2835	1.2813
27.04			1.2768
T0.63	5	1.2790	1.2790

* - reflections observed only in
Guinier-Lenné patterns

Carlosturanite
a = 36.70   β = 101.1°
b = 9.41   Cm
c = 7.291
Sampeyre, Italy; Fe/Mn. diffractometer
and Guinier-Lenné; Compagnoni et al
(1985)

Caryopilite

1T(calc) d Å	hkl	Observed d Å	I	1M(calc) d Å	hkl
7.273	001	7.30	80	7.270	001
3.636	002	3.639	50	3.635	002
		2.825	60	2.822	20$\bar{1}$,130
2.650	201,131	2.648	5		
		2.521	100	2.523	20$\bar{2}$,131
2.424	003	2.427	7	2.423	003
		2.382	20	2.384	201,13$\bar{2}$
2.241	202,132	2.239	4		
		2.101	35	2.101	20$\bar{1}$,132
		1.968	20	1.969	202,13$\bar{1}$
1.846	203,133	1.850	2	1.817	004+
1.818	004	1.820		1.731	20$\bar{1}$,133
		1.731	25		060,33$\bar{1}$
1.643	060,330	1.643	35	1.643	060,33$\bar{1}$
		1.626	20	1.627	203,13$\bar{1}$
1.603	061,331	1.603	25	1.603	061,330,33$\bar{2}$
1.533	204,134	1.535	3		
1.497	062,332	1.497	10	1.497	062,331,33$\bar{2}$
1.455	005	1.453	10	1.454	005
		1.445		1.445	205,13$\bar{4}$
		1.420	10	1.420	40$\bar{1}$,26$\bar{1}$
		1.377	15	1.377	400,26$\bar{2}$
		1.367	20	1.366	204,135
1.295	205,135	1.295	10	1.294	40$\bar{1}$,26$\bar{2}$
		1.260	2	1.261	40$\bar{1}$,262
1.219	064,334	1.215	2	1.219	064,333,335
1.212	006			1.212	006
		1.192	10	1.192	402,26$\bar{4}$
		1.154	5	1.156	405,263
1.089	065,335	1.087	10	1.089	065,334,336
				1.085	403,265

1T: a = 5.693, b = 9.854, c = 7.273, β = 90°

1M: a = 5.692, b = 9.860, c = 7.513, β = 104.6°

FeKα radiation, 114.6mm camera; intensities estimated visually. cell dimensions determined by least-squares fit; (Guggenheim et al (1982))

Eggletonite A-centred supercell

hkl	I	d(obs) Å	d(calc) Å
004	>100	12	12.5
040	10	6.8	6.8
008	10	6	6.24
320	25	5.11	5.12
32$\bar{4}$	15	4.59	4.63
062	15	4.42	4.43
0012	40	4.17	4.16
066	40	3.96	3.96
346	15	3.722	3.726
02.1$\bar{4}$	35	3.449	3.449
00.16	50.		3.122
382	45	2.850	2.852
06.1$\bar{4}$	15	2.794	2.796
202,32.16,60$\bar{4}$	30	2.741	2.741
386	100	2.694	2.694
608	55	2.598	2.599
38.10	55	2.458	2.458
60.12	30	2.385	2.383
38.1$\bar{4}$	15	2.28$\bar{4}$	2.283
60.12	20	2.23$\bar{4}$	2.234
38.1$\bar{4}$	30	2.203	2.203
38.18	15	2.036	2.036
38.22,36.2$\bar{4}$	15	1.75$\bar{4}$	
016.0	20	1.689	1.688
016.$\bar{4}$	45	1.673	1.673
982	30	1.622	1.622
986		1.608	

* - obscured by Si standard; I calculated from single crystal data

a = 16.647 (3) b = 27.012 (4) c = 50.06 (2)

β = 94.02° (3) Aa, A2/a

Little Rock, Arkansas, U.S.A.: Fe/Mn.
114.6 mm Debye-Scherrer; modified from Guggenheim and Eggleton (1986)

Friedelite

hkl	I	d(obs) Å	d(calc) Å
110*	10	11.4	11.51
200*			11.26
001	90	7.17	7.19
020*			6.70
11$\bar{1}$*			6.50
201*			5.45
002	70	3.60	3.59
$\bar{4}$40,$\bar{3}$01	60	2.88	2.88
$\bar{4}$41,$\bar{8}$02	100	2.56	2.56
$\bar{7}$42,801	30	2.408	2.408
$\bar{4}$42,$\bar{8}$03	40	2.115	2.113
$\bar{7}$43,$\bar{8}$02	20	1.974	1.976
$\bar{4}$43,$\bar{8}$04	30	1.731	1.730
080	60	1.676	1.676
081	20	1.632	1.632
$\bar{7}$44,803	10	1.625	1.625
082	10	1.520	1.519
$\bar{8}$81,$\bar{1}$6.01	10	1.449	1.448
$\bar{4}$44,$\bar{8}$05	10	1.439	1.441
005			1.438
$\bar{8}$82,16.00	20	1.402	1.402
881,$\bar{1}$6.03	10	1.374	1.376
083			1.374
$\bar{7}$45,804	30	1.359	1.361
$\bar{8}$83,16.01	10	1.313	1.313
$\bar{5}$91*			1.289
$\bar{8}$84,16.02	20	1.204	1.204
006	10	1.200	1.198
883,$\bar{1}$6.05	10	1.167	1.167
$\bar{8}$85,16.03	10	1.093	1.092
085	10	1.090	1.091
4.12.$\bar{2}$	10	1.065	1.063

* - superstructure reflections

a = 23.33 (5) b = 13.396 (8) c = 7.447 (4)

β = 105.08° (8) C2/m

Kimberley, B.C., Canada : Fe/Mn.
Debye-Scherrer; pattern from Donnay et al (1980); indexing from Ozawa et al (1983). 6.50 Å line re-indexed by authors.

Ganophyllite

hkl	I	d Å
004	100	12.5
022	2	11.9
026	2	7.2
040	2	6.8
008	4	6.3
140	1	6.2
04̄4	6	5.94*
02.10	1	5.1̄+
32̄4	2	4.71
062	3	4.63
34̄2	1	4.42
00.12	9	4.25
32̄8	2	4.18
066	7	4.10
34̄6	2	3.97
328	15	3.90
346	4	3.835
02.1̄4	3	3.724
34̄.1̄0	25	3.464
00.16	6	3.349
368	4	3.136
38̄2	5	2.990
382	5	2.901
06.1̄4	5	2.857
60̄4	17	2.806
386	1	2.737
36̄.12	10	2.696
60̄8	2	2.618
38̄.10	2	2.596
00.20	10	2.543
60̄8	5	2.506
60.1̄2	4	2.463
38.1̄4	3	2.383
60.1̄2	3	2.291
38.1̄4	6	2.232
60.16	3	2.208
62.1̄4	3	2.148
38.1̄8	4	2.086
	3	2.043
	3	1.970
38.22,36.2̄4	7	1.927
	1	1.759
0160	3	1.725
0164	1	1.698
		1.676
		1.631

Ganophyllite

982	d Å
3	1.621
5	1.607
2	1.578
3	1.567
1	1.554*
3	1.455
3	1.419
2	1.323
2	1.314

a = 16.60   β = 94.2
b = 27.06   Aa or A2/a
c = 50.35

Harstig, Sweden (Fe/Mn), and Franklin, U.S.A. (Cr/V); 114.6 mm camera, averaged; Smith and Frondel (1968)

Gonyerite

hkl	I	d Å
001*	2	28.5
002	30	14.6
012	5	8.0
004	100	7.23
005	5	5.79
101	10	5.33
006	50	4.79
008	80	3.61
123	10	3.33
009	5	3.22
030	5	3.13
033	5	3.03
00.10	20	2.888
	20	2.801
202	30	2.697
00.11	10	2.610
204	5	2.550
	10	2.441
00.14	5	2.056
	30	1.634
	50	1.574
00.20	10	1.437

a = 5.47   d_{001} = 28.8
b = 9.46

Långban, Sweden; Fe/Mn powder photography (* supplemented by author's diffractometry); Frondel (1955)

Greenalite

1T(calc)		Observed		1M(calc)	
d Å	hkl	d Å	I	d Å	hkl
7.212	001	7.20	100	7.211	001
3.606	002	3.61	50	3.605	002
2.796	200	2.78	30	2.77	20̄1,130
2.607	201,131	2.605	60	2.486	20̄2,131
		2.486	30	2.404	003
2.404	003	2.408	10	2.350	201,13̄2
		2.355	10		
2.209	202,132	2.210	40	2.075	20̄3,132
		2.078	15	1.944	202,13̄3
		1.944	5		
1.823	203,133	1.823	15	1.712	20̄3,133
1.614	060,330	1.715	10	1.616	060,33̄1
1.575	061,331	1.616	60	1.576	061,330,33̄2
1.515	204,134	1.577	45		
1.473	062,332	1.519	10	1.474	062,331,33̄3
1.442	005	1.474	15	1.442	005
1.398	400,260	1.442	5	1.355	400,26̄2
		1.397	10	1.341	063,332,33̄7
		1.373	5		
1.340	063,333	1.340	10	1.202	006,064,333
1.303	402,262	1.302	5		
1.203	006,064,334	1.202	5		

1T: a = 5.591
b = 9.685
c = 7.212
β = 90°

1M: a = 5.601
b = 9.691
c = 7.453
β = 104.6°

FeKα radiation, 114.6mm camera; intensities estimated visually, cell dimensions determined by least-squares fit; Guggenheim et al (1982)

mcGillite

hkl	I	d(obs) Å	d(calc) Å
110•	20	11.67	11.55
200•			11.25
001	70	7.16	7.16
020•	10	6.75	6.73
11T̄•	10	6.50	6.49
201•	10	5.44	5.43
002	40	3.570	3.582
440,8̄01	60	2.888	2.888
441,8̄02	100	2.560	2.562
7̄42,801	20	2.409	2.412
442,8̄03	40	2.112	2.114
7̄43,802	20	1.971	1.975
443,8̄04	20	1.727	1.731
080	40	1.683	1.682
081	20	1.638	1.638
7̄44,803	20	1.619	1.623
082	10	1.524	1.523
8̄81,16.01	10	1.454	1.454
4̄44,805	20	1.420	1.438
005			1.433
8̄82,16.00			1.406
8̄81,16.03			1.380
083			1.375
7̄45,804	20	1.356	1.358
8̄83,16.01	10	1.317	1.316
9̄91•	10	1.292	1.294
8̄84,16.02	10	1.205	1.206
006			1.194
8̄83,16.05	10	1.167	1.168
8̄85,16.03			1.093
085	10	1.0897	1.0908
4̄36•	10	1.0795	1.0780

•- superstructure reflections

a = 23.312   c = 7.423
b = 13.459   β = 105.17°
Kimberley, B.C., Canada; Fe/Mn,
Debye-Scherrer; pattern from Donnay
et al (1980), indexing from Ozawa
et al (1983). 6.50 Å line re-indexed
by authors.

Minnesotaite

hkl	I	d(obs) Å	d(calc) Å
001	100	9.54	9.54
002	15	4.78	4.77
	5.b	4.62	
003	50	3.18	3.18
2̄04	2	2.796	2.798
203	25	2.759	2.759
13T̄	15	2.721	2.721
T̄32	35	2.655	2.652
T̄34	25	2.538	2.537
202	45	2.528	2.525
T̄33	32	2.405	2.403
131	5	2.356	2.351
206	12	2.318	2.316
T̄35	15	2.248	2.250
20T̄	20	2.212	2.212
T̄34	10	2.103	2.103
132	5	2.050	2.051
2̄07	10	2.007	2.005
T̄36	5	1.956	1.954
131	10	1.913	1.912
			1.911
200	4	1.822	1.822
T̄35	5	1.695	1.695
T̄3T̄	5	1.656	1.655
201	15	1.609	1.608
336			1.608
335			1.597
33T̄	6.b	1.597	1.596
337			1.596
336	8	1.576	1.575
06T̄	13	1.565	1.565
062			1.564
13T̄			1.548
333	2	1.541	1.543
338			1.543
060	2	1.524	1.525
063			1.523
2̄09			1.509
33T̄	5.b	1.509	1.506
337			1.505
133	2	1.450	1.452
06T̄			1.451

Minnesotaite

hkl	I	d(obs) Å	d(calc) Å
T̄09			1.365
007	8.b	1.362	1.363
062			1.362
26T̄	8.b	1.351	1.352
26T̄			1.352
20.10			1.328
T̄64	8.b	1.325	1.326
T̄67			1.325
339			1.308
2T̄0	5.b	1.307	1.308
T̄0.10			1.308
T̄39			1.306

b - broad peak

a = 5.623 (2)   α = 99.44° (3)
b = 9.419 (2)   β = 136.35° (2)
c = 14.234 (5)   γ = 90.00° (2)
Cuyuna, Minnesota, U.S.A.; FeKα,
114.6 mm Gandolfi; Guggenheim & Eggleton
(1986)

Nelenite

hkl	I	d(obs) Å	d(calc) Å
001	40	7.10	7.12
002	60	3.55	3.56
80T̄	70	2.878	2.879
T̄40			2.879
802	100	2.552	2.552
441			2.552
80T̄	40	2.402	2.403
T̄42			2.403
803			2.105
442	40	2.104	2.105
152			2.102
802			1.966
T̄43	10	1.962	1.966
931			1.960
114			1.725
462			1.723
T̄43	50	1.723	1.721
80T̄			1.721
024			1.721
13T̄			1.680
12.T̄1	60	1.677	1.677
080			1.677
10.0T̄			1.616
803	10	1.616	1.616
T̄4T̄			1.615
	10	1.518	
	5	1.449	
	5	1.429	
	10	1.401	
	20	1.373	
	5	1.202	

a = 23.24   β = 105.21°
b = 13.418   C2/m
c = 7.382
Gandolfi; Dunn and Peacor (1984)

Parsettensite

hkl	I	d(obs) Å	d(calc) Å
001	200	12.62	12.56
020	40	11.37	11.39
40T	25	9.62	9.70
400	15	8.62	8.48
+2T	10	7.34	7.38
002	30	6.27	6.28
150,73T	50	4.51	4.50
91Z	30	4.28	4.28
003	50	4.195	4.190
932,550	25.b	3.78	3.78
several	50.b	3.70	
004	40	3.145	3.142
several	15	3.006	
several	20	2.923	
11.0Z	70	2.791	2.789
	20	2.708	
77Z	100	2.639	2.639
005	20	2.512	2.512
970?	15	2.464	2.463
77Z	60	2.417	2.418
77Z	70	2.174	2.176
0.14.0,21.7Z	70	1.628	1.627
0.14.1	45	1.615	1.614
0.14.14	35	1.575	1.575
008	1	1.570	1.570
0.14.21	20	1.516	1.517

b - broad peak

a = 39.47   B = 120.8°
b = 22.78
c = 14.63

Data from authors using 114.6 mm
Debye-Scherrer photograph by
I. M. Threadgold, Fe Kα /Mn filtered radiation.
Sample from Parsettens Alp, Switzerland,
supplemented by diffractometer data
for 00l.

Fe-pyrosmalite

tr - trace, b - broad peak

a = 13.33   c = 7.11
Pegmont, Queensland, Australia ; 57.3mm
camera ; Vaughan (1986)

Fe-pyrosmalite

hkl	I	d(obs) Å	d(calc) Å
10T0	20	11.58	11.55
0001	80	7.13	7.11
11Z0	tr	6.64	6.70
10T1	20	6.08	6.06
20Z0	tr	5.78	5.78
11Z1	tr.b	4.871	4.865
20Z1	5	4.493	4.483
2130	10	4.359	4.367
21Z1	10	3.728	3.721
0002	60	3.564	3.555
10T2	20	3.399	3.398
30Z1			3.386
22Z0	5	3.326	3.335
20Z2	5	3.012	3.028
22Z1			3.019
40Z0	10	2.889	2.888
21Z2	tr	2.762	2.757
40T1	100		2.676
30Z2	tr	2.609	2.612
31T2			2.380
0003	10	2.374	2.370
+15T	60	2.243	2.242
11Z3			2.333
+0T3	40	1.833	1.832
+37T			1.835
++8T	40	1.667	1.667
33Z3	30	1.622	1.622
++8T	40	1.513	1.514
0005	5	1.423	1.422
63ZT			1.426
10TZ	10	1.413	1.411
80ZT	10	1.365	1.366
6173	10	1.336	1.334
50Z+	10	1.276	1.276
5+9Z	5	1.233	1.233
++8Z	5.b	1.121	
30Z5	10.b	1.803	
+0ZT	10.b	1.078	
73.TÖ1	10.b	1.043	
80ZZ	10.b	0.991	

Fe - Mn Pyrosmalite

hkl	I	d Å
100	20	11.5
001	80	7.16
110	5	6.65
101	30	6.06
200	5	5.74
111	20	4.87
201	10	4.50
210	20	4.35
211	20	3.73
002	50	3.58
102,301	40	3.41
220	5	3.34
112	20	3.15
202,221	10	3.03
400	10	2.89
212	10	2.77
401	100	2.69
302	5	2.61
+10	10	2.52
312,+11+	10	2.38
+02,113	70	2.25
203,501+	tr	2.20
322,331	5	2.13
+21,510+	tr	2.09
303	tr	2.02
600	tr	1.930
313	tr	1.912
332	50	1.892
+03,+31	tr	1.840
512,521	tr	1.797
10+,610	tr	1.766
+13,11+	tr	1.734
++0	50	1.672
333,++1	+0	1.629
531,701	tr	1.608
612	10	1.588
513,621+	tr	1.566
+0+	50	1.521
622,523+	5.b	1.459

Plus 30 lines to 0.9953

tr - trace, b - broad peak, + - additional
indices possible

a = 13.36   P 3m1
c = 7.16
Broken Hill, N.S.W., Australia : Fe/ Mn.
114.6mm Debye-Scherrer ; Stillwell and
McAndrew (1957)

Mn - pyrosmalite

hkl	I	d Å
100	30	11.6
001	100	7.16
110	10	6.71
101	30	6.09
200	10	5.77
111	20	4.886
201	10	4.509
210	20	4.376
211	20	3.736
002	80	3.583
102	40	3.419
220	30	3.338
221,202	20	3.035
400	20	2.882
212	20	2.770
+01	90	2.683
+10	20	2.549
003	20	2.385
103	10	2.334
+02,113	70	2.251
213	10.b	2.102
+03	+0	1.843
10+	10	1.768
11+	10.b	1.733
++0	50	1.672
30+	+0	1.627
+0+	50	1.523
005	10	1.432
105,50+	20	1.419
+2+,205	20	1.371
305	20	1.342
+05	20	1.285
820	10	1.266
633	10	1.238
006	10	1.194
	10.bb	1.790
	10.bb	1.406
	20.b	1.126
	10.bb	1.106

b - broad peak

a = 13.36   P 3m1
c = 7.16
Frondel and Bauer (1953)

724

Schallerite I		
hkl	I	d Å
002	90	7.37
004	100	3.59
104	20	3.45
2207	10	3.30
ni	10	3.11
401	50	2.848
402	70	2.687
410	30	2.546
403	60	2.478
006	10	2.395
323	10	2.321
330	30	2.253
331	10	2.208
	20	2.109
	50	2.037
	30	1.842
	20	1.727
	90	1.674
	30	1.631
	40	1.521
	10	1.422
	30	1.394
	20	1.373
	10	1.294
	10	1.286
	20	1.185

ni - not indexed

a = 13.43
c = 14.31
Bauer and Berman (1928) ;
JCPDS file # 12-248

Schallerite II		
hkl	I	d Å
002	10	7.02
004	40	3.547
104	5	3.411
220	5	3.313
401	30	2.825
402	60	2.673
403	50	2.466
006	5	2.370
323	5	2.309
330	10	2.222
332	5	2.105
422	5	2.087
405	50	2.022
406	40	1.975
440	100	1.688
442	20	1.6275
408	60	1.5108
445,801	5	1.4420
00.10	10	1.4256
802	10	1.4176
338,409	30	1.3906
544,446	20	1.3687
723,21.10	5	1.3546
617,812+	5	1.3432
552,22.10	20	1.2913
813,438+	20	1.2796
821	5	1.2580

+ - additional indices possible

a = 13.36
c = 14.24
FeKα , 114.6 mm camera ;
McConnell (1954)

Zussmanite		
hkl	I	d Å
003	100	9.60
104,110	2	5.82
105,113+	4	4.98
006	45	4.78
107,205+	10	3.78
116,212	8	3.69
108,214	4	3.38
009,207	25	3.19
208,220	2	2.92
119,217	4	2.79
306,312	10	2.74
314,218+	2	2.61
315,401	16	2.51
00.12	6	2.39
309,317+	4	2.31
410,324+	10	2.20
413	2	2.14
31.10,416	6	2.00
ni	2	1.938
31,11,21.13	8	1.908
21.14	4.d	1.806
31.13,32.11	4	1.731
31.14	8	1.653
434,520	10	1.616
00.18,345+	10	1.594
526,612	2	1.531
	6	1.510
	2	1.489
	4	1.444
	4	1.393
	4	1.367
	6	1.360
	6	1.339
	8	1.329
	2	1.304
	2	1.289
	2.d	1.260
	4	1.233
	6	1.196

ni - not indexed , d - doublet ,
+ - additional indices possible

a = 11.65   R3 or R3̄
c = 28.668
Laytonville , California , U.S.A. ; Fe/Mn ,
114.6 mm ; JCPDS File # 19-500

X-ray powder patterns of stilpnomelanes and lennilenapeite

Triclinic hkl	Ortho-hexagonal hkl	1 Mn-ferro-stilpnomelane I	d Å	2 Lennilenapeite I	d Å	3 Mg-ferro-stilpnomelane I	d Å	4 Mg-ferri-stilpnomelane I	d Å	5 Ferri-stilpnomelane I	d Å
001	003	10	12.+	100	12.11	10	12.1	10	12.+	10	12.6
102		4	7.21			4	6.96				
002	006	2	6.07	2	6.09	2	6.06	2	6.16	+	6.31
222,242		3	5.51	2	5.50	5	5.44	3	5.42	2	5.40
441		2	4.77	5	4.76	5	4.74	3	4.70	2	4.69
232		2	4.38	2	4.39	4	4.34	2	4.32	1	4.29
003	009	7	4.05	20	4.07	10	4.02	6	4.11	7	4.20
502,053 / 551		1	3.82			2	3.79	1	3.76	1	3.73
333,363 / 014		2.b	3.6	5	3.67	5.b	3.6	3.b	3.6	1	3.6
004	00.12	6	3.035	20	3.04	8	3.025	3	3.082	*	3.15
445	202	6	2.735	30	2.734	7	2.703	4	2.691	7	2.681
443	204	1	2.646					1	2.61		
446	205	10	2.583	40	2.582	10	2.558	10	2.552	10	2.550
442	207	3	2.442	2	2.439	6	2.424	1	2.415		
447	208	8	2.362	30	2.365	6	2.344	5	2.345	5	2.350
441	20.10	4	2.202	2	2.204	3	2.186	5	2.190	5	2.202
448	20.11	6	2.121	10	2.125	5	2.107	5	2.114	4	2.126
440	20.13	3	1.968		1.965	1	1.952	1	1.967	1	1.981
449	20.14	3	1.894			4	1.885	2	1.895		
441.10	20.17	4	1.693	1	1.695	5	1.686	5	1.699		
12.05	060	8	1.596	30	1.593	8	1.580	5	1.570	+	1.564
12.04	063	7	1.582	30	1.578	7	1.566	1	1.558	3	1.551
12.03	066	3	1.543	10	1.542	3	1.529	1	1.523	1	1.517
441.11	20.20	3	1.520			3	1.516	1	1.530		
12.02	069	1	1.485			1	1.471	1	1.467		
443	20.22	2	1.420			3	1.415	1	1.431		
12.01	06.12	2	1.412	1	1.413	1	1.401	1	1.399		
888	402	2	1.377	1	1.373	2	1.364	2	1.356		
887	405					4	1.345	2	1.377		
12.00	06.15	1	1.33			2	1.324			1	1.333

b - broad peak, + - additional indices possible, * - obscured by Si standard

1 Grythyte, Sweden   a = 22.11, d_{001} = 12.14
2 Franklin, N.J., U.S.A.   a = 22.05, d_{001} = 12.19
3 Auburn Mine, Michigan, U.S.A.   a = 21.88, d_{001} = 12.14
4 Matukuki R., New Zealand   a = 21.76, d_{001} = 12.33
5 Canberra, A.C.T., Australia   a = 21.66, d_{001} = 12.60

Samples 1, 3-5 : Fe/Mn 114.6mm Debye-Scherrer, Eggleton and Chappell (1978)
Sample 2 : Cu/Fe 114.6mm Gandolfi, Dunn et al (1984)
(cell dimensions assume b = a√3, calculated by least squares by authors)